Electron Beams and Microwave Vacuum Electronics

Electron Beams and Microwave Vacuum Electronics

SHULIM E. TSIMRING

WILEY-
INTERSCIENCE
A JOHN WILEY & SONS, INC., PUBLICATION

Published by John Wiley & Sons, Inc., Hoboken, New Jersey.
Published simultaneously in Canada.

Library of Congress Cataloging-in-Publication Data:

Tsimring, Shulim E., 1924–
Electron beams and microwave vacuum electronics / by Shulim E. Tsimring.
p. cm.
"A Wiley-Interscience publication."
Includes bibliographical references and index.
ISBN-13: 978-0-470-04816-0
ISBN-10: 0-470-04816-6
1. Vacuum microelectronics. 2. Electron beams. I. Title.

TK7874.9.T75 2006
621.3815′12–dc22 2006044263

10 9 8 7 6 5 4 3 2 1

Contents

Electron Beams and Microwave Vacuum Electronics

SHULIM E. TSIMRING

WILEY-INTERSCIENCE
A JOHN WILEY & SONS, INC., PUBLICATION

Published by John Wiley & Sons, Inc., Hoboken, New Jersey.
Published simultaneously in Canada.

Library of Congress Cataloging-in-Publication Data:

Tsimring, Shulim E., 1924–
Electron beams and microwave vacuum electronics / by Shulim E. Tsimring.
p. cm.
"A Wiley-Interscience publication."
Includes bibliographical references and index.
ISBN-13: 978-0-470-04816-0
ISBN-10: 0-470-04816-6
1. Vacuum microelectronics. 2. Electron beams. I. Title.

TK7874.9.T75 2006
621.3815′12–dc22 2006044263

10 9 8 7 6 5 4 3 2 1

To my dear wife, Miroslava Sverdlova

Preface

One of the most important features of vacuum electronics is the strong interaction of two subjects: the physics of electron beams, and vacuum microwave electronics, including millimeter-wave electronics. This is evident in the statement of electron beam problems and in the process of creating practically all microwave electron devices. Comprehension of these subjects is important for the development and application of devices; in general, it determines the professional level of a researcher or an engineer in this field.

There are a number of books devoted to the physics of electron beams and to microwave electronics. However, in books on the physics of electron beams, the problems of microwave electronics are usually treated briefly. Similarly, in books on microwave electronics, the theory of electron beams usually occupies a modest place. Two books (Barker and Schamiloglu, 2001; Barker et al., 2005) are a notable exception. These books give detailed coverage of most problems of physics, engineering, technology, and the use of power microwave electronics, combined with the physics of electron beams and a detailed historical survey. However, with the exception of the klystron analysis, these books do not include a systematic exposition of the theory of different devices; instead, they assume the reader's familiarity with the theory.

The primary goal of this book is discussion of the foundations of the physics and theory of electron beams and microwave electronics. The structure is dictated by a historical sequence from classical vacuum electronics of the twentieth century, to the impressive achievements of recent years. A similar presentation was offered by the remarkable books of Chodorow and Susskind (1964), Granatstein and Alexeff (1987), Kirstein et al. (1967), Kleen (1958), Lawson (1988), and Szilagyi (1988). The sense of historical perspective gives each new generation of researchers and teachers confidence that the path chosen is correct, and it protects against repeating previous errors.

The book consists of two parts. In Part I, the motion of charged particles in static fields, the theory of electron lenses, and problems of the formation and transport of intense electron beams are considered. Part II covers the principles and theory of the interaction of electron beams with electromagnetic waves in quasistationary systems (e.g., diodes, klystrons), systems with continuous interactions (e.g., traveling-wave tubes and backward-wave oscillators), crossed-field systems (e.g., traveling-wave and backward-wave tubes of M type, magnetrons, crossed-field amplifiers), and finally, the extensive class of systems based on the stimulated radiation of classical electron oscillators: classical electron masers, including gyrotrons, classical autoresonance masers, and free electron lasers. Prominent space is given to the relativistic beams and corresponding powerful relativistic devices that occupy center stage in contemporary vacuum electronics. The book also provides significant coverage of theoretical methods. Statements of problems, discussion of models and approximations, and derivations of formulas are shown as completely as possible. Simultaneously, my goals are to represent the material in a form accessible to students as well as to engineers and researchers working in adjacent fields of science and technology. Careful attention is given to explanations of the physical principles underlying the operation of devices and the representation of related background topics. Consequently, the book could be used as a textbook. The work is based on my lectures at Saratov and Nizhny Novgorod (Gorky) State Universities in Russia, and research work at the Institute of Applied Physics of the Russian Academy of Sciences and at General Atomics (San Diego, California).

The book would not be possible without my close contacts with pioneers of the physics of electron masers and, in particular, gyrotrons: Profs. Andrey Gaponov-Grekhov and Michael Petelin [Institute of Applied Physics of the Russian Academy of Science (IAP RAS)]. I wish to express my sincere gratitude to them for their collaboration, attention, and support. I am also indebted to the director of IAP RAS, Prof. Alexandr Litvak, and all the friends and associates with whom I worked for decades at IAP RAS and Nizhny Novgorod State University, for their cooperation and invaluable help in research. I am grateful to Prof. Jay Hirshfield and Dr. Vyacheslav Yakovlev (Yale University), Prof. Richard Temkin and Dr. Michael Shapiro (MIT), Prof. Edl Schamiloglu (New Mexico University), Prof. Michail Rabinovich (UCSD), Prof. Manfred Thumm (University Karlsruhe, Germany), and Prof. Olgiert Dumbrajs (Helsinki University of Technology, Finland), for their support and helpful advice. I am very thankful to my sons, Dr. Michael Tsimring and especially to Dr. Lev Tsimring (UCSD), for their considerable help in the preparation of the manuscript.

Finally, I would like to thank George Telecki, Rachel Witmer, and Angioline Loredo, of Wiley-Interscience, for their great editorial support in the production of the book.

S.E. Tsimring

Introduction

I.1 OUTLINE OF THE BOOK

Recent trends in the development of vacuum microwave electronics and the physics of electron beams have been shaped in part by competition with solid-state high-frequency electronics. So practically all information technology and microwave devices of small power and limited frequency are based on solid-state electronics. Contemporary vacuum microwave electronics and the physics of charged-particle beams include the formation and transport of intense and relativistic electron beams, electron optics, powerful microwave devices together with millimeter- and submillimeter-wave devices, charged-particle accelerators, material procession, and free electron lasers. This narrowing of focus has led to considerable progress in the following aspects of theory and engineering:

- Theory of electromagnetic fields
- Dynamics of charged-particle beams
- Interaction of electron beams with high-frequency fields
- Electron emission and optics
- Development and application of vacuum electron devices
- Methods of computer simulation
- Technology and powerful and high-voltage experimental techniques and equipment

It would thus be very difficult to embrace within a single volume a complete description of the state of the art in this field. There are many books on the subject of vacuum electronics (many of which we cite in the book). They do not, however, provide a thorough treatment of the theory of both electron beams and microwave

Electron Beams and Microwave Vacuum Electronics. By Shulim E. Tsimring

electronics. Two books, *Modern Microwave and Millimeter-Wave Power Electronics* (Barker et al., 2005) and *High Power Microwave Sources and Technologies* (Barker and Schamiloglu, 2001), are the exception. These volumes are characterized by an exceptionally wide scope of information. They do not, however, include a systematic exposition of the theory of basic processes assuming the reader's familiarity with the theory. In this book I strike a compromise, providing the reader with a foundation in the physics and theory of electron beams and vacuum microwave electronics. The material is presented in historical sequence, and classical results and concepts are treated alongside contemporary issues.

The book is divided into two parts: Part I, Electron Beams (Chapters 1 to 5), and Part II, Vacuum Microwave Electronics (Chapters 6 to 10). Auxiliary information (e.g., equations of motion, Maxwell's equations, Hamiltonian formalism, the Liouville theorem) is presented in the Introduction. This material cannot, however, replace corresponding background fundamental guides. It serves a reference function and provides notation and definitions.

Part I begins in Chapter 1 with a discussion of the motion of charged particles in static electric and magnetic fields. Special attention is devoted to an analysis of relativistic beams and the motion of charged particles in weakly inhomogeneous fields (e.g., adiabatic invariants, drift equations).

In addition to classical paraxial electron optics, in Chapter 2 we describe the theory and applications of quadrupole lenses, which are important elements of accelerators and effective correctors of aberration in paraxial electron-optical systems. Principles of electronic image construction are an important element in electron beam formation in microwave devices and accelerators.

In Chapters 3 and 4, an extensive area of the physics of intense electron beams that are used in most high-frequency electron devices is considered. The self-consistent equation of steady-state space-charge beams is derived. Self-consistent solutions for certain space-charge curvilinear flows as well as gun synthesis methods are described. A number of electron guns with compressed electron beams are discussed. The theory of noncongruent space-charge beams and its application to the design of magnetron-injected guns are considered.

Electron guns that use explosive electron emission, making it possible to obtain electron beams with energy on the order of MeV and currents of hundreds of kiloamperes, have acquired great significance in powerful high-frequency electronics and electron beam technology. Guns using planar explosive emission and magnetically insulated diodes are considered.

Transport problems of lengthy intense electron beams that are key problems for microwave devices are discussed in Chapter 5. A group of relevant problems is connected with the transport of nonrelativistic and relativistic Brillouin beams of various configurations. The transport of intense beams in an infinite magnetic field approximation and centrifugal focusing is also discussed. A theory of intense axially symmetric paraxial electron beams with arbitrary shielding of the cathode magnetic field is described. A criterion for stiffness beam formation is formulated.

Finally, the transport of intense electron beams in spatially periodic fields is considered. A theory of periodic magnetic focusing, which has the most practical value for beam-type tubes, is expounded.

Part II opens (Chapter 6) with an analysis of quasistationary microwave devices in which the electric field is potential but the energy integral is not conserved. Analysis of the simplest element of these systems, a planar electron gap, demonstrates two principal effects: bunching of electrons and phasing of bunches. The latter, and also the effects of velocity and energy modulation, are crucial for all vacuum microwave devices. All these effects in the electron gap are not optimal, however. Their implementation led to the first truly microwave amplifiers and oscillators: klystrons based on electron-stimulated transition radiation. In Chapter 7 a number of klystron systems, including reflex and relativistic klystrons, are considered.

Linear and nonlinear theories of traveling-wave tubes of O type (TWTOs) based on the synchronous radiation of rectilinearly moved electrons in the field of a slow electromagnetic wave are discussed in Chapter 8. These tubes and backward-wave oscillators (BWOs), in which an electron beam interacts with an electromagnetic wave whose phase and group velocities are opposite, possess unique properties as wideband oscillators. Relativistic TWTOs are considered there as well. These tubes have output power on the order of gigawatts and should provide very high gain, because only in this case can conventional low-power input sources be used. Powerful relativistic TWTs have spatially extended electromagnetic structures. Therefore, mode selection is an important problem in these tubes.

The energy of the electromagnetic field in TWTs and BWOs of O type is fed by electron kinetic energy. Decrease in electron velocities in the process of interaction violates the synchronism. So the efficiency of these tubes, especially of BWOs, is comparatively low. An essentially different mechanism is implemented when electron beams interact with electromagnetic fields in crossed static electric and magnetic fields (M-type systems). In this case the energy of the electromagnetic field is extracted from the potential energy of particles. As a result, synchronism is maintained along a deep conversion of the electron energy. M-type systems can have an efficiency close to 100%. In Chapter 9, typical devices of M type are considered: magnetrons, injected-beam traveling-wave and backward-wave amplifiers and oscillators, and amplifiers of magnetron type. The very high efficiency and high pulse power of the latter allow them to be used as basic high-frequency sources in radar systems and electronic countermeasure devices. Also, the high efficiency, compactness, and low cost of low-power magnetrons explain their exceptional use in domestic microwave ovens. Relativistic magnetrons that use explosive emission cathodes are also considered in Chapter 9. These oscillators are very promising high-frequency sources in radar systems and countermeasure means.

A very interesting power oscillator that utilizes crossed fields is the magnetically insulated line oscillator (MILO). In this tube, the magnetic field of the electron beam replaces the external magnetic field of a conventional magnetron. This requires a very high beam current that can be provided only by explosive electron emission. The constructive simplicity of such systems provides potential advantages with respect to other pulse sources of electromagnetic oscillation with power on the order of gigawatts in the L and S frequency bands.

The microwave amplifiers and oscillators mentioned above exploit radiation of electrons executing rectilinear or close to rectilinear particle motion: transition and Cerenkov radiation. In the latter, synchronous radiation of particles is possible, due to their interaction with slow and therefore surface electromagnetic fields. The output power and efficiency of corresponding devices inevitably drop with the frequency. So the shortest nonrelativistic BWOs have a maximum output power on the order of milliwatts in the submillimeter-wavelength range. Relativistic devices of O type are an exception, but the possibilities for their practical application, especially in continuous-wave (CW) regimes, are limited.

New ideas were put forward at the end of the 1950s and the beginning of the 1960s. The natural attenuation was turned to electron beams with curvilinear periodical trajectories of particles in which electrons radiate at an arbitrary ratio of their velocity to the phase velocity of a wave in a given medium. This concept is the idea underlying classical electron masers (CEMs), where stimulated radiation of oscillating electrons takes place.

Chapter 10 is devoted to the mechanism, theory, and sources of stimulated radiation of classical electron oscillators. This area of vacuum electronics reflects perhaps the most significant tendencies in modern high-frequency electronic developments. The analysis of an ensemble of classical electron oscillators in electromagnetic fields displays two important mechanisms: linear and quadratic bunching. The latter is the result of the nonisochronism of oscillators. Among the examples of subrelativistic classical electron masers considered in the book, the gyrotron and the ubitron are notable, in which takes places the stimulated bremsstralung of electrons in uniform and spatially periodic magnetic fields, respectively. The surprising property of the gyrotron is the existence of a strong essentially relativistic quadratic bunching for subrelativistic energies of electrons (on the order of tens of keV). Another important property of a gyrotron is the possibility of using spatially developed electrodynamic and electron-optical systems, due to the existence of effective mode selection methods for gyrotrons. That allows one to obtain record average output power in the millimeter- and submillimeter-wave ranges. Unique gyromonotrons have been developed that deliver CW output power up to 1 MW in a 2-mm wavelength. Similar gyromonotrons find wide application in controlled fusion experiments (e.g., electron–cyclotron resonance heating and electron–cyclotron current drive in tokamak–stellarator plasmas). A substantial part of Chapter 10 covers an analysis of the gyrotron mechanism, gyrotron electron-optical systems, methods of mode selection in gyrotrons, and various gyrotron applications.

The efficiency of gyrotrons drops, however, as the electron energy approaches the relativistic energy, because the decrease in the relativistic electron mass in the process of radiation violates the synchronism between oscillating electrons and the electromagnetic field. In this case, cyclotron autoresonance masers (CARMs) and free-electron masers (FELs) are alternatives. In a CARM, synchronism is supported when the phase velocity is closed to the light velocity c, due to compensation of the electron relativistic gyrofrequency shift and the Doppler shift stipulated by a change in the electron drift velocity. In a FEL, an ultrarelativistic version of the ubitron, the stimulated radiation of electrons in wiggler (undulator) devices with a spatially periodic magnetic field, is used. A very important property of a FEL is the bremsstrahlung Doppler frequency up-conversion, according to which the radiation frequency in the laboratory frame of reference increases approximately proportionally to the square of the relativistic mass of the moving radiating particle. This property, together with those specific for the FEL pondermotive bunching effect, allows one to obtain powerful coherent radiation in the infrared, optical, ultraviolet, and potentially, even hard x-ray ranges.

CARMs and FELs are considered in the book comparatively briefly. FELs were invented in 1971, and at present the number of published papers dedicated to FELs is on the order of 10^4. Due to its wavelengths, coherent properties, frequency tunability, and high output power, FELs open an unprecedented range of applications. A list that is very far from complete includes such topics as biology, biomedicine, surgery, solid-state physics, chemistry and the chemical industries, defense, micromachining, photophysics of polyatomic molecules, and military and domestic applications.

Unfortunately, a reader will find very little material on electron emission, stochastic oscillations in electron beams and microwave tubes, and charged-particle beam problems in accelerators. That restriction certainly narrows the scope of the book but opens additional possibilities for a more detailed discussion of theory and methods in the main topics mentioned above. Also, there was no room for numerous constructive implementations of devices. This is a very complicated problem taking into account the modern dynamics of high-power vacuum electronics development.

Finally, computation algorithms and softwares that have reached a very high level of sophistication are not considered in this book. Surely, they should now be treated as an independent area of vacuum electronics. Certainly, numerical methods are a necessary component of the design of all electron devices. However, the application of the numerical simulation can turn out to be useless without a clear understanding of theoretical foundations.

1.2 LIST OF SYMBOLS

Vector values are denoted in bold face. MKS units are used.

$\mathbf{A}$	magnetic vector potential
$\mathbf{B}, B$	magnetic induction
$c = 2.997925 \times 10^8$ m/s	light velocity
$\mathbf{E}, E$	electric field
$e_0 = 1.602177 \times 10^{-19}$C	absolute value of an electron charge
$h = 6.626076 \times 10^{-34}$ J · s	Planck's constant
$I_0 = \dfrac{4\pi\varepsilon_0 c^3}{\eta} \approx 17$ kA	relativistic current
i	imaginary unity
$\mathbf{j}, j$	current density
$k = 1.38066 \times 10^{-23} J/K$	Boltzmann's constant
m	electron relativistic mass
$m_0 = 9.109390 \times 10^{-31}$ kg	electron rest mass
n	particle density
$\mathbf{p}, p$	momentum

$\mathbf{v}, v$	velocity of a particle
v_g	group velocity
v_{ph}	phase velocity
$w_0 = m_0c^2 = 8.187111 \times 10^{-14}$ J	electron rest energy
$\beta = \dfrac{v}{c}$	dimensionless velocity
β	propagation constant
$\gamma = \dfrac{m}{m_0} = 1/\sqrt{1-\beta^2}$	relativistic factor
$\varepsilon_0 = (\mu_0 c^2)^{-1} = 8.854188 \times 10^{-12}$F/m	permittivity of free space
$\eta = \dfrac{e_0}{m_0} = 1.758820 \times 10^{11}$ C/kg	electron specific charge
$\mu_0 = 4\pi \times 10^{-7}$ H/m	permeability of free space
ρ	space-charge density
φ	electric potential
$\varphi_0 = \dfrac{m_0c^2}{e_0} \approx 511 \times 10^3$ V	reduced electron rest energy
$\omega_g = \dfrac{e_0B}{m} = \dfrac{e_0B}{m_0\gamma}$	gyrofrequency (electron–cyclotron frequency)
$\omega_p = \left(\dfrac{n_e e_0^2}{m_0\varepsilon_0}\right)^{\frac{1}{2}}$	electron plasma frequency
ω_q	reduced plasma frequency

I.3 ELECTROMAGNETIC FIELDS AND POTENTIALS

Considered below are electromagnetic fields acting on a *moving particle* with location $\mathbf{r}(t)$; therefore, the fields and potentials are expressed as

$$\mathbf{E} = \mathbf{E}[\mathbf{r}(t),t], \qquad \mathbf{B} = \mathbf{B}[\mathbf{r}(t),t], \qquad \varphi[\mathbf{r}(t),t], \qquad \mathbf{A} = \mathbf{A}[\mathbf{r}(t),t] \tag{I.1}$$

For static fields, $\partial/\partial t = 0$ and

$$\mathbf{E} = \mathbf{E}[\mathbf{r}(t)], \qquad \mathbf{B} = \mathbf{B}[\mathbf{r}(t)], \qquad \varphi[\mathbf{r}(t)], \qquad \mathbf{A} = \mathbf{A}[\mathbf{r}(t)] \tag{I.2}$$

Field–potential relations according to Maxwell's equations are

$$\mathbf{E} = -\text{grad}\,\phi - \frac{\partial A}{\partial t}, \qquad \mathbf{B} = \text{curl}\,\mathbf{A} \tag{I.3}$$

For static fields:

$$\mathbf{E} = -\text{grad}\,\varphi, \qquad \mathbf{B} = \text{curl}\,\mathbf{A} \tag{I.4}$$

Maxwell's equations (in free space):

$$\text{curl}\,\mathbf{B} = \mu_0\mathbf{j} + \frac{1}{c^2}\frac{\partial \mathbf{E}}{\partial t} \tag{I.5}$$

$$\text{curl}\,\mathbf{E} = -\frac{\partial \mathbf{B}}{\partial t} \tag{I.6}$$

$$\operatorname{div} \mathbf{E} = \frac{\rho}{\varepsilon_0} \tag{I.7}$$

$$\operatorname{div} \mathbf{B} = 0 \tag{I.8}$$

For static fields according to Eqs. (I.4) and (I.7), Poisson's equation is valid:

$$\operatorname{div} \operatorname{grad} \varphi = -\frac{\rho}{\varepsilon_0} \tag{I.9}$$

This equation is reduced to the following forms in Cartesian and cylindrical coordinates, respectively:

$$\frac{\partial^2 \varphi}{\partial x^2} + \frac{\partial^2 \varphi}{\partial y^2} + \frac{\partial^2 \varphi}{\partial z^2} = -\frac{1}{\varepsilon_0} \rho(x,y,z) \tag{I.10}$$

$$\frac{1}{r}\frac{\partial}{\partial r}\left(r\frac{\partial \varphi}{\partial r}\right) + \frac{1}{r^2}\frac{\partial^2 \varphi}{\partial \theta^2} + \frac{\partial^2 \varphi}{\partial z^2} = -\frac{1}{\varepsilon_0} \rho(r,\theta,z) \tag{I.11}$$

I.4 PRINCIPLE OF LEAST ACTION. LAGRANGIAN. GENERALIZED MOMENTUM. LAGRANGIAN EQUATIONS

The *principle of least action* (*Hamilton's principle*), is stated: For each mechanical system, the functional (the action integral of specific function L) exists as

$$S = \int_{t_1}^{t_2} L(q_1, q_2, \ldots, q_N, \dot{q}_1, \dot{q}_2, \ldots, \dot{q}_N, t)\, dt \tag{I.12}$$

For real trajectories, $q_i = [q_i(t)]_{\text{real}}$ S has the least (in general, extreme) value. Here L is the Lagrangian, q_i and $\dot{q}_i (i = 1, 2, \ldots, N)$ are generalized coordinates and velocities, and N is the number of degrees of freedom. A system of n particles has $N = 3n$ degrees of freedom. All possible (comparable) trajectories belong to the class

$$q_i(t_1) = q_i^{(1)}, \qquad q_i(t_2) = q_i^{(2)}, \qquad i = 1, 2, \ldots, N \tag{I.13}$$

It can be shown that a necessary condition for realization of the extreme of the functional S is the system of equations

$$\frac{d}{dt}\left(\frac{\partial L}{\partial \dot{q}_i}\right) - \frac{\partial L}{\partial q_i} = 0, \qquad i = 1, 2, \ldots, N \tag{I.14}$$

which determines the real trajectories. These are the famous *Lagrangian equations*. *Generalized momenta* are defined as

$$P_i = \frac{\partial L}{\partial \dot{q}_i}, \qquad i = 1, 2, \ldots, N \tag{I.15}$$

It is convenient for the particle ($N = 3$) to use the vector notation $\mathbf{r}(q_1, q_2, q_3)$, $\mathbf{v}(\dot{q}_1, \dot{q}_2, \dot{q}_3)$. Then the generalized momentum is

$$\mathbf{P} = \frac{\partial L}{\partial \mathbf{v}} \tag{I.16}$$

An alternative *Lagrangian equation* to (I.14) is

$$\frac{d}{dt}\left(\frac{\partial L}{\partial \mathbf{v}}\right) - \frac{\partial L}{\partial \mathbf{r}} = \frac{d\mathbf{P}}{dt} - \frac{\partial L}{\partial \mathbf{r}} = 0 \tag{I.17}$$

The *Lagrangian* for the electron in the electromagnetic field is given by (see, e.g., Landau and Lifshitz, 1987)

$$L = -m_0c^2\sqrt{1-\beta^2} - e_0\,\mathbf{A}\mathbf{v} + e_0\varphi \tag{I.18}$$

The *electron generalized momentum* according to Eqs. (I.16) and (I.18) is

$$\mathbf{P} = \frac{\partial L}{\partial \mathbf{v}} = \frac{m_0\mathbf{v}}{\sqrt{1-\beta^2}} - e_0\mathbf{A} = \mathbf{p} - e_0\mathbf{A} \tag{I.19}$$

where the mechanical momentum $\mathbf{p} = m_0\gamma\mathbf{v}$. Then the *electron Lagrangian* may be written as

$$L = -\frac{m_0c^2}{\sqrt{1-\beta^2}}(1-\beta^2) - e_0\mathbf{A}\mathbf{v} + e_0\phi = \mathbf{P}\mathbf{v} - w \tag{I.20}$$

where the quantity

$$w = mc^2 - e_0\varphi \tag{I.21}$$

is the electron energy [see Eq. (1.3)] and $m = m_0/\sqrt{1-\beta^2}$ is the relativistic electron mass. The equation of motion according to Eqs. (I.17) and (I.19) is

$$\frac{d\mathbf{p}}{dt} - e_0\frac{d\mathbf{A}}{dt} + e_0\frac{\partial(\mathbf{A}\mathbf{v})}{\partial \mathbf{r}} - e_0\frac{\partial\varphi}{\partial \mathbf{r}} = 0 \tag{I.22}$$

Using a known relation of vector analysis,

$$\frac{\partial(\mathbf{A}\mathbf{v})}{\partial \mathbf{r}} = (\mathbf{v}\nabla)\mathbf{A} + \mathbf{v} \times \operatorname{curl}\mathbf{A} \tag{I.23}$$

we obtain

$$\frac{d\mathbf{p}}{dt} - e_0\frac{\partial \mathbf{A}}{\partial t} - e_0(\mathbf{v}\nabla)\mathbf{A} + e_0(\mathbf{v}\nabla)\mathbf{A} + e_0\mathbf{v}\times\mathbf{B} - e_0 \operatorname{grad}\varphi = 0 \tag{I.24}$$

So, corresponding to Eqs. (I.3), Eq. (I.24) becomes

$$\frac{d\mathbf{p}}{dt} = -e_0\mathbf{E} - e_0\mathbf{v}\times\mathbf{B} \tag{I.25}$$

This equation can also be expressed in terms of the relativistic factor $\gamma = m/m_0$:

$$\frac{d(\gamma\mathbf{v})}{dt} = -\eta(\mathbf{E} + \mathbf{v}\times\mathbf{B}) \tag{I.26}$$

The equation of motion in the nonrelativistic approximation ($\gamma = 1$) is

$$\frac{d\mathbf{v}}{dt} = -\eta(\mathbf{E} + \mathbf{v}\times\mathbf{B}) \tag{I.27}$$

I.5 HAMILTONIAN. HAMILTONIAN EQUATIONS (e.g., Landau and Lifshitz, 1987)

According to Eq. (I.20), the energy in terms of the momentum and the Lagrangian is

$$w = \mathbf{P}\mathbf{v} - L \tag{I.28}$$

Here L and w are functions of **r**, **v**, and t. Using the formula for generalized momentum (I.19), we can express **v** in terms of **P**, **r**, and **t**. The energy as a function of these variables is called the *Hamiltonian*:

$$H = w(\mathbf{P}, \mathbf{r}, t) = \mathbf{P}\mathbf{v} - L \tag{I.29}$$

We can express the equation of motion via the Hamiltonian. The total differential of H is

$$dH = \mathbf{P}\,dv + \mathbf{v}\,d\mathbf{P} - \frac{\partial L}{d\mathbf{r}}d\mathbf{r} - \frac{\partial L}{\partial \mathbf{v}}d\mathbf{v} - \frac{\partial L}{\partial t}dt \tag{I.30}$$

Taking the Lagrangian equation [Eq. (I.17)] and the Lagrangian momentum definition [Eq. (I.16)], we find that

$$dH = \frac{\partial H}{\partial \mathbf{r}}d\mathbf{r} + \frac{\partial H}{\partial \mathbf{P}}d\mathbf{P} + \frac{\partial H}{\partial t}dt = \mathbf{v}\,d\mathbf{P} - \frac{d\mathbf{P}}{dt}d\mathbf{r} - \frac{\partial L}{\partial t}dt \tag{I.31}$$

Equating corresponding terms in Eq. (I.31), we obtain the *Hamiltonian equations*

$$\frac{d\mathbf{P}}{dt} = -\frac{\partial H}{\partial \mathbf{r}}, \qquad \frac{d\mathbf{r}}{dt} = \frac{\partial H}{\partial \mathbf{P}} \tag{I.32}$$

Also, we find that

$$\frac{dH}{dt} = \frac{\partial H}{\partial t} = -\frac{\partial L}{\partial t} \tag{I.33}$$

The Hamiltonian equations are the equations of motion in the variables **P** and **r**. Note that for static fields according to Eq. (I.33), $dH/dt = 0$. This relation is equivalent to the conservation of energy.

The explicit expression of the Hamiltonian for an electron in the electromagnetic field as a function of **P, r** and t according to Eq. (I.19) is

$$H = mc^2 - e_0\varphi = c\sqrt{m_0^2c^2 + (\mathbf{P} + e_0\mathbf{A})^2} - e_0\varphi \tag{I.34}$$

Here **A** and φ are functions of **r** and t. We also used the obvious relationship for the relativistic mass:

$$m^2c^2 = m_0^2c^2 + p^2$$

If one differentiates the first of Eqs. (I.32) with respect to **P** and the second with respect to **r** and add them, we obtain

$$\frac{\partial\dot{\mathbf{P}}}{\partial\mathbf{P}} + \frac{\partial\dot{\mathbf{r}}}{\partial\mathbf{r}} = 0 \tag{I.35}$$

I.6 LIOUVILLE THEOREM

I.6.1 Liouville Theorem for Interaction Particles

When $F(P, q, t)$ is the distribution function in the phase space (i.e., a number of particles in the unity phase volume),

$$F = \frac{dn}{dP\,dq}$$

where dn is the number of particles in the phase volume $dP\,dq$. The value F denotes density in the phase space. In general, if there are N interacting particles, the Lagrangian for each particle depends on the coordinates and velocities of all particles, and the phase space has $6N$ dimensions.

Each element of $dP_i\,dq_i$ is really a six-dimensional element in Euclidean space. The $6N$ dimension's current density of these particles is

$$\mathbf{j} = \mathbf{v}\frac{dn}{dP\,dq} = \mathbf{v}\mathrm{F}$$

where the 6N velocity $\mathbf{v}$ has the components $\dot{P}$ and $\dot{q}$. According to a continuity theorem (conservation of the particle number),

$$\operatorname{div}\mathbf{j} + \frac{\partial F}{\partial t} = 0 \quad \text{or} \quad \frac{\partial j_P}{\partial P} + \frac{\partial j_q}{\partial q} + \frac{\partial F}{\partial t} = 0$$

Then

$$\frac{\partial(F\dot{P})}{\partial P} + \frac{\partial(F\dot{q})}{\partial q} + \frac{\partial F}{\partial t} = 0 \qquad \text{or}$$

$$\dot{P}\frac{\partial F}{\partial P} + F\frac{\partial \dot{P}}{\partial P} + \dot{q}\frac{\partial F}{\partial q} + F\frac{\partial \dot{q}}{\partial q} + \frac{\partial F}{\partial t} = 0 \tag{I.36}$$

By letting the particles move along trajectories, Eq. (I.35) becomes valid, and the second and fourth terms in Eq. (I.36) cancel. We obtain

$$\frac{\partial F}{\partial P}\dot{P} + \frac{\partial F}{\partial q}\dot{q} + \frac{\partial F}{\partial t} = 0 \quad \text{or} \quad \frac{DF}{Dt} = 0 \tag{I.37}$$

This is a version of the *Liouville theorem*: The density F of particles moving along trajectories is constant.

Consider a group of N particles that move along their trajectories. The number of the particles in the group is constant:

$$N = \int_{\Omega} F\, dP\, dq = \text{const.} \tag{I.38}$$

Because $F = \text{const.}$, we obtain

$$\int_{\Omega} dP\, dq = \text{const.} \tag{I.39}$$

The volume of the 6N phase space that encloses a chosen group of the particles is constant. This is the second version of the Liouville theorem.

Note that the Hamiltonian equations (I.32) are valid if the forces acting on the particles have a potential. These forces are called *Hamiltonian forces*. For *non-Hamiltonian forces* (i.e., radiation losses, frictional forces, etc.) the first of Eqs. (I.32) must be changed (Landau and Lifshitz, 1987). In this case, the Liouville theorem is not valid, and the phase density as well as the phase volume of the group of particles chosen has no more invariants.

I.6.2 Liouville Theorem for Noninteraction Identical Particles

In this case, the Lagrangian of each particle will depend on the coordinate and momentum of the particle. Then we can write the Hamiltonian equations for each particle in the same six-dimensional phase space and use a distributive function in this space for identical particles:

$$f = \frac{dn}{dP\,dq}$$

Furthermore, formulas (I.37)–(I.39) are repeated and we obtain Liouville's theorem for six-dimensional phase space. In particular,

$$f = \text{const.} \quad \text{or} \quad \frac{\partial f}{\partial \mathbf{P}}\dot{\mathbf{P}} + \frac{\partial f}{\partial \mathbf{r}}\dot{\mathbf{r}} + \frac{\partial f}{\partial t} = 0, \qquad \int_{V_6} dP\,dq = \text{const.} \tag{I.40}$$

where V_6 is the six-dimensional phase volume.

I.6.3 Liouville Theorem for a Phase Space of Lesser Dimensions

The invariant phase volume in the Liouville theorem may be shortened even more if some degree of freedom of the noninteraction particles is independent of the other(s). In this case, the corresponding Hamiltonian may be represented as the sum of independent components, and for each component we obtain invariants corresponding to Eq. (I.40) (Moss, 1968; Reiser, 1994). For example, if the motion in the transverse plane ($p_\perp$, $q_\perp$) and the longitudinal motion (p_z, q_z) are independent, the invariants are

$$f(P_\perp, q_\perp) = \text{const.}, \qquad \int_{V_4} d\dot{P}_\perp\, dq_\perp = \text{const.} \tag{I.41}$$

$$f(P_z, q_z) = \text{const.}, \qquad \int_{V_2} dP_z\, dq_z = \text{const.} \tag{I.42}$$

Finally, if motions in the x and y directions are independent, the invariant of Eq. (1.41) is split to

$$\begin{aligned} &f(P_x, x) = \text{const.}, \qquad f(P_y, y) = \text{const.} \\ &\int_{V_2} dP_x\, dx = \text{const.}, \qquad \int_{V_2} dP_y\, dy = \text{const.} \end{aligned} \tag{I.43}$$

I.7 EMITTANCE. BRIGHTNESS (Humphries, 1990; Lawson, 1988; Lejuene and Aubert, 1980)

I.7.1 Emittance in a Zero Magnetic Field

In this case, $\mathbf{P} - \mathbf{p} + e_0\mathbf{A} = \mathbf{p}$. Consider the first of the integrals in Eq. (I.43):

$$\int dp_x\, dx = \int \frac{dp_x}{p_z} p_z\, dx = \int p_z\, dx\, dx' = \text{const.} \tag{I.44}$$

Suppose that p_z is constant in each transverse section of the beam. Then

$$p_z \int dx\, dx' = p_z V_2 = \text{const.} \tag{I.45}$$

If the particles move with constant energy in the z-direction ($p_z = \text{const.}$),

$$\Xi_x = \int dx\, dx' = V_2 = \text{const.} \tag{I.46}$$

The (x, x') phase space is called a *trace space*. The integral Ξ_x is called an *emittance*. In the literature, the emittance is often defined as $(1/\pi)\Xi_x$. According to Eq. (I.40), the area of the trace space for some groups of the particles with constant axial momentum along a trajectory is *invariant*. Note that the configuration of the contour enclosing a chosen group of particles in the trace space can vary, but the area into the contour is conserved. A reader can find many interesting pictures displaying contour transformations in the literature (see, e.g., Humphries, 1990; Lawson, 1988; Lejuene and Aubert, 1980).

If the beam is axially symmetric, the single emittance of Eq. (I.46) may be utilized. For the nonsymmetrical paraxial beams with constant p_z, only the invariance of a *hyperemittance*

$$\Xi_h = \int dx\, dx' dy\, dy' = V_4 \tag{I.47}$$

is correct. If the beam has two planes of symmetry x and y, the hyperemittance is

$$\Xi_h = \int_{V_2} dx\, dx' \int_{V_2} dy\, dy' = \Xi_x \Xi_y$$

For a beam with variable p_z momentum, the invariant is

$$p_z^2 \Xi_h \sim \beta^2\gamma^2\Xi_h = \beta^2\gamma V_4 \tag{I.48}$$

In nonrelativistic approximation, that is equivalent to

$$\varphi(z)\Xi_h = \text{const.} \tag{I.49}$$

I.7.2 Brightness

Microscopic brightness is, by definition,

$$B_m = \frac{dI}{dV_4} = \frac{dI}{dx\, dx' dy\, dy'} \qquad \text{A/rad}^2 \cdot \text{m}^2 \tag{I.50}$$

where I is the beam current. Usually, the average brightness value is utilized:

$$B = \frac{I}{V_4} = \frac{I}{\Xi_h} \tag{I.51}$$

For the small beam angles, this definition is equivalent to

$$B \approx \frac{j}{\Omega} = \frac{j}{\pi\alpha^2(z)} \qquad \text{A/m}^2 \cdot \text{st} \tag{I.52}$$

where j is a current density, Ω is a solid angle enclosing all the rays emerging from a point on the z-axis, and $\alpha(z)$ is the angle between the z-axis and the beam edging rays. According to Eq. (I.47), the brightness is invariant if $\beta\gamma$ or $\varphi(z)$ is a constant. Otherwise, invariants are

$$B_{\text{inv}} = \frac{B\overline{\beta}^2\overline{\gamma}^2}{\beta^2\gamma^2} \ (\text{relativistic}), \qquad B_{\text{inv}} = \frac{B\overline{\varphi}}{\varphi(z)} \ (\text{nonrelativistic}) \tag{I.53}$$

where $\overline{\beta}$, $\overline{\gamma}$, and $\overline{\varphi}$ are average values.

I.7.3 Maximum Langmuir Brightness for Thermionic Emitters

Assuming a Maxwellian distribution of initial velocities and neglecting all other factors that can limit the beam current density, Langmuir (1937) obtained the following formula for a theoretical maximum of current density:

$$j(z) = j_c \frac{e_0\varphi(z)}{kT} \sin^2\alpha(z) \tag{I.54}$$

where j_c is the cathode current density, T is the cathode temperature, and k is Boltzmann's constant. This formula is valid in a nonrelativistic approximation and for $\varphi(z)/kT \gg 1$.

Substituting Eq. (I.54) for small $\alpha(z)[\sin\alpha(z) \approx \alpha(z)]$ into Eq. (I.52), we obtain the *maximum brightness*:

$$B_{\max} = j_c \frac{e_0\varphi(z)}{\pi kT} \tag{I.55}$$

Correspondingly, the *invariance maximum brightness* [Eq. (I.53)] is

$$B_{\text{inv.max}} = j_c \frac{e_0}{\pi kT} \overline{\varphi} \tag{I.56}$$

According to Eq. (I.54), for small α the maximum current density is

$$j_{\max} = j_c \frac{e_0\overline{\varphi}}{kT} \alpha(z)^2 \tag{I.57}$$

PART I

ELECTRON BEAMS

Electron beams are flows of free electrons moving in the direction chosen. This is called the *axis* of the beam. The trajectories of the beam particles often come close to being rectilinear, although recently, specific devices with periodic curvilinear beams, in particular classical electron masers (CEMs) and free electron lasers (FELs), are attracting more and more attention. A beam's axis may be either straight or curved. Finally, beams may have different symmetry: for example, a cylindrical, a sheet, or a strip.

Let us list some properties of electrons as elementary particles. These make electron beams very important for creating and controlling images, material processing, transferring energy with high density, generating other particles, and generating and transforming high-frequency signals.

- Electrons are charged particles that provide effective control of a particle's movement by means of electric and magnetic fields. At the same time, interaction between the particles through self fields (space-charge and self magnetic fields) raises many specific effects and problems connected with the control and stability of intense electron beams.
- Electrons are stable (long-life) particles.
- Electrons are very light particles. They are the lightest among all long-living elementary charged particles and the longest living of the light charged particles. The small mass and correspondingly, inertia of electrons forms the functioning base of high-frequency systems. The transfer of an electron's beams is not accompanied by a large mass transport. So a 100-kA beam transfers only 2 g of electrons during 1 hour. Self fields are relatively weak, due to their high level of electron mobility.
- Electrons are chemically neutral particles. Their transfer does not change the chemical content of electrode surfaces.

Electron Beams and Microwave Vacuum Electronics. By Shulim E. Tsimring

- Simple and effective methods are available to extract electrons from solid, liquid, or gaseous media. It is essential that each type of electron emission provide many important applications of corresponding electronic devices.

Usually, beams are thought of as thin flows with transverse dimensions (at least, a single transverse dimension for sheet beams) and are quite small compared to beam length.

The physics of electron beams certainly does not exhaust the topic of vacuum electronics. Nevertheless, the sphere of electron beam applications is great. A list of some important applications follows:

- Electron optics (e.g., lenses, electron guns, beam deflection systems)
- Microwave vacuum electronics, including relativistic microwave electronics and free-electron lasers
- Electron microscopy, including electron-probe microanalysis
- Electron beam tubes in television, radar, and radio-meter systems
- Electron beam technology (e.g., precise cutting, drilling, welding, melting, high-resolution lithography, manufacture of integrated contours)
- Electron accelerators
- Formation and transport of intense electron beams

To stay within the scope of the book, we will not give a detailed exposition of the huge family of microwave devices and in general, devices using electron beams, or catalogue their numerous specific properties. Only a description of key devices and clarification of their design schemes are offered.

CHAPTER ONE

Motion of Electrons in External Electric and Magnetic Static Fields

1.1 INTRODUCTION

Taking external fields into account—the sources of fields that are independent of beam electrons—is the simplest approximation. In general, the motion of charged particles is also determined by interaction with other particles in the beams. For pure electron beams, this interaction is realized through macroscopic (collective) forces created by electrons themselves (self fields, i.e., space-charged electric and magnetic fields) and by microscopic fields (in practice, through collisions). Considering motion and ignoring space-charge fields is sufficient for rarefied beams (e.g., for beams in electron lenses). But to study dense beams, knowledge of motion principles in external fields is also necessary. Beams with the essential effect of self fields are considered in Chapters 3 to 5.

In this chapter we limit ourselves to a study of static fields. Aspects of electron motion in nonstationary fields are considered in Chapters 6 to 10.

1.2 ENERGY OF A CHARGED PARTICLE

The important principle of charged-particle motion in static fields is energy conservation. Let us multiply Eq. (I.25) by $\mathbf{v}$:

$$\mathbf{v}\frac{d\mathbf{p}}{dt} = -e_0\,\mathbf{v}\mathbf{E} \tag{1.1}$$

Electron Beams and Microwave Vacuum Electronics. By Shulim E. Tsimring

The transformation of Eq. (1.1) for the static fields ($\partial\varphi/\partial t = 0$) gives

$$\frac{d\varphi[\mathbf{r}(t)]}{dt} = \frac{\partial\varphi}{\partial\mathbf{r}}\frac{d\mathbf{r}}{dt} = -\mathbf{vE}$$

$$\mathbf{v}\frac{d\mathbf{p}}{dt} = \frac{1}{m}\mathbf{p}\frac{d\mathbf{p}}{dt} = \frac{1}{2m}\frac{dp^2}{dt} = \frac{1}{2m}\frac{d}{dt}(m^2c^2 - m_0^2c^2) = \frac{d}{dt}(mc^2)$$

Here we used the relation

$$m^2c^2 = p^2 + m_0^2c^2 \tag{1.2}$$

following the expression for the relativistic mass $m = m_0/\sqrt{1 - v^2/c^2}$. Substituting these formulas in Eq. (1.1), we obtain the *conservation of energy*:

$$\frac{d}{dt}(mc^2 - e_0\varphi) = 0, \qquad w = mc^2 - e_0\varphi = \text{const.} \tag{1.3}$$

The energy w can be called a *full energy*, one that is equal to the sum of the electron potential $e_0\varphi$ and kinetic mc^2 energies. Often, the kinetic energy is assumed as the difference $w_k = w - w_0 = (m - m_0)c^2 = m_0c^2(\gamma - 1)$. Note that in the important *quasistatic approximation*, where the equation $\mathbf{E} = -\text{grad}\,\varphi$ is still correct but the electric field depends on the time explicitly, the following relation is true:

$$\frac{dw}{dt} = \frac{\partial\varphi}{\partial t} \tag{1.4}$$

In *nonrelativistic approximation*, the energy is $mc^2 = (m_0v^2/2) + m_0c^2$. Omitting m_0c^2, we obtain the conservation of nonrelativistic energy:

$$w_n = \frac{m_0v^2}{2} - e_0\varphi = \text{const.} \tag{1.5}$$

1.3 POTENTIAL–VELOCITY RELATION (STATIC FIELDS)

Let us assume that an electron leaves a cathode with velocity v_c and mass m_c correspondingly. We believe that the cathode potential $\varphi_{\text{cath}} = 0$. Below m and φ are the electron mass at an arbitrary point $\mathbf{r}$; $e_0\varphi_c$ is the initial energy of the electron at the cathode. Then, according to energy conservation,

$$mc^2 - e_0\varphi = m_0c^2 + e_0\varphi_c \tag{1.6}$$

Let us divide Eq. (1.6) by m_0c^2:

$$\gamma = \frac{m}{m_0} = 1 + \frac{e_0(\varphi + \varphi_c)}{m_0c^2} = 1 + \frac{e_0\varphi^*}{m_0c^2} = 1 + \frac{\varphi^*}{\varphi_0} = 1 + \frac{\eta\varphi^*}{c^2} \tag{1.7}$$

where $\varphi_0 = m_0c^2/e_0 = c^2/\eta$ is the reduced electron rest energy in volts; $\varphi^* \equiv \varphi + \varphi_c$. Taking 1 MV as a unit of potential, we obtain $\varphi_0 \approx 0.51100$ MV, and

$$\gamma \approx 1 + 1.9569\phi^*_{\text{MV}} \approx 1 + 2\phi^*_{\text{MV}} \tag{1.8}$$

For zero initial velocity, $\varphi^* = \varphi$ and

$$\gamma = 1 + \frac{\varphi}{\varphi_0} \approx 1 + 1.9569\varphi_{\text{MV}} \approx 1 + 2\varphi_{\text{MV}} \tag{1.9}$$

where φ_{MV} is the potential in megavolts. In nonrelativistic approximation and for $\varphi_{\text{cath}} = 0$, we obtain from Eq. (1.5) the nonrelativistic dimensionless velocity

$$\beta_n = \frac{v}{c} = \sqrt{\frac{2\varphi^*}{\varphi_0}} \approx \frac{1}{16}\sqrt{\phi^*_{\text{kV}}} \tag{1.10}$$

where φ^*_{kV} is the reduced potential in kilovolts. Using the dependence $\gamma = 1/\sqrt{1-\beta^2}$ and Eq. (1.7), it is easy to find that

$$\beta = \frac{\sqrt{\varphi^*(\varphi^* + 2\varphi_0)}}{\varphi^* + \varphi_0}, \qquad \beta_{\varphi_c=0} = \frac{\sqrt{\varphi(\varphi + 2\varphi_0)}}{\varphi + \varphi_0} \tag{1.11}$$

In the extreme relativistic limit $\varphi_0/\varphi \ll 1$, according to Eq. (1.11),

$$\beta_{\text{ext}} \approx 1 - \frac{1}{2}\left(\frac{\varphi_0}{\varphi}\right)^2 \tag{1.12}$$

The values of β_n, β, and β_{ext} are given in Table 1.1 for different values of φ. According to the table, nonrelativistic approximation "works" until $\varphi \sim 10$ kV. Extreme relativistic approximation is acceptable after $\varphi = 5$ MV.

TABLE 1.1 Reduced Nonrelativistic, Relativistic, and Extreme Relativistic Velocities as Functions of the Potential

φ	1.0 V	100 V	1 kV	10 kV	100 kV	300 kV	1 MV	5 MV	25 MV
β_n	0.001978	0.01978	0.06256	0.1976	0.625	1.083	1.976	4.42	9.9
β	0.001978	0.01978	0.06247	0.1950	0.5482	0.7253	0.9411	0.9957	0.9998
β_{ext}	-13×10^{10}	-13×10^6	-13×10^4	-1300	-12	-0.451	0.8694	0.9948	0.9998

1.4 ELECTRONS IN A LINEAR ELECTRIC FIELD $e_0E = kx$

Equations of motion:

$$\frac{dp}{dt} = -e_0E_x = -kx \tag{1.13}$$

$$\frac{dx}{dt} = \frac{p}{m} = \frac{pc}{\sqrt{p^2 + m_0^2c^2}} \tag{1.14}$$

1.4.1 Nonrelativistic Approximation

Equation (1.14) becomes

$$\frac{dx}{dt} = \frac{p}{m_0} \tag{1.14a}$$

Equations (1.13) and (1.14a) are the equations of a harmonic oscillator. Their solution is

$$\begin{aligned} x(t) &= x_0 \cos(\omega t + \alpha) \\ p(t) &= -x_0 m_0 \omega \sin(\omega t + \alpha) \end{aligned} \tag{1.15}$$

where $\omega = \sqrt{k/m_0}$ is the oscillation frequency. The trajectory in a phase space for this oscillator is the ellipse:

$$\frac{x^2}{x_0^2} + \frac{p^2}{x_0^2 km_0} = 1 \tag{1.16}$$

1.4.2 Relativistic Oscillator

Return to the relativistic system of equations. Dividing Eq. (1.14) by Eq. (1.13), we obtain

$$-kx\,dx = \frac{cp\,dp}{\sqrt{p^2 + m_0^2c^2}} \tag{1.17}$$

Integration of this equation yields the trajectory in phase space:

$$a^2 - \frac{kx^2}{2} = c\sqrt{p^2 + m_0^2c^2} \tag{1.18}$$

where a^2 is an arbitrary constant. It is readily verified that this trajectory describes a finite (periodic) motion. Actually, the variables x and p have maximum values (amplitudes):

$$\begin{aligned} x_{\max} &= x_{p=0} = c\sqrt{\frac{2\,m_0}{k}(u-1)} \\ p_{\max} &= p_{x=0} = m_0c\sqrt{u^2 - 1} \end{aligned} \tag{1.19}$$

where $u = a^2/m_0c^2 > 0$ is a reduced arbitrary constant. The results of a numerical integration of Eqs. (1.13) and (1.14) have been given by Humphries (1990). It was shown that phase trajectories have the form of distorted ellipses. According to Eq. (1.19), they are transformed to correct nonrelativistic ellipses when $u - 1 \approx kx^2/2\,m_0c^2 \ll 1$. This inequality means that the potential energy of an oscillator is negligible compared with the particle rest energy.

1.5 MOTION OF ELECTRONS IN HOMOGENEOUS STATIC FIELDS

Homogeneous fields certainly are idealizations. However, the principles and results of this theory are a basis for the solution of much more complicated problems.

1.5.1 Electric Field

We assume that the electric field is opposite the y-axis (Fig. 1.1), so that $E_y = -E$. Consider two cases:

1. *Initial momentum p_0 of the electron turns toward the x-axis* (Fig. 1.1). The equations of motion are

$$\frac{d\mathbf{p}}{dt} = -e_0\,\mathbf{E} \tag{1.20}$$

$$\frac{dp_x}{dt} = 0, \qquad \frac{dp_y}{dt} = e_0E \tag{1.21}$$

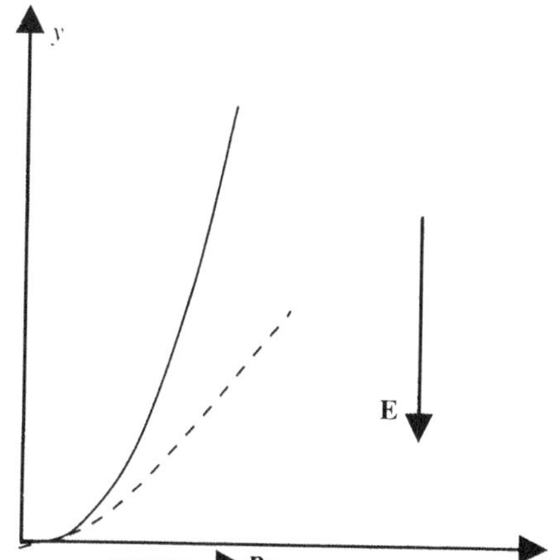

FIGURE 1.1 Motion of an electron in a homogeneous electric field. Solid curve, relativistic trajectory; dashed curve, nonrelativistic trajectory.

Integrating the first equation in Eq. (1.21), we obtain

$$p_x = mv_x = p_0 \tag{1.22}$$

As follows from the second equation in Eq. (1.21),

$$\frac{dp_y}{dt} = \frac{dp_y}{dx}\frac{dx}{dt} = \frac{p_0}{m}\frac{dp_y}{dx} = e_0 E \tag{1.23}$$

According to Eq. (1.2), the mass may be written as

$$m = \frac{1}{c}\sqrt{p^2 + m_0^2 c^2} = \frac{1}{c}\sqrt{p_0^2 + m_0^2 c^2 + p_y^2} = \frac{1}{c}\sqrt{\frac{W_0^2}{c^2} + p_y^2} \tag{1.24}$$

where $W_0 = c\sqrt{m_0^2 c^2 + p_0^2}$ is the initial kinetic energy of the particle. Substituting m into Eq. (1.16), we obtain

$$\frac{dp_y}{\sqrt{(W_0^2/c^2) + p_y^2}} = \frac{e_0 E}{cp_0} dx$$

Integration of this equation gives

$$\sinh^{-1}\frac{p_y c}{W_0} = \frac{e_0 E}{cp_0} x \tag{1.25}$$

Note that

$$p_y = m\frac{dy}{dt} = m\frac{dy}{dx}\frac{dx}{dt} = p_x\frac{dy}{dx} = p_0\frac{dy}{dx}$$

After a corresponding transformation (1.25) and integration, we obtain

$$y = \frac{W_0}{e_0 E}\left(\cosh\frac{e_0 E x}{cp_0} - 1\right) \tag{1.26}$$

This is the equation of a catenary curve. In the nonrelativistic approximation ($e_0 Ex/cp_0 \ll 1$), the energy $W_0 \approx m_0 c^2$. The first term of the Taylor expansion of cosh in Eq. (1.19) gives the well-known parabolic trajectory

$$y = \frac{e_0 E}{2 m_0 v_0^2} x^2$$

In Fig. 1.1 the catenary curve is situated above the parabola because of the more rapid (almost exponential) change of the relativistic y-coordinate as a function of x [Eq. (1.26)].

2. *Initial momentum p of the particle is parallel to the y-axis* (Fig. 1.1). According to Eqs. (1.21),

$$p_x = 0$$
$$p_y = m\frac{dy}{dt} = e_0 Et + p \tag{1.27}$$

Then using relation (1.2) for m, we find that

$$dy = \frac{c(e_0 Et + p)}{\sqrt{m_0^2c^2 + (e_0 Et + p)^2}}\,dt$$

Integrating this equating with the initial condition $y_0 = 0$, we obtain

$$y = \frac{c}{e_0 E}\left(\sqrt{m_0^2c^2 + (e_0 Et + p)^2} - \sqrt{m_0^2c^2 + p^2}\right)$$

1.5.2 Magnetic Field

The equation of motion

$$\frac{dm\,\mathbf{v}}{dt} = -e_0\,\mathbf{v} \times \mathbf{B} \tag{1.28}$$

Since the force is perpendicular to the velocity, the modulus of $\mathbf{v}$ and the electron mass m are constants:

$$v = |\mathbf{v}| = \text{const.}, \qquad m = \text{const.} \tag{1.29}$$

The same result follows from the energy integral (1.3) because the electric field is zero. Then the potential $\varphi = \text{const.}$ and $m = \text{const.}$ Equation (1.28) becomes

$$\frac{d\mathbf{v}}{dt} = -\frac{e_0}{m}\,\mathbf{v} \times \mathbf{B} \tag{1.30}$$

Let us assume that v_B and $\mathbf{v}_\perp$ are velocity components parallel and perpendicular to $\mathbf{B}$, respectively. We obtain from Eq. (1.30)

$$v_B = \text{const.} \tag{1.31}$$

$$\frac{d\mathbf{v}_\perp}{dt} = -\frac{e_0}{m}\,\mathbf{v}_\perp \times \mathbf{B} \tag{1.32}$$

From Eqs. (1.29) and (1.31) follow

$$v_\perp \equiv |\mathbf{v}_\perp| = \sqrt{v^2 - v_B^2} = \text{const.} \tag{1.33}$$

So in a static magnetic field, the values $m, v, v_\perp$, and v_B are constants. The quantity $\mathbf{a}_\perp = d\mathbf{v}_\perp/dt$ is the acceleration of a plane motion. In this case, it may be represented as a sum of the tangential and the centripetal components:

$$a_n = \frac{v_\perp^2}{R}, \qquad a_\tau = \frac{dv_\perp}{dt} \tag{1.34}$$

where R is the radius of curvature in the given point of the trajectory. According to Eq. (1.33), $a_\tau = 0$. Then

$$a_n = \frac{v_\perp^2}{R} = \frac{e_0}{m}|\mathbf{v}_\perp \times \mathbf{B}| = \frac{e_0}{m} v_\perp B \tag{1.35}$$

The radius of the curvature

$$R = r_\perp = \frac{v_\perp}{(e_0/m)B} = \text{const.} \tag{1.36}$$

Hence, the electron trajectory in the plane perpendicular to **B** is a circle. The quantity

$$\omega_g \equiv \frac{e_0}{m} B = \frac{\eta B}{\gamma} \tag{1.37}$$

is the angular velocity of gyration.

We find that the motion of the electron (in general, of a charged particle) is the superposition of two motions: uniform drift along the magnetic field with velocity v_B and uniform gyration on a circle of radius $r_\perp$ [Eq. (1.36)] with frequency ω_g [Eq. (1.37)]. Therefore, the spatial trajectory of the particle is a helical line (Fig. 1.2) with pitch $h = (2\pi/\omega_g)v_B$. It is evident that equations of the trajectory for $x_0 = y_0 = z_0 = 0$ are

$$\begin{aligned} x &= r_\perp \sin(\omega_g t) \\ y &= r_\perp[\cos(\omega_g t) - 1] \\ z &= v_B t \end{aligned} \tag{1.38}$$

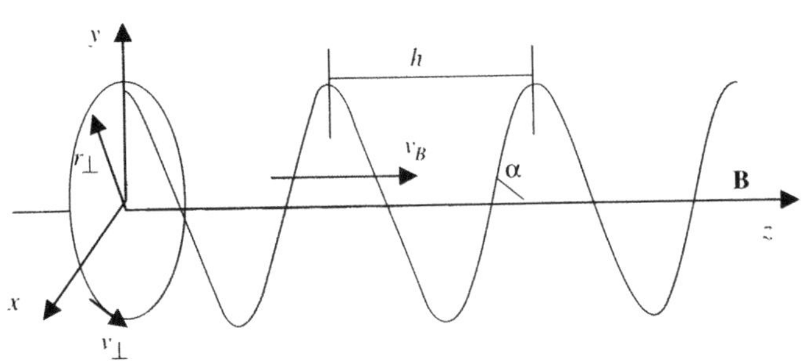

FIGURE 1.2 Motion of an electron in a homogeneous magnetic field.

The circle is called a *Larmor circle*, the radius is a *Larmor radius*, the center of the Larmor orbit is a *guiding center*, and ω_g is the *gyrofrequency* or *cyclotron frequency*. It is essential that the relativistic gyrofrequency depends on the particle's kinetic energy:

$$\omega_g = \frac{e_0 c^2}{w} B = \frac{\eta B}{\gamma} \tag{1.39}$$

Therefore, electrons in magnetic fields behave as nonisochronous oscillators. This property of electrons has very important applications in electron masers (Chapter 10).

In the nonrelativistic approximation $\gamma = 1$, the gyrofrequency is $\omega_g = \eta B$. In this case, the electrons are isochronous oscillators.

Example 1.1 Find the trajectory parameters of an electron that is injected into the uniform magnetic field $B = 1$ T at an angle $\alpha = 30^\circ$ (Fig. 1.2). The electron energy is 1 MeV.

For $w = 1$ MeV, $\gamma = 1.957$ [Eq. (1.12)], and $\beta = 0.941$ (Table 1.1). The gyrofrequency is $\omega_g = \eta B/\gamma = 8.99 \times 10^{10}$ rad/s. The velocity components (Fig. 2.1) are $v = \beta c = 2.82 \times 10^8$ m/s, $v_B = v \cos\alpha = 2.43 \times 10^8$ m/s, and $v_\perp = v \sin\alpha = 1.41 \times 10^8$ m/s. The pitch of the helical trajectory is $h = (2\pi/w_g) v_B = 1.7 \times 10^{-2}$ m. The radius of the helix is $r_\perp = v_\perp/\omega_g = 1.6 \times 10^{-3}$ m.

1.5.3 Parallel Electric and Magnetic Fields

Now assume that the electric and magnetic fields are oriented along the z-axis. Let $E_z = -E$ and $B_z = B$ (Fig. 1.3). The equations of motion are

$$\frac{d\mathbf{p}_\perp}{dt} = -e_0 \mathbf{v}_\perp \times \mathbf{B} \tag{1.40}$$

$$\frac{dp_z}{dt} = e_0 E \tag{1.41}$$

Nonrelativistic Approximation *($m = m_0$)* Equations (1.40) and (1.41) become

$$\frac{d\mathbf{v}_\perp}{dt} = -\eta \mathbf{v}_\perp \times \mathbf{B} \tag{1.42}$$

$$\frac{dv_z}{dt} = \eta E \tag{1.43}$$

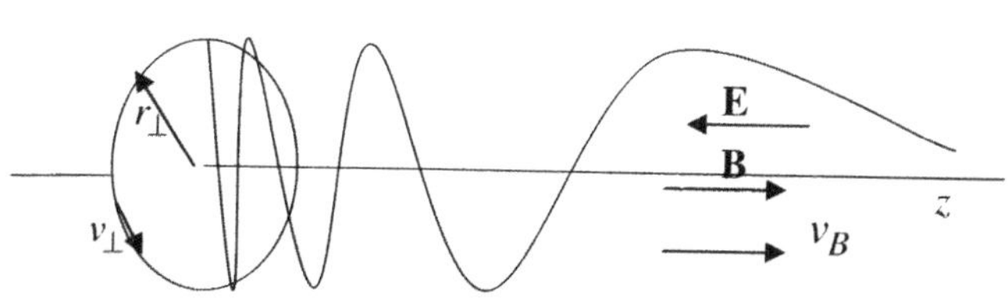

FIGURE 1.3 Motion of an electron in parallel fields **E** and **B**.

Comparison of Eq. (1.42) with Eq. (1.32) shows that motion of the electron perpendicular to the z-axis plane is the same as motion without an electric field; the electron gyrates on the Larmor circle with gyrofrequency $\omega_g = \eta B$. However, the velocity v_B (the velocity of the guiding center along the magnetic field) is now proportional to t. As a result, the trajectory is a helix with variable pitch h. Note that for large t the velocity v_B reaches arbitrarily large values that contradict relativity theory. In fact, for a velocity of order c, the mass becomes variable and the motion in the transversal plane is not described by Eq. (1.42).

Relativistic Motion Let us turn to Eqs. (1.40) and (1.41). According to Eq. (1.41),

$$p_z = e_0 E t \tag{1.44}$$

(for simplicity we take $p_{z0} = 0$). Since $\mathbf{p}_\perp$ is perpendicular to $\mathbf{v}_\perp \times \mathbf{B}$ and according to Eq. (1.40), the magnitude of the perpendicular momentum

$$p_\perp \equiv |\,\mathbf{p}_\perp| = \text{const.} \tag{1.45}$$

Then the full momentum of the particle is

$$p = \sqrt{p_\perp^2 + p_z^2} = \sqrt{p_\perp^2 + (e_0 E t)^2} \tag{1.46}$$

The mass is [Eq. (1.2)]

$$m = \frac{1}{c}\sqrt{p^2 + m_0^2 c^2} = \frac{1}{c}\sqrt{p_\perp^2 + m_0^2 c^2 + (e_0 E t)^2} = \frac{1}{c}\sqrt{\frac{w_0^2}{c^2} + (e_0 E t)^2} \tag{1.47}$$

where w_0 is the initial kinetic energy of the particle.

The perpendicular velocity now decreases with time:

$$v_\perp = \frac{p_\perp}{m} = \frac{p_\perp c}{\sqrt{(w_0^2/c^2) + (e_0 E t)^2}} \tag{1.48}$$

The magnitude of the full acceleration of the particle in the plane perpendicular to $\mathbf{B}$ is

$$a = \left|\frac{d\mathbf{v}_\perp}{dt}\right| \tag{1.49}$$

This value may be found from Eq. (1.4):

$$\frac{d}{dt}(m\mathbf{v}_\perp) = m\frac{d\,\mathbf{v}_\perp}{dt} + \mathbf{v}_\perp\frac{dm}{dt} = e_0 \mathbf{v}_\perp \times \mathbf{B} \tag{1.50}$$

and

$$a = \frac{1}{m}\left|e_0 \mathbf{v}_\perp \times \mathbf{B} - \mathbf{v}_\perp \frac{dm}{dt}\right| = \sqrt{\frac{(e_0 v_\perp B)^2}{m^2} + \frac{v_\perp^2}{m^2}\left(\frac{dm}{dt}\right)^2} \tag{1.51}$$

It is readily verified that the second term under the square root in Eq. (1.51) is a_τ^2, where a_τ is the tangential acceleration in the transverse plane:

$$a_\tau = \frac{dv_\perp}{dt} = \frac{d}{dt}\left(\frac{p_\perp}{m}\right) = -\frac{p_\perp}{m^2}\frac{dm}{dt} = -\frac{v_\perp}{m}\frac{dm}{dt} \tag{1.52}$$

Therefore, the first term equals a_n^2, where a_n is the centripetal acceleration. Thus,

$$a_n = \frac{v_\perp^2}{R} = \frac{e_0 v_\perp B}{m} \tag{1.53}$$

where R is the curvature radius of the trajectory in the perpendicular plane. We obtain

$$R = \frac{mv_\perp}{e_0 B} = \frac{p_\perp}{e_0 B} = \text{const.} \tag{1.54}$$

The angular velocity is

$$\omega = \frac{v_\perp}{R} = \frac{e_0}{m}B \tag{1.55}$$

This value is analogous to the gyrofrequency ω_g [Eq. (1.33)], but m increases with t [Eq. (1.47)]. We find that the trajectory of the particle is a helical line of constant radius, variable pitch, and variable angular velocity (Fig. 1.3). It may be seen from Eqs. (1.44), (1.47), and (1.55) that $\omega \sim 0, v_z \sim c$ for very large t and the trajectory becomes a straight line parallel to the z-axis at a distance from it of $R = p_\perp / e_0 B$.

1.5.4 Perpendicular Fields E and B

Nonrelativistic Approximation The coordinate system and directions of fields are shown in Fig. 1.4. For this configuration

$$E_y = -E = \text{const.}, \qquad B_x = B = \text{const.} \tag{1.56}$$

The nonrelativistic equations of motion are

$$\frac{dv_x}{dt} = 0 \tag{1.57}$$

$$\frac{d\mathbf{v}_\perp}{dt} + \eta \mathbf{v}_\perp \times \mathbf{B} = -\eta \mathbf{E} \tag{1.58}$$

Equation (1.57) corresponds to uniform drift in the x direction with velocity

$$v_x \equiv v_B = \text{const.} \tag{1.59}$$

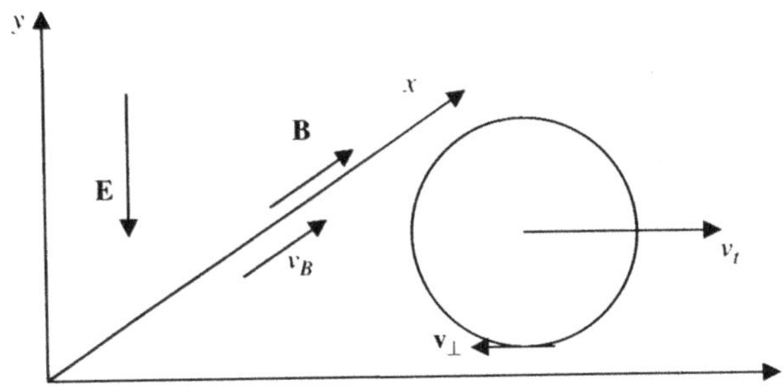

FIGURE 1.4 Motion of an electron in perpendicular fields **E** and **B**.

The drift velocity v_B is determined by the initial condition. Equation (1.58) describes motion in the plane (Y, Z) perpendicular to **B**. This is an ordinary linear inhomogeneous differential equation of first order in $\mathbf{v}_\perp$. The general solution of this equation is the sum of a general solution of the homogeneous equation and a partial solution of the inhomogeneous equation.

The *homogeneous equation is*

$$\frac{d\mathbf{v}_\perp}{dt} + \eta \mathbf{v}_\perp \times \mathbf{B} = 0 \tag{1.60}$$

which we considered in Section 1.4.2. As has been shown, its general solution describes the electron motion as gyration along the Larmor circle in the (Y, Z) plane with gyrofrequency $\omega_g = \eta B$. The magnitude of the gyratic velocity $v_\perp$ is determined by the initial conditions.

The *nonhomogeneous equation* is Eq. (1.58). The right side of this equation is a constant. Because we need a partial solution, we can use any solution of the type $\mathbf{v}_\perp = \mathbf{v}_t = \text{const}$. Then Eq. (1.58) is reduced to

$$\mathbf{v}_t \times \mathbf{B} = -\mathbf{E} \tag{1.61}$$

Vector multiplication of Eq. (1.61) by **B** yields

$$\mathbf{B} \times (\mathbf{v}_t \times \mathbf{B}) = -\mathbf{B} \times \mathbf{E} = \mathbf{E} \times \mathbf{B} \tag{1.62}$$

Using the rule of vector algebra for a double cross product, we have

$$\mathbf{v}_t B^2 - \mathbf{B}(\mathbf{B}\mathbf{v}_t) = \mathbf{E} \times \mathbf{B} \tag{1.63}$$

We can further specialize the partial solution setting $\mathbf{v}_t$ perpendicular to **B**. Then

$$\mathbf{v}_t = \frac{\mathbf{E} \times \mathbf{B}}{B^2}, \qquad v_t = |\mathbf{v}_t| = \frac{E}{B} \tag{1.64}$$

Velocity $\mathbf{v}_t$ is perpendicular to both **E** and **B**. It is the *transversal* (*perpendicular*) *drift*. Its value does not depend on initial conditions. The full velocity can be represented as a sum:

$$\mathbf{v} = \mathbf{v}_\omega + \mathbf{v}_d \tag{1.65}$$

where $\mathbf{v}_\omega$ is the velocity of gyration and $\mathbf{v}_d = \mathbf{v}_t + \mathbf{v}_B$ is the velocity of the guiding center. Its trajectory is evidently a straight line in the (X, Z) plane with slope v_B/v_t. The trajectory in the plane perpendicular to **B** is the superposition of the drift $\mathbf{v}_t$ and gyration $\mathbf{v}_\omega$. The character of the trajectory depends on the velocities v_t and v_ω. In the lower part of the trajectory, the resulting velocity is (see Fig. 1.4).

$$v_r = v_t - v_\omega$$

The following are possible types of trajetories:

- *Extended trochoid*: $v_r > 0$, $v_t > v_\omega$
- *Cycloid*: $v_r = 0$, $v_t = v_\omega$
- *Contracted trochoid*: $v_r < 0$, $v_t < v_\omega$

These trajectories are shown in Fig. 1.5 and can be described in kinematic terms as trajectories of the point on the rim of rolling wheel.

Note that for the electric field direction chosen, the magnitude of the electron velocity on the lower point of the trajectory is less than that on the upper point, where the potential is higher.

Perpendicular Fields* E *and* B*: Relativistic Version One method of solving this problem is based on the transformation of electromagnetic fields **E** and **B** by transition to another inertial frame of reference. Below we limit ourselves to a

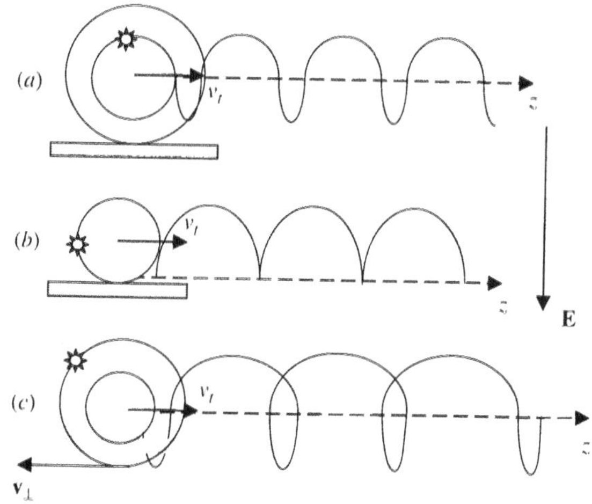

FIGURE 1.5 Electron trajectories in perpendicular **E** and **B** fields for different values of $v_t/v_\perp$: (*a*) extended trochoid; (*b*) cycloid; (*c*) contracted trochoid.

qualitative discussion. According to the properties of the electromagnetic field tensor (see, e.g., Landau and Lifshitz, 1987; Lehnert, 1964), there are two invariants that are conserved with the transition:

$$E^2 - c^2B^2 = \text{invariant}, \qquad \mathbf{EB} = \text{invariant} \tag{1.66}$$

In the case of perpendicular **E** and **B**, the second invariant is $\mathbf{EB} = 0$. Therefore, it is possible to find an inertial frame of reference in which either the E or the B field is zero.

A field that is not zero depends on the first invariant. Let's assume that $E/B > c$, so the first invariant is positive $E^2 - c^2B^2 > 0$. Hence, it is possible to find an inertial frame of reference with a purely electric field where the first invariant is equal to E'^2 and is also positive. According to Eq. (1.62), $E' = \sqrt{E^2 - c^2B^2}$. As can be shown, the velocity of the reference frame must be $V/c = cB/E$. In the new frame the particle velocity as a function of time would not be periodical. It is clear that after returning to the old frame of reference, the motion will not be periodical either.

In the alternative case, $E^2 - c^2B^2 < 0$, a transition is possible to an inertial frame with a purely magnetic field. In the new inertial frame of reference, the motion evidently would be a superposition of the gyration and the uniform drift along the new magnetic field (it is equal to $B' = \sqrt{B^2 - E^2/c^2}$ and has the same direction as the initial magnetic field). However, the coordinates are functions of time within a moving inertial frame of reference. It is readily verified that this time is

$$dt' = \frac{1 - Ev/Bc^2}{\sqrt{1 - E^2/B^2c^2}}\,dt \tag{1.67}$$

It is not the proper time (the time in the frame of reference linked with the moving particle). We can return to the old system by recalculating coordinates and time simultaneously using the Lorentz transformation. Numerically, this operation is very simple. However, the result computed for the analysis does not have a simple physical interpretation such as that for superposition of the gyration and uniform drift as in nonrelativistic approximation. Therefore, we do not discuss this topic in greater detail. Another possible solution of this problem is given by Landau and Lifshitz (1987), but their result is also not very obvious.

1.5.5 Arbitrary Orientation of Fields E and B. Nonrelativistic Approximation

Let us draw an (X, Y) plane through fields **E** and **B** and decompose vector **E** on the components $\mathbf{E}_B$ and $\mathbf{E}_\perp$ parallel and perpendicular to the magnetic field (Fig. 1.6). The nonrelativistic equations of motion are

$$\frac{dv_B}{dt} = \eta E_B \tag{1.68}$$

$$\frac{d\mathbf{v}_\perp}{dt} + \eta \mathbf{v}_\perp \times \mathbf{B} = -\eta \mathbf{E}_\perp \tag{1.69}$$

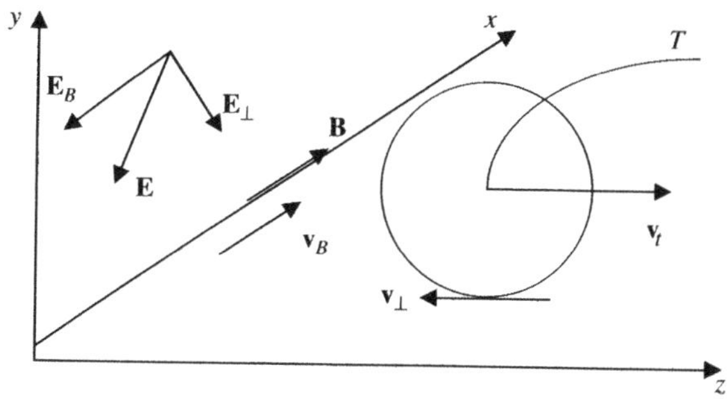

FIGURE 1.6 Electron velocities for arbitrary orientation of fields **E** and **B**. Trajectory T of the guiding center in the (X, Z) plane is parabolic.

Equation (1.68) describes uniformly accelerated drift in the direction of the magnetic field:

$$v_B = \eta E_B t + v_{B0} \tag{1.70}$$

The only difference between Eqs. (1.69) and (1.58) is the replacement of field **E** by $\mathbf{E}_\perp$; therefore, all characteristics of motion perpendicular to the **B** plane coincide in both cases except for the substitution of transversal drift velocity [Eq. (1.64)]:

$$\mathbf{v}_t = \frac{1}{B^2}\mathbf{E}_\perp \times \mathbf{B} \tag{1.71}$$

The full velocity of the guiding center is

$$\mathbf{v}_d = \mathbf{v}_B + \mathbf{v}_t = v_B\frac{\mathbf{B}}{B} + \frac{1}{B^2}\mathbf{E}_\perp \times \mathbf{B} \tag{1.72}$$

The electron gyrates on the Larmor circle in the plane perpendicular to **B** with gyrofrequency $\omega_g = \eta B$, and the guiding center moves with uniform acceleration along **B**. Evidently, the trajectory of the guiding center is a parabola in the (X, Z) plane (Fig. 1.6).

1.6 MOTION OF ELECTRONS IN WEAKLY INHOMOGENEOUS STATIC FIELDS (Lehnert, 1964; Northrop, 1963; Vandervoort, 1960)

The motion of electrons in weakly inhomogeneous fields is the next approximation in the dynamics of charged particles. A physical system in general may be considered as slowly varying when change in its properties is small on a characteristically finite scale (temporal or spatial) for the system. Small varying systems often display important specific properties. Typical are dynamic oscillatory systems with slowly varying parameters. Here the finite scale is a period T of oscillations. The condition of slow changes is

$$T\left|\frac{dX}{dt}\right| \ll |X| \tag{1.73}$$

or

$$T\left|\frac{dX}{dt}\right| = \varepsilon|X| \tag{1.74}$$

where X is some parameter of the system and $\varepsilon \ll 1$ is the smallness parameter.

1.6.1 Small Variations in Electromagnetic Fields Acting on Moving Charged Particles

If one passes from a laboratory frame where a particle moves in inhomogeneous static fields to a frame moving with a guiding center, the particle will experience action of the variables in time fields. The conditions of slow change in fields (adiabatic approximation) according to Eq. (1.74) are

$$T_g\left|\frac{d\mathbf{B}}{dt}\right| = \varepsilon_B|\mathbf{B}|, \qquad T_g\left|\frac{d\mathbf{E}}{dt}\right| = \varepsilon_E|\mathbf{E}| \tag{1.75}$$

where $T_g = 2\pi/\omega_g$ is the cyclotron period (gyroperiod), and ε_B and ε_E are the smallness parameters. Below we assume for simplicity that $\varepsilon_{B,E} = \varepsilon$ and $\mathbf{B},\mathbf{E} = \mathbf{F}$.

Let us return to the laboratory frame. The velocity magnitude is $|\mathbf{v}| = |d\mathbf{r}/dt|$ and $dt = |d\mathbf{r}|/|\mathbf{v}|$. Substituting in Eq. (1.75) gives conditions for a small inhomogeneity of

$$T_g|\mathbf{v}|\left|\frac{d\mathbf{F}}{d\mathbf{r}}\right| = \varepsilon|\mathbf{F}| \tag{1.76}$$

The full velocity of a particle in homogeneous fields [Eq. (1.65)] is

$$\mathbf{v} = \mathbf{v}_\omega + \mathbf{v}_B + \mathbf{v}_t \tag{1.77}$$

Assume that the gyratic velocity magnitude is

$$|\mathbf{v}_\omega| \gg |\mathbf{v}_B| + |\mathbf{v}_t| \tag{1.78}$$

This means that $|\mathbf{v}| \approx |\mathbf{v}_\omega| = v_\perp$. From Eq. (1.76) we obtain

$$2\pi r_\perp\left|\frac{d\mathbf{F}}{d\mathbf{r}}\right| = \varepsilon|\mathbf{F}| \tag{1.79}$$

where $r_\perp = (1/2\pi)T_g v_\perp$ is the Larmor radius. Usually, these conditions are written without a factor 2π. This means that changes in field magnitudes on the Larmor radius scale must be much smaller than full field magnitudes. The alternative condition takes place when

$$|\mathbf{v}_B| \gg |\mathbf{v}_\omega| + |\mathbf{v}_t| \quad \text{and hence} \quad |\mathbf{v}| \approx v_B \tag{1.80}$$

In this case, we obtain from Eq. (1.72)

$$h\left|\frac{d\mathbf{F}}{d\mathbf{r}}\right| = \varepsilon|\mathbf{F}| \tag{1.81}$$

where $h = v_B T_g$ is the pitch of the helical trajectory. Hence, Eq. (1.79) must be fulfilled as a condition of small field changes on the helical trajectory pitch. Note that this is possible only if the component of the electric field E_B parallel to the magnetic field is notzero in the limited time interval, and $E_B \ll E_\perp$. Thus, in the presence of magnetic fields, two natural spatial scales of the weak inhomogeneous electromagnetic field appear: the Larmor radius and the pitch of the helical trajectory.

Example 1.2 It is instructive to calculate the parameter ε_B for electrons in a near-Earth space magnetic field $B_e \sim 2 \times 10^{-5}$ T. The characteristic scale of change of the magnetic field is Earth's radius, $\sim 10^4$ km. Therefore, the value $|\nabla B_e|/B_e$ is on the order of $\sim 10^{-7}\text{m}^{-1}$. Let us assume that the electron moves toward the Earth perpendicular to the magnetic field. Find the value of ε_B for various electron energies.

1. *Nonrelativistic Electron Velocity*. $\beta = 0.2$, $\gamma \approx 1$. The gyrofrequency is $\omega_g = \eta B_e/\gamma \approx 3.5 \times 10^6$ rad/s. The Larmor radius is $R = \beta c/\omega_g \approx 17$ m. $\varepsilon_B = R(|\nabla B_e|/B_e) \approx 1.7 \times 10^{-6}$. As shown in this case, Earth's magnetic field is essentially adiabatic.

2. *Relativistic Velocity*. $\beta = 0.995$, $\gamma = 1/\sqrt{1-\beta^2} \approx 10$, the electron energy is $\sim$5 MeV, $\omega_g \approx 3.5 \times 10^5$ rad/s, $R \sim 800$ m, and $\varepsilon_B \sim 8 \times 10^{-5}$. The magnetic field is again adiabatic.

3. *Supreme Relativistic Energy*. $\sim$50 GeV $(\gamma = 1 \times 10^5)$, $\omega_g \sim 35$ rad/s, $R \sim 10^7$ m, and $\varepsilon_B \sim 1$. The magnetic field is nonadiabatic.

1.6.2 Adiabatic Invariants (Landau and Lifshitz, 1987; Northrop, 1963)

Consider an oscillatory system with a slowly varying parameter X so that condition (1.74) is being fulfilled. Let us assume that only one of the system's degrees of freedom is oscillatory with the coordinate q and generalized momentum P. The Lagrangian equation of the system is

$$\frac{dP}{dt} - \frac{\partial L}{\partial q} = 0 \tag{1.82}$$

Because the Lagrangian L depends on X as a parameter,

$$\frac{\partial L}{\partial q} = \left(\frac{\partial L}{\partial q}\right)_{X=\text{const.}} + \frac{\partial L}{\partial X}\frac{dX}{dq} \tag{1.83}$$

The first term in Eq. (1.83) can be found by integrating Eq. (1.82) with $X = \text{const}$. Then we obtain a nonperturbed trajectory, and this term must be a strictly periodic function of q. Evidently, the second term is nonperiodic if the function $X = X(t)$ does not have periodicities commensurable with the system oscillation. We can represent this term as

$$\frac{\partial L}{\partial X}\frac{dX}{dq} = \frac{\partial L}{\partial X}\frac{dX}{dt}\frac{dt}{dq} \tag{1.84}$$

Now let us integrate Eq. (1.82) along the closed contour corresponding to the nonperturbed particle trajectory:

$$\frac{d}{dt}\oint P\,dq = \oint\left(\frac{\partial L}{\partial q}\right)_{X=\text{const}} dq + \oint \frac{\partial L}{\partial X}\frac{dX}{dt}\frac{dt}{dq}\,dq \tag{1.85}$$

The first term on the right-hand side equals zero as an integral over the period of a periodic function. The second term is

$$\int_0^T \frac{\partial L}{\partial X}\frac{dX}{dt}\,dt = \overline{\frac{\partial L}{\partial X}}\,\overline{\frac{dX}{dt}}\int_0^T dt = \overline{\frac{\partial L}{\partial X}}\,T\,\overline{\frac{dX}{dt}}$$

So taking Eq. (1.74) into account, we obtain

$$\frac{d}{dt}\oint P\,dq = \varepsilon\,\overline{\frac{\partial L}{\partial X}}\,\overline{X} \tag{1.86}$$

This equality means that at least in the first order of the smallness parameter ε, the integral

$$I = \oint P\,dq = \text{const.} \tag{1.87}$$

Remember that the integration in Eq. (1.87) is performed along a nondisturbed oscillating trajectory. This integral is called the *adiabatic invariant*. The number of adiabatic invariants for any system is equal to the number of its oscillatory degrees of freedom. For one particle the maximum possible number of degrees is 3. Particle motion described by three adiabatic invariants is an uncommon event. For example, oscillations of charged particles captured in Earth's Van Allen radiation belt have three invariants, corresponding to three types of oscillating motions (see, e.g., Northrop, 1963): gyration on the Larmor orbits, north–south oscillations, and precession about the Earth.

Transversal Adiabatic Invariant of an Electron in an Electromagnetic Field Let us consider the integral from Eq. (1.87) within the frame of reference of the guiding center. In this frame the contour in Eq. (1.87) is the Larmor circle of the electron gyrating in the static electromagnetic field if $E/cB < 1$. Then, taking into account the fact that the momentum on the circle is $P = P_\perp = \text{const.}$, we obtain

$$\oint P\,dl = \oint P_\perp dl = 2\pi p_\perp r_\perp - e_0 2\pi r_\perp A_\perp \qquad (1.88)$$

Application of Stokes' theorem to Eq. (I.3) yields

$$\int_S B_n dS = \int_S \text{curl}_n\,\mathbf{A}\,dS = \oint A_\perp dl = 2\pi r_\perp A_\perp \qquad (1.89)$$

The integral (1.89) is taken over by the Larmor circle, where normal components of a weakly inhomogeneous magnetic field $B_n = B = \text{const.}$ Therefore, $\int_S B_n\,dS = \pi r_\perp^2 B$. By substituting this in Eqs. (1.88) and (1.89), we obtain

$$\oint P\,dl = \pi(2p_\perp r_\perp - e_0 r_\perp^2 B) \qquad (1.90)$$

The Larmor radius is equal to

$$r_\perp = \frac{v_\perp}{\omega_g} = \frac{v_\perp m}{e_0 B} = \frac{p_\perp}{e_0 B} \qquad (1.91)$$

Then we obtain from Eqs. (1.87) and (1.90),

$$I = \frac{\pi}{e_0}\frac{p_\perp^2}{B} \qquad (1.92)$$

This invariant refers to the gyrating degree of freedom and is called the *transversal (perpendicular) adiabatic invariant*. Usually, I is written

$$I_\perp = \frac{p_\perp^2}{B} = \text{const.} \qquad (1.93)$$

In nonrelativistic approximation we can replace $p_\perp$ by $v_\perp$, and

$$I_{\perp n} = \frac{v_\perp^2}{B} \qquad (1.94)$$

It is instructive to calculate the invariant $I_\perp$ directly, considering the gyration of an electron in weakly inhomogeneous magnetic fields. Assume that the guiding center drifts along an increasingly magnetic field (Fig. 1.7). The magnetic field lines are shown in Fig. 1.7. The magnetic field in the moving frame would increase in time. In this frame of reference, the electron experiences the action of the inducing

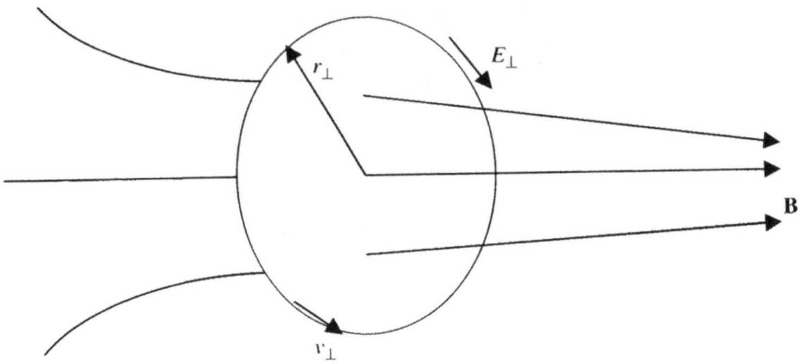

FIGURE 1.7 Perpendicular adiabatic invariant.

electric field $E_\perp$ directed along the Larmor circle clockwise (looking from the magnetic field side; Fig. 1.7). The field $E_\perp$ can be found by integrating Maxwell's equation (I.6) over the Larmor circle and using Stokes' theorem:

$$\int_S \frac{\partial B_n}{\partial t} dS = -\int_S \text{curl}_n\, \mathbf{E}\, ds = -\oint E_\perp\, dl \tag{1.95}$$

Obviously, $B_n = B$, and as a result of Eq. (1.79), fields B and $E_\perp$ are almost uniform on the Larmor circle. We now obtain

$$\frac{dB}{dt}\pi r_\perp^2 = -2\pi r_\perp E_\perp \tag{1.96}$$

$$E_\perp = -\frac{r_\perp}{2}\frac{dB}{dt} = -\frac{p_\perp}{2e_0 B}\frac{dB}{dt} \tag{1.97}$$

This electric field (according to Fig. 1.7) accelerates the electron along the Larmor orbit (the particle has a negative charge). The corresponding change in momentum is

$$\frac{dp_\perp}{dt} = -e_0 E = \frac{p_\perp}{2B}\frac{dB}{dt} \tag{1.98}$$

By integrating this equation, we obtain conservation of the perpendicular adiabatic invariant:

$$I_\perp = \frac{p_\perp^2}{B} = \text{const.} \tag{1.99}$$

The electron motion on the Larmor circle forms the elementary current

$$I = \frac{e_0}{T} = \frac{e_0 \omega_g}{2\pi} \tag{1.100}$$

The magnetic moment of this circular current is $M = (\mu_0/4\pi)IS$ (Northrop, 1963), where S is the area of the circle. Thus, the magnetic moment of the electron is

$$M = \frac{\mu_0}{4\pi}\frac{e_0\omega_g}{2\pi}\pi r_\perp^2 = \frac{\mu_0}{8\pi m}I_\perp \tag{1.101}$$

We see that for $E = 0$ ($m = \text{const.}$) in the nonrelativistic approximation, the magnetic moment of the electron is the adiabatic invariant. The transversal adiabatic invariant has the meaning of a magnetic flux through the Larmor circle:

$$\Psi = \pi r_\perp^2 B = \frac{\pi}{e_0^2}I_\perp \tag{1.102}$$

Accuracy of Conservation of an Adiabatic Invariant According to numerical calculations in many specific problems, the accuracy of the conservation of the transversal adiabatic invariant generally is higher than the first power of ε [see Eq. (1.86)]. But a rigorous consideration of this problem is complicated (Arnold et al., 1988; Kruskal, 1960; Littlejohn, 1980). Let us suppose, following Arnold et al. (1988), that the state of some system depends on a slow parameter λ, which quickly pushes the adiabatic invariant I to its limits: $I(t = -\infty)$ and $I(t = +\infty)$. Then one can introduce an increment of I: $\Delta I = I(+\infty) - I(-\infty)$. Although for finite t, oscillations of I are of order ε, the increment ΔI is much smaller than ε. If λ depends analytically on εt, then $\Delta I \sim O(\exp(-c/\varepsilon))(c > 0)$. *So the increment of an adiabatic invariant decays faster than any power of* ε *as* $\varepsilon \to 0$.

1.6.3 Motion of the Guiding Center

Drift Equations The conditions (1.74) and (1.81) of the small inhomogeneity allow us to set the following scaling orders for the particle trajectory parameters:

$$r_\perp \sim \varepsilon, \qquad T_g \sim \varepsilon, \qquad \omega_g = \frac{2\pi}{T_g} \sim \frac{1}{\varepsilon} \tag{1.103}$$

In the inhomogeneous fields considered, the particles [as seen from Eq. (1.103)] perform oscillations of very high frequency but small amplitude. As a result, the vibrations do not perturb slow movement of the guiding center (evolution of the system). This effect is typical for oscillatory systems with slowly varying parameters. The key method in the theory is the averaging procedure, which separates small oscillations from the drift. This procedure, applying the motion of particles in inhomogeneous static electromagnetic fields, is described briefly below. We use nonrelativistic approximation as well as the conditions

$$\frac{E}{B} < v \qquad E_B \ll E \tag{1.104}$$

Let us write the instantaneous position of the particle (see, e.g., Morozov and Solov'ev, 1960; Northrop, 1963) as

$$\mathbf{r}(t) = \mathbf{R}_d(t) + \mathbf{r}_\perp(t) \tag{1.105}$$

where $\mathbf{R}_d$ is the position of the guiding center and $\mathbf{r}_\perp$ is a radius vector of the particle relative to the guiding center (Fig. 1.8). Thus, the particle velocity is

$$\mathbf{v} = \frac{d\mathbf{R}_d}{dt} + \mathbf{v}_\perp \tag{1.106}$$

Substituting Eq. (1.104) in the nonrelativistic equation of motion, we obtain

$$\ddot{\mathbf{R}} + \ddot{\mathbf{r}}_\perp = -\eta[\mathbf{E}(\mathbf{R} + \mathbf{r}_\perp) + (\dot{\mathbf{R}} + \dot{\mathbf{r}}_\perp) \times \mathbf{B}(\mathbf{R} + \mathbf{r}_\perp)]$$

Expand the fields in a Taylor series about $\mathbf{R}$ up to the first degree of $\mathbf{r}_\perp$. In this case the error is of order $O(\varepsilon)$. We obtain

$$\ddot{\mathbf{R}} + \ddot{\mathbf{r}}_\perp = -\eta\{\mathbf{E}(\mathbf{R}) + \mathbf{r}_\perp \nabla \mathbf{E}(\mathbf{R}) + (\dot{\mathbf{R}} + \dot{\mathbf{r}}_\perp) \times [\mathbf{B}(\mathbf{R}) + \mathbf{r}_\perp \nabla \mathbf{B}(\mathbf{R})]\} + O(\varepsilon) \tag{1.107}$$

Now we must express the gyration of vector $\mathbf{r}_\perp$ explicitly. Let us introduce three orthogonal unit vectors $\boldsymbol{\tau}_1, \boldsymbol{\tau}_2$, and $\boldsymbol{\tau}_3$, where $\boldsymbol{\tau}_1 = \mathbf{B}/B$ is parallel to the magnetic field and $\boldsymbol{\tau}_2$ and $\boldsymbol{\tau}_3$ are perpendicular to the $\mathbf{B}$ plane (Fig. 1.8). The particles gyrate in the plane perpendicular to $\mathbf{B}$. Therefore, we can write

$$\mathbf{r}_\perp = r_\perp(\tau_2 \sin\theta + \tau_3 \cos\theta), \qquad \theta = \int \omega_g \, dt \tag{1.108}$$

Repeated differentiation of Eq. (1.108) gives

$$\dot{\mathbf{r}}_\perp = \omega r_\perp(\tau_2 \cos\theta - \tau_3 \sin\theta) + (\mathbf{r}_\perp \tau_2) \sin\theta + (\mathbf{r}_\perp \tau_3) \cos\theta \tag{1.109}$$

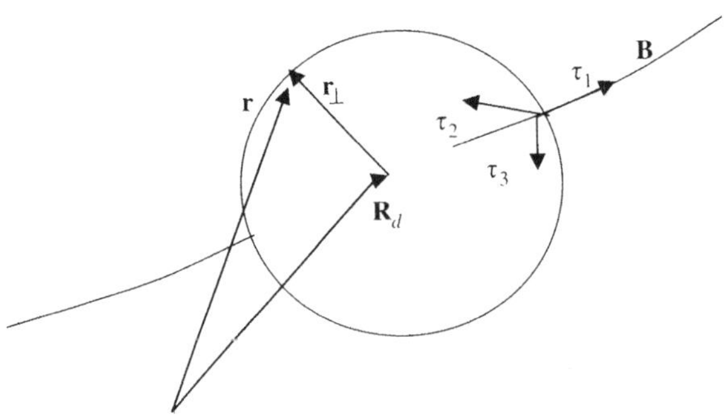

FIGURE 1.8 Motion of a particle as a superposition of the guiding center on a gyrating cyclotron.

We have omitted the corresponding expression of $\ddot{\mathbf{r}}_\perp$ (Northrop, 1963). The next step is the substitution of Eqs. (1.108) and (1.109) and a similar expression for $\ddot{\mathbf{r}}_\perp$ into Eq. (1.107) and integration of the resulting expression:

$$\ddot{\mathbf{R}} = F[\mathbf{R}, \dot{\mathbf{R}}, \mathbf{E}(\mathbf{R}), \mathbf{B}(\mathbf{R}), r_\perp, \theta]$$

over θ with a 2π period $\int_0^{2\pi}(\cdots)d\theta$. After a complicated iterative procedure which takes into account the estimations of Eq. (1.103) and the conditions of Eq. (1.104), the following basic equation of the guiding center's motion is given by Morozov and Solov'ev (1960) and Northrop (1963):

$$\frac{d\mathbf{R}_d}{dt} = v_B\frac{\mathbf{B}}{B} + \frac{\mathbf{E}\times\mathbf{B}}{B^2} - \frac{v_\perp^2 + 2v_B^2}{2\eta B^3}\mathbf{B}\times\nabla B \tag{1.110}$$

Let us augment the energy relation to this equation:

$$W = \frac{m_0}{2}v^2 - e_0 U = \frac{m_0}{2}(v_B^2 + v_\perp^2) - e_0\varphi \tag{1.111}$$

where v_B is the velocity component parallel to $\mathbf{B}$. Recall that the *drift equation* (1.110) is correct in nonrelativistic approximation. Relativistic drift equations may be found in an article by Vandervoort (1960).

We see that the guiding center velocity consists of three components: (1) drift along the curvilinear line of the magnetic field, (2) transversal (perpendicular) drift, and (3) gradient drift. The first two components are similar to the drifts in uniform fields. The third component exists only in an inhomogeneous magnetic field.

Equation (1.10) is a first-order differential equation, a significant simplification of the original second-order equation of motion because Eq. (1.110) gives the velocity of the particle directly. It is interesting to compare the accuracy of the numerical calculation of the velocity using the original equation of motion in second order and Eqs. (1.110) and (1.111). The latter gives acceptable accuracy for the times $t \sim \varepsilon^{-1}$, whereas the accuracy of the numerical methods is reduced with time.

Integration of the Drift Equation The right-hand side of Eq. (1.110) can be expressed explicitly in terms of the potential $U = U(\mathbf{r})$ and the magnetic field $B = B(\mathbf{r})$. Let us suppose that the cathode potential and the initial velocity are zeros and the oscillatory velocity $v_{\perp 0}$ is given at the point where $B = B_0$. The full velocity is

$$v = \sqrt{2\eta U(\mathbf{r})} \tag{1.112}$$

The oscillatory velocity can be found from the nonrelativistic adiabatic invariant

$$I = I_n = \frac{v_\perp^2}{B(\mathbf{r})} = \text{const.} \tag{1.113}$$

The value of I is determined by the initial condition

$$I = I_0 = \frac{v_{\perp 0}^2}{B_0}$$

The oscillatory velocity then is $v_\perp = v_{\perp 0}\sqrt{B(\mathbf{r})/B_0}$. The longitudinal velocity according to the relation of the energy [Eq. (1.11)] is

$$v_B = \sqrt{v^2 - v_\perp^2} = \sqrt{v^2 - IB(\mathbf{r})} \tag{1.114}$$

The electric field $\mathbf{E}(\mathbf{r}) = -\nabla U(\mathbf{r})$. Thus, all values on the right-hand side of Eq. (1.110) have been expressed as functions of $\mathbf{r}$.

Example 1.3: Drift Motion of Electrons in an Adiabatic Magnetic Trap The scheme of a *magnetic trap* (elementary mirror machine or magnetic bottle) is shown in Fig. 1.9. The particle moves in the magnetic field, which is assumed to be weakly inhomogeneous, so that Eq. (1.113) is valid. As can be seen, the magnetic field in the bottle ends $B = B_M$ is maximum. According to Eq. (1.14), the velocity component v_B is reduced when the particle moves in the direction of increasing magnetic field. This effect can be interpreted as an action of the radial component B_r existing in the region of an inhomogeneous magnetic field. The corresponding force $\mathbf{F}_B = -e_0\mathbf{v}_\perp \times \mathbf{B}_r$ is directed along magnetic lines. In increasing magnetic field, one is opposite to the velocity component v_B. When the oscillatory velocity $v_\perp$ achieves the full velocity v, the value $v_B = 0$ and the particle is reflected. The corresponding planes are named ***magnetic mirrors***. Regions where $v_\perp > v$, called *magnetic corks*, are unattainable for the particle.

Let us suppose that the particle is injected into the magnetic field B at the angle α with the magnetic line (Fig. 1.9). The adiabatic invariant

$$I = \frac{v_\perp^2}{B} = \frac{v^2 \sin^2\alpha}{B} \tag{1.115}$$

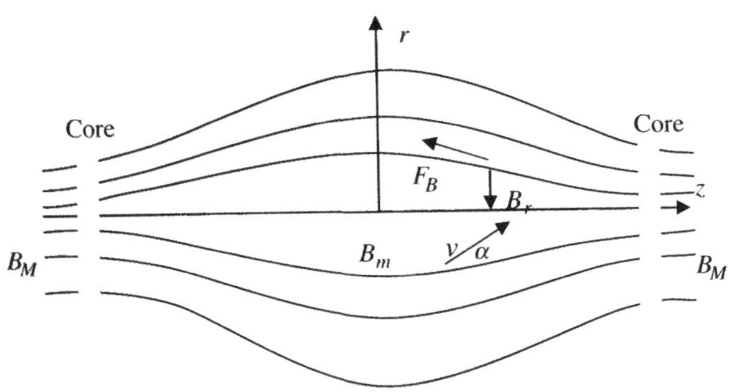

FIGURE 1.9 Adiabatic magnetic trap (mirror machine).

The velocity v_B of the particle in the maximum of the magnetic field B_M according to Eq. (1.114) is

$$v_B = \sqrt{v^2 - IB_M} = \sqrt{v^2\left(1 - \frac{B_M}{B}\sin^2\alpha\right)} \tag{1.116}$$

Particles injected with $\sin\alpha > \sqrt{B/B_M}$ are reflected and captured in the adiabatic trap. When the particles emerge in the medium plane of the trap, where the magnetic field B_n is minimal, the critical angle is also minimal.

$$\sin\alpha_{\min} = \sqrt{\frac{B_n}{B_M}} \tag{1.117}$$

1.7 MOTION OF ELECTRONS IN FIELDS WITH AXIAL AND PLANE SYMMETRY. BUSCH'S THEOREM

Similar to the way that field independence (Lagrangian independence) of time leads to a conservation of energy, any symmetry of fields (i.e., independence of the Lagrangian of some other variables) leads to conservation of the corresponding momenta. These laws of conservation are the foundation of mechanics.

Assume that the Lagrangian L does not depend on the generalized coordinate q (i.e., $\partial L/\partial q = 0$), which is a cyclic variable. The Lagrangian equation (I.17) for this case is

$$\frac{dP_q}{dt} = \frac{\partial L}{dq} = 0 \tag{1.118}$$

$$P_q = \text{const.} \tag{1.119}$$

1.7.1 Systems with Axial Symmetry. Busch's Theorem

Consider a cylindrical coordinate system (Fig. 1.10). For axially symmetric fields, the cyclic variable is θ. This tells us that a turning does not change an electromagnetic field in the system. We can find corresponding generalized momentum by writing the Lagrangian as a function of θ and $\dot{\theta}$. According to Eq. (I.18), the Lagrangian is

$$\begin{aligned} L &= -m_0c^2\sqrt{1 - \frac{v^2}{c^2}} - e_0\mathbf{A}\mathbf{v} + e_0U \\ &= -m_0c^2\sqrt{1 - \frac{1}{c^2}(\dot{r}^2 + r^2\dot{\theta}^2 + \dot{z}^2)} - e_0(\dot{r}A_r + r\dot{\theta}A_\theta + \dot{z}A_z) + e_0\varphi \end{aligned} \tag{1.120}$$

where $\partial\mathbf{A}/\partial\theta = \partial\varphi/\partial\theta = 0$. Then the angular generalized momentum is

$$P_\theta = \frac{\partial L}{\partial\dot{\theta}} = \frac{m_0c^2}{\sqrt{1 - v^2/c^2}}\frac{r^2\dot{\theta}}{c^2} - e_0rA_\theta = mr^2\dot{\theta} - e_0rA_\theta = \text{const.} \tag{1.121}$$

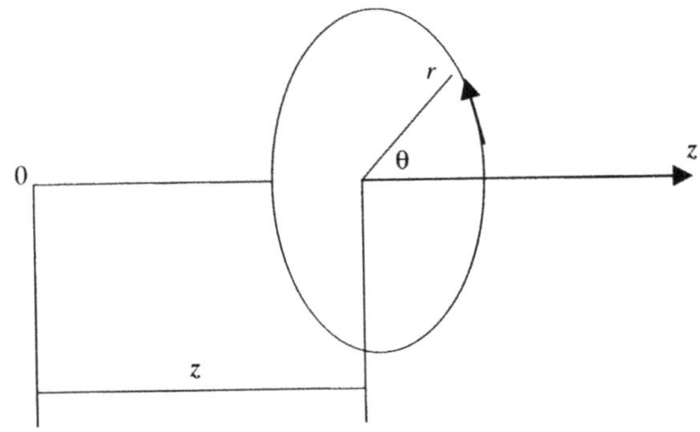

FIGURE 1.10 Cylindrical coordinate system.

To obtain A_θ, let us apply Stokes' theorem to Maxwell's equation (I.3):

$$\int B_n\, dS = \int \text{curl}_n\, \mathbf{A}\, dS = \oint A_\theta\, r d\theta \tag{1.122}$$

Integration is performed along the circle of radius r shown in Fig. 1.1. Here $B_n = B_z$. The value rA_θ is constant on the contour; therefore,

$$rA_\theta = \frac{1}{2\pi}\Psi = \frac{1}{2\pi}\int B_z\, dS \tag{1.123}$$

So the equation of angular momentum conservation is

$$P_\theta = mr^2\dot{\theta} - \frac{e_0}{2\pi}\Psi = \text{const.} \tag{1.124}$$

where $\Psi = 2\pi \int_0^r B_z(r, z) r\, dr$ is the magnetic flux through a circle of radius r. Suppose that the electron leaves the cathode at a distance r_c from the axis of symmetry with angular velocity $\dot{\theta}_c$. Then, according to Eq. (1.124),

$$mr^2\,\dot{\theta} - \frac{e_0}{2\pi}\Psi = m_c r_c^2\,\dot{\theta}_c - \frac{e_0}{2\pi}\Psi_c \tag{1.125}$$

We obtain the formula for angular velocity:

$$\dot{\theta} = \frac{m_c r_c^2}{mr^2}\,\dot{\theta}_c + \frac{e_0}{2\pi m r^2}(\Psi - \Psi_c) = \frac{\gamma_c}{\gamma}\left(\frac{r_c}{r}\right)^2\,\dot{\theta}_c + \frac{\eta}{2\pi\gamma r^2}(\Psi - \Psi_c) \tag{1.126}$$

This relation is known as *Busch's theorem*. In nonrelativistic approximation,

$$\dot{\theta} = \left(\frac{r_c}{r}\right)^2\,\dot{\theta}_c + \frac{\eta}{2\pi r^2}(\Psi - \Psi_c) \tag{1.127}$$

Busch's theorem for zero cathode velocity is

$$\dot{\theta} = \frac{\eta}{2\pi r^2}(\Psi - \Psi_c) \tag{1.128}$$

1.7.2 Formation of Helical Trajectories at a Jump in a Magnetic Field

The electron in an axially symmetric system is injected into a jump in the magnetic field so that in the plane $z = 0$ the magnetic field changes from $B = B_1$ to $B = B_2$ (Fig. 1.11). Before the jump ($z < 0$) the electron moves parallel to the z-axis. In the plane of the jump, the magnetic field is radial. One rotates the particle, and the latter continues to gyrate after the jump in the magnetic field B_2 along the helical trajectory. The magnetic fields on both sides of the jump are assumed uniform.

Let us find the parameters of the electron trajectory. For $z < 0$, $\dot{\theta} = 0$ and we can write Busch's theorem similar to Eq. (1.128):

$$\dot{\theta} = \frac{\eta}{2\pi r^2}(\Psi_2 - \Psi_1) \tag{1.129}$$

Ψ_1 and Ψ_2 are the magnetic fluxes through circles of radii r_1 and r_2, which are the distances of the electron from the z-axis before and just after the jump. We assume that the radial position of the particle does not change across the jump. Therefore, $r_1 = r_2 = r$ and

$$\Psi_1 = \pi r^2 B_1, \qquad \Psi_2 = \pi r^2 B_2 \tag{1.130}$$

The angular velocity according to Eqs. (1.129) and (1.130) is

$$\dot{\theta} = \frac{\eta}{2}(B_2 - B_1) \tag{1.131}$$

The azimuthal velocity is

$$v_\theta = r\dot{\theta} = \frac{\eta r}{2}(B_2 - B_1) = v_\perp \tag{1.132}$$

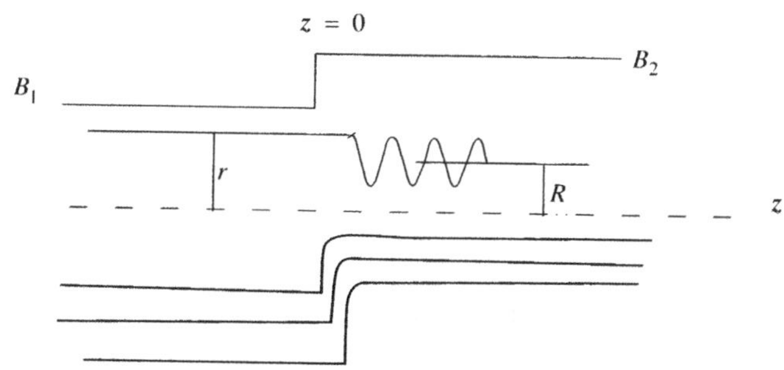

FIGURE 1.11 Jump of a magnetic field as the formation system of a helical trajectory.

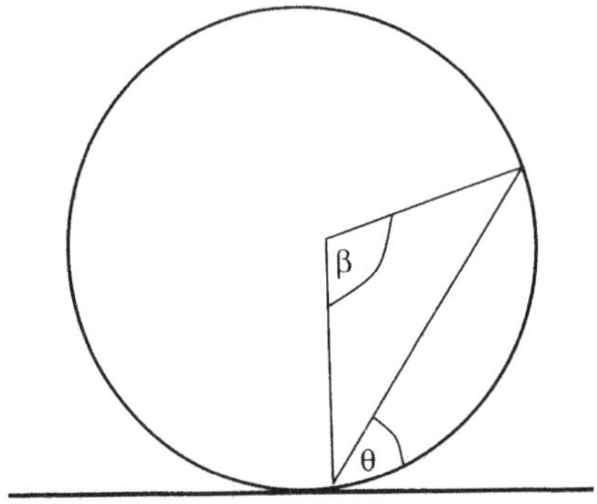

FIGURE 1.12 Motion of an electron injected from a region with zero magnetic field.

where $v_\perp$ is the velocity of the cyclotron gyration. The Larmor radius is

$$r_\perp = \frac{v_\perp}{\omega_{g2}} = \frac{v_\perp}{\eta B_2} = \frac{r}{2}\left(1 - \frac{B_1}{B_2}\right) \tag{1.133}$$

The guiding center radius (see Fig. 1.11) is

$$R = r - r_\perp = \frac{r}{2}\left(1 + \frac{B_1}{B_2}\right) \tag{1.134}$$

Thus, Busch's theorem allows us to solve the problem.

Let us consider two particular cases:

1. *Injection from a Region with Zero Magnetic Field.* $B_1 = 0$. We find from Eqs. (1.131)–(1.134) that

$$\dot{\theta} = \frac{\eta B_2}{2}, \qquad r_\perp = \frac{r}{2}, \qquad R = \frac{r}{2} \tag{1.135}$$

We see that the Larmor circle touches the z-axis. According to Fig. 1.12, angle θ of the turn of the radius vector about the axis equals the half-angle β of the Larmor radius turn. Therefore, $\dot{\theta} = \frac{1}{2}\dot{\beta} = \frac{1}{2}\omega_g$.

2. *Reverse (cusp) of the Magnetic Field.* $B_1 = -B_2$. We find from Eqs. (1.131), (1.133), and (1.134) that $\dot{\theta} = \eta B_2$, $r_\perp = r$, and $R = 0$. So in this case, after the jump the electron gyrates at cyclotron frequency around the guiding center located on the z-axis. The Larmor radius after the cusp is equal to the initial distance of the electron from the axis (Fig. 1.13). Axis-encircling helical electron beams that are formatted in the magnetic cusp are used last time in high-harmonic gyrotrons (Appendix 9, Idehara et al., 2004).

1.7.3 Systems with Plane Symmetry

Consider the Cartesian coordinate system in Fig. 1.14. By definition the plane symmetric system is the electromagnetic system that is uniform along one

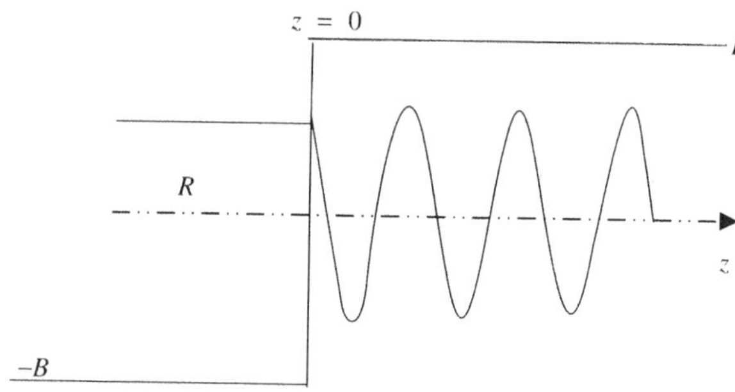

FIGURE 1.13 Motion of an electron through a magnetic cusp.

coordinate (e.g., in the y direction) and symmetrical about a plane parallel to this direction [e.g., the (Y, Z) plane]. It means that the Lagrangian is independent of y; that is,

$$\frac{\partial L}{\partial y} = 0 \tag{1.136}$$

and the potential and z-components of the fields are symmetrical functions of x:

$$\begin{aligned}
&\varphi(x) = \varphi(-x), \qquad && E_z(x) = E_z(-x), \qquad && B_z(x) = B_z(-x) \\
&A_y(x) = -A_y(-x), \qquad && E_x(x) = -E_x(-x), \qquad && B_x(x) = -B_x(-x) \\
&A_x = A_z = E_y = B_y = 0
\end{aligned} \tag{1.137}$$

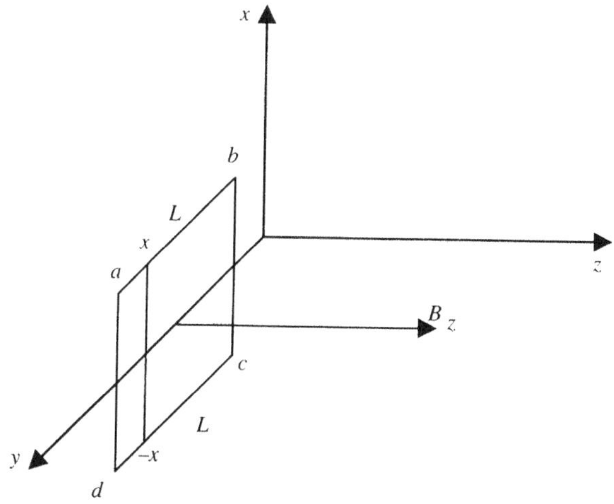

FIGURE 1.14 Derivation of Busch's theorem for a plane symmetrical system.

The generalized momentum for a plane symmetrical system according to Eq. (1.136) is

$$P_y = p_y - e_0 A_y = m\dot{y} - e_0 A_y = \text{const.} \tag{1.138}$$

Applying Eq. (1.3) and Stokes' theorem to the contour *abcd* (Fig. 1.14), we find that

$$\int B_z \, ds = \oint_{abcd} A_l \, dl \tag{1.139}$$

Taking into account the relations (1.137), we obtain $2L \int_0^x B_z \, dx = 2LA_y$; that is,

$$A_y = \Psi = \int_0^x B_z \, dx \tag{1.140}$$

where Ψ is the magnetic flux through a strip of width x and unit length in the y direction. So

$$m\dot{y} - e_0 \Psi = \text{const.} \tag{1.141}$$

Let us assume that the electron leaves the cathode at distance x_c from the plane of symmetry with the velocity $\dot{y}_c$. Then the velocity for the particle in the question is

$$\dot{y} = \dot{y}_c + \frac{\eta}{\gamma}(\Psi - \Psi_c) = \dot{y}_c + \frac{\eta}{\gamma}\left[\int_0^x B_z(x, z) \, dx - \int_0^{x_c} B_z(x, z_c) \, dx\right] \tag{1.142}$$

Equation (1.142) is a version of Busch's theorem for a plane symmetric system.

CHAPTER TWO

Electron Lenses

2.1 INTRODUCTION

The origin of electron optics dates back to the first quarter of the twentieth century, when it was discovered that electric and magnetic fields could be used as lenses for electron images. The term *electron optics* is not casual. The proximity of the fundamental Fermat principle in optics and the Maupertuis principle for particle trajectories, basic ideas of image formation, lens classification, the theory of aberrations, and even the terminology are thoroughly analogous in electron and light optics. It is significant that publication of de Broglie's famous paper which established wave–particle duality, and the first steps of electron optics, were almost simultaneous.

Generally speaking, many electron-optical devices are systems that produce spatial separation and focusing of charged particles in a beam according to a chosen criterion. So mass spectrometers separate the particles according to their masses (or the ratio of the mass to the charge) and bring them into focus on a screen. Beta-spectrometers handle an analogous problem for particles according to their energy. The main function of electron lenses is to prepare images of objects: to sort particles according to their points of exit from the object and their convergence to corresponding points of the image. Problems such as the transport of electron beams and beam concentration at small targets or channels cannot be solved without electron lenses. Naturally, electron lenses are components of almost all other electron-optical devices.

Two of the most important qualities of lenses must be provided.

1. *Stigmatic Imaging:* Rays emanating from any point of an "object" on plane O in Fig. 2.1 that are perpendicular to some line z called an *optical axis* must intersect after the lens at one point of the image in plane I, perpendicular to the z-axis. The beam

Electron Beams and Microwave Vacuum Electronics. By Shulim E. Tsimring

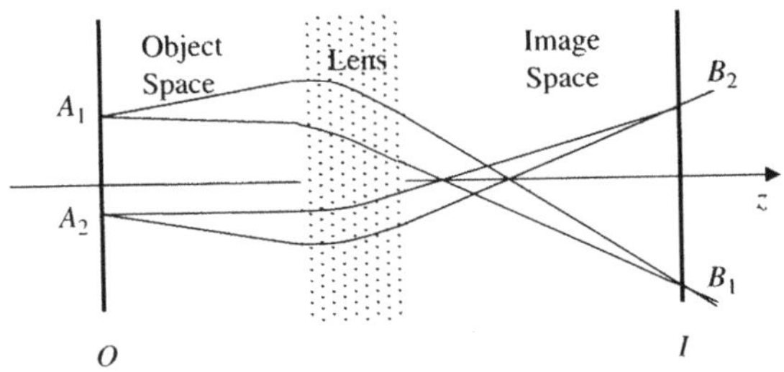

FIGURE 2.1 Stigmatic imaging.

formed by rays diverging from or converting to some point is called a *homocentric beam*. The lenses form stigmatic images if they support the homocentricity of beams. The stigmatic image ensures the sharpness of a picture.

2. *Geometric Similarity of an Image to an Object:* The distances between any points of an image are proportional to the distances between corresponding points of an object.

Below, a differential equation of electron trajectories is derived and discussed. The greatest attention is devoted to axially symmetric lenses. These lenses, among other extensive systems, have the maximum degree of symmetry. The majority of light optical systems are also axially symmetrical.

2.2 MAUPERTUIS'S PRINCIPLE. ELECTRON-OPTICAL REFRACTIVE INDEX. DIFFERENTIAL EQUATIONS OF TRAJECTORIES

2.2.1 Maupertuis's Principle. Differential Equations of Trajectories

Rewrite Hamilton's principle [Eq. (I.12)] using the Lagrangian form [Eq. (I.20)]

$$S_{\text{min}} = \int_{t_1}^{t_2} (\mathbf{Pv} - w)\, dt \tag{2.1}$$

Take into account the time independence of the energy for a static field. We can then narrow the class of comparable trajectories. Let us assume *equal* and *constant* particle energy. Then we obtain

$$S_{\text{min}} = \int_{t_1}^{t} \mathbf{Pv}\, dt - w(t_2 - t_1) \tag{2.2}$$

The second term is constant and may be omitted because all trajectories compared have the same energy. We obtain *Hamilton's principle for shortened action*:

$$S^*_{\text{sh}} = \int_{t_1}^{t_2} \mathbf{Pv}\, dt \tag{2.3}$$

In this integral, **P** and **v** depend on time only via the time-dependent coordinates of the particles [see. Eq. (I.2)]. The time may be excluded completely. Substituting $\mathbf{Pv}\,dt = \mathbf{P}\,d\mathbf{l}$, we obtain

$$S_M = \int_{\mathbf{r}_1}^{\mathbf{r}_2} \mathbf{P}\,d\mathbf{l} = \int_{\mathbf{r}_1}^{\mathbf{r}_2} (p\,dl - e_0\mathbf{A}\,d\mathbf{l}) = \int_{\mathbf{r}_1}^{\mathbf{r}_2} (p - e_0\mathbf{A}\mathbf{i}_l)\,dl \tag{2.4}$$

where $\mathbf{i}_l$ is the unit vector directed along the trajectory. We come to the variational Maupertuis principle: *Among all trajectories with given constant energy that connect points* $\mathbf{r}_1$ and $\mathbf{r}_2$, *only that trajectory which is real imparts to the integral* (2.4) *the least* (*in general, extreme*) *value*. Compare this principle with Fermat's principle that in light optics,

$$S_F = \int_{\mathbf{r}_1}^{\mathbf{r}_2} n\,dl = \text{extr.} \tag{2.5}$$

According to this principle, the optical path length S_F (the *eikonal*) is extreme for a real ray. Thus, the rays in the optical medium with an index of refraction $n(\mathbf{r}, \mathbf{r}') = n_{e_0}(\mathbf{r}, \mathbf{r}')$ have the same geometry as the trajectories in corresponding electromagnetic fields. The quantity

$$n_{e_0}(\mathbf{r}, \mathbf{r}') = p - e_0\mathbf{A}\mathbf{i}_l \tag{2.6}$$

is called the *electron-optical index of refraction*.

Let us express the momentum p in terms of the potential and electron initial energy. Take the cathode potential $\varphi_{\text{cath}} = 0$. Then, from the integral energy equation (1.3),

$$mc^2 = e_0\varphi + m_0c^2 + e_0\varphi_c = e_0(\varphi_0 + \varphi^*) \tag{2.7}$$

where

$$\varphi^* = \varphi + \varphi_c \tag{2.8}$$

and φ_c (V) is the reduced electron kinetic energy at the cathode:

$$\varphi_c = \frac{m_0c^2}{e_0}(\gamma_c - 1) = \frac{c^2}{\eta}\left(\frac{1}{\sqrt{1-\beta_c^2}} - 1\right)$$

Using Eq. (1.2), we obtain

$$p = m_0c\sqrt{\left(\frac{mc^2}{m_0c^2}\right)^2 - 1} = m_0c\sqrt{\frac{(\varphi_0 + \varphi^*)^2}{\varphi_0^2} - 1} = \frac{e_0}{c}\sqrt{\varphi^*(\varphi^* + 2\varphi_0)}$$
$$= \frac{e_0}{c}\Phi \tag{2.9}$$

where $e_0\varphi_0$ is the electron rest energy in electron volts, $\Phi = \sqrt{\varphi^*(\varphi^* + 2\varphi_0)}$. Then we have

$$n_{e_0} = e_0\left(\frac{1}{c}\Phi - \mathbf{A}\mathbf{i}_l\right) \tag{2.10}$$

A nonrelativistic index of refraction can be obtained by setting $\varphi_0 \gg \varphi^*$ and omitting the coefficient $\sqrt{2m_0e_0}$ in n_{e_0}:

$$n_{e_0}^{\text{nrel}} = \sqrt{\varphi^*} - \sqrt{\frac{\eta}{2}}\mathbf{A}\mathbf{i}_l \tag{2.11}$$

Differential Equations of Trajectories Let us select particle coordinate z as an independent variable. The other two coordinates are denoted q_1 and q_2. Then the trajectory can be written as

$$q_1 = q_1(z) \qquad q_2 = q_2(z)$$

The integral (2.4) is equal to

$$\begin{aligned} S_M &= \int_{z_1}^{z_2}\left(p\frac{dl}{dz} - e_0\frac{\mathbf{A}\,d\mathbf{l}}{dz}\right)dz = \int_{z_1}^{z_2}\left(\frac{e_0}{c}\Phi\sqrt{1 + q_1'^2 + q_2'^2} - e_0\frac{\mathbf{A}\,d\mathbf{l}}{dz}\right)dz \\ &= \int_{z_1}^{z_2} N\,dz \end{aligned} \tag{2.12}$$

where $d\mathbf{l} = \mathbf{i}_l\,dl$ and

$$N(q_1, q_2, q_1', q_2', z) = e_0\left(\frac{1}{c}\Phi\sqrt{1 + q_1'^2 + q_2'^2} - \frac{\mathbf{A}\,d\mathbf{l}}{dz}\right)$$

The necessary conditions of the extreme functional (2.7) are the Euler–Lagrange equations:

$$\begin{aligned} \frac{d}{dz}\left(\frac{\partial N}{\partial q_1'}\right) - \frac{\partial N}{\partial q_1} &= 0 \\ \frac{d}{dz}\left(\frac{\partial N}{\partial q_2'}\right) - \frac{\partial N}{\partial q_2} &= 0 \end{aligned} \tag{2.13}$$

These are differential equations of trajectories. N may be called a *reduced index of refraction*.

2.2.2 General Properties of Charged-Particle Trajectories in Electromagnetic Fields

Assume a static electromagnetic field in space with charged particles as an electron-optical medium. Consider some properties of the electron trajectories in this medium.

1. *Anisotropy:* The refraction index [Eqs. (2.10) and (2.11)] in the presence of a magnetic field depends not only on the coordinates, but also on the ray direction at a given point.

2. *Conservation of Trajectories:* If the refraction indicated in two media are linear dependent, the charged-particle trajectories in these media are the same. Assume that the initial energy of nonrelativistic particles on the cathode equals zero (i.e., $\varphi^* = \varphi$). Increasing the potentials in each point n times and the magnetic fields $\sqrt{n}$ times does not change the trajectories. Also, increasing the zero initial energy of the particles' proportional change of all electrode potentials n times and the currents in coils creating the magnetic field $\sqrt{n}$ times does not change the trajectories.

3. *Dependence of Trajectories on e_0 and m_0:* Nonrelativistic charged particles with equal specific charge η move on the same trajectories. The trajectories of nonrelativistic particles in an electrostatic field do not depend at all on e_0 and m_0.

The theory of electron lenses developed below does not take space-charge fields into account.

2.3 DIFFERENTIAL EQUATIONS OF TRAJECTORIES IN AXIALLY SYMMETRIC FIELDS

A cylindrical coordinate system is shown in Fig. 1.10. For axially symmetric fields, the coordinate θ is a cyclic variable. This means that φ^* and $\mathbf{A}$ are independent of θ; that is,

$$\varphi^* = \varphi^*(r, z), \qquad \mathbf{A} = \mathbf{A}(r, z) \tag{2.14}$$

For an axially symmetric system without space charge, the magnetic potential has only the component A_θ (Appendix 1), and according to Eq. (1.123),

$$rA_\theta = \frac{1}{2\pi}\Psi = \int_0^r rB_z(r, z)\,dr \tag{2.15}$$

where Ψ is the magnetic flux through a circle of radius r. The circle center is on the z-axis (Fig. 1.1). Then

$$\frac{\mathbf{A}\,d\mathbf{l}}{dz} = A_\theta r\theta' = \frac{\Psi}{2\pi}\theta' \tag{2.16}$$

The reduced index of refraction N is [see Eq. (2.12)]

$$N = e_0\left(\frac{1}{c}\Phi\sqrt{1 + r'^2 + r^2\theta'^2} - \frac{\Psi}{2\pi}\theta'\right) \tag{2.17}$$

where

$$\Phi = \Phi(r, z) = \sqrt{\varphi^*(\varphi^* + 2\varphi_0)} = \frac{c}{e_0}p \tag{2.18}$$

Because $\theta \equiv q_2$ is a cyclic variable, we obtain, according to the second of Eqs. (2.13),

$$\frac{\partial n_{e_0}}{\partial \theta'} = e_0\left(\frac{\Phi r^2}{c\sqrt{1 + r'^2 + r^2\theta'^2}}\theta' - \frac{\Psi}{2\pi}\right) = \text{const.} \tag{2.19}$$

Assume that a subscript c denotes cathode variables. Then, where all variables are supplied by c (taken at the cathode), the constant on the right-hand side of Eq. (2.19) will equal the left-hand side and we obtain the following relation for θ':

$$\frac{\theta'}{\sqrt{1 + r'^2 + r^2\theta'^2}} = \frac{1}{\sqrt{1 + r_c'^2 + r_c^2\theta_c'^2}}\frac{\Phi_c}{\Phi}\left(\frac{r_c}{r}\right)^2\theta_c' + c\frac{\Psi - \Psi_c}{2\pi\Phi r^2} \tag{2.20}$$

where

$$\Phi_c = \Phi_{\varphi=0} = \sqrt{\varphi_c(\varphi_c + 2\varphi_0)} \tag{2.21}$$

[see Eqs. (2.18) and (2.8)]. Equation (2.20) is a version of Busch's theorem [Eq. (1.126)] for trajectories. Substitution of n_{e_0} from Eq. (2.17) in the first of Eqs. (2.13), where $r \equiv q_1$ gives

$$\frac{d}{dz}\left(\frac{\Phi r'}{\sqrt{1+r'^2+r^2\theta'^2}}\right) - \sqrt{1+r'^2+r^2\theta'^2}\frac{\partial\Phi}{\partial r} - \frac{\Phi r\theta'^2}{\sqrt{1+r'^2+r^2\theta'^2}} + crB_z\theta' = 0 \tag{2.22}$$

Equations (2.20) and (2.22) are *differential equations of trajectories in axially symmetric fields*. Substituting θ' from Eq. (2.20) in Eq. (2.22), we obtain an ordinary second-order differential equation for the function $r = r(z)$. Substitution of $r(z)$ in Eq. (2.20) gives a first-order differential equation for $\theta = \theta(z)$. The following initial conditions determine the solution:

$$r_c = r_{z=0}, r_c', \theta_c, \theta_c', \varphi_c, \Psi_c = 2\pi\int_0^{r_c} rB_z(r, 0)\,dr \tag{2.23}$$

These equations are too complicated and we will not write them out explicitly. Without further strong approximations, only their numerical solutions can be found. Fortunately, the beams considered are not suitable for obtaining images. The light optics gives clear proof of this assertion: Almost all optical tools have essentially bound-beam apertures. Expansion of the apertures always increases aberrations, whose minimization is very labor-intensive.

In principle, only paraxial (close to the optical axis) trajectories (rays) can give stigmatic and similar images. Therefore, the theory of image construction in electron (light) optics in the first approximation is formulated as paraxial theory. Next, approximations lead to aberration theory.

2.4 DIFFERENTIAL EQUATIONS OF PARAXIAL TRAJECTORIES IN AXIALLY SYMMETRIC FIELDS WITHOUT A SPACE CHARGE

The following conditions define the paraxial trajectory.

1. *Small Transversal Velocities:* (i.e., a small slope of the trajectory to the z-axis)

$$\frac{dr}{dt} \ll \frac{dz}{dt} \quad \frac{dr}{dz} = r' \ll 1, \qquad r\frac{d\theta}{dt} \ll \frac{dz}{dt} \quad r\frac{d\theta}{dz} = r\theta' \ll 1 \tag{2.24}$$

Then

$$\sqrt{1 + r'^2 + r^2\theta'^2} \approx 1 \tag{2.25}$$

2. *Small Distances r of Trajectories from the Axis:* This means that trajectories deviate from the z-axis by no more than the distance at which the potential and magnetic field are equal to their axis values:

$$\varphi(r, z) \approx \varphi(0, z) \equiv \varphi_z(z) \tag{2.26}$$

Equation (2.26) is also applicable to the *axis components of electric and magnetic fields*:

$$\begin{aligned} E_z(r, z) &= -\frac{\partial\varphi(r, z)}{\partial z} \approx -\frac{d\varphi(0, z)}{dz} \equiv -\varphi_z'(z) = E_z(0, z) \\ B_z(r, z) &\approx B_z(0, z) \equiv B(z) \end{aligned} \tag{2.27}$$

The conditions of Eq. (2.27) are not applicable to radial components of fields because these components are zero on the axis. The radial components in the paraxial approximation are followed from Maxwell's equation for an axially symmetric electric field [Eq. (I.7)] with $\rho = 0$:

$$\operatorname{div}\mathbf{E} = \frac{1}{r}\frac{\partial(rE_r)}{\partial r} + \frac{dE_z(0, z)}{dz} = 0 \tag{2.28}$$

Integrating Eq. (2.28), we obtain $rE_r = -(r^2/2)E_z'(z) + f(z)$, where $f(z)$ is an arbitrary function and

$$E_r = -\frac{r}{2}E_z'(z) + \frac{f(z)}{r}$$

For $f(z) \neq 0$, the field has a singularity on the axis. That is possible only if there are line charges on the z-axis. So $f(z) = 0$, and the *radial electric field* in the paraxial approximation is

$$E_r = -\frac{\partial \varphi}{\partial r} = -\frac{r}{2} E_z'(z) = \frac{r}{2} \varphi_z''(z) \tag{2.29}$$

Then, according to Eqs. (2.18) and (2.26), we obtain

$$\frac{\partial \Phi}{\partial r} = \frac{\varphi^* + \varphi_0}{\sqrt{\varphi^*(\varphi^* + 2\varphi_0)}} \frac{\partial \varphi}{\partial r} = -\frac{r}{2} \frac{\varphi_z^* + \varphi_0}{\sqrt{\varphi_z^*(\varphi_z^* + 2\varphi_0)}} \varphi_z'' = -\frac{r}{2\beta_z} \varphi_z'' \tag{2.29a}$$

where $\beta_z = v_z/c$.

Using Maxwell's equation (I.8) and assuming the absence of magnetic charges on the axis, we obtain an analogous expression for the *radial magnetic field*:

$$B_r = -\frac{r}{2} B'(z) \tag{2.30}$$

Magnetic fluxes in the paraxial approximation are obviously

$$\Psi = \pi r^2 B(z), \qquad \Psi_c = \pi r_c^2 B(0) \tag{2.31}$$

Equation (2.31) is valid for a cathode located on the plane perpendicular z-axis. Substitution of Eqs. (2.25)–(2.27) and (2.31) in Eqs. (2.20) and (2.22) gives

$$\theta' = \frac{c}{2\Phi_z} \left[Q \frac{r_c^2}{r^2} + B(z) \right] \tag{2.32}$$

$$\frac{d}{dz}(\Phi_z r') - \frac{\partial \Phi}{\partial r} - \Phi_z r \theta'^2 + crB(z)\theta' = 0 \tag{2.33}$$

where, taking into account Eqs. (1.7), (1.11), and (2.18), we find that

$$\Phi_z \equiv \sqrt{\varphi_z^*(\varphi_z^* + 2\varphi_0)} = \beta_z(\varphi_z^* + \varphi_0) = \beta_z \gamma \varphi_0 = \beta_z \gamma \frac{c^2}{\eta} = \frac{c}{e_0} p_z \tag{2.34}$$

$$Q = \frac{2}{c} \Phi_c \theta_c' - B(0) \tag{2.35}$$

It is readily verified that according to Eqs. (1.124) and (2.32),

$$Q = \frac{2P_\theta}{e_0 r_c^2} = -\frac{\Psi_0}{\pi r_c^2} \tag{2.35a}$$

where P_θ is the azimuthal generalized momentum and Ψ_0 is the magnetic flux in the plane where $\dot{\theta} = 0$. Substitution of Eq. (2.32) for the angular velocity, and Eqs. (2.29a) and (2.34) into Eq. (2.33), gives a radial paraxial equation

$$\frac{\Phi_z^2}{\varphi_z^* + \varphi_0}\frac{r''}{r_c} + \varphi_z'\frac{r'}{r_c} + \left[\frac{\varphi_z''}{2} + \frac{c^2 B(z)^2}{4(\varphi_z^* + \varphi_0)}\right]\frac{r}{r_c} - \frac{c^2 Q^2}{4(\varphi_z^* + \varphi_0)}\frac{r_c^3}{r^3} = 0 \qquad (2.36a)$$

Equations (2.32) and (2.36a) in terms $\gamma = (\varphi_z/\varphi_0) + 1$ and $\Psi_0 = -\pi r_c^2 Q$ can be written as

$$\theta' = \frac{\eta B}{2\beta_z c\gamma} - \frac{\eta\Psi_0}{2\pi\beta_z c\gamma}r''$$
$$+\frac{\gamma'}{\beta_z^2\gamma}r' + \left[\frac{\gamma''}{2\beta_z^2\gamma} + \left(\frac{\eta B}{2\beta_z c\gamma}\right)^2\right]r - \left(\frac{\eta\Psi_0}{2\pi\beta_z c\gamma}\right)^2\frac{1}{r^3} = 0 \qquad (2.36b)$$

These equations, according to Eq. (2.35a) in the term P_θ, are

$$\theta' = \frac{\eta B}{2\beta_z c\gamma} + \frac{P_\theta}{\beta_z c\gamma m_0}\frac{1}{r^2}r''$$
$$+\frac{\gamma'}{\beta_z^2\gamma}r' + \left[\frac{\gamma''}{2\beta_z^2\gamma} + \left(\frac{\eta B}{2\beta_z c\gamma}\right)^2\right]r - \left(\frac{P_\theta}{\beta_z c\gamma m_0}\right)^2\frac{1}{r^3} = 0 \qquad (2.36c)$$

Equations (2.32) and (2.36) are the *fundamental equations of paraxial electron optics* without a space charge for axially symmetric systems.

If the cathode is disposed at a plane perpendicular to the z-axis, the following initial conditions at the cathode ($z = 0$) must be set: $r_c = r(0)$, $\theta_c = \theta(0)$, $r_c' = r'(0)$, θ_c', and φ_c. The initial angular velocity θ_c' and the energy φ_c enter correspondingly in Q and in φ_z or γ.

Nonrelativistic paraxial equations can be obtained by setting $\varphi^* \ll \varphi_0 = m_0 c^2/e_0 = c^2/\eta$ in Eqs. (2.32) and (2.36):

$$\theta' = \sqrt{\frac{\eta}{8\varphi_z^*}}\left[Q_n\frac{r_c^2}{r^2} + B(z)\right], \qquad Q_n = 2\sqrt{\frac{\varphi_z^*}{\eta}}\theta_c' - B(0) \qquad (2.37)$$

$$\frac{r''}{r_c} + \frac{\varphi_z'}{2\varphi_z^*}\frac{r'}{r_c} + \left[\frac{\varphi''}{4\varphi_z^*} + \frac{\eta B(z)^2}{8\varphi_z^*}\right]\frac{r}{r_c} - \frac{\eta Q_n^2}{8\varphi_z}\frac{r_c^3}{r^3} = 0 \qquad (2.38)$$

2.5 FORMATION OF IMAGES BY PARAXIAL TRAJECTORIES

2.5.1 Linearization of Trajectory Equations

Later it will be shown that the linearity of trajectory differential equations is a *sufficient condition* for the formation of similar and stigmatic images. From a formal mathematical standpoint, derivation of the paraxial ray equation is a linearization procedure for the original system. Unfortunately, this goal is not achieved! According to Eqs. (2.32) and (2.36)–(2.38), the nonlinear terms remain. What is the reason? Nonlinearities disappear only for $Q = 0$. According to Eq. (2.35), it is possible when the cathode magnetic field $B(0)$ and the cathode angular velocity θ'_c are zero simultaneously. But the initial energy φ_c can be nonzero. In this case (if $C = 0$), the linearity of the equation is not violated. It means that only the r and z components of the initial velocities are admissible (i.e., the trajectory has to lie in a meridional plane).

However, the linearity of the original trajectory equations and of meridional trajectories are *not necessary conditions*. It is shown (see, e.g., Glazer, 1952; Kelman and Yavor, 1963; Klemperer and Barsnett, 1971; Lawson, 1988; Rusterholz, 1950; Szilagyi, 1988) that the proper transformation of variables can linearize the equations. After this transformation, the trajectories become more complicated and do not lie in the meridional plane. Nevertheless, the fundamental possibility of obtaining stigmatic and similar images is restored. This result is very important because it means that arbitrary axial symmetric fields in the approximation of paraxial optics are lenses. It is interesting that after the transformation, the new equation has the same form (in reduced variables) as the original but with $Q = 0$. Therefore, below we choose the condition $Q = 0$ as a simple approximation.

The corresponding equations are now *relativistic equations*:

$$\theta' = \frac{cB(z)}{2\Phi_z} \tag{2.39}$$

$$\frac{\Phi_z^2}{\varphi_z^* + \varphi_0}\frac{r''}{r_c} + \varphi'_z\frac{r'}{r_c} + \left[\varphi''_z + \frac{c^2B^2(z)}{4(\varphi_z^* + \varphi_0)}\right]\frac{r}{r_c} = 0 \tag{2.40}$$

Nonrelativistic equations will be

$$\theta' = \sqrt{\frac{\eta}{8\varphi_z^*}}B(z) \tag{2.41}$$

$$\frac{r''}{r_c} + \frac{\varphi'_z}{2\varphi_z^*}\frac{r'}{r_c} + \left[\frac{\varphi''_z}{4\varphi_z^*} + \frac{\eta B^2(z)}{8\varphi_z^*}\right]\frac{r}{r_c} = 0 \tag{2.42}$$

2.5.2 Rotation of an Image. Stigmatic Imaging. Image Similarity

Rotation of an Image Assume that a trajectory came to point $r(b)$ from point $r(a)$. According to Eqs. (2.39) and (2.41), the rotation angle is

$$(\theta_b - \theta_a)_{\text{rel}} = \frac{c}{2}\int_a^b \frac{B(z)\,dz}{\Phi_z} \tag{2.43}$$

$$(\theta_a - \theta_b)_{\text{nrel}} = \int_a^b \sqrt{\frac{\eta}{8\varphi_z^*}}B(z)\,dz \tag{2.44}$$

Because Φ_z and φ_z^* are functions of z alone, we have an important result: Any trajectory going between two planes perpendicular to the optical z-axis rotates by the same angle. Assume that the stigmatic image of the object has gotten into a plane. Then the entire image will be turned about the object without angular distortion. The rotation angle is the linear function of B, and reversing the sign of B changes the rotation to the opposite direction. The rotation angle depends only on the specific charge $\eta = e_0/m_0$ for the same initial energy of the trajectories. This property is also valid for relativistic particles because $\varphi_0 = c^2/\eta$ [see Eq. (2.34)].

Stigmatic Imaging and Image Similarity Turn to the radial equations (2.40) and (2.42). Note that the radial variable in these equations is the ratio r/r_c. It means that trajectories emanating from different points of the cathode are similar geometrically if their initial angles r' are the same (e.g., $r' = 0$ when trajectories start normally to the cathode plane). All of these trajectories can be crossed only at the axis ($r = 0$).

Both equations are ordinary linear homogeneous differential equations of the second order. As is known, the general solution of these equations may be represented as

$$r(z) = Pr_1(z) + Qr_2(z) \tag{2.45}$$

where P and Q are arbitrary constants. Functions r_1 and r_2 are linearly independent partial solutions of the trajectory equations; that is, the relation

$$r_1(z) = kr_2(z) \tag{2.46}$$

is prohibited. Geometrically, it means that any trajectories expressed by the partial solutions must not be similar to each other.

Assume that a trajectory $r_1(z)$ crosses the z-axis in two points, a and b (Fig. 2.2). Consider the bundle of trajectories

$$r(z) = Pr_1(z) \tag{2.47}$$

with an arbitrary P. All of these trajectories pass through points a and b. These points differ only by the slope [i.e., by values of the derivative $r'(z)$]. If point a is

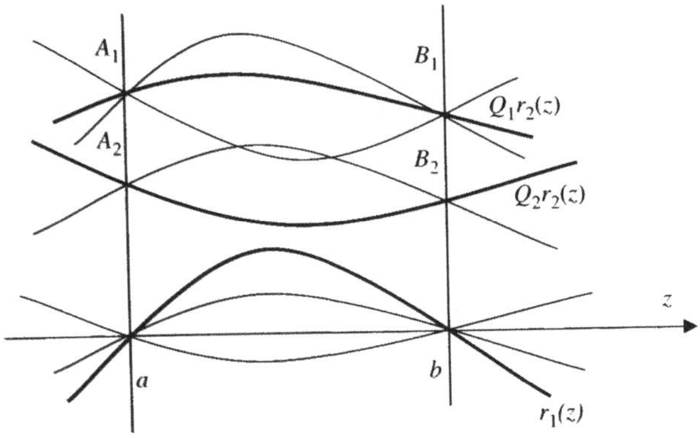

FIGURE 2.2 Stigmatic and similar images.

an object, point b will be the stigmatic image of this object. Can it be that any trajectory that crosses the z-axis only at point a does not cross it at point b? It is easy to see that such a solution would contradict the Cauchy uniqueness theorem (by changing the constant P, we sample all possible values of the first derivative of the function in point a).

Let us take another solution of the trajectory equation $r_2(z)$ that does not pass through the points $z = a$ and $z = b$. Obviously, the functions $r_2(z)$ and $r_1(z)$ are linearly independent. Now add to the trajectory $Q_1r_2(z)$ passing through points A_1 and B_1 the bundle of the trajectories [Eq. (2.47)]. We obtain the bundle of trajectories that intersect the planes $z = a$ and $z = b$ at the points A_1 and B_1, respectively (Fig. 2.2):

$$R_1(z) = Pr_1(z) + Q_1r_2(z) \tag{2.48}$$

Repeating the argumentation above, we conclude that point B_1 is a stigmatic image of point A_1. Now take the bundle of trajectories

$$R_2(z) = Pr_1(z) + Q_2r_2(z) \tag{2.49}$$

where $Q_2r_2(z)$ is a trajectory that passes through points A_2 and B_2. We see that point B_2 will be the stigmatic image of point A_2. As a result, trajectories that are described by Eq. (2.40) or (2.42) produce in plane $z = b$ the stigmatic image of the object in plane $z = a$.

Now let us show the geometrical similarity of the image to the object. Because $r_1(a) = r_1(b) = 0$, we obtain

$$R_1(a) = Q_1r_2(a), \quad R_1(b) = Q_1r_2(b), \quad R_2(a) = Q_2r_2(a), \quad R_2(b) = Q_2r_2(b) \tag{2.50}$$

Dividing the first relation from Eqs. (2.50) by the third, and the second relation by the fourth, we obtain

$$\frac{R_1(a)}{R_2(a)} = \frac{R_1(b)}{R_2(b)} = \frac{Q_1}{Q_2} \tag{2.51}$$

The ratio of the radial positions of any two points of the object in the plane perpendicular to the z-axis is equal to that for corresponding points of the image. Previously, it was shown that all trajectories in the image plane rotate by the same angle around the object plane. So the similarity of the image and the object has been proven. Note that the first formulation of these electron-optical relations belongs to Busch (1926).

2.5.3 Magnifications

According to Eq. (2.51), the quantity

$$M = \frac{R_1(b)}{R_1(a)} = \frac{R_2(b)}{R_2(a)} = \text{const.} \tag{2.52}$$

The constant M determines a similar change in radial sizes and is called *magnification*. It is the image parameter for a given position of the object around the lens.

Note that when a trajectory crosses the axis, we change the sign of the variable r, although as the radial coordinate of the cylindrical system, it must be positive. This difficulty may be overcome in the following way. As we saw, trajectories rotate in the rotating meridional planes. Motion in the meridional plane is described by the linear homogenous equation (2.40) or (2.42), which is satisfied by any sign of r. If we hadn't used the paraxial approximation, we could have taken $r < 0$ and defined the electromagnetic fields for $r < 0$ as the analytical continuation of the fields at $r > 0$. The axis potential in the paraxial equations is independent of r. Therefore, we can use the trajectory equations with any sign of r, assuming that we operate with analytical continuation of the fields. In this case, M may by either positive or negative. For $M < 0$ the image will be inverted, respectively, to the object in the meridian plane.

Angular Magnification Assume that the trajectory $r_1(z)$ intersects the z-axis in the points $z = a$ and $z = b$. Any trajectory that intersects the z-axis in these points is expressed as

$$r(z) = Pr_1(z) \tag{2.53}$$

where P is an arbitrary constant. We find from Eq. (2.53) that the quantity

$$G = \frac{r'(b)}{r'(a)} = \frac{r_1'(b)}{r_1'(a)} \tag{2.54}$$

is the same for all trajectories passing through the points a and b on the optical axis. It is called *angular magnification* and is also the image parameter for a given position of the object.

Correlations M and G Represent the differential radial equation of the trajectory [Eq. (2.40)] as

$$\Theta[r(z)] = F(z)r'' + \tfrac{1}{2}F'(z)r' + \Psi(z)r = 0 \tag{2.55}$$

where

$$F(z) = \varphi_z^*(\varphi_z^* + 2\varphi_0) = \Phi_z^2, \qquad \Psi(z) = \frac{\varphi_z''(\varphi_z^* + \varphi_0)}{2} + \frac{c^2B^2(z)}{4} \tag{2.55a}$$

We now form the combination

$$\begin{aligned} r_1(z)\Theta[r_2(z)] - r_2(z)\Theta[r_1(z)] &= F(z)(r_2''r_1 - r_1''r_2) + \tfrac{1}{2}F'(z)(r_2'r_1 - r_1'r_2) \\ &= 0 \end{aligned} \tag{2.56}$$

Taking it into account that

$$r_2''r_1 - r_1''r_2 = \frac{d}{dz}(r_2'r_1 - r_1'r_2)$$

we obtain from Eq. (2.56),

$$\frac{\dfrac{d}{dz}(r_2'r_1 - r_1'r_2)}{r_2'r_1 - r_1'r_2} + \frac{\dfrac{d}{dz}F(z)}{2F(z)} = 0 \tag{2.57}$$

Integrating this equation gives

$$\sqrt{F(z)}\,(r_2'r_1 - r_1'r_2) = \Phi_z(r_2'r_1 - r_1'r_2) = \text{const.} \tag{2.58}$$

Assume that the trajectory $r_2(z)$ crosses the z-axis in the points $z = a, b$. The trajectory $r_1(z)$ does not pass through these points. Applying Eq. (2.58) to planes $z = a$ and $z = b$, we obtain

$$\Phi_z(b)r_2'(b)r_1(b) = \Phi_z(a)r_2'(a)r_1(a) \tag{2.59}$$

so that

$$\frac{r_2'(b)\,r_1(b)}{r_2'(a)\,r_1(a)} = \frac{\Phi_z(a)}{\Phi_z(b)} \tag{2.60}$$

According to the definitions of M [Eq. (2.52)] and G [Eq. (2.54)], Eq. (2.60) may be represented as

$$GM = \frac{\Phi_z(a)}{\Phi_z(b)} = \frac{n_{e_0}(a)}{n_{e_0}(b)} \tag{2.61}$$

For nonrelativistic beams,

$$GM = \sqrt{\frac{\varphi_z^*(a)}{\varphi_z^*(b)}} = \frac{n_{e_0}(a)}{n_{e_0}(b)} \tag{2.62}$$

Equations (2.61) and (2.62) are known as the *Helmholtz and Lagrange formulas*. The original light optics versions of the formulas are the *Helmholtz and Lagrange theorems*. It is obvious that if the trajectory crossed the z-axis twice, G would be negative. In this case, the magnification M is also negative (i.e., the image is inverted in the meridional plane). We note that the right-hand sides of formulas (2.60) and (2.61) are equal to the ratios of corresponding electrostatic paraxial indexes of refraction $n_{e_0}(a)/n_{e_0}(b)$ [see Eqs. (2.10) and (2.11) for $r = 0$ and $\mathbf{A} = 0$].

2.6 ELECTROSTATIC AXIALLY SYMMETRIC LENSES

2.6.1 Classification of Electrostatic Lenses

We have written relativistic and nonrelativistic equations of paraxial trajectories in the electrostatic field, setting $B = 0$ in Eqs. (2.39), (2.40), and (2.42):

$$\theta' = 0 \tag{2.63}$$

$$\Phi_z^2 \frac{r''}{r_c} + \varphi_z'(\varphi_z^* + \varphi_0)\frac{r'}{r_c} + \frac{\varphi_z''}{2}(\varphi_z^* + \varphi_0)\frac{r}{r_c} = 0 \qquad \text{(relativistic)} \tag{2.64}$$

$$\frac{r''}{r_c} + \frac{\varphi'}{2\varphi_z^*}\frac{r'}{r_c} + \frac{\varphi_z''}{4\varphi_z^*}\frac{r}{r_c} = 0 \qquad \text{(nonrelativistic)} \tag{2.65}$$

As shown, the nonrelativistic equation has two important features that are related to the absence of the reduced rest energy $\varphi_0 = c^2/\eta$ in Eq. (2.65), unlike Eq. (2.64).

1. The trajectories geometry of any charged particle is independent of its charge and mass.

2. Proportional variation of the potential in all points of the lens space does not affect the geometry of the trajectories. It means that a proportional variation of the electrode potentials of all the lens does not influence the trajectory.

According to Eq. (2.63), $\theta = \text{const.}$ (i.e., particles move in nonrotational meridional planes).

It may readily be shown that the *lens effect* is possible only in an inhomogeneous electric field. The lens effect is displayed by the rays emerging in the lens parallel to the optical axis. The real lens necessarily deflects the rays to or out of the axis. Let us rewrite Eq. (2.64) in the following form [see Eq. (2.40)]:

$$\Phi_z \frac{d}{dz}\left(\Phi_z \frac{r'}{r_c}\right) + \frac{\varphi_z''(\varphi_z^* + \varphi_0)}{2}\frac{r}{r_c} = 0 \tag{2.66}$$

Assume that the lens electric field is homogeneous [i.e., $\varphi'(z) = \text{const.}$]. Then $\varphi_z'' = 0$ and we obtain from Eq. (2.66),

$$\Phi_z \frac{r'}{r_c} = \text{const.} \tag{2.67}$$

Thus, if the trajectory in a plane is parallel to the optical axis ($r' = 0$), it would not be deflected at all. So the electrostatic lens space is localized in the region of the inhomogeneous electric field.

Let us suppose that the object and image are located outside the lens (i.e., in the region of the homogeneous or zero electric field). It is impossible to detect strict boundaries of any electron lens (unlike optic lenses) because fields, including inhomogeneous fields in open structures, have, in principle, infinite extent. But this issue is rather abstract because it is always possible to use as a criterion the existence of almost straight trajectories near the object and the image. Besides, for typical axially symmetric electrode configurations, edged fields decrease abruptly. We specify the boundary planes of the lens as $z = z_a$ and $z = z_b$ (Fig. 2.3). The regions $z < z_a$ and $z > z_b$ will be called an *object space* and an *image space*, respectively. All quantities in the object space will be denoted by a subscript a and accordingly in the image space by a subscript b. Note that inside the lens, the potential distribution as a function of z is displayed by lines with a nonzero curvature.

The electron optics interact with a huge variety of electrostatic lens devices. The choice of a particular device depends on many factors: purpose, construction

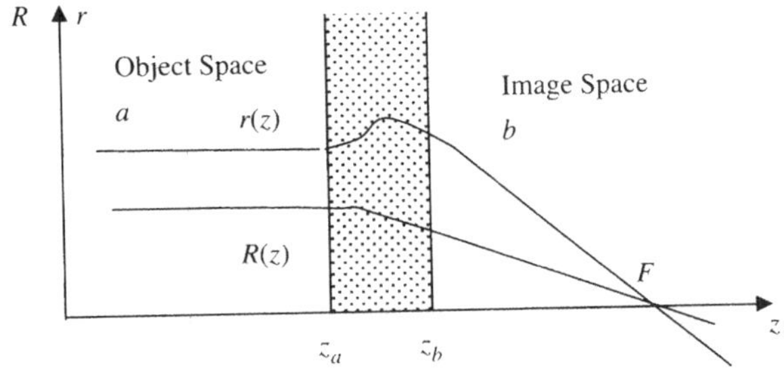

FIGURE 2.3 Object and image spaces in a lens. Trajectories $r(z)$ and $R(z)$ display a convergent effect of the immersion lens [see Eqs. (2.68) and (2.69)].

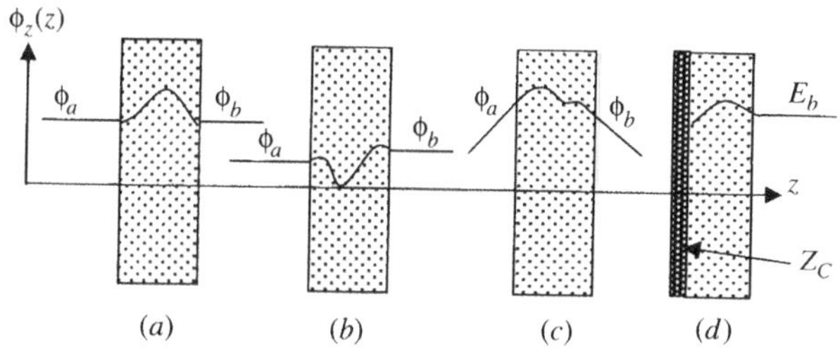

FIGURE 2.4 Axial distributions of the potential in lenses: (*a*) unipotential lens ($\varphi_a = \varphi_b$); (*b*) immersion lens ($\varphi_a \neq \varphi_b$); (*c*) aperture lens; (*d*) cathode lens.

requirements, power supply, electrical breakdown, aberrations, and so on. It is difficult to classify lenses according to all these factors. As the main criterion, the structure of the axis potential distribution in the object and image spaces is usually taken. Following are the basic types of electrostatic lenses:

1. *Lenses with a Limited Field Extent*: The potential outside the lenses is constant (the field is zero). These lenses can be divided into two categories: (a) *unipotential* (*Einzel*) *lenses*, which have equal constant potentials in the object and image spaces, and (b) *immersion* (*bipotential*) *lenses*, which have different constant potentials in these spaces.

2. *Aperture Lenses*: A nonzero (homogeneous) field exists in at least one space (the object space or image space).

3. *Cathode Lenses*: The inhomogeneous field of the lens begins directly from the cathode surface.

In Fig. 2.4, the axis potential distributions for these lenses are shown. The basic properties of the lenses are considered briefly below.

2.6.2 Immersion and Unipotential Lenses

Note first that immersion and unipotential lenses are focusing (i.e., the trajectories that parallel the optical axis in the object space are collected at the axis in the image space, and vice versa). This effect can be demonstrated by the example of the simplest immersion lens formed by two half-cylinders with different potentials. In Fig. 2.5, the equipotential surfaces are shown by thin lines. The electric fields (arrows in Fig. 2.5) are directed normally to equipotential surfaces and point to the right.

The radial field in the open area of the left half-cylinder increases the radial velocity of the electron. This part of the electron lens acts as a diverging light lens. The field on the right-hand side of the lens produces a contrary effect. This part is analogous to a convergent light lens. But the focusing effect prevails over the defocusing effect because the axial velocity of the electron at the right-hand side is less than that

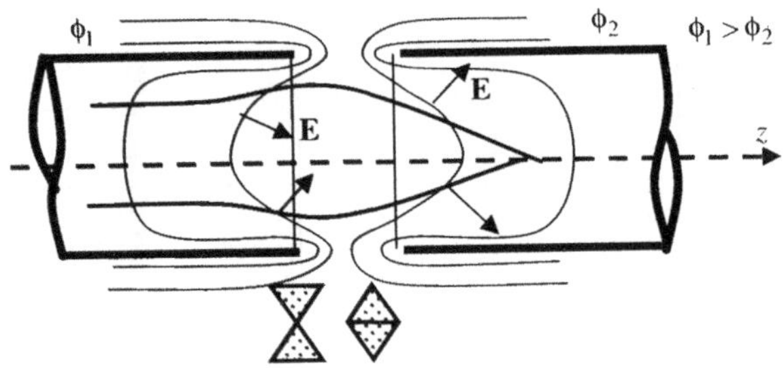

FIGURE 2.5 Focusing effect of an immersion lens.

at the left-hand side (the potential $\varphi_1 > \varphi_2$). The particle spends more time in the right half-cylinder than in the left one. Note that for this reason, the final radial velocity is proportional to the square of the field (second-order focusing).

Below it is shown that *any* electrostatic lens with limited field extent is convergent. Let us introduce the following variable in Eq. (2.64):

$$r = R\Phi_z^{1/2} = R[\varphi^*(\varphi^* + 2\varphi_0)]^{-1/4} \tag{2.68}$$

We obtain the equation of the paraxial trajectory in the following form:

$$R'' + \frac{1}{4}\left[\frac{\varphi_z'^2[(\varphi_z^* + \varphi_0)^2 + 2\varphi_0^2]}{(\varphi_z^{*2} + 2\varphi_z^*\varphi_0)^2}\right]R = 0 \tag{2.69}$$

Let us take the ray $r(z)$ parallel to the z-axis in the object space (Fig. 2.3). Because the potential $\varphi_z^*(z)$ is constant in the object space, the curve $R(z)$ is parallel to the z-axis also. In the lens, the curvature of $R(z)$ is negative because the coefficient in brackets in Eq. (2.69) is essentially positive and the curve $R(z)$ is convex. Therefore, after the lens, the straight line $R(z)$ intersects the z-axis (point F in Fig. 2.3). According to Eq. (2.68), the trajectory $r(z)$ is also a straight line in the image space and crosses the z-axis at the same point, F. In Fig. 2.6, the simplest specific schemes of unipotential and immersion lenses are presented.

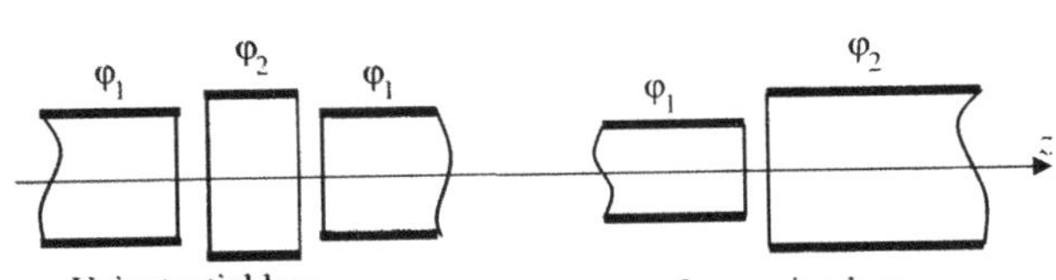

FIGURE 2.6 Lenses with limited field extent.

2.6.3 Cardinal Elements of a Lens with Limited Field Extent

The fundamental idea of studying the image properties of light lenses on the basis of cardinal elements was suggested by Gauss. He introduced the idea that cardinal elements could describe a lens (in a paraxial rays approximation) by a small number of parameters and lead to a unified and clear theory of image formation for lens systems of any degree of complexity. Below we define the cardinal elements for electrostatic lenses with a limited field extent when an object and an image are on the outside of the lens field. Consider a trajectory $r_1(z)$ that is parallel to the z-axis in the object's space and intersects the z-axis in point F (Fig. 2.7). Obviously, the equation $r_c(z) = cr_1(z)$ represents a bundle of trajectories that is also parallel to the z-axis and crosses it at point F. The trajectories are straight lines outside the lens ($z < z_1$ and $z > z_2$). Let us take the trajectory $r_2(z) = c_2 r_1(z)$ and move it backward from plane $z = z_b$ as a straight line (dashed line in Fig. 2.7). The distance Z_b from the plane $z = z_b$ to point F_b, where this line crosses the z-axis, is

$$Z_b = z_b - z_{F_b} = \frac{r_2(z_b)}{\tan\alpha} = \frac{r_2(z_b)}{r_2'(z_b)} = \frac{c_2 r_1(z_b)}{c_2 r_1(z_b)} = \frac{r_1(z_b)}{r_1'(z_b)} \tag{2.70}$$

As we see, Z_b does not depend on c_1. So any trajectory of the bundle $r_c(z)$ after continuing backward from the image space intersects the z-axis at the same point, F_b. This point is called the *principal focus of the image space*. In the same way, it can readily be shown that these trajectories intersect their continuations from the object space on the common plane H_b. This plane, $z = z_{H_b}$, is called the *principal plane of the image space*. The distance $f_b = z_{F_b} - z_{H_b}$ is called the *principal focal length of the image space*. By drawing the trajectories parallel to the z-axis in the image space, we can obtain analogous cardinal elements of the object space: F_a, z_{F_a}, z_{H_a}, and f_a.

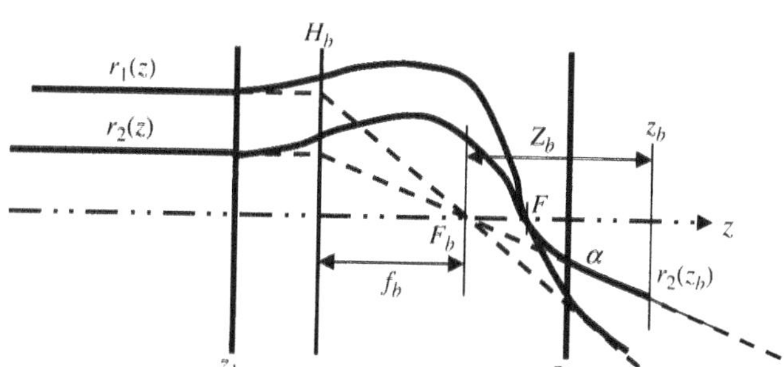

FIGURE 2.7 Cardinal elements of lenses with limited field extent. z_1 and z_2, lens boundaries; z_{H_b}, position of the principal plane in the image space; z_{F_b}, focal coordinate in the image space; $f_b = z_{F_b} - z_{H_b}$, focal length in the image space.

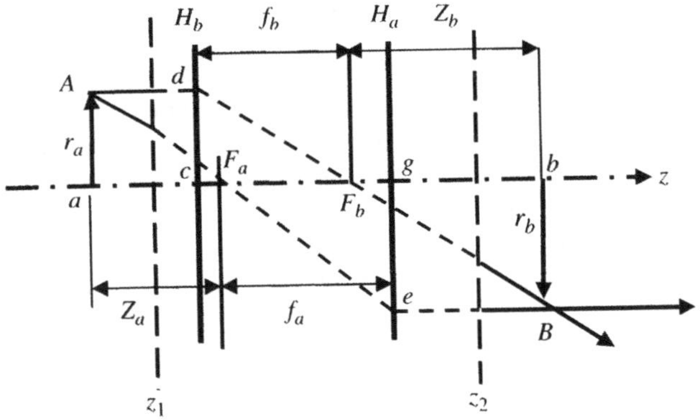

FIGURE 2.8 Construction of an image with cardinal elements.

Construction of an Image A scheme for image construction is given in Fig. 2.8. It is sufficient to know the positions of the principal planes H_a and H_b of the principal foci F_a and F_b and the position of the object. Planes z_1 and z_2 in Fig. 2.8 are the lens boundaries, and a and b are the coordinates of the corresponding points A and B on the object and the image. We can very easily form the image shown in Fig. 2.8. Let us draw a ray parallel to the z-axis through object A. This is the *first principal ray*. The continuation of this ray is the dashed line inside the lens ($z_1 < z < z_2$). It runs parallel to the z-axis until it reaches the principal plane H_b and further through the principal focus F_b. The real ray is in $z > z_2$. The *second principal ray* is drawn through object A and the focus through F_a until the intersection with plane H_a. After that happens, this ray must be drawn parallel to the z-axis.

The intersection of both principal rays creates point B of the image. Let us find two basic parameters of the image: the position and the magnification. From the geometrical similarity of triangles BbF_b, cdF_b and egF_a, aAF_a, we find that

$$M = \frac{r_b}{r_a} = \frac{Z_b}{f_b} = \frac{f_a}{Z_a} \quad \text{or} \quad Z_a Z_b = f_a f_b \tag{2.71}$$

The distances from the object and image to the corresponding principal planes are (see Fig. 2.8) $d_a = Z_a + f_a$ and $d_b = Z_b + f_b$. Then we obtain from Eq. (2.71),

$$\frac{f_a}{d_a} + \frac{f_b}{d_b} = 1 \tag{2.72}$$

Relation (2.72) is known in light optics as *Newton's formula*. If the object position (Z_a) is given, the magnification M can be found from the first of Eqs. (2.71) and the image position (Z_b) from the second of Eqs. (2.71). The angular magnification can be obtained from Eq. (2.61) or (2.62). Note that for unipotential lenses, $G = 1/M$.

Thus, image construction only requires knowing the cardinal elements. If the distribution of the axial potential φ_z^* is known, it is sufficient to calculate two principal rays (trajectories) that are parallel to the z-axis in the object and image spaces. In general, it is necessary to find two solutions of Eq. (2.64) or (2.65) with initial conditions $r_{z=a} = r_a$ and $r'_{z=a} = 0$ (the first principal ray) and $r_{z=b} = r_b$ and $r'_{z=b} = 0$ (the second principal ray). Here, r_a and r_b are arbitrary constants; a and b are z-coordinates in the object and image spaces accordingly.

2.6.4 Focal Length of Thin Unipotential and Immersion Lenses

A *thin lens* is defined as a lens whose thickness is small in comparison with both focal lengths, $z_2 - z_1 \ll f_a, f_b$ (Fig. 2.9). This lens may be also called *weak*. It means that $r \approx$ const. inside the lens ($z_1 < z < z_2$). We can assume that the principal planes coincide and are disposed of in the middle of the lens. Let us take the trajectory parallel z-axis in the object space. Integrating the reduced trajectory equation (2.69) over an interval $z_1 < z < z_2$ yields

$$\int_{z_1}^{z_2} R''\,dz = -\frac{1}{4}\int_{z_1}^{z_2} R\left[\frac{\varphi_z'^2[(\varphi_z^* + \varphi_0)^2 + 2\varphi_0^2]}{(\varphi_z^{*2} + 2\varphi_z^*\varphi_0)^2}\right]dz \tag{2.73}$$

where

$$R = r\Phi_z^{1/2} \tag{2.74}$$

Taking into account the boundary conditions

$$r'(z_1) = 0, \qquad r'(z_2) = -\tan\alpha, \qquad \varphi_z'(z_1) = 0, \qquad r(z_2) = r_a \tag{2.75}$$

we obtain

$$\varphi_z^*(z_1) = \varphi_{za}^*, \quad \varphi_z^*(z_2) = \varphi_{zb}^*, \quad R(z_2) = r_a\Phi_{za}^{1/2}, \quad R'(z_2) = -\Phi_{zb}^{1/2}\tan\alpha \tag{2.76}$$

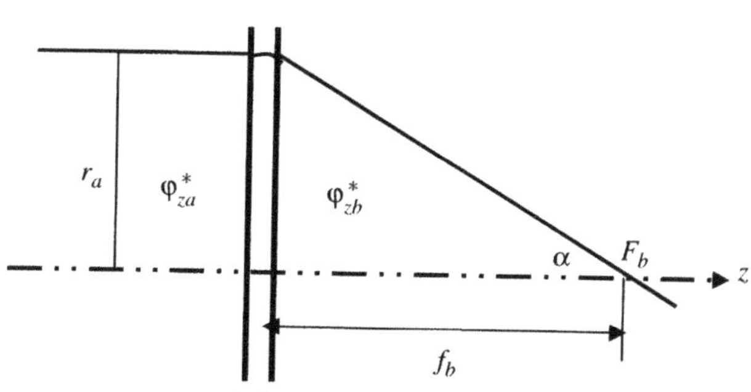

FIGURE 2.9 Thin lens with limited field extent.

Subscripts a and b in Φ_z indicate values of Φ calculated respectively in the object and image spaces. Taking into account both these relations and Eq. (2.34), we obtain

$$\frac{1}{f_b}=\frac{\tan\alpha}{r_a}=\frac{\Phi_{z_a}^{1/2}}{4\Phi_{z_b}^{1/2}}\int_{z_1}^{z_2}\left[\frac{\varphi_z'^2[(\varphi_z^*+\varphi_0)^2+2\varphi_0^2]}{(\varphi_z^{*2}+2\varphi_z^*\varphi_0)^2}\right]dz$$

$$=\frac{\eta^2(\beta_{z_a}\gamma_a)^{1/2}}{4(\beta_{z_b}\gamma_b)^{1/2}c^4}\int_{z_1}^{z_2}\frac{E_z^2(\gamma^2+2)}{\beta_z^4\gamma^4}dz \tag{2.77}$$

Note that the values z_1 and z_2 are not given. Formally, we can replace z_1 and z_2 by $-\infty$ and ∞ because $\varphi_z'=0$ for $z<z_1$ and $z>z_2$. In reality, this substitution is based on the assumption that the decrease in the fields is sufficiently abrupt. The focal length in the object space may be found analogously:

$$\frac{1}{f_a}=\frac{\Phi_{z_b}^{1/2}}{4\Phi_{z_a}^{1/2}}\int_{-\infty}^{\infty}\left[\frac{\varphi_z'^2[(\varphi_z^*+\varphi_0)^2+2\varphi_0^2]}{(\varphi_z^{*2}+2\varphi_z^*\varphi_0)^2}\right]dz \tag{2.78}$$

Nonrelativistic focal lengths are found from Eqs. (2.34) and (2.78) for $\varphi_0\gg\varphi_z^*$ and equal, respectively, to

$$\frac{1}{f_b}=\frac{3}{16}\left(\frac{\varphi_{z_a}^*}{\varphi_{z_b}^*}\right)^{1/4}\int_{-\infty}^{\infty}\left(\frac{\varphi_z'}{\varphi_z^*}\right)^2dz,\qquad \frac{1}{f_a}=\frac{3}{16}\left(\frac{\varphi_{z_b}^*}{\varphi_{z_a}^*}\right)^{1/4}\int_{-\infty}^{\infty}\left(\frac{\varphi_z'}{\varphi_z^*}\right)^2dz \tag{2.79}$$

The ratio of the focus lengths is

$$\frac{f_a}{f_b}=\begin{cases}\dfrac{\Phi_{z_a}}{\Phi_{z_b}}=\dfrac{\beta_{z_a}\gamma_a}{\beta_{z_b}\gamma_b} & \text{(relativistic)}\\[2ex] \dfrac{\sqrt{\phi_{z_a}^*}}{\sqrt{\phi_{z_b}^*}} & \text{(nonrelativistic)}\end{cases} \tag{2.80}$$

Compare Eqs. (2.80) with Eqs. (2.61) and (2.62). We see that in both (relativistic and nonrelativistic) approximations,

$$\frac{f_a}{f_b}=\frac{n_{e_0}(a)}{n_{e_0}(b)}=GM \tag{2.81}$$

For the same angular magnification G, the magnification M increases with $n_{e_0}(a)$. This property is used in light microscopy by immersing the object into a optically denser medium (e.g., oil). Similar lenses in light optics are called *immersion lenses*. Another reason for object immersion in a dense medium is related to shortening the wavelength in a dielectric. This effect increases the diffraction resolution of light microscopes.

2.6.5 Aperture Lenses

The characteristic distribution of the axis potential in the aperture lenses is shown in Fig. 2.4*d*. A nonzero homogeneous field must be in at least one space (the object or image space). In Fig. 2.10, the simplest electrode configuration is shown: three parallel electrodes with potentials φ_a, φ_d, and φ_b. The middle electrode (a diaphragm) has a hole (an aperture). Its radius R is usually supposed to be small compared with interelectrode distances. In this case, the inhomogeneous field of the lens is localized in the R region of the aperture.

Two versions of the equipotential pattern are shown in Fig. 2.10. In Fig. 2.10*a*, the potential of the diaphragm φ_d is greater than φ_a and φ_b; therefore, the directions of the electric fields in the object and image spaces are opposite and a saddle point is formed on the z-axis. It is easy to see that in the vicinity of the saddle point, the universal form of the equipotential surface is a cone with an angle in the vertex of $2\alpha \approx 108^\circ$ ($\tan\alpha = \sqrt{2}$). In Fig. 2.10*b*, the value of the potential φ_d is intermediate between φ_a and φ_b. The inhomogeneous field of the aperture in this case appears only if the values of the electric fields in the object and image spaces are different ($E_a \neq E_b$).

If the field in the image space is nonzero, the trajectory of the particle is a parabola. Usually, the focus is defined as the point of intersection of the tangent to a trajectory with the z-axis on the boundary of the inhomogeneous field (Fig. 2.11). Let us find the focal length in the thin (weak) lens approximation: the width of the inhomogeneous lens field $z_2 - z_1 \ll f$ (Fig. 2.11).

We begin with the integration of Eq. (2.65) over the interval z_1 and z_2:

$$\int_{z_1}^{z_2} \frac{d}{dz}(\Phi_z r')\,dz = -\frac{1}{2}\int_{z_1}^{z_2} \frac{\varphi_z''(\varphi_z^* + \varphi_0)}{\Phi_z} r\,dz \tag{2.82}$$

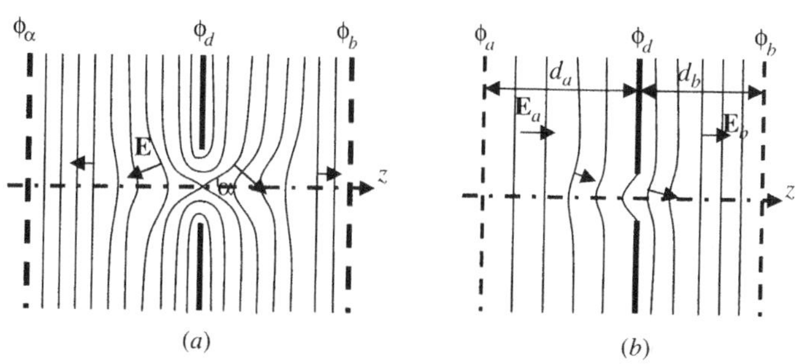

FIGURE 2.10 Equipotential patterns in an aperture lens: (*a*) $\varphi_a < \varphi_d, \varphi_b < \varphi_d$; (*b*) $E_b > E_a$.

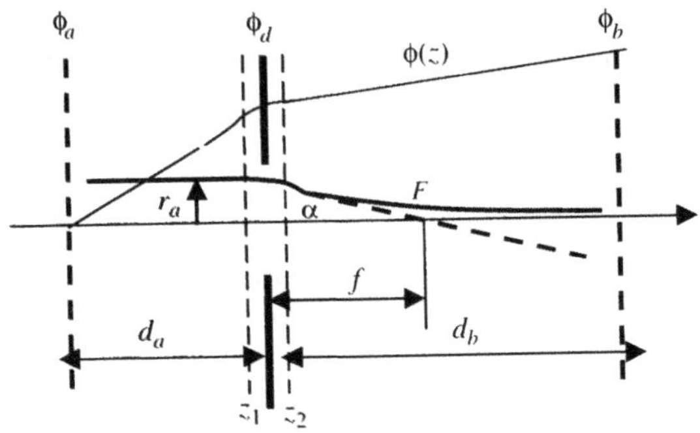

FIGURE 2.11 Focal length of a thin aperture lens.

The axis potential in the thin lens is almost constant and equal to the diaphragm potential; that is,[1]

$$\varphi_z^*(z) = \varphi_d^*, \qquad \Phi_z = \sqrt{\varphi_z^*(\varphi_z^* + 2\varphi_0)} = \sqrt{\varphi_d^*(\varphi_d^* + 2\varphi_0)}, \qquad z_1 < z < z_2 \quad (2.83)$$

Taking Eq. (2.83) into account and writing

$$\begin{aligned} r &= r_a; \qquad \varphi_z'' = -E'; \qquad r'(z_1) = 0, \\ r'(z_2) &= -\tan\alpha; \quad E(z_1) = E_a; \qquad E(z_2) = E_b \end{aligned} \quad (2.84)$$

we obtain from Eq. (2.82),

$$-\tan\alpha = \frac{r_a(\varphi_d^* + \varphi_0)}{2\varphi_d^*(\varphi_d^* + 2\varphi_0)}(E_b - E_a) \quad (2.85)$$

so the focal length is

$$f = \frac{r_a}{\tan\alpha} = \frac{2\varphi_d^*(\varphi_d^* + 2\varphi_0)}{(\varphi_d^* + \varphi_0)(E_a - E_b)} \quad (2.86)$$

In the nonrelativistic approximation, this is

$$f = \frac{4\varphi_d^*}{E_a - E_b} \quad (2.87)$$

The approximate values of fields E_a and E_b in these formulas are easily calculated. It is sufficient to notice that cutting a hole in the diaphragm perturbs the field only in the R region of the aperture. If we close the hole with metal, the fields will not

[1] We remind readers that a superscript asterisk indicates that the corresponding variable is taken with the initial energy φ_ε [see Eq. (2.8)].

change in remote regions; therefore, E_a and E_b can be found as the fields in the corresponding flat capacitive gaps:

$$E_a = \frac{\varphi_a - \varphi_d}{d_a}, \qquad E_b = \frac{\varphi_d - \varphi_b}{d_b} \tag{2.88}$$

These fields can have any sign; therefore, both convergent and divergent aperture lenses can be realized, as in the following particular cases:

- *Convergent lens:* $E_a > E_b$
- *Divergent lens:* $E_a < E_b$
- *Saddle, convergent lens:* $E_a > 0,\ E_b < 0$
- *Saddle, divergent lens:* $E_a < 0,\ E_b > 0$
- *No lens:* $E_a = E_b,\ f = \infty$

Example 2.1: Anode Aperture as a Divergent Lens In Fig. 2.12, the simplest version of the electron gun based on a planar diode is shown. Use of an electron gun together with a transportation system makes it possible to get extended, often intense, electron beams. After transportation, the electron beam emerges in a working space, where its energy is used for interaction with electromagnetic fields, other particles, or for material processing. The open end of the beam tube at the diode is an aperture lens.

Let us find its focal length using Eqs. (2.87) and (2.88). Here $\varphi_d = \varphi_a$, $E_a = -\varphi_a/d$, and $E_b = 0$. According to Eq. (2.87), nonrelativistic and extreme relativistic focal lengths are, respectively,

$$f_{\text{nrel}} = -4d, \qquad f_{\text{ext}} = -2d \tag{2.89}$$

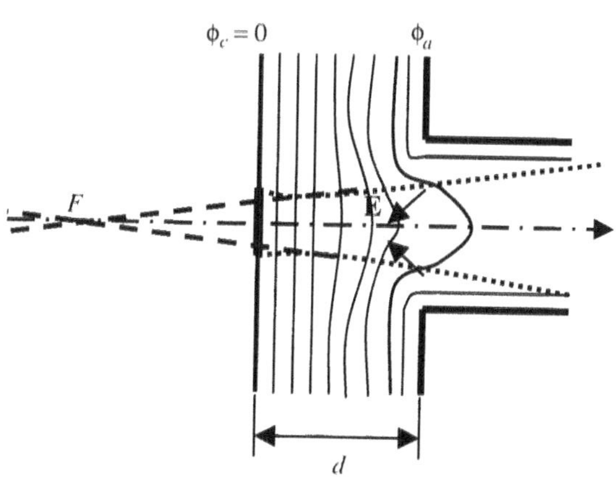

FIGURE 2.12 Equipotential pattern in an anode aperture lens.

As shown, the lens is divergent. The virtual focuses in the nonrelativistic and relativistic cases are located behind the cathode at a distance of $3d$ and d, respectively. The divergent effect may also be seen on the pattern equipotentials: the electrical forces defocus the beam (Fig. 2.12). In effective guns, various methods of minimization of this defocusing effect are used. This topic is considered in detail in Chapter 3.

Cathode Lens (Immersion Objective) According to Eqs. (2.86) and (2.87) and Example 2.1, the near-cathode aperture lens may be convergent only if the electric field in the object space (where the field is homogeneous) $E_a > 0$. But in this case, the electric field in the cathode is retarding and the electron beam current is practically zero. This situation can be changed if the inhomogeneous field of the lens reaches the cathode. Similar fields can be realized in a three-electrode system. In Fig. 2.13 a typical scheme for a three-electrode cathode lens is shown. The field nearest the cathode electrode *Wehnelt* has a small negative potential (the cathode potential is zero). But for a sufficiently positive anode potential, the electric field in the zone of the central cathode region is accelerating. According to Fig. 2.13, the electric field in this region focuses the beam and a convergent beam will be formed.

Note that the current to the negative Wehnelt is practically zero. Variation of the Wehnelt potential φ_W provides effective control of the beam current. For thermionic cathodes in the space-charge regime, the beam current I_b changes approximately as

$$I_b = C(\varphi_W + D\varphi_a)^{3/2} \tag{2.90}$$

where φ_a is the anode potential; D is the penetration factor of the Wehnelt, which is analogous to the penetration factor of a thermionic triode control grid; and C is a constant proportional to the effective cathode area S_c, where the electric field accelerates. Note that S_c increases with φ_W; therefore, the dependence $I_b = I_b(\varphi_W)$ is sharper than the $3/2$ law. Electrons in the cathode lens leave the cathode at arbitrary angles to the axis (if one takes the initial velocities into account), and the paraxial approximation is not directly applicable. This difficulty is overcome by the

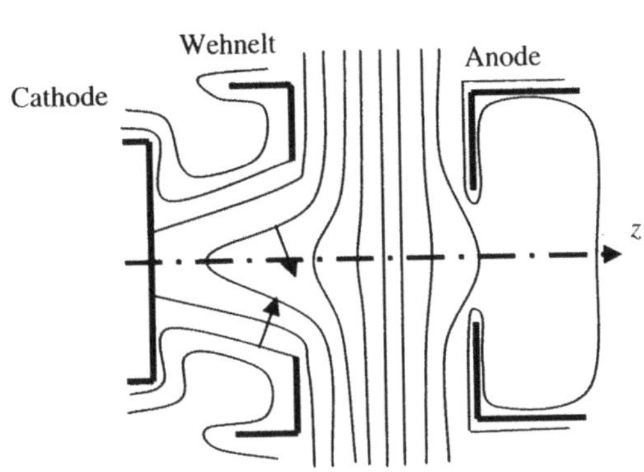

FIGURE 2.13 Equipotential pattern of a cathode lens (immersion objective).

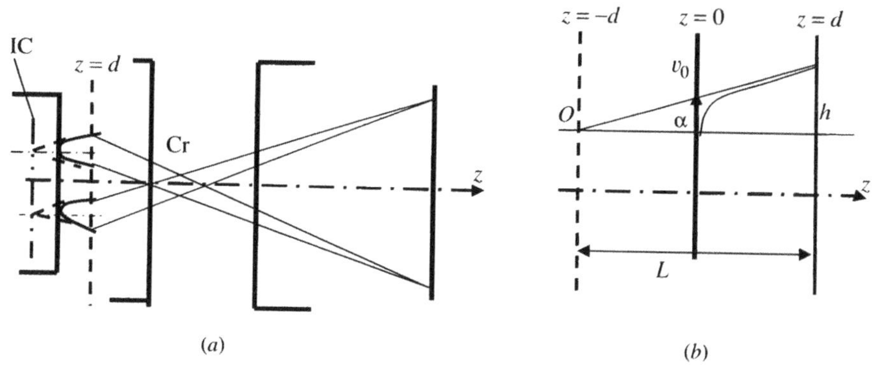

FIGURE 2.14 (*a*) Image formatting by a cathode lens (IC, imaginary cathode; Cr, crossover); (*b*) formation of an imaginary cathode.

introduction of an imaginary cathode (Rusterholz, 1950). At the small distance d from the cathode, the electric field may be considered uniform. Then the electron trajectory that goes into the region $0 \leq z \leq d$ is a parabola. Let us draw a tangent to the trajectory in the plane $z = d$ (Fig. 2.14*b*) and find the intersection point of the tangent with the straight line that passes through the output point parallel to the z-axis (Fig. 2.14*b*). Assume that the electron leaves the cathode ($z = 0$) parallel to the cathode surface with a velocity of v_0. Then

$$L = \frac{h}{tg\alpha} = \frac{v_0 t}{v_0/v_z} = v_z t = 2d \tag{2.91}$$

because v_z is the velocity of the uniformly accelerated motion in the plane $z = d$. We see that L does not depend on the initial velocity v_0. Now we can assume the trajectories in the region $-d < z < d$ to be straight lines that leave the imaginary source O at arbitrarily small angles with the z-axis. The inhomogeneous lens field begins in the plane $z = d$, and the paraxial approximation is restored. The plane $z = -d$ can be called an imaginary cathode.

2.6.6 Applications of Cathode Lenses

Emission Microscopes Figure 2.14*a* shows an emission microscope that can be used to investigate the emitting structure's details. As we can see, a magnified image of the cathode surface is formed on the screen. The contrast of the image is determined by the emission currents from different elements on the cathode. Note that to obtain a sharp cathode image, it is necessary to operate in the regime of a temperature-limited emission.

Electron Guns These guns are used to concentrate the electron beam on small targets. The scheme of the electron gun can be obtained by adding a second anode to the schemes shown in Figs. 2.13 and 2.14*a* (see Fig. 2.15). The rays are

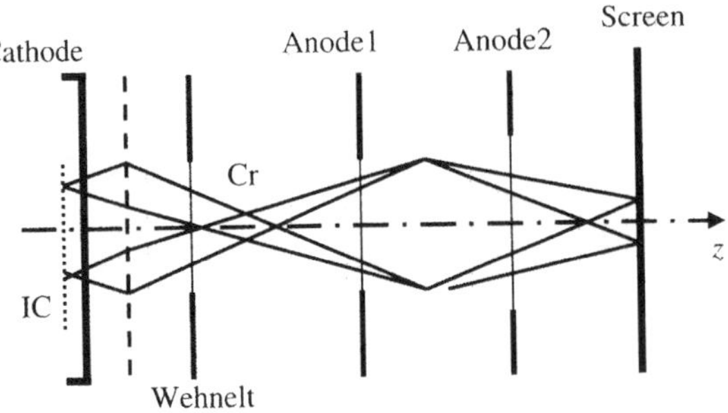

FIGURE 2.15 Two-anode electron gun. IC, imaginary cathode; Cr, crossover.

also refracted, and with the proper electrode geometry and potentials, they form a little circle on the screen that may be considered the crossover image. One disadvantage of this gun is the impossibility of independent control of the beam current and image focusing. Variations in the Wehnelt potential or the potential of anode 1 change these parameters simultaneously.

Usually, a version of the electron gun with sectioned anode 2 (Fig. 2.16) is used. Here, anode 2 is divided into two sections between which anode 1 is placed. Anode 1 is called a *focusing electrode*; variations in its potential affect focusing only. A variation in the Wehnelt potential basically influences only the beam current. As a whole, two sections of anodes 2 and 1 form a unipotential lens (Fig. 2.6) that projects the crossover on the screen.

The electron guns described are used to obtain very small spots on various targets. These guns usually operate with small currents and high voltages; therefore, the space-charge fields on the targets are very small. Properties of electron guns are essentially different from high-current applications in microwave and technological devices (e.g., klystrons, traveling-wave tubes, free electron lasers and masers, electron beam welding and machining systems). In the latter the space-charge field is significant, and it is important to use the maximum part of the emission current (considered in Chapter 4).

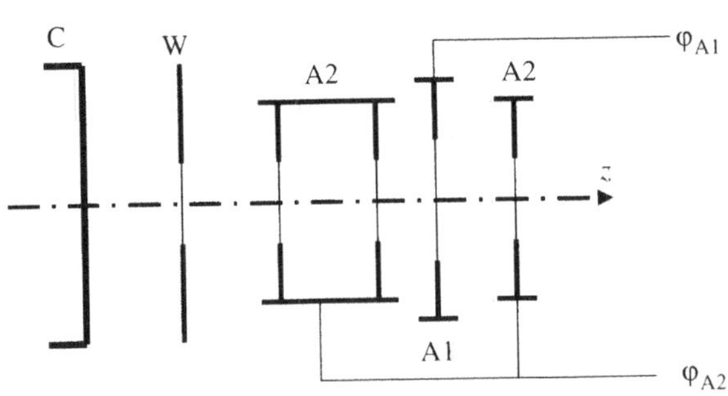

FIGURE 2.16 Electron gun with sectioned anode (2). C, cathode; W, Wehnelt; A1, first anode (focusing electrode); A2, second anode.

The following devices are known (see, e.g., Goldstein et al., 1992; Moss, 1968; Reimer, 1985; Reiser, 1994; Whitaker, 2000):

1. *Cathode-Ray Tubes:* The distance between the lens and the screen in these tubes is usually more than the distance between the lens and the crossover. The electron beam spot on the phosphorescent screen is a magnified image of the crossover. It is essential to have the minimal crossover size, which is limited basically by the initial electron velocities (Langmuir, 1937). The formula for estimating the crossover diameter of thermionic cathodes is (see, e.g., Glazer, 1952)

$$d_{\mathrm{cr}} = f_a \sqrt{\frac{kT}{e_0 \varphi_d}} \tag{2.92}$$

where f_a is the focal length of the cathode lens in the object space, T is the cathode temperature, $k = 1.3807 \times 10^{-23}$ J/K is Boltzmann's constant, and φ_d is the potential of the boundary plane $z = d$ of the uniform field (Fig. 2.14*a*). Usually, $d_{\mathrm{cr}} \sim 0.1$ mm. For example, let us take $f_a = 1$ mm, $T = 1100$ K ($kT/e_0 \approx 0.1$ V), and $\varphi_d = 5$ V. We obtain $d_{\mathrm{cr}} \approx 0.14$ mm.

2. *Scanning Electron Microscopes:* These devices utilize highly demagnified images of the crossover as electron spots on the specimens. The typical size of the spot that determines the microscope resolution is on the order of 10 nm. The required demagnification factors $M = d_{\mathrm{cr}}/d_s$ (d_s is the diameter of the electron spot), on the order of 1000, are realized in systems with one or two intermediate condenser lenses (see, e.g., Goldstein et al., 1992; Reimer, 1985). In each lens (and also in a final lens) $M_i \approx S_i/f_i \gg 1$, where M_i is the ith lens demagnification factor, S_i is the distance from the intermediate crossover image to the ith lens, and f_i is the focal length of the ith lens.

For the given size of the electron spot, the maximum probe current is determined by the maximum beam brightness [Eq. (I.57)], and it increases with α, j_c, and $\bar{\varphi}$. The aperture lens angle α is limited primarily by the spherical aberration (see Section 2.8). It is $\sim$5 to 10 mrad. The voltage is usually 10 to 30 kV. The maximum values of φ are determined by the increase in parasitic X-ray radiation. Reducing φ decreases the brightness and increases the chromatic aberration (Goldstein et al., 1992). The cathode current density depends on the type of emitter used. The best results for thermionic cathodes are those of the LB_6 cathode ($j_{c,\max} \sim 100$ A/sm^2). The probe current for the spot size $\sim$1 to 10 nm is about 0.01 to 1 nA. That is enough to form a picture. In general, field emission cathodes are best. They improve all parameters by an order of magnitude. The disadvantage of these cathodes is the necessity of working with ultrahigh vacuum ($\sim 10^{-9}$ torr).

3. *Electron-Probe Analyzers* (Goldstein et al., 1992): These analyzers are often provided with scanning microscopes in a common installation. They are applied in X-ray spectrometers for the analysis of characteristic X-ray radiation and of the chemical content of specimens irradiated by scanning electrons.

4. *Electron beam lithography* (e.g., Valiev, 1992): Electron beam lithography is provided by scanning systems similar to an electron-optical ESM system. A sample is covered by a thin layer of specific matter that changes its properties after being irradiated by scanning electrons. In particular, the exposed layer changes its solubility toward certain chemicals. The motion of a scanning electron beam is usually controlled by a computer. As a result, the hyperfine microstructures required can be obtained.

2.7 MAGNETIC AXIALLY SYMMETRIC LENSES

A magnetic lens is a focusing system with a purely magnetic field. The electric field found in the lens (i.e., in the object and image space) is equal to zero, and the potential is constant. The electric field differs from zero only in the region where the electron beam is formed. We do not consider that region here.

2.7.1 Equations of Paraxial Trajectories. Classification of Magnetic Lenses

Consider the equations of paraxial trajectories [Eqs. (2.39)–(2.42)] with $\varphi'_z = \varphi''_z = 0$:

$$\left.\begin{aligned} \theta' &= \frac{cB(z)}{2\Phi_z} \\ r'' + \left[\frac{cB(z)}{2\Phi_z}\right]^2 r &= 0 \end{aligned}\right\} \quad \text{relativistic} \tag{2.93}$$

$$\left.\begin{aligned} \theta' &= \sqrt{\frac{\eta}{8\varphi_z^*}}B(z) \\ r'' + \frac{\eta B^2(z)}{8\varphi_z^*} r &= 0 \end{aligned}\right\} \quad \text{nonrelativistic} \tag{2.94}$$

Equations (2.93) and (2.94) may be applied directly for trajectories in meridional planes and for zero cathode magnetic fields. Nevertheless, as mentioned in Section 2.5.1, these equations can be written in the identical form without the restrictions indicated but with other variables instead of θ and r (see, e.g., Lawson, 1988; Szilagyi, 1988).

Consider the following properties of magnetic lenses (from Eqs. (2.93) and (2.94)):

1. The rotational angle θ of any trajectory is a linear function of $B(z)$ and does not depend on r directly. Therefore, the entire image turns without angular distortion. The direction of the rotation is determined by the sign of B_z.

2. For any $B(z) \neq 0$, the second derivative $d^2r/dz^2 \neq 0$. Therefore, a ray entering a lens parallel to the z-axis will always be deflected. This means that any magnetic field is a lens. We are reminded that only inhomogeneous electric fields form electrostatic lenses.

3. Because $r'' < 0$, all magnetic lenses are convergent (rays entering the lens parallel to the z-axis approach this axis up to the first intersection with the axis).

Taking Eq. (2.34) into account for the function Φ_z, we can rewrite Eqs. (2.93) and (2.94) as follows:

$$\theta = \int_{z_1}^{z_2} p\,dz \tag{2.95}$$

$$r'' + p^2 r = 0 \tag{2.96}$$

where

$$p = \begin{cases} \dfrac{cB(z)}{2\Phi_z} = \dfrac{\eta B(z)}{2\beta\gamma c} = \dfrac{\omega_{gr}(z)}{2\beta c} & \text{(relativistic)} \\[2ex] \dfrac{\eta B(z)}{\sqrt{8\eta\phi_z^*}} = \dfrac{\omega_{gn}}{\sqrt{8\eta\phi_z^*}} & \text{(nonrelativistic)} \end{cases} \tag{2.97}$$

ω_{gr} and ω_{gn} are the relativistic and nonrelativistic gyrofrequencies, respectively, and z_1 and z_2 are coordinates of the lens boundary planes (we assume that the magnetic field in the lens has sharp boundaries). If the magnetic field is decreased fast enough at $z < z_1$ and $z > z_2$, we can change z_1, z_2 to $(-\infty, \infty)$.

Equation (2.97) can also be written as

$$p = \frac{\pi}{h} \tag{2.98}$$

where $h = h(z)$ is an instantaneous pitch of the helical trajectory (accordingly, relativistic or nonrelativistic) computed for the magnetic field $B(z)$.

The principle of electrostatic lenses classification (see Section 2.6.1) cannot be utilized here because unlike the latter, the field in the magnetic lens is zero outside the lens. Instead, let us use the value of h for classification in the natural spatial scale of the magnetic field. Then the following types of lenses can be defined:

$$\begin{aligned} \chi &\equiv \bar{p}L \sim \frac{L}{h} \ll 1 && \text{short lenses} \\ \chi &\equiv \bar{p}L \sim \frac{L}{h} \geq 1 && \text{strong lenses} \\ \chi &\equiv \bar{p}L \sim \frac{L}{h} \gg 1 && \text{long lenses} \end{aligned} \tag{2.99}$$

Here $L = z_2 - z_1$ is the lens length and $\bar{p}$ is an average of p in the lens. Note that the inequality defining the long lenses is similar to condition in Eq. (1.81) of the small magnetic field inhomogeneity.

2.7.2 Short Magnetic Lenses

Let us rewrite Eq. (2.96) as

$$\frac{d^2 r}{d\zeta^2} + \chi^2 r = 0 \tag{2.100}$$

where $\zeta = z/L$ and the lens region is

$$\frac{z_1}{L} = \zeta_1 \leq \zeta \leq \zeta_1 + 1 \tag{2.101}$$

It can readily be shown from Eq. (2.100) that for $\chi \ll 1$, the deflection of the ray entering the lens parallel to the z-axis $[(dr/d\zeta)_{\zeta_1} = 0]$ is small; that is,

$$\Delta r = r_2 - r_1 \ll r_1 \tag{2.102}$$

The inequality (2.102) can also be used as a definition of *short lens*.

Focal Length of a Short Lens Assume that the trajectory enters the lens parallel to the z-axis at the radius r_1; that is,

$$r'(z_1) = 0, \qquad r(z_1) = r_1 \tag{2.103}$$

Integration of the second of Eqs. (2.97) over the interval $z_1 \leq z \leq z_2$ gives

$$\int_{z_1}^{z_2} \frac{d^2 r}{dz^2}\, dz = -\int_{z_1}^{z_2} \left[\frac{cB(z)}{2\Phi_z}\right]^2 r\, dz = -\left(\frac{c}{2\Phi_z}\right)^2 \int_{z_1}^{z_2} B(z)^2\, dz \tag{2.104}$$

According to Eqs. (2.102) and (2.103), we obtain

$$r'(z_2) = -\tan\alpha = -r_1 \left(\frac{c}{2\Phi_z}\right)^2 \int_{z_1}^{z_2} B(z)^2\, dz \tag{2.105}$$

where α is the slope of the trajectory with respect to the z-axis in the plane $z = z_2$. For $z > z_2$, the trajectory is a straight line lying in the nonrotating meridional plane. The focal length of the image space is the distance between $z = z_2$ and the intersection point of the trajectory with the z-axis: $f_b = r_2/\tan\alpha$. Obviously, the focal length for the object space is the same: $f_b = f_a$. Thus, according to Eq. (2.105), we obtain the focal lens power:

$$\frac{1}{f_a} = \frac{1}{f_b} = \left(\frac{c}{2\Phi_z}\right)^2 \int_{z_1}^{z_2} B(z)^2\, dz = \frac{\eta^2}{4\beta^2 c^2 \gamma^2} \int_{z_1}^{z_2} B_z^2\, dz \approx \left(\frac{c}{2\Phi_z}\right)^2 \int_{-\infty}^{\infty} B(z)^2\, dz \tag{2.106}$$

The focal length of a nonrelativistic short lens is obtained from Eq. (2.106) by setting $\gamma = 1$ and $\beta^2 c^2 = v^2 = 2\eta\phi_z^*$:

$$\frac{1}{f_a} = \frac{1}{f_b} = \frac{\eta}{8\varphi_z^*} \int_{-\infty}^{\infty} B(z)^2 dz \tag{2.107}$$

Taking Eq. (2.97) into account, formulas (2.106) and (2.107) can be written as

$$\frac{1}{f} = \int_{z_1}^{z_2} p^2\, dz = L\overline{p^2} \tag{2.108}$$

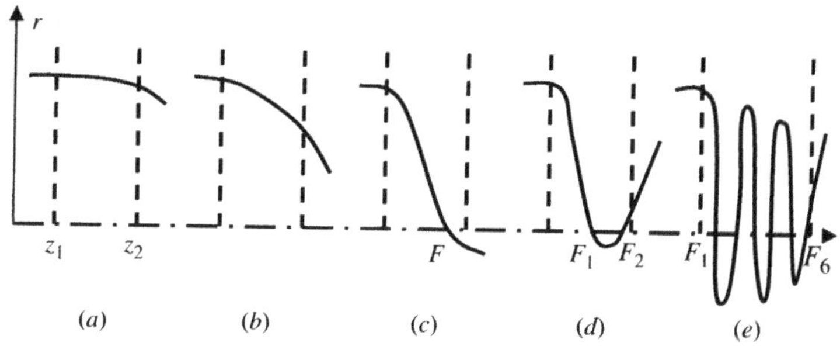

FIGURE 2.17 Change of shape of lens trajectories with increasing focal power: (*a*) short lens; (*b*) strong lens; (*c*) the focus is inside the lens (the latter is a divergent lens); (*d*) two focuses are into the lens; (*e*) multifocusing lens.

where $\overline{p^2}$ is the average of p^2 over the length of the lens. The lens's length is small in compared with the focal length. Then, according to Eq. (2.108),

$$\frac{L}{f} = L^2\overline{p^2} \ll 1 \tag{2.109}$$

This inequality coincides with the definition of a short lens [Eq. (2.99)].

According to Eqs. (2.106) and (2.107), the focal lens power $1/f$ grows with the magnetic field B and/or with a decrease in energy (the potential) $\Phi_z(\phi_z^*)$. Is it possible to increase the optical power indefinitely? It is easy to see that Eq. (2.106) or (2.107) becomes inapplicable when the focal length decreases beyond the range of inequality (2.109) because (1) inequality (2.102), which is the basis of Eq. (2.105), is not valid, and (2) knowledge of focal lengths in general is insufficient for construction of the image. It is necessary to determine the cardinal elements of the lens. Essentially, for $f \sim L$ the form of the trajectories is changed. In particular, an effect of multifocusing appears. The qualitative shape of the trajectories for different values of $B(z)^2/\Phi_z$ is shown in Fig. 2.17.

It is instructive to consider qualitatively the focusing effect of the magnetic lens. We draw a magnetic field line pattern of a solenoid lens in Fig. 2.18. This pattern

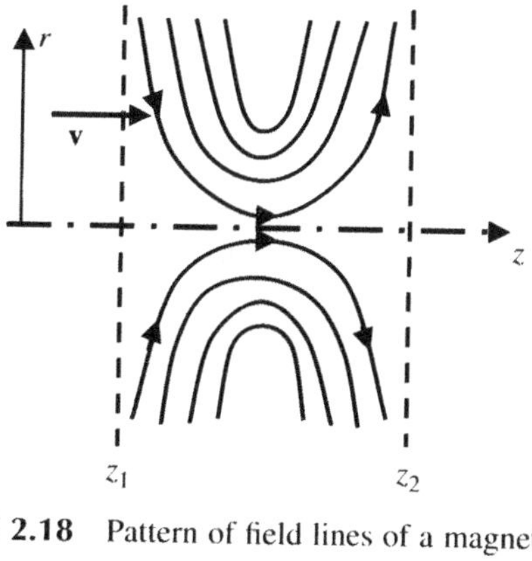

FIGURE 2.18 Pattern of field lines of a magnetic lens.

satisfies the following requirements: (1) axial symmetry is present; (2) the sources of the magnetic field (currents) are arranged on the outside of the operation (near-axis) region of the lens; and (3) the magnetic field is localized in the closed region $(-z_1 \leq z \leq z_2)$.

Assume that an electron enters the lens parallel to the z-axis. Initially, it is accelerated in the azimuthal direction by the radial component of the magnetic field. The azimuthal velocity increases up to the middle plane of the lens, and afterward, when B_r changes the sign of rotation, it is retarded. But according to Eqs. (2.93) and (2.94), the sign of the azimuthal velocity does not change. When looking from the z-axis, the rotation direction for the pattern shown in Fig. 2.18 is counterclockwise. The presence of the azimuthal velocity in combination with the axial component of the magnet field induces radial movement of the electrons directed toward the axis. Changing the sign of the magnetic field in Fig. 2.18 does not affect the sign of the radial velocity, because the direction of the azimuthal velocity and that of the axial magnetic field change simultaneously. This proves that any magnetic lens is convergent. The radial motion is a quadratic function on the magnetic field because the radial force is proportional to $v_\theta B_z$, where v_θ is also proportional to B.

2.7.3 Strong Magnetic Lenses

The optical properties of short lenses are determined by one parameter: the focal length. In turn, according to Eqs. (2.109) and (2.97), the focal length is determined by the average of $B^2(z)$ over the lens length. Therefore, details of the magnetic field distribution, as well as analysis of the trajectories in short lenses, are not important. But when we turn to strong lenses, these problems become essential.

Glazer's Bell-Shaped Distribution The field distribution in real magnetic lenses does not have well-defined boundaries. The magnetic field decreases more or less smoothly from the middle lens plane and almost reaches the object and image spaces. Some models of the distribution $B(z)$ are, on the one hand, similar to the real distribution, and on the other hand, permit a closed solution of the paraxial trajectory equation (2.93) or (2.94) (Glazer, 1952; Szilagyi, 1988). Below we consider the bell-shaped distribution proposed by Glazer (1952):

$$B(z) = \frac{B_0}{1 + (z/d)^2} \tag{2.110}$$

where B_0 is the magnetic field in the lens center and d is a half-width of the distribution (Fig. 2.19). Substituting Eq. (2.110) for Eqs. (2.96) and (2.97), we obtain the following equation for the paraxial trajectories:

$$\frac{d^2 r}{dz^2} + \frac{k^2}{[1 + (z/d)^2]^2} r = 0 \tag{2.111}$$

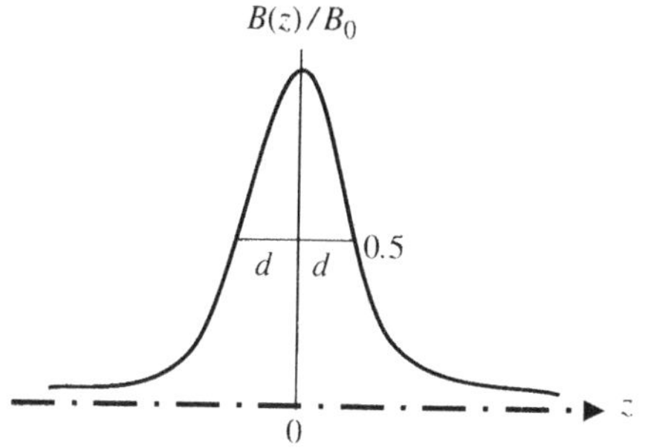

FIGURE 2.19 Bell-shaped distribution of an axis magnetic field.

where

$$k = \begin{cases} \dfrac{cB_0}{2\Phi_z} & \text{(relativistic)} \\ \dfrac{\eta B_0}{\sqrt{8\eta\phi_z^*}} & \text{(nonrelativistic)} \end{cases} \tag{2.112}$$

We see that the magnetic field in this model extends over the entire space, and the trajectories are curvilinear everywhere. That makes this theory significantly different from the immersion electrostatic lenses theory (Section 2.6.2). In particular, the definition of cardinal elements must be changed.

It is easy to verify that the solution of Eq. (2.110) is (Glazer, 1952)

$$r = \frac{d[C_1 \cos(\omega\psi) + C_2 \sin(\omega\psi)]}{\sin\psi} \tag{2.113}$$

where

$$\psi = a\cot\frac{z}{d}, \qquad \omega = \sqrt{1 + k^2 d^2} \tag{2.114}$$

and C_1 and C_2 are constants of the integration. The interval $(\pi, 0)$ of the independent variable ψ corresponds to the domain $(-\infty, \infty)$ of the variable z. The lens center is at the point $\psi = \pi/2$. We can introduce two functions,

$$\begin{aligned} r_1(\psi) &= \frac{d[A_1 \cos(\omega\psi) + B_1 \sin(\omega\psi)]}{\sin\psi} \\ r_2(\psi) &= \frac{d[A_2 \cos(\omega\psi) + B_2 \sin(\omega\psi)]}{\sin\psi} \end{aligned} \tag{2.115}$$

as linear independent solutions of Eq. (2.111). Let them intersect in points of the object $\psi = \psi_a$ and an image $\psi = \psi_b$ [i.e., $r_1(\psi_a) = r_2(\psi_a)$, $r_1(\psi_b) = r_2(\psi_b)$]. Then we obtain a system of two homogeneous linear equations for $A_{1,2}$ and $B_{1,2}$: The condition of compatibility of this system yields the following relation between coordinates of the object and image:

$$\sin[\omega(\psi_a - \psi_b)] = 0 \quad \text{or} \quad \psi_a - \psi_b = \frac{n\pi}{\omega}, \qquad n = 1, 2, \ldots \tag{2.116}$$

According to Eqs. (2.116) and for the given position of the object ψ_a can be n images in the planes

$$\psi_{bn} = \frac{\omega\psi_a - n\pi}{\omega}, \qquad n = 1, 2, \ldots \tag{2.117}$$

The maximum number of images depends on ω because $0 < \psi < \pi$. Therefore, we have n images if $n < \omega < n + 1$. Using the definition of ω [Eq. (2.114)], we obtain the following condition of having n images:

$$n^2 - 1 < k^2d^2 < (n+1)^2 - 1 \tag{2.118}$$

So if $0 < k^2d^2 < 3$, we have one image; if $3 < k^2d^2 < 8$, we have two images; and so on.

Foci of Glazer's Lens Let us define the focus of the image space as the intersection point of the image's principal ray with the z-axis. This ray is parallel to the z-axis in the zero magnetic field [i.e., in the plane $\psi = 0$ ($z = -\infty$)]. It can readily be shown that the equation of the corresponding trajectory is the second term of the general solution [Eq. (2.113)]. In particular, the equation for the ray that passes $\psi = 0$ parallel to the z-axis at the distance r_b is[2]

$$r = \frac{r_b}{\omega}\frac{\sin\omega\psi}{\sin\psi} \tag{2.119}$$

Positions of foci according to Eq. (2.119) are $\omega\psi_{F_{b,n}} = n\pi$ or [see Eq. (2.114)]

$$z_{F_{b,n}} = d\cot\frac{n\pi}{\omega} = d\cot\frac{n\pi}{\sqrt{1 + k^2d^2}} \tag{2.120}$$

Evidently, coordinates of the focal points shift to the left with increased n. The number of foci is equal to the number of images for each object. In the strong magnetic field when the number of foci is more than two [according to Eq. (2.118), $1 + k^2d^2 > 4$], the foci with small n can lie within the right (image) side of the lens. However, if there is only one focus ($1 + k^2d^2 < 4$), it always lies on the left (object) side. The principal ray of the object space also forms n symmetric foci,

[2]Reminder: Subscripts a and b refer to the object and image spaces, respectively.

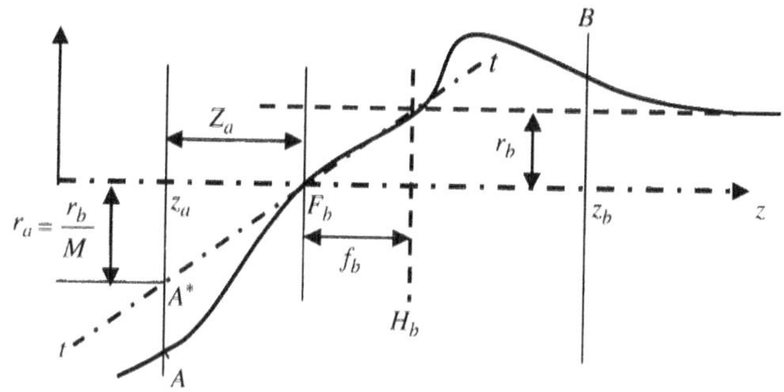

FIGURE 2.20 Cardinal elements of an image space.

which accords with the symmetry of the magnetic field distribution [Eq. (2.110)]. So

$$z_{F_{a,n}} = -d \cot \frac{n\pi}{\omega} = -d \cot \frac{n\pi}{\sqrt{1+k^2 d^2}} \tag{2.121}$$

Principal Planes Let us define the principal plane of the image space H_b as the plane drawn through the intersection point of the tangent to the image's principal ray in the focus and the straight line that is a continuation of this trajectory from $z = \infty$ ($\psi = 0$) (Fig. 2.20). Then

$$z_{H_b} = z_{F_b} + \frac{r_b}{(dr/dz)F_b} = z_{F_b} + \frac{r_b}{(dr/d\psi)_{F_b}} \frac{dz}{d\psi} \tag{2.122}$$

where $r(\psi)$, $z(\psi)$, and z_{F_b} are given by Eqs. (2.119), (2.114), and (2.120). Substituting these functions into Eq. (2.121) for $\psi = \psi_{F_{b,n}} = n\pi/\omega$, we obtain

$$\begin{aligned} z_{H_{b,n}} &= d \cot \frac{n\pi}{\omega} - \frac{d}{\sin(n\pi/\omega) \cos n\pi} \\ &= \frac{d}{\sin(n\pi/\omega)} \left[(-1)^{n+1} + \cos \frac{n\pi}{\omega} \right] \end{aligned} \tag{2.123}$$

Analogously, the location of the principal plane of the object space is

$$\begin{aligned} z_{H_{a,n}} &= -d \cot \frac{n\pi}{\omega} - \frac{d}{\sin(n\pi/\omega) \cos n\pi} \\ &= \frac{d}{\sin(n\pi/\omega)} \left[(-1)^{n+1} - \cos \frac{n\pi}{\omega} \right] \end{aligned} \tag{2.124}$$

Focal Length By definition, the focal length is the distance between the principal plane and the focus:

$$f_{b,n} = z_{H_{b,n}} - z_{F_{b,n}} = \frac{(-1)^{n+1} d}{\sin(n\pi/\omega)} \tag{2.125}$$

It is easy to verify that

$$f_{a,n} = z_{H_{a,n}} - z_{F_{a,n}} = f_{b,n} \tag{2.126}$$

Magnification Assume that the position of the object is $z_a(\psi_a)$. The image positions $z_{b,n}$ can be obtained from Eq. (2.117). Because the ray equation is known [Eq. (2.119)], we can find the values of the radii $r(z_a)$ and $r(z_{b,n})$ on this trajectory. The ratio of these values is the magnification

$$M = \frac{r(z_{b,n})}{r(z_a)} = \frac{\sin\omega(\psi_a + n\pi/\omega)}{\sin(\psi_a + n\pi/\omega)} \frac{\sin\psi_a}{\sin\omega\psi_a} = \frac{(-1)^n \sin\psi_a}{\sin(\psi_a + n\pi/\omega)} \tag{2.127}$$

Sign inversion of M for subsequent images shows that periodically, the images move to another side of the axis. For the minimum focal length ($\omega = 2, n = 1$) $M = -\tan g\psi_a$.

Construction of Images. Newton's Relationship Consider the tangent $t - t$ to the trajectory at the focal point F_b (see Fig. 2.20) and find the length of the segment z_aA^* cut off by the tangent in the object plane:

$$\begin{aligned} r_a &= -(z_{F_b} - z_a)(r')_{F_b} = -\left(d\cot\frac{n\pi}{\omega} - d\cot\psi_a\right)\frac{r_b}{d}(-1)^{n+1}\sin\frac{n\pi}{\omega} \\ &= r_b(-1)^n \frac{\sin(\psi_a - n\pi/\omega)}{\sin\psi_a} = \frac{1}{M} r_b \end{aligned} \tag{2.128}$$

We have obtained an important result: The straight line parallel to the axis in the image space after refraction on the cardinal plane H_b and crossing the focus F_b crosses the object. We call this line the *first auxiliary principal ray*. An analogous line in the object space after refraction in the principal plane H_a and crossing the focal point F_a hits the image. This is the *second auxiliary principal ray*. Thus, if the positions of the principal planes and the foci are known, the parameters of images are obtained by drawing the foregoing two auxiliary principal rays through the object. Their intersection gives the position of the image and magnification. Although the definitions of principal elements in both cases are different, this method of the image's construction is completely identical to the method in the theory of electrostatic immersion lenses (Section 2.6.3 and Fig. 2.8). Evidently, the relationship (2.71) between the magnification of focal length and distances of the object and image from the corresponding foci, $M = r_b/r_a = Z_b/f_b = f_a/Z_a$, and Newton's formula,

$$Z_a Z_b = f_a f_b \tag{2.129}$$

are valid also.

Image Rotation The angle of the image rotation according to Eqs. (2.95), (2.97), (2.110), and (2.116) is

$$\theta = k\int_{z_a}^{z_b} \frac{dz}{1 + (z/d)^2} = kd\left(a\tan\frac{z_b}{d} - a\tan\frac{z_a}{d}\right) = kd(\psi_a - \psi_b) = n\frac{kd\pi}{\omega}$$

or, taking Eqs. (2.114) into account, we obtain

$$\theta = n\frac{\pi kd}{\sqrt{1+k^2d^2}} \tag{2.130}$$

where k is given by Eq. (2.112).

General Characteristics of Glazer's Lens It is interesting to determine the dependence of the focal length and focal point position of a magnetic field. Let us consider the half-width d. The coefficient k is proportional to the value of the maximum magnetic field B_0 [see Eq. (2.112)]. When B_0 is small, $\omega \approx 1$, the focal length $f \gg d$, and $z_{F_a} = -z_{F_b} \gg d$. This is the case for a short lens. For a larger B_0, the focus of the object space moves to the left and the focal length decreases. For $\omega = 2$, $n = 1$, and we obtain the minimum focal length $f_a = f_b = d$. Both foci move to the lens center $z = 0$. For $\omega = 2$, the parameter $kd = \sqrt{3}$; that is, according to Eq. (2.112),

$$B_0 d = \begin{cases} \dfrac{2\sqrt{3}\,\Phi_z}{c} & \text{(relativistic)} \\[2ex] \sqrt{\dfrac{24\varphi_z^*}{\eta}} & \text{(nonrelativistic)} \end{cases} \tag{2.131}$$

Further increase in B_0 removes the object focus to the left of the lens, the focal length increases, a second focus ($n = 2$) appears, and so on. Relations (2.131) determine the optimal regime of Glazer's lens. The maximum focal-lens power is obtained for minimum d and corresponds to a large value of B_0. On the contrary, if we increase B_0, we must provide an adequate decrease in d; otherwise, the lens power would be reduced. In practical versions of strong magnetic lenses, the foregoing problems (increase in B_0 and reduction in d) are solved simultaneously by the application of suitable systems for magnetic field concentration. Figure 2.21 is a schematic of a magnetic lens with an axially symmetric magnetic iron circuit. Similar systems are typical in electron microscopes. It is a very promising application of superconducting lenses and lenses with permanent-magnet circuits in connection with progress in high-temperature superconductors and magnetic alloys.

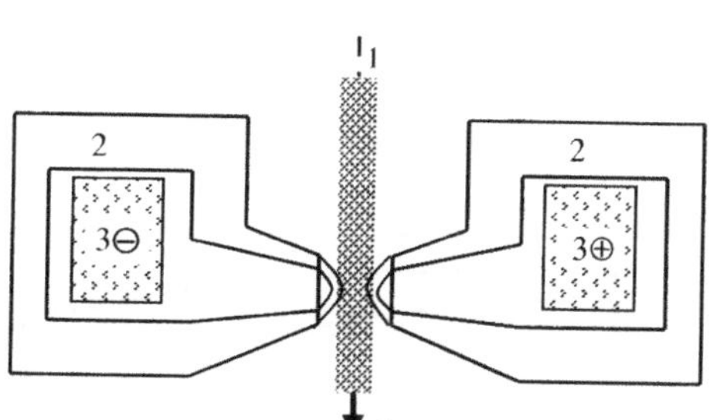

FIGURE 2.21 Magnetic lens with an iron circuit. 1, Electron beam; 2, iron shield; 3, winding.

2.7.4 Long Magnetic Lenses

When the lens length is much more than the pitch of the helical trajectory, there are many foci in the lens. The simplest case of a long lens is a homogeneous magnetic field. The electron-optical properties of a homogeneous magnetic field can easily be analyzed without using the equations of paraxial trajectories. In this case, the condition of the paraxial approximation $B_z(r, z) \approx B_z(0, z)$ holds automatically because the magnetic field is homogeneous. It remains to provide the condition $\alpha \ll 1$, where α is the slope of the trajectory to the axis. Then the axial velocity is $v_z = v\cos\alpha \approx v =$ const., where v is the total velocity, and the pitch $h = v_g T =$ const., where T is the gyroperiod.

Let us consider a bundle of trajectories emanating from point A_1 in Fig. 2.22 with different angles to the z-axes. They are helical trajectories with equal pitch that intersected at point B_1 on the same magnetic line. This is also valid for points A_2 and B_2. Therefore, if points A_1 and A_2 lie in a plane perpendicular **B**, the stigmatic image of object A_1A_2 in plane B_1B_2 will have a magnification $M = 1$. Thus, lenses with a homogeneous magnetic field transfer the image. Note that the screen may be placed at any distance nh, $n = 1, 2, \ldots,$ from the object. This lens is a typical multifocusing element.

A more general class of long magnetic lenses are axially symmetric lenses with a small inhomogeneous field. In this case and according to Eq. (1.110), guiding centers move along the curvilinear magnetic field lines with a constant velocity (Fig. 2.23) and drift slowly in the plane perpendicular to the magnetic field. The pitch of helical trajectories does not depend on the guiding center radius (in the paraxial approximation) or on the input angle of the trajectories. Therefore, in the plane

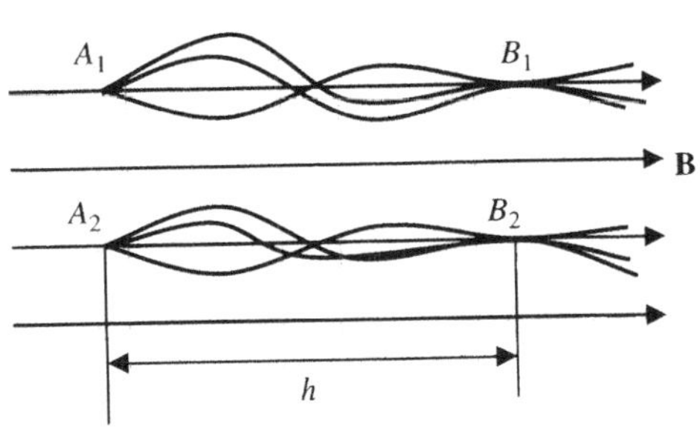

FIGURE 2.22 Focusing in an homogeneous magnetic field.

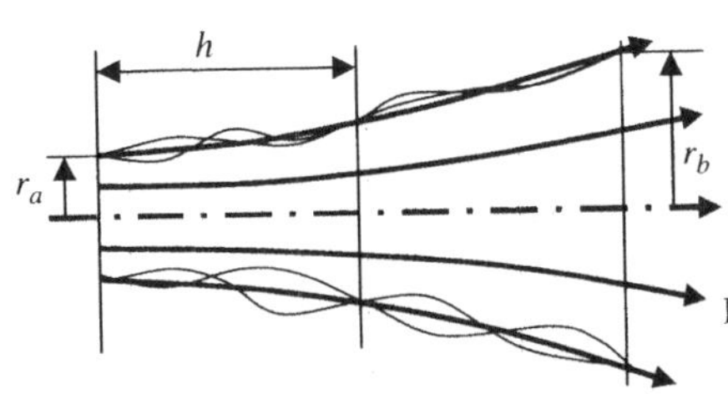

FIGURE 2.23 Long magnetic lens with an inhomogeneous field.

perpendicular to the magnet field, a stigmatic image of the object is formed. Obviously, corresponding points of the object and image lie on the same magnetic field lines. We see that these lenses give a magnified or reduced image of the object, depending on the sign of $B_a/B_b - 1$. It is easy to see that the condition of constant flux into the magnetic field tube leads to the following formula for lens magnification:

$$M = \sqrt{\frac{B_a}{B_b}} \tag{2.132}$$

We see that if $B_b < B_a$, the magnetic field lines form a widening tube and $M > 1$.

2.8 ABERRATIONS OF AXIALLY SYMMETRIC LENSES

The theory of electrostatic and magnetic lenses described earlier in the chapter was based on a number of assumptions. Their implementation provides the creation of an ideal lens that forms stigmatic and similar images. Let us recall these assumptions: (1) rigorous axial symmetry; (2) paraxial trajectory approximation; (3) energy homogeneity, including the absence of time-dependent processes; and (4) negligible space-charge fields and small effects of electron diffraction. Violation of at least one of these conditions leads to *aberrations* that are responsible for blurred or distorted images and complicate beam transport problems. The theory of aberrations is the most extensively studied area in both light optics and electron optics. Intensive investigations of aberrations in axially symmetric electron-optical systems were carried out in the 1930s in connection with studies of the electron transmission microscope problem. Detailed analyses of aberrations may be found in books by Glazer (1952), Grivet (1972), Klemperer and Barnett (1971), Sturrock (1955), and Szilagyi (1988), and in a number of related journal papers. Below, we consider briefly classification and basic notions of aberrations in axially symmetric lenses.

2.8.1 Geometric Aberrations

Geometric aberrations result from violation of paraxial trajectory approximation. The beginning of the geometric aberration theory in light optics dates from the middle of the nineteenth century, when the classification of geometric aberrations and corresponding aberration figures were developed. Note that the basic ideas in aberration theory regarding light and electron optics are very similar.

Strictly speaking, a paraxial ray is a ray infinitesimally close to the axis with infinitesimal transverse velocities of particles. Perturbation theory is used to estimate the influence of small but *finite* disturbances. According to this theory, a current radius vector of the particle can be represented as a perturbation of the paraxial approximation:

$$\rho = \rho_G + \Delta\rho \tag{2.133}$$

where $\rho(z) = re^{i\theta}$ is the complex radius vector and r and θ are cylindrical coordinates. $\rho_G(z)$ is the solution of paraxial equations (2.39) and (2.40), or (2.41) and

(2.42), and can be called a *paraxial trajectory* because in these equations, the axial potential φ_z^* is implied, although $\rho(z)$ and $\rho'(z)$ are finite. $\Delta\rho(z)$ is the perturbation. To find $\Delta\rho$, the fields and the potential are represented as a power series of ρ. In third-order aberration theory, expansions of ρ up to third order are used.

Consider a possible scheme for aberration analysis of electrostatic lenses in a nonrelativistic approximation. Let us take the trajectory equations (2.22) and (2.18), where we assume, according to these conditions, $\theta' = 0$ and $\Phi = \sqrt{2\varphi_0\varphi^*}$, that

$$\frac{d}{dz}\frac{\sqrt{\varphi^*}r'}{\sqrt{1+r'^2}} - \frac{\sqrt{1+r'^2}}{2\sqrt{\varphi^*}}\frac{\partial\varphi^*}{\partial r} = 0 \tag{2.134}$$

[We are reminded that according to Eq. (2.8), $\varphi^*(z) = \varphi(z) + \varphi_c$.] Expansions of the potential in terms of r up to $O(r^3)$ are

$$\varphi^*(r,z) = \varphi_z^* - \frac{r^2}{4}\varphi_z'', \qquad \frac{\partial\varphi^*}{\partial r} = -\frac{r}{2}\varphi_z'' + \frac{r^3}{16}\varphi_z^{\mathrm{IV}}$$

where $\varphi_z^* = \varphi^*(0,z)$. Also, $\sqrt{1+r'^2} \approx 1 + r'^2/2$. Substituting these equations into Eq. (2.134), we obtain up to third-order terms in the equation for the perturbation:

$$\Delta r'' + \Delta r'\frac{\varphi_z'}{2\varphi_z^*} + \Delta r\frac{\varphi_z''}{4\varphi_z^*} = \frac{r_G^3}{16}\left[\frac{\varphi_z^{\mathrm{IV}}}{2\varphi_z^*} - \left(\frac{\varphi_z''}{\varphi_z^*}\right)^2\right] + \frac{r_G^2 r_G'}{8}\left[\left(\frac{\varphi_z'''}{\varphi_z^*} - \frac{\varphi_z'\varphi_z''}{\varphi_z^{*2}}\right)\right]$$
$$- \frac{r_G r_G'^2}{4}\frac{\varphi_z''}{\varphi_z^*} - \frac{r_G'^3}{2}\frac{\varphi_z'}{\varphi_z^*} \tag{2.135}$$

Here $r_G = r_G(z)$ is a solution of the paraxial equation

$$r_G'' + \frac{\varphi_z'}{2\varphi_z^*}r_G' + \frac{\varphi_z''}{4\varphi_z^*}r_G = 0 \tag{2.136}$$

which is obtained using initial values $r_G(z_a)$ and $r_G'(z_a)$. Equation (2.135) is an ordinary linear differential equation with a right-hand side that is a known function of z. The solution of Eq. (2.135) can be represented as a linear combination of two linearly independent solutions $r_1(z)$ and $r_2(z)$ of the homogeneous equation that is the left-hand side of Eq. (2.135):

$$\Delta r = C_1(z)r_1(z) + C_2(z)r_2(z)$$

According to the method using parameter variation, functions $C_1(z)$ and $C_2(z)$ can be expressed as definite integrals along the z-axis of the function on the right-hand side of Eq. (2.135), and a combination of the functions $r_1(z)$ and $r_2(z)$ (i.e., integrals of the axial potential and its derivatives up to the fourth order). Note that the homogeneous equation on the left-hand side of Eq. (2.135) coincides with the equation of paraxial trajectories [Eq. (2.136)].

The perturbation in the image plane can be described by the function $[\Delta\rho(z_b)]_{r_a,\theta_a,r'_a} = e^{-i\theta_a}[\Delta(z_b)]_{r_a,r'_a}$, where r_a, θ_a, and r'_a are cylindrical coordinates and the slope of the trajectory to the axis in the object plane, and $z = z_b$ is the position of the image plane. According to Eq. (2.135), $\Delta r(z_b)$ is a polynomial of third degree in terms such as $r_a^3, r_a'^3, r_a^2 r'_a, r_a r_a'^2$. The coefficients of this expansion, $A_i = A_i(r_a, r'_a, \theta_a)$, found after integration, are called *geometrical aberration coefficients*. Note that in reality, realization of this simple procedure is very complicated. Other methods are used more often, in particular, the method of characteristic functions (see, e.g., Glazer, 1952; Sturrock, 1955; Szilagyi, 1988).

Let the paraxial trajectories give a stigmatic image of the object in the plane $z = z_b$, called a *Gauss plane*. According to Eq. (2.133), in the Gauss plane

$$\rho_b = \rho_{G,b} + (\Delta\rho)_b = M\rho_a + \Delta\rho_b \tag{2.137}$$

where M is the magnification. For the electrostatic lenses, M is a real value. If a cone of nonparaxial rays emerges from point ρ_a, it forms after intersecting with the image plane of some figure (an aberration figure). Classification of the geometrical aberrations is based on the characteristic features of aberration figures.

Below we use an analysis of aberration figures that is based on the introduction of an auxiliary *aperture plane* perpendicular to the z-axis. This plane has a circular aperture with the center on the axis at the point $z = z_d$. Previously, we determined the particular trajectory by setting the values ρ_a and ρ'_a. Let us replace ρ'_a by the complex radius vector ρ_d of the trajectory in the aperture plane. The particular trajectory is determined unambiguously by the complex values $\rho_a = r(z_a)e^{i\theta_a}$ and $\rho_d = r(z_d)e^{i\theta_d}$. The perturbation $\Delta\rho_b$ at the Gauss plane is now

$$\Delta\rho_b = f(\rho_a, \rho_d) \tag{2.138}$$

It is convenient to arrange the aperture plane after the lens so that the trajectories in the region $z_d < z < z_b$ will be straight lines. A diagram of aberration figure formation is shown in Fig. 2.24. Here $\rho_G(z)$ is one of the paraxial trajectories that satisfies the paraxial equation (2.136). These trajectories emerge from point ρ_a with different ρ'_a values, but they pass through the Gauss plane at the same point, $M\rho_a$.

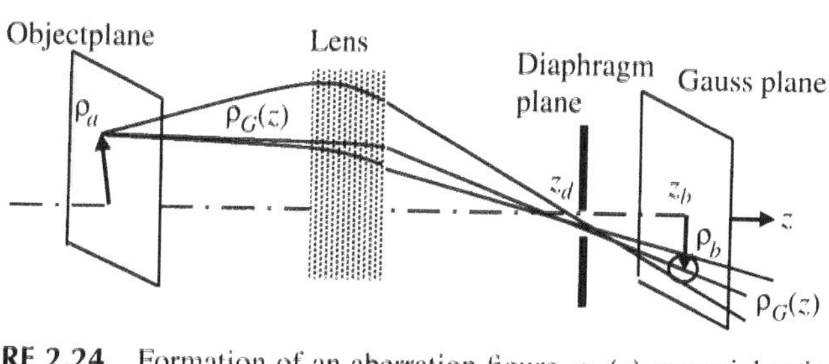

FIGURE 2.24 Formation of an aberration figure. $\rho_G(z)$, paraxial trajectory.

Take into account the small size of ρ_a and ρ_d. Let us to expand $\Delta\rho_b$ into a power series of the products $\rho_a, \rho_a^*, \rho_d,$ and ρ_d^* up to third order:

$$\Delta\rho_b = a_1\rho_a^2 + a_2\rho_a\rho_a^* + a_3\rho_a\rho_d + \cdots + b_1\rho_a^3 + b_2\rho_a^2\rho_a^*$$
$$+ b_3\rho_a\rho_a^{*2} + b_4\rho_a^2\rho_d + \cdots \tag{2.138a}$$

The number of terms in this sum can be reduced considerably if we take into account the axial symmetry of the system. In particular, if the system is turned as a whole by an angle α, the vectors ρ_a, ρ_d, and $\Delta\rho_b$ must be multiplied on $e^{i\alpha}$ and ρ^* on $e^{-i\alpha}$. But then the term $\rho_1\rho_2$ will be multiplied by e^{2ia} and the term $\rho_1\rho_2^*$ by 1. So all quadratic terms must be dropped. As far as the cubic terms are concerned, only terms like $\rho_1^2\rho_2^*$ yield multiplication by $e^{i\alpha}$. Therefore, the sum (2.138*a*) transforms into a thirdorder sum:

$$\Delta\rho_b = \mathbf{A}\rho_d^2\rho_d^* + \mathbf{B}\rho_a^*\rho_d^2 + \mathbf{C}\rho_a\rho_d\rho_d^* + \mathbf{D}\rho_a^2\rho_d^* + \mathbf{E}\rho_a\rho_a^*\rho_d + \mathbf{F}\rho_a^2\rho_a^* \tag{2.139}$$

Here the values **A**, **B**, **C**, **D**, **E**, and **F** are, *in general*, complex aberration coefficients. Altogether, we have 12 real coefficients. This number may be reduced further if one takes into account that after the lens, trajectories are straight lines. Therefore, the vector $\Delta\rho_b$ must not be rotated when $r_d = |\rho_d|$ is changed; that is,

$$\operatorname{Im}\left(\frac{\partial\Delta\rho_b}{\partial r_d}\right) = 0 \tag{2.140}$$

It is possible to show (Born and Wolf, 1965) that

$$\operatorname{Im}\mathbf{A} = \operatorname{Im}\mathbf{E} = 0, \qquad \mathbf{C} = 2\mathbf{B}^* \tag{2.141}$$

Equation (2.139) becomes

$$\Delta\rho_b = A\rho_d^2\rho_d^* + \mathbf{B}\rho_a^*\rho_d^2 + 2\mathbf{B}^*\rho_a\rho_d\rho_d^* + \mathbf{D}\rho_a^2\rho_d^* + \mathbf{E}\rho_a\rho_a^*\rho_d + \mathbf{F}\rho_a^2\rho_a^* \tag{2.142}$$

The number of independent aberration coefficients now is eight. For electrostatic lenses, the coefficients are real and there are five. Note that in obtaining Eq. (2.142), the concrete form of the paraxial equations has not been used. Therefore, properties of the aberrations, including aberration figures, must be the same in light and electrostatic electron optics. Let us consider these aberrations very briefly.

Spherical Aberration The problem of spherical aberration is fundamental in electron microscopy, in which objects are very small and arranged close to the optical axis. Let us arrange the point object exactly at the axes. According to Eq. (2.42), for $\rho_a = 0$ only the first term remains:

$$\Delta\rho_b = Ar_d^3e^{i\theta d} = Ar_d^3e^{i\theta_b} \tag{2.143}$$

We see that the trajectories leaving the point on the z-axis at different angles form in the Gauss plane a circle with radius Δr_b that is proportional to r_d^3. Note, however, that

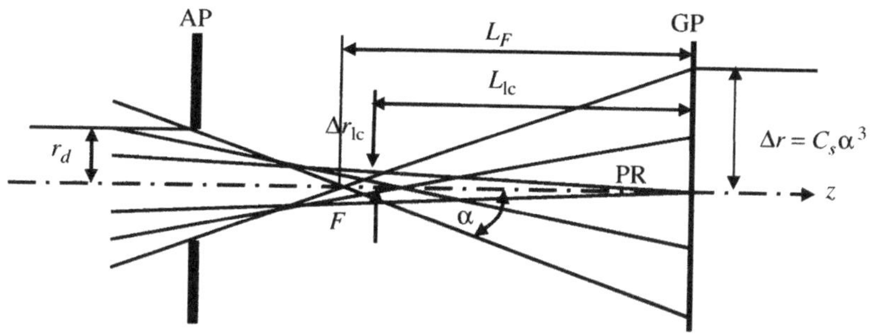

FIGURE 2.25 Formation of a spherical aberration. AP, aperture plane; GP, Gauss plane; PR, paraxial rays; Δr_{lc}, radius of least confusion.

the coefficient A in Eq. (2.142) depends on the aberration plane position. Therefore, it is difficult to compare different lenses using A as a criterion. One can show that r_d is proportional to the angle α between the ray that passes through an aperture edge and the z-axis in the Gauss plane, and we can write instead of Eq. (2.143),

$$\Delta r_b = C_s \alpha^3 \tag{2.144}$$

Here C_s is a spherical aberration coefficient that does not depend on the aperture plane position.

In Fig. 2.25, the formation of an aberration circle is shown. The boundary rays cross the z-axis at point F, which is to the left of the Gauss plane. In light optics, there are lenses with different signs of C_s, so aberration rays intersect the z-axis on different sides of the Gauss plane. Combining similar lenses allows one to design optical objectives with a very large aperture, practically without spherical aberration. The problem of spherical aberration in the electron optics is much more complicated. As determined by Scherzer (1936a,b), all electrostatic and magnetic lenses with electric and magnetic Laplace potential fields have the same sign of C_s, and the aberration cannot be eliminated. It is possible to decrease only the coefficient C_s in choosing a proper lens configuration. Therefore, the radical way is to use very small apertures. Fortunately, de Broglie's electron wavelength is sufficiently small to neglect (in certain limits) the effects of electron diffraction.

If the image plane moves to the left of the Gauss plane position, the radius of the aberration disk can be decreased. The radius $\Delta r_{lc} = \frac{1}{4}\Delta r_b$ (the *radius of least confusion*) is minimal when the displacement $L_{lc} = \frac{3}{4}L_F$, where L_F is the distance from the Gauss plane to the point where the boundary rays intersect with the z-axis (Fig. 2.25).

Coma This aberration is described by the sum of the second and third terms in Eq. (2.142). For the electrostatic lenses (the coefficient **B** is real), it reads

$$\Delta\rho_b = B(\rho_a^* \rho_d^2 + 2\rho_a \rho_d \rho_d^*) \tag{2.145}$$

Let rays enter from the point $r_a, \theta_a = 0$ on the object plane, and $\rho_d = r_d e^{i\theta_d}$. Then, taking into account that $\theta_d = \theta_b$, we obtain

$$\Delta\rho_b = Br_a r_d^2 (2 + e^{2i\theta_b}) \tag{2.146}$$

We see that the aberration figure is a circle. The center of this circle is located at a distance $2Br_a r_d^2$ from the image point on the Gauss plane, and the radius of the circle is $Br_a r_d^2$. The radius and the displacement of the center increase linearly with the object point's distance from the axis, but the radius of the circle and the displacement of the center in the image plane increase as a square of the aperture radius. Note that the shift of image points for other aberrations increases according to Eq. (2.142) as r_a^2. The image of the objects may be strongly disguised, as they are being covered by the coma when large apertures are used. When the object is close to the axis, spherical aberration dominates. It is also necessary to keep in mind that spherical aberration is the only type of aberration that cannot be suppressed by the proper lens configuration or combination of lenses (electrostatic and magnetic).

Astigmatism The fourth term in Eq. (2.142) describes astigmatism. For electrostatic lenses, the coefficient $\mathbf{D}$ is real ($\mathbf{D} = D$). The aberration figures are formed by rays emerging from the object point, O. Let us take $\rho_a = r_a$, and $\theta_d = \theta_b$. Then

$$\Delta\rho_b = Dr_a^2 r_d e^{-i\theta_b} \tag{2.147}$$

The aberration figure in the Gauss plane is a circle, but the direction of rotation of any point on this circle is opposite the rotation of the corresponding point on the aperture plane. It can be shown that horizontal and vertical rays passing through the aperture have different foci. In Fig. 2.26, the focus of the horizontal rays

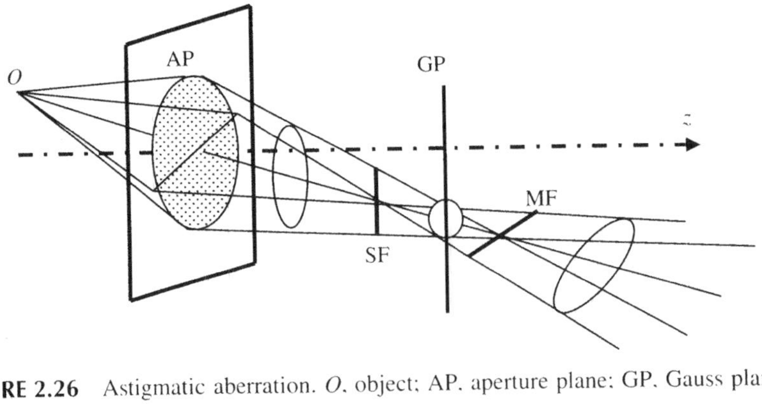

FIGURE 2.26 Astigmatic aberration. O, object; AP, aperture plane; GP, Gauss plane; SF, sagittal focus; MF, meridional focus.

(the sagittal focus) locates the left of the Gauss plane and the focus of the vertical rays (the meridional focus) locates the right.

If the image plane is moved from the Gauss plane to the left or right, the circle is transformed to ellipses. These ellipses degenerate into straight lines when the image plane is at the sagittal or meridional foci (see Fig. 2.26). Suppression of astigmatic aberration is possible in principle (unlike spherical aberration). According to Eq. (2.147), it becomes very important when wide bundles of rays, especially large objects or objects distant from the axis, are used.

Field Curvature This is the fifth term in Eq. (2.142). Because the coefficient E is real, aberrations are the same for electrostatic and magnetic lenses. Taking $\rho_a = r_a\varphi$ and $\varphi\theta_d = \theta_b$, we obtain

$$\Delta\rho_b = E r_a^2 r_d \, e^{i\theta_b} \tag{2.148}$$

The aberration figure is also a circle with the image point in the Gauss plane as the center. Now, the rotation direction of any point in the aperture and the corresponding point on the circle in the Gauss plane are the same. According to Eq. (2.148), the image of the axial point $r_a = 0$ is a point.

It can be shown (see, e.g., Rusterholz, 1950) that if the image plane is moved to the left of the Gauss plane, a circle in the object plane of definite radius r_a will be imaged as a sharp circle with the center on the axis. For each r_a, there is a definite position of the image plane with a stigmatic image of the corresponding circle. Therefore, a curved surface can be found that touches the Gauss plane at the axis with the stigmatic image of a wide object. It can be shown that this surface is a paraboloid for an electron lens that is concave toward the lens. The stigmatic image can be located in the plane if the object is disposed on the paraboloid surface.

Distortion The distortion is described by the sixth term in Eq. (2.142). For electrostatic lenses and $\rho_a = r_a$,

$$\Delta\rho_b = F r_a^3 \tag{2.149}$$

Because $\Delta\rho_b$ does not depend on r_d, the image in the Gauss plane is stigmatic. But the image points are displaced by $\sim r_a^3$ from the paraxial image. As a result, horizontal or vertical lines in the object plane will be imaged as curved on the Gauss plane. Specific features of distortions (i.e., barrel or cushion distortions) are determined by the sign of F (Klemperer and Barnett, 1971).

Anisotropic Aberrations Anisotropic aberrations are aberrations of magnetic lenses. They are described by the terms in Eq. (2.142), with complex aberration coefficients **B**, **D**, and **F**. These aberrations, called *anisotropic coma*, *anisotropic astigmatism*, and *anisotropic distortion*, do not have analogs in light optics. It can be shown (see, e.g., Glazer, 1950; Rusterholz, 1950) that unlike corresponding aberrations in electrostatic lenses, these aberration figures are twisted as a result of azimuthal shift.

2.8.2 Chromatic Aberration

The cause of chromatic aberration in light optics is the dependence on the refraction index of the wavelength. The use of nonmonochromatic lightning is therefore accompanied by blurring of the image. In electron optics, the particle energy is the analog of the wavelength in light optics. The formulas for the electron-optical index of refraction [Eq. (2.10) or (2.11)] include the reduced potential $\varphi^* = \varphi + \varphi_c$, where φ_c is the initial electron energy. The spread in values of φ_c is one of the causes of chromatic aberration of electron lenses. Another cause is the instability of voltage or current power supplies of electric and magnetic lenses.

The spread in the initial energy is determined by various factors. For thermionic cathodes, the energy spread is about 0.1 to 0.2 eV, and the potentials of electrodes in electric and magnetic lenses are usually higher than 1 kV; therefore, this factor is not important. For photocathodes, the energy spread is greater (on the order of 1 eV), and the chromatic aberration can be significant. Secondary electrons can have a much greater energy spread, so the energy spread of secondary particles reflected from the target can be comparable to the primary energy. In transmission electron microscopes, the factor that determines the spread is the inelastic scattering of electrons into a specimen. In this case the energy losses can reach ~10 eV. The instability of modern power supply sources is not large (it can be better than 10^{-5}). Therefore, this factor is not considerable, especially for electrostatic lenses if the potentials of all electrodes are set from a common source. The chromatic aberration does not disappear, even for nearby axis objects. Therefore, similar to spherical aberration, it is the most important limiting factor for electron microscopy.

Let us consider as an example the chromatic aberration of thin nonrelativistic (unipotential electrostatic or magnetic) lenses when the point object is on the axis (axial chromatic aberration). Assume that rays leave an axial object O into the cone at the angle α_a (Fig. 2.27). According to Eq. (2.54) and the Helmholtz–Lagrange formula [Eq. (2.62)], the edge ray in the image plane is

$$\tan \alpha_b = G \tan \alpha_a = \frac{1}{M} \tan \alpha_a \tag{2.150}$$

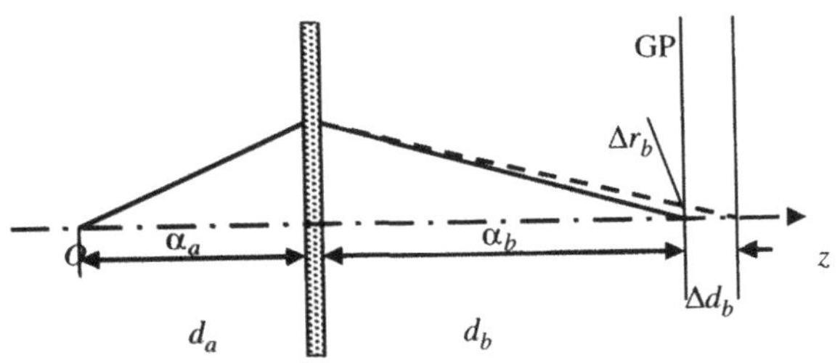

FIGURE 2.27 Axial chromatic aberration. GP, Gauss plane for $\Delta f = 0$; Δr_b, radius of least confusion.

because for the lenses being considered, $n_{e_0}(a) = n_{e_0}(b)$. The radius of least confusion of a disk is

$$\Delta r_b = \Delta d_b \tan\alpha_b = \frac{1}{M}\Delta d_b \tan\alpha_a \tag{2.151}$$

where Δd_b is the shift in the Gauss plane for perturbed trajectories. This value can be found from Newton's formula [Eq. (2.72)], which can be written as

$$\frac{1}{d_a} + \frac{1}{d_b} = \frac{1}{f} \tag{2.152}$$

(d_a and d_b are the distance between the lens and the object and the image, respectively). Taking into account that d_a = const., we obtain

$$\Delta d_b = \Delta f\left(\frac{d_b}{f}\right)^2 \approx M^2 \Delta f \tag{2.153}$$

Equation (2.153) is valid for $M \gg 1$. Substituting Eq. (2.153) in Eq. (2.51), we find that

$$\Delta r_b = M \tan\alpha_a \,\Delta f \tag{2.154}$$

Consider the two following cases.

1. *Initial energy spread in a thin unipotential electrostatic lens.* According to Eq. (2.79),

$$\frac{1}{f} = \frac{3}{16}\int_{-\infty}^{\infty}\left(\frac{\varphi_z'}{\varphi_z^*}\right)^2 dz \tag{2.155}$$

In this formula only the value φ_z^* depends on the initial energy. Therefore, it can readily be shown that

$$\Delta f = \frac{2\Delta\varphi}{\bar{\varphi}} f, \qquad \Delta r_b = M\frac{2\Delta\varphi}{\bar{\varphi}} f \tan\alpha_a \tag{2.156}$$

where $\bar{\varphi}$ is an average potential in the lens.

2. *Nonstability of the power supply of thin magnetic lenses.* In Eq. (2.107) for the focal length of a thin magnetic lens, the perturbation ΔB of the value $B(z)$ does not depend on z. Therefore,

$$\begin{aligned} \Delta f &= \left(\frac{\Delta\varphi}{\varphi} - \frac{2\Delta B}{B}\right) f \\ \Delta r_b &= M\left(\frac{\Delta\varphi}{\varphi} - \frac{2\Delta B}{B}\right) f \tan\alpha_a \end{aligned} \tag{2.157}$$

Note that the value of the radius of least confusion Δr_b is the linear function of the aperture angle, whereas for spherical aberration, $\Delta r_b \sim \alpha^3$. Therefore, at low aperture angles the performance of the lenses is limited by the chromatic aberration. For larger apertures, spherical aberration prevails.

A comprehensive analysis of chromatic aberration problems can be found, for example, in a book by Szilagyi (1988).

2.8.3 Disturbances of Axial Symmetry

Axially symmetric aberrations are the result of unavoidable errors in the manufacture of parts or assembly of systems. Errors in assembly, or *alignment aberration*, can be excluded almost entirely by proper adjustment. It is also possible to correct a finished system. If displacements from the axis are small in terms of the aperture radius, the errors have little effect on image quality.

Errors defined by the asymmetry of some parts of a system are more serious. Very often, it is impossible to either avert or eliminate them. For example, asymmetric magnetization of the iron circuits in magnetic lenses may be a result of the structural inhomogeneity of the magnet. A simple analytical model of asymmetric perturbations is presented by the quasielliptic distribution of the electric or magnetic potential. This model gives two different stigmatic images of the axial point object: one image for rays in a vertical plane and another for horizontal rays. As a whole, the picture of image formation is similar to that of astigmatic aberration (Fig. 2.26) and is called *axial astigmatism*. These aberrations in electron optics (especially in electron microscopy) are corrected by special devices, *stigmators*, that add the same type of asymmetry as that of the original device but with the opposite sign. Often, multipole lenses are used as stigmators (see, e.g., Szilagyi, 1988).

2.8.4 Space-Charge Fields

The effects of space-charge field can create problems, even in such low-current electron-optical topics as electron microscopy (especially, scanning microscopy) and electron beam lithography. As the dimensions of an electron beam spot in corresponding devices shrink, the importance of the influence of space-charge fields increases. There are two types of effects. The first is collective interaction, when every particle participates in the formation of the average field of a beam. If we neglect the dynamic effects in the beam (instabilities), this effect becomes the classical problem in space-charge beam theory. Similar problems are considered in Chapter 3. The space-charge fields determine the waist of the beam near the spot that defines the minimal thickness of the beam. This effect is added to the broadening of the beam upon the action of spherical and chromatic aberrations.

The second effect is related to particles collisions. As discovered by Boersh, individual interactions can lead not only to direct spatial broadening of the beam, but also to the effect of energy broadening with subsequent deterioration of beam quality (the *Boersh effect*) (see, e.g., Arnold et al., 1988).

2.8.5 Electron Diffraction

The impossibility of creating electron lenses without spherical aberration dictates the narrowing of apertures as a radical way of obtaining adequate high resolution in electron microscopy, which inevitably lowers the maximum current density on the image [Eq. (I.57)], although this difficulty can be overcome by sensitive modern methods of recording the optical information. But the principal limiting factor in both transmission and scanning microscopes is the effect of electron diffraction. According to the de Broglie formula, the electron wavelength is

$$\lambda = \frac{h}{p} = \frac{hc}{e_0\Phi} = \frac{hc}{e_0\sqrt{\varphi^*(\varphi^* + 2\varphi_0)}} \tag{2.158}$$

where $h = 6.6261 \times 10^{-34}$ J/s is Planck's constant. For $\varphi^* = 10$ kV, $\lambda \approx 1.2 \times 10^{-11}$ m $= 0.12$ Å; for $\phi^* = 500$ kV, $\lambda \approx 1.4 \times 12^{-12}$ m $= 0.014$ Å. These small wavelengths do not exclude, however, large sizes of the Airy disk for small apertures. So for object aperture angle α_a, the image aperture angle is $a_b = \alpha_a/M$ [see Eq. (2.62)]. The radius of the first Fresnel zone in the image plane is

$$\Delta r_F = \frac{0.6\lambda}{\alpha_b} = \frac{M0.6\lambda}{\alpha_a} \tag{2.159}$$

Then the radius of the Airy disk in the object plane is

$$\Delta r_{da} = \frac{\Delta r_F}{M} = \frac{0.6\lambda}{\alpha_a} \tag{2.160}$$

So for $\alpha_a = 1 \times 10^{-3}$ and for $\varphi^* = 10$ kV and 500 kV, $\Delta r_{da} = 72$ Å and 8 Å, correspondingly. *The last value essentially limits the resolution of an electron transmission microscope.* If α_a is increased, problems related to chromatic and spherical aberrations will arise.

2.9 COMPARISON OF ELECTROSTATIC AND MAGNETIC LENSES. TRANSFER MATRIX OF LENSES

2.9.1 Comparison of the Optical Power of Electrostatic and Magnetic Lenses

Averaging over the lens length fields and potentials in Eqs. (2.77) and (2.106), and for the focal lengths of electrostatic and magnetic lenses, we obtain

$$\frac{1}{f_E} = \frac{\eta^2\overline{E_z^2}(\overline{\gamma^2} + 2)}{4c^4\overline{\beta_z^4}\overline{\gamma^4}}L, \qquad \frac{1}{f_B} = \frac{\eta^2\overline{B_z^2}}{4c^2\overline{\beta_z^2}\overline{\gamma^2}}L \tag{2.161}$$

In a nonrelativistic approximation, these relations become

$$\frac{1}{f_E} = \frac{3}{16}\frac{\overline{E_z^2}}{\varphi_z^2}L, \qquad \frac{1}{f_B}\frac{\eta\overline{B_z^2}}{8\varphi_z}L \tag{2.162}$$

where L is the lens length. Then, omitting the signs of the averages, we find that

$$F_{BE} \equiv \frac{f_B}{f_E} = \frac{E_z^2(\gamma^2+2)}{B_z^2\beta_z^2c^2\gamma^2} \quad \text{(relativistic)}$$

$$F_{BE} = \frac{3}{2\eta}\left(\frac{E_z}{B_z}\right)^2\frac{1}{\varphi_z} \quad \text{(nonrelativistic)} \tag{2.163}$$

Let us take intermediate values of E_z and B_z: $E_z = 3 \times 10^6$ V/m and $B_z = 0.2$ T. We find in the nonrelativistic approximation that $F_{BE} \approx 2000\ /\varphi_z$. So the optical power of magnetic lenses is higher than that of electrostatic lenses for all $\varphi_z > 2$ kV. If one takes the limiting values $E_z = 10^7$ V/m (a breakdown strength) and $B_z = 3$ T (a saturation of the ferromagnetic), we obtain $F_{BE} \approx 100\ \varphi_z$. In this case, the optical power of magnetic lenses is already higher for $\varphi_z > 100$ V.

Obviously, for relativistic energies, magnetic lenses have a clear advantage. Furthermore, the relativistic electrostatic lenses for values of electrical fields less than the breakdown values have unacceptably low optical power. One advantage of magnetic lenses is the possibility to place parts of the lens outside the vacuum chamber. That essentially simplifies the correction of lens parameters. The advantages of electrostatic lenses are small mass, compactness, and a simpler power supply.

2.9.2 Second-Order Focusing of Axially Symmetric Lenses

For thin electrostatic lenses according to the first of Eqs. (2.162),

$$\frac{L}{F_E} = \frac{3}{16}\left(\frac{E_zL}{\varphi_z}\right)^2 \ll 1$$

The parameter $(E_zL/\varphi_z)^2$ is the square of the ratio of the axial voltage on the lens to the average energy of the electron. The smallness of this parameter leads to weak focusing by axially symmetric electrostatic lenses. The same conclusion is valid for magnetic lenses [Eq. (2.162)]. The cause of this effect is the absence in both types of lenses of fields that could provide constant radial acceleration of electrons along the lens. The radial component of an electric field according to Eq. (2.29) is proportional to E_z', which changes the sign in electrostatic lenses with a limited field extent.

For magnetic lenses, the weak (second-order) focusing [Eq. (2.161)] is the result of the zero azimuthal component of the magnetic field B_θ in axially symmetric systems. When particles moving parallel to the axis enter the lens, the radial acceleration is zero. Acceleration begins only after the initiation of rotational motion whose velocity is proportional to the axial component of the magnetic field B_z. This component further determines the radial acceleration that is proportional to B_z^2.

2.9.3 Transfer Matrix of Lenses

Consider the general solution of the linear paraxial equation

$$r(z) = Pu(z) + Qv(z)$$

where u and v are linearly independent partial solutions of the equation. Let the values $r(z_1) = r_1$, $r'(z_1) = r_1'$ belong to some trajectory in the plane $z = z_1$. We obtain

$$\begin{aligned} r_1 &= Au(z_1) + Bv(z_1) \\ r_1' &= Au'(z_1) + Bv'(z_1) \end{aligned} \tag{2.164}$$

The solution of Eq. (2.164) is

$$A = \frac{r_1 v'(z_1) - r_1' v(z_1)}{W_1}, \qquad B = \frac{r_1' u(z_1) - r_1 u'(z_1)}{W_1} \tag{2.165}$$

where $W_1 = u(z_1)v'(z_1) - u'(z_1)v(z_1)$ is the Wronskian of the paraxial equation, which is nonzero if u and v are linearly independent. Substitute Eq. (2.165) into Eq. (2.164) written for another point, z_2, of the same trajectory. We obtain the linear equation

$$\begin{aligned} r_2(z_2) &= m_{11} r_1 + m_{12} r_1' \\ r_2'(z_2) &= m_{21} r_1 + m_{22} r_1' \end{aligned} \tag{2.166}$$

where the m_{ij} are known functions of z_1 and z_2. We can write Eqs. (2.166) in matrix form as

$$\begin{pmatrix} r_2 \\ r_2' \end{pmatrix} = \begin{pmatrix} m_{11} & m_{12} \\ m_{21} & m_{22} \end{pmatrix} \begin{pmatrix} r_1 \\ r_1' \end{pmatrix} = M_{21} \begin{pmatrix} r_1 \\ r_1' \end{pmatrix} \tag{2.167}$$

The matrix M_{21}, called the *transfer matrix*, sets the linear relation between vectors $\begin{pmatrix} r_1 \\ r_1' \end{pmatrix}$ and $\begin{pmatrix} r_2 \\ r_2' \end{pmatrix}$, which belong to the same trajectory in two arbitrary sections of the electron-optical system. The matrix elements m_{ij} depend only on z_1 and z_2. Therefore, any pair of conjugated vectors in the z_1 and z_2 planes will be transferred by the same matrix, M_{21}. Obviously, successive transfer through a number of the sections is performed by multiplication of the transfer matrix for each section. Consider some examples.

1. *Transfer matrix of a drift space.* Let the vectors (r_1/r_1') and (r_2/r_2') belong to the planes z_1 and $z_2 = z_1 + l$ in the drift space where the trajectory is rectilinear. We obtain

$$r_2 = r_1 + lr_1'$$
$$r_2' = r_1' = 0r_1 + r_1'$$

so the transfer matrix of this section is

$$M_{21} = \begin{pmatrix} 1 & l \\ 0 & 1 \end{pmatrix} \tag{2.168}$$

2. *Transfer matrix of a thin immersion lens.* Take planes z_1 and z_2 at the left and right boundaries of the lens in Fig. 2.28. Placing the object A and the image B on the axis, we obtain

$$r_2 = r_1 = r_1 + 0r_1' \tag{2.169}$$

$$d_1 = \frac{r_1}{\tan\alpha_1} = \frac{r_1}{r_1'}, \quad d_2 = \frac{-r_2}{r_2'} \tag{2.170}$$

Substituting Eqs. (2.170) into Newton's formula $f_1/d_1 + f_2/d_2 = 1$, we obtain

$$r_2' = -\frac{1}{f_2}r_1 + \frac{f_1}{f_2}r_1' \tag{2.171}$$

Let us combine Eqs. (2.169) and (2.171) into a transfer matrix:

$$M_{21} = \begin{pmatrix} 1 & 0 \\ -\dfrac{1}{f_2} & \dfrac{f_1}{f_2} \end{pmatrix} \tag{2.172}$$

For a unipotential immersion lens, $f_1 = f_2 = f$ and

$$M_{21} = \begin{pmatrix} 1 & 0 \\ -\dfrac{1}{f} & 1 \end{pmatrix} \tag{2.173}$$

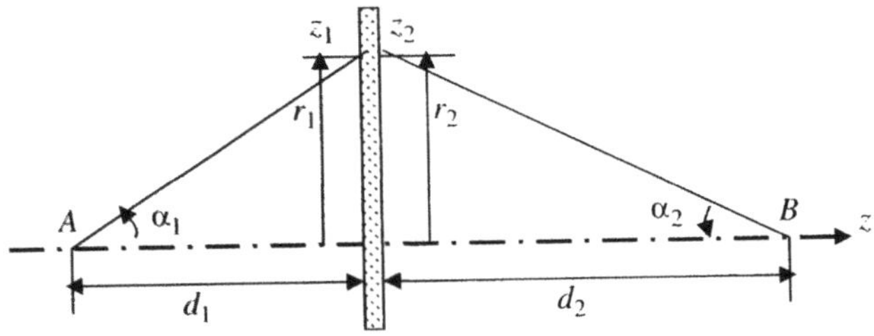

FIGURE 2.28 Transfer matrix of a thin electron lens.

3. *Two thin unipotential lenses separated by a drift space of length l.* The transfer matrix is equal to the product of three matrices [Eqs. (2.173) and (2.168)]:

$$M = \begin{pmatrix} 1 & 0 \\ -\dfrac{1}{f_2} & 1 \end{pmatrix} \begin{pmatrix} 1 & l \\ 0 & 1 \end{pmatrix} \begin{pmatrix} 1 & 0 \\ -\dfrac{1}{f_1} & 1 \end{pmatrix} \tag{2.174}$$

Performing the matrix multiplication, we find that

$$M = \begin{pmatrix} 1 - \dfrac{l}{f_1} & l \\ -\dfrac{1}{f_1} - \dfrac{1}{f_2} + \dfrac{l}{f_1 f_2} & 1 - \dfrac{l}{f_2} \end{pmatrix} \tag{2.175}$$

Let the distance between the lenses be small: $l/f \ll 1$. Then we can write Eq. (2.75) as

$$M = \begin{pmatrix} 1 & 0 \\ -\dfrac{1}{F} & 1 \end{pmatrix} \tag{2.176}$$

where

$$\frac{1}{F} = \frac{1}{f_1} + \frac{1}{f_2} - \frac{l}{f_1 f_2} \tag{2.177}$$

So a pair of nearly unipotential thin lenses is equivalent to a single thin lens with the focal length F. It is interesting to note that the combination of a divergent and a convergent lens with equal focal lengths ($f_1 = -f_2$) always forms a convergent lens ($F > 0$).

2.10 QUADRUPOLE LENSES

2.10.1 Introduction

As we have seen, the main disadvantage of axially symmetric lenses is the second order of focusing, which is the result of a comparative weakness of the net radial acceleration in axially symmetric systems with a simple connective cross section. To avoid this problem, one possibility is to apply coaxial electrostatic or magnetic configurations. But these systems obviously are useless in electron microscopy, where the objects and images must be arranged near the axis. Abandoning the axial symmetry changes the problem (although it brings new problems). The simplest system of this type is a flat capacitor oriented along the z-axis. In this system, the transverse electric field is constant along the axis, and excluding the end fields, the axial field is zero. Unfortunately, this device can be used only as a beam deflecting system, because in this case, the function of image formation is absent. However, there are systems without axial symmetry but with a higher degree of symmetry than that of this flat capacitor. The corresponding fields are obtained as solutions of the two-dimensional Laplace equation

$$r\frac{\partial}{\partial r}\left(r\frac{\partial\varphi}{\partial r}\right) - \frac{\partial^2\varphi}{\partial\theta^2} = 0 \tag{2.178}$$

Consider a system of $2N$ identical electrodes placed equally in the azimuthal direction. The system has N planes of symmetry. If one of these planes coincides with the (x, z) plane, the solution of Eq. (2.178) can be written as (Hawkes, 1970; Szilagyi, 1988)

$$\varphi(r, \theta) = \sum_n a_{nN} r^{nN} \cos(nN\theta) \tag{2.179}$$

Naturally, this representation is valid for a potential independent of z. An analogous formula can be written for the magnetic potential. Every term of this sum satisfies Eq. (2.178). In general, a sum of the terms in Eq. (2.179), including $n = \infty$, is required to satisfy the boundary conditions. However, it is possible to use single terms of the sum as solutions of boundary value problem, setting the electrode configuration as an equipotential surface of the type

$$r^{nN} \cos nN\theta = \text{const.}, \qquad nN = 1, 2, \ldots \tag{2.180}$$

In general, these equations represent the electrode geometry of multipoles (Hawkes, 1970). The quadrupole distribution corresponds to the term $n = 1$ and two symmetry planes (i.e., $nN = 2$):

$$\varphi(r, \theta) = a_2 r^2 \cos 2\theta$$

In Cartesian coordinates ($x = r\cos\theta$, $y = r\sin\theta$), the quadrupole potential is

$$\varphi(x, y) = a_2(x^2 - y^2) = \frac{\varphi_0}{a^2}(x^2 - y^2) \tag{2.181}$$

Let the potential of the electrode be φ_0. Then the equation of the electrode surface is

$$x^2 - y^2 = a^2 \tag{2.182}$$

These are two horizontal hyperboles with a distance between vertices of $2a$. The other two surfaces are symmetrical vertical hyperboles corresponding to the potential $-\varphi_0$ (Fig. 2.29).

The components of the electrical field according to Eq. (2.181) are

$$E_x = E_0\frac{x}{a}, \qquad E_y = -E_0\frac{y}{a} \tag{2.183}$$

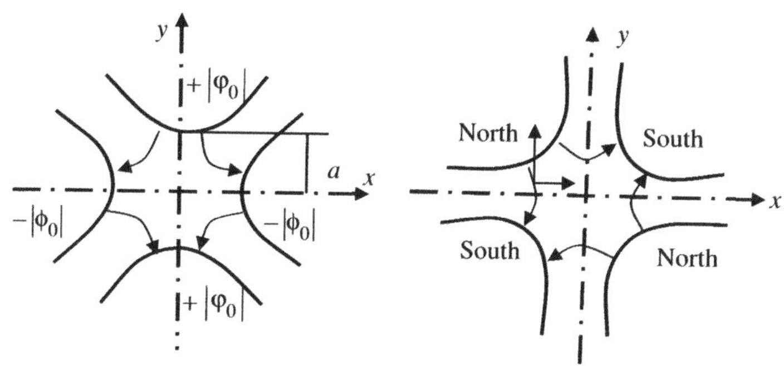

FIGURE 2.29 Electric and magnetic quadrupole lenses. The arrows show the field lines and the directions of the magnetic force acting on electrons that move "under the paper."

where $E_0 = -2\varphi_0/a$ is the electric field in the hyperbole vertices. So if $\varphi_0 < 0$, the vertical electrodes have a positive potential, and $E_0 > 0$. Correspondingly, the components of the magnetic field are

$$B_y = B_0\frac{x}{a}, \qquad B_x = B_0\frac{y}{a} \tag{2.184}$$

These simple relations are very significant. They mean that the field components are linear functions of the transverse coordinates in the *entire volume* of the system. In the practice lens, lengths are limited. A structure of fields at the lens' ends (fringe fields) is like a multipole and is essentially more complicated than one described by Eq. (2.183). Obviously, the increase in lens length weakens the influence of the fringe fields. A detailed discussion of this topic and of the analysis of the influence of nonhyperbolic perturbations on electrode configurations may be found in a book by Hawkes (1970).

2.10.2 Equation of Paraxial Trajectories

In general, trajectory equations for the field [Eqs. (2.183) and (2.184)] contain nonlinear terms. However, the equations of the paraxial trajectories are linear (see, e.g., Szilagyi, 1988):

$$\begin{aligned} x'' &= -\frac{\eta E_0}{\beta^2 c^2 \gamma a}x + \frac{\eta B_0}{\beta c \gamma a}x \\ y'' &= \frac{\eta E_0}{\beta^2 c^2 \gamma a}y - \frac{\eta B_0}{\beta c \gamma a}y \end{aligned} \tag{2.185}$$

Let us write the equations for the electric and the magnetic quadrupoles separately:

$$x'' + k_e^2 x = 0, \qquad y'' - k_e^2 y = 0 \tag{2.186}$$

$$x'' - k_b^2 x = 0, \qquad y'' + k_b^2 y = 0 \tag{2.187}$$

where

$$k_e^2 = \frac{\eta E_0}{\beta^2 c^2 \gamma a} = \frac{\eta(\varphi^* + \varphi_0)^2 E_0}{c^2 \varphi^* (\varphi^* + 2\varphi_0)\gamma a} \tag{2.188}$$

$$k_b^2 = \frac{\eta B_0}{\beta c \gamma a} = \frac{\eta(\varphi^* + \varphi_0) B_0}{c\sqrt{\varphi^*(\varphi^* + 2\varphi_0)}\gamma a} \tag{2.189}$$

In a nonrelativistic approximation,

$$k_e^2 = \frac{E_0}{2a\varphi^*}, \qquad k_b^2 = \frac{\eta B_0}{a\sqrt{2\eta\varphi^*}} \tag{2.190}$$

Equations (2.186) and (2.187) are the equations of the harmonic oscillator. The solution of these equations is a linear combination of harmonic or hyperbolic functions:

$$x = a\cos k_e z + b\sin k_e z, \qquad y = c\cosh k_e z + d\sinh k_e z \tag{2.191}$$

Also for the magnetic quadrupole,

$$x_b = a_b \cosh k_b z + b_b \sinh k_b z, \qquad y_b = c_b \cos k_b z + d_b \sin k_b z \tag{2.192}$$

Analysis of the image properties of the lenses is better performed using transfer matrices.

2.10.3 Transfer Matrix

Consider an electric quadrupole lens. Let both the object and image be situated outside the lens. Assume that the x-coordinate of the ray at the left lens boundary ($z_1 = 0$) is x_1 and the ray slope is x_1'. According to the first of Eqs. (2.191), $x_1 = a$ and $x_1' = bk_e$. Then the ray equation (2.191) in the (x, z) plane can be written as

$$x = x_1 \cos k_e z + \frac{x_1'}{k_e}\sin k_e z \tag{2.193}$$

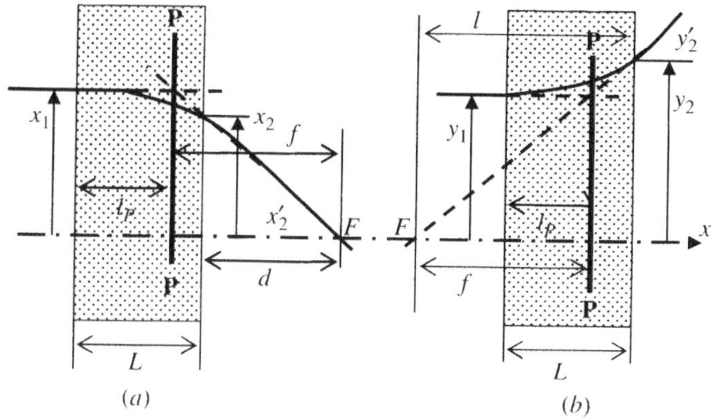

FIGURE 2.30 Cardinal elements of (*a*) convergent and (*b*) divergent quadrupole lenses. P–P, principal plane; F, focus.

The coordinate and the ray slope on the right-hand boundary of the lens can be obtained using Eq. (2.191) in the plane $z = L$ (Fig. 2.30):

$$
\begin{aligned}
x_2 &= x_1 \cos k_e L + x_1' \frac{\sin (k_e L)}{k_e} \\
x_2' &= -x_1 k_e \sin k_e L + x_1' \cos k_e L
\end{aligned}
\tag{2.194}
$$

Thus, the transfer matrix in the (x, z) plane is

$$
\begin{vmatrix} m_{11} & m_{12} \\ m_{21} & m_{22} \end{vmatrix} = \begin{vmatrix} \cos k_e L & \dfrac{\sin k_e L}{k_e} \\ -k_e \sin k_e L & \cos (k_e L) \end{vmatrix}
\tag{2.195}
$$

Using the second of Eqs. (2.192), we obtain the transfer matrix in the (y, z) plane:

$$
\begin{vmatrix} \cosh k_e L & \dfrac{\sinh (k_e L)}{k_e} \\ k_e \sinh k_e L & \cosh k_e L \end{vmatrix}
\tag{2.196}
$$

We are reminded that these formulas are valid in neglecting the fringe fields at the ends of the lens.

2.10.4 Cardinal Elements

Let us find the focal lengths and position of the principal planes. We assume that the lens foci are located in the image (the convergent lens) or object (the divergent lens) spaces. The principal planes are defined as intersection planes of the ray continuations from the object and image spaces (dashed lines in Fig. 2.30).

Assume that the trajectory is in the (x, z) plane (Fig. 2.30a) and that the electron enters the lens parallel to the optical axis. Hence, the beginning vector of the particle is $\begin{vmatrix} x_1 \\ 0 \end{vmatrix}$. The vector in the output plane of the lens is

$$\begin{vmatrix} x_2 \\ x_2' \end{vmatrix} = \begin{vmatrix} m_{11} & m_{12} \\ m_{21} & m_{22} \end{vmatrix} \begin{vmatrix} x_1 \\ 0 \end{vmatrix} \tag{2.197}$$

or $x_2 = m_{11}x_1$ and $x_2' = m_{21}x_1$

The focal length f and distance l_p between the left lens boundary and principal planes, respectively, are equal:

$$f = -\frac{x_1}{x_2'} = \frac{-1}{m_{21}}, \qquad l_P = L - \frac{x_1 - x_2}{-x_2'} = L + \frac{1 - m_{11}}{m_{21}} \tag{2.198}$$

Substituting m_{ij} for Eq. (2.195), we obtain

$$f = \frac{1}{k_e \sin k_e L} \tag{2.199}$$

$$l_P = L\left[1 - \frac{\tan(k_e L/2)}{k_e L}\right] \tag{2.200}$$

For the trajectory in the (y, z) plane, we obtain, according to Fig. 2.30b and Eqs. (2.196) and (2.198).

$$f = -\frac{1}{k_e \sinh k_e L}, \qquad l_P = L\left[1 - \frac{\tanh(k_e L/2)}{k_e L}\right] \tag{2.201}$$

The parameters of the magnetic quadrupole lens are obtained by substitution of k_e for k_b in Eqs. (2.199)–(2.201).

We assume that the foci are situated outside the lens. Therefore, we require that $L < |f|$ or, according to Eqs. (2.199) and (2.201),

$$k_e L \sin k_e L < 1, \qquad k_e L \sinh k_e L < 1 \tag{2.202}$$

According to these conditions, $L_{\max} k_{e,b} = 1$ approximately. Let us also assume that $L_{\min} = a$, because neglecting the fringe fields for $L < a$ is impossible. So we obtain the following proper range for L:

$$a < L < 1/k_{e,b} \tag{2.203}$$

Note that the criterion $k_{e,b}L < 1$ leads to the conclusion that quadrupole lenses, especially magnetic quadrupole lenses, must be relativistic devices. For example, for $B_0 = 1$ T, $L = 2$ cm, and $a = 0.5$ cm, the criterion indicated is satisfied when $\gamma > 50$.

In Table 2.1, values of the parameters for different values of $k_{e,b}L$ are presented. The transfer matrix of a *thin quadrupole lens* can be found from Eqs. (2.195) and (2.196), assuming that $k_e L \ll 1$. Then, $m_{11} = 1$, $m_{12} = L$, $m_{21} = -1/f$, and $m_{22} = 1$. Note that $m_{12} = 0$ is usually assumed also. This change hardly influences the value of x_2. So the transfer matrix is

$$M_{\text{th}} = \begin{vmatrix} 1 & 0 \\ -\dfrac{1}{f} & 1 \end{vmatrix} \tag{2.204}$$

where

$$\frac{1}{f_e} \approx \pm k_e^2 L = \pm \frac{\eta E_0}{\beta^2 c^2 \gamma a} L, \qquad \frac{1}{f_b} \approx \pm k_b^2 L = \pm \frac{\eta B_0}{\beta c \gamma a} L \tag{2.205}$$

The $\pm$ signs refer to convergent and divergent lenses, respectively. Assume that the object and image are located at distances p and q from the lens, respectively. The transfer matrix of the system is equal to the product of matrix (2.204) and the two matrices of the drift object and drift image spaces:

$$M = \begin{vmatrix} 1 & p \\ 0 & 1 \end{vmatrix} M_{\text{th}} \begin{vmatrix} 1 & q \\ 0 & 1 \end{vmatrix} = \begin{vmatrix} 1 - \dfrac{p}{f} & p + q - \dfrac{pq}{f} \\ -\dfrac{1}{f} & 1 - \dfrac{q}{f} \end{vmatrix} \tag{2.206}$$

The condition of the stigmatic image formation is equal to zero of element m_{12} of matrix M. Indeed, coordinate x_2 of the image point in this case will not depend on the slope x_1' of the rays leaving point x_1 of the object [see Eq. (2.162)]. So

$$p + q - \frac{pq}{f} = 0 \quad \text{or} \quad \frac{1}{p} + \frac{1}{q} = \frac{1}{f}$$

This relation coincides with Newton's formula [Eq. (2.72)] for a thin unipotential lens. We see that a thin quadrupole lens forms a real stigmatic image in the converging

TABLE 2.1 Parameters of Quadrupole Lenses for Different Values of $k_{e,b}L$

	$k_{e,b}L$					
	0.1		0.5		1.0	
	$\dfrac{f}{L}$	$\dfrac{l_p}{L}$	$\dfrac{f}{L}$	$\dfrac{l_p}{L}$	$\dfrac{f}{L}$	$\dfrac{l_p}{L}$
Convergent lens	100	0.50	4.2	0.49	1.2	0.45
Divergent lens	−100	0.50	−3.8	0.51	−0.85	0.54

plane. In the diverging plane, the stigmatic image is absent. Therefore, the image of the point in the plane determined by Eq. (2.206) is a line.

2.10.5 Quadrupole Doublets

The quadrupole doublet is a system of two quadrupole lenses turned by $\pi/2$ with respect to each other, so that in each plane of symmetry there are both convergent and divergent lenses. Consider the properties of doublets consisting of two thin quadrupole lenses separated by a drift section of length d. The transfer matrix of this doublet can be written as

$$M_d = \begin{vmatrix} 1 & 0 \\ -\dfrac{1}{f_1} & 1 \end{vmatrix} \begin{vmatrix} 1 & d \\ 0 & 1 \end{vmatrix} \begin{vmatrix} 1 & 0 \\ -\dfrac{1}{f_2} & 1 \end{vmatrix} = \begin{vmatrix} 1 - \dfrac{d}{f_2} & d \\ -\dfrac{1}{f_1} - \dfrac{1}{f_2} + \dfrac{d}{f_1 f_2} & 1 - \dfrac{d}{f_1} \end{vmatrix} \tag{2.207}$$

The focal power of the doublets is

$$\frac{1}{F} = -m_{21} = \frac{1}{f_1} + \frac{1}{f_2} - \frac{d}{f_1 f_2} \tag{2.208}$$

Let us consider the version of the doublet consisting of lenses with equal focal lengths $f_1 = -f_2 = f$. So

$$\frac{1}{F} = \frac{d}{f^2}, \qquad m_{11} = 1 + \frac{d}{f}, \qquad m_{21} = -\frac{d}{f^2}, \qquad m_{22} = 1 - \frac{d}{f} \tag{2.209}$$

We see that the doublet is likened to a thick convergent lens with equal focal lengths in both planes. Exchanging lenses 1 and 2 is equivalent to a change of sign of f. Let $f = f_1 > 0$. It means that the ray described by the transfer matrix M_d in Eq. (2.207) is in the converging plane of lens 1 (the lens nearest the object). According to Eqs. (2.198) and (2.209), the distance from the left boundary of lens 1 to the principal plane of the doublet is

$$l_P = L + \frac{1 - m_{11}}{m_{21}} = L + f \tag{2.210}$$

Obviously, in the diverging plane of lens 1, $l_P = L - f$, so the positions of the principal planes and the foci in both planes are sufficiently different (i.e., the doublet is a strong astigmatic lens). According to Eqs. (2.209) and (2.205), the optical powers of the thin electrostatic and magnetic quadrupole doublets are

$$\frac{1}{F_e} = \frac{\eta^2 d L^2}{\beta^4 c^4 \gamma^2 a^2} E_0^2, \qquad \frac{1}{F_b} = \frac{\eta^2 d L^2}{\beta^2 c^2 \gamma^2 a^2} B_0^2 \tag{2.211}$$

It is seen that the doublet focusing is of second order. The reader may recall that in the qualitative discussion of the focusing mechanism of an immersion lens (Section 2.6.2), the focusing effect was interpreted as a result of the consecutive action of both lens halves, one-half a divergent lens and one-half a convergent lens.

Compare the optical power of an axially symmetric magnetic lens (superscript "as") and a quadrupole magnetic doublet (superscript "d"). According to Eqs. (2.161) and (2.211), we have

$$\frac{F_b^{(\mathrm{as})}}{F_b^{(\mathrm{d})}} = \frac{4dL^2}{a^2\Lambda} \tag{2.212}$$

where Λ is the length of an axially symmetric lens. Assume for purposes of estimation that $\Lambda = 2L$ and $d = L$. Then we obtain $F_b^{(\mathrm{as})}/F_b^{(\mathrm{d})} = 2(L/a)^2$. Only for $L/a \sim 1$ do an axially symmetric lens and a quadrupole doublet have approximately equal optic powers. But in real quadrupole doublets, $L/a \gg 1$ (up to 10), and therefore the superiority of the doublets is beyond doubt.

2.10.6 Quadrupole Triplets

A quadrupole triplet is a system of three quadrupole lenses. The first and third ("outer") quadrupoles are oriented so that their focusing planes coincide. The second ("central") quadrupole is turned by $\pi/2$. In symmetric triplets, the outer quadrupoles are identical and the drift spaces are equal.

It has been shown (Regenstreif, 1967) that the triplet could provide a stigmatic image with equal magnifications in both planes, although the corresponding distance between an object and the first quadrupole is not arbitrary but depends on the focal lengths of the triplet's elements. It is interesting that the triplet can be considered as a thin lens because the principal planes for the rays in both symmetric planes for a symmetric triplet are closed to the center of the triplet.

2.10.7 Applications of Quadrupole Lenses

Transport of Charged Particles in Accelerators A periodic lattice of quadrupoles has been shown to be the most accurate system to describe the precision transport of relativistic beams at large distances. The properties of quadrupole lenses, such as strong focusing and the linearity of the focusing (transverse) forces, are most important in this respect. The latter property is most important in the problem of transport of real nonlaminar beams. Let us touch briefly on this problem by neglecting self fields (space-charge and self magnetic forces).

Let us write the solution of the paraxial trajectory [Eq. (2.186)] in the converging plane as

$$x = a\cos(kz + \beta) \tag{2.213}$$

where the phase β of any trajectory is arbitrary for different particles in a nonlaminar beam. Find the equation of a trajectory in the trace space (see Section I.7).

$$x' = -ak\sin(kz + \beta) \tag{2.214}$$

$$\frac{x^2}{a^2} + \frac{x'^2}{a^2k^2} = 1 \tag{2.215}$$

We obtain the equation of an ellipse with semiaxes a and ak. The trajectory equation (2.215) does not include the phase β. It means that all particles with equal amplitude move in the trace space along the common elliptic trajectory. The particles of the beam with amplitudes $a_i < a$ move with the same frequency along similar ellipses that are inserted into an ellipse [Eq. (2.215)]. According to Eq. (I.46), the emittance in this case is equal to the area of the ellipse that is constant if motions in the (x, z) and (y, z) planes are independent and the momentum $p_z = \text{const}$. In an arbitrary field, the configuration of the trace contours can be modified, but for noninteracting particles the area of the corresponding figures and the particle density are invariants according to the Liouville theorem. The elliptical form of trace trajectories is conserved when conversion of the particle coordinates in the trace space is determined by transfer matrices, because of the linearity of the transformation. Note that the determinant of the transfer matrix must be equal to 1 because it coincides with a Jacobian of the transformation. A Jacobian value of 1 is a condition of trace figure area conservation for the conversion of variables.

The transport of a nonlaminar beam with a given emittance depends on the solution of a *matching problem* (Humphries, 1990; Lawson, 1988; Regenstreif, 1967; Reiser, 1994). Here, the concept of *acceptance* (defined as a contour in the trace space determined by limiting coordinates of the particles that can pass through a given section of the system without interception by the walls) is essential. If the emittance is less than the acceptance, the beam can pass through the system. However, the form and orientation of the emittance contour must be properly modified. This is the problem of *matching*, which is best approached using the concept of a beam envelope. If the envelope of a nonlaminar beam is known, the matching consists of joining the waist of the beam envelope with the acceptance contour. Figure 2.31 depicts passage of a nonlaminar beam through a thin lens. In the lower part of the figure, the positions of the trace ellipses in various cross sections of the beam are shown. The horizontal position of the big semiaxis of the ellipse in the lens plane corresponds to the maximum width of the envelope. The ellipse upright position in the right part of the scheme corresponds to the envelope waist. It can be shown (Arnold et al., 1988) that the envelope equation for the drift space in the vicinity of the waist is a hyperbola:

$$a^2 = a_0^2 + \frac{\Xi^2 z^2}{a_0^2} \tag{2.216}$$

where a_0 is the radius of the waist (Fig. 2.31) and Ξ is the emittance.

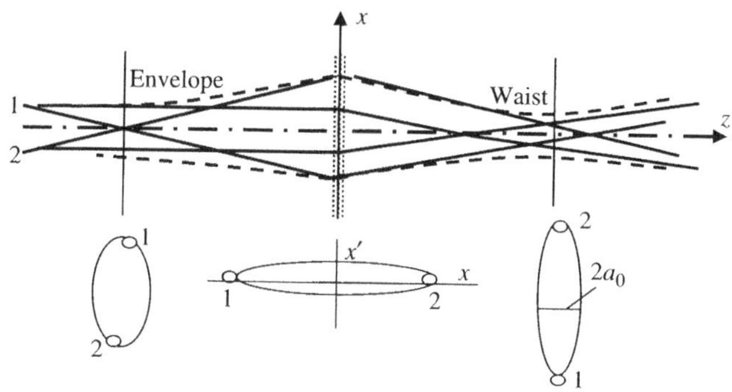

FIGURE 2.31 Passage of a nonlaminar beam with straight trajectories through a thin lens. 1, 2, any two trajectories of the beam. In the lower part of the scheme, 1 and 2 are projections of these trajectories in the trace space. The dashed lines are the beam envelope.

An analysis of the envelope problem for both nonlaminar beams and beams with self-fields can be found in monographs by Humphries (1990), Lawson (1988), and Reiser (1994). The transport and matching of the beam in accelerator systems are reached by using adjustable lenses and drift spaces. In general, this is a complicated and specialized procedure. The periodic lattice of the quadrupoles, the FODO lattice, is used in accelerators. The simplest version of a FODO cell is a magnetic doublet with two-sided symmetric drift sections (see, e.g., Reiser, 1994). Large accelerators contain hundreds of FODO cells, each several meters in length. So, for example, the length of Fermilab's proton–antiproton collider Tevatron, so far the world's largest collider, is approximately 6 miles. Quadrupoles in the Tevatron are formed by superconducting magnets with field 4.2 T. CERN (Switzerland) is building a proton–proton collider LHC 17 miles in length. The field projected for the superconducting magnets is 8.4 T.

Correction of Aberration As we have seen, all axially symmetric lenses have nonvanishing spherical and chromatic aberrations. Unlike axially symmetric elements, multipole elements have no fixed sign of spherical aberration and therefore may be used as compensation devices. Also, quadrupole lenses may be used to compensate for astigmatism. Effective compensation spherical aberration can be realized with octupole lenses. Octupole lenses can also compensate the aperture aberrations of quadrupoles. The combination of electrostatic and magnetic quadrupoles can form an achromatic system. The main limiting factor for the use of similar systems is the problem of their alignment (see, e.g., Szilagyi, 1988).

CHAPTER THREE

Electron Beams with Self Fields

3.1 INTRODUCTION

So far, our consideration of electron beams has been based on neglecting self fields, which is valid for small beam currents. This approximation provides both high-quality images and simplicity of electron-optical theory. However, there is extensive area of charge-particle physics in which disregarding space-charge fields in general deprives the theory of intense charge beams of physical content. The space-charge effects often are the main limiting factors in powerful beam systems, especially of microwave devices.

A criterion for a space-charge beam[1] is revealed in any model of intense electron beams. Consider, for example, an entire cylindrical electron beam in a conductive tube. The beam is supposed homogeneous with space-charge density $\rho = \text{const.}$, and the trajectories are supposed parallel to the z-axis (Fig. 3.1). The potential φ depends only on the radial coordinate, and the electric field E_ρ of the electron space charge is directed radial to the cylinder axis.

The potential satisfies Poisson's equation [Eq. (I.11)], which in this case is

$$\frac{1}{r}\frac{d}{dr}\left(r\frac{d\varphi}{dr}\right) = -\frac{\rho}{\varepsilon_0} \tag{3.1}$$

The integral of this equation is

$$\varphi = \varphi_a - \frac{\rho}{4\varepsilon_0}(r^2 - r_a^2) \tag{3.2}$$

[1]Below we use the term *space-charge field* in the sense of *self fields* (i.e., space charge and self magnetic fields).

Electron Beams and Microwave Vacuum Electronics. By Shulim E. Tsimring

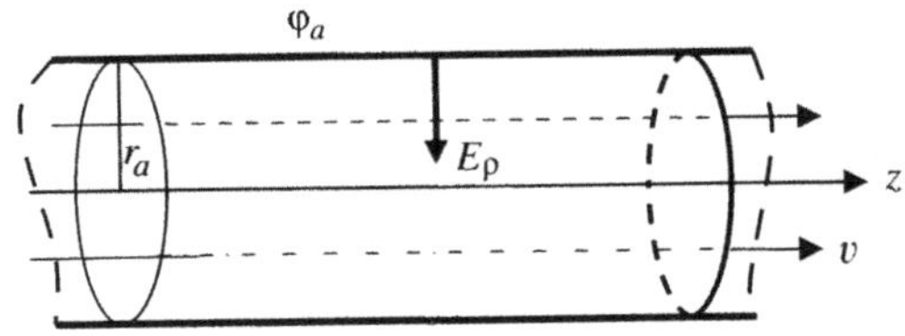

FIGURE 3.1 Potential depression in a cylindrical electron beam. φ_a, potential of the conductive tube.

Let us estimate the depression $\delta\varphi$ of the axis potential for the given tube potential φ_a:

$$\delta\varphi \equiv \frac{\varphi_a - \varphi_{r=0}}{\varphi_a} = -\frac{\rho r_a^2}{4\varepsilon_0 \varphi_a} = \frac{j_a r_a^2}{4\varepsilon_0 \varphi_a v_a} \tag{3.3}$$

where j_a and v_a are correspondingly a current density and the electron velocity for $r = r_a$.

Assume that

$$j_a r_a^2 = \frac{j_a}{\bar{j}} \frac{I}{\pi} \sim \frac{I}{\pi} \tag{3.4}$$

where I and $\bar{j}$ are the beam current and average current density, respectively. Taking Eq. (I.11) into account, we obtain

$$\delta\varphi \sim \frac{1}{4\pi\varepsilon_0} \frac{I}{\varphi_a v_a} = \frac{1}{4\pi c\varepsilon_0} \frac{I}{\varphi_a^{3/2}} \frac{\varphi_a + \varphi_0}{\sqrt{\varphi_a + 2\varphi_0}} \tag{3.5}$$

The parameter $P = I/(\varphi_a)^{3/2}$ is called a *perveance*. It is the base parameter of intense nonrelativistic electron beams because according to Eq. (3.5), the corresponding potential depression is

$$\delta\varphi = \frac{1}{4\pi\varepsilon_0 \sqrt{2\eta}} P \tag{3.6}$$

We can estimate the limiting values of the perveance by setting $\delta\varphi = 1$ in Eq. (3.6). Then

$$P_{\max} = 4\pi\varepsilon_0 \sqrt{2\eta} \approx 6.6 \times 10^{-5} \mathrm{A/V^{3/2}} \tag{3.7}$$

Usually, a special unity of the perveance, mkp $\equiv 1 \times 10^{-6} \mathrm{A/V^{3/2}}$, is used, and instead of Eq. (3.7) we can write $P_{\max} = 66$ mkp, so nonrelativistic space-charge beams can be defined as beams with a perveance of order 10 mkp. The perveance of corresponding ion beams is less. So for a proton beam, $\eta_{\mathrm{pr}} = \frac{1}{1840}\eta$ and $P_{\mathrm{pr.max}} \approx \frac{1}{43} P_{\max}$. For example, the limiting electron current for

$\varphi = 10$ kV is $6.6 \times 10^{-5}(10^4)^{3/2} = 66$ A. The proton limiting current for the same voltage is $66/43 \approx 1.5$ A. For $\varphi = 1$ kV, the limiting electron and ion currents are 2 A and 47 mA, respectively.

Considering the proper coefficient in Eq. (3.5), we obtain the following correction for moderate relativistic beams:

$$\delta\varphi = (\delta\varphi)_{\text{nrel}}\sqrt{2\eta}\frac{\varphi_a + \varphi_0}{c\sqrt{\varphi_a + 2\varphi_0}} \approx (\delta\varphi)_{\text{nrel}}\left[1 + \frac{3}{4}\frac{\varphi_a}{\varphi_0} - \frac{5}{32}\left(\frac{\varphi_a}{\varphi_0}\right)^2 + \cdots\right]$$

For extreme relativistic beams, in Eq. (3.5) the velocity $v_a \approx c$. Therefore, a measure of the extreme relativistic beam intensity is a *conductance*:

$$P_{\text{extr}} = \frac{I}{\phi_a} \qquad \text{A/V} \tag{3.8}$$

The limiting value of P_{extr} for $\delta\varphi = 1$ is

$$(P_{\text{extr}})_{\max} = 4\pi\varepsilon_0 c \approx 3.3 \times 10^{-2}\ \text{A/V} \tag{3.9}$$

Principal Problems of Space-Charge Beam Theory In Fig. 3.2 the scheme for an electron-optical system with a space-charge beam is shown. We can define the following basic regions and detect the corresponding problems as basic:

1. *Gun region.* This is a region of beam formation. The basic problems are (a) gathering an electron beam from the cathode (sometimes from the maximum area of the cathode with the maximum current density) and guiding it into a transport channel, and (b) formation of the appointed shape of the trajectories.

2. *Regions of beam transport, working space, and collector.* The beam must be guided to a working space for interaction with the electromagnetic field or with matter. The transport region itself can also be used as the working space. Problems are the conservation or additional transformation of trajectory shapes. The specific problem is ensuring depressed collector operation when the efficiency of the system can be increased by applying retarded potential to the collector. In this case the principal problem is adequate distribution of secondary electrons.

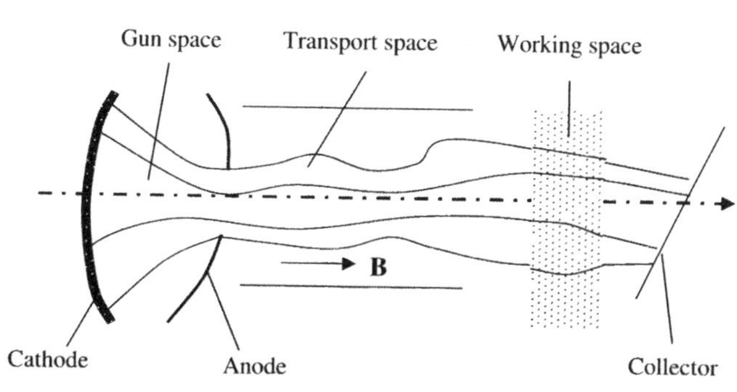

FIGURE 3.2 Electron-optical system with an intense electron beam.

3. *Beam stability*. In principle, an intense beam is always unstable: The particle interaction leads to excitation and the growth of spontaneous oscillatory processes. The theory of beam instabilities and their suppression is the most complicated aspect of space-charge beam physics.

In this chapter, we consider only the theory of electron space-charge beams in static fields.

3.2 SELF-CONSISTENT EQUATIONS OF STEADY-STATE SPACE-CHARGE ELECTRON BEAMS

The particles in space-charge beams move in the fields created both by external sources (charges and currents) and by charged particles themselves, so that the resulting fields that determine electron motion are not given and must be found simultaneously with the solution of equations of motion. This leads to self-consistent equations of the electron beams. Three versions of these equations are considered below.

3.2.1 Single-Flow Approximation (Laminar Beams). Pinch Effect

In laminar beams trajectories do not cross. Therefore, the velocities of particles at any points of the electron flow are single-valued, and the vector $\mathbf{v} = \mathbf{v}(\mathbf{r}) = \mathbf{v}(\mathbf{r}(t))$ is an unambiguous function of the coordinates. The particle velocities in the flow form a velocity field. A vector of the electron current density $\mathbf{j}(\mathbf{r})$ is

$$\mathbf{j} = -\rho\,\mathbf{v} \tag{3.10}$$

where $\rho > 0$ is the volume charge density. This approximation is typical for the mechanics of fluids and is called *hydrodynamic approximation*. The self-consistent equations in this case are

$$\frac{d\gamma\mathbf{v}}{dt} = -\eta\mathbf{E} - \eta\mathbf{v} \times \mathbf{B} \tag{3.11}$$

$$\gamma = \frac{1}{\sqrt{1 - (v^2/c^2)}} \tag{3.12}$$

$$\mathbf{E} = -\text{grad}\ \varphi \tag{3.13}$$

$$\Delta\varphi = \frac{\rho}{\varepsilon_0} + \frac{\rho_{\text{ext}}}{\varepsilon_0} \tag{3.14}$$

$$\text{curl}\,\mathbf{B} = \mu_0\,\mathbf{j} + \mu_0\,\mathbf{j}_{\text{ext}} \tag{3.15}$$

$$\mathbf{j} = -\rho\mathbf{v} \tag{3.16}$$

$$\text{div}\,\mathbf{j} = 0 \tag{3.17}$$

$$\text{div}\,\mathbf{B} = 0 \tag{3.18}$$

where ρ_{ext} and $\mathbf{j}_{\text{ext}}$ are the external values of the electron charge density and the current density. This is a system of 15 scalar equations; the number of unknown scalar functions is also 15, so this is really a self-consistent system.

If we neglect the self magnetic field, the system (3.11)–(3.18) is simplified to 12 equations (below we omit the term ρ_{ext}/ε_0):

$$\frac{d\gamma\mathbf{v}}{dt} = -\eta\mathbf{E} - \eta\mathbf{v} \times \mathbf{B}_{ext} \tag{3.19}$$

$$\gamma = \frac{1}{\sqrt{1 - v^2/c^2}} \tag{3.20}$$

$$\mathbf{E} = -\text{grad}\ \varphi \tag{3.21}$$

$$\Delta\varphi = \frac{\rho}{\varepsilon_0} \tag{3.22}$$

$$\mathbf{j} = -\rho\mathbf{v} \tag{3.23}$$

$$\text{div}\,\mathbf{j} = 0 \tag{3.24}$$

Neglecting the self magnetic field is usually correct in nonrelativistic approximation. Let us consider, for example, a section of a round solid electron beam (Fig. 3.3) and compare the forces of the self electric and magnetic fields acting on peripheral electrons. The electric (Coulomb) force according to Gauss's theorem is given as

$$\int E_r d\sigma = \frac{-Q}{\varepsilon_0}, \qquad E_r 2\pi r L = \frac{-\pi r L\rho}{\varepsilon_0}, \qquad E_r = \frac{-\rho r}{2\varepsilon_0} \tag{3.25}$$

$$F_E = -e_0 E_r = \frac{e_0 \rho r}{2\varepsilon_0}$$

The magnetic (Lorentz) force is given as

$$\oint B_\theta dl = \mu_0 I, \qquad B_\theta 2\pi r = \mu_0 \pi r^2 j, \qquad B_\theta = -\frac{r}{2}\mu_0 \rho v$$

$$F_B = e_0 v B_\theta = -\frac{r}{2}\mu_0 e_0 \rho v^2 \tag{3.26}$$

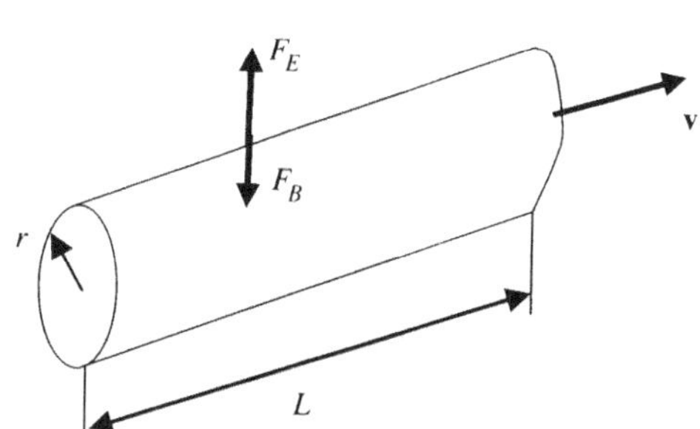

FIGURE 3.3 Comparison of self electric and magnetic forces in an electron beam.

The ratio

$$k_B \equiv \left|\frac{F_B}{F_E}\right| = \varepsilon_0\mu_0 v^2 = \frac{v^2}{c^2} = 1 - \gamma^{-2} < 1 \tag{3.27}$$

We see that the resulting self force for the electron beam model being considered is always repulsive (although for relativistic beams, $k_B \approx 1$). Note if there is a positive ion background, the resulting force can be attractive and the electron beam be convergent. That is the case for the *pinch effect*. The criterion for the pinch effect is

$$f - \gamma^{-2} > 0 \tag{3.28}$$

where f is the ratio of positive ion charge to electron beam charge (Arnold et al., 1988).

Estimation of the self magnetic field for other models of nonrelativistic beams can give different results. Consider, for example, an electron beam in a wide planar nonrelativistic diode (Fig. 3.4). Here, according to the boundary condition, the radial component of the electric field $E_r \approx 0$, unlike components of the electric field E_z and the self magnetic force B_θ. Let us compare the electric and magnetic forces $F_E = -e_0E_z$ and $F_B = e_0vB_\theta$. The magnetic force according to Eq. (3.26) is

$$F_B = -\frac{r}{2}\mu_0 e_0 \rho\bar{v}^2 = \frac{\mu_0 e_0 I\bar{v}}{2\pi r} \tag{3.29}$$

where $I = -\pi r^2\rho\bar{v}$ is the beam current. The velocity $\bar{v}$ is assumed equal to an average electron velocity in the cathode–anode gap. The electric field E_z can be expressed in terms of the beam current using the Child–Langmuir formula

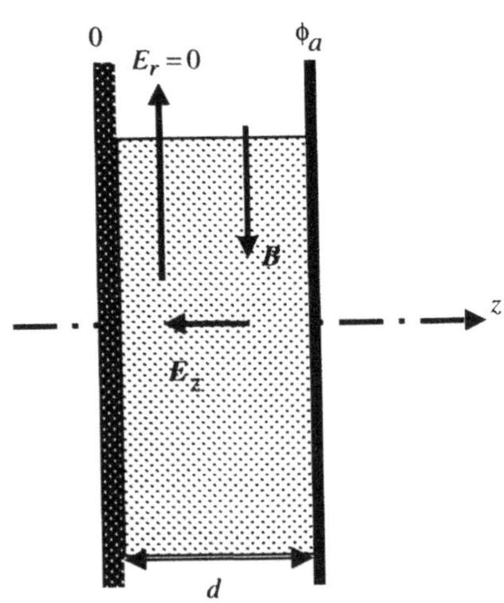

FIGURE 3.4 Electric and magnetic forces in a planar diode.

(see Section 3.4.4) for a planar diode in the space-charge regime:

$$I = P\varphi_a^{3/2} \approx \frac{4}{9}\varepsilon_0\sqrt{2\eta}\frac{\pi r^2}{d^2}\varphi_a^{3/2} \approx \frac{4}{9}\varepsilon_0\frac{v_a\pi r^2}{d}E_z = \frac{8}{9}\varepsilon_0\frac{\bar{v}\pi r^2}{d}E_z \tag{3.30}$$

Then $F_E = -(9e_0dI/8\varepsilon_0\bar{v}\pi r^2)$.

Here we assume that $E_z = \varphi_a/d$ is equal to the average field in the diode gap. The ratio of magnetic to electric forces is

$$k_B = \frac{F_B}{F_E} = \frac{\mu_0 e_0 I\bar{v}8\varepsilon_0\bar{v}\pi r^2}{2\pi r 9e_0 dI} = \frac{4r}{9d}\frac{\bar{v}^2}{c^2} \tag{3.31}$$

We see that in principle the pinch effect ($k_B > 1$) is possible in nonrelativistic very narrow diodes ($r/d \gg 1$) even without a positive ion background.

3.2.2 Multistream Flows

The next approximation in the theory of space-charge beams suppose that an electron beam can be represented as a superposition of N laminar flows (streams). Therefore, for each partial flow the continuity equation

$$\text{div}\,\mathbf{j}_i = \text{div}(\rho_i\mathbf{v}_i) = 0, \qquad i = 1, 2, \ldots, N \tag{3.32}$$

where values with a subscript i refer to the ith partial flow, must be satisfied. Generally, the flow is not laminar if partial flows intersect. Such a beam can be called *quasilaminar* (Kirstein et al., 1967). Equations for a multistream flow can be written in the same way as Eqs. (3.11)–(3.18):

$$\frac{d(\gamma_i\mathbf{v}_i)}{dt} = -\eta\mathbf{E} - \eta\mathbf{v}_i \times \mathbf{B}, \qquad i = 1, 2, \ldots, N \tag{3.33}$$

$$\gamma_i = \frac{1}{\sqrt{1 - v_i^2/c^2}}, \qquad i = 1, 2, \ldots, N \tag{3.34}$$

$$\mathbf{E} = -\text{grad}\,\varphi \tag{3.35}$$

$$\Delta\varphi = \sum_{i=1}^{N}\frac{\rho_i}{\varepsilon_0} \tag{3.36}$$

$$\text{curl}\,\mathbf{B} = \mu_0\sum_{i=1}^{N}\mathbf{j}_i \tag{3.37}$$

$$\mathbf{j}_i = -\rho_i\mathbf{v}_i, \qquad i = 1, 2\ldots, N \tag{3.38}$$

$$\text{div}\,\mathbf{j}_i = 0, \qquad i = 1, 2, \ldots, N \tag{3.39}$$

$$\text{div}\,\mathbf{B} = 0 \tag{3.40}$$

This is the self-consistent system of $8N + 7$ scalar equations with $8N + 7$ unknown scalar functions.

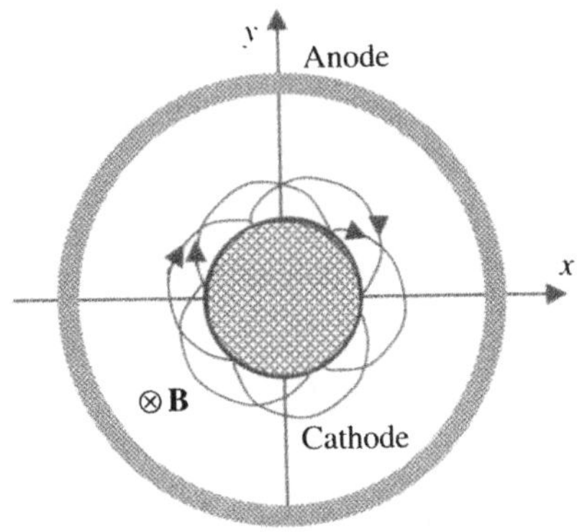

FIGURE 3.5 Two-stream flow in a static cylindrical magnetron.

The magnetically insulated electron beam in a cylindrical magnetron (the cylindrical diode in the axial magnetic field **B**) is a good example of multistream electron flow. The electrons are emitted from the cathode and accelerated by the electric field in the radial direction, but the magnetic field bends the trajectories, and for a sufficiently strong magnetic field, the electrons return to the cathode (Fig. 3.5). As a result, a two-stream electron flow arises: In any point of the beam there are particles with two different (on the direction) velocities.

3.2.3 Kinetic Description

Now we consider total nonlaminar beams that cannot be represented as a sum of a finite number of laminar flows. In this case, individual treatment of particle motion is impossible. Instead, statistic description and corresponding distribution functions must be used.

We have seen (Section I.6) that a distribution function in phase space satisfies the Liouville equation, which for a steady state has the form

$$\frac{\partial f}{\partial \mathbf{r}}\mathbf{v} + \dot{\mathbf{P}}\frac{\partial f}{\partial \mathbf{P}} = 0 \tag{3.41}$$

Equation (3.41) is applied rigorously for an ensemble of the noninteracting particles. However, interaction can be included in this equation if we take into account only collective forces through average electric and magnetic fields of the particles and neglect short-range interactions that are described as interparticle collisions.

Let us take the charge density as an integral over all values of the momentum of particles in the unit volume; that is,

$$\rho = e_0 \int \frac{dn}{d^3\mathbf{r}} \tag{3.42}$$

where dn is the particle number in the phase volume $d^3\mathbf{r}\,d^3\mathbf{P}$. The distribution function in six-dimensional space is (Section I.6) $f(\mathbf{r}, \mathbf{P}) = dn/d^3\mathbf{r}\,d^3\mathbf{P}$. Then the electron charge density as the source of a self electric field is

$$\rho = -e_0 \int f(\mathbf{r}, \mathbf{P})\, d^3\mathbf{P} \tag{3.43}$$

Analogously, the electron current density is

$$\mathbf{j} = -e_0 \int \mathbf{v} f(\mathbf{r}, \mathbf{P})\, d^3\mathbf{P} \tag{3.44}$$

The self electric and magnetic fields must satisfy the following equations:

$$\mathbf{E} = -\text{grad}\ \varphi \tag{3.45}$$

$$\Delta\varphi = \frac{e_0}{\varepsilon_0} \int f(\mathbf{r}, \mathbf{P})\, d^3\mathbf{P} \tag{3.46}$$

$$\text{curl}\ \mathbf{B} = -\mu_0 e_0 \int \mathbf{v} f(\mathbf{r}, \mathbf{P})\, d^3\mathbf{P} \tag{3.47}$$

$$\text{div}\ \mathbf{B} = 0 \tag{3.48}$$

Equations (3.41) and (3.45)–(3.48) describe a self-consistent steady-state (equilibrium) of the electron system in the space $\mathbf{r}, \mathbf{P}$.

Usually, an alternative and more convenient form of the distribution function $f = f(\mathbf{r}, \mathbf{p})$, where $\mathbf{p}$ is the mechanical momentum is used. A modified Liouville equation for $f(\mathbf{r}, \mathbf{p})$ is similar to Eq. (3.41):

$$\frac{\partial f}{\partial \mathbf{r}} \mathbf{v} + \dot{\mathbf{p}} \frac{\partial f}{\partial \mathbf{p}} = 0 \tag{3.49}$$

This statement is founded on the equality of the volumes in phase space:

$$\int D\, d^3\mathbf{r}\, d^3\mathbf{p} = \int d^3\mathbf{r}\, d^3\mathbf{P}$$

where

$$D = \frac{\partial(q_1, q_2, q_3, p_1, p_2, p_3)}{\partial(q_1, q_2, q_3, P_1, P_2, P_3)} \tag{3.50}$$

is a Jacobian of the transformation from the variables $(\mathbf{r}, \mathbf{P})$ to the variables $(\mathbf{r}, \mathbf{p})$ (here s_1, s_2, and s_3 denote the components of the vector $\mathbf{s}$). It is readily shown that $D = 1$, and

$$\frac{\partial q_i}{\partial q_k} = \delta_{ik}, \quad \frac{\partial p_i}{\partial q_k} = 0, \quad \frac{\partial q_i}{\partial P_k} = 0, \quad \frac{\partial p_i}{\partial P_k} = \frac{\partial p_i}{\partial (p_k - e_0 A_k)} = \delta_{ik} \tag{3.51}$$

We see that only the elements of determinant D on the main diagonal $= 1$ are nonzero, and $D = 1$. Therefore, the volume of six-dimensional phase space $(\mathbf{r}, \mathbf{p})$ enclosing a chosen group of particles that move along the trajectories is constant, and the distribution function of the particles is conserved:

$$f = \frac{dn}{d^3\mathbf{r}\, d^3\mathbf{p}} = \text{const.} \tag{3.52}$$

Thus, the version of the Liouville equation in (3.49) is proven. Substitution of the motion equation [Eq. (I.25)] in Eq. (3.49) yields the *Vlasov–Maxwell equation* (see, e.g., Davidson, 1990):

$$\frac{\partial f}{\partial \mathbf{r}}\mathbf{v} - e_0(\mathbf{E} + \mathbf{v} \times \mathbf{B})\frac{\partial f}{\partial \mathbf{p}} = 0 \tag{3.53}$$

where the velocity

$$\mathbf{v} = \frac{\mathbf{p}}{m} = \frac{\mathbf{p}c}{\sqrt{p^2 + m_0^2 c^2}} \tag{3.54}$$

Together with Eqs. (3.45)–(3.48) (we must replace $\mathbf{p}$ in these equations by $\mathbf{P}$), we obtain a self-consistent system of equations (Vlasov–Maxwell equations).

A nonrelativistic equation is

$$\frac{\partial f}{\partial \mathbf{r}}\mathbf{v} - \eta(\mathbf{E} + \mathbf{v} \times \mathbf{B})\frac{\partial \mathrm{f}}{\partial \mathbf{v}} = 0 \tag{3.55}$$

The entire system of Vlasov–Maxwell equations obviously is formed by supplementing Eq. (3.55) with the field eqations

$$\mathbf{E} = -\text{grad}\ \varphi \tag{3.56}$$

$$\Delta\varphi = \frac{e_0}{\varepsilon_0}\int f(\mathbf{r}, \mathbf{v})\, d^3\mathbf{v} \tag{3.57}$$

$$\text{curl}\,\mathbf{B} = -\mu_0 e_0 \int \mathbf{v} f(\mathbf{r}, \mathbf{v})\, d^3\mathbf{v} \tag{3.58}$$

$$\text{div}\,\mathbf{B} = 0 \tag{3.59}$$

Solution of the self-consistent system (3.41) and (3.45)–(3.48) gives the desired distribution function $f(\mathbf{r}, \mathbf{p})$ and any macroscopic quantities [$\mathbf{E}(\mathbf{r})$, $\mathbf{B}(\mathbf{r})$, $\bar{\rho}(\mathbf{r})$, etc.] that describe the equilibrium state. Finding solutions of that system of integrodifferential equations is the difficult problem.

Essential simplification of the problem gives knowledge of the integrals of motion $I_k(\mathbf{r}(t), \mathbf{p}(t))$ (i.e., an energy, components of the momentum, etc.). Now we can seek the distribution function as a function of I_k:

$$\frac{df}{dt} = \sum_k \frac{\partial f}{\partial I_k}\frac{dI_k}{dt} = 0 \tag{3.60}$$

and this distribution function satisfies the Vlasov–Maxwell equation. Note that the maximum number of independent integrals of motion is $2s - 1$, where s is the number of degrees of freedom (Landau and Lifshitz, 1976, 1987). If some integrals of motion are known, the problem is reduced to finding a concrete form of the function $f(I_1, I_2, \ldots, I_n)$ and then to a solution of Eqs. (3.45)–(3.48) or (3.56)–(3.59). This solution is far from unambiguous and depends on many geometric and physical conditions, approximations, and so on. In general, one can note the absence of regular analytic methods of problem solution. Each known solution is something of a work of art (see, e.g., Arnold et al., 1988; Davidson, 1990; Reiser, 1994).

An example of a distribution function is the Boltzmann–Maxwell distribution of a nonrelativistic electron gas. In this case, the integral of motion is the energy

$$w = \frac{m_0 v^2}{2} - e_0\varphi = \text{const.} \tag{3.61}$$

The Boltzmann–Maxwell function

$$f(w) = f_0 \exp\left(\frac{-w}{kT}\right) = f_0 \exp\left[-\left(\frac{m_0 v^2}{2} - e_0\varphi\right)\Big/ kT\right] \tag{3.62}$$

obviously satisfies the Vlasov–Maxwell equation. Substituting Eq. (3.62) in Eq. (3.57) allows us in principle (for corresponding boundary conditions) to get the potential distribution. Note that the right side of Eq. (3.57) gives the density of particles:

$$n = \int f d^3\mathbf{v} = f_0 \int \exp\left(-\frac{m_0 v^2}{2}\right) d^3\mathbf{v} \exp\left(\frac{e_0\varphi}{kT}\right) \tag{3.63}$$

Denoting the integral in Eq. (3.63) as n_0, we obtain a known barometric formula,

$$n = n_0 \exp\left(\frac{e_0\varphi}{kT}\right) \tag{3.64}$$

3.3 EULER'S FORM OF A MOTION EQUATION. LAGRANGE AND POINCARÉ INVARIANTS OF LAMINAR FLOWS

3.3.1 Equation of Motion in Euler Form. Regular Beams

As we noted in Section 3.2.1, the particle velocities in laminar flow form a velocity field. The corresponding description of motion is typical for mechanics of fluids and can be called an *Euler description*. Consider the velocity of particles as the velocity of the stream for a stationary observer. Assuming that the flow is steady, we find that the derivative of the momentum $\partial\mathbf{p}/\partial t = 0$. Then the equation of motion can be expressed as

$$\begin{aligned}\frac{\partial\mathbf{p}}{\partial\mathbf{r}}\dot{\mathbf{r}} + e_0 \operatorname{grad}\varphi - e_0\mathbf{v} \times \operatorname{curl}\mathbf{A} &= \frac{1}{m}(\mathbf{p}\nabla)\mathbf{p} + e_0 \operatorname{grad}\varphi - e_0\mathbf{v} \times \operatorname{curl}\mathbf{A}\\ &= 0\end{aligned} \tag{3.65}$$

Apply the formula from vector analysis $(\mathbf{a}\nabla)\mathbf{a} = \frac{1}{2}\,\text{grad}\,a^2 - \mathbf{a} \times \text{curl}\,\mathbf{a}$. Then

$$\frac{1}{2m}\,\text{grad}\,p^2 - \frac{1}{m}\,\mathbf{p} \times \text{curl}\,\mathbf{p} = e_0\,\text{grad}\,\varphi - e_0\mathbf{v} \times \text{curl}\,\mathbf{A} \tag{3.66}$$

Finally, using the relation (1.2), we obtain the equation of motion in Euler form:

$$\text{grad}\,w = \mathbf{v} \times \text{curl}\,\mathbf{P} \tag{3.67}$$

In this equation the quantity $w = mc^2 - e_0\varphi$ is no longer the energy of a separate particle but the energy in the stream as a function of the coordinates $w = w(\mathbf{r})$. Equation (3.67) contains an energy integral because $\text{grad}_t w = 0$ (i.e., w is constant along the trajectories). Note that if the beam is quasilaminar, Eq. (3.67) is true for the each partial flow.

Among the various types of charged-particle beams, *regular beams*, beams with uniform energy within the flow, are very important. According to Eq. (3.67), for a regular beam

$$\mathbf{v} \times \text{curl}\,\mathbf{P} = 0 \tag{3.68}$$

So the vortex of the generalized momentum in the regular bean is parallel to the trajectories. Electrons are emitted from an equipotential cathode with equal velocities. In this case the electrons have equal energy at all cathode points. Because the energy is constant along each trajectory, it is constant within the entire beam. In general, regular beams are a useful idealization. Very often, real electron beams in a state of equilibrium can be considered regular.

3.3.2 Lagrange and Poincaré Invariants

Lagrange's and Poincaré's invariants are applicable for an arbitrary laminar beam in the steady state or separately for each partial flow in a quasilaminar beam. Consider an arbitrary contour L_1 (Fig. 3.6). Assume that the particles move from points of this contour along corresponding nonintersecting trajectories. Through a time interval dt, these particles shift on a distance $\mathbf{v}\,dt$, and form the contour L_2. Take a vortex vector $\mathbf{G} = \text{curl}\,(\mathbf{v} \times \text{curl}\,\mathbf{P})$. As follows from Eq. (3.67), this vector is

$$\mathbf{G} = \text{curl}\,(\mathbf{v} \times \text{curl}\,\mathbf{P}) = 0 \tag{3.69}$$

Represent the flux of $\mathbf{G}$ through the surface S_2 that is carried by contour L_2 and apply Stokes' theorem:

$$\int_S G_n\,d\sigma = \int_S \text{curl}_n(\mathbf{v} \times \text{curl}\,\mathbf{P})\,d\sigma = \oint_{L_2} (\mathbf{v} \times \text{curl}\,\mathbf{P})_l\,dl = 0 \tag{3.70}$$

Transform the latter equality:

$$dt \oint_{L_2} (\mathbf{v} \times \text{curl}\,\mathbf{P})\,d\mathbf{l} = \oint_{L_2} \text{curl}\,\mathbf{P}(d\mathbf{l} \times \mathbf{v}\,dt) = 0 \tag{3.71}$$

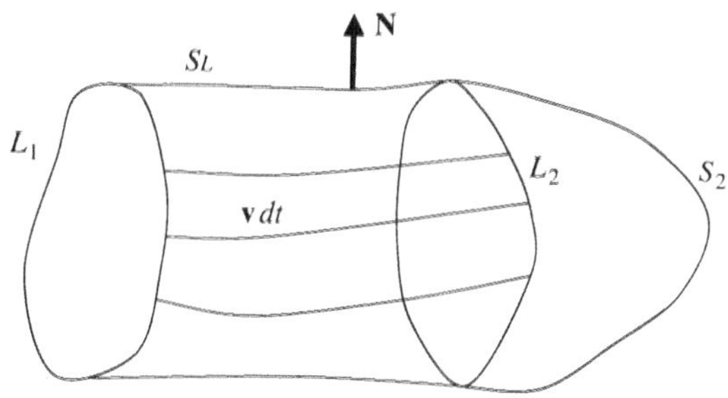

FIGURE 3.6 Derivation of the Lagrange and Poincaré invariants.

The vector product $d\tau = d\mathbf{l} \times \mathbf{v}\,dt$ in Eq. (3.71) is directed along the normal **N** to the lateral surface S_L that is formed by trajectories and passes through contours L_1 and L_2. The value of $|d\tau|$ is equal to the area dS_L of the elementary strip on the surface S_L, so the integral in Eq. (3.71) is equal to

$$\int_{S_L} \mathrm{curl}_N \mathbf{P}\, dS_L = 0 \tag{3.72}$$

Now let us take the flux of the vector curl **P** through the surface $S_1 = S_L + S_2$ that is carried by the contour L_1:

$$\int_{S_1} \mathrm{curl}_N \mathbf{P}\, dS = \int_{S_L} \mathrm{curl}_N \mathbf{P}\, dS + \int_{S_2} \mathrm{curl}\, \mathbf{P}_N\, dS \tag{3.73}$$

Taking Eq. (3.72) into account, we obtain *Lagrange's invariant*:

$$\Lambda \equiv \int_{S} \mathrm{curl}_N \mathbf{P}\, dS = \text{const.} \tag{3.74}$$

The flux of the momentum vortex through the surface that is carried by the contour attached to the particles of the beam is conserved in particle motion and the contour.

Applying Stokes' theorem to the integral in Eq. (3.74),

$$\int_{S} \mathrm{curl}_N \mathbf{P}\, dS = \oint_{L} P_l\, dl \tag{3.75}$$

we obtain *Poincaré's invariant*:

$$\Pi = \oint_{L} P_l\, dl = \text{const.} \tag{3.76}$$

Circulation of the generalized momentum around a contour attached to the moving particles of the beam is conserved.

The Case of a Regular Beam As we have seen [Eq. (3.68)], the vector curl **P** in the regular beam is parallel to the trajectories. Therefore, the flux of this vector through the lateral surface of a current tube is zero. Take two arbitrary contours L_1 and L_2 at the current tube. Because the integral $\int_{S_L}$ curl$_N$**P** is zero, the integrals

$$\int_{S_1} \text{curl}_N \mathbf{P}\, dS = \int_{S_2} \text{curl}_N \mathbf{P}\, dS \quad \text{or} \quad \Lambda = \text{const.}, \qquad \Pi = \text{const.} \tag{3.77}$$

As we can see, for regular beam the Lagrange and Poincaré invariants are applicable for arbitrary contours L_1 and L_2 on a current tube, not only for contours that move together as particles.

3.3.3 Generalized Busch Theorem

Apply Poincaré's invariant [Eq. (3.76)] to contours L_1 and L_2 on two arbitrary cross sections of a current tube:

$$\oint_{L_1} P_l\, dl = \oint_{L_2} P_l\, dl \tag{3.78}$$

It is readily seen that

$$\oint_L P_l\, dl = \oint_L p_l\, dl - e_0 \oint_L A_l\, dl = \oint_L p_l\, dl - e_0 \int_S B_n\, dS = \oint_L p_l\, dl - \Psi \tag{3.79}$$

where Ψ is a flux of the magnetic field through the surface S that is carried by the contour L. We obtain the *generalized Busch theorem*

$$\oint_{L_1} p_l\, dl - \oint_{L_2} p_l\, dl = e_0(\Psi_1 - \Psi_2) \tag{3.80}$$

For axially symmetric beams, let us take contours L_1 and L_2 as circles with centers on the z-axis. Then

$$\oint_L p_l\, dl = 2\pi m r^2 \dot{\theta} = 2\pi m_0 \gamma r^2 \dot{\theta} \tag{3.81}$$

which gives the following version of the Busch theorem:

$$\dot{\theta}_2 = \frac{\gamma_1}{\gamma_2}\left(\frac{r_1}{r_2}\right)^2 \dot{\theta}_1 + \frac{\eta}{2\pi\gamma_2 r_2^2}(\Psi_2 - \Psi_1) \tag{3.82}$$

This formula coincides with Eq. (1.127). However, Eq. (3.82) has been derived directly only for axially symmetric flows. Equation (1.127) is more universal and applicable for separate particles in axially symmetric fields.

3.4 NONVORTEX BEAMS. ACTION FUNCTION. PLANAR NONRELATIVISTIC DIODE. PERVEANCE. CHILD–LANGMUIR FORMULA. ρ- AND T-MODES OF ELECTRON BEAMS

Nonvortex beams are flows for which

$$\text{curl}\,\mathbf{P} = 0 \tag{3.83}$$

Such a beam is called a *normal congruence* (Sturrock, 1955). We use the term *congruent beams* for these beams (Kirstein et al., 1967), reserving the term *normal congruent beam* for a nonmagnetic nonvortex beam (see below).

As we see from Eq. (3.68), these beams are a particular case of regular beams. According to this definition, the momentum in the congruent beam is

$$\mathbf{P} = \text{grad}\,\Psi \tag{3.84}$$

where the function of the coordinates $\Psi = \Psi(\mathbf{r})$ is called an *action function*. This function, which is similar to the function $S = S(\mathbf{r}, t)$ in Hamiltonian mechanics (see, e.g., Jackson, 1999; Landau and Lifshitz, 1976), is of fundamental importance for all theoretical physics. The action function $\Psi(\mathbf{r})$ is sometimes an effective tool for congruent beam analysis.

3.4.1 Indications of Congruent Beams

1. A beam is congruent if it is regular and $\text{curl}_n\,\mathbf{P} = 0$ in some cross section of the beam. According to the version of the Lagrange invariant in Eq. (3.77), the integrals $\int_S \text{curl}_n \mathbf{P}\, dS$ for a regular beam are the same on any two cross sections of a current tube. But then it is obvious that $\text{curl}\,\mathbf{P} = 0$ always.

2. A beam is congruent if it is emitted from a cathode that is not intersected by a magnetic field (zero normal component $B_{c,n}$ of the cathode magnetic field) and the tangential initial velocities in any point of the cathode are zero.

Take an arbitrary contour L_c on a cathode surface and find the flux of rot $\mathbf{P}$ through the cathode element restrained by this contour:

$$\int_{S_c} \text{curl}_n \mathbf{P}\, dS = \int_{S_c} \text{curl}_n \mathbf{p}\, dS + \int_{S_c} \text{curl}_n \mathbf{A}\, dS = \oint_{L_c} p_\tau dl + \int_{S_c} B_n\, dS \tag{3.85}$$

Both integrals in Eq. (3.85) are zero. Because the contour is arbitrary, it means that $\text{curl}_n \mathbf{P} = 0$ on the cathode. Therefore, according to the first indication, the beam is congruent.

3.4.2 Differential Equation of a Congruent Beam

Relation (3.84), which expresses generalized momentum in terms of the action function Ψ leads to a differential equation that describes completely the self-consistent state of a congruent beam. According to Eq. (3.84), the mechanical momentum is equal to

$$\mathbf{p} = \text{grad}\,\Psi + e_0\mathbf{A} \tag{3.86}$$

Write the remainder of the self-consistent equations of the congruent electron beam:

$$w = mc^2 - e_0\varphi = \text{const.} \tag{3.87}$$

$$mc^2 = c\sqrt{p^2 + m_0^2c^2} \tag{3.88}$$

$$\Delta\varphi = \frac{\rho}{\varepsilon_0} \tag{3.89}$$

$$\mathbf{j} = -\rho\mathbf{v} \tag{3.90}$$

$$\operatorname{div}\mathbf{j} = 0 \tag{3.91}$$

We have from Eqs. (3.86)–(3.89),

$$\rho = \varepsilon_0\Delta\varphi = \frac{\varepsilon_0}{e_0}\Delta mc^2 = \frac{\varepsilon_0 c}{e_0}\Delta\sqrt{(\operatorname{grad}\Psi + e_0\mathbf{A})^2 + m_0^2c^2} \tag{3.92}$$

It follows from Eqs. (3.88) and (3.86) that

$$\mathbf{v} = \frac{\mathbf{p}}{m} = c\frac{\operatorname{grad}\Psi + e_0\mathbf{A}}{\sqrt{(\operatorname{grad}\Psi + e_0\mathbf{A})^2 + m_0^2c^2}} \tag{3.93}$$

We obtain the following differential equation from Eqs. (3.90)–(3.93):

$$\operatorname{div}\left[\frac{(\operatorname{grad}\Psi + e_0\mathbf{A})\Delta\sqrt{(\operatorname{grad}\Psi + e_0\mathbf{A})^2 + m_0^2c^2}}{\sqrt{(\operatorname{grad}\Psi + e_0\mathbf{A})^2 + m_0^2c^2}}\right] = 0 \tag{3.94}$$

In the nonrelativistic approximation $p^2 = (\operatorname{grad}\Psi + e_0\mathbf{A})^2 \ll m_0^2c^2$, the differential equation (3.94) becomes

$$\operatorname{div}\left[(\operatorname{grad}\Psi + e_0\mathbf{A})\Delta(\operatorname{grad}\Psi + e_0\mathbf{A})^2\right] = 0 \tag{3.95}$$

We now have only one equation, a very complicated one: the boundary problem for a fourth-order nonlinear partial differential equation. Solutions of this equation succeeded in finding rare cases. Therefore, it is often preferable to use the original equations (3.86)–(3.91).

3.4.3 Nonmagnetic Congruent Beams

This is a case when $\mathbf{B} = 0$ everywhere in the beam. For these beams, according to (3.86),

$$\mathbf{p} = \operatorname{grad}\Psi \tag{3.96}$$

that is, the particle trajectories are perpendicular to surfaces of the constant action function. According to indication 2 in Section 3.4.1, the velocities of a normal congruent beam are perpendicular to the cathode surface. These beams also called *normal congruent beams*, are the full analog of the normal congruence in light optics (Kirstein et al., 1967; Sturrock, 1955). Congruent beams with $\mathbf{B} \neq 0$ are called *skew congruent beams* (Kirstein et al., 1967).

The differential equation of a relativistic normal congruent beam is

$$\operatorname{div}\left[\frac{\operatorname{grad}\Psi\Delta\sqrt{(\operatorname{grad}\Psi)^2 + m_0^2c^2}}{\sqrt{(\operatorname{grad}\Psi)^2 + m_0^2c^2}}\right] = 0 \tag{3.97}$$

The corresponding equation in a nonrelativistic approximation is

$$\operatorname{div}\left\lfloor \operatorname{grad}\Psi\Delta(\operatorname{grad}\Psi)^2 \right\rfloor = 0 \tag{3.98}$$

3.4.4 Application of an Action Function to Analysis of a Planar Nonrelativistic Diode. Child–Langmuir Formula. ρ- and T-Modes

In the planar diode (Fig. 3.7) all quantities depend only on x. Assume that the initial velocities of the electrons are zero. According to Section 3.4.3, the nonmagnetic electron beam is normal. Therefore, we can apply Eq. (3.98), which can be written in a one-dimensional version as

$$\frac{d}{dx}\left[\frac{d\Psi}{dx}\frac{d^2}{dx^2}\left(\frac{d\Psi}{dx}\right)^2\right] = 0 \tag{3.99}$$

The nonrelativistic velocity for zero initial velocity and zero cathode potential according to Eqs. (3.96) and (1.19) is given as

$$p_x = \frac{d\Psi}{dx} = m_0 v = m_0\sqrt{2\eta\phi} \tag{3.100}$$

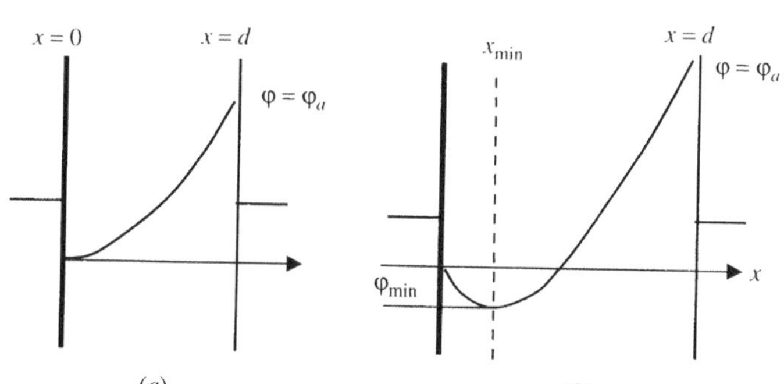

FIGURE 3.7 Potential distribution in a planar diode: (*a*) zero initial velocities; (*b*) nonzero initial velocities.

Then Eq. (3.99) in terms of φ will be

$$\sqrt{\varphi}\frac{d^2\varphi}{dx^2} = \text{const.} \tag{3.101}$$

where (Fig. 3.7)

$$\varphi(0) = 0, \qquad \varphi(d) = \varphi_a \tag{3.102}$$

Let us represent the solution of Eq. (3.101) as $\varphi(x) = Ax^n$. It satisfies the first of Eqs. (3.102). Substitution of $\varphi(x)$ in Eq. (3.101) gives the relation $A^{3/2}(n-1)/(n-2)x^{n/2+(n-2)} = \text{const.}$ We obtain $n = 4/3$ and then $\varphi = Ax^{4/3}$. According to boundary condition $\varphi(d) = \varphi_a$,

$$\varphi(x) = \varphi_a\left(\frac{x}{d}\right)^{4/3} \tag{3.103}$$

The beam current corresponding to Eqs. (3.89) and (3.100) is

$$I = Sj = S\rho v = S\varepsilon_0\frac{d^2\varphi}{dx^2}\sqrt{2\eta\varphi} \tag{3.104}$$

where S is the area of the diode electrodes. Substituting Eq. (3.103) in Eq. (3.104), we obtain the *Child–Langmuir formula* (Childs, 1911; Langmuir and Blogett, 1923, 1924):

$$I_{\text{CL}} = \frac{4}{9}\frac{S}{d^2}\varepsilon_0\sqrt{2\eta}\,\varphi_a^{3/2} = P\varphi_a^{3/2} \approx 2.33 \times 10^{-6}\frac{S}{d^2}\varphi_a^{3/2} \tag{3.105}$$

where P is the perveance.

The relation $\varphi(x) = Ax^n$ that was used for derivation of the Child–Langmuir formula implies that the cathode field $E_c = -[d\varphi(x)/dx]_{x=0} = 0$ for $n > 1$. If $E_c = E_{c0} - E_{c\rho}$ ($E_{c0} = \varphi_a/d$ is the cathode field in the diode without the beam and $E_{c\rho}$ is the cathode electric field that is created by the space charge of the beam), we obtain $E_{c\rho} = E_{c0}$. In fact, it means that for a given a φ_a, the current I_{CL} is the *limiting current*. If we tried to raise the current more than I_{CL} by, for example, increasing the cathode temperature, $E_{c\rho}$ would be more than E_{c0} and then $E_c < 0$. But in this case, electrons with zero initial energy could not leave the cathode and the current would not depend on the cathode emission. We call this current I_{CL} a *current limited by the space charge* and the corresponding diode regime the ρ-*mode*. Note that a power dependence $I(\varphi_a)$ of the type of Eq. (3.105) is called the *3/2 law*. It is shown in Appendix 1 that this law is applied to any non-relativistic diode in the ρ-mode.

On the contrary, if the current is decreased up to $I < I_{\text{CL}}$ (e.g., by cooling the cathode), the cathode field E_c will be accelerating and this current will not longer depend on φ_a. We call this current a *current limited by the emission* and the diode regime the *T-mode*.

3.4.5 Influence of Initial Velocities

Assume that the initial electron velocity is not zero. For simplicity, suppose that initial velocities of all the electrons are the same (in reality, they have a Boltzmann distribution) and their energy is equal to $e_0\varphi_0$. For a current limited by the space charge (the ρ-mode), the cathode electric field must be retarded (negative). Therefore, the curve of the potential distribution has a minimum (Fig. 3.7*b*). It is easy to see that the value of the potential minimum for the ρ-mode is $\varphi_{\min} = \varphi_0$. Then the electron velocities at the plane $x = x_{\min}$ are equal to zero, and this plane has properties of the cathode in an idealized plane diode for the ρ-mode (zero initial velocity and zero electric field). This cathode is called a virtual cathode. This diode has the anode potential $\varphi_a + \varphi_0$ and an anode–cathode distance $d - x_{\min}$. According to Eq. (3.105), the current in the ρ-mode is

$$I = 2.33 \times 10^{-6} \frac{S}{(d - x_{\min})^2} (\varphi_a + \varphi_0)^{3/2}$$

In this equation the virtual cathode location $x_{\min}$ is unknown. To determine $x_{\min}$ we should consider the region $0 \le x \le x_{\min}$, that is, a diode with negative anode potential. We do not consider this problem. Note only that in general (for velocity distribution) there are two flows (a forward and a reflected flow) in the left diode.

3.4.6 Optical Definition of a Congruent Beam

Assume that a laminar beam is focused to a point. Circumscribe the flow at this point by a circle of indefinitely small radius. It is obvious that the Lagrange and Poincaré integrals on this circle are zero if the generalized momentum **P** is finite. But if we take into account the invariance of these integrals, we find that invariants Λ and P [Eqs. (3.74) and (3.76)] are both zero in a whole beam. According to indication 1 in Section 3.4.1, this beam is congruent (i.e., $\operatorname{rot}\mathbf{P} = 0$). Thus, a congruent beam can be focused to a point. If the beam is not congruent, it obviously cannot be focused to a point. In Chapter 2 we called beams formed by rays diverging from or converting to a point *homocentric*. Thus, the term *normal congruence* from classical geometrical optics is equivalent to the term *homocentric beam.*

3.5 SOLUTIONS OF SELF-CONSISTENT EQUATIONS FOR CURVILINEAR SPACE-CHARGE LAMINAR BEAMS. MELTZER FLOW. PLANAR MAGNETRON WITH AN INCLINED MAGNETIC FIELD. DRYDEN FLOW

3.5.1 Forms of Representation of Solutions for Curvilinear Space-Charge Beams

The self-consistent equations of space-charge flows belong to the nonlinear partial differential equations of mathematical physics. Even if we limit ourselves to equilibrium states, solution of these equations is very complicated, especially if we want to obtain analytical results. Certainly, modern numerical methods and computer resources create innumerable possibilities for investigation of concrete problems. However, numerical methods need a high degree of detailed elaboration of the structures analyzed: geometry of the beams, electrodes and magnets, field distribution, and so on. These problems cannot be set properly if real common characteristics of flows (there can be many types) are not given. That requires certain analytical work.

We can arrange the types of curvilinear space-charge beams in the order of complexity of analysis:

1. Normal (nonmagnetic) congruent beams
2. Skew congruent beams
3. Laminar noncongruent beams
4. Nonlaminar beams

For types 1 and 2, differential equations of the action function (3.94) or (3.95) can be used. Also, the original system of self-consistent differential equations (3.11)–(3.18) or (3.19)–(3.24) for types 1 to 3 is applicable, although it is not so easy in these cases to control beam laminarity. Finally, the Vlasov–Maxwell equations [Eqs. (3.55)–(3.59)] can be used in the analysis of beams of type 4.

For linear partial differential equations, the most universal method is perhaps the separation of variables. A partial differential equation is split after a functional representation to a set of independent ordinary differential equations for certain variables or even for each variable. Due to the linearity of the original equation, a general solution is sought as a linear combination of the partial solutions obtained. This method is not applicable for finding general solutions of nonlinear equations. However, the idea of the separation of variables is used to reduce the original equation to the ordinary differential equations that determine a certain class of helpful solutions (Kirstein and Kino, 1958; Kirstein et al., 1967). In particular, they can be solved numerically.

The additive and product forms of the method of the separation of variables are known. The action function in *additive form* is represented as

$$\Psi(q_1, q_2, q_3) = \Psi_1(q_1) + \Psi_2(q_2) + \Psi_3(q_3) \tag{3.106}$$

This method was used successfully for solution of the Hamilton–Jacobi equation in classical dynamics (Landau and Lifshitz, 1976, 1987). However, in the theory of normal congruent flows, the method was not of great interest because, as shown by Kirstein et al. (1967), the cathode in corresponding devices can only be a point source or a finite number of discrete points.

More important applications are connected with the *product form* of the separation of variables:

$$\Psi(q_1, q_2, q_3) = \Psi_1(q_1)\Psi_2(q_2)\Psi_3(q_3) \tag{3.107}$$

Note that this form is used almost exclusively for the solution of boundary problems in Maxwell's theory. In a paper by Kirstein and Kino (1958) and a book by Kirsten et al. (1967), important examples are given of the use of this form in the analysis of congruent and even noncongruent beams. Single-component normal congruent flow and some cases of noncongruent beams are considered next.

3.5.2 Normal Congruent Nonrelativistic Beams. Single-Component Flows

The single-component version is the simplest form of separation of variables. In this case the action function is represented as a function of only one coordinate:

$$\Psi(q_1, q_2, q_3) = \Psi(q_1) \tag{3.108}$$

According to Eq. (3.96), the velocity components $\mathbf{v}(v_1,v_2,v_3)$ are equal:

$$v_1 = \frac{1}{m_0 h_1}\frac{d\Psi}{dq_1}, \qquad v_2 = v_3 = 0 \tag{3.109}$$

where h_1 is a metric coefficient. In this flow the particle trajectories coincide with coordinate lines q_1. Let us obtain an ordinary differential equation of a beam in an orthogonal coordinate system. For this system the length element is represented as

$$dl^2 = h_1^2\, dq_1^2 + h_2^2\, dq_2^2 + h_3^2\, dq_3^2 \tag{3.110}$$

where $h_i(q_1, q_2, q_3)$ $(i = 1, 2, 3)$ are metric coefficients. Differential operators in an orthogonal coordinate system are

$$\text{grad}\,\Psi = \frac{1}{h_1}\frac{d\Psi}{dq_1}\mathbf{e}_1 \tag{3.111}$$

$$\text{div}\,\mathbf{b} = \frac{1}{h_1 h_2 h_3}\left(\frac{\partial}{\partial q_1}(h_2 h_3 b_1) + \frac{\partial}{\partial q_2}(h_1 h_3 b_2) + \frac{\partial}{\partial q_3}(h_1 h_2 b_3)\right) \tag{3.112}$$

$$\Delta\Phi = \frac{1}{h_1 h_2 h_3}\left[\frac{\partial}{\partial q_1}\left(\frac{h_2 h_3}{h_1}\frac{\partial\Phi}{\partial q_1}\right) + \frac{\partial}{\partial q_2}\left(\frac{h_1 h_3}{h_2}\frac{\partial\Phi}{\partial q_2}\right) + \frac{\partial}{\partial q_3}\left(\frac{h_1 h_2}{h_3}\frac{\partial\Phi}{\partial q_3}\right)\right] \tag{3.113}$$

where $\mathbf{e}_1$ is the unit vector in the q_1 direction. $\mathbf{b}(b_1, b_2, b_3)$ and $\Phi(q_1, q_2, q_3)$ are an arbitrary vector and function.

Let us turn to Eq. (3.98) of normal congruent flow:

$$\text{div}\left\lfloor \text{grad}\,\Psi\Delta(\text{grad}\,\Psi)^2 \right\rfloor = 0 \tag{3.114}$$

Denote the derivative $d\Psi/dq_1 \equiv Q(q_1)$. Then

$$\text{grad}\,\Psi = \frac{Q}{h_1}\mathbf{e}_1 \tag{3.115}$$

The vector into div for Eq. (3.114) has only one nonzero component:

$$b_1 = \frac{Q}{h_1}\Delta\left(\frac{1}{h_1^2}Q^2\right) \tag{3.116}$$

Then, according to Eqs. (3.112) and (3.114),

$$\operatorname{div} \mathbf{b} = \frac{1}{h_1 h_2 h_3}\frac{\partial}{\partial q_1}\left[\frac{h_2 h_3}{h_1} Q\Delta\left(\frac{1}{h_1^2}Q^2\right)\right] = 0 \tag{3.117}$$

So the function in brackets does not depend on q_1:

$$\frac{h_2 h_3}{h_1} Q\Delta\left(\frac{1}{h_1^2}Q^2\right) = f(q_2, q_3) \tag{3.118}$$

Substituting the operator Δ according to Eq. (3.113), we obtain

$$\begin{aligned}\frac{Q}{h_1^2}\Bigg\{\frac{\partial}{\partial q_1}\left[\frac{h_2 h_3}{h_1}\frac{\partial}{\partial q_1}\left(\frac{Q^2}{h_1^2}\right)\right] + Q^2\Bigg[\frac{\partial}{\partial q_2}\left(\frac{h_1 h_3}{h_2}\frac{\partial}{\partial q_2}\left(\frac{1}{h_1^2}\right)\right)\\ + \frac{\partial}{\partial q_3}\left(\frac{h_1 h_2}{h_3}\frac{\partial}{\partial q_3}\left(\frac{1}{h_1^2}\right)\right)\Bigg]\Bigg\} = f(q_2, q_3)\end{aligned} \tag{3.119}$$

This is an ordinary differential equation about the function $Q(q_1)$. The right side of Eq. (3.119) is an arbitrary function of q_2 and q_3. Equation (3.119) is considerably complicated; nevertheless, it can be solved in some cases.

Before considering a concrete example, we note the following. In principle, we have two approaches to our problem:

1. We can consider the metric coefficients h_1, h_2, and h_3 of a coordinate system as given functions of the coordinates q_1, q_2, and q_3. In this case, integration of Eq. (3.119) gives a result (analytical or numerical): trajectories and other parameters of the beam. Certainly, the problem of determining the corresponding electrode configuration must still be solved. Unfortunately, use of this method is very limited because no adequate orthogonal system is available for arbitrary beams.

2. The metric coefficients are not given. It was shown by V. A. Syrovoy that trajectories of a spatial electron beam in an external magnetic field in general may be included as the coordinate lines only in essentially nonorthogonal system comprising at least five nonzero elements of the metric tensor. In this case, *geometrized beam equations* are developed that include equations of motion and of the electromagnetic field, and equations for components of the metric tensor (see, e.g., Syrovoy, 1999).

3.5.3 Meltzer Flow (Meltzer, 1956)

Meltzer flow is infinite flow formed by electrons rotating on circular trajectories in planes perpendicular to the z-axis with centers on this axis (Fig. 3.8). For this flow

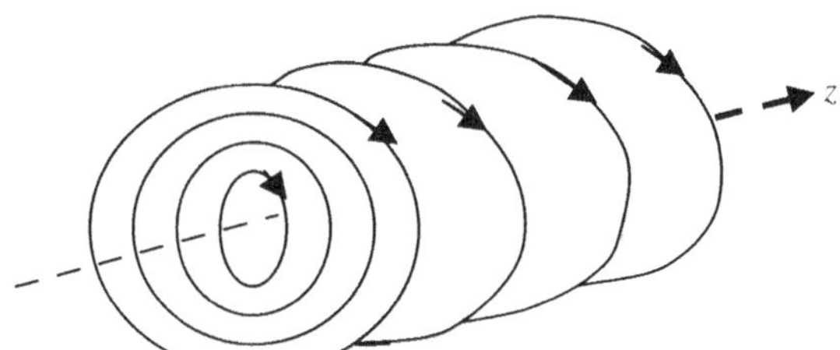

FIGURE 3.8 Meltzer flow.

the coordinates are $q_1 = \theta$, $q_2 = r$, and $q_3 = z$. The length element and metric coefficient in this coordinate system are

$$dl^2 = r^2 d\theta^2 + dr^2 + dz^2$$
$$h_1 = r, \qquad h_2 = h_3 = 1 \tag{3.120}$$

Substituting h_1 from Eq. (3.120) in Eq. (3.119), we obtain the following differential equation for the function $Q(\theta)$:

$$Q\left(\frac{d^2Q^2}{d\theta^2} + 4Q^2\right) = r^5 f(r, z) \tag{3.121}$$

The left and right sides of this equation depend on different variables (θ and r, z). Therefore, each side is constant. Thus, Eq. (3.121) is reduced to

$$\frac{d^2Q^2}{d\theta^2} + 4Q^2 = \frac{C}{Q} \tag{3.122}$$

where $C = \text{const}$. It is readily verified that a solution of this equation is

$$Q = C^{1/3}\left(\sin\tfrac{3}{2}\theta\right)^{2/3} \tag{3.123}$$

Let us now find the beam parameters. According to Eq. (3.109), the velocity is

$$v = \frac{1}{m_0 r}\frac{d\Psi}{d\theta} = \frac{Q}{m_0 r} = \frac{C^{1/3}}{m_0 r}\left(\sin\tfrac{3}{2}\theta\right)^{2/3} \tag{3.124}$$

The angular velocity is

$$\omega = \frac{v}{r} = \frac{C^{1/3}}{m_0 r^2}\left(\sin\tfrac{3}{2}\theta\right)^{2/3} \tag{3.125}$$

One is not uniform as the function of the radius and azimuth. The nonrelativistic potential is

$$\varphi = \frac{v^2}{2\eta} = \frac{C^{2/3}}{2r^2 e_0 m_0}\left(\sin\tfrac{3}{2}\theta\right)^{4/3} \tag{3.126}$$

The space-charge density $\rho = \varepsilon_0 \Delta\varphi$. Using the expression of the operator Δ from Eq. (3.113), we readily obtain

$$\rho = \frac{\varepsilon_0 C^{2/3}}{2r^4 e_0 m_0}\left(\sin\tfrac{3}{2}\theta\right)^{-2/3} \tag{3.127}$$

Then the current density

$$j = \rho v = \frac{\varepsilon_0 C^{2/3}}{2r^4 e_0 m_0}\left(\sin\tfrac{3}{2}\theta\right)^{-2/3}\frac{C^{1/3}}{m_0 r}\left(\sin\tfrac{3}{2}\theta\right)^{2/3} = \frac{\varepsilon_0 C}{2r^5 e_0 m_0^2} \tag{3.128}$$

We have obtained a result that agrees with Eq. (3.91): The current density is constant along the beam (one does not depend on θ). The interesting features of this flow are the very strong dependence on the current density of the radius and the nonuniform rotation of particles. The problem of electrode synthesis for a gun with Meltzer flow is considered in Section 4.3.2.

3.5.4 Laminar Noncongruent Beams

The vortex of the momentum rot **P** in this case is nonzero, and it is impossible to replace the original system of equations (3.11)–(3.18) by the sole equation for the action function. Nevertheless, some important solutions are known (see, e.g., Dryden, 1962; Kirstein et al., 1967). Below we consider two cases using the theory of magnetron injection guns as developed by Dryden (1962), Manuilov and Tsimring (1978), and Tsimring (1977).

Nonrelativistic Electron Flow in a Planar Magnetron with an Inclined Magnetic Field A scheme for a planar static magnetron with an inclined magnetic field in a near cathode region is depicted in Fig. 3.9. It is a uniform infinite system

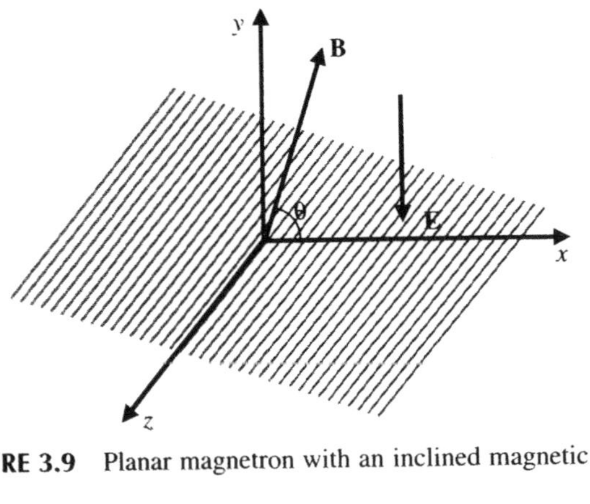

FIGURE 3.9 Planar magnetron with an inclined magnetic field.

along the x and z directions. Therefore, all quantities depend on y only, and the electric field has only one nonzero component, $E \equiv -E_y = d\varphi/dy$. The cathode is arranged in the (X,Z) plane. The magnetic field is assumed uniform, parallel to the (X,Y) plane and inclined to the cathode plane with angle θ. The distinguishing feature of this problem is taking into account versions with a nonzero cathode electric field [Eq. (3.141)]. The system of self-consistent flow equations is reduced to a system of 11 scalar equations with 11 unknown functions because the self magnetic field can be neglected in a nonrelativistic approximation (see Section 3.2.1):

$$\begin{aligned} \frac{d\mathbf{v}}{dt} &= -\eta\mathbf{E} - \eta\mathbf{v} \times \mathbf{B} \\ \mathbf{E} &= -\text{grad}\ \varphi \\ \Delta\varphi &= \frac{\rho}{\varepsilon_0} \\ \mathbf{j} &= -\rho\mathbf{v} \\ \text{div}\ \mathbf{j} &= 0 \end{aligned} \tag{3.129}$$

Let us write Eqs. (3.129) in the terms of the coordinates:

$$\frac{dv_x}{dt} = -\eta(\mathbf{v} \times \mathbf{B})_x = \eta B v_z \sin\theta = \omega_g v_z \sin\theta \tag{3.130}$$

$$\frac{dv_y}{dt} = -\eta E_y - \eta(\mathbf{v} \times \mathbf{B})_y = \eta E - \omega_g v_z \cos\theta \tag{3.131}$$

$$\frac{dv_z}{dt} = -\eta(\mathbf{v} \times \mathbf{B})_z = \omega_g(v_y \cos\theta - v_x \sin\theta) \tag{3.132}$$

$$E = \frac{d\varphi}{dy} \tag{3.133}$$

$$\frac{d^2\varphi}{dy^2} = \frac{\rho}{\varepsilon_0} \tag{3.134}$$

$$j_x = -\rho v_x \tag{3.135}$$

$$j_y = -\rho v_y \tag{3.136}$$

$$j_z = -\rho v_z \tag{3.137}$$

$$\frac{dj_y}{dy} = 0 \tag{3.138}$$

where ω_g is the gyrofrequency. Here equations $E_x = E_z = 0$ are omitted and there are nine equations. The number of independent self-consistent equations can be further reduced to six because Eqs. (3.135) and (3.137) are necessary only for the determination of the current density components j_x and j_z, and instead of Eqs. (3.133) and (3.134), one can write

$$\frac{dE}{dy} = \frac{\rho}{\varepsilon_0} \tag{3.139}$$

Integration of the System Eqs. (3.130)–(3.132), (3.136), (3.138), and (3.139) This problem has an analytic solution in elementary functions. According to continuity equation (3.138),

$$j_y = -j_0 = \text{const.} \tag{3.140}$$

Find $E(t) \equiv E[y(t)]$ using Eqs. (3.137), (3.139), and (3.140):

$$\frac{dE}{dt} = \frac{dE}{dy}\frac{dy}{dt} = \frac{\rho}{\varepsilon_0}v_y = -\frac{j_y}{\varepsilon_0} = \frac{j_0}{\varepsilon_0}$$

After integration, we obtain

$$E = \frac{j_0}{\varepsilon}t + E_c \tag{3.141}$$

where E_c is an electric field for $t = 0$ (i.e., the electric field on the cathode). Note that the time in Eq. (3.141) is determined by the position of the particle on the trajectory. To obtain the field as an explicit function of the coordinates, it is necessary to know the function $t = t(\mathbf{r})$. According to Eq. (3.141), for the ρ-mode (see Section 3.4.4) we must assign $E_c = 0$ and set the potential $\varphi = \varphi_a$ at a plane $y = d$. So we will obtain a current j_0 limited by the space charge. For the T-mode the field E_c is unknown but the emission current j_0 must be given.

For determination of the velocity components, let us differentiate Eq. (3.132) with respect to t and substitute dv_y/dt and dv_z/dt from Eqs. (3.130) and (3.131). We obtain

$$\frac{d^2v_z}{dt^2} + \omega_g^2 v_z = \eta\omega_g \cos\theta\left(\frac{j_0}{\varepsilon_0}t + E_c\right) \tag{3.142}$$

This is the equation of a harmonic oscillator, with the right side a linear function of time. Assume that the component of the initial velocity $v_{x0} = 0$. The solution of the equation is

$$v_z = M(\Phi - \sin\Phi) + N(1 - \cos\Phi) \tag{3.143}$$

where $\Phi = \omega_g t$ is a transit angle. The quantities $M = (\eta j_0/\varepsilon_0\omega_g^2)\cos\theta$ and $N = (\eta E_c/\omega_g)\cos\theta$ are current and field parameters, respectively. The other two components of the velocity can be found by integration of Eqs. (3.131) and (3.132) with initial conditions $v_{y0} = v_{z0} = 0$ after substitution of Eq. (3.142).

Finally, integration of the functions $v_x = dx/dt$, dy/dt, and dz/dt with initial conditions $(x)_{t=0} = x_0$, $(y)_{t=0} = 0$, and $v\,(z)_{t=0} = z_0$ gives the equations of the trajectory:

$$X(\Phi) = \omega_g(x - x_0) = M\sin\theta\left(\frac{\Phi^3}{6} - \Phi + \sin\Phi\right) + N\sin\theta\left(\frac{\Phi^2}{2} - 1 + \cos\Phi\right) \tag{3.144}$$

$$Y(\Phi) = \omega_g y = M\cos\theta\left(\frac{\Phi^3}{6}\tan^2\theta + \Phi - \sin\Phi\right) + N\cos\theta\left(\frac{\Phi^2}{2}\tan^2\theta + 1 - \cos\Phi\right) \tag{3.145}$$

$$Z(\Phi) = \omega_g(z - z_0) = M\left(\frac{\Phi^2}{2} - 1 + \cos\Phi\right) + N(\Phi - \sin\Phi) \tag{3.146}$$

The potential according to Eq. (3.133) is $\varphi = \int E\,dy = \int_0^t Ev_y\,dt$, which gives

$$U(\Phi) = \eta\varphi = M^2\left(\frac{\Phi^4}{8}\tan^2\theta + \frac{\Phi^2}{2} - \Phi\sin\Phi + 1 - \cos\Phi\right) + MN\left(\frac{\Phi^3}{2}\tan^2\theta + \Phi - \Phi\cos\Phi\right) + N^2\left(\frac{\Phi^2}{2}\tan^2\theta + 1 - \cos\Phi\right) \tag{3.147}$$

For different values of x_0 and z_0 we find a bundle of trajectories. Note that in general the trajectories obtained intersect each other and form a two-stream flow. This effect contradicts the primary condition of beam laminarity. In particular, Eqs. (3.135)–(3.137) are not applicable. These equations are true only if returning points on the trajectories are absent (i.e., for the conditions $dy/d\Phi \geq 0$ and $d^2y/d\Phi^2 \geq 0$). The signs of equality determine the critical angle θ_c; exceeding that provides a laminar beam. Using conditions (3.145), (3.147) the critical angle θ_c is readily obtained as a function of the parameter $\kappa = M/N = j_0/\varepsilon_0\omega_g E_c$. The function $\theta_c = \theta_c(\gamma)$ calculated is shown in Fig. 3.10. As we see, the angle $\theta_c \approx 24^\circ$ for

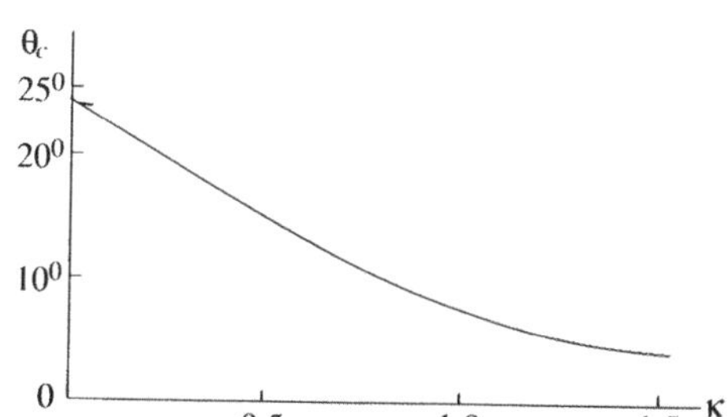

FIGURE 3.10 Critical angle of inclination of a magnetic field to a cathode as a function of the parameter $\kappa = j_0/\varepsilon_0\omega_g E_c$.

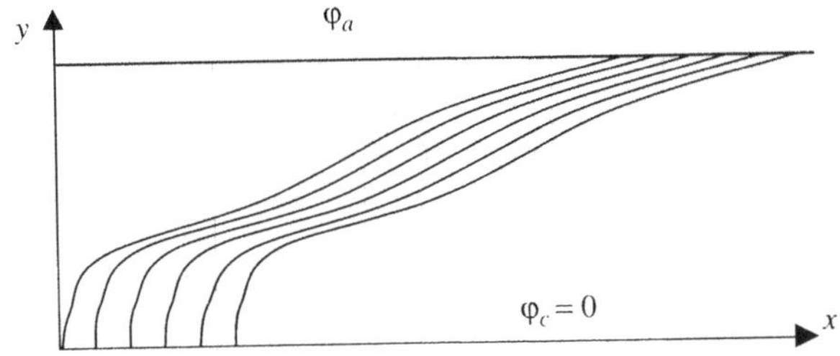

FIGURE 3.11 Trajectories of laminar flow in a planar magnetron with an inclined magnetic field.

the "cold" diode ($\kappa = j_0 = 0$). It is readily seen that $\theta_c = 0$ for the ρ-mode ($E_c = 0$, $\kappa = \infty$). Thus, an electron beam in the ρ-mode is laminar for any θ. Typical trajectories of the laminar beam ($\theta > \theta_c$) are shown in Fig. 3.11.

Dryden Flow Dryden flow (Dryden, 1962) is a laminar electron flow from a conic cathode in an axially symmetric magnetic field. Similitude relationships in this flow correspond to transferring any trajectory to another by a proportional variation of the radius (in a spherical coordinate system). Then a power law dependence on the flow dynamical variables of the radius is ensured (Dryden, 1962). Any variable of the axially symmetric flow is represented in product form as

$$f(r,\theta) = r^m f(1,\theta) \tag{3.148}$$

where r and θ are the radius and polar angle in a spherical coordinate system. Taking into account the relationships between the basic flow variables, one can obtain the following radial variations of these quantities:

$$\begin{aligned}
&\text{Velocity:}\quad v \sim r^n\\
&\text{Electric field:}\quad E \sim r^{2n-1}\\
&\text{Magnetic field:}\quad B \sim r^{n-1}\\
&\text{Potential:}\quad \varphi \sim r^{2n}\\
&\text{Charge density:}\quad \rho \sim r^{2n-2}\\
&\text{Current density:}\quad j \sim r^{3n-2}
\end{aligned} \tag{3.149}$$

where n is an arbitrary constant (integer or fractional). These relationships can be considered as the product form of the separation variables. As a result, the nonrelativistic self-consistent equations of flow [Eqs. (3.129)] are transformed into a system of ordinary differential equations. Equations in a spherical coordinate system are given by Dryden (1962). Note that according to Eq. (3.149), the case $n = 1$ corresponds to a uniform magnetic field. The cylindrical coordinate system in this important case is more suitable because unlike the spherical system, the system of differential equations does not contain equations for the magnetic field. Equations in the cylindrical coordinate system r,θ,z are given by Manuilov and Tsimring (1978).

$$
\begin{aligned}
&\frac{dr}{dt} = v_r, \qquad \frac{d\theta}{dt} = \frac{v_\theta}{r}, \qquad \frac{dv_z}{dt} = v_z \\
&\frac{dv_r}{dt} = \frac{v_\theta^2}{r} - \eta E_r - \eta v_\theta B, \qquad \frac{dv_\theta}{dt} = -\frac{v_r v_\theta}{r} + \eta v_r B, \qquad \frac{dv_z}{dz} = -\eta E_z \\
&\frac{dE_r}{dt} = \left\{ \left[E_r v_r \left(1 - \frac{z^2}{r^2}\right) + E_z v_z \right] r - (E_z v_r - 2E_r v_z) z \right. \\
&\qquad \left. + \frac{j_c r_c^2 z}{\varepsilon_0 r \sin\alpha_c} \right\} \frac{1}{r^2 + z^2} \\
&\frac{dE_z}{dt} = \left[(2E_r v_r + E_z v_z) z + (E_z v_r - 2E_r v_z) r - \frac{j_c r_c^2}{\varepsilon_0 \sin\alpha_c} \right] \frac{1}{r^2 + z^2} \\
&\frac{d\varphi}{dt} = -(E_r v_r + E_z v_z)
\end{aligned}
\tag{3.150}
$$

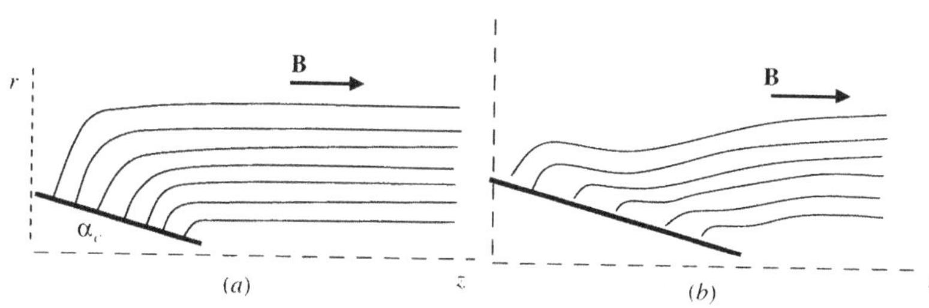

FIGURE 3.12 Trajectories in Dryden flow: (*a*) $G \geq 10$; (*b*) $G \leq 5$.

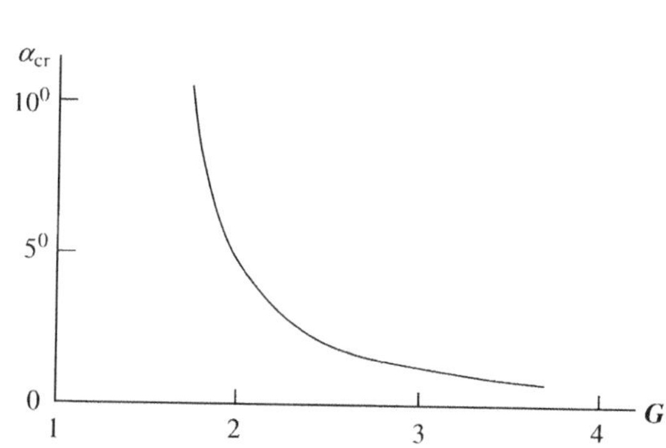

FIGURE 3.13 Critical angle of a cathode cone as a function of the parameter $G = (1/\eta B)(2\eta j/\varepsilon_0 r_0)^{1/3}$. (From Manuilov and Tsimring, 1978.)

where j_c is the cathode current density, r_c is the initial radius of the trajectory on the cathode, and α_c is an angle of the cathode cone (Fig. 3.12).

So we have a system of nine ordinary differential equations of the first order. The numerical solution of these equations was obtained by Dryden (1962) for a regime of current limited by the space charge (the ρ-mode). As it turned out, the trajectories oscillate, in general, into a near-cathode region. However, for a strong magnetic field [the coefficient $G^{-1} = \eta B(\varepsilon_0 r_c/2\eta j)^{1/3} \sim 10$], trajectories quickly transform to straight lines parallel to the z-axis (Fig. 3.12*a*). For less B ($G \leq 5$), the extent of the region of oscillating trajectories increases notably (Fig. 3.12*b*). The frequency of the oscillations is close to the gyrofrequency ηB.

Note that some solutions of the equations (3.150) can give two-stream flows even for the ρ-mode (unlike the planar magnetron; Section 3.5.2). These solutions must be excluded. In Fig. 3.13 the critical cone angle α_{cr} curve is shown (Manuilov and Tsimring, 1978).[2] The region of the laminar beams corresponds to $\alpha_c > \alpha_{cr}$.

[2]In a paper by Manuilov and Tsimring (1978), another parameter of the magnetic field $\lambda = G^3$ was used.

CHAPTER FOUR

Electron Guns

4.1 INTRODUCTION

An electron gun is one of the basic components of an electron beam system, which includes microwave tubes, electron accelerators, and technology procession devices. The gun transforms the electron flow emitted from a cathode in a finite (dense) electron beam with rectilinear or curvilinear electron trajectories. In this chapter, the principles of finite electron beam formation and basic types of electron guns (excluding electron maser guns) (see Chapter 10) are considered.

It is necessary to emphasize that an important element of gun design is the numerical simulation of systems. However, as mentioned in the Introduction, computation analysis problems are beyond the scope of this book.

4.2 PIERCE'S SYNTHESIS METHOD FOR GUN DESIGN

So far, our analysis of space-charge electron beams was connected with two fundamental conditions: (1) single-flow approximation (laminar beams), and (2) beams that fill all space (boundless beams). The second condition allows us to ignore no-cathode boundary value problems. Essentially, this simplifies solution of self-consistent equations and gives in closed form, beam quantities (i.e., trajectories, velocities, space-charge and current densities, distribution of the potential, etc.) as a function of coordinates or other parameters. However, this consideration does not solve the entire problem of electron gun design with bound, sharply outlined beams and with the given beam trajectories. Is it possible to use the solutions of boundless electron beam equations for this problem? A statement and method for solving the problem of electrostatic beams with rectilinear electron trajectory were given by J. R. Pierce (see, e.g., Pierce, 1954).

Electron Beams and Microwave Vacuum Electronics. By Shulim E. Tsimring

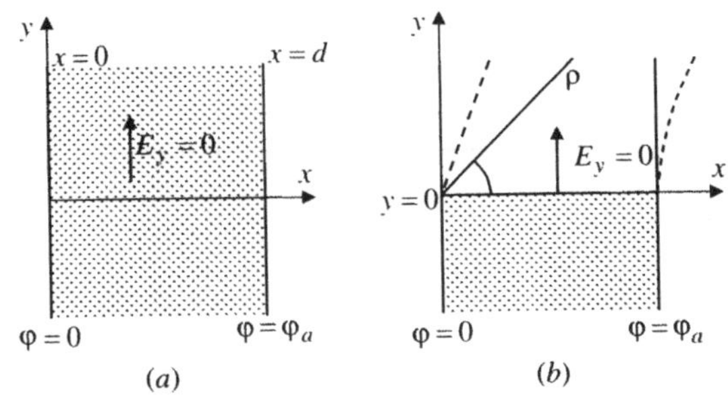

FIGURE 4.1 Pierce's synthesis of electrodes for a planar diode: (*a*) boundless beam; (*b*) prepared beam.

Pierce's method consists of the following: Consider, for example, a nonrelativistic electron beam in a planar diode. The flow fills the entire diode space (Fig. 4.1*a*). According to Eq. (3.103), the potential distribution in this diode for the ρ-mode is $\varphi(x, y) = \varphi_a(x/d)^{4/3}$, and the electric field perpendicular to the trajectories is $E = E_y = -\partial\varphi/\partial y = 0$.

Let us eliminate the beam section from the region $y > 0$ (Fig. 4.1*b*). This reduces the problem in this region to that of obtaining a potential distribution that does not change the electric field at the beam boundary from that of the original beam (Fig. 4.1*a*). In this case, after beam "preparation," the rectilinear electron trajectories will not be disturbed. In the region $y > 0$, Poisson's equation is transformed to Laplace's equation $\Delta\varphi = 0$, and we arrive at the problem of solving Laplace's equation with boundary conditions

$$\left(\frac{\partial\varphi(x,y)}{\partial y}\right)_{y=0} = 0$$
$$(\varphi(x,y))_{y=0} = \varphi_a\left(\frac{x}{d}\right)^{4/3} \tag{4.1}$$

Equipotential surfaces in this region determine the configuration desired for electrodes outside the beam. In Fig. 4.1*b*, the dashed and dash-dotted lines show the shapes of two electrodes for potentials $\varphi = 0$ and φ_a, which follow an elegant Pierce solution (see Section 4.4.2). Afterward, the Pierce method was extended to beams with various boundary surface forms. We examine similar problems in subsequent sections.

The general idea behind the Pierce synthesis method is contained in the division of a single problem into two particular subproblems:

1. *Internal Problem of Synthesis:* Solution of the self-consistent equation for a boundless beam
2. *External Problem:* Preparation of a beam by eliminating a section of it lying outside the envelope of trajectories, which is the given boundary of the beam, solution of the Laplace equation in the volume selected, and determination of the electrode configuration

4.3 INTERNAL PROBLEMS OF SYNTHESIS. RELATIVISTIC PLANAR DIODE. CYLINDRICAL AND SPHERICAL DIODES

So far, we have considered the following internal problems:

1. Flow in planar nonrelativistic diodes (Section 3.4.4)
2. Meltzer flow (Section 3.5.3)
3. Nonrelativistic electron flow in a planar magnetron with an inclined magnetic field (Section 3.5.4)
4. Dryden flow (Section 3.5.4)

Next we study some other important cases.

4.3.1 Relativistic Planar Diode in the ρ-Mode

The scheme of a planar diode is depicted in Fig. 3.7. The problem is reduced to a solution of the one-dimensional Poisson equation

$$\frac{d^2\varphi}{dx^2} = -\frac{\rho}{\varepsilon_0} \tag{4.2}$$

with boundary conditions

$$\varphi(0) = 0, \qquad \varphi(d) = \varphi_a, \qquad (v)_{x=0} = 0, \left(\frac{d\varphi}{dx}\right)_{x=0} = 0 \tag{4.3}$$

The final condition in Eq. (4.3) is referred to the ρ-regime. Using the energy integral [Eq. (1.7)], $\gamma = 1 + \eta\varphi/c^2$, expressions for space-charge density, $\rho = -I/Sv$, and velocity, $v = c\left(\sqrt{\gamma^2 - 1}/\gamma\right)$, we rewrite Eq. (4.2) as

$$\frac{d^2\gamma}{dx^2} = \frac{I}{I_r S}\frac{\gamma}{\sqrt{\gamma^2 - 1}} \tag{4.4}$$

where I is the beam current, S is the square of the electrodes, and the current parameter I_r is

$$I_r = \frac{c^3\varepsilon_0}{\eta} \approx 1356 \text{ A} \tag{4.5}$$

We express the charge density as $\rho = j/v = I\gamma/Sc\sqrt{\gamma^2 - 1}$.

Boundary conditions [Eq. (4.3)] in the term γ is now

$$\gamma_{x=0} = 1, \qquad \gamma_{x=d} = 1 + \frac{\eta\varphi_a}{c^2}, \qquad \left(\frac{d\gamma}{dx}\right)_{x=0} = 0 \tag{4.6}$$

Let us multiply both sides of Eq. (4.4) by the factor $d\gamma/dx$. After integration and taking Eq. (4.6) into account we obtain

$$\frac{d\gamma}{dx} = \sqrt{\frac{2I}{I_r S}}\,\sqrt[4]{\gamma^2 - 1} \tag{4.7}$$

Repeated integration gives the diode volt–ampere characteristic:

$$I = \frac{I_r S}{2d^2}\left(\int_1^{1+\eta\varphi_a/c^2} \frac{d\gamma}{\sqrt[4]{\gamma^2 - 1}}\right)^2 \tag{4.8}$$

The integral in Eq. (4.8) can be reduced to elliptic integrals of the first and second type (Miller, 1982). The simplest way, however, is numerical computation of the integral. Let us consider two extreme cases when the integral can be expressed through elementary functions.

1. *Nonrelativistic Diode*, $\eta\varphi/c^2 \ll 1$. In this case, $\sqrt[4]{\gamma^2 - 1} \approx \sqrt[4]{2(\gamma - 1)}$. Then

$$I \approx \frac{8 I_r S}{9\sqrt{2}\, d^2}(\gamma - 1)^{3/2} = \frac{4}{9}\varepsilon_0\sqrt{2\eta}\,\frac{S}{d^2}\varphi_a^{3/2}$$

$$\approx 2.334 \times 10^{-6}\,\frac{S}{d^2}\varphi_a^{3/2} \qquad \text{A/V}^{3/2} \tag{4.9}$$

The formula obviously agrees with Eq. (3.105).

2. *Supreme Relativistic Diode*, $\gamma \gg 1$. Then $\sqrt[4]{\gamma^2 - 1} \approx \sqrt{\gamma}$, and the current is

$$I \approx \frac{I_r S}{2d^2}\int_1^{\gamma_a}\left(\frac{d\gamma}{\sqrt{\gamma}}\right)^2 \approx \frac{2 I_r S}{d^2}\gamma_a \approx \frac{2 I_r S}{d^2}\frac{\eta\varphi_a}{c^2} = 2c\varepsilon_0\frac{S}{d^2}\varphi_a \tag{4.10}$$

Thus, the supreme relativistic diode is equivalent to a linear admittance:

$$Y = 2c\varepsilon_0\frac{S}{d^2} \approx \frac{1}{188}\frac{S}{d^2} \qquad \Omega^{-1} \tag{4.11}$$

For a diode with circular cross section ($S = \pi r^2$), the admittance is

$$Y \approx \frac{1}{60}\frac{r^2}{d^2} \qquad \Omega^{-1} \tag{4.12}$$

The normalized admittance of the diode for various approximations is depicted in Fig. 4.2. According to the plot, nonrelativistic approximation is applicable up to $\varphi \sim 500$ kV; supreme relativistic approximation is acceptable for values above $\varphi \sim 50$ MV.

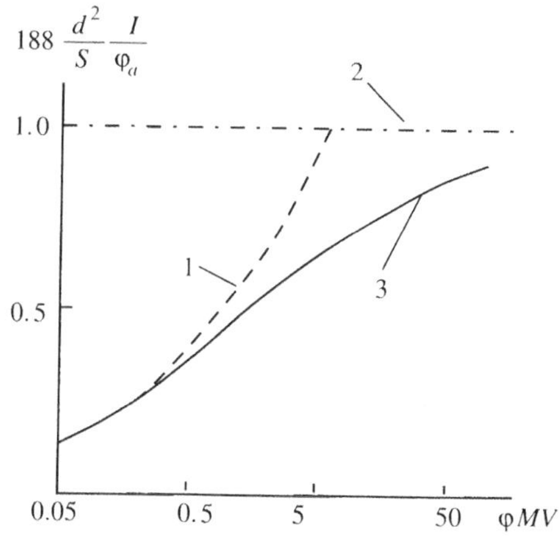

FIGURE 4.2 Diode admittance for different approximations. 1, Nonrelativistic diode; 2, supreme relativistic diode; 3, exact solution.

4.3.2 Nonrelativistic Cylindrical and Spherical Diodes in the ρ-Mode

A diode scheme is shown in Fig. 4.3. Let us confine ourselves to consideration of diodes with an external cylinder as the cathode (i.e., $r_c/r_a > 1$). Corresponding flows are generally used in electron guns that compress the beam following an emission. Assume that the fringe effects in the cylindrical diode are absent. The potential in both diodes will depend only on the radial coordinate, and the electric field will have only a radial component. Assume that the initial velocities are zero. In this case, the velocities and current densities are radial and the trajectories are rectilinear. So we can write

$$\varphi(\mathbf{r}) = \varphi(r), \qquad \mathbf{v} = v_r \equiv v, \qquad \mathbf{j} = j_r \equiv j \tag{4.13}$$

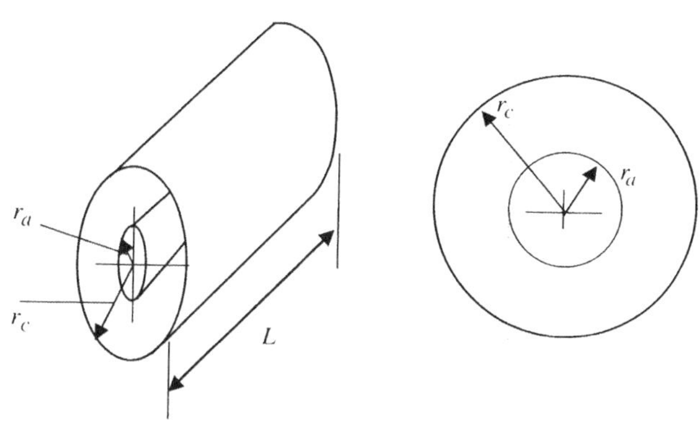

FIGURE 4.3 Cylindrical and the spherical diodes with an external cathode.

A system of self-consistent equations for both diodes is

$$\begin{aligned} \Delta_r \varphi &= -\frac{\rho}{\varepsilon_0} \\ j &= \frac{I}{S} = \rho v \\ v &= \sqrt{2\eta\varphi} \\ \text{div}\, \mathbf{j} &= \frac{dI}{dr} = 0 \end{aligned} \tag{4.14}$$

where Δ_r is the radial component of the Laplace operator and I is the beam current, which is the same in any cross section S of the flow. Consider each diode separately.

Cylindrical Diode According to Fig. 4.3 and Eqs. (4.14),

$$S = 2\pi r L, \qquad \rho = \frac{I}{Sv} = \frac{I}{2\pi r L\sqrt{2\eta\varphi}} \tag{4.15}$$

Using the known expression for Δ_r in the cylindrical coordinate system, we obtain after substituting Eq. (4.15) in the first of Eqs. (4.14), the equation

$$\sqrt{\varphi}\frac{d}{dr}\left(r\frac{d\varphi}{dr}\right) = \frac{I}{2\pi L\varepsilon_0\sqrt{2\eta}} \tag{4.16}$$

and the boundary conditions

$$\varphi(r_c) = 0, \qquad \varphi(r_a) = \varphi_a, \qquad \left(\frac{d\varphi}{dr}\right)_{r=r_c} = 0 \tag{4.17}$$

As was shown (Langmuir and Blogett, 1923), it is convenient to express the current through an auxiliary function $\beta(r/r_c)$:

$$I = \frac{8\pi}{9}\varepsilon_0 L\sqrt{2\eta}\frac{\varphi^{3/2}}{r\beta^2(r/r_c)} \tag{4.18}$$

Note that this representation is valid for $L \gg r_c - r$. The potential as a function of r according to Eq. (4.18) is

$$\varphi(r) = Cr^{2/3}\left[\beta\left(\frac{r}{r_c}\right)\right]^{4/3} \tag{4.19}$$

where the constant C and taking the boundary conditions (4.17) into account is

$$C = \frac{\varphi_a}{r_a^{2/3}[\beta(r_a/r_c)]^{4/3}} \tag{4.20}$$

Then Eq. (4.16) for the term $\beta = \beta(r/r_c)$ is

$$C^{3/2} r^{1/3} \beta^{2/3} \frac{d}{dr}\left[r \frac{d}{dr}(r^{2/3}\beta^{4/3}) \right] = \frac{I}{2\pi L \varepsilon_0 \sqrt{2\eta}} \tag{4.21}$$

Langmuir obtained the solution of Eq. (4.21) by expanding $\beta(r/r_c)$ in a quickly convergent power series of the variable $\omega = \ln(r_c/r)$ (Langmuir and Blogett, 1923):

$$\beta = \omega + \frac{2\omega^2}{5} + \frac{11\omega^3}{120} + \frac{47\omega^4}{3300} + \cdots \tag{4.22}$$

Spherical Diode According to Fig. 4.3,

$$S = 4\pi r^2, \qquad \rho = \frac{I}{Sv} = \frac{I}{4\pi r^2 \sqrt{2\eta\varphi}} \tag{4.23}$$

Using an expression for Δ_r in the spherical coordinate system, after substituting Eq. (4.13) in the first of Eqs. (4.14), we obtain

$$\sqrt{\varphi}\frac{d}{dr}\left(r^2 \frac{d\varphi}{dr} \right) = \frac{I}{4\pi\varepsilon_0 \sqrt{2\eta}} \tag{4.24}$$

with boundary conditions (4.17).

According to Langmuir and Blogett (1924), the current can be represented through a function $\alpha(r/r_c)$:

$$I = \frac{16\pi}{9} \varepsilon_0 \sqrt{2\eta} \frac{\varphi^{3/2}}{\alpha^2(r/r_c)} \tag{4.25}$$

Note that the current depends on neither the anode nor the cathode radius separately but only on the ratio of these values. The potential as the function of r according to Eq. (4.25) is

$$\varphi(r) = D\left[\alpha\left(\frac{r}{r_c} \right) \right]^{4/3} \tag{4.26}$$

where the constant D, taking the boundary conditions (4.17) into account, is

$$D = \frac{\varphi_a}{[\alpha(r_a/r_c)]^{4/3}} \tag{4.27}$$

So Eq. (4.24) in the term α is

$$D^{3/2} \alpha^{2/3} \frac{d}{dr}\left[r^2 \frac{d}{dr}(\alpha^{4/3}) \right] = \frac{I}{4\pi\varepsilon_0 \sqrt{2\eta}} \tag{4.28}$$

The function $\alpha(r/r_c)$ can be expanded (Langmuir and Blogett, 1924) in a quickly convergent power series of the variable $\omega = \ln(r_c/r)$:

$$\alpha = \omega + \frac{3\omega^2}{10} + \frac{3\omega^3}{40} + 0.00143\omega^4 + 0.0021\omega^5 + \cdots \tag{4.29}$$

In conclusion, let us rewrite the expressions for distribution of the potential in the diodes:

$$\varphi(r) = \begin{cases} \varphi_a \left(\dfrac{r}{r_a}\right)^{2/3} \left(\dfrac{\beta^2(r/r_c)}{\beta^2(r_a/r_c)}\right)^{2/3} & \text{(cylindrical diode)} \quad (4.30) \\ \varphi_a \left(\dfrac{\alpha^2(r/r_c)}{\alpha^2(r_a/r_c)}\right)^{2/3} & \text{(spherical diode)} \quad (4.31) \end{cases}$$

4.4 EXTERNAL PROBLEMS OF SYNTHESIS. CAUCHY PROBLEM

Next we turn to the second half of Pierce's synthesis, the external subproblem.

4.4.1 Cauchy Problem for Laplace's Equation

As seen earlier, the external problem can be reduced to solution of the Laplace equation in the domain outside the beam with the electric field on the beam boundary, which in turn was found in the solution of the internal problem. It can readily be proven that assignment of the boundary electric field for the Laplace equation is equivalent to setting the boundary potential and its normal derivative. In this case, the boundary value problem is called the *Cauchy problem*. Note that the boundary potential and its normal derivative (Cauchy data) are given on the part of the domain boundary that is common with the beam. This problem of the potential theory is distinguished from classic Dirichlet–Neiman problems. In the latter case, the potential or its normal derivative is given on a closed boundary of the domain. The main peculiarity of the Cauchy problem when used in Laplace's equation is its incorrect setting. According to Hadamard's definition (see, e.g., Fattoriny, 1983), the conditions of the correct setting involve small variations of the solution that occurs when variations of the boundary values are small. It can be shown that, in general, the setting of the Cauchy problem is correct for hyperbolic equations and is incorrect for elliptic equations. In the latter case, the Cauchy problem is correct only if the boundary values are analytical functions of arguments. This condition is not fulfilled in direct numerical solution of boundary value problems.

Let us illustrate this effect by using the example of an electric field in the triode. Consider a planar model of the grid–cathode gap. The grid is carried out from thin equidistant rectilinear wires arranged perpendicular to the plane of the

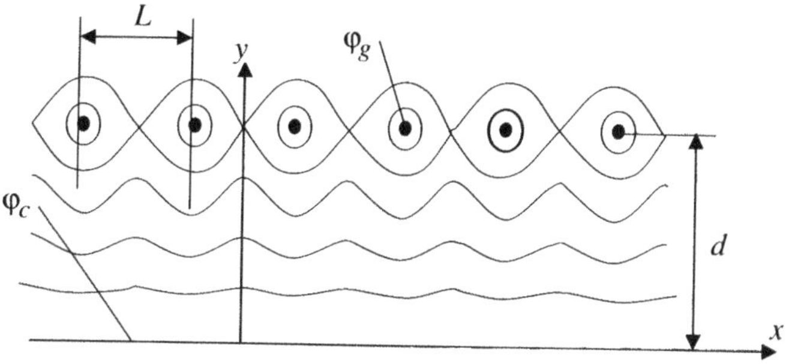

FIGURE 4.4 Potential pattern in the grid–cathode gap of a triode.

drawing (Fig. 4.4). Assume that we have solved the Dirichlet problem for the given boundary potentials φ_g of the grid and φ_c of the cathode. We obtain a pattern of periodic equipotential lines of the type in Fig. 4.4 that appears along the coordinate, x. It can be shown that the amplitude of the potential variations decreases exponentially with distance from the grid. The potential on the cathode as a function x will be constant, but the electric field (the normal derivative of the potential) changes as

$$E_y = -\left[\frac{\partial\varphi(x,y)}{\partial y}\right]_{y=0} \sim a\sin\left(\frac{2\pi}{L}x\right) \tag{4.32}$$

where the amplitude a is an exponentially small value as a function of d.

Consider the same problem as the Cauchy problem with the Cauchy data on the plane $y = 0$:

$$[\varphi(x,y)]_{y=0} = \varphi_c = \text{const.}, \qquad \left[\frac{\partial\varphi(x,y)}{\partial y}\right]_{y=0} = a\sin\left(\frac{2\pi}{L}x\right) \tag{4.33}$$

If we found a rigorous solution to this problem, we would obviously obtain the pattern in Fig. 4.4. Now let us replace the small oscillating normal derivative with one of uniform value: for example, $\partial\varphi/\partial y = a$. Then the pattern corresponding to a solution of the Cauchy problem is changed drastically. We would get a uniform potential distribution, and the shape of the equipotential on the plane of the grid would be plane. Thus, the small nonanalytic variation of the Cauchy data led to a qualitative reformation of the solution. Essentially, the amplitude a of the periodic variations of the cathode field is decreased with L. Therefore, the instability of the Cauchy problem is intensified when the scale L decreases. If the Cauchy problem is solved by a finite-difference method, small rounding errors and machine noises make the solution untrue. Below we consider some important examples of Cauchy problem solutions when analytical Cauchy data are given.

4.4.2 Synthesis of a Pierce Gun

The Cauchy data for a Pierce gun are given by Eqs. (4.1). Following Pierce (1954), let us continue the potential $\varphi(x)$ analytically into the upper half-plane (Fig. 4.1):

$$\varphi(z) = \varphi_a \left(\frac{z}{d}\right)^{4/3} \tag{4.34}$$

where $z = x + iy$ and $y > 0$. The real function

$$\varphi(x, y) = \text{Re } \varphi(z) \tag{4.35}$$

is the desired solution of our problem.

Indeed, as is generally known, real and imaginary parts of any analytical function satisfy a Laplace equation. It is readily verified that Eq. (4.35) satisfies the boundary conditions (4.1) also:

$$\begin{aligned} \Phi(x,0) &= [\text{Re } \varphi(z)]_{y=0} = \text{Re}[\varphi_a(x+iy)^{4/3}]_{y=0} = \varphi_a x^{4/3} \\ \left[\frac{\partial\Phi(x,y)}{\partial y}\right]_{y=0} &= \{\text{Re}[\varphi_a i \tfrac{4}{3}(x+iy)^{1/3}]\}_{y=0} = \text{Re}(i\tfrac{4}{3}\varphi_a x^{1/3}) = 0 \end{aligned} \tag{4.36}$$

The equation of a equipotentials in a Pierce gun with potential φ_i is obviously

$$\text{Re}[\varphi_a(x+iy)^{4/3}] = \varphi_i = \text{const.} \tag{4.37}$$

It is convenient to use polar coordinates ρ and θ (Fig. 4.1*b*). Then $x + iy = \rho \exp(i\theta)$ and the equation of the equipotentials is

$$\varphi_a \rho \cos(4/3\theta) = \varphi_i \tag{4.38}$$

In particular, the equation of the cathode equipotential ($\varphi_i = 0$) is $\theta = \frac{3}{8}\pi$. This straight line inclines to the beam edge an angle of $\frac{3}{8}\pi = 67^\circ 30'$. According to Eq. (4.38), the equation of the anode equipotential ($\varphi_i = \varphi_a$) is $\rho = 1/\cos(\frac{3}{8}\theta)$. It can readily be shown that equipotentials with $\varphi_i \neq 0$ cut the beam at an angle of $\pi/2$. The equipotentials are shown in Fig. 4.1*b* by dashed lines.

It is interesting to note that Pierce's angle $67^\circ 30'$ is *universal*: The cathode equipotentials in any gun in the space-charge regime (the ρ-mode) and for zero initial velocities cut the beam at the same angle. Indeed, trajectories and a potential distribution for any flows (including relativistic ones) in the nearest vicinity to the cathode are the same as in a planar nonrelativistic diode. Therefore, a local geometry of cathode equipotentials near the point of its intersection with a beam must coincide with the geometry of Pierce's planar gun.

The difficult problem is the exit of the beam from the gun considered. If an anode aperture is used to achieve this goal, it will act as a divergent lens (Section 2.6.5) and the beam will not be rectilinear. Use of the grid as an anode electrode leads to partial interception of the beam and to the disturbance of rectilinear trajectories near grid wires.

4.5 SYNTHESIS OF ELECTRODE SYSTEMS FOR TWO-DIMENSIONAL CURVILINEAR BEAMS WITH TRANSLATION SYMMETRY (LOMAX–KIRSTEIN METHOD). MAGNETRON INJECTION GUN

Let us consider an infinite laminar electron flow with curvelinear trajectories that are identical in the z-direction. The translation of the flow in the stated direction does not change the beam. In general, the trajectories can be three-dimensional curves of the laminar flow. We can consider the latter as a system of the curvilinear surfaces with a straight generatrix parallel z-axis. The beam quantities would not, however, depend on the variable z.

Let us choose one surface within the flow as the beam boundary. The equation of the surface coincides with the equation of a projection of the boundary trajectory on the (x,y) plane. We write this equation in parametric form,

$$x = X(t), \qquad y = Y(t) \tag{4.39}$$

where X and Y are analytic functions. The parameter t cannot be obligatory to the *time*. Remove parts of the beam above (or under) the boundary surface. We come to the Cauchy problem for the Laplace equation. It is supposed that the internal problem has been solved and that the Cauchy data on the beam boundary are known:

$$\varphi[x(t), y(t)] = \varphi(t) \tag{4.40}$$

$$\frac{\partial \varphi}{\partial n} = -E_n(t) \tag{4.41}$$

Here n is used as a direction of a normal-to-boundary surface. We can, instead [Eq. (4.40)], use the tangential component of the electric field:

$$\frac{\partial \varphi(t)}{\partial \tau} = -E_\tau(t) \tag{4.42}$$

4.5.1 Lomax–Kirstein Method of Synthesis of Two-Dimensional Systems (Kirstein, 1958; Lomax, 1957)

Let us continue analytically the functions X and Y from Eq. (4.39) into the upper half-plane $w = t + iu\ (u > 0)$, and consider the following conformal transformation of the plane $v = x + iy$ to the plane $w = t + iu$:

$$x + iy = X(t + iu) + iY(t + iu) \tag{4.43}$$

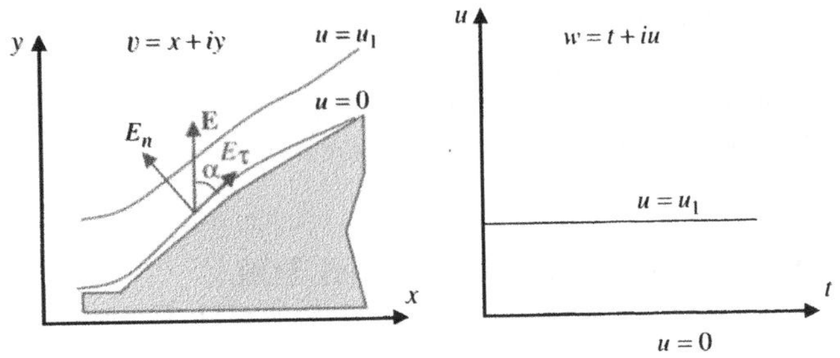

FIGURE 4.5 Conformal transformation of the external domain of a beam.

It is readily seen from Eq. (4.41) what the transformation corresponding to Eq. (4.43) transfers from the boundary trajectory of the beam into the t-axis of the plane w (Fig. 4.5). Because of invariance of Laplace's equation under conformal transformation, the transformed potential $\varphi(t,u)$ must satisfy the equation

$$\frac{\partial^2\varphi}{\partial t^2}+\frac{\partial^2\varphi}{\partial u^2}=0 \tag{4.44}$$

To obtain the Cauchy data in the (t,u) plane, we take into account the correspondence of the lines for the conformal transformation. It means that the translation along τ and n in the (x,y) plane corresponds to translation along t and u in the (t,u) plane. Then the fields on the line $u=0$ are

$$E_{t0}(t)=-\left[\frac{\partial\varphi(t,u)}{\partial t}\right]_{u=0}=-\left[\frac{\partial\varphi(t,u)}{\partial\tau}\right]_{u=0}\frac{d\tau}{dt}=E_\tau(t)\frac{d\tau}{dt} \tag{4.45}$$

$$E_{u0}(t)=-\left[\frac{\partial\varphi(t,u)}{\partial u}\right]_{u=0}=-\left[\frac{\partial\varphi(t,u)}{\partial n}\right]_{u=0}\frac{dn}{du}=E_n(t)\frac{dn}{du} \tag{4.46}$$

The geometric meaning of the derivatives $d\tau/dt$ and dn/du is an extensions of the (x,y) plane in the directions τ and l for transfer to the (t,u) plane. By virtue of conformal transformation, these extensions are the same. Then according to Eqs. (4.45) and (4.46),

$$E_{u0}=E_{t0}\frac{E_n(t)}{E_\tau(t)}=E_{t0}\tan\alpha \tag{4.47}$$

where α is an angle contained by the boundary electric field and the trajectory (see Fig. 4.1). But the quantity E_{t0} according to Eq. (4.45) is

$$E_{t0}=-\frac{d\varphi}{dt} \tag{4.48}$$

Analytic continuation of the fields from the line $u = 0$ into the upper half-plane gives

$$E_{t0}(t) \longrightarrow E_{t0}(t + iu) \equiv E_{t0}(w), \quad E_{u0}(t) \longrightarrow E_{u0}(t + iu) \equiv E_{u0}(w) \tag{4.49}$$

Now let us form a specific complex field and a corresponding complex potential:

$$\Xi(w) = E_{t0}(w) - iE_{u0}(w) \tag{4.50}$$

$$\varphi(w) = -\int_0^w \Xi(w)\, dw \tag{4.51}$$

It is readily verified that the solution desired for the Cauchy problem in the (t,u) plane is

$$\varphi(t,u) = \operatorname{Re} \varphi(w) = -\operatorname{Re} \int_0^w [E_{t0}(w) - E_{u0}(w)]\, dw \tag{4.52}$$

$\varphi(t,u)$ obviously satisfies Laplace's equation [Eq. (4.44)], One task remains—check the boundary conditions (4.4). Let us try $E_{u0}(t)$:

$$E_{u0}(t) = -\left[\frac{\partial \varphi(t,u)}{\partial u}\right]_{u=0} = -\left[\frac{\partial}{\partial u} \operatorname{Re}[\varphi(w)]\right]_{u=0} = -\operatorname{Re}\left[\frac{d\varphi(w)}{dw}\frac{\partial w}{\partial u}\right]_{u=0}$$

$$= \operatorname{Re}\{i[E_{t0}(w) - iE_{u0}(w)]_{u=0} = \operatorname{Re}[iE_{t0}(t) + E_{u0}(t)] = E_{u0}(t) \tag{4.53}$$

The condition for $E_{t0}(t)$ could be checked analogously.

The equipotentials (electrode shape) can be calculated by integrating Eq. (4.52) numerically. If we separate the real and imaginary parts in Eq. (4.43), we obtain the functions

$$x = x(t, u), \qquad y = y(t, u) \tag{4.54}$$

While integrating Eq. (4.52), we replace t,u on x,y in every point according to Eqs. (4.54) and obtain $\varphi = \varphi(x,y)$. As a result, we must find the surfaces of an equal potential.

A simpler procedure is based on direct numerical integration of the differential equation of equipotentials. The equation of the equipotentials in the (t,u) plane is obviously

$$\frac{du}{dt} = -\frac{\partial\varphi/\partial t}{\partial\varphi/\partial u} = -\frac{\dfrac{\partial}{\partial t}\left[\operatorname{Re}\displaystyle\int_0^w \Xi(w)\, dw\right]}{\dfrac{\partial}{\partial u}\left[\operatorname{Re}\displaystyle\int_0^w \Xi(w)\, dw\right]} = \frac{\operatorname{Re}[\Xi(w)]}{\operatorname{Im}[\Xi(w)]} \tag{4.55}$$

Taking Eq. (4.50) into account, we obtain the following ordinary differential equation:

$$\frac{du}{dt} = \frac{\mathrm{Re}[E_{t0}(w) - iE_{u0}(w)]}{\mathrm{Im}[E_{t0}(w) - iE_{u0}(w)]} \tag{4.56}$$

While integrating this equation we are moving along the equipotential in the (u,v) plane. If we replace u,v in each point by x,y, we immediately obtain an equipotential in the real (x,y) plane. This problem can be solved easily with a computer, but it is necessary to have a solution of the internal problem in analytic form. This is, in essence, the method's limitation.

4.5.2 Design of Electrodes for a Meltzer Flow Gun (Lomax, 1957)

Let us take as the boundaries of a beam circles of radius b and a (Fig. 4.6; see also Section 3.5.3). According to Eq. (3.126), the potential and the components of the electric field on the outer boundary $r = b$ are

$$\varphi_{r=b} = \frac{A}{b^2}\left(\sin\frac{3}{2}t\right)^{4/3}, \qquad E_n = -\left(\frac{\partial\varphi}{\partial r}\right)_{r=b} = \frac{2A}{b^3}\left(\sin\frac{3}{2}t\right)^{4/3}$$
$$E_\tau = -\left(\frac{\partial\varphi}{r\partial\theta}\right)_{r=b} = -\frac{2A}{b^3}\left(\sin\frac{3}{2}t\right)^{1/3}\cos\left(\frac{3}{2}t\right) \tag{4.57}$$

[We formally replace the argument θ in Eq. (3.26), by the parameter t.] The equation of the outer boundary trajectory is

$$x = b\cos t, \qquad y = b\sin t \tag{4.58}$$

Let us find the angle α (Fig. 4.5):

$$\tan\alpha = \frac{E_n}{E_\tau} = -\frac{\left(\sin\frac{3}{2}t\right)^{4/3}}{\left(\sin\frac{3}{2}t\right)^{1/3}\cos\frac{3}{2}t} = -\tan\frac{3}{2}t, \qquad \alpha = -\frac{3}{2}t \tag{4.59}$$

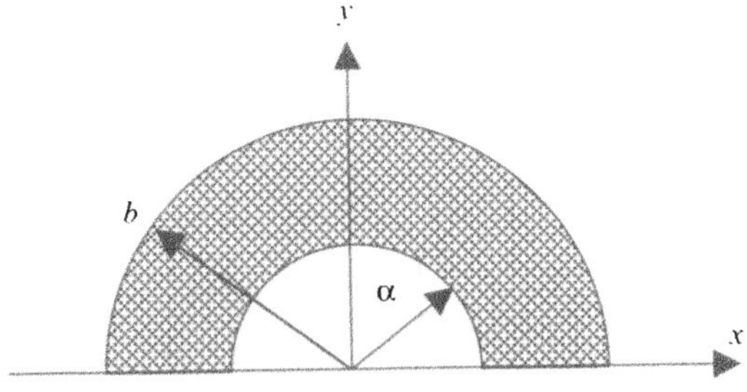

FIGURE 4.6 Meltzer flow.

The conformal transformation according to Eqs. (4.43) and (4.58) is

$$x + iy = b\cos(t + iu) + ib\sin(t + iu) = b\exp[i(t + iu)]$$

Then

$$x = b\cos t\exp(-u), \qquad y = b\sin t\exp(-u) \tag{4.60}$$

The differential equation of the equipotential is Eq. (4.60), where according to Eqs. (4.47), (4.48), (4.57), and (4.59),

$$\begin{aligned}\Xi = E_{t0}(w) - iE_{u0}(w) &= -\frac{2A}{b^2}\left(\sin\frac{3}{2}w\right)^{1/3}\cos\frac{3}{2}w - i\frac{2A}{b^2}\left(\sin\frac{3}{2}w\right)^{4/3}\\ &= -\frac{2A}{b^2}\left(\sin\frac{3}{2}w\right)^{1/3}\exp\left(i\frac{3}{2}w\right)\end{aligned} \tag{4.61}$$

Let us now transform this equation to polar coordinates. According to Eq. (4.60),

$$t = \theta, \qquad r = \sqrt{x^2 + y^2} = b\exp(-u), \qquad u = \ln\left(\frac{b}{r}\right) \tag{4.62}$$

Then it is readily verified that the differential equation of the equipotentials [Eq. (4.58)] is

$$\frac{du}{dt} = \cot[\arg \Xi(w)] \tag{4.63}$$

In polar coordinates it becomes

$$\frac{du}{dt} = -\frac{1}{r}\frac{dr}{d\theta} = \cot\left\{\frac{1}{3}a\tan\left[\frac{1 - (r/b)^3}{1 + (r/b)^3}\cot\frac{3}{2}\theta\right] + \frac{3}{2}\theta\right\} \tag{4.64}$$

The equation is an ordinary first-order differential equation. It can easily be integrated numerically. The equipotentials corresponding to the inner part of the gun is obtained from Eq. (4.64) by replacing the ratio $r/b > 1$ by $r/a < 1$.

The pattern of the calculated equipotentials is shown in Fig. 4.7. The system is symmetric near the line $\theta = \pi/3$. The gun's important feature is the presence of equipotentials that don't intercept the beams. Thus, for this gun, unlike that of Pierce guns, there is no anode hole problem. The peculiar feature of the gun is not having uniform rotation of the particles in the flow. The velocity is zero on the cathode ($\theta = 0$), attains a maximum at $\theta = \pi/3$, and again is zero at $\theta = 2\pi/3$. If one sets two electrodes as planes at positions $\theta = 0$ and $2\pi/3$, it is possible to pass current between them without transferring beam energy to these electrodes. As one would expect, the cathode equipotentials make Pierce's angle 67.5° with the beam.

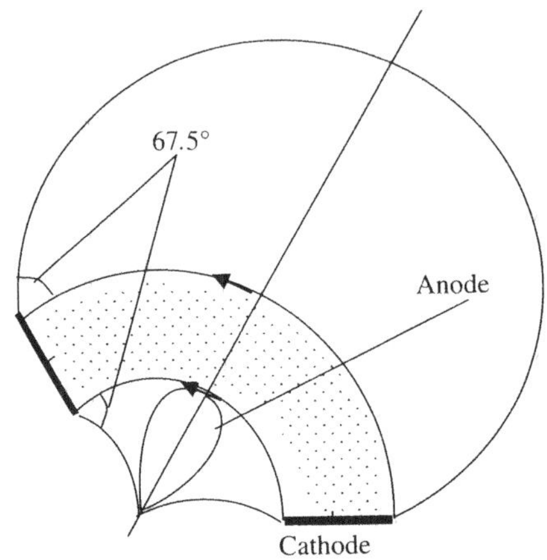

FIGURE 4.7 Meltzer gun.

4.5.3 Design of Electrodes for a Magnetron Injection Gun (Tsimring, 1977)

Let us take a nonrelativistic electron flow in the planar magnetron with an inclined magnetic field as the basis for a magnetron injection gun (MIG) design (see Section 4.7.2). Corresponding flow was used first by Kino (1960) in his theory of the MIG . The model used in that work was of a MIG with a zero cathode electric field (ρ-mode). As a simplified condition, Kino has also introduced a small initial electron velocity that allows us to exclude from the flow equations oscillatory terms of type $\sin\Phi$ [see Eqs. (3.144)–(3.147)]. This approximation simplifies the solution of the external problem significantly. Disregarded oscillating trajectories can represent independent interest in electron maser gyrotrons in that interaction of a high-frequency electromagnetic field with oscillating electrons is used. The same reasoning also requires using a T-mode with a nonzero cathode electric field. Corresponding topics are considered in detail in Chapter 10.

The solution based on the full system of the equations (3.144)–(3.147) is described below. The system takes into account an arbitrary cathode electric field [Eq. (3.141)] and zero initial velocity. According to the Lomax–Kirstein method, we continue the boundary beam trajectory of Eqs. (3.144), (3.145) analytically:

$$X = X_b(\Phi), \qquad Y = Y_b(\Phi) \tag{4.65}$$

into a complex plane $\Theta = \Phi + i\Psi$ by a conformal mapping with the function

$$W = X + iY = X_b(\Theta) + iY_b(\Theta) \tag{4.66}$$

This mapping transforms the trajectory T in the real axis $\Psi = 0$ (Fig. 4.8). After separation of the real and imaginary parts in Eq. (4.66) and using Eqs. (3.144) and

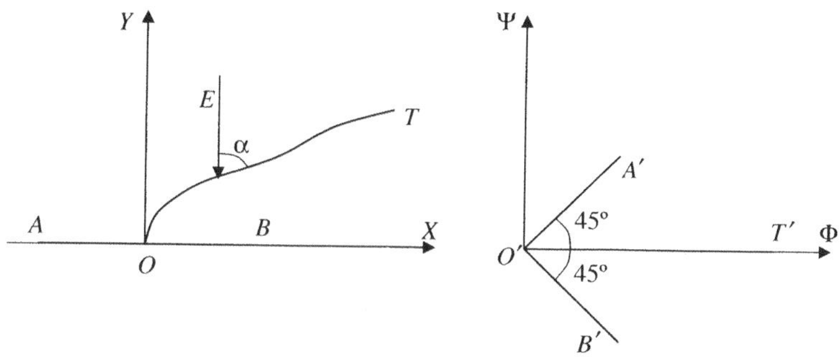

FIGURE 4.8 Conformal mapping of a (X,Y) plane to Φ,Ψ. The rays OA′ and OB′ are the mapping of the rays OA and OB in the vicinity of the origin of the coordinates.

(3.145), we obtain

$$X = \tan\theta\left[\frac{\Phi^2-\Psi^2}{2} + \cos\Phi\cosh\Psi - 1 + \gamma\left(\frac{\Phi^3-3\Phi\Psi^2}{6} - \Phi + \sin\Phi\cosh\Psi\right)\right]$$
$$-\left[\Phi\Psi\tan^2\theta + \sin\Phi\sinh\Psi + \gamma\left(\frac{3\Phi^2\Psi-\Psi^3}{6}\tan^2\theta + \Psi + \cos\Phi\sinh\Psi\right)\right] \tag{4.67}$$

$$Y = \tan\theta\left[\Phi\Psi - \sin\Phi\sinh\Psi + \gamma\left(\frac{3\Phi^2\Psi-\Psi^3}{6} - \Psi + \cos\Phi\sinh\Psi\right)\right]$$
$$+\left[\frac{\Phi^2-\Psi^2}{2}\tan^2\theta - \cos\Phi\cosh\Psi\right.$$
$$\left.+\gamma\left(\frac{\Phi^3-3\Phi\Psi^2}{6}\tan^2\theta + \Phi - \sin\Phi\cosh\Psi\right) + 1\right] \tag{4.68}$$

The shape of electrodes is determined from a differential equation of equipotentials [Eq. (4.56)] in the (Φ,Ψ) plane, which in the given case is

$$\frac{d\Psi}{d\Phi} = \frac{\mathrm{Re}[E_\Phi(\Theta) - iE_\Psi(\Theta)]}{\mathrm{Im}[E_\Phi(\Theta) - iE_\Psi(\Theta)]} \tag{4.69}$$

where $E_\Phi(\Theta)$ and $E_\Psi(\Theta)$ are analytically continued electric fields:

$$E_\Phi = -\frac{dU(\Phi)}{d\Phi}, \qquad E_\Psi = -\frac{dU(\Phi)}{d\Phi}\tan\alpha \tag{4.70}$$

Here $U(\Phi)$ is determined by Eq. (3.147) and α is the angle between the boundary electric field and the trajectory. The electric field is directed along the Y-axis; therefore, $\tan\alpha = dx/dy = X'(\Phi)/Y'(\Phi)$.

With the aid of Eqs. (4.67)–(4.70) and some algebra, we obtain the following differential equation of the electrodes in the (Φ, Ψ) plane:

$$\frac{d\Psi}{d\Phi} = \frac{H[(C^2 + D^2)\cot\theta + BC - AD] + F(AC + BD)}{F[(C^2 + D^2)\cot\theta + BC - AD] - H(AC + AD)} \tag{4.71}$$

where

$$\begin{aligned} H = {} & \Phi\tan^2\theta + \sin\Phi\cosh\Psi + \gamma\left[\frac{3(\Phi^2 - \Psi^2)}{2}\tan^2\theta + \Phi\sin\Phi\cosh\Psi \right. \\ & \left. - \Psi\cos\Phi\sinh\Psi - \cos\Phi\cosh\Psi + 1\right] \\ & + \gamma^2\left(\frac{\Phi^3 - 3\Phi\Psi^2}{2}\tan^2\theta + \Phi - \Phi\cos\Phi\cosh\Psi - \Psi\sin\Phi\sinh\Psi\right) \end{aligned} \tag{4.72}$$

$$\begin{aligned} F = {} & \Psi\tan^2\theta + \cos\Phi\sinh\Psi + \gamma(3\Phi\Psi\tan^2\theta + \Phi\cos\Phi\sinh\Psi \\ & + \Psi\sin\Phi\cosh\Psi + \sin\Phi\sinh\Psi) \\ & + \gamma^2\left(\frac{3\Phi^2\Psi - \Psi^3}{2}\tan^2\theta + \Psi + \Phi\sin\Phi\sinh\Psi - \Psi\cos\Phi\cosh\Psi\right) \end{aligned} \tag{4.73}$$

$$A = \Phi - \sin\Phi\cosh\Psi + \gamma\left(\frac{\Phi^2 - \Psi^2}{2} + \cos\Phi\cosh\Psi - 1\right) \tag{4.74}$$

$$B = \Psi - \cos\Phi\sinh\Psi + \gamma(\Phi\Psi - \sin\Phi\sinh\Psi) \tag{4.75}$$

$$C = \Phi\tan^2\theta + \sin\Phi\cosh\Psi + \gamma\left(\frac{\Phi^2 - \Psi^2}{2}\tan^2\theta - \cos\Phi\cosh\Psi + 1\right) \tag{4.76}$$

$$D = \cos\Phi\sinh\Psi - \Psi\tan^2\theta + \gamma(\Phi\Psi\tan^2\theta + \sin\Phi\sinh\Psi) \tag{4.77}$$

The point $\Phi = 0$ and $\Psi = 0$ is the singular point for the right side of Eq. (4.71). It is a branch point for the function $\Theta = \Theta(W)$ that maps the (X, Y) plane into the (Φ, Ψ) plane. Therefore, for the integration of Eq. (4.71), calculating the zero equipotential must not start from zero but from some adjacent point identified by the trend in the integral curves and by conformal mapping in the vicinity of the origin. Disregarding terms of higher orders than the first, we obtain from Eqs. (4.71)–(4.77) the equation

$$\frac{d\Psi}{d\Phi} = \frac{\Phi}{\Psi} \tag{4.78}$$

according to which the zero is a saddle point. The integral curves passing through it are inclined to the Φ-axis at angles $\pm 45^\circ$ (Fig. 4.8). Let us expand the right-hand sides of Eqs. (3.144) and (3.145) into series in Φ up to second-order terms inclusively. We then find from Eq. (4.66) the conformal mapping in the vicinity of the origin of coordinates:

$$W = X + iY = j\frac{\Theta^2}{2\cos^2\theta} \tag{4.79}$$

Near the coordinates' origin, the electrodes lie in the plane $Y = 0$: the rear one at $X < 0$ and the front electrode at $X > 0$ (respectively, lines OA and OB in Fig. 4.8). In Fig. 4.9*a*, the trajectory calculated for $\gamma = 0.5$ and $\theta = 15^\circ$ is shown as a result of numerical integration of Eq. (4.71) and the transfer from Φ,Ψ to X,Y with Eqs. (4.67) and (4.68).

According to Fig. 4.9*a*, the anode electrode found by the synthesis method described intercepts the electron beam. We meet a problem that is similar to the anode-hole problem of Pierce planar guns. But this task is more difficult because it is necessary to conserve the complicated oscillating trajectories. As a guide, it is useful to apply the method of a correction of the anode shape based on the requirement for minimal perturbations of the electrostatic field at the cathode and its vicinity. The solution is reduced to an electrostatic problem for a system without the beam with subsequent testing of the trajectories by a numerical simulation. This approach is close to the superposition method developed by Kirstein et al. (1967). Note the possibility that considerable variation of the anode shape without substantial perturbation of the cathode field can be considered as a consequence of the Cauchy problem setting: Strong perturbations of the potential distribution in a remote space produce small variations in the boundary (cathode) conditions. To illustrate, the corrected configuration of the electrodes is depicted in Fig. 4.9*b*, showing small perturbations of the trajectories and beam current.

The special case $\theta = 0$ requires a separate consideration (Appendix 2). In this case the electron trajectories according to Eqs. (3.144)–(3.146) are disposed in the plane perpendicular to the magnetic field. The corresponding scheme is used in electron guns of crossed-field traveling-wave tubes and backward-wave oscillators (Chapter 9).

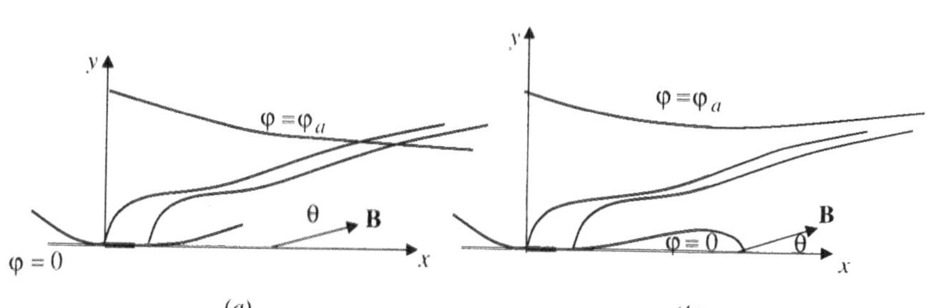

FIGURE 4.9 (*a*) Equipotentials and trajectories in an electron beam with $\gamma = 0.5$, $\theta = 15^\circ$; (*b*) electrode configuration and trajectories after correction.

4.6 SYNTHESIS OF AXIALLY SYMMETRIC ELECTRODE SYSTEMS (Harker, 1960, 1963)

Axially symmetric electron beam devices have obvious advantages: There are no edge effect problems, and technologically, these systems are significantly simpler. A general method for the synthesis of axially symmetric systems was developed by D. Radley (Radley, 1958). His method can be used in cases where Laplace's equation into outer domains of the beams is separable. However, realization of the Radley method encounters difficulties that are connected to the need to operate with analytical representations of the internal problem solutions. Harker's method is free of these limitations.

4.6.1 Statement of the Problem. Harker's Method. Conformal Transformation of a Boundary Trajectory. Field Equations in a Transformed Plane

Consider an infinite laminar electron beam with trajectories that are identical in the azimuthal direction. As in the two-dimensional case, the trajectories can be spatial (three-dimensional) curves (i.e., they could not be arranged in the meridional planes). But the laminarity and the axial symmetry of the flow allow us to consider the latter as a system of cylindrical corrugated surfaces, $r_i = r_i(z)$. Each surface is formed by rotating one of the trajectories around the z-axis. In the meridional (r,z) plane these surfaces are depicted as trajectories of the two-dimensional laminar beam. Let us choose two trajectories as boundaries of the beam (Fig. 4.10). Owing to the axial symmetry, the Cauchy data can be given on the curves in the meridional (r,z) plane.

Assume that the equation of the upper boundary in parametric form is

$$r = R(t), \qquad z = Z(t) \tag{4.80}$$

The functions $R(t)$ and $Z(t)$ are known from a solution of the internal problem. These functions are supposedly analytical, although knowledge of their analytical expressions is not obligatory. Harker's method provides the possibility of an operation with numerical values of these functions. That also concerns the Cauchy data

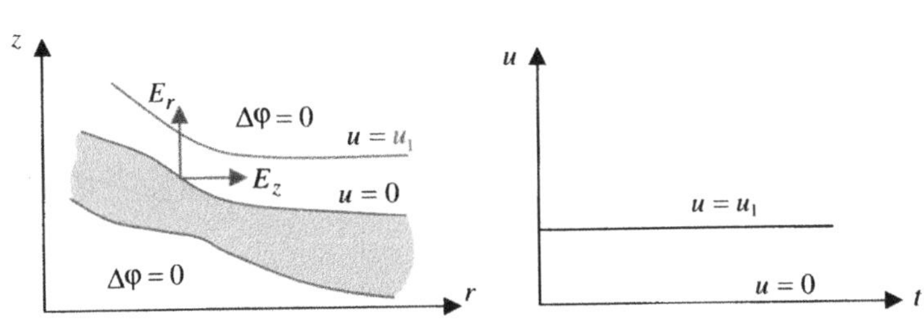

FIGURE 4.10 Conformal transformation of an external domain of an axially symmetric beam.

on the beam boundary [Eq. (4.80)]:

$$\varphi = \varphi(t) \tag{4.81}$$

$$E_r(t) = -\frac{\partial\varphi(t)}{\partial r}, \qquad E_z(t) = -\frac{\partial\varphi(t)}{\partial z} \tag{4.82}$$

The next step is analogous to the theory of plane-symmetric beams. Let us continue Eqs. (4.82) analytically into the upper half-plane of the (t,u) plane and consider the conformal transformation of the (z,r) plane to the (t,u) plane (Fig. 4.10):

$$z + ir = Z(t + iu) + iR(t + iu) \tag{4.83}$$

It is seen that the boundary trajectory in the (z,r) plane is transformed on the line $u = 0$ of the plane (t,u).

At this point, the similarity between the axially symmetric and plane symmetric cases ends because the Laplace equation in the cylindrical coordinate system,

$$\frac{\partial^2\varphi}{\partial r^2} + \frac{1}{r}\frac{\partial\varphi}{\partial r} + \frac{\partial^2\varphi}{\partial z^2} = 0 \tag{4.84}$$

is not invariant under the conformal transformation. To obtain a field equation in the (t,u) plane we will use Maxwell's equations

$$\operatorname{div}\mathbf{E} = 0, \qquad \operatorname{curl}_\theta \mathbf{E} = 0 \tag{4.85}$$

and the Cauchy–Riemann equations for a conformal transformation [Eq. (4.83)]:

$$\frac{\partial z}{\partial t} = \frac{\partial r}{\partial u}, \qquad \frac{\partial r}{\partial t} = -\frac{\partial z}{\partial u} \tag{4.86}$$

Let us write Eq. (4.85) in coordinate form:

$$\frac{\partial E_r}{\partial r} + \frac{E_r}{r} + \frac{\partial E_z}{\partial z} = 0 \tag{4.87}$$

$$\frac{\partial E_r}{\partial z} - \frac{\partial E_z}{\partial r} = 0 \tag{4.88}$$

Derivatives of E_z and E_r as functions of t and u, taking Eqs. (4.86)–(4.88) into account, are

$$\frac{\partial E_z}{\partial u} = \frac{\partial E_z}{\partial r}\frac{\partial r}{\partial u} + \frac{\partial E_z}{\partial z}\frac{\partial z}{\partial u} = \frac{\partial E_z}{\partial r}\frac{\partial z}{\partial t} - \frac{\partial E_z}{\partial z}\frac{\partial r}{\partial t} \tag{4.89}$$

$$\frac{\partial E_r}{\partial t} = \frac{\partial E_r}{\partial r}\frac{\partial r}{\partial t} + \frac{\partial E_r}{\partial z}\frac{\partial z}{\partial t} = -\frac{\partial E_z}{\partial z}\frac{\partial r}{\partial t} - \frac{E_r}{r}\frac{\partial r}{\partial t} + \frac{\partial E_z}{\partial r}\frac{\partial z}{\partial t} \tag{4.90}$$

A combination of these equations gives

$$\frac{\partial E_z}{\partial u} = \frac{\partial E_r}{\partial t} + \frac{E_r}{r}\frac{\partial r}{\partial t} \tag{4.91}$$

This is the first equation for the fields in the (t,u) plane. The second equation,

$$\frac{\partial E_r}{\partial u} = -\frac{\partial E_z}{\partial t} - \frac{E_r}{r}\frac{\partial z}{\partial t} \tag{4.92}$$

is obtained if we express the derivatives $\partial E_z/\partial t$ and $\partial E_r/\partial u$ analogously to Eqs. (4.89) and (4.90). Equations (4.91) and (4.92) together with the Cauchy–Riemann equations [Eqs. (4.86)] form a complete system of equations for the functions $r(t,u)$, $z(t,u)$, $E_z(t,u)$, and $E_r(t,u)$. These equations must be solved with the boundary conditions

$$(r)_{u=0} = R(t), \qquad (z)_{u=0} = Z(t), \qquad (E_z)_{u=0} = E_{z0}(t),$$
$$(E_r)_{u=0} = E_{r0}(t) \tag{4.93}$$

which are known from the solution of the internal problem. This system is simpler than the original one because the boundary conditions are now given on the straight line $u = 0$. Unfortunately, it can be shown that this system belongs to the elliptic type also. For example, Eqs (4.86) follows the elliptic Laplace equation,

$$\frac{\partial^2 z}{\partial t^2} + \frac{\partial^2 z}{\partial u^2} = 0 \tag{4.94}$$

and for that, the Cauchy problem is not correct. Therefore, at the stage of the theory given, it is very difficult to construct a stable numerical realization of the problem.

4.6.2 Transformation of Beam Equations to a Hyperbolic Type. Numerical Realization of Harker's Method

The method pertaining to the transformation of elliptic equations to the hyperbolic type was first worked out by Garabedian and Lieberstein (1958) and applied to electron optics by K. Harker. Let us introduce an additional axis s perpendicular to the (t,u) plane; then a three-dimensional (t,u,s) space is formed (Fig. 4.11). Further, let us continue the functions $E_z(t,u)$, $E_r(t,u)$, $r(t,u)$ and $z(t,u)$ analytically along the variable t to an upper half-plane (t,s) by replacing $t \rightarrow t + is$. The quantities are comprised of the functions of three arguments: $E_z(t,u,s)$, $E_r(t,u,s)$ and so on. Now, if we fix the argument $t = t_1$, the variables will be the functions of arguments (u,s) in the plane $t = t_1$ (Fig. 4.11).

Which will be the field equation in this plane? We must take into account that the differentiation of functions with respect to the complex quantity $t \rightarrow t + is$ for the

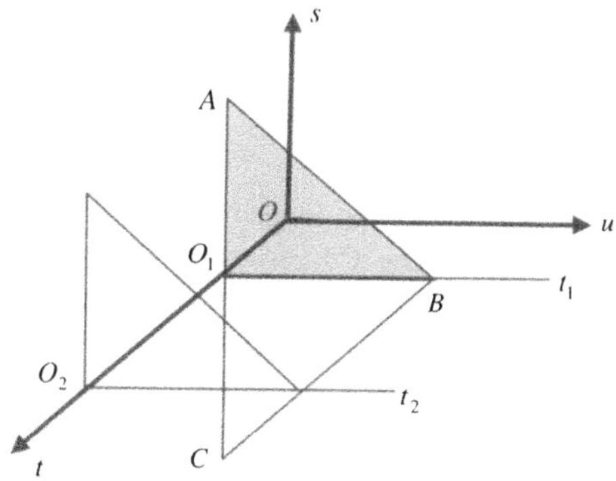

FIGURE 4.11 Complex space (t,u,s) for transformation of beam equations to hyperbolic equations.

constant $t = t_1$ can be written as

$$\frac{\partial}{\partial(t+is)_{t=t_1}} = \frac{\partial}{\partial is} = -i\frac{\partial}{\partial s} \tag{4.95}$$

Then Eqs. (4.86), (4.90), and (4.91) take the form

$$\frac{\partial E_z}{\partial u} = -i\left(\frac{\partial E_r}{\partial s} + \frac{E_r}{r}\frac{\partial r}{\partial s}\right) \tag{4.96}$$

$$\frac{\partial E_r}{\partial u} = i\left(\frac{\partial E_z}{\partial s} + \frac{E_r}{r}\frac{\partial z}{\partial s}\right) \tag{4.97}$$

$$\frac{\partial r}{\partial u} = -i\frac{\partial z}{\partial s}, \qquad \frac{\partial z}{\partial u} = i\frac{\partial r}{\partial s} \tag{4.98}$$

Let us add to these equations the equation for the potential in the (u,s) plane:

$$\frac{\partial \varphi}{\partial u} = \frac{\partial \varphi}{\partial r}\frac{\partial r}{\partial u} + \frac{\partial \varphi}{\partial z}\frac{\partial z}{\partial u}$$

Taking Eqs. (4.98) into account, this gives a differential equation for the potential in the (s,u) plane:

$$\frac{\partial \varphi}{\partial u} = iE_r\frac{\partial z}{\partial s} - iE_z\frac{\partial r}{\partial s} \tag{4.99}$$

A remarkable distinction of these equations from the original equations (4.91), (4.92), and (4.86) is that they are of hyperbolic type (Garabedian and Lieberstein, 1958; Harker, 1960). For example, Eqs. (4.98) give direct hyperbolic (wave) equations in any (u,s) plane:

$$\frac{\partial^2 r}{\partial u^2} - \frac{\partial^2 r}{\partial s^2} = 0, \qquad \frac{\partial^2 z}{\partial u^2} - \frac{\partial^2 z}{\partial s^2} = 0 \tag{4.100}$$

The Cauchy problem of integrating the hyperbolic system [Eqs. (4.96)–(4.99)] in the (u,s) plane, is properly set. The Cauchy data must be given on the line AC (Fig. 4.11). They can be obtained as the analytic continuation of the Cauchy data on the boundary $u = 0$ of the real (t,u) plane.

According to the theory of partial differential equations (see, e.g., Garabedian, 1998), the solution is defined inside the characteristic domain. Here, the latter is the triangle ABC bounded by the characteristics $s = \pm u + \text{const}$. The numerical integration of the system in the (u,s) plane can easily be executed by a finite-difference method (Harker, 1960; Soluyanova et al., 1986) when the solution of the internal problem can be described by a system of ordinary differential equations.

Here the next Harker step is important: to continue analytically not the solution of the internal problem but the ordinary differential equations of the original flow. In this case, the internal and external problems should be solved simultaneously. The numerical procedure of the synthesis consists of the following steps (Fig. 4.11):

1. Integrating the self-consistent ordinary differential equations of the flow containing boundary conditions that correspond to chosen boundary trajectories: first along the real axis t from $t = 0$ *to* $t = t_1$ and then along the imaginary axis s (changing $t \to is$) until point A. We obtain the complex Cauchy data on the interval O_1A. The Cauchy data on the interval O_1C are the complex conjugation of data on the interval O_1C.

2. Integrating the hyperbolic equations (4.96)–(4.99) in the characteristic triangle ABC, with the Cauchy data obtained on the first step of the interval AC. The values of all functions desired are real on the line O_1B that belongs to the (t,u) plane.

3. Translating the triangle ABC to other positions along the t-axis (e.g., to $t = t_2$) and repeating steps 1 and 2 allows coverage of the (t,u) plane by a net of real values of the potential $\varphi(t,u)$ and the functions $z(t,u)$ and $r(t,u)$, which give a mapping of the conformal transformation (4.83).

4. Connecting the points with the given potential yields equipotentials. Using the functions $z(t,u)$ and $r(t,u)$ determines desired electrode shapes in the (z,r) plane.

This process can be automatized as a computer code (see, e.g., Manuilov and Tsimring, 1978; Soluyanova et al., 1986). For its realization as applied to any axially symmetric beam, it is sufficient to give the right sides of the flow ordinary differential equations and boundary conditions. It is supposed that the flow equations can be reduced to system ordinary differential equations of first order with analytical right sides. It is also possible to use a method of the Riemann function (Garabedian, 1998; Harker, 1963; Syrovoy, 1994a), although that is more laborious because it involves transformation and a tabulation of hypergeometric functions and elliptic integrals in the complex domain. However, if the flow equations are not reduced to a system of ordinary differential equations, use of Riemann's method can be inevitable, for example, in the theory of three-dimensional electron beam formation (Syrovoy, 1994a). Note that analytic continuation of the Cauchy data can be complicated and is determined essentially by the specific character of the problem. This method can easily be transferred to two-dimensional systems with translation symmetry if $|r - \bar{r}|/\bar{r} \ll 1$($\bar{r}$ is the average radius of the trajectories in an axially symmetric system).

4.7 ELECTRON GUNS WITH COMPRESSED BEAMS. MAGNETRON INJECTION GUN

4.7.1 Electron Guns with Wedge-Shaped and Conic Beams

Electron flows in the cylindrical and spherical diodes can be used. In Fig. 4.12 (x,z) is the symmetry plane for the wedge-shaped gun. For the conic gun, z is the axis of the axial symmetry. Following the Pierce method, let us cut a sector from the cylindrical or conic flows involving an angle greater than θ_0 (Fig. 4.12). We come to Laplace's equation in the section of the beam that has been eliminated, with Cauchy data on the beam boundary: the potential distribution [Eq. (4.30) or (4.31)] and the normal derivative of the potential

$$\frac{\partial\varphi(r,\theta)}{\partial\theta} = 0 \tag{4.101}$$

That is the condition that of the trajectories will obtain a rectilinear shape.

A comprehensive theory of these guns has been developed by Radley (1958, 1963). The Cauchy problem was reduced to a pair of integral equations that were solved analytically. The general pattern of the equipotentials is similar for both guns and is shown schematically in Fig. 4.12. The Radley method is applicable in principle not only to planar and axially symmetric systems but also to any flows with the boundary surfaces in that Laplace's equation is separable.

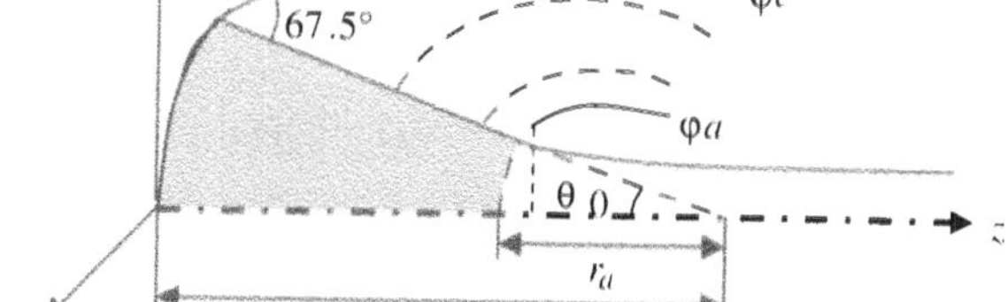

FIGURE 4.12 Electron gun with wedge-shaped and conic beams.

The beam currents in both guns for a given anode potential according to Eqs. (4.18) and (4.25) and Fig. 4.12 are

$$I = \begin{cases} \dfrac{8}{9}\varepsilon_0\sqrt{2\eta}\dfrac{\theta_0 L}{r_a}\dfrac{\phi_a^{3/2}}{\beta^2(r_a/r_c)} \\ \quad = 4.668\times 10^{-6}\dfrac{\theta_0 L}{r_a}\dfrac{\phi_a^{3/2}}{\beta^2(r_a/r_c)} \quad \text{A} \quad \text{(wedge-shaped beam)} & (4.102) \\ \dfrac{16\pi}{9}\varepsilon_0\sqrt{2\eta}\sin^2\dfrac{\theta_0}{2}\dfrac{\varphi_a^{3/2}}{\alpha^2(r_a/r_c)} \\ \quad = 29.33\times 10^{-6}\sin^2\dfrac{\theta_0}{2}\dfrac{\varphi_a^{3/2}}{\alpha^2(r_a/r_c)} \quad \text{A} \quad \text{(conic beam)} & (4.103) \end{cases}$$

According to Eq. (4.102), the current in the wedge-shaped gun for a given ratio r_a/r_c increases inversely proportional to the anode (cathode) radius. The current in the conic gun depends only on the ratio r_a/r_c.

Let us concentrate on the problem of an anode aperture. As we have seen, the anode hole in a planar Pierce gun leads to beam divergence. This effect impedes use of this gun as an electrostatic device (without an additional strong focusing magnetic field) for the formation of dense, lengthy electron beams. In compression guns, the initial convergence of beams allows beams to become parallel or even convergent after the anode (see, e.g., Fig. 4.12).

An analysis of the anode-hole action can be executed considering the anode aperture as a divergent lens. So for the axially symmetric case (Section 2.6.5) it can be shown according to Eq. (2.82) that in the nonrelativistic approximation the angle γ of the beam after the lens is a function of the ratio r_c/r_a:

$$\tan\gamma = \left[1 - \frac{1}{4(r_c/r_a - 1)}\right]\tan\theta_0 \qquad (4.104)$$

As we see, a parallel beam after aperture ($\gamma = 0$) is achieved for $r_c/r_a = 1.25$. It can be seen that for the wedge-shaped gun, the corresponding ratio is $r_c/r_a = 1.5$. Note that similar considerations give rough guide results. Real designs are usually based on available computer codes of the trajectory analysis.

The use of compressed guns allows a significant increase in beam currents for limited cathode emission density. By neglecting variation in a cross-section beam beyond the anode space, a compression coefficient

$$K_c = \begin{cases} \dfrac{r_c}{r_a} & \text{(wedge-shaped gun)} \\ \left(\dfrac{r_c}{r_a}\right)^2 & \text{(spherical gun)} \end{cases}$$

The disadvantages of guns with an anode aperture are irregularities of the emission current distribution and reduction in the cathode current.

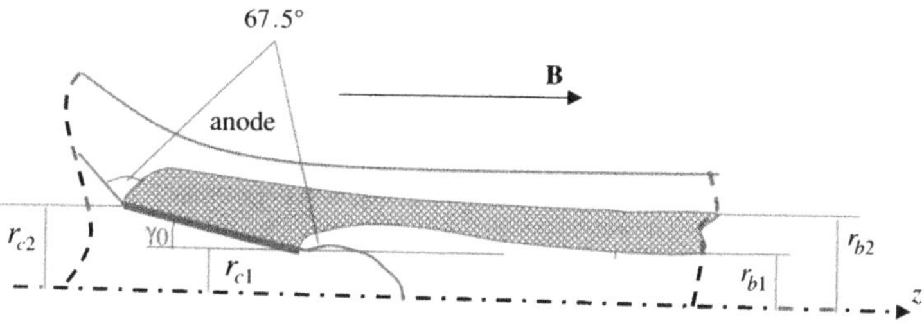

FIGURE 4.13 Magnetron injection gun.

4.7.2 Magnetron Injection Electron Guns (Kino–Taylor Guns)

Magnetron injection electron guns (Kino and Taylor, 1962; Midford and Kino, 1962) use laminar flow in the planar magnetron with an inclined magnetic field or Dryden flow (Sections 3.1 and 3.5). After cutting off the upper and adjoining axis section of the electron flow, we obtain a hollow beam. In Fig. 4.13 a diagram of an axially symmetric magnetron injection gun (MIG) situated on the base of Dryden flow is shown. The Harker method (Harker, 1960, 1963), which gives a solution of the Cauchy problem for similar electrode systems, was considered in Section 4.6. Assuming that the magnetic field is uniform, let us estimate the dimensions of the hollow beam and the compression coefficient. We use the Busch theorem [Section 1.7, Eqs. (1.126) and (1.127)]:

$$\dot{\theta}_b = \frac{\eta B}{2}\left[1 - \left(\frac{r_c}{r_b}\right)^2\right] \tag{4.105}$$

where r_c and r_b are corresponding radiuses of the trajectory on the cathode and after the gun (Fig. 4.13); $\dot{\theta}_b$ is the angular velocity of the beam on the radius $r = r_b$. The beam is supposedly laminar; therefore, the beam envelope coincides with the trajectories of the flow.

Assume rectilinear motion of the electrons at the internal boundary of the beam appearing after the gun. According to Eq. (4.105), the necessary condition of nonrotational motion ($\dot{\theta}_{b1} = 0$) is $r_{b1} = r_{c1}$. This condition is sufficient for equilibrium of the internal boundary because the space-charge field for $r \leq r_{b1}$ is zero. Because the space-charge field is not zero at $r > r_{b1}$, equilibrium is possible if the outer layers of the beam rotate counterclockwise ($\dot{\theta}_b > 0$). So for the chosen direction of the magnetic field in Fig. 4.13, the Lorentz force is directed to the axis. According to Eq. (4.105), the equilibrium radius $r_{b2} > r_{c2}$. But for the small space-charge field, we can take $r_{b2} \approx r_{c2}$. The thickness of the beam in this case is $r_{b2} - r_{b1} \approx r_{c2} - r_{c1}$. Then the coefficient of compression is

$$K = \frac{S_c}{S_b} \approx \frac{1}{\sin \gamma_0} \tag{4.106}$$

Note two important advantages of the MIG:

1. Formation of a hollow beam improves the performance of some microwave devices with distributed interaction of the electron beam and an electromagnetic field (i.e., traveling-wave tubes, backward-wave tubes, etc.) (see Chapter 8).
2. No problem connected with the anode hole.

A disadvantage of use of the MIG is the comparatively high noise level. Among different mechanisms of hollow electron beams instability in crossed magnetic and space-charge electric fields is the very likelihood of diochotron instability. According to Davidson (1990), oscillations on a *diochotron frequency* $\omega_p^2/2\omega_g$ and its harmonics (ω_p and ω_g are the plasma and gyrofrequencies, respectively) could be in the unstable regime. However, one should take into account that diochotron instability belongs to a convective type, and therefore in a near-cathode domain must be instability of another origin. Experiments with MIGs (Midford and Kino, 1962) registered considerable currents on sole electrodes that were negatively biased with respect to the cathode—evidence of the presence of alternating electromagnetic fields.

4.7.3 High-Convergence Electron Gun with a Magnetic Accompaniment (Baryshev et al., 1994; Yakovlev and Nezhevenko, 1999)

Relativistic electron guns have been applied in power magnicon microwave amplifiers (Nezhevenko et al., 2002; Yakovlev and Nezhevenko, 1999) (see also Section 10.4.5) as a prototype of power pulse sources for future electron–positron colliders. The gun is depicted schematically in Fig. 4.14. The gun's base is the conic beam described in Section 4.5.1. But this gun had important peculiarities that together with comprehensive computer-aided design, and taking a self magnetic field into account, have made it possible to obtain a beam with very high compression and low emittance. In particular, magnetic accompaniment of the beam was used. Following the anode hole, the electrons enter into a nonuniform magnetic field with magnetic force lines parallel to the electron trajectories. As a result, additional magnetic compression of the beam (i.e., ~2) has been obtained. Magnetic screening

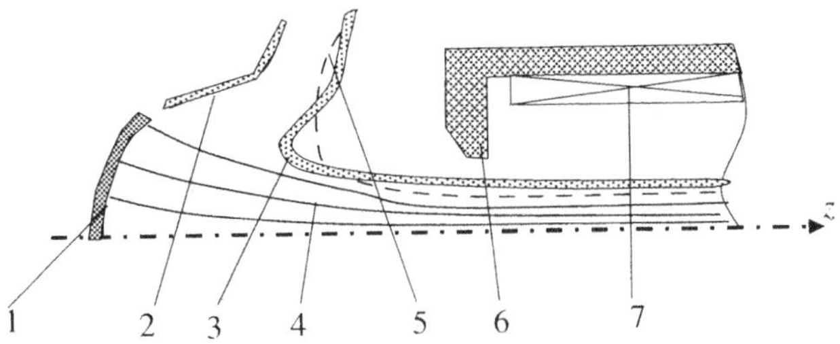

FIGURE 4.14 High-convergence magnicon electron gun. 1, Cathode; 2, focusing electrode; 3, anode; 4, electron trajectory; 5, magnetic force line; 6, ferromagnetic yoke; 7, solenoid.

TABLE 4.1 Design and Experimental Parameters of the High-Convergence Gun

Parameters	Design	Experiment
Current (A)	210	225
Voltage (kV)	500	480
Cathode radius (mm)	37	37
Beam radius (mm)	0.75	1.0
Beam compression	2500	1400
Beam transverse emittance (mrad·cm)	1.6π	

close to a Brillouin structure ensured electron beam transport (see Section 5.2.3). Finally, use of an electrically isolated focusing electrode biased negative with respect to the cathode has allowed a decrease in the beam emittance and reduction or elimination of the beam halo.

In Table 4.1, some design and experimental parameters of this gun are presented (Nezhevenko et al., 2002). The beam radius was close to a Brillouin radius. Note that in an earlier version of a 7-GHz magnicon, a beam compression of 2300 was measured for a design compression of 2400.

4.7.4 Centrifugal Electrostatic Guns

The central idea of a centrifugal electrostatic electron beam formation with high compression is based on the space separating the region where current is drawn from the cathode and the region of beam acceleration and compression. The guns were proposed and studied by Chernov et al. (1978).

In Fig. 4.15, a centrifugal electrostatic gun (CEG) is represented. An annular-shaped Pierce gun with a cathode (1) and first anode (2) forms a circular disk–shaped electron beam with trajectories in meridional planes. Further, the beam enters into a deflection system consisting of two electrodes: an anode (3) and a

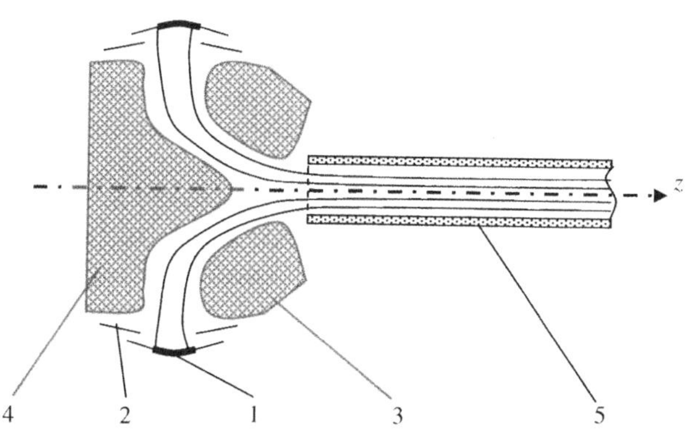

FIGURE 4.15 Centrifugal-electrostatic gun. 1, Cathode; 2, anode 1; 3, anode 2; 4, mirrors; 5, drift tube.

conic mirror (4). A potential of the mirror is higher than a potential of an anode (2). A corresponding electric field balances the centrifugal force that arises when the deflected electrons gain curvilinear motion. After leaving the deflecting system and entering a drift tube, the disk-shaped electron beam is transformed into an axially symmetric hollow electron beam. In principle, the beam can be a solid cylindrical beam. If the magnetic field does not penetrate the gun, the structure of this beam will be close to that of the Brillouin structure. In this case, the maximum perveance of the beam is $\sim 25 \times 10^{-6} \mathrm{A/V^{3/2}}$ (see Section 5.2.3) and does not depend on the drift tube radius. In principle the beam compression in this gun is not limited. Therefore, these guns can be used to shape thin electron beams with very high current density. The perveance can be significantly greater for thin hollow electron beams (see Section 5.2.4).

Note that a compression in the Pierce gun is based on the use of initial wedge-shaped or conic convergent beams (Section 4.5.1). Further transformation of parallel beams is a result of the repulsive action of the space charge and lens fields. The action of the space-charge field can be altered by the buildup of positive ions. So in many cases, beam transformation can be unstable. In a CEG, the action of the space-charge field in the deflected region is accomplished on the basis of two strong opposed stable centrifugal and electrostatic interelectrode forces. Furthermore, the cathode in a CEG is protected from bombardment by positive ions that arise in ionization processes in residual gases. The positive ions from the drift tube are deflected in the vicinity of the opening in electrodes 4 and 3, strike the low potential anode (3), and cannot reach the thermoionic cathode (1). The latter can be subjected to the positive ions formed only in a Pierce gun. The number of these particles is small.

According to Chernov et al. (1978, 1983), additional optimization of electrode geometry is necessary to obtain a laminar beam in the drift tube. That is attained by a numerical simulation method. A relativistic version of a CEG is considered in Chernov et al. (1983).

4.8 EXPLOSIVE EMISSION GUNS

4.8.1 Introduction

The discovery of explosive electron emission (EEE) and related effects (Dyke and Trolan, 1953; Fursei and Vorontsov-Vel'yaminov, 1968; Mesyats, 1995; Mesyats and Proskurovskii, 1968) marked a new age in high-current and high-power electronics. EEE and progress in powerful pulse equipment allow us to reach pulse beam currents and the electron energy up to 10^6 A and 10^7 eV for a pulse duration of 10^{-10} to 10^{-5} s. A beam current in EEE diodes is emitted from the cathode plasma that is generated by explosive processes on the cathode surface. A brief qualitative scheme of EEE evolution follows.

According to Fowler–Nordheim field emission theory (FNT), the values of an electric field supporting the emission developed from a metal with a smooth surface are on the order of 10^7 V/cm. However, appreciable currents appear for essentially smaller electric fields (about two orders of magnitude). The explanation of this effect is reduced by taking into account the many micropoints on the real metal surface. The electric field near these micropoints increases the magnitude by up to two to three orders with a comparatively smooth surface. This effect develops irreversible features with further increases in the electric field. When the emission current density reaches $\sim 10^7$ A/cm^2, the Joule heat of the points leads to increased thermionc emission. A corresponding increase in the temperature and current density (up to 10^9 A/cm^2) is terminated by destruction of emission centers. The transient time of this process is about 10^{-9} s, and it has the characteristics of an explosion. Dense cathode flare plasma—a composition of the ions, neutral atoms, and free electrons—is formed close to the metal surface. This plasma determines further processing in two ways (Mesyats, 1991). On the one hand, a high temperature in the cathode area and a potential difference between the metal surface and the cathode flare plasma support intense electron emission from the metal to the plasma. On the other hand, the potential difference between the plasma and the anode gives a current of free electrons from the plasma to the anode. So the plasma layer plays the role of an electron cathode. This cathode has a small effective work function, and the electron current in the plasma cathode–anode gap is limited by the space charge of the electrons in the ρ-mode and determined by the Child–Langmuir law (Sections 3.4.4 and 4.3.1).

The cathode plasma expands during explosive emission with an average velocity of about 10^6 to 10^7 cm/s. The velocity is essentially dependent on a value and a distribution of the magnetic field in the gap. It determines the rate at which the accelerating gap is filled and closed by the plasma.

Below, two basic types of explosive emission diodes as parts of high-current accelerators are considered: planar and magnetically insulated diodes.

4.8.2 Planar Explosive Emission Diodes

Foiled Anode Diodes A diagram of a diode is shown in Fig. 4.16*a*. The accelerating gap is formed by a planar plasma cathode and a planar foil anode. The electrons emitting from the cathode pierce the thin anode foil ($\sim$0.01 cm) with small energy losses because the interaction of high-energy electrons with atoms of the foil is practically elastic. After escape from the foil, the electrons move in a drift

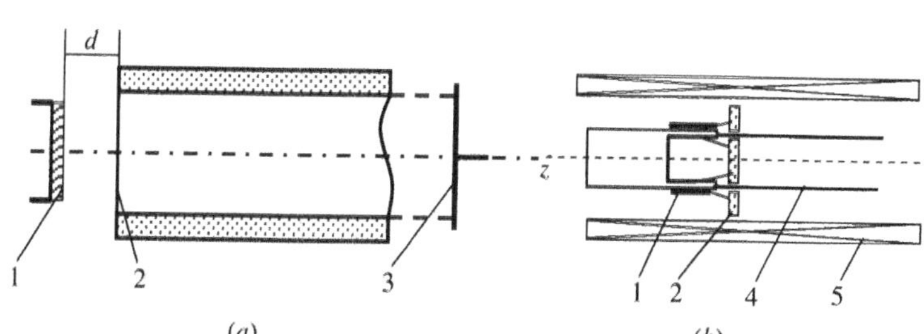

FIGURE 4.16 Planar explosive emission diodes: (*a*) foiled anode diode; (*b*) foil-free annular diode. 1, Plasma cathode; 2, anode; 3, collector; 4, electron beam; 5, solenoid.

space. In general, the electron beam is nonuniform and nonlaminar; as a result, there is strong inhomogeneity of the cathode plasma due to the chaotic distribution of emission centers and the random nature of the explosions. The electron beam is also scattered in the foil. Nevertheless, the motion of electrons in a drift tube is approximately rectilinear because of the focusing action of the self magnetic field in the relativistic beams (see Graybill and Nable, 1966; see also Section 3.2.1).

Two types of such planar diodes are used. The first type is used in the production of dense electron beams with a current density of 10^4 to 10^7 A/sm^2 and a pulse duration of 10 to 100 ns. They are used in nuclear fusion investigations, microwave electronics, X-ray generation, and so on. The second type of diode produces wide beams with a comparatively low current density ($j \sim 1$ to $10\,\mathrm{A/cm^2}$), a large area ($S \sim 10^2$ to 10^4 cm^2), and a pulse duration up to 1000 ns. These diodes are used for pumping lasers and for technological purposes. The maximum pulse duration equals approximately

$$\tau_{\max} = \frac{d}{v_p} \tag{4.107}$$

where v_p is the velocity of expansion of the cathode plasma. For $v_p = 10^7$ cm/s, $d = 1$ cm and $\tau_{\max} = 100$ ns. Real $\tau < \tau_{\max}$.

Among the various causes for shortening of the real pulse duration, one essential cause arises from the anode flare plasma, its expansion to the cathode, and closure of the cathode–anode gap, an effect called *gap closure*. The current pulse duration can be increased by extension of the acceleration gap, slowing down of the cathode plasma (e.g., by using a transverse magnetic field), and elimination of the anode flare (Bugaev et al., 1975).

Foil-Free Annular Diode A diode is depicted in Fig. 4.16*b*. A hollow ringlike cathode is placed close to an anode plate with annular cut. The annulus is interrupted by radial struts to support the central region of the anode. Emission takes place from the edge of a hollow cathode. The gun is inserted in a strong axial magnetic field, and the electrons in the annular electron beam outside the anode plate have low transverse momentum. Only a small fraction of the total current is passed through the annular slot. However, the diode can be considered as an effective low-emittance filter. This diode has been used as a part of an electron-optical system, forming a helical electron beam in relativistic gyrotrons (Black et al., 1990; see also Section 10.7.1).

4.8.3 Magnetically Insulated Diodes

The idea of increasing pulse duration by using the transverse to an acceleration gap magnetic field, together with the production of annual high current electron beams, has been realized in magnetically insulated diodes. The first experimental results were obtained by Friedman and Uri (1970).

A diagram of a magnetically insulated diode (MID) is shown in Fig. 4.17. A tubular cathode and an anode form a coaxial diode. The drift tube is a continuation of the anode. In the simplest version of an MID, the magnetic field is assumed to be

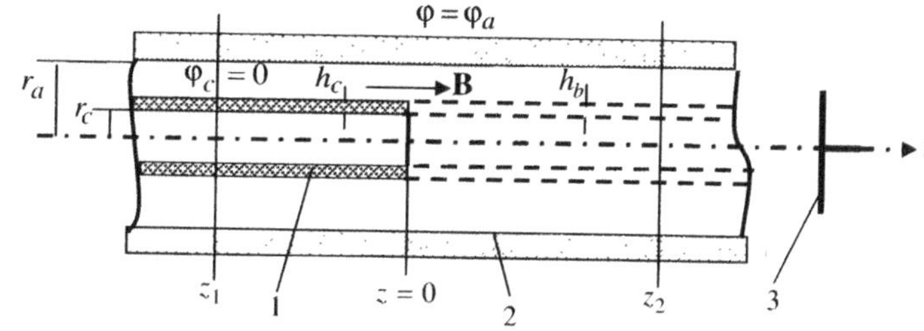

FIGURE 4.17 Magnetically insulated diode. 1, Cathode; 2, anode; 3, collector.

parallel to the axis of axial symmetry and is infinitely large, providing rectilinearity to the electron trajectories and their parallelism to the z-axis. Taking into account the specific character of explosive emission, including nonstationary states even in an approximation of an infinite magnetic field, comprehensive analysis of the MID is complicated and requires computer simulation.

An instructive simplified analytic MID theory developed by Fedosov et al. (1977) is considered below. The theory allows us to obtain MID volt–ampere characteristics without consideration of electron motion in the acceleration region. Let us write the beam equations and boundary conditions:

$$\text{div}\,\mathbf{E} = \frac{\rho}{\varepsilon_0} \tag{4.108}$$

$$\text{curl}_\theta\,\mathbf{E} = \frac{\partial E_z}{\partial r} - \frac{\partial E_r}{\partial z} = 0 \tag{4.109}$$

$$\frac{dp}{dt} = -e_0 E_z \tag{4.110}$$

$$j = -\rho v \tag{4.111}$$

$$(E_z)_{r=r_c} = 0 \tag{4.112}$$

$$(p)_{r=r_c,\,z=0} = 0 \tag{4.113}$$

$$(\varphi)_{r=r_c,\,z<0} = 0, \qquad (\varphi)_{r=r_a} = \varphi_a \tag{4.114}$$

Here p, v, and j are nonzero z-components of a momentum, a velocity, and a current density, respectively. Equation (4.112) expresses the assumed regime of the current limitation by the space charge.

Multiply Eqs. (4.108) and (4.110). Then using Eq. (4.111), we obtain

$$E_z\,\text{div}\,\mathbf{E} = -\frac{\rho}{e_0\varepsilon_0}\frac{dp}{dt} = \frac{j}{e_0\varepsilon_0}\frac{dp}{dz} \tag{4.115}$$

Let us represent $E_z\,\text{div}\,\mathbf{E}$ as div of a vector $\mathbf{A}$:

$$\begin{aligned} E_z\,\text{div}\,\mathbf{E} &= \text{div}(E_z\mathbf{E}) - \mathbf{E}\,\text{grad}\,E_z \\ &= \frac{1}{r}\frac{\partial}{\partial r}(rE_zE_r) + \frac{\partial}{\partial z}(E_z^2) - E_z\frac{\partial E_z}{\partial z} - E_r\frac{\partial E_z}{\partial r} \end{aligned} \tag{4.116}$$

According to Eq. (4.109), replace $\partial E_z/\partial r$ by $\partial E_r/\partial z$. We obtain

$$E_z \operatorname{div} \mathbf{E} = \frac{1}{r}\frac{\partial}{\partial r}(rE_zE_r) + \frac{\partial}{\partial z}\left(\frac{E_z^2 - E_r^2}{2}\right) = \operatorname{div} \mathbf{A} \tag{4.117}$$

where

$$A_r = E_zE_r, \qquad A_z = \frac{E_z^2 - E_r^2}{2} \tag{4.118}$$

so

$$\operatorname{div} A = \frac{j}{e_0\varepsilon_0}\frac{dp}{dz} \tag{4.119}$$

Let us to integrate Eq. (4.119) over a volume Ω of the diode between radiuses r_c and r_a and the cross sections z_1 and z_2, including the volume of the beam (Fig. 4.17), and use Gauss's theorem:

$$\oint A_n d\sigma = \frac{1}{e_0\varepsilon_0}\int_\Omega j\frac{dp}{dz}d\tau \tag{4.120}$$

The surface integral is the sum of four integrals:

$$\oint A_n\, d\sigma = -\int_{z=z_1} A_z\, d\sigma + \int_{z=z_2} A_z\, d\sigma - \int_{r=r_c} A_r\, d\sigma + \int_{r=r_a} A_r\, d\sigma \tag{4.121}$$

Take into account that $E_z = 0$ for $z = z_1$ and $z = z_2$ if we choose z_1 and z_2 sufficiently far from the plane $z = 0$. In these planes, $A_z = -A_r^2/2$. Also, $E_z = 0$ for $r = r_a$. Therefore, the third integral in Eq. (4.121) is missing. Then Eq. (4.121) becomes

$$\oint A_n\, d\sigma = \int_{z_1} \frac{E_r^2}{2}2\pi r\, dr - \int_{z_2} \frac{E_r^2}{2}2\pi r\, dr - \int_0^{z_2} 2\pi r_cE_zE_r\, dz \tag{4.122}$$

The last integral in Eq. (4.122) must be taken over the inner boundary of the beam. The lower limit in this integral is zero because $(E_z)_{r=r_c, z<0} = 0$. In general, one can see (Fedosov et al., 1977) that this integral is small for narrow beams (see Fig. 4.17):

$$h_c \approx h_b \ll r_c, r_a, r_a - r_c \tag{4.123}$$

The volume integral in Eq. (4.120) in this approximation can be represented as

$$\int_\Omega j\frac{dp}{dz}\, d\tau = \int_0^{z_2}\frac{dp}{dz}\, dz\int_S j\, d\sigma = I(p_2 - p_0) = Ip_b \tag{4.124}$$

where $p_b = p_2$ is the electron momentum in the far region $z_2 \gg r_c,\ r_a$. We assumed the initial momentum $p_0 = (p)_{z=0} = 0$. Neglecting the integral $\int_0^{z_2} 2\pi r_c E_z E_r\, dz$ in Eq. (4.122), we can write Eq. (4.120) as

$$\int_{r_c}^{r_a} E_{r1}^2 r\, dr - \int_{r_c}^{r_a} E_{r2}^2 r\, dr = \frac{Ip_b}{\pi e_0 \varepsilon_0} \tag{4.125}$$

It can be shown that Eq. (4.125) expresses conservation of the z-axis momentum of a system electron beam–electrostatic field (Barker et al., 2005).

The configuration of the system in the cross sections $z = z_1$ and $z = z_2$ is geometrically similar to a uniform coaxial line. Then the electric field in the plane $z = z_1$ is

$$E_{r1} = \frac{\varphi_a}{r \ln (r_a/r_c)} \tag{4.126}$$

The electric field in the plane $z = z_2$ is

$$E_{r2} = \frac{q}{2\pi r \varepsilon_0} = -\frac{I}{2\pi r v_b \varepsilon_0} \tag{4.127}$$

where q is a beam charge on a length unit and v_b is the beam electron velocity. The potential difference across the beam–tube gap is

$$\varphi_a - \varphi_b = -\int_{r_b}^{r_a} E_{r2} dr = \frac{I}{2\pi \varepsilon_0 v_b} \ln \frac{r_a}{r_b} \tag{4.128}$$

This relationship gives the beam current

$$I = \frac{(\varphi_a - \varphi_b) 2\pi \varepsilon_0 v_b}{\ln (r_a/r_b)} = I_0 \frac{(\gamma_a - \gamma_b)\sqrt{\gamma_b^2 - 1}}{2\gamma_b \ln (r_a/r_b)} \tag{4.129}$$

where γ_a and γ_b are relativistic factors, corresponding to the potentials φ_a and φ_b, and

$$I_0 = 4\pi \varepsilon_0 \frac{c^3}{\eta} \approx 17\,\text{kA} \tag{4.130}$$

is a relativistic unity of the current (*relativistic current*).

Substitution of Eqs. (4.126), (4.127), and (4.129) in Eq. (4.125), and assuming that $r_c = r_b$, we obtain, after integrating,

$$(\gamma_a - 1)^2 - (\gamma_a - \gamma_b)^2 = \frac{2(\gamma_a - \gamma_b)(\gamma_b^2 - 1)}{\gamma_b} \tag{4.131}$$

It is readily verified that Eq. (4.131) is reduced to

$$\gamma_b^2 + \gamma_b - 2\gamma_a = 0 \tag{4.132}$$

to give

$$\gamma_b = \left(\tfrac{1}{4} + 2\gamma_a\right)^{\frac{1}{2}} - \tfrac{1}{2} \tag{4.133}$$

Equations (4.129), (4.133) determine the volt–ampere characteristics of the diode. Consider the following examples:

1. $\varphi_a = 1$ MV, $r_a = 1.1r_b$, $\gamma_a \approx 3$, $\gamma_b = \sqrt{\frac{1}{4} + 6} - \frac{1}{2} = 2$, and $I = I_0(\sqrt{3}/4 \ln 1.1) \approx 4.5I_r \approx 77$ kA.
2. $\varphi_a = 10$ MV, $r_a = 1.1r_b$, $\gamma_a \approx 21$, $\gamma_b = 6$, and $I \approx 78I_0 \approx 1.3$ MA.
3. Nonrelativistic approximation ($\gamma - 1 = e_0\varphi/m_0c^2 \ll 1$).

According to Eqs. (4.129) and (4.131), we obtain

$$\varphi_a = \frac{2}{3}\varphi_b, \qquad I = I_0\frac{(\gamma_a - 1)^{3/2}}{3\sqrt{3}\,\ln(r_a/r_b)} \approx \frac{0.5}{\ln(r_a/r_b)} I_0\varphi a^{3/2} \tag{4.134}$$

As we see, the perveance of the MID, $P = [0.5/\ln(r_a/r_b)]I_0$, is enormous with respect to ordinary planar and cylindrical diodes [see, e.g., Eqs. (3.105) and (4.18)]. This effect can be explained in the frame using a diode model (i.e., a narrow cathode, an infinite magnetic field) and anode screening of the beam space-charge field (Fig. 4.18).

According to Fig. 4.18, the space-charge field on the cathode E_{sc} is formed by the small part of the beam length, on the order of the cathode thickness, h_c. The majority of the space-charge lines are completed through the anode. Therefore, the field balance in the ρ-mode $E_0 - E_{sc} = 0$ (E_0 is the cathode field without the beam) is attained with large beam currents.

Some refinement of the theory (Fedosov et al., 1977) was obtained by Belomytsev et al. (1981). In particular, these authors have extended the theory up to a finite magnetic field. In this case, it is necessary to take into account the self magnetic field and related effects: the expansion and radial displacement of the beam, the spread of particle momentum, and so on. Estimation of the external finite magnetic field, providing validity of the approximation $B = \infty$, has been given by Belomytsev et al. (1981).

The principal advantages of magnetically insulated coaxial diodes (MICDs) is the absence of a fast-exploding anode foil and considerable weakening of the gap

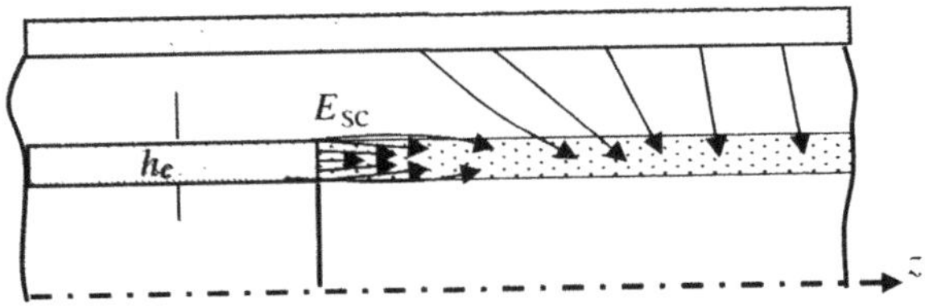

FIGURE 4.18 Pattern of space-charge field lines in an idealized model of an MID.

closure effect, which is determined by a lower rate of filling in the accelerating gap by the cathode plasma. MICDs are widely used to generate high-current beams in pulsed electron accelerators of direct action. These accelerators are applied in thermonuclear fusion experiments for collective ion acceleration, especially in high-power microwave electronics. In this connection, a very important problem is further increase in the pulse duration. According to Nikonov et al. (1983), a definite rise in the pulse duration pulse in an MICD that is determined by the time of filling of the drift space by the cathode plasmas can be achieved by using an increasing magnetic field from the cathode to the drift tube. Data reported by Nikonov et al. (1983) include a magnetic field ratio of $\sim$4, an expanding cathode plasma velocity of $\sim$(1 to 2) $\times 10^6$ cm/s, and a collector plasma of $\sim 7 \times 10^5$ cm/s, whereas the expanding velocity of the cathode plasma in the uniform magnetic field exceeds 10^7 cm/s. The stable pulse duration attained was 10 μs, which is approximately three orders of magnitude above the pulse duration in planar explosive emission diodes.

Increasingly interesting is the use of fiber cathodes, which in comparatively lower electric fields (i.e., $\sim$10 to 30 kV/cm produce current densities of 1 to 10 A/cm^2. Using cesium iodide (CsI)–coated carbon fiber reduces closure velocity up to 0.46×10^6 cm/s (see, e.g., Barker and Schamiloglu, 2001, and Section 9.3.1).

CHAPTER FIVE

Transport of Space-Charge Beams

5.1 INTRODUCTION

Following the scheme shown in Fig. 3.2, we pass to the drift space, where the electron beam formed in a gun should be transferred with minimal distortions in a work space or interact directly after the gun with a high-frequency electromagnetic field or substance. It is clear that the transport of electron beams is one of the most important problems in determining the effective operation of microwave devices and electron accelerators. In this chapter transport problems of smooth (unrippled) or small rippled beams are considered. These problems are essential for klystrons, especially for multicavity and multibeam klystron amplifiers (Chapter 7) and for beam-type devices: TWTOs, MTWTs, BWOs, and MBWOs (Chapters 8 and 9). The transport of electron beams with oscillating electron trajectories is a specific problem. One is considered in Chapter 10: dedicated, in particular, to gyrotrons. The transport of nonlaminar beams with constant or growing emittance and with Maxwell–Boltzmann velocity distribution is a key problem in the theory of high-energy accelerators with extensive charged-particle beams. Consideration of these topics is, however, beyond the scope of this book. An interested reader can find an analysis of these problems in Davidson (1990), Humphries (1990), Lawson (1988), and Reiser (1994).

In Section 5.2, transport of unrippled nonrelativistic axially symmetric laminar electron beams in an external magnetic field is considered. It is the easiest problem in the theory of the propagation of space-charge electron beams, although even in this case the idealized scheme of Fig. 3.2 is insufficient because it is not possible to achieve equilibrium of nonrotating space-charge beams in a finite magnetic field. For proper rotation of a beam, it is necessary either to change the gun itself or to introduce a transition section between the gun and the drift space with

Electron Beams and Microwave Vacuum Electronics. By Shulim E. Tsimring

corresponding distribution of electric and magnetic fields. It is a problem of matching the gun and the drift beam. Naturally, this problem is more complicated for relativistic beams when it is necessary to take into account the self magnetic field (Section 5.3). The matching problem is solved automatically only in an infinite magnetic field. A similar case was considered in the theory of magnetically insulated guns (Section 4.8). Note that the stability of beams in an infinite magnetic field can be disturbed by transverse potential depression with a rising problem of limiting currents (Section 5.4).

A number of factors make the unrippled beam model unsatisfactory. Among these we can mention the nonlaminarity of a beam, dismatching of the gun and the beam, constructive disturbances, inevitable velocity spread, and dynamic effects. Finally, there is a wide range of problems where the use of unrippled electron beams is not optimal or may even be impossible (e.g., helical beams in gyrotrons, free electron lasers, or electron beams in periodic focusing systems).

As a first approximation to the theory, a paraxial-ray approximation based on a model of pulsatory laminar beams (Kirstein et al., 1967) can be used. The envelope of these beams is formed by boundary trajectories or by the rotation of any boundary trajectory for an axially symmetric beam with $dr/dz \neq 0$. The derivation of paraxial equations for solid and annular beams and some of their applications are considered in Section 5.5. It will be seen that paraxial approximation is very effective for the optimization of beam parameters in the presence of disturbances.

In Section 5.6, electrostatic focusing of electron beams is considered. Note that stable equilibrium of these beams is also impossible without rotation. The theory of periodic focusing of nonrelativistic electron beams based on paraxial approximation is given in Section 5.7.

5.2 UNRIPPLED AXIALLY SYMMETRIC NONRELATIVISTIC BEAMS IN A UNIFORM MAGNETIC FIELD

5.2.1 Statement of the Problem. Equations of an Equilibrium Beam

Consider propagation of a round unrippled space-charge beam in vacuum. Note that in the literature, transport systems are often called *focusing systems*. A beam in a conducting drift tube is depicted in Fig. 5.1. It is supposed that after the transition space ($z > z_{tr}$), an electron beam propagates in a uniform magnetic field, and all beam quantities do not depend on z; that is, $\varphi = \varphi(r)$, $\mathbf{v} = \mathbf{v}(r)$, and $\rho = \rho(r)$. The electric field at $z > z_{tr}$ has only the radial component $E_r = -d\varphi/dr$. It can readily be shown that the external magnetic field at $z > z_{tr}$ has only axial component $B_0 = \text{const}$. Let us assume that the magnetic field in the gun B_c is uniform. In general, $B_c \neq B_0$. Below we obtain the basic system of equilibrium beam equations. This system takes into account the following relationships that imply a stationary state and axial symmetry of the space-charge beam: (1) the rotary beam equilibrium, (2) the energy integral, (3) Poisson's equation; and (4) conservation of the azimuthal generalized momentum (Busch's theorem).

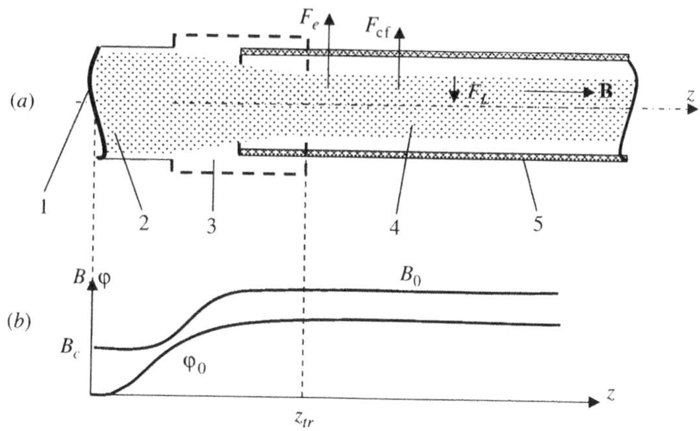

FIGURE 5.1 (*a*) Space-charge electron beam in a drift tube (1, cathode; 2, gun; 3, transition space; 4, electron beam; 5, conducting drift tube); (*b*) distribution of the axial component of a magnetic field and axis potential.

1. The condition of the rotary beam equilibrium is

$$F_r + F_{\mathrm{cf}} + F_L = 0 \tag{5.1}$$

where (Fig. 5.1*a*)

$$\begin{aligned} F_e &= -e_0 E_r \\ F_{\mathrm{cf}} &= \frac{m_0 v_\theta^2}{r} = m_0 \omega^2 r \\ F_L &= -e_0 v_\theta B_0 = -m_0 r \omega \omega_c \end{aligned} \tag{5.2}$$

are the space-charge, centrifugal, and Lorentz forces, respectively. Here $\omega_c = \eta B_0$ is the nonrelativistic cyclotron frequency. Substituting Eq. (5.2) in Eq. (5.1), we can write the equilibrium condition as

$$\eta E_r = \omega^2 r - \omega \omega_c r \tag{5.3}$$

2. The energy integral in the nonrelativistic approximation for zero cathode potential and zero initial electron velocities is

$$2\eta\varphi = v_z^2 + v_\theta^2 = v_z^2 + \omega^2 r^2 \tag{5.4}$$

Differentiation of Eq. (5.4) with respect to r and taking $\varphi' \equiv d\varphi/dr = -E_r$ into account gives

$$\eta E_r = -v_z v_z' - \omega\omega' r^2 - \omega^2 r \tag{5.5}$$

Combining Eqs. (5.3) and (5.5), we obtain the equation

$$v_z v_z' + \omega \frac{d}{dr}\left[r^2\left(\omega - \frac{\omega_c}{2}\right)\right] = 0 \tag{5.6}$$

3. Poisson's equation can be written as

$$\text{div } \mathbf{E} = \frac{1}{r}\frac{d}{dr}(r\eta E_r) = -\omega_p^2 \tag{5.7}$$

where $\omega_p^2 = -\rho\eta/\varepsilon_0 = e_0^2 n/m_0 \varepsilon_0$ is the electron plasma frequency (n is an electron density). Substituting the electric field from Eq. (5.3) into Eq. (5.7), we obtain

$$2\omega(\omega_c - \omega) + r\omega'(\omega_c - 2\omega) = \omega_p^2 \tag{5.8}$$

4. Busch's theorem provides matching the beam parameters in the gun and in the drift space. The nonrelativistic Busch theorem for zero cathode velocity according to Eq. (1.128) is

$$\dot{\theta} = \omega = \frac{\eta}{2\pi r^2}(\Psi - \Psi_c) \tag{5.9}$$

The magnetic fields B_c and B_0 in the gun and the drift space, respectively, are assumed to be uniform. Then the magnetic fluxes are

$$\Psi = \pi r^2 B_0, \qquad \Psi_c = \pi r_c^2 B_c \tag{5.10}$$

where r and r_c are the radii of the same trajectory in the drift space and on the cathode. Letting $\eta B_c \equiv \omega_{ck}$, we can write Busch's theorem as

$$\omega = \frac{\omega_c}{2}\left(1 - \frac{\omega_{ck}}{\omega_c}\frac{r_c^2}{r^2}\right) \tag{5.11}$$

Equations (5.6), (5.8), and (5.11) are the basic equations of an equilibrium beam. Some particular cases are considered below.

5.2.2 Isovelocity Beams

The longitudinal velocity of the isovelocity beam does not depend on the radius (i.e., $v_z' = 0$). These beams are active components of microwave electron devices with distributed interaction, similar to traveling-wave tubes of O-type (TWTOs) or backward-wave oscillators (BWOs) based on a synchronism of the beam with an electromagnetic wave. We obtain from Eq. (5.6)

$$r^2\left(\omega - \frac{\omega_c}{2}\right) = K = \text{const.} \tag{5.12}$$

Assume that the angular velocity at some radius r_0 is

$$(\omega)_{r=r_0} = \omega_0 \tag{5.13}$$

Then according to Eq. (5.12),

$$\omega = \frac{\omega_c}{2}\left[1 - \left(1 - \frac{2\omega_0}{\omega_c}\right)\frac{r_0^2}{r^2}\right] \tag{5.14}$$

Taking Busch's theorem [Eq. (5.11)] into account, we obtain

$$\frac{\omega_c}{2}\left[1 - \left(1 - \frac{2\omega_0}{\omega_c}\frac{r_0^2}{r^2}\right)\right] = \frac{\omega_c}{2}\left(1 - \frac{\omega_{ck}}{\omega_c}\frac{r_c^2}{r^2}\right) \tag{5.15}$$

which gives us

$$r_c = r_0\left[\frac{\omega_c}{\omega_{ck}}\left(1 - \frac{2\omega_0}{\omega_c}\right)^{1/2}\right] \tag{5.16}$$

We see that the cathode radius r_c in general does not depend on r (i.e., all trajectories of the beam should be emitted from a cathode of constant radius; Fig. 5.2). Note that $\omega_0 < \omega_c/2 = \omega_L$, where ω_L is the Larmor frequency. An exception is the case

$$\omega_0 = \omega_L; \qquad \omega_{ck} = 0 \tag{5.17}$$

This case corresponds to that of a Brillouin beam (considered in the next section). According to Eq. (5.14), the condition of the positive frequency ω for $\omega_0 < \omega_L$ is

$$r > r_0\sqrt{1 - \frac{\omega_0}{\omega_L}} \tag{5.18}$$

This unequality means that isovelocity beams, with the exception of Brillouin beams [Eq. (5.17)], are always hollow beams. In Fig. 5.2 two possible schemes of isovelocity beam formation with inner and external cathodes are depicted. As we can see,

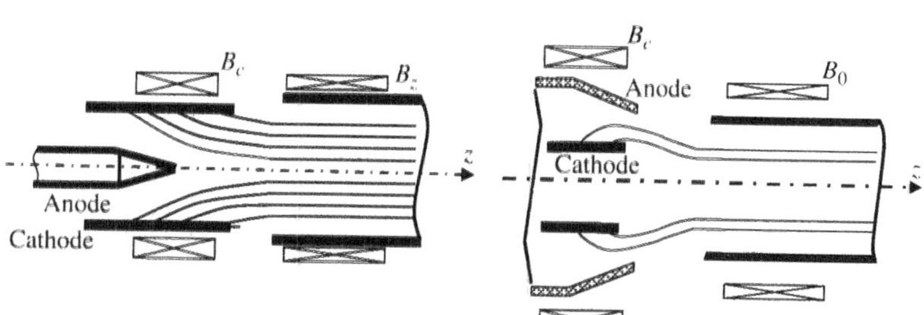

FIGURE 5.2 Two versions of guns with isovelocity beams.

these devices are versions of magnetron injection guns (Section 4.5.2). However, these schemes are only approximate. Their design is possible by a numerical method on the basis of adequate simulation codes with subsequent experimental adjustment.

5.2.3 Solid Brillouin Beams (Brillouin, 1945)

Earlier we defined a Brillouin beam (BB) as a beam for which the angular velocity at a particular radius is equal to the Larmor frequency [Eq. (5.17)]. It can readily be seen after substituting $\omega_0 = \omega_L = \omega_c/2$ into Eq. (5.12) that constant $K = 0$. This means that at any radius,

$$\omega = \frac{\omega_c}{2} = \omega L \tag{5.19}$$

that is, BBs are not only isovelocity but also isorotational beams. According to the Busch theorem [Eq. (5.11)] for BB formation, a zero cathode magnetic field $\omega_{ck} = 0$ must be provided. In this case, Eq. (5.15) is satisfied for any r_c and the limitations of Eq. (5.16) disappear. Thus, in contrast to other isovelocity beams, BBs are solid beams. These beams rotate similar to a rigid rotator. A diagram of Brillouin beam formation is shown in Fig. 5.3. The iron shield should provide a zero magnetic field near the cathode. According to the theory above, in some domain there must be a transition from a zero cathode magnetic field to a uniform field B_0 in the drift space. An idealized Brillouin beam is formed when the radial component of the magnetic field is nonzero only in the transition space. As electrons are emitted they move along straight trajectories up to the transition space. After the transition, all trajectories become helical lines with centers on the z-axis and with equal pitch. Note that the formation of helical trajectories encircling the z-axis in rare space-charge beams is provided by the cusp of the magnetic field (Section 1.7).

Real systems certainly do not provide ideal cathode shielding. Therefore, the characteristics of experimental beams are different from those of idealized beams. This topic is discussed in Section 5.7, where the paraxial theory of the space-charge beams is described.

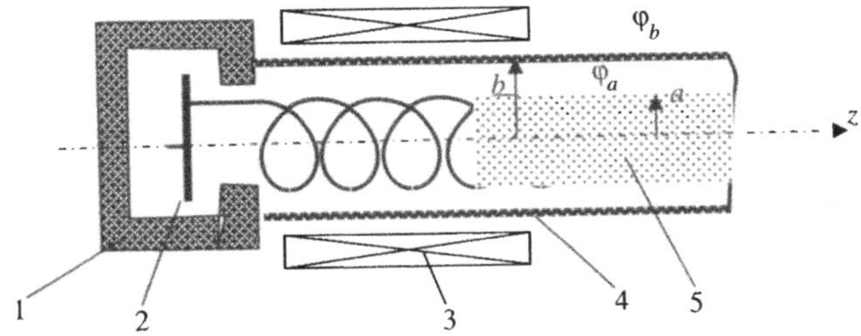

FIGURE 5.3 Brillouin beam formation. 1, Iron shield; 2, cathode; 3, solenoid; 4, drift tube; 5, electron beam.

Space-Charge Density and Brillouin Beam Current The plasma frequency in BBs is constant according to Poisson's equation [Eqs. (5.8) and (5.19)]:

$$\omega_p = \frac{\omega_c}{\sqrt{2}} = \sqrt{2}\,\omega_L = \text{const.} \tag{5.20}$$

The electron charge density and current density are also constant:

$$\rho = -\frac{\omega_p^2 \varepsilon_0}{\eta} = -\frac{\omega_c^2 \varepsilon_0}{2\eta} = \text{const.} \tag{5.21}$$

$$j = -\rho v_z = \frac{\omega_c^2 \varepsilon_0}{2\eta} v_z = \text{const.} \tag{5.22}$$

The current in the beam of radius a is

$$I = \pi a^2 j = \frac{\pi a^2 \omega_c^2 \varepsilon_0}{2\eta} v_z \tag{5.23}$$

The z-component of the electron velocity according to the energy integral, Eq. (5.4), is

$$v_z = \sqrt{2\eta\varphi_a - \frac{a^2\omega_c^2}{4}} \tag{5.24}$$

where φ_a is the beam potential at the radius $r = a$. We see that velocity v_z in the beam decreases with increased a, due to the increasing azimuthal linear speed. Since the rotational velocity $v_\theta = \omega r$ is zero on the axis, the velocity axial component is

$$v_z = \sqrt{2\eta\varphi_0} \tag{5.25}$$

where φ_0 is the potential on the axis. Then

$$a^2\omega_c^2 = 8\eta(\varphi_a - \varphi_0) \tag{5.26}$$

The beam current according to Eqs. (5.23), (5.24), and (5.26) is

$$I = \frac{\pi a^2 \omega_c^2 \varepsilon_0}{2\eta}\sqrt{2\eta\varphi_a - \frac{a^2\omega_c^2}{4}} = 4\pi\varepsilon_0\sqrt{2\eta\varphi_0}\,(\varphi_a - \varphi_0) \tag{5.27}$$

It is straightforward to express the potential φ_a through the potential of the drift tube φ_b as the potential of a uniform charged cylinder. Then we obtain

$$I = \frac{\pi a^2 \omega_c^2 \varepsilon_0}{2\eta}\sqrt{2\eta\varphi_b - \frac{a^2\omega_c^2}{4}\left(1 + 2\ln\frac{b}{a}\right)} \tag{5.28}$$

where b is the drift tube radius.

According to Eq. (5.27), the current first increases with increased $\varphi_a - \varphi_0$ (or $a\omega_c$), but after reaching a maximum begins to decrease as a result of lowering v_z. The maximum of the current is determined from the condition

$$\frac{\partial I}{\partial(a\omega_c)} = 0 \tag{5.29}$$

Let us find the current maximum assuming the ratio $b/a = \text{const}$. From Eq. (5.28) we obtain

$$(a^2\omega_c^2)_{\max} = \frac{16\eta\varphi_b}{3[1 + 2\ln(b/a)]} \tag{5.30}$$

After substitution of Eq. (5.30) in Eq. (5.28), we find that

$$I_{\max} = \frac{8\sqrt{6}\,\pi\varepsilon_0\sqrt{\eta}}{9[1 + 2\ln(b/a)]}(\varphi_b)^{3/2} = P_{\max}(\varphi_b)^{3/2} \tag{5.31}$$

where P is the perveance. For a solid Brillouin beam that fills the drift tube ($b/a = 1$) completely, we obtain from Eq. (5.30)

$$(a^2\omega_c^2)_{\max} = \tfrac{16}{3}\eta\varphi_a = \tfrac{16}{3}\mu\varphi_b \tag{5.32}$$

or according to Eq. (5.26), the axis potential for the maximum current is

$$\varphi_{0,\max} = \tfrac{1}{3}\varphi_a = \tfrac{1}{3}\varphi_b \tag{5.33}$$

The maximum perveance is

$$P_{\max} = \frac{8\sqrt{6}\,\pi\varepsilon_0\sqrt{\eta}}{9} \approx 25.4 \times 10^{-6}\,\text{A/V}^{3/2} \tag{5.34}$$

The magnetic field $B_{0,\max}$ corresponding to the maximum of the BB current can be found from Eq. (5.30):

$$B_{0,\max} = \frac{4}{a\sqrt{3}}\sqrt{\frac{\varphi_b}{\eta[1 + 2\ln(b/a)]}} \tag{5.35}$$

Note that we can search the maximum current as the extreme of the function $I(\varphi_0)$ from Eq. (5.27). For $a/b = 1$ we obtain the same Eq. (5.33) and the same maximum current [Eq. (5.31)].

It is instructive to depict the function $I = I(\varphi_0)$ corresponding to Eq. (5.27) (Fig. 5.4). Let us increase the cathode injection current from zero. When the beam current is zero, the potential depression is absent and $\varphi_0 = \varphi_a$. Increase in

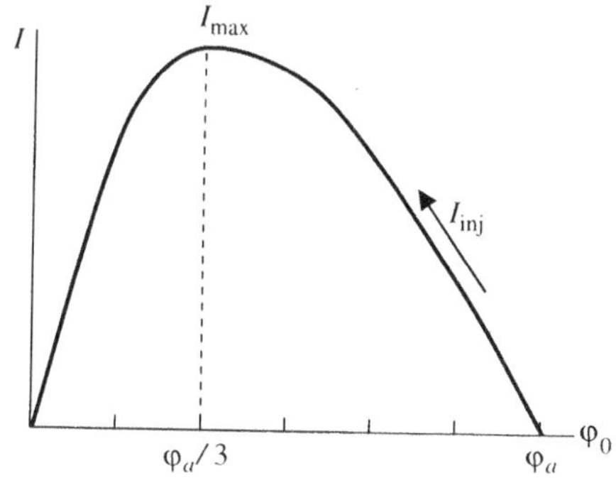

FIGURE 5.4 Current of a solid Brillouin beam as a function of axis potential.

the injection current reduces the axis potential. The point $\varphi_0 = \varphi_a/3$ corresponds to the beam that completely fills the drift tube. When the injection current attains the values I_{max}, determined by Eq. (5.31) with $a/b = 1$, the axis potential drops to $\varphi_a/3$. Further increase in the injection current (supercritical current) cannot raise the beam current above I_{max}. The descending (left) branch of the curve $I(\varphi_0)$ for $\varphi_0 < \varphi_a/3$ is unstable and not realizable experimentally. In the static version of a theory with $I_{inj} > I_{max}$, a virtual cathode occurs and a fraction of the injected current is reflected back toward to the cathode. The corresponding solution should take into account two beams (straight and backward) in the domain between real and virtual cathodes. However, according to numerical simulation, solution of the steady-state problem is physical unrealizable. The beam is unstable, and for $I_{inj} > I_{max}$, oscillations of various types can be developed (see, e.g., Haeff, 1939). In any case, the equilibrium will be disturbed and an unrippled beam structure is not possible.

5.2.4 Hollow Brillouin Beams

According to Section 5.2.3, all isovelocity beams with the exception of solid Brillouin beams are hollow. *Brillouin hollow beams* (BHBs), for which by definition the angular velocity on the inner boundary beam is zero, have the most practical interest. Let us take r_0 in Eq. (5.14) as the inner radius of a BHB. Then $\omega_0 = \omega(r_0) = 0$, and we obtain from Eq. (5.14),

$$\omega = \frac{\omega_c}{2}\left(1 - \frac{r_0^2}{r^2}\right) \tag{5.36}$$

The charge density of the BHB is found by substituting Eq. (5.31) into Eq. (5.36):

$$\omega_p^2 = -\frac{\rho\eta}{\varepsilon_0} = \frac{\omega_c^2}{2}\left(1 + \frac{r_0^4}{r^4}\right) \tag{5.37}$$

Then the current of the beam with the outer radius a is

$$I = -\int_{r_0}^{a} \rho v_z \cdot 2\pi r dr = -2\pi v_z \int_{r_0}^{a} \rho r\, dr = \frac{\pi\varepsilon_0\omega_c^2 a^2}{2\eta} v_z \left(1 - \frac{r_0^4}{a^4}\right) \tag{5.38}$$

Here $v_z = \text{const}$. Expressing $(v_z)_{r=a}$ from the energy integral (5.4) and substituting it into Eq. (5.38), we obtain

$$I = \frac{\pi\varepsilon_0\omega_c^2 a^2}{2\eta}\left(1 - \frac{r_0^4}{a^4}\right)\sqrt{2\eta\varphi_a - \frac{\omega_c^2 a^2}{4}\left(1 - \frac{r_0^2}{a^2}\right)^2} \tag{5.39}$$

Let us find the maximum BHB current as a function of the product $\omega_c a$, assuming the ratio $r_0/a = \text{const}$. Then application of the condition $\partial I/\partial(\omega_c a) = 0$ to the current of Eq. (5.39) with fixed values φ_a is analogous to application of Eq. (5.29) to the current $I(a)$ from Eq. (5.28). So we obtain

$$\left[a^2\omega_c^2\left(1 - \frac{r_0^2}{a^2}\right)^2\right]_{\max} = \frac{16\eta\varphi_a}{3} \tag{5.40}$$

Substitution of Eq. (5.40) in Eq. (5.39) gives the maximum current of the BHB:

$$I_{\max} = \frac{8\sqrt{6}\,\pi\varepsilon_0\sqrt{\eta}}{9}\frac{1 + r_0^2/a^2}{1 - r_0^2/a^2}(\varphi_a)^{3/2} \tag{5.41}$$

The corresponding perveance of the BHB is

$$P_{\max} = 25.4 \times 10^{-6}\frac{1 + r_0^2/a^2}{1 - r_0^2/a^2} \quad \text{A/V}^{3/2} \tag{5.42}$$

Comparison of Eqs. (5.42) and (5.34) shows that the maximum perveance of hollow Brillouin beams is greater than the corresponding perveance for solid Brillouin beams. For thin beams, the ratio of the perveances is $\sim r_0/\Delta r$, where Δr is the beam thickness. The cathode of BHB as cathodes in all other hollow isoivelocity beams must be a surface of constant radius.

The important practical advantage of hollow Brillouin beams is the lower requirement of magnetic field distribution than in solid BBs, where ideal magnetic shielding of the cathode must be provided. In particular, a uniform magnetic field over the entire system length can be used. In this case ($\omega_{ck} = \omega_c$), and according to Eq. (5.16), the inner radius of the beam r_0 $[\omega_0(r_0) = 0]$ is equal to the cathode radius r_c. Note that the use of hollow Brillouin beams increases the efficiency of microwave high-frequency devices with slow-wave structures, similar to nonrelativistic TWTOs and BWOs, due to the proximity of the surface structure of slow waves and the structure of BHBs.

5.3 UNRIPPLED RELATIVISTIC BEAMS IN A UNIFORM EXTERNAL MAGNETIC FIELD

5.3.1 Introduction

Formulation of a problem for relativistic beams is somewhat similar to the foregoing. In particular, the beam is supposed to be uniform in the z-direction and axially symmetrical. Also, an equilibrium state of flow is assumed. We can therefore write the following equalities:

$$v_r = 0, \qquad \frac{\partial Q_i}{\partial z} = \frac{\partial Q_i}{\partial \theta} = \frac{\partial Q_i}{\partial t} = 0 \tag{5.43}$$

where Q_i are any beam quantities. So beam variables are a function of r only. The beam is injected from a gun into a conducting cylindrical drift tube with radius $r = b$ (Fig. 5.3). The drift tube is placed inside a uniform round solenoid that produces a magnetic field B_0. The natural condition that provides the laminarity of the beam is the zero initial (cathode) velocities of the particles.

The relativistic formulation of the problem suggests the following distinctions from the nonrelativistic formulation, which complicates the theory. First, it is the necessity to take into account the self magnetic field of the beam. According to Eqs. (5.43), the self magnetic field has two nonzero components: axial B_{sz} and azimuthal $B_{s\theta}$. Another problem with the theory is related to the use of short pulse high-current relativistic beams when the effects of self magnetic field shielding by the drift tube are relevant. Then the equilibrium state of the beam is possible in two extreme cases: either during an initial short period of time when the self magnetic field does not penetrate the metal drift tube and is localized inside the drift tube (e.g., when using very short pulses, on the order of nanoseconds), or when a long pulse duration is possible because the drift tube does not disturb the self magnetic field. We postulate the latter case below. An additional complication of the theory is determined by the use of relativistic equations of motion.

Development of the theory of equilibrium intense relativistic beams began the 1960s. Here we should refer a paper by DePackh and Ulrich (1960), in which a fully relativistic theory of space-charge cylindrical Brillouin beams was developed. Reiser (1977) has given a comprehensive fully relativistic theory of space-charge cylindrical flow in the magnetic field, taking into account the effect of locking the self magnetic field into the drift tube. Presented below is an analysis that follows works of DePackh and Ulrich (1960) and Reiser (1977) in the approximation of a long-pulse self magnetic field.

5.3.2 Equations of Relativistic Equilibrium Electron Flow

Relativistic equilibrium of radial forces is

$$\frac{\gamma(r) m_0 v_\theta^2(r)}{r} - e_0 E_r - e_0 v_\theta(r) B_z(r) + e_0 v_z(r) B_{s\theta}(r) = 0 \tag{5.44}$$

where $B_z(r) = B_0 + B_{sz}(r)$ and B_0 is the uniform external magnetic field given. The relativistic factor is

$$\gamma(r) = \left(1 - \frac{v_\theta^2}{c^2} - \frac{v_z^2}{c^2}\right)^{-1/2} \tag{5.45}$$

In suggesting zero cathode potential and zero initial electron velocities, the energy integral can be written as

$$\gamma(r) = 1 - \frac{e_0\varphi(r)}{m_0c^2} \tag{5.46}$$

Poisson's equation is

$$\frac{1}{r}\frac{d}{dr}\left[r\frac{d\varphi(r)}{dr}\right] = \frac{e_0 n(r)}{\varepsilon_0} \tag{5.47}$$

The condition of conservation of azimuthal generalized momentum $P_\theta(r)$ [Eq. (1.124)] can be written as

$$P_\theta(r) = P_{\theta c}(r_c) = m_0 r\gamma(r)v_\theta(r) - \frac{e_0}{2\pi}\Psi(r) = m_0 r\gamma(r)v_\theta(r) - e_0\int_0^r rB_z(r)\,dr \tag{5.48}$$

Here $P_{\theta c}(r_c)$ is the cathode azimuthal momentum. Expressing this quantity through the cathode magnetic field and cathode electron velocity leads to Busch's theorem. Below, $P_\theta(r)$ will be assumed as a given function.

Let us add to these equations for electric and magnetic self fields. The electric field is

$$E_r(r) = -\frac{d\varphi(r)}{dr} \tag{5.49}$$

The magnetic field is determined by Maxwell's equation, curl $\mathbf{B} = \mu_0\mathbf{j} = -\mu_0 e_0 n\mathbf{v}$. That leads to the equations

$$\frac{d}{dr}[rB_{s\theta}(r)] = -\mu_0 e_0 r n(r) v_z(r) \tag{5.50}$$

$$\frac{dB_{sz}(r)}{dr} = \mu_0 e_0 n(r) v_\theta(r) \tag{5.51}$$

Here we took into account $dB_0/dr = 0$, and then $dB_z/dr = dB_{sz}/dr$. Note that initiation of the self magnetic field B_{sz}, which is always directed opposite B_0, characterizes the electron beam as a gyrotropic diamagnetic medium.

Thus, we have got a set of eight self-consistent equations (5.44)–(5.51) for eight functions: $\gamma(r)$, $v_z(r)$, $v_\theta(r)$, $E_r(r)$, $B_z(r)$, $B_{s\theta}(r)$, $\varphi(r)$, and $n(r)$. As one has been obtained, this system can be reduced to three equations for the functions $\gamma(r)$, $v_z(r)$, and $v_\theta(r)$ by elimination of the remaining functions from the original set [Eqs. (5.44)–(5.51)]. These are the following nonlinear ordinary differential equations:

$$[r(\gamma v_z)']' - v_z(r\gamma')' + \frac{1}{m_0}\left(\frac{v_\theta P'_\theta}{v_z}\right)' = 0 \tag{5.52}$$

$$\frac{v_\theta}{r}\gamma' + (r\gamma)'\left(\frac{v_\theta}{r}\right)' + (\gamma v'_\theta)' + \frac{1}{m_0}\left(\frac{P'_\theta}{r}\right)' = 0 \tag{5.53}$$

and Eq. (5.45) for the relativistic factor γ. Here all primed symbols signify differentiation, d/dr. The azimuthal momentum $P_\theta(r)$ is determined by the initial cathode conditions and can be assumed as a given function. Solving these equations in general is a difficult problem. Below, the simplest case (although an important one), $P_0 = \text{const.}$, is considered. In this case Eqs. (5.52) and (5.53) are reduced to

$$\frac{1}{\gamma}(r\gamma^2 v'_z)' = 0 \tag{5.54}$$

$$\frac{v_\theta}{r}\gamma' + (r\gamma)'\left(\frac{v_\theta}{r}\right)' + (\gamma v'_\theta)' = 0 \tag{5.55}$$

The obvious first integral of Eq. (5.54) is

$$r\gamma^2 v'_z = K = \text{const.} \tag{5.56}$$

5.3.3 Solid Relativistic Brillouin Beams

Assume that similarly to nonrelativistic solid Brillouin flow (Section 5.2.3), the cathode magnetic field in relativistic flow is zero, and correspondingly, $P_\theta = 0$. This is the simplest way to realize a distribution of the cathode magnetic field with $P_\theta = \text{const.}$

According to Eq. (5.56), for solid beams, $K = 0$; otherwise, we obtain $(v'_z)_{r=0} = \infty$. Then $v'_z = 0$, and we obtain the isovelocity beam:

$$v_z = v_0 = \text{const.} \tag{5.57}$$

The azimuthal velocity is expressed through γ from the energy integral:

$$v_\theta = \frac{c}{\gamma\gamma_0}\sqrt{\gamma^2 - \gamma_0^2} \tag{5.58}$$

where

$$\frac{1}{\gamma_0^2} = 1 - \frac{v_0^2}{c^2} \tag{5.59}$$

Because the azimuthal velocity on the axis is zero, we see, according to Eq. (5.58), that for solid beams, Eq. (5.59) is equivalent to

$$(\gamma)_{r=0} = \gamma_0 \tag{5.60}$$

Substitution of Eq. (5.58) in Eq. (5.55) gives the following differential equation for γ:

$$\gamma'' - \frac{\gamma\gamma'^2}{\gamma^2 - \gamma_0^2} + \frac{\gamma'}{r} - \frac{\gamma}{r^2\gamma_0^2}(\gamma^2 - \gamma_0^2) = 0 \tag{5.61}$$

The first integral of this equation can be presented as (see DePackh and Ulrich, 1960)

$$\frac{r^2\gamma'^2}{\gamma^2 - \gamma_0^2} - \frac{\gamma^2}{\gamma_0^2} = C = \text{const.} \tag{5.62}$$

A solution of this equation for an arbitrary C is expressed through the elliptic integral of the first kind. According to Eq. (5.60), for a solid Brillouin beam in the vicinity of a point $r = 0$, the ratio γ/γ_0 can be represented as

$$\frac{\gamma}{\gamma_0} = 1 + kr^2 \tag{5.63}$$

Substituting Eq. (5.63) in Eq. (5.62), we obtain for the solid beams, $C = -1$. In this case, Eq. (5.62) becomes

$$\frac{\gamma_0\gamma'}{\gamma^2 - \gamma_0^2} = \frac{1}{r} \tag{5.64}$$

It is readily verified that the solution of this equation is

$$\gamma = \gamma_0 \frac{r_0^2 + r^2}{r_0^2 - r^2} \tag{5.65}$$

where r_0 is a constant of integration. The constant r_0 depends on the electron energy and the beam dimensions. For nonrelativistic energies the factors $\gamma \approx \gamma_0 \approx 1$, and we obtain

$$r_{\text{nrel}} \ll r_0 \tag{5.66}$$

Next, we obtain the remaining quantities of the Brillouin flow.

The charge density is determined by Poisson's equation [Eq. (5.47)]. The corresponding relationship in terms of γ is

$$n = \frac{\varepsilon_0 m_0 c^2}{e_0^2} \frac{1}{r} (r\gamma')' \tag{5.67}$$

Substituting Eq. (5.65) in Eq. (5.67), we find that

$$n = \frac{8\varepsilon_0\gamma_0 m_0 c^2 r_0^2(r_0^2 + r^2)}{e_0^2(r_0^2 - r^2)^3} \tag{5.68}$$

We see that unlike the nonrelativistic solid Brillouin beams, the charge particle density here is essentially nonuniform. However, for nonrelativistic beams correspondingly to the inequality (5.66) and $\gamma_0 \approx 1$, we obtain

$$n_{\text{nrel}} \approx \frac{8\varepsilon_0 m_0 c^2}{r_0^2 e_0^2} = \text{const.} \tag{5.69}$$

which agrees with Eq. (5.21) for a nonrelativistic solid Brillouin beam. Comparison of this formula with Eq. (5.21) gives us

$$(r_0)_{\text{nrel}} = \frac{4c}{\omega_c} \tag{5.70}$$

It is readily seen that this quantity is really much greater than r.

The angular velocity is found from Eqs. (5.58) and (5.65):

$$\omega = \frac{v_\theta}{r} = \frac{c}{r\gamma\gamma_0}\sqrt{\gamma^2 - \gamma_0^2} = \frac{2r_0 c}{\gamma_0(r_0^2 + r^2)} \tag{5.71}$$

Thus, the relativistic Brillouin solid beam does not represent a rigid rotor. In nonrelativistic approximation, neglecting r in Eq. (5.71) and taking Eq. (5.70) into account, we obtain $\omega = \omega_c/2$ (i.e., we return to the beam model as a rigid rotor). According to Eqs. (5.57) and (5.67), a beam current that fills a tube of radius $r = a$

$$I = \int_0^a 2\pi r v_z e_0 n(r)\, dr = 2\pi v_0 e_0 \frac{\varepsilon_0 m_0 c^2}{e_0^2}\int_0^a r\frac{1}{r}(r\gamma')' dr = I_0 \frac{v_0}{2c} a\gamma_a' \tag{5.72}$$

where $I_0 = 4\pi\varepsilon_0 m_0 c^3/e_0 \approx 17$ kA is the relativistic current [see Eq. (4.129)], $\gamma_a \equiv \gamma(a)$. Using Eqs. (5.64) and (5.72), we obtain the current in terms of γ/γ_0:

$$I = I_0 \frac{v_0}{2c}\frac{\gamma_a^2 - \gamma_0^2}{\gamma_0} = I_0 \frac{\sqrt{\gamma_0^2 - 1}}{2}\left(\frac{\gamma_a^2}{\gamma_0^2} - 1\right) \tag{5.73}$$

This is a volt–ampere characteristic of a diode with a solid Brillouin beam. Consider the current as a function of the axis potential $\varphi_0 = (m_0c^2/e_0)(\gamma_0 - 1)$ and the beam potential on the surface $r = a$, $\varphi_a = (m_0c^2/e_0)(\gamma_a - 1)$. According to Eq. (5.73), the current is zero for $\gamma_0 = 1$ and $\gamma_0 = \gamma_a$. In the first case it indicates a zero axis potential. In the second case we have zero potential depression on the beam, which requires zero current. The maximum current is found from the condition $dI/d\gamma_0 = 0$, where I is determined by Eq. (5.73). That yields

$$\gamma_{0,\max}^2 = \frac{\gamma_a^2}{2}\left[\left(1 + \frac{8}{\gamma_a^2}\right)^{1/2} - 1\right] \tag{5.74}$$

Substitution of $\gamma_{0,\max}$ into Eq. (5.73) gives the maximum current:

$$I_{\max} = \frac{I_0}{2}\left\{\frac{\gamma_a^2}{2}\left[\left(1 + \frac{8}{\gamma_a^2}\right)^{1/2} - 1\right] - 1\right\}^{1/2}\left[\frac{2}{[1 + (8/\gamma_a^2)]^{1/2} - 1} - 1\right] \tag{5.75}$$

Let us find the beam current as a function of γ_b, which corresponds to the drift tube potential φ_b (Fig. 5.3). Considering the electric field in the gap $r_a \le r \le r_b$ as the field of a charged cylinder, we obtain, according to Eq. (5.73),

$$E_r \approx -\frac{I}{2\pi r v_0 \varepsilon_0} = \frac{I_0}{4\pi\varepsilon_0 c r}\frac{\gamma_a^2 - \gamma_0^2}{\gamma_0} \tag{5.76}$$

Then the potential difference $\varphi_b - \varphi_a$ and, respectively, the difference $\gamma_b - \gamma_a$ is equal to

$$\gamma_b - \gamma_a = \frac{\gamma_a^2 - \gamma_0^2}{\gamma_0}\ln\frac{b}{a} \tag{5.77}$$

For computation of the maximum current as a function γ_b, it is necessary to solve Eqs. (5.74) and (5.77) to find a function $\gamma_a = \gamma_a(\gamma_b)$ and substitute it into Eq. (5.75). If the beam fills the drift tube completely then $r_a = r_b$, $\gamma_a = \gamma_b$, and the current is obtained directly from Eq. (5.75).

An external magnetic field B_0, focusing the maximum current, can be found from Eq. (5.48) for $P_\theta = 0$:

$$m_0 r\gamma(r) v_\theta(r) - e_0\int_0^r r[B_0 + B_{sz}(r)]\,dr = 0 \tag{5.78}$$

where $B_{sz}(r)$ is determined by Eq. (5.51). Taking into account that in the long-pulse approximation, the self magnetic field on the beam surface $B_{sz}(a) = 0$, we obtain from Eq. (5.51),

$$B_{sz}(r) = \int_a^r \mu_0 e_0 n(r) v_\theta(r)\,dr \tag{5.79}$$

Substituting the functions $n(r)$ and $v_\theta(r)$ from Eqs. (5.68) and (5.71), respectively, and integrating, we find the axial component of the self magnetic field:

$$B_{sz} = \frac{4cr_0^3}{\eta}\left[\frac{1}{(r_0^2 - r^2)^2} - \frac{1}{(r_0^2 - a^2)^2}\right] \tag{5.80}$$

Now substitute the functions $\gamma(r)$, $v_\theta(r)$, and $B_{sz}(r)$ from Eqs. (5.65), (5.71), and (5.80), respectively, into Eq. (5.78). After integration, we obtain

$$B_0 = \frac{4c}{\eta r_0}\frac{1}{(1 - a^2/r_0^2)^2} \tag{5.81}$$

Note that this formula distinguishes it from the relation (see Reiser, 1977)

$$B_0 = \frac{4c}{\eta r_0}\frac{(1 - a^4/r_0^2 b^2)}{(1 - a^2/r_0^2)^2} \tag{5.82}$$

obtained by taking into account the shielding effect of the drift tube. Equation (5.82) is reduced to Eq. (5.81) in $b \to \infty$.

B_0 is readily expressed as a function of γ_0 and γ_a. According to Eq. (5.65),

$$\frac{r_0}{a} = \sqrt{\frac{\gamma_a + \gamma_0}{\gamma_a - \gamma_0}} \tag{5.83}$$

Then we find from Eq. (5.81),

$$B_0 = \frac{c}{\eta a}\left(\frac{\gamma_a}{\gamma_0} + 1\right)\sqrt{\frac{\gamma_a^2}{\gamma_0^2} - 1} \tag{5.84}$$

We can find the external magnetic field that corresponds to the maximum of the beam current by substituting the ratio γ_a/γ_0 from Eq. (5.74) in Eq. (5.84).

The curves of the current for the beams, which fill the drift tube ($b/a = 1$) completely, and the corresponding values of the magnetic field as functions of the ratio γ_0/γ_a for different values of γ_a, are shown in Fig. 5.5. As we see, the current reaches a maximum near small values of γ_0/γ_a, which correspond to large values of the potential depression. Here, just as in Fig. 5.4, the left branch or curve $I(\gamma_0/\gamma_a)$ is unstable and not realizable physically. Also, by exceeding the injection current, the value $I_{\max}$ forms a virtual cathode and the growth of beam instabilities.

In the extreme relativistic limit ($\gamma_a \gg 1$) according to Eq. (5.74), $\gamma_{0,\max}^2 = 2$ and does not depend on γ_a. Substitution of this quantity into Eqs. (5.73) and (5.84) gives

$$I_{\text{erel}} = I_0 \frac{\gamma_a^2}{4} \tag{5.85}$$

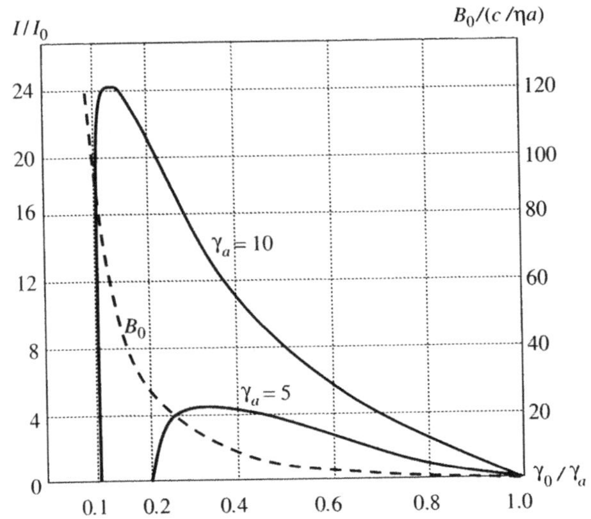

FIGURE 5.5 Dimensionless maximum beam current and external magnetic field as functions of the ratio γ_0/γ_a for different values of the relativistic factor γ_a.

$$(B_0)_{\text{erel}} = \frac{c}{\eta a}\frac{\gamma_a^2}{2} \tag{5.86}$$

Note that the extreme relativistic magnetic field in a quasistationary short-pulse approximation, taking into account the effect of self magnetic field shielding, is (Reiser, 1994)

$$(B_0)_{\text{erel}} = \frac{c}{\eta a}\sqrt{2}\gamma_a \tag{5.87}$$

We see that the external magnetic field required in the long-pulse approximation in an extreme relativistic limit essentially exceeds the magnetic field for the model of locking the self magnetic field when short-pulse processes are used. Thus, the effect of magnetic shielding in this case softens considerable the requirements for an external magnetic field. Let us consider as an example an electron beam with $b = a = 1$ cm and $\gamma_b = \gamma_a = 5(\varphi_b = 4$ MV). We obtain from Eq. (5.74), $\gamma_{0,\max}/\gamma_a = 0.27$. Then according to Fig. 5.5, we find in dimensionless units $I_{\max} = 4$ and $B_0 = 20$, which correspond to $I_{\max} = 68$ kA and $B_0 = 3.4$ T.

The formulas for maximum current $I_{\max}$ and corresponding external magnetic field B_0 in nonrelativistic approximation follow from Eq. (5.75) and Eqs. (5.84) and (5.74) after substituting $\gamma = 1 + \varphi/\eta c^2$ for $\varphi/\eta c^2 \ll 1$ in these relations. It is readily verified that the same current is obtained from corresponding nonrelativistic formulas (5.31) and (5.35) (for $b/a = 1$).

Note that the formation of equilibrium solid Brillouin beams with maximum currents is not always possible or even necessary. In this case there are wide possibilities for choosing the beam parameters because only three equations [Eqs. (5.68),

(5.72), and (5.79)] must be satisfied. As we can see, the beam is described by the five experimental quantities I, B_0, γ_b, a, and b and by the two theoretical quantities γ_0 and γ_a. In principle it is possible to specify any four quantities and calculate the remaining three. This is a comparatively simple computational problem, although its experimental realization can be complicated.

An analysis of equilibrium hollow relativistic beams with distributions of momentum $P_\theta(r) \neq 0$ (different from the solid Brillouin beams) is a significantly more difficult problem. Consideration of the corresponding theory exceeds the scope of this book however. A case involving a cathode immersed in an external magnetic field with $P_\theta(r) \propto r^2$ (neglecting the self magnetic field) has been analyzed by Miller (1982). The reader can find a useful qualitative discussion of this problem in a book by Reiser (1994). An exception is he analysis of relativistic beams in an infinite magnetic field (Section 5.4).

5.4 CYLINDRICAL BEAMS IN AN INFINITE MAGNETIC FIELD

The solid Brillouin beams (SBBs) considered in Section 5.3 are particular types of the flows for a full set of self-consistent equations that has a rigorous analytical solution. Simulation methods provide a considerable expansion of a class of solvable problems connected with self-consistent flows. However, the limits of applicability of these methods are often indefinite. In this regard, the SBB solution can be considered a standard for the control of simulation software.

Practically, SBBs are not the best type of beams, due to the complexity of obtaining adequate magnetic shielding of the cathode. In addition, it could be necessary to obtain structures of relativistic flows that differ from the SBBs. Note that design difficulties of self-consistent relativistic beams are connected basically with an account of the self magnetic fields. However, if experimental conditions do not limit values of the external magnetic field, self-consistent relativistic flows are available and the self magnetic fields can be neglected. A strong magnetic field, together with adequate distribution of cathode emission, will allow the transversal structure of the electron flows to vary within wide limits. In these cases, however, a problem of limiting currents arises. The theory of beams with rectilinear trajectories in the infinite magnetic field is a theoretical idealization of this problem. In this section we consider hollow and solid flows in an infinite magnetic field. The first works on the theory of space-charge rectilinear beams in the nonrelativistic limit are those of Haeff (1939) and Smith and Hartman (1940). Thin annular and solid relativistic beams are considered next.

5.4.1 Thin Annular Flows in an Infinite Magnetic Field

Let us consider an axially symmetric beam formed by electrons injected from a thin annular cathode into a strong uniform magnetic field $B_0 \to \infty$ (Fig. 5.6). The electrons emitted move parallel to the magnetic field and pass into a long metal drift tube with the potential φ_b and the radius $r = b$. It is suggested that the beam is thin

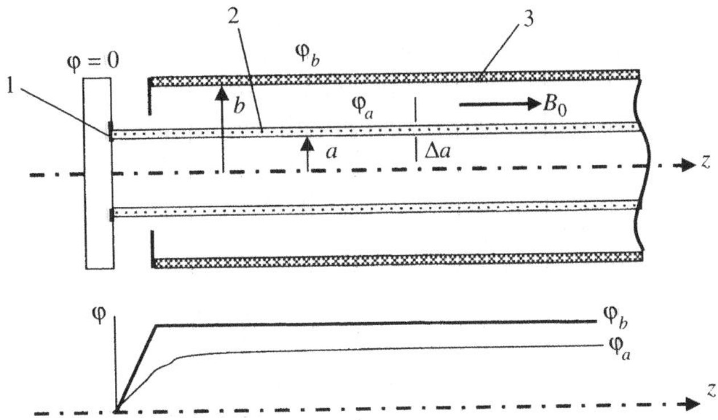

FIGURE 5.6 Thin electron beam transport in a strong magnetic field, and distribution of the potential. 1, Emitter; 2, electron beam; 3, drift tube.

[i.e., $\Delta a \ll a(b-a)$]. Therefore, the potential into the beam is constant, $\varphi = \varphi_a$. The radial equilibrium of the beam is supported by the strong magnetic field, and the electron velocity has only the z-axis component v. The beam equations are

$$\gamma(r) = 1 + \frac{\eta\varphi(r)}{c^2} \tag{5.88}$$

$$\gamma_a = \left(\frac{1-v^2}{c}\right)^{-1/2} = 1 + \frac{\eta\varphi_a}{c^2} \tag{5.89}$$

$$\varphi_b - \varphi(r) = \begin{cases} \dfrac{I}{2\pi\varepsilon_0 v}\ln\dfrac{b}{r}, & a \le r \le b \\ \dfrac{I}{2\pi\varepsilon_0 v_a}\ln\dfrac{b}{a}, & 0 < r \le a \end{cases} \tag{5.90}$$

Equations (5.88) and (5.89) are consequences of the energy integral for zero electron initial velocity and zero potential of the cathode; Eq. (5.90) is the potential of the charged cylinder outside the annular beam with charge density $j = -\rho v$. From Eqs. (5.88) and (5.90) we obtain

$$\gamma_b - \gamma_0 = \frac{2I}{I_0 v_a/c}\ln\frac{b}{a} \tag{5.91}$$

where the relativistic unity of a current is

$$I_0 = \frac{4\pi\varepsilon_0 c^3}{\eta} \approx 17\,\text{kA} \tag{5.92}$$

In the nonrelativistic limit

$$\varphi_b - \varphi_a = \frac{I}{2\pi\varepsilon_0 v_a}\ln\frac{b}{a} \tag{5.93}$$

Let us express current I as a function of γ_a. Finding the electron velocity v from Eq. (5.89) and substituting it in Eq. (5.91), we obtain

$$I = \frac{I_0}{2\ln(b/a)}\left(\frac{\gamma_b}{\gamma_a} - 1\right)\sqrt{\gamma_a^2 - 1} \tag{5.94}$$

According to this equation, the current is zero in two cases: $\gamma_a = 1$ (full drop of the depressed potential) and $\gamma_a = \gamma_b$ (zero potential depression). The maximum current according to the condition $\partial I/\partial\gamma_a = 0$ corresponds to the point $\gamma_a = \gamma_b^{1/3}$. Then this current is

$$I_{\max} = \frac{I_0}{2\ln(b/a)}(\gamma_b^{2/3} - 1)^{3/2} \tag{5.96}$$

It is readily verified that in the nonrelativist limit,

$$\varphi_{a\max} = \frac{1}{3}\varphi_b \tag{5.97}$$

$$I_{\max} = \frac{4\pi\varepsilon_0\sqrt{6\eta}}{9\ln(b/a)}\varphi_b^{3/2} \approx 12.7\times 10^{-6}\frac{\varphi_b^{3/2}}{\ln(b/a)} \quad \text{A} \tag{5.98}$$

The beam current as a function of the relativistic factor γ_a is shown in Fig. 5.7. For small injection currents the beam potential is closed to the drift tube potential (i.e., $\gamma_a \approx \gamma_b$). When the injection current increases, the corresponding point moves along the right branch of the curve. The maximum current $I_{\max}$ is the limiting beam current. When $I_{\text{inj}} > I_{\max}$, both a virtual cathode and beam instabilities in the input domain of the beam arise. A state of the beam corresponding to the left branch of the curve $I(\gamma_b)$ (here, $1 < \gamma_a < \gamma_b^{1/3}$; in the relativistic limit it corresponds to $0 < \varphi_a < \frac{1}{3}\varphi_b$) is unstable and physically unrealizable, similar to solid Brillouin beams (see Figs. 5.4 and 5.5).

According to Eq. (5.93), the potential depression depends on two factors that determine the charge density of the beam: the beam current I and the electron velocity v. When the injection current increases, the potential depression also increases and

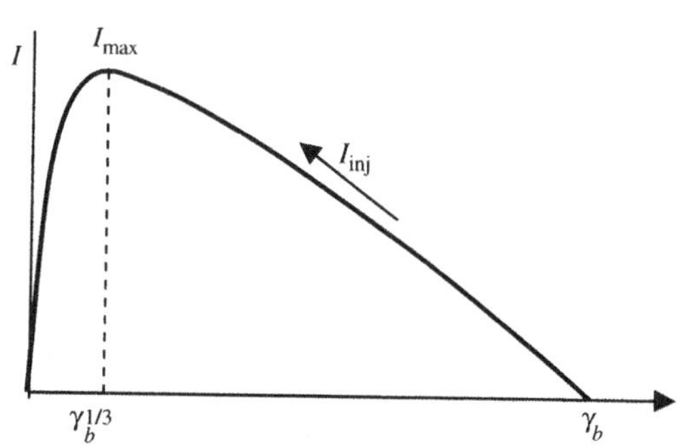

FIGURE 5.7 Current of a thin relativistic beam in the infinity magnetic field as a function of the relativistic factor.

the velocity decreases; that leads to increased charge density, further growth of the potential depression, and so on. Thus, charged particles determine a specific positive (regenerative) feedback in the beam that finally (for $I_{\text{inj}} > I_{\max}$) makes the beam unstable.

Note that it is possible to interpret the instability of the beam on the left branch of the curve $I = I(\gamma_a)$ from the energy minimum principle (Breizman and Ryutov, 1974). It is seen that for each value of the injection current $I_{\text{inj}} < I_{\max}$, there are two states with different beam energies. Because the energy of the electrons is equal to $-e_0\varphi_a$, the state of the beam on the right branch, which answers to a higher potential, has a lower energy. Therefore, of the two branches, only the right branch will be stable.

5.4.2 Cylindrical Solid Beams in an Infinite Magnetic Field

In this case the electrons are injected from an entire circular emitter. In infinite magnetic field they form a cylindrical electron flow (Fig. 5.8). The electron velocity has only the z-component v. The beam is described by the following equations:

$$\gamma(r) = 1 + \frac{\eta\varphi(r)}{c^2} = \left(1 + \frac{v^2}{c^2}\right)^{-1/2} \tag{5.99}$$

$$\Delta\varphi = \frac{1}{r}\frac{d}{dr}\left(r\frac{d\varphi}{dr}\right) = \frac{c^2}{\eta r}\frac{d}{dr}\left(r\frac{d\gamma}{dr}\right) = -\frac{\rho}{\varepsilon_0} \qquad 0 \le r \le a \tag{5.100}$$

$$j = -\rho v \tag{5.101}$$

Equation (5.99) is the energy integral written for zero cathode potential and zero initial velocity. Equations (5.100) and (5.101) are Poisson's equation and the equation for electron current. Assume that the current density $j(r) = \text{const}$. That is a natural sequence of a uniform distribution of the emission current. Then the potential in the region between the beam and the drift tube is [similar to Eq. (5.90) for the tubular beam]

$$\varphi(r) = \varphi_b - \frac{I}{\pi a^2 \varepsilon_0}\ln\frac{b}{r}\int_0^a \frac{r\,dr}{v}, \qquad a \le r \le b \tag{5.102}$$

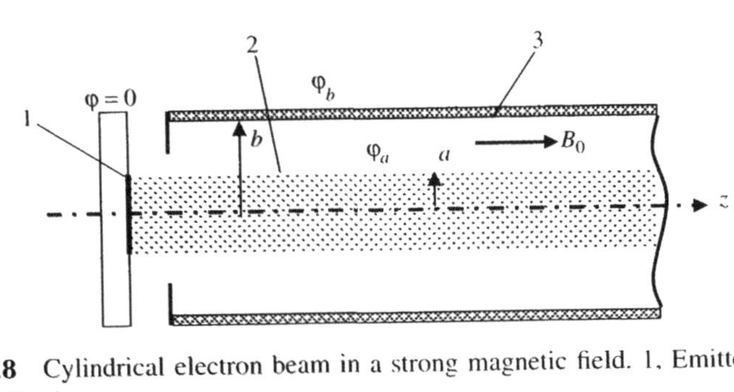

FIGURE 5.8 Cylindrical electron beam in a strong magnetic field. 1, Emitter; 2, electron beam; 3, drift tube.

Equations (5.99)–(5.102) are self-consistent equations for the set of functions $\gamma(r)$, $\rho(r)$, and $v(r)$, or in terms of φ, for the set $\varphi(r)$, $\rho(r)$, and $v(r)$. The current density j and current $I = \pi a^2 j$ in general are arbitrary parameters. Eliminating v and ρ from Eqs. (5.99)–(5.102) reduces these equations to

$$\frac{1}{r}\frac{d}{dr}\left(r\frac{d\gamma}{dr}\right) = \frac{4I}{I_0 a^2}\frac{\gamma}{\sqrt{\gamma^2 - 1}}, \qquad 0 \le r \le a \tag{5.103}$$

$$\gamma(r) = \gamma(b) - \frac{4I}{I_0}\frac{c}{a^2}\ln\frac{b}{r}\int_0^a \frac{r\,dr}{v}, \qquad a \le r \le b \tag{5.104}$$

where $I = \pi a^2 j$. These equations in the nonrelativistic limit ($\rho\varphi/c^2 \ll 1$) are reduced to

$$\begin{aligned} &\frac{1}{r}\frac{d}{dr}\left(r\frac{d\varphi}{dr}\right) = \frac{I}{\pi a^2 \varepsilon_0\sqrt{2\eta}}\varphi^{-3/2}, && 0 \le r \le a \\ &\varphi(r) = \varphi(b) - \frac{I}{\pi a^2 \varepsilon_0\sqrt{2\eta}}\ln\frac{b}{r}\int_0^a \frac{r\,dr}{\varphi^{1/2}}, && a \le r \le b \end{aligned} \tag{5.105}$$

Analytic solutions of Eqs. (5.103) and (5.104) or Eqs. (5.105) in general are complicated, and only numerical integration can be used. The nonrelativistic beams were treated by Smith and Hartman (1940). The typical curves $I = I(\varphi_0/\varphi_b)\,[\varphi_0 \equiv \varphi(0)]$ for different values of b/a are depicted in Fig. 5.9. As we see, these functions have maximums for some $\varphi_0 = \varphi_{0,\max}$. The left branches of the curves $\varphi_0 < \varphi_{0,\max}$ correspond to unstable regimes. In this region, increase in the beam current leads to

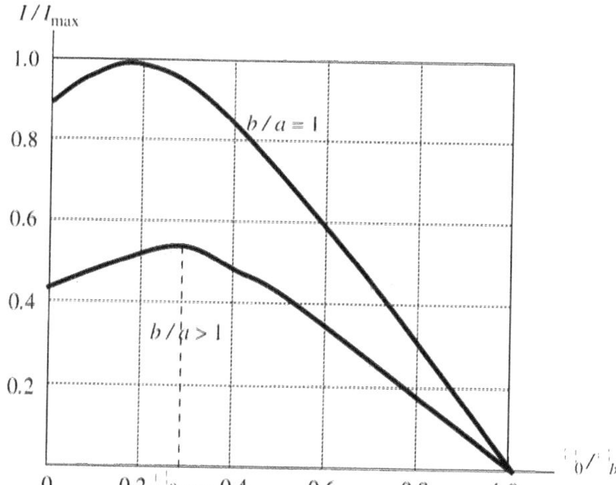

FIGURE 5.9 Currents of a nonrelativistic beam in the infinity magnetic field as functions of an axis potential.

reduction in the potential depression $\varphi_b - \varphi_0$. When an injection current I_{inj} is greater than I_{max}, the axis potential drops to zero and a virtual cathode arises. Between the cathode and the virtual cathode, a reflected beam appears. According to Pierce (1944), in the case of complete filling by the beam of the drift tube, the current for zero axis potential ($\varphi_0 = 0$) is

$$I_{\varphi_0=0} = \frac{16\pi\varepsilon_0\sqrt{2\eta}}{9}\varphi_b^{3/2} \approx 29.33 \times 10^{-6}\varphi_b^{3/2} \quad \text{A} \tag{5.106}$$

The maximum current for complete filling is

$$I_{max} = 1.963\pi\varepsilon_0\sqrt{2\eta}\varphi_b^{3/2} \approx 32.4 \times 10^{-6}\varphi_b^{3/2} \quad \text{A} \tag{5.107}$$

There are situations in which relativistic equation (5.103) has simple solutions: low filling of the drift tube and supreme relativistic beams. In the first case the beam is localized in the small domain near the axis $a/b \ll 1$. Therefore, the potential in the beam $[\varphi(r) \approx \varphi_0]$ is constant, and the right side of Eq. (5.103) is also constant. The integrand into Eq. (5.104) is also constant and that is true for Eqs. (5.105). It is readily seen that formulation of the problem and a solution will exactly coincide with the problem of the thin annular beam (Section 5.4.1). In particular, the limiting currents are determined for relativistic and nonrelativistic beams by Eqs. (5.96) and (5.98), respectively.

For extreme relativistic beams ($\gamma \to \infty$, $v \to c$) the right side of Eq. (5.103) and an integrand into Eq. (5.104) are constant, and the equations become

$$\frac{1}{r}\frac{d}{dr}\left(r\frac{d\gamma}{dr}\right) = \frac{4I}{I_0 a^2} \tag{5.108}$$

$$\gamma_a = \gamma_b - \frac{2I}{I_0}\ln\frac{b}{a} \tag{5.109}$$

It is readily verified that the solution of Eq. (5.108) is

$$\gamma = \frac{Ir^2}{I_0 a^2} + C \tag{5.110}$$

Finding the integration constant C from Eq. (5.109) and substituting a constant into Eq. (5.109), we obtain the solution of Eq. (5.107):

$$\gamma = \frac{Ir^2}{I_0 a^2} + \gamma_b - \frac{I}{I_0}\left(1 + 2\ln\frac{b}{a}\right) \tag{5.111}$$

Then the potential minimum is obtained on the axis:

$$\gamma_0 = \gamma_b - \frac{I}{I_0}\left(1 + 2\ln\frac{b}{a}\right) \tag{5.112}$$

The limiting current corresponds obviously to $\gamma_0 = 1$, which gives

$$I_{\max} = I_0 \frac{\gamma_b - 1}{1 + 2\ln(b/a)} \approx I_0 \frac{\gamma_b}{1 + 2\ln(b/a)} \tag{5.113}$$

It is interesting to note that in the supreme relativistic limit the unstable branch of the current characteristic $I = I(\varphi_0)$ is absent. The maximum of the current is attained for $\gamma_0 = 1$. Equation (5.113) for the near axis beams [$\ln(b/a) \gg 1$] follows directly from Eq. (5.96) if we assume that in the latter, $\gamma_b \gg 1$.

Using an iteration procedure, Bogdankevitch and Rukhadze (1971) have obtained the following formula as a good approximation for the limiting current in all cases:

$$I_{\max} = \frac{I_0}{1 + 2\ln(b/a)} (\gamma_b^{2/3} - 1)^{3/2} \tag{5.114}$$

As is seen, this formula differs from Eq. (5.96) in replacing the coefficient $2\ln(b/a)$ by $1 + 2\ln(b/a)$. Remember that Eq. (5.96) concerns annular flows, whereas Eq. (5.114) is applied entirely to cylindrical flows.

5.5 CENTRIFUGAL ELECTROSTATIC FOCUSING

5.5.1 Introduction

Centrifugal electrostatic focusing of electron beams is based on the existence of stable helical electron orbits in a coaxial cylindrical tube with a nonzero potential difference between coaxial cylinders. Assume that a particle moves on a circular trajectory in the crosssection of a coaxial system (Fig. 5.10). The condition of orbital equilibrium is equal to a sum of radial electric and centrifugal forces of zero:

$$F(r) = F_r + F_c = 0 \tag{5.115}$$

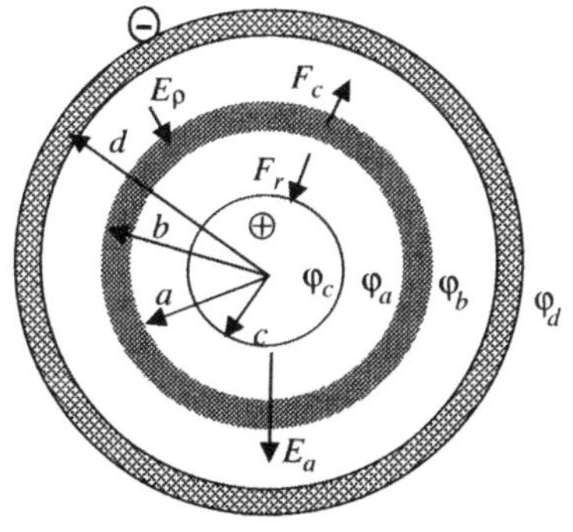

FIGURE 5.10 Cross section of a centrifugal electrostatic focusing system.

The radial electric force acting on an electron in neglecting the space charge is

$$F_r = -\frac{e_0 V}{r \ln(c/a)} = -\frac{N}{r} \tag{5.116}$$

where $V = \varphi_d - \varphi_c$ is the potential difference between electrodes (a voltage). If $\varphi_d > \varphi_c$, $N > 0$ and equilibrium is possible. The centrifugal force is $F_c = m_0\omega^2 r$. The condition of angular momentum conservation in the axially symmetrical system $P_\theta = \omega r^2 = \text{const.}$, leads to the equality $F_c = M/r^3$. So a full radial force is

$$F(r) = \frac{M}{r^3} - \frac{N}{r} \tag{5.117}$$

The orbital equilibrium [Eq. (5.115)] on the radius r_0 requires that $M = Nr_0^2$. Therefore, the radial force is equal to

$$F(r) = N\left(\frac{r_0^2}{r^3} - \frac{1}{r}\right) \tag{5.118}$$

The condition of stability of the equilibrium orbit is

$$\left[\frac{dF(r)}{dr}\right]_{r_0} = -\frac{2N}{r_0^2} < 0 \tag{5.119}$$

We see that since $N > 0$, the equilibrium is always stable.

A method of electron beam formation based on the centrifugal focusing effect was proposed by Harris (1952). The first theory and experimental study of a centrifugal focusing system (the *Harris flow*) are also credited to Harris. A diagram of the Harris flow formation is shown in Fig. 5.11. An electron gun (2) is enclosed in a ferromagnetic yoke (1). The magnetic flow, created by the drive winding (3),

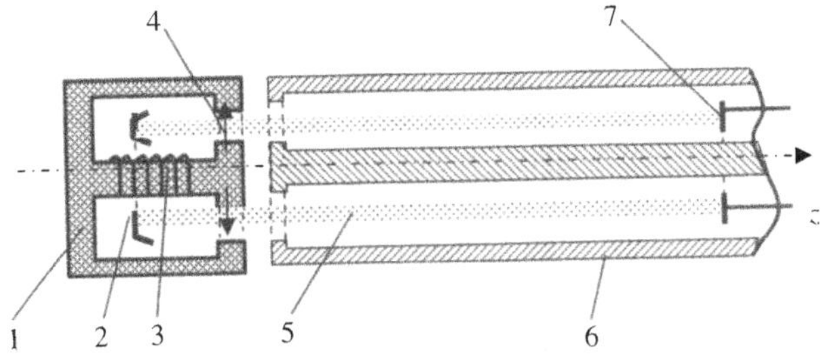

FIGURE 5.11 System forming Harris flow. 1, Ferromagnetic yoke; 2, gun cathode; 3, drive winding; 4, radial magnetic field; 5, electron beam; 6, drift tube; 7, collector.

forms a radial magnetic field (4) in an annular gap of the yoke. The electrons crossing the gap acquire angular velocity and form a rotating electron beam (5) into a drift tube (6). The angular velocity according to Busch's theorem [Eq. (1.128)] is

$$\omega = \dot{\theta} = \frac{\eta}{2\pi r^2}\Psi_c = \frac{A}{r^2} \qquad (5.120)$$

where Ψ_c is a magnetic flux into the central rod of the yoke. We use nonrelativistic approximation because the specific constructive features of a Harris gun do not allow us to use relativistic voltage.

5.5.2 Self-Consistent Centrifugal Focusing of an Annular Electron Beam

The condition of orbital equilibrium according to Eq. (5.118) can be fulfilled only for a single trajectory of a given radius r_0. If the beam has finite thickness, orbits with $r \neq r_0$ would not be circles in the cross section, and the beam inevitably would be rippled. Kirstein et al. (1967) have shown that it is possible for nonrelativistic beams to take the space-charge field into account to obtain self-consistent flow of the finite thickness, with rigorously concentric helical trajectories and with equal axial velocities on any radius.

Let us take for simplicity a beam that fills the system completely. Then according to Fig. 5.10, $c = a$, $d = b$, $\varphi_a = \varphi_c$, and $\varphi_b = \varphi_d$. A self-consistent set of equations for an equilibrium beam is given as

$$e_0 E_r = m_0 \omega^2 r \qquad (5.121)$$

$$2\eta\varphi = v_z^2 + \omega^2 r^2 \qquad (5.122)$$

$$\frac{1}{r}\frac{d}{dr}(r\eta E_r) = \frac{\rho\eta}{\varepsilon_0} = -\omega_p^2 \qquad (5.123)$$

$$P_\theta = m_0 \omega r^2 = \text{const.} \qquad (5.124)$$

Here Eq. (5.121) is the condition of radial equilibrium, Eq. (5.122) is the energy integral, Eq. (5.123) is Poisson's equation, and Eq. (5.124) is the condition of angular momentum conservation. According to Eq. (5.124),

$$\omega = \frac{P_\theta}{m_0 r^2} = \frac{A}{r^2} \qquad (5.125)$$

where the coefficient A is determined according to Eq. (5.120) by the value of the magnetic flux Ψ_c into the ferromagnetic yoke (Fig. 5.11).

Let us limit ourselves to consideration of *isovelocity beams* ($v_z = v_0 = \text{const.}$), which are most important for microwave devices. It can readily be shown that in this case the condition of radial equilibrium is fulfilled on any radius in the flow.

Differentiation of Eq. (5.122) with respect to r, taking Eq. (5.125) into account, gives

$$-2\eta E_r = \frac{d}{dr}(\omega^2 r^2) = -\frac{2A^2}{r^3} = -2\omega^2 r \tag{5.126}$$

for which Eq. (5.121) follows. Note that radial equilibrium is possible only with specific distribution of the space-charge density. Substituting the electric field from Eq. (5.121) into Poisson's equation, and taking Eq. (5.125), into account, we obtain

$$\rho = -\frac{2\varepsilon_0 A^2}{\eta r^4} \tag{5.127}$$

Thus, formation of the equilibrium space-charge beam must provide a distribution of the space-charge density inversely proportional to the fourth power of the radius.

Find a beam current

$$\begin{aligned} I &= \int_a^b 2\pi rj\,dr = -2\pi \int_a^b r\rho v_0\,dr = \frac{4\pi\varepsilon_0 A^2 v_0}{\eta}\int_a^b \frac{dr}{r^3} \\ &= \frac{2\pi\varepsilon_0 A^2 v_0}{\eta}\left(\frac{1}{a^2} - \frac{1}{b^2}\right) \end{aligned} \tag{5.128}$$

Accordingly to the energy integral (5.122) and Eq. (5.125), the axial velocity is

$$v_0 = \sqrt{2\eta\varphi_a - \frac{A^2}{a^2}} \tag{5.129}$$

Then the beam current is

$$I = \frac{2\pi\varepsilon_0}{\eta}\left(\frac{1}{a^2} - \frac{1}{b^2}\right)A^2\sqrt{2\eta\varphi_a - \frac{A^2}{a^2}} \tag{5.130}$$

As we see, the beam current as the function of A that according to Eq. (5.120) determines the angular velocity ω has a maximum. For small ω the electric field that supports the radial equilibrium is also small. It leads to smallness of the space-charge density [Eq. (5.127)] and beam current. To the contrary, for a large value of ω (coefficient A), the radial energy increases, and the axial velocity decreases with the corresponding drop in beam current. The optimal value of A corresponding to the condition $\partial I/\partial A = 0$ is

$$A_{\text{opt}}^2 = \tfrac{4}{3}\eta\varphi_a a^2 \tag{5.131}$$

Substitution of this value into Eq. (5.130) gives the maximum current:

$$I_{\max} = \frac{8\pi\varepsilon_0\sqrt{6\eta}}{9}\left(1 - \frac{a^2}{b^2}\right)\varphi_a^{3/2} \equiv P_{\max}\varphi_a^{3/2} \approx 25.4 \times 10^{-6}\varphi_a^{3/2} \quad \text{A} \tag{5.132}$$

where P_{max} is a corresponding perveance. In this case the electron rotational energy on the inner cylinder is equal to $(m_0/2)\omega_a^2 a^2 = \frac{2}{3}e_0\varphi_a$ (i.e., twice as much as the axial energy of the electrons). On most radiuses the energy of rotation decreases as a^2/r^2, and a fraction of the axial energy grows. It can readily be shown that the ratio of the axial and rotational energy in the regime $I = I_{max}$ is $r^2/2a^2$.

5.5.3 Electron Guns Formatting Harris Flow

As we have seen, a specific feature of Harris flow is the very heterogeneous radial distribution of the space-charge density $\sim 1/r^4$ [Eq. (5.127)]. If a gun in the ferromagnetic yoke does not provide a similar distribution of charge density, the unrippled structure of the flow is inevitably disturbed.

Harris–Waters Gun Waters (1959) has developed a gun with a tubular azimuthal electron flow in a spherical diode. This gun has a conical cathode similar to that of a magnetron injection gun. But this gun is purely electrostatic. The gun provides adequate distribution of the space-charge density $\sim 1/r^4$. A diagram of this Harris–Waters gun is depicted in Fig. 5.12. The system forms a hollow beam. In principle, Eq. (5.132) is used to estimate the limiting current in the hollow beam, at although the inner potential φ_a of the beam is not given directly. However, it is not difficult to express the potential φ_a through given values of φ_c and φ_d for the maximum current. Let us write the voltage as

$$V = \varphi_c - \varphi_d = \frac{Q}{2\pi\varepsilon_0}\ln\frac{a}{c} + \Delta\varphi + \frac{Q}{2\pi\varepsilon_0}\ln\frac{d}{b} - \frac{I_{max}}{2\pi\varepsilon_0 v_0}\ln\frac{d}{b} \tag{5.133}$$

Here Q is the surface charge on an inner electrode per unit of length. According to Eqs. (5.129), (5.131), and (5.125), for the maximum beam current

$$v_0^2 = 2\eta\varphi - \frac{A_{opt}^2}{a^2} = \frac{2}{3}\eta\varphi_a \tag{5.134}$$

The potential difference on the beam

$$\Delta\varphi = \varphi_a - \varphi_b = \frac{A_{opt}^2}{a^2}\left(1 - \frac{a^2}{b^2}\right) = \frac{2}{3}\varphi_a\left(1 - \frac{a^2}{b^2}\right) \tag{5.135}$$

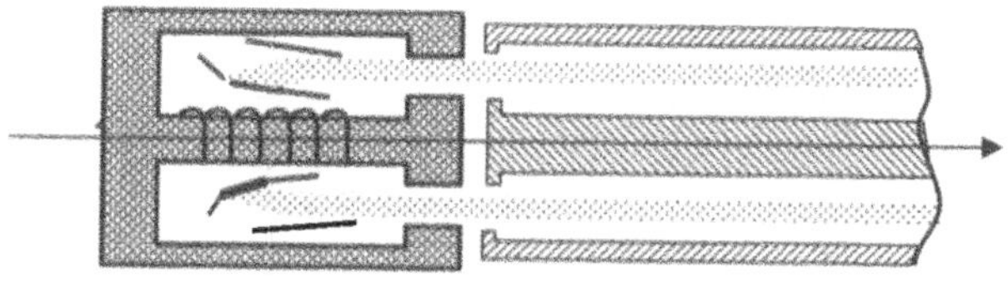

FIGURE 5.12 Harris–Waters gun formatting a centrifugal focusing electron beam.

Substituting v_0, $\Delta\varphi$, and I_{max} from Eqs. (5.134), (5.135), and (5.132), respectively, into Eq. (5.133), we obtain

$$V = \frac{Q}{2\pi\varepsilon_0}\ln\frac{ad}{bc} + \varphi_a\frac{2}{3}\left(1 - \frac{a^2}{b^2}\right)\left(1 - 2\ln\frac{d}{b}\right) \tag{5.136}$$

Let us express $Q/2\pi\varepsilon_0$ from Eq. (5.136) through V and φ_a. Then we can write

$$\varphi_a = \varphi_c - \frac{Q}{2\pi\varepsilon_0}\ln\frac{a}{c}$$

Finally,

$$\varphi_a = \left[\varphi_c - V\frac{\ln(c/a)}{\ln(ad/bc)}\right]\frac{1}{K} \tag{5.137}$$

where

$$K = 1 - \frac{2}{3}\left(1 - \frac{a^2}{b^2}\right)\left(1 - 2\ln\frac{d}{b}\right)\frac{\ln(a/c)}{\ln(ad/bc)}$$

Using potentials of the electrodes φ_c and φ_d and dimensions of the system a,b,c,d, we find the potential φ_a and current I_{max} according to Eqs. (5.137) and (5.132). Applying Eq. (5.131) and Busch's theorem [Eq. (5.120)], we also obtain a necessary magnetic flux in the ferromagnetic yoke:

$$\Psi_c = 4a\sqrt{\frac{\varphi_a}{3\eta}} \tag{5.138}$$

Chernov Gun Chernov (1956) developed an electrostatic screw electron gun for formation of Harris flow without a ferromagnetic yoke. A diagram of this gun is shown in Fig. 5.13. Harris flow with helical electron trajectories into a coaxial system is formatted. However, unlike a Harris gun, rotary flow is created by an electrostatic diode system of screw electrodes. The faces of the electrodes have spiral sections (Fig. 5.13), so their configuration is similar to the geometry of single-threated turbine screws. The advantage of a Chernov gun is the absence of the complicated magnetic system that must be mounted close to hot electrodes inside

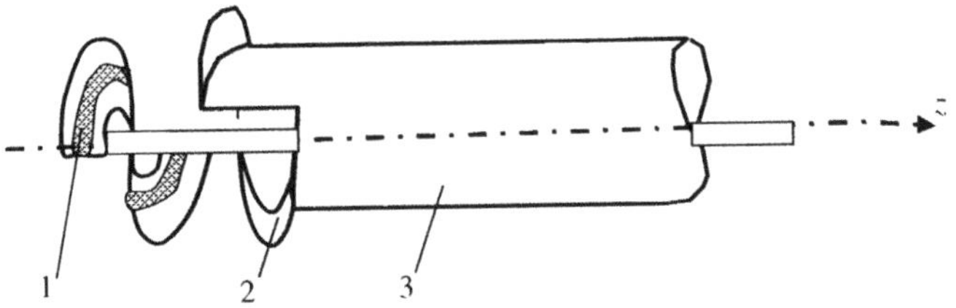

FIGURE 5.13 Chernov gun formatting a centrifugal focusing electron beam. 1, Cathode; 2, anode; 3, low potential cylinder.

a Harris–Webster gun. Experimentally, Chernov (1956) obtained sharply bounded electron beams with a current density up to 0.4 A/cm^2 and diode and focusing voltages of 800 V and 300 V, respectively. The system was used in TWTO microwave amplifiers (spiratrons) in the frequency range 100 to 4500 MGz.

It is important to note that thermionic cathodes of guns in any centrifugal focusing electron beam devices are protected from bombardment by positive ions that arise during the ionization of residual gases. The radial electric field in a transportation area makes it possible to draw off the ions going to the low-potential cylinder (3) (Fig. 5.13). Bombardment of the cathode is carried out by ions only created inside the gun.

5.6 PARAXIAL-RAY EQUATIONS OF AXIALLY SYMMETRIC LAMINAR BEAMS

As we have seen, the formation of unrippled self-consistent beams is difficult. For non-Brillouin beams, only cylindrical cathodes are available. The formation of Brillouin beams requires ideal magnetic isolation of the cathode. And finally, unrippled beams have to be launched into a uniform magnetic field rigorously along cylindrical surfaces with assumed distribution of the current or space-charge densities. Obviously, the violation of ideal conditions disturb the radial equilibrium, and oscillating trajectories are developed. However, if the amplitudes of the oscillations are small, these mismatched beams can be used successfully.

The small amplitude of trajectory oscillations allows us to employ the paraxial theory of beams. However, that is only one of the conditions of paraxial theory validity. Another condition is small thickness of the beams (Section 2.3). Earlier (Chapter 2), paraxial equations were obtained by negelcting the selffields. Below, we derive paraxial equations by taking into account self-electric and self-magnetic fields. Examples of application of these equations are also considered.

5.6.1 Paraxial-Ray Equations of Axially Symmetric Hollow Beams with Rectilinear Axis[1]

The differential equation of relativistic trajectories in an axially symmetric field [Eq. (2.22)] taking the self magnetic field into account must include in general all components of the magnetic vector potential. Therefore, in Eq. (2.12) we must assume that

$$\frac{\mathbf{A}\,\mathbf{dl}}{dz} = A_\theta r\theta' + A_r r' + A_z \tag{5.139}$$

So the trajectory differential equation (2.22) will include an additional term:

$$c\frac{\partial A_r}{\partial r}r' + c\frac{\partial A_z}{\partial r} \tag{5.140}$$

[1]General theory of paraxial space-charge beams with curvilinear axis has been developed in the book of Syrovoy, 2004 (in Russian).

For the paraxial trajectory we can neglect A_r, and according to Eq. (A3.8) from Appendix 3, we can write the additional component as

$$c\frac{\partial A_z}{\partial r} = -\frac{\mu_0 c I_r}{2\pi r} \tag{5.141}$$

where I_r is the beam current within the tube of the radius $r = r(z)$. It is also necessary to change Eq. (2.29) in the paraxial equation, because in taking the self electric field into account we must use instead of Eq. (2.28) the equation

$$\operatorname{div}\mathbf{E} = \frac{1}{r}\frac{\partial(rE_r)}{\partial r} + \frac{dE_z(0,z)}{dz} = \frac{\rho}{\varepsilon_0} \tag{5.142}$$

Integration of this equation on the cross section of a current tube of a hollow beam (Fig. 5.14) gives

$$\int_{r_0}^{r} \frac{\partial}{\partial r}(rE_r)\,dr = -\int_{r_0}^{r} r\frac{dE_z}{dz}\,dr + \frac{1}{\varepsilon_0}\int_{r_0}^{r} \rho r\,dr \tag{5.143}$$

Here $dE_z/dz = -\varphi''(0,z) \equiv -\varphi''$. After integration we find that

$$E_r = -\frac{r_0}{r}E_{\text{in}} + \frac{r^2 - r_0^2}{2r}\varphi'' + \frac{1}{\varepsilon_0 r}\int_{r_0}^{r} \rho r\,dr \tag{5.144}$$

where $E_{\text{in}} = E_r(r_0)$ is the electric field into the beam, which can be nonzero if there is a charged cylinder inside the beam (Fig. 5.14); $r_0 = r_0(z)$ is an inner radius of the beam. The integral on the right side of Eq. (5.144) is expressed in paraxial approximation through the beam current:

$$I_r = -\int_{r_0}^{r} \rho v_z \cdot 2\pi r\,dr = -2\pi c\beta_z \int_{r_0}^{r} \rho r\,dr \tag{5.145}$$

so

$$E_r = -\frac{r_0}{r}E_{\text{in}} + \frac{r^2 - r_0^2}{2r}\varphi'' - \frac{I_r}{2\pi c\varepsilon_0 r\beta_z} \tag{5.146}$$

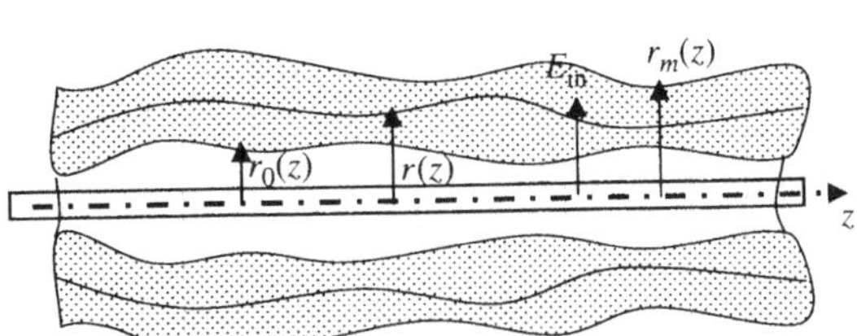

FIGURE 5.14 Hollow axially symmetrical electron beam.

Using Eq. (5.146), the term $\partial\Phi/\partial r$ can be written instead of Eq. (2.29*a*) as

$$\frac{\partial\Phi}{\partial r} = \frac{r_0}{r\beta_z}E_{\text{in}} - \left(1 - \frac{r_0^2}{r^2}\right)\frac{r}{2\beta_z}\varphi'' + \frac{I_r}{2\pi c\varepsilon_0 r\beta_z^2} \tag{5.147}$$

Then the radial paraxial equation (2.36*a*) becomes

$$\frac{\Phi_z^2}{\varphi_z^* + \varphi_0}r'' + \varphi' r' - \frac{r_0}{r}E_{\text{in}} + \left[\left(1 - \frac{r_0^2}{r^2}\right)\frac{\varphi''}{2} + \frac{c^2 B(z)^2}{4(\varphi_z^* + \varphi_0)}\right]r$$
$$- \frac{c^2 Q_1^2}{4(\varphi_z^* + \varphi_0)}\frac{1}{r^3} - \Xi\beta_z = 0 \tag{5.148}$$

where $Q_1 = 2P_\theta/e_0$ and Ξ is the sum of the additional components determined by the self-fields [Eqs. (5.141), (5.146)]:

$$\Xi = \frac{I_r}{2\pi c r\varepsilon_0\beta_z^2} - \frac{\mu_0 c I_r}{2\pi r} = \frac{I_r}{I_0}\frac{2\varphi_0}{\beta_z^2\gamma^2 r} \tag{5.149}$$

where I_0 is the relativistic current. The azimuthal paraxial equation does not depend on the self fields and coincides with Eq. (2.32). The latter and the terms γ and P_θ in Eq. (5.149) [see also Eq. (2.34)] are

$$\theta' = \frac{\eta B}{2\beta_z c\gamma} + \frac{P_\theta}{\beta_z c\gamma m_0}\frac{1}{r^2}$$

$$r'' + \frac{\gamma'}{\beta_z^2\gamma}r' + \left[\left(1 - \frac{r_0^2}{r^2}\right)\frac{\gamma''}{2\beta_z^2\gamma} + \left(\frac{\eta B}{2\beta_z c\gamma}\right)^2\right]r - \frac{K_r}{r} - \left(\frac{P_\theta}{\beta_z c\gamma m_0}\right)^2\frac{1}{r^3} = 0 \tag{5.150}$$

where

$$K_r = \frac{2I_r}{I_0\beta_z^3\gamma^3} \tag{5.151}$$

In nonrelativistic approximation ($\beta_z = \sqrt{2\eta\varphi_z}/c$, $\gamma = 1$), the coefficient K_r is reduced to the perveance term:

$$K_r^{(nr)} = \frac{I_r}{4\pi\varepsilon_0\sqrt{2\eta}\,\varphi_z^{3/2}} = \frac{P_r}{4\pi\varepsilon_0\sqrt{2\eta}} \approx 1.515 \times 10^4 \frac{I_r}{(\varphi_z^*)^{3/2}} \quad \text{A/V}^{3/2} \tag{5.152}$$

The coefficient K_r is called a *generalized perveance*. According to Lawson (1988), if a background of neutralizing ions is presented, the generalized perveance K_r can be expressed as

$$K_r = \frac{I_r}{I_0}\frac{2}{\beta_z^3\gamma^3}(1-\gamma^2 f) \tag{5.153}$$

where f is the ratio of the neutralizing charge to the beam charge. Note we can replace a relativistic factor γ in the paraxial equations on the axial factor $\gamma_0 = (\gamma)_{r=0}$. This follows from Eq. (5.65).

A nonrelativistic radial paraxial equation is obtained from Eq. (5.150) and differs from Eq. (2.38) by additional terms which take into account the space-charge field and the hollow structure of the beam:

$$r'' + \frac{\varphi_z'}{2\varphi_z^*}r' + \left[\left(1-\frac{r_0^2}{r^2}\right)\frac{\varphi_z''}{4\varphi_z} + \frac{\eta B^2}{8\varphi_z^*}\right]r - \frac{P_r}{4\pi\varepsilon_0\sqrt{2\eta}}\frac{1}{r}$$
$$-\frac{P_\theta^2}{2e_0 m_0 \varphi_z^*}\frac{1}{r^3} = 0 \tag{5.154}$$

Assume that the cathode angular velocity $\theta_c = 0$ and that the cathode magnetic field is uniform. Then according to Eq. (1.124), the azimuthal generalized momentum is given as

$$P_\theta = \frac{e_0}{2\pi}\Psi_c = \frac{e_0 r_c^2}{2}B_c \tag{5.155}$$

In this case the last terms in Eqs. (5.150), (5.154) are, respectively,

$$\left(\frac{r_c^2\eta B_c}{2\beta_z c\gamma}\right)^2\frac{1}{r^3}, \qquad \frac{\eta r_c^4 B_c^2}{8\varphi_z^*}\frac{1}{r^3} \tag{5.156}$$

It is important that for a laminar beam, the trajectory $r = r(z)$ is the envelope equation for a beam of radius r. Therefore, I_r as a magnitude of the beam current within a tube of radius r is constant (i.e., does not depend on z).

5.6.2 Paraxial-Ray Equations of Axially Symmetric Solid Beams

These equations are obtained from Eqs. (5.150) and (5.154) setting $r_0 = 0$. If we assume that $\theta_c = 0$ and uniformity of the cathode magnetic field, the radial paraxial equations according to Eq. (5.156) would be

$$r'' + \frac{\gamma_0'}{\beta_z^2\gamma_0}r' + \left[\frac{\gamma_0''}{2\beta_z^2\gamma_0} + \left(\frac{\eta B}{2\beta_z c\gamma_0}\right)^2\right]r$$
$$-\frac{K_r}{r} - \left(\frac{r_c^2\eta B_c}{2\beta_z c\gamma_0}\right)^2\frac{1}{r^3} = 0 \qquad \text{(relativistic)} \tag{5.157}$$

$$r'' + \frac{\varphi_z'}{2\varphi_z^*} r' + \left(\frac{\varphi''}{4\varphi_z^*} + \frac{\eta B^2}{8\varphi_z^*} \right) r$$
$$- \frac{P_r}{4\pi\varepsilon_0 \sqrt{2\eta}} \frac{1}{r} - \frac{r_c^4 \eta B_c^2}{8\varphi_z^*} \frac{1}{r^3} = 0 \quad \text{(nonrelativistic)} \tag{5.158}$$

Let us consider the interesting case of a laminar beam with uniform charge density $\rho = \text{const}$. In this case, according to Eq. (5.145),

$$I_r = I \frac{r^2}{r_m^2} \tag{5.159}$$

where I is the full beam current and $r_m = r_m(z)$ is an external radius of the beam (Fig. 5.14). Then the perveance and the generalized perveance are

$$P_r = P \frac{r^2}{r_m^2}, \qquad K_r = K \frac{r^2}{r_m^2} \tag{5.160}$$

where

$$P = \frac{I}{(\varphi_z^*)^{3/2}}, \qquad K = \frac{I}{I_0} \frac{2}{\beta_z^3 \gamma_0^3} (1 - \gamma^2 f) \tag{5.161}$$

It is can readily be shown that in nonrelativistic approximation,

$$K = \frac{P}{4\pi\varepsilon_0 \sqrt{2\eta}} (1 - f) \tag{5.161a}$$

The values I and $r_m = r_m(z)$ are the same for all trajectories of the beam. Substituting Eq. (5.160) into Eq. (5.157) and Eq. (5.155) into the first of Eqs. (5.150), we can write the paraxial relativistic equations as

$$\theta' = \frac{\eta B}{2\beta_z c\gamma_0} + \frac{\eta B_c}{2\beta_z c\gamma_0} \frac{r_c^2}{r^2} \tag{5.162}$$

$$\frac{r''}{r_c} + \frac{\gamma_0'}{\beta_z^2 \gamma_0} \frac{r'}{r_c} + \left[\frac{\gamma_0''}{2\beta_z^2 \gamma_0} + \left(\frac{\eta B}{2\beta_z c\gamma_0} \right)^2 \right] \frac{r}{r_c} - \frac{K}{r_m^2} \frac{r}{r_c} - \left(\frac{\eta B_c}{2\beta_z c\gamma_0} \right)^2 \frac{r_c^3}{r^3} = 0 \tag{5.163}$$

An analogous system can be obtained for the nonrelativistic equation (5.158). We see that the trajectories determined by these equations are functions of the ratio r/r_c, which means that all paraxial trajectories in a laminar beam with uniform space-charge density distribution are similar, beginning with the cathode.

5.6.3 Expansion of a Laminar Beam in a Uniform Drift Tube

Let us consider the expansion of a free electron beam in a drift tube as an example of the use of paraxial-ray equations. The electron beam enters the tube with beginning radius $(r)_{z=0} = r_0$ and slope $(r')_{z=0} = 0$. Assume that the external electric and magnetic fields are absent, so that in Eq. (5.157), $\gamma' = \gamma'' = B = B_c = 0$. We obtain the following equation for any trajectory in the laminar beam:

$$r'' - \frac{K_r}{r} = 0 \tag{5.164}$$

The generalized perveance K_r is determined by Eq. (5.153). Since $\gamma = \text{const.}$, the perveance is constant. In principle, K_r can have any sign, depending on the magnitudes of γ and f. Assume for simplicity that $f = 0$. Then $K_r > 0$. The trajectory $r = r(z)$ can be considered as a boundary of the beam. Note that the equation of a beam envelope, $r_m = r_m(z)$, can be obtained if to take $r = r_m$ in Eq. (5.163). We would see that this equation coincides with Eq. (5.164) if we replace r by r_m.

Before integrating Eq. (5.164), let us introduce the dimensionless variables

$$R = r/r_0, \qquad Z = \sqrt{2K_r}\frac{z}{r_0} \tag{5.165}$$

Equation (5.164) becomes

$$\frac{d^2}{dZ^2} = R'' = \frac{1}{2R} \tag{5.166}$$

The first integral of this equation with $(R)_{Z=0} = 1$ and $(R')_{z=0} = 0$ after multiplying on R' is equal to

$$\frac{dR}{dZ} = R'^2 = \ln R \tag{5.167}$$

Let us enter the auxiliary function

$$\Phi = \sqrt{\ln R} \tag{5.168}$$

Then $R' = \Phi$. According to Eq. (5.168), $R = e^{\Phi^2}$. Differentiating this equation with respect to Z, we obtain

$$R' = \Phi = 2\Phi\Phi' e^{\Phi^2} \tag{5.169}$$

Canceling Φ out of Eq. (5.169) and integrating with respect to Z and Φ gives

$$Z = 2\int_0^{\Phi} e^{\Phi^2}\, d\Phi = 2\int_0^{\sqrt{\ln R}} e^{\Phi^2}\, d\Phi \tag{5.170}$$

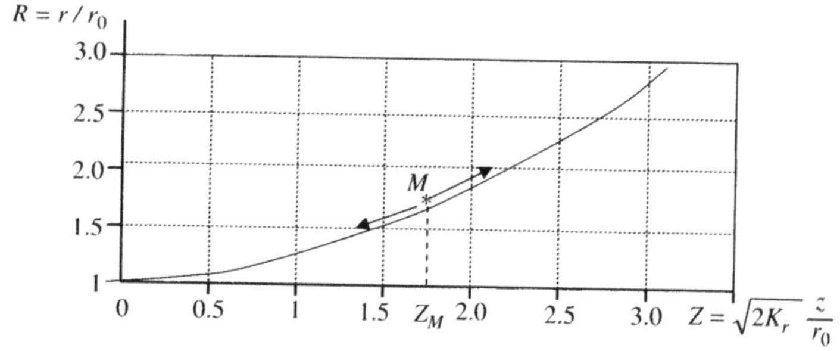

FIGURE 5.15 Universal beam extension curve.

The low limit of the integral for $R = 1$ is found from Eq. (5.168). In the literature the integral

$$D(x) = e^{-x^2}\int_0^x e^{t^2}\,dt \tag{5.171}$$

is known as the *Doson integral*. Tabulated values of the integral can be found in Abramowitz and Stegun (1972).

A function $R = R(Z)$ covers the dependence $r = r(z)$ for all possible values of the beam parameter. The corresponding universal beam extension curve is shown in Fig. 5.15. Note that in the vicinity of the point $Z = 0$, $R \approx 1$ and $R' \ll 1$. In this region the upper limits of the integral in Eq. (5.170) are $\sqrt{\ln R} \approx \sqrt{R-1}$ and $\Phi \ll 1$. Therefore, the approximation of Eq. (5.170) is $Z \approx 2\sqrt{R-1}$, and the function $R(z)$ is the parabola

$$R = 1 + \frac{1}{4}Z^2 = 1 + \frac{K_r}{2}\left(\frac{z}{r_0}\right)^2 \tag{5.172}$$

The curve in Fig. 5.15 allows us to determine the profile of the beam for any initial conditions and any perveances. Let us, for example, assign the perveance, initial beam slope r', and radius r_0. We obtain $R' = r'/r_0$ and according to Eq. (5.167), $R = e^{R'^2}$ and $r = Rr_0$. Using R we can find a corresponding point M on the curve in Fig. 5.15. The abscissa Z_M for the given perveance determines the coordinate $z = r_0 Z_M/\sqrt{2K_r}$. Now, moving point M in any direction along the curve, we obtain the current values of Z and R (i.e., a beam profile with a positive or negative slope). For a negative slope the curve determines the profile until the point $z = 0$, where the beam radius is minimal. The profile of the beam in $Z < 0$ is obtained as a symmetric continuation of the curve from $Z > 0$.

5.6.4 Transfer of a Maximal Beam Current Through a Drift Tube Without External Fields

In Fig. 5.16 a drift tube of length L and radius r_T is depicted. Trajectories of the particles with initial slope too small and too large are shown. As we see in both cases, the particles are quickly intercepted by the tube wall.

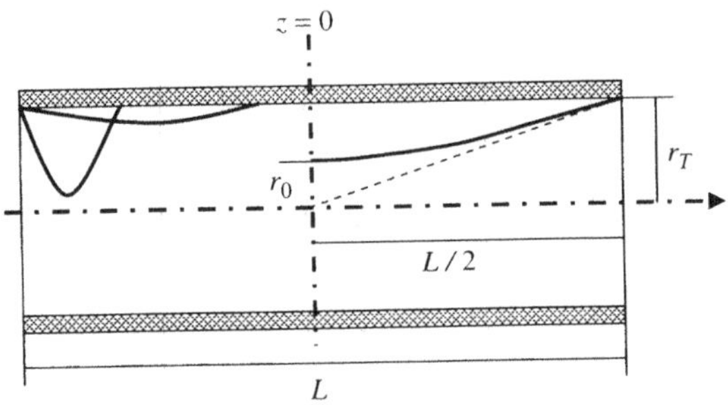

FIGURE 5.16 Transfer of a beam through a drift tube.

The following setting of the problem is convenient. Place the origin of coordinates $z = 0$ in the middle plane of the tube. It can be assumed that the beam profile is symmetric about a middle plane of the tube. Then we can choose the origin coordinate in this plane and require that the boundary trajectory touch the tube in the plane $z = \pm L/2$, (i.e., $r_{z=\pm L/2} = r_T$; Fig. 5.16). According to Eqs. (5. 170) and (5.165), we have

$$\frac{L}{2r_0}\sqrt{2K_r} = 2\int_0^{\Phi_T} e^{t^2}\,dt \tag{5.173}$$

where $\Phi_T = \sqrt{\ln(r_T/r_0)}$. r_0 is the beam radius in the plane $z = 0$. Let us write Eq. (5.173) as

$$\frac{L}{4r_T}\sqrt{2K_r} = e^{-\Phi_T^2}\int_0^{\Phi_T} e^{t^2}\,dt \tag{5.174}$$

This formula allows us to find the beam perveance for any given parameters L, r_0, and r_T.

The condition of the maximum K_r is

$$\frac{d}{d\Phi_T}\left(e^{-\Phi_T^2}\int_0^{\Phi_T} e^{t^2}\,dt\right) = 0$$

This condition gives

$$e^{-\Phi_T^2}\int_0^{\Phi_T} e^{t^2}\,dt = \frac{1}{2}\Phi_T \tag{5.175}$$

The numerical solution of this equation is $(\Phi_T)_{\text{opt}} = 0.924$. That gives us $r_T/r_0 = e^{(0.924)^2} \approx 2.35$. The extreme generalized perveance according to Eqs. (5.174) and (5.175) is

$$(K_r)_{\text{ext}} = \frac{2}{\Phi_T^2}\left(\frac{r_T}{L}\right)^2 \approx 2.34\left(\frac{r_T}{L}\right)^2 \tag{5.176}$$

The nonrelativistic perveance is [see Eq. (5.161*a*)]

$$P_{\text{ext}} = \frac{8\pi\varepsilon_0\sqrt{2\eta}}{\Phi_T^2}\left(\frac{r_T}{L}\right)^2 \approx 155\times 10^{-6}\left(\frac{r_T}{L}\right)^2 \tag{5.177}$$

Corresponding values of the beam current are

$$I_{\text{ext}} = \begin{cases} 1.17 I_0 \beta^3\gamma^3\left(\dfrac{r_T}{L}\right)^2 \quad \text{kA} & \text{(relativistic)} \quad (5.178) \\ 155\times 10^{-6}(\phi_{\_}^*)^{3/2}\left(\dfrac{r_T}{L}\right)^2 \quad \text{A} & \text{(nonrelativistic)} \quad (5.179) \end{cases}$$

As we see, the current diminishes very quickly with increased drift length L. So for $L/r_T = 10$, $\phi_{\_}^* = 1000$ V and $I_{\text{ext}} \approx 50$ mA.

A slope of the beam boundary trajectory in the plane $z = L/2$ according to Eqs. (5.165) and (5.169) is

$$\frac{dr}{dz} = \sqrt{2K_r}\left(\frac{dR}{dZ}\right)_{z=L/2} = \sqrt{2K_r}\,\Phi_T \tag{5.180}$$

Substitution $\sqrt{2K_r}$ from Eq. (5.174) gives

$$\frac{dr}{dz} = \frac{4r_T}{L}\Phi_T e^{-\Phi_T^2}\int_0^{\Phi_T} e^{t^2}\,dt \tag{5.181}$$

Taking Eq. (5.175) into account, we find that

$$\frac{dr}{dz} = \frac{r_T}{L/2} \tag{5.182}$$

Thus, to provide passage of the beam through the drift tube with maximum perveance, a tangent to the beam boundary trajectory in an input plane of the tube must be directed to the tube center (Fig. 5.16).

5.7 AXIALLY SYMMETRIC PARAXIAL BEAMS IN A UNIFORM MAGNETIC FIELD WITH ARBITRARY SHIELDING OF A CATHODE MAGNETIC FIELD

In Section 5.6 we analyzed the transport of a paraxial beam in a drift tube without external fields as an example of the use of paraxial equations. Below we add to this model an external uniform magnetic field. In general, trajectories of those beams, unlike the foregoing, will oscillate and the beam will be rippled. The problem is to find the conditions for sufficiently small rippling of the beams for a given initial perturbation. Solution of this problem is always important in designs and experiments.

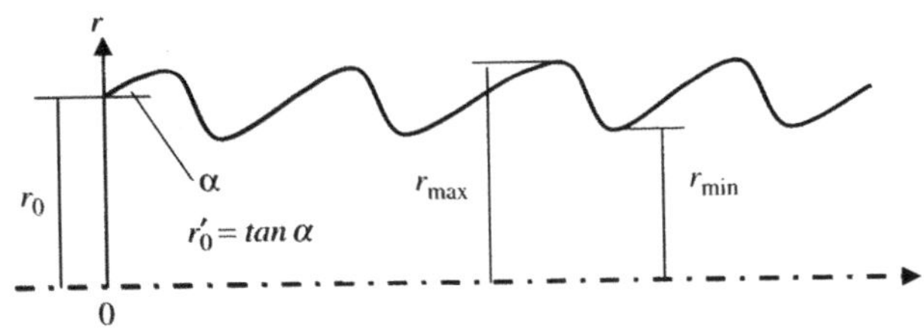

FIGURE 5.17 Electron trajectory in a uniform magnetic field.

5.7.1 Paraxial-Ray Equation of an Axially Symmetric Laminar Beam in a Uniform Magnetic Field and Its First Integral

For a zero external electric field we can assume that $\gamma = \gamma_0 = \text{const.}$, in the relativistic equation (5.157). We obtain the following paraxial-ray equation:

$$r'' + \left(\frac{\eta B}{2\beta_z c\gamma_0}\right)^2 r - \frac{K_r}{r} - \left(\frac{r_c^2 \eta B_c}{2\beta_z c\gamma_0}\right)^2 \frac{1}{r^3} = 0 \qquad (5.183)$$

The nonrelativistic equation differs from Eq. (5.183) only by the coefficients:

$$\gamma \rightarrow 1, \qquad \beta_z^2 c^2 \rightarrow 2\eta\varphi_z^*, \qquad K_r \rightarrow \frac{P_r}{4\pi\varepsilon_0\sqrt{2\eta}} \qquad (5.184)$$

After multiplication of Eq. (5.183) by $2r'$, we obtain the following first integral:

$$r'^2 = r_0'^2 + \left(\frac{\eta B}{2\beta_z c\gamma_0}\right)^2 (r_0^2 - r^2) + 2K_r \ln\frac{r}{r_0} + \left(\frac{r_c^2 \eta B_c}{2\beta_z c\gamma_0}\right)^2 \left(\frac{1}{r_0^2} - \frac{1}{r^2}\right) \qquad (5.185)$$

where r_0 and r'_0 are initial values of r and r', respectively. The motion determined by Eq. (5.185) is finite. As is readily verified, an increase in the radius by more than some value of r_{max} or a decrease by less than some value of r_{min} results in $r'^2 < 0$, so that $r_{min} \le r \le r_{max}$ (Fig. 5.17).

5.7.2 Equilibrium Radius of an Electron Beam

The equilibrium radius is a beam radius r_e on which $r''_e = 0$ (i.e., all forces acting on the beam are balanced). According to Eq. (5.183), the following equation determines the equilibrium radius:

$$r_e^4 - K_r \left(\frac{2\beta_z c\gamma_0}{\eta B}\right)^2 r_e^2 - r_c^4 \left(\frac{B_c}{B}\right)^2 = 0 \qquad (5.186)$$

The corresponding equation for a nonrelativistic beam differs by a coefficient less than r_e^2, which according to Eq. (5.184) is

$$\frac{\sqrt{2}\,P_r\varphi_z^*}{\pi\varepsilon_0\eta^{3/2}B^2} \tag{5.187}$$

Let us consider two particular cases:

1. *Zero current*, $I_r = K_r = P_r = 0$. Denoting the equilibrium radius for this case as $r_e = r_a$, we obtain from Eq. (5.186)

$$r_a = r_c\left(\frac{B_c}{B}\right)^{1/2} \tag{5.188}$$

According to this relation, magnetic fluxes through the beam cross section on the cathode and on a circle with radius r_a are equal (i.e., a particle on the equilibrium radius beginning at the cathode moves along the magnetic flux surface).

2. *The cathode is shielded from the magnetic field* $B_c = 0$, but $I_r \neq 0$. Denote the equilibrium radius as $r_e = r_b$. According to Eq. (5.186),

$$r_b^2 = K_r\left(\frac{2\beta_z c\gamma_0}{\eta B}\right)^2 = \frac{I_r}{I_0}\frac{8c^2}{\eta^2B^2\beta_z\gamma_0} \tag{5.189}$$

In nonrelativistic approximation using Eq. (5.184), we obtain

$$r_b^2 = P_r\frac{\sqrt{2}}{\pi\varepsilon_0\eta^{3/2}B^2}\varphi_z^* = I_r\frac{\sqrt{2}}{\pi\varepsilon_0\eta^{3/2}B^2\sqrt{\varphi_z^*}} \tag{5.190}$$

Let us compare the equilibrium radius [Eq. (5.189)] with the radius of the equilibrium (nonrippled) solid Brillouin beam. According to Eq. (5.189),

$$I_r = I_0\frac{\eta^2B^2r_b^2\beta_z\gamma_0}{8c^2} \tag{5.191}$$

On the other hand, according to Eq. (5.73), the current of a solid equilibrium relativistic beam is

$$I_r = I_0\frac{\beta_z\gamma_0}{2}\left(\frac{\gamma_r^2}{\gamma_0^2} - 1\right)$$

Substituting

$$\frac{\gamma_r^2}{\gamma_0^2} - 1 = \frac{\omega^2r^2\gamma_r^2}{c^2}$$

from Eq. (5.71), we obtain

$$I_r = I_0 \frac{\omega^2 r^2 \beta_z \gamma_0 \gamma_r^2}{2c^2} \tag{5.192}$$

where the angular velocity ω in general is a function of the radius (unlike the angular velocity in a nonrelativistic beam). If we assume as in the nonrelativistic case that $\omega = \omega_c/2 = \eta B/2\gamma_r$, Eq. (5.192) would coincide with Eq. (5.191). But this assumption in general is not correct. Thus, the equilibrium radius of a magnetic shielded cathode for relativistic beams in general does not coincide with the radius of an unrippled Brillouine beam. For nonrelativistic beams, as we have seen, the two radiuses are equal.

Let us estimate a change in the equilibrium radius as a function of the magnetic field and beam current. Replacing coefficients in Eq. (5.186) through the radiuses r_a and r_b, we obtain the equation

$$r_e^4 - r_b^2 r_e^2 - r_a^4 = 0 \tag{5.193}$$

Let us write the solution of this equation in two forms:

$$r_e = \begin{cases} r_b\left[\dfrac{1 + (1 + 4r_a^4/r_b^4)^{1/2}}{2}\right]^{1/2} & (5.194) \\[2ex] r_a\left[\left(1 + \dfrac{r_b^4}{4r_a^4}\right)^{1/2} + \dfrac{r_b^2}{2r_a^2}\right]^{1/2} & (5.195) \end{cases}$$

According to Eqs. (5.194) and (5.195), the equilibrium radius $r_e \approx r_b$ when $r_a/r_b \ll 1$, but $r_e \approx r_a$ when $r_a/r_b \gg 1$. Consider some particular cases.

1. *Small Magnetic Field*. Because $r_b \sim 1/B$ and $r_a \sim 1/\sqrt{B}$, the ratio $r_a/r_b \ll 1$, so that when the magnetic field decreases, the equilibrium radius r_e, being greater than r_b, approaches r_b. Note that for a small magnetic field and a large current, the radius r_b is large. However, the current density is small. Therefore, too great a diminution of the magnetic field is not acceptable.

2. *Small Beam Current*. The radius r_b decreases but r_a does not change. Therefore, r_e, being greater than r_a, approaches r_a.

3. *Large Cathode Magnetic Field*. According to Eq. (5.188), the radius r_a increases but r_b does not change. So for large B_c, r_e, being more than r_a, approaches r_a.

5.7.3 Stiffness of an Electron Beam. Frequency and Wavelength of Small Ripples

Let us rewrite Eq. (5.185) assuming that the initial radius equals the equilibrium radius (i.e., $r_0 = r_e$). According to Eqs. (5.188) and (5.193) we have

$$r_c^4 B_c^2 = r_a^4 B^2 = B^2(r_e^4 - r_e^2 r_b^2) \tag{5.196}$$

Replacing r_0 in Eq. (5.185) with r_e and using Eqs. (5.196) and (5.190), we obtain

$$R'^2 = R_e'^2 + K_B\left(2 - R^2 - \frac{1}{R^2}\right) + \frac{K_r}{r_e^2}\left(\ln R^2 + \frac{1}{R^2} - 1\right) \tag{5.197}$$

where $R = r/r_e$, $R_e' = r_e'/r_e$, and $K_B = (\eta B/2\beta_z c\gamma)^2$. The extreme values of R are found from the condition $R' = 0$. That gives a transcendental equation determining $R_{\max}$ and $R_{\min}$ as functions of R_e', K_B, and K_r. Note that these parameters do not depend on z. It proves that the rippled beam is periodical. Integration of Eq. (5.197) is possible in general only by numerical methods. However, assuming small ripples ($\rho \equiv \Delta r/r_e \ll 1$), it is easy to obtain the solution in closed form. Expansion of Eq. (5.197) in a power series on ρ and neglecting quantities of order greater than 2 gives the equation

$$R'^2 = R_e'^2 - 4K_b\left(1 - \frac{r_b^2}{2r_e^2}\right)\rho^2 = R_e'^2 - \left(\frac{\omega_g}{v_z}\right)^2\left(1 - \frac{r_b^2}{2r_e^2}\right)\rho^2 \tag{5.198}$$

where $\omega_g = \eta B/\gamma$ is a gyrofrequency. The extreme amplitude of the ripples is

$$\rho_{\max} = \frac{R_e'}{S} \quad \text{or} \quad \Delta r_{\max} = \frac{r_e'}{S} \tag{5.199}$$

A quantity $S = r'/\Delta r_{\max}$ in general is called *beam stiffness*. One characterizes a capacity of the beam to conserve its form in the presence of perturbations. In our case the stiffness is

$$S = \frac{\omega_g}{v_z}\left(1 - \frac{r_b^2}{2r_e^2}\right)^{1/2} \tag{5.200}$$

This quantity has dimensions of inverse length. Note that for the shielding cathode, $r_e = r_b$ and $S = \frac{\sqrt{2}}{2}\frac{\omega_g}{v_z}$. That is a minimal beam stiffness because in a nonzero cathode field according to Eq. (5.194), $r_b/r_e < 1$. The maximum stiffness takes place for zero current ($r_b = 0$), $S_{\max} = \omega_g/v_z$.

Equation (5.198) shows that the beam envelope accomplishes harmonic oscillations. Differentiation of this equation with respect to z and taking $\rho \ll 1$ into account gives the differential equation of a linear harmonic oscillator:

$$\ddot{\rho} + S^2 v_z^2 \rho = 0 \tag{5.201}$$

The frequency of the small envelope oscillation is

$$\omega = S v_z = \omega_g \left(1 - \frac{r_b^2}{2r_e^2}\right)^{1/2} \tag{5.202}$$

The frequency in the absence of space charge ($r_b = 0$) is equal to the relativistic gyrofrequency. The ripple wavelength is

$$\lambda = \frac{2\pi}{\omega} v_z = \frac{2\pi v_z}{\omega_g (1 - r_b^2/2r_e^2)^{1/2}} \tag{5.203}$$

5.8 TRANSPORT OF SPACE-CHARGE BEAMS IN SPATIAL PERIODIC FIELDS

5.8.1 Introduction

The methods of beam focusing considered earlier in fact do not solve the transport problems of space-charge beams over a long distance, a vital problem in microwave electronics and accelerator physics. The maximum beam current that can be transferred due to inertial focusing (i.e., focusing without external fields) is diminished inversely proportional to the square of the transport length (Section 5.6.4). Also in this case, too high precision of the beam launching must be provided. In principle, transport of a space-charge beam in a uniform magnetic field is possible over any distance. However, implementation of the corresponding magnetic system is very difficult in practice. In Fig. 5.18 the magnetic force lines of a long solenoid are depicted. As can be seen, the lines diverge in a radial direction over a distance on the order of the solenoid length L. Hence, a volume occupied by the magnetic field and correspondingly, the magnetic field energy, are quantities of order L^3. The solenoid weight is also proportional to L^3, and there is large external leakage of the field.

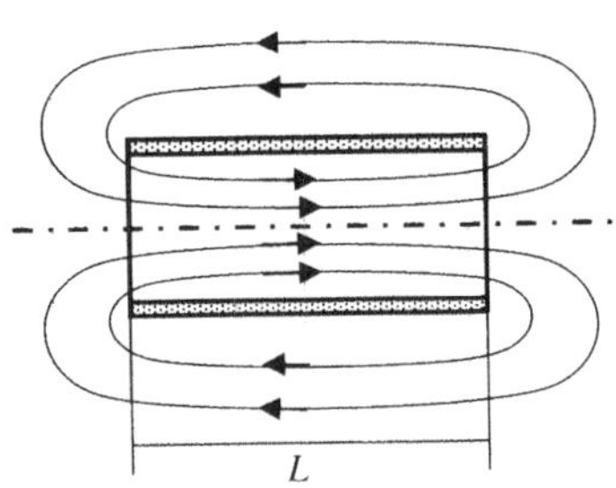

FIGURE 5.18 Magnetic force lines.

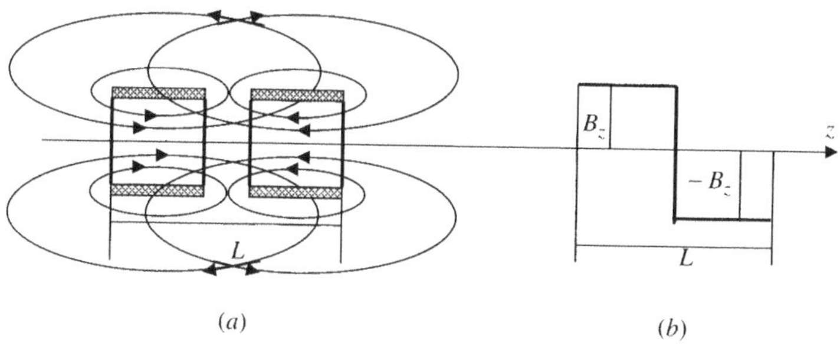

FIGURE 5.19 (*a*) Two solenoids with opposite directions of the magnetic field; (*b*) distribution of an axial magnetic field in an idealized reverse system.

Now let us consider a case of magnetic field reverse: two closely placed solenoids with an opposite the magnetic directions of the axial magnetic field (Fig. 5.19). As we see, the directions of the magnetic force lines on the periphery of the solenoids are opposite and cancel each other out. Therefore, a summary volume of the magnetic field is approximately equal to the doubled volume of a single solenoid field with length $L/2$ [i.e., proportional to $2(L/2)^3 = \frac{1}{4}L^3$]. On the other hand, since the magnetic term in the equations of paraxial trajectories is proportional to B^2 [see, e.g., Eq. (5.183)], the trajectories in the reverse system are the same as in a uniform magnetic field with an equal total solenoid length. So use of the reverse system gives fourfold saving, on the order of the magnitude of the solenoid weight and the magnetic energy. This result is also true for a reverse system of permanent magnets.

Let us continue this concept and take a system of N identical short close-placed solenoids with alternately opposite magnetization (Fig. 5.20). We again conclude that the paraxial trajectories in this periodical system would not differ from the trajectories in a single solenoid of the same total length. Assume that the magnetic energy and weight of each solenoid are both $P = k(L/N)^3$, where L is the total length of the system. The total magnetic energy and the weight of the system would be $PN = k(1/N^2)L^3$. Thus, the use of a periodic system of N magnets gives an approximate N^2-fold saving in weight and energy consumption of the system.

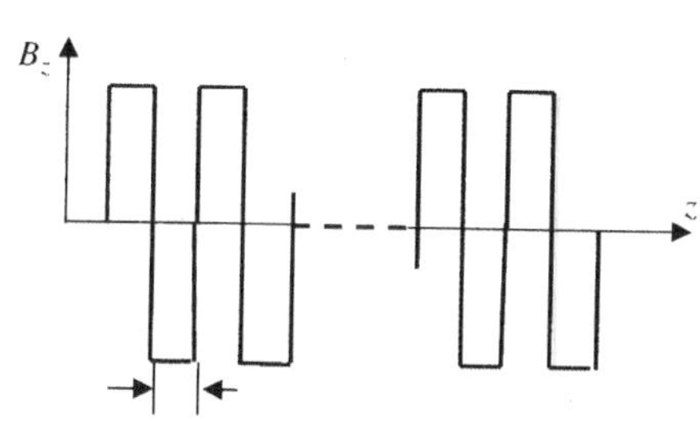

FIGURE 5.20 Idealized distribution of an axial magnetic field in a magnetic periodical system.

The real distribution of a magnetic field is essentially different from the idealized distribution and correspondingly, the transport characteristics of a periodic system are much more complicated. A periodic magnetic system can be considered as a lattice of short magnetic lenses. According to Section 2.6, each lens is convergent. The focusing effect balances the effect of the space-charge field. But, in general, this system is unstable: The ripples can grow indefinitely as a function of distance. Nevertheless, analysis of the beam stability allows us to choose the proper range of initial and system parameters, providing beam stability. As result, transport of a stable rippled beam is attained.

Related to the magnetic focusing is *periodic electrostatic focusing*. In this case the beam in general is transferred by a periodic system of electrostatic immersion or unipotential lenses. As shown in Section 2.5.2, all these lenses are convergent, and their electric field can balance the space-charge force. As a result, transport of long space-charge beams is possible. Many electrostatic periodic system versions have been proposed, and corresponding theory has been developed (basically in the 1950s; see, e.g., Kirstein et al., 1967; Tien, 1954; Cook et al., 1957; Hoggs, 1958; Waters, 1960). Outlines of some systems are given below.

In Fig. 5.21, two electrostatic focusing systems with axially symmetric electron beams are represented. The system shown in Fig. 5.21*a* is a periodic sequence of immersion lenses. Each lens is formatted by two metal rings with different potentials φ_a and φ_b. According to Kirstein et al. (1967), the equilibrium perveance of the beam is

$$P \approx 110 \frac{(\varphi_b - \varphi_a)^2}{(\varphi_b + \varphi_a)^2} \left(\frac{r_0}{L}\right)^2 \frac{1}{I_0^2(2\pi b/L)} \left[\frac{\sin(2\pi d/L)}{2\pi d/L}\right]^2 \quad \text{micropervs} \qquad (5.204)$$

where r_0 is a beam radius and I_0 is a modified Bessel function.

A bifilar helix spiral system is shown in Fig. 5.21*b*. The potential difference between the spirals forms a periodical electric field similar to a lens system. This system was proposed by Tien (1954). The interesting property of bifilar helixes is its applicability simultaneously as a focusing system and as a slow-wave electrodynamic TWT or BWO structure.

Two electrostatic periodic systems for the transport of sheet beams are illustrated in Fig. 5.22. The system shown in Fig. 5.22*a* is formed by the lattice of tape lines

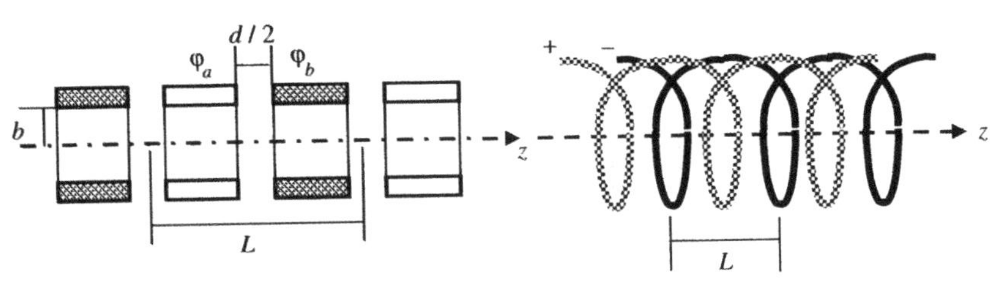

FIGURE 5.21 Electrostatic periodic axially symmetric focusing systems: (*a*) rings with alternatively changed potential; (*b*) bifilar bipotential helix spiral.

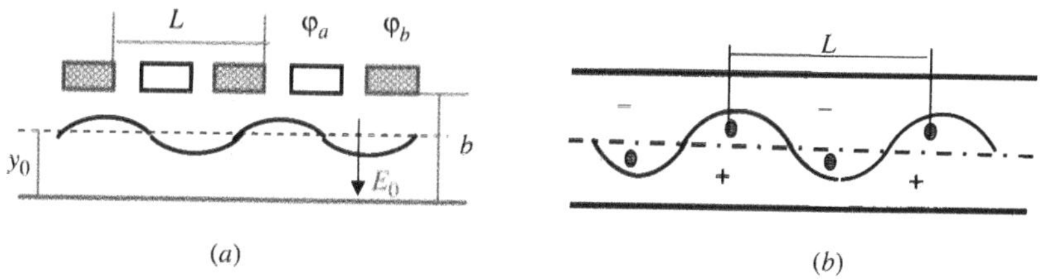

FIGURE 5.22 Electrostatic periodic systems focusing sheet beams: (*a*) asymmetrical electrode system; (*b*) slalom focusing system.

above the plane base (Waters, 1960). The tapes alternating potentials φ_a and φ_b. The equilibrium position of the rippled electron beam is determined by the relation (Kirstein et al., 1967).

$$\frac{(\varphi_b - \varphi_a)^2}{|(\varphi_b + \varphi_a)E_0|\pi L}\frac{\sinh(4\pi y_0/L)}{\cosh^2(2\pi b/L)}\left[\frac{\sin(2\pi d/L)}{2\pi d/L}\right]^2 \approx 1 \qquad (5.205)$$

where E_0 is the average electric field between electrodes. An important virtue of this system is the possibility of using the unipotential lower electrode as an electrodynamic structure in microwave devices with distributed parameters similar to TWTOs and BWOs. Disadvantages of the system are the requirement for three different electrode potentials and the very critical beam adjustment conditions.

The structure in Fig. 5.22*b* is called *Slalom focusing*. A deflection type of focusing system (Kirstein et al., 1967; Cook et al., 1957; Hoggs, 1958) was invented by Compfner et al. The system contains a series of equidistant-spaced wires with equal potentials between two parallel plates with another potential. It is interesting that the electron trajectory is close to one of the equipotential surfaces. As can be shown (Kirstein et al., 1967), the perveance of the slalom system is

$$P \sim 200\frac{S}{L^2} \qquad \mu\text{perv} \qquad (5.206)$$

were S is the area of a beam cross section. The basic difficulty of slalom focusing is very strong criticality of a trajectory with respect to injection conditions.

A common disadvantage of electrostatic periodic systems is the impossibility of their use in focusing of high-voltage beams. Even in a nonrelativistic range of energy the small distance between electrodes in a periodic system and the nearness of the system itself to the beam increase the probability of an electric breakdown and limit the energy and power of the beam. Magnetic focusing has far wider applications, especially periodic permanent magnetic focusing on the highly effective permanent-magnet base materials.

Periodic magnetic systems have found exceptional use in the transport of relativistic beams. In the range of extreme relativistic energies, in particular for accelerators of charged particles, the systems are used on the base of quadrupole and multipole lenses (see Section 2.9). In this aspect, such properties of these lenses as strong focusing and linearity of the focusing (transversal) forces to provide support for beams with very low emittance are the most important. In large accelerators, basically FODO lattices are used, containing up to hundreds of sections. Below we limit ourselves to an analysis of focusing only in a periodic axially symmetric magnetic field.

5.8.2 Magnetic Periodic Focusing

Periodic Focusing System with Permanent Magnets A typical axially symmetrical periodic permanent-magnet system (PPM) is depicted in Fig. 5.23. Permanent magnetic rings are usually manufactured from an effective magnetic alloy such as SmCo or NdFeB. The rings are magnetized in the axial direction with alternately changing polarity, and separated by ferromagnetic inserts from soft-magnetic material. This construction provides some regularization of axial magnetic field distribution. Note that a distribution of the axial magnetic field created by a magnetic ring is essentially irregular. In particular, there are large backswings where the field changes sign (Fig. 5.24*a*). The presence of backswings is explained by the following. A schematic pattern of molecular currents in the volume of the magnetized ring is shown in Fig. 5.24*b*. As we see, the adjacent molecular currents into a volume of the magnets are concealed from each other. Two unbalanced currents remain that flow in opposite directions along the boundaries of the ring. These currents can

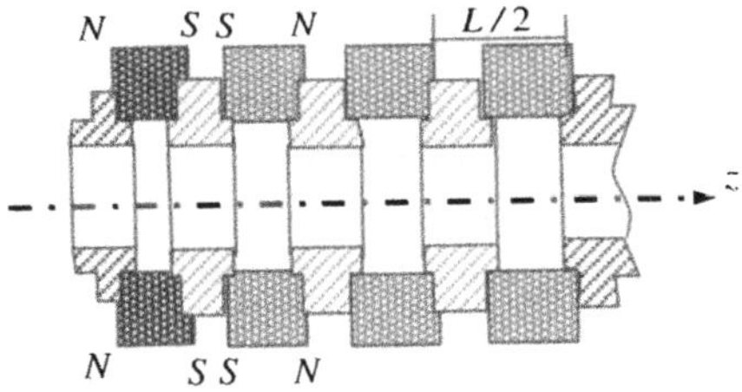

FIGURE 5.23 Periodic permanent magnetic system.

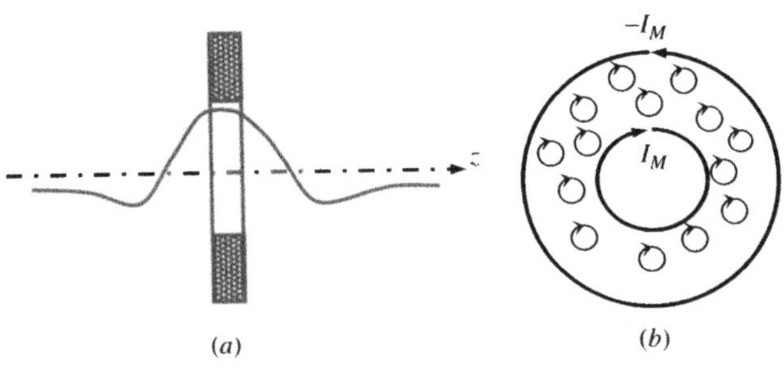

FIGURE 5.24 (*a*) Axial magnetic field distribution of a magnetized element of a PPM system; (*b*) pattern of PPM micromagnetization.

be considered as a source of the magnetic field. The magnetic field of an interior current on a contour of smaller radius is stronger in the central plane than the field of the outer current. However, the field of the interior current decreases faster with distance from the central plane. Therefore, the sign of the axial field is changed within some distance. The soft-magnetic inserts regularize the field distribution. It can be shown that the distribution of the axial magnetic field is approximately sinusoidal:

$$B = B_m \cos\frac{2\pi z}{L} \tag{5.207}$$

Periodic Focusing in a Sinusoidal Magnetic Field A relativistic equation of the paraxial trajectory can be obtained by substituting Eq. (5.207) in Eq. (5.183):

$$r'' + \left(\frac{\eta B_m}{2\beta_z c\gamma_0}\right)^2 r\cos^2\frac{2\pi z}{L} - \left(\frac{r_c^2\eta B_c}{2\beta_c c\gamma_0}\right)^2\frac{1}{r^3} - \frac{K_r}{r} = 0 \tag{5.208}$$

A nonrelativistic equation follows Eq. (5.208) using Eqs. (5.184). It is convenient to transform Eq. (5.208) to dimensionless form. Let us define the following dimensionless variables (the corresponding coefficients for the nonrelativistic equation are given if they are different from the relativistic ones):

$$Z = \frac{2\pi}{L}z \quad \text{(a length)} \tag{5.209}$$

$$R = \frac{r}{r_e} \quad \text{(a radius)} \tag{5.210}$$

where r_e is the equilibrium radius.

$$\alpha = \frac{1}{2}\left(\frac{\eta B_m L}{2\pi\beta_z c\gamma_0}\right)^2, \quad \left(\frac{\eta B_m^2 L^2}{64\pi^2\varphi_z^*}\right) \quad \text{(parameter of the periodic magnetic field)} \tag{5.211}$$

$$K_B = \frac{r_c^4 B_c^2}{r_e^4 B_m^2} \quad \text{(parameter of the cathode shielding)} \tag{5.212}$$

$$\beta = \frac{K_r L^2}{4\pi^2 r_e^2} \quad \left(\frac{P_r L^2}{16\pi^3\varepsilon_0\sqrt{2\eta}\, r_e^2}\right) \quad \text{space-charge parameter} \tag{5.213}$$

Equation (5.208) then becomes

$$\frac{d^2R}{dZ^2} + \alpha(1 + \cos 2Z)R - \frac{2\alpha K_B}{R^3} - \frac{\beta}{R} = 0 \tag{5.214}$$

This is a nonlinear second-order differential equation with a periodic coefficient. Let us linearize the equation mean about the small ripples. Assume that

$$R = \frac{r}{r_e} = \frac{r_e + \Delta r}{r_e} = 1 + \rho \tag{5.215}$$

where $\rho = \Delta r / r_e \ll 1$. Substituting R into Eq. (5.214) and omitting the terms with ρ'' of order $n > 1$, we obtain the following linear differential equation:

$$\frac{d^2\rho}{dZ^2} + (\alpha + 6\alpha K_B + \beta + \alpha \cos 2Z)\rho = 2\alpha K_B + \beta - \alpha(1 + \cos 2Z) \tag{5.216}$$

A general solution of this inhomogeneous equation can be represented as a sum of a general solution ρ_h of the homogeneous equation and a partial solution ρ_i of the inhomogeneous equation:

$$\rho = \rho_h + \rho_i \tag{5.217}$$

Let us write the homogeneous equation in the form

$$\frac{d^2 y}{dx^2} + (a - 2q \cos x) = 0 \tag{5.218}$$

where

$$y = \rho, \qquad x = Z, \qquad a = \alpha + 6\alpha K_B + \beta, \qquad q = -\frac{\alpha}{2} \tag{5.219}$$

Equation (5.218) is a standard form of a *Mathieu equation* (see, e.g., Abramovitz and Stegan, 1972). A general solution of the Mathieu equation according to the Floquet theorem (a general theorem for differential equations with periodical coefficients) is

$$y = C_1 e^{\mu x} P(x) + C_2 e^{-\mu x} P(-x) \tag{5.220}$$

where $P(x)$ is a function with period π. That corresponds to a period of the envelope rippling equal to a half-period L of the magnetic system. The parameter μ, in general, is a complex constant. If Re $\mu \neq 0$, the function $y(x)$ is increased infinitely for $x \to \infty$, which indicates beam instability. In this case the amplitudes of the ripples increase but the period remains $L/2$. In principle, it is possible to eliminate of from the coefficients C_i by proper choice of an initial condition and to obtain a bounded solution. However, this solution will in the end be unstable because an arbitrarily small addition of the coefficient that is eliminated again leads to an exponential grow of ripples. If Re $\mu = 0$, the solutions are stable; that is, the functions $y(x)$ are bounded for $x \to \infty$ and the envelope oscillates with a period L/μ that depends on the values of the constants a and q. Let us call these oscillations *free ripples*. Values of the parameters a and q corresponding to stable (periodic) or unstable solutions form connective regions in the (q,a) plane. In Fig. 5.25, the corresponding Mathieu stability diagram is shown. The shaded parts mark unstable regions. Note that solutions of the Mathieu equation that correspond to boundaries between stable and unstable regions are functions with period π or 2π (Mathieu functions).

First, let us consider a case of ideal cathode magnetic shielding and neglect by space charge (i.e., assume that the parameters $K_B = K_r = \beta = 0$) [see Eqs. (5.212) and (5.213)]. Then according to Eq. (5.219),

$$a = \alpha + 6\alpha K_B + \beta = \alpha = -2q \qquad (5.221)$$

In this case Eq. (5.214) is linear for any R. Let us trace the stability of a beam with changing amplitude B_m of the magnetic field. According to Eqs. (5.221) and (5.211), the parameter q is negative, and when we change B_m we move along the straight line $a = -2q$ in the left half-plane of the stability diagram (dashed line in Fig. 5.25). As can be seen from Fig. 5.25, we are in the first stability zone, beginning with $\alpha = q = 0$ until $q \approx -0.3$ (i.e., $a = -2q \approx 0.6$). For better accuracy one can use an equation for the upper boundary of the first stability zone 86]:

$$a = 1 + q - \frac{q^2}{8} - \frac{q^3}{64} - \cdots \qquad (q < 0) \qquad (5.222)$$

Then taking Eq. (5.221) into account, we obtain $a \approx 0.66$. For $a > 0.66$ we find ourselves in an unstable zone, and the trajectories diverge. That effect is similar to the phenomenon of multifocusing in magnetic lenses (Section 2.7.2). For too large a magnetic field, the trajectory can leave the first lens under an angle greater than an

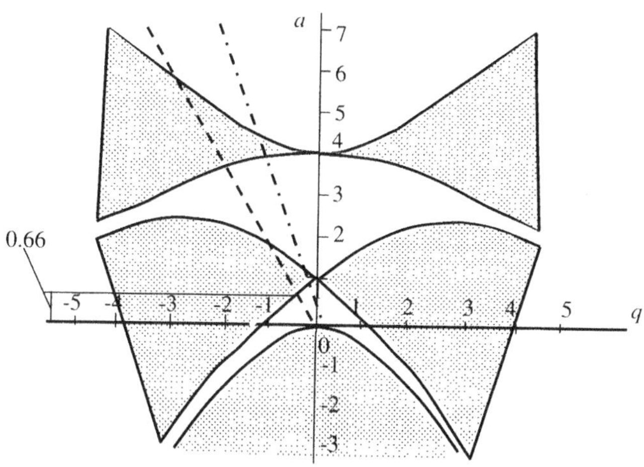

FIGURE 5.25 Mathieu stability diagram. Shaded parts are unstable zones. Dashed curve, $K_r = \beta = 0$; dash-dotted curve, $K_r > 0$, $\beta > 0$.

entrance angle (Fig. 2.17); in the second lens this effect is repeated; and so on. Recall that it was a case of zero space charge. For $\beta > 0$ the value of the parameter a can be positive even for $B_m = \alpha = 0$, and we can skip the first stable zone. Then it is necessary to use higher stability zones, and considerable more values B_m can be required.

For a space-charge beam and nonzero cathode magnetic field ($\beta \neq 0$, $B_r \neq 0$), we obtain from Eq. (5.219),

$$a = -2q(1 + 6K_B) + \beta \tag{5.223}$$

The corresponding straight line is shown in Fig. 5.25 as a dash-dotted line. As we see, the upper boundary of stability shifts to values $|q| < 0.33$ and $\alpha > 0.66$ (i.e., maximum permission amplitudes B_m are reduced with an increase in the cathode magnetic field).

It is necessary to take this assertion into account as applied to the solution of the homogeneous linear equation. For a final conclusion about the stability of the beam, it is necessary first, to estimate a contribution of the right side in the solution of Eq. (5.216) (i.e., the particular solutions of the inhomogeneous equation), and second, to evaluate the influence of the nonlinearity in the original equation [Eq. (5.208)].

The right side of Eq. (5.216) contains two terms:

$$\rho_i = A_1 + A_2 = -\alpha \cos 2Z + (2\alpha K_B + \beta - \alpha) \tag{5.224}$$

It is readily seen that the influence of the term A_1 in the stable zone is equivalent to nonresonance excitation of the oscillatory circuit. This effect combines with free beam rippling and increases total beam rippling. Therefore, it is advisable to use not too large values of α. That simplifies the magnetic system and removes parameter a from the boundary of an unstable zone.

The component A_2 in Eq. (5.224) plays a part in the constant force that strives to widen the beam. This effect is minimal for the equilibrium condition when

$$2\alpha K_B + \beta - \alpha = 0 \quad \text{or} \quad \alpha = \frac{\beta}{1 - 2K_B} \tag{5.225}$$

Since the space-charge parameter $\beta > 0$, the equilibrium parameter of the magnetic field α will be positive if the shielding parameter $K_B < \frac{1}{2}$. Note that a decrease in K_B reduces the required equilibrium value α (i.e., the amplitude B_m).

Let us write the equilibrium condition (5.225) in an explicit form using Eqs. (5.211)–(5.213):

$$\left(\frac{\eta B_m L}{4\pi\beta_z c\gamma_0}\right)^2 \frac{r_c^4 B_c^2}{r_e^4 B_m^2} + \frac{K_r L^2}{4\pi^2 r_e^2} - \frac{1}{2}\left(\frac{\eta B_m L}{4\pi\beta_z c\gamma_0}\right)^2 = 0 \tag{5.226}$$

This is the equation for an equilibrium radius r_e that we can write as

$$r_e^4 - r_{bm}^2 r_e^2 - r_{am}^4 = 0 \tag{5.227}$$

where

$$r_{bm}^2 = K_r \left(\frac{2\beta_z c \gamma_0}{\eta B_m / \sqrt{2}} \right)^2 \tag{5.228}$$

$$r_{am} = r_c \left(\frac{B_c}{B_m / \sqrt{2}} \right)^{1/2} \tag{5.229}$$

We see that the equilibrium condition for a periodic magnetic field coincides with the equilibrium condition for the beam in a uniform magnetic field $B = B_m/\sqrt{2}$ [Eq. (5.193)]. In particular, for a zero cathode magnetic field ($r_{am} = 0$), the equilibrium radius r_{bm} is equal to the radius of the Brillouin beam in the uniform magnetic field $B = B_m/\sqrt{2}$.

Nonlinear Effects Analysis of a linearized equation with the exception of the case $B_c = \beta = 0$ can give only trends for changing the ripples. In fact, even unbounded growth of the ripples in an unstable zone as a consequence of a linear Mathieu equation does not yet mean real loss of stability (i.e., does not exclude the possibility to establish the stabilization with finite values of R). Taking nonlinearity into account requires numerical computations. This design and an approximate analytical solution for the search of minimal ripples were undertaken by Harker (1955). These investigations have given not only general results but also concrete values for the amplitudes of rippling for different values of the system parameters.

Numerical solutions of nonlinear equations have been found by Harker directly on an analog computer with initial condition $(dR/dZ)_{Z=0} = 0$. Particular values α, K_B, and β were determined from the need to obtain minimal ripples. Analytical solutions were close to computed solutions when the trial solution was taken as

$$R = 1 - a + a \cos 2Z \tag{5.230}$$

where the parameter a is the ripple amplitude. Note that this form of the beam corresponds to the nonresonance excitation of the envelope oscillations by the force on the right side of Eq. (5.216). Some results obtained (Harker, 1955) are the following.

1. Stable solutions of Eq. (5.214) continue to exist when the parameter of a periodic magnetic field α becomes greater than 0.66. In general, that is a consequence of the nonlinearity because deleting the nonlinear terms in Eq. (5.214) returns this equation to Mathieu form with Eq. (5.221) or (5.223).

2. Ripples of the beam are decreased with increased cathode shielding. However, even for full magnetic shielding of the cathode, minimal ripples can be too large, and

the parameter α of the cathode magnetic field must be chosen considerably lower than the critical value 0.66.

3. The presence of a cathode magnetic field in beams with a small space charge significantly reduces the maximum permissible values of α.

4. No minimal ripple solutions were obtained for $K_B > \frac{1}{2}$, which corresponds to Eq. (5.225).

The theory of PPM focusing described here may be used as a starting design point. The final versions are usually obtained as a result of a numerical design and empirical experimental adjustment.

In concluding this section, let's compare some properties of focusing using a uniform magnetic field and permanent periodic magnets for the same lengths of both systems and the same root mean square value of the average magnetic field.

1. PPM systems provide values of a perveance and a beam stiffness comparable with those in a uniform magnetic field.

2. PPM systems are much lighter and have considerable lower external magnetic field leakage.

3. PPM systems do not require a power supply. Permanent magnets with a uniform magnetic field are also independent of the power supply, but their cost and weight is much higher.

One of the substantial advantages of PPM systems over electrostatic focusing systems in microwave devices is the possibility of design and optimization of the parameters (including experimental adjusting) independent of electrodynamic properties of systems.

PART II

MICROWAVE VACUUM ELECTRONICS

The history of radio electronics from the very beginning (late nineteenth to early twentieth centuries) was to a certain extent a history of the search for sources with increasing frequencies and electromagnetic radiation power. The first step in the foundation of vacuum radio electronics was undoubtedly the discovery by Thomas Edison in 1880 of the effect of unipolar conductance: Evacuated light bulbs with a second electrode would allow current to flow in one direction. Using the Edison effect in 1904, John Fleming invented a vacuum diode as a detector of electromagnetic oscillations. By 1907, Lee de Forest had created a three-electrode vacuum tube. This device, called the Audion, was the first successful electron amplifier. Since 1907, arrays of different electron tubes with permanently increased frequency and power of electromagnetic radiation have been invented. However, until the 1920s, vacuum electronics was of low-frequency type. At that time, principles of the electron tubes were founded implicitly with the following restrictions.

1. *Quasistationarity of a working space where an electron beam interacts with a high-frequency field.* The electromagnetic field does not change while light is passing through an interaction space: that is,

$$t_c \ll \widetilde{T} \tag{PII.1}$$

where t_c is the light transit time and $\widetilde{T}$ is a specific time of field change. For a periodic signal, $\widetilde{T} = T$, where T is the period. Then criterion (PII.1) is equivalent to the condition

$$l_i \ll \lambda \tag{PII.2}$$

Electron Beams and Microwave Vacuum Electronics. By Shulim E. Tsimring

where l_i and λ are a length of the working space and a wavelength of the radiation in the free space, respectively. It is clear that shortening the wavelength leads at the end to an unacceptable reduction in the working space.

2. *Quasistationarity of device leads*. Lengths of open sections l_{out} in the devices outside the electron beam are small compared to the wavelength:

$$l_{\text{out}} \ll \lambda \tag{PII.3}$$

In the opposite case, radiation emerges, and the leads work as antennas.

3. *Noninertia of electrons*. The electromagnetic field interacting with the electrons is not changed for the time t_e of the electron transit through the working space:

$$t_e \ll \widetilde{T} \tag{PII.4}$$

This condition can be written analogous to Eq. (PII.2) as

$$l_i \ll \lambda\beta \tag{PII.5}$$

where $\beta = v/c$. Effects connected with violation of the criterion (PII.4) are called *transit-time effects*.

Microwave electronics originates as a result of overcoming at least one of these restrictions. Note that conditions (PII.4) and (PII.5) are stronger than (PII.1) and (PII.2) if the microwave devices are not relativistic. Thus, in general, transit-time effects determine a low-frequency boundary f_L of the microwave range. For $f > f_L$, the high-frequency electric field is not static, and the energy integral is not conserved.

Example PII.1 Let $\varphi = 3$ kV, $\lambda = 10$ cm ($f = 3$ GHz), $l_i = 1$ cm, $v = \sqrt{2\eta\varphi} \approx 3.2 \times 10^7$ m/s.

Then $t_c = l_i/c \approx 3 \times 10^{-11}$ s, $T = \lambda/c \approx 3 \times 10^{-10}$ s, and $t_e = l_i/v \approx 3 \times 10^{-10}$ s. We see that the condition (PII.1) is fulfilled, whereas the condition (PII.4) is not true, and we find that we are in the microwave range. Note that the restriction (PII.4) is important for vacuum electronics. Therefore, strictly speaking, the term *microwave electronics* is inherent to vacuum electronics. For solid-state electronics, the term *ultrahigh-frequency electronics* would perhaps be more appropriate.

The low-frequency boundary of the microwave range is rather relative and is determined by the specific dimensions of the devices and by the electron energy. Conventionally, the microwave range is 300 MHz to 3000 GHz. The upper limit of the microwave is the boundary between the submillimeter-wave and infrared spectrums. Thus, the frequency overlap in the microwave range is 10^4. Note that in the acoustic range, the frequency overlap is 10^3, and in the optical range it is 2.

TABLE PII.1 Frequency Bands in the Microwave Range (Old Designation)

Band	Range (GHz)	Average Wavelength (cm)
L	1–2	20
S	2–4	10
C	4–8	5
X	8–12	3
K_u	12–18	2
K	18–26	1.4
K_a	26–40	0.9
Q	30–50	0.75
U	40–60	0.5
V	46–56	0.5
W	56–100	0.38

The standard notation of frequency bands for the microwave spectrum is presented in Table PII.1 and Fig. PII.1 (see, e.g., Gilmour, 1986). Here:

- Bands B and C (300 MHz to 1 GHz) form the ultrahigh-frequency (UHF) range.
- Bands D to I (1 to 10 GHz) form the superhigh-frequency (SHF) range.
- Bands J to M (10 to 100 GHz) form the extra-high-frequency (EHF) range.
- Band N (100 GHz to 3 THz) forms the short millimeter and submillimeter range.

The wide frequency range of microwaves results in the essential difference in generating and amplification methods, and in general, in a difference in signal processing and a layout of devices at different bands of the microwave spectrum. Gridded tubes (triodes, tetrodes, etc.) were the first amplifier vacuum electron devices that worked in the microwave range. Construction of these tubes provided conservation of the fundamental principle of acceptable functioning, the classic low-frequency gridded tubes, and gridded control of the electron current, excluding velocity modulation and bunching of the electron beam. In fact, that requires the carrying out of criterion (PII.5), where $l_i = l_{g1-c}$ is the distance between the first grid and the cathode. Therefore, l_{g1-c} and the grid structure must be increasingly smaller with increasing frequency

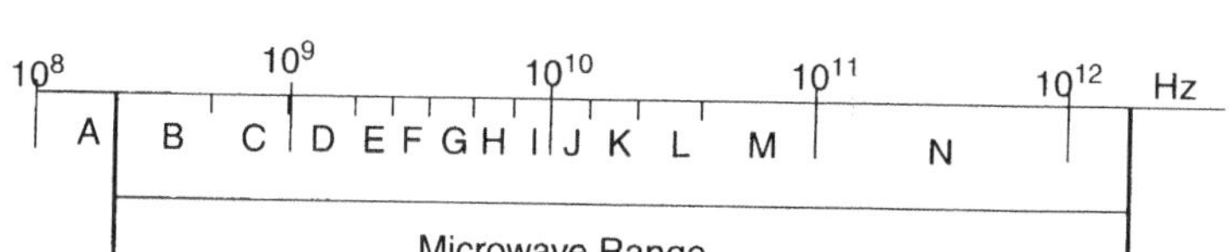

FIGURE PII.1 Frequency bands in the microwave range (new designation).

and with a corresponding reduction in the output power. These tubes have been applied in low microwave bands (300 to 3000 MHz). They may be called *microwave* only because non-quasistationary electromagnetic structures are used in them.

The first true microwave tubes were klystrons in the sense that velocity modulation, space bunching of the electron beam, and transfer of the energy to the electromagnetic field in the output cavity that provides the stimulated electron transition radiation are the fundamental working factors. The velocity modulation and energy extraction in the input and output gaps, respectively, are accomplished on the full electron velocities. Therefore, these gaps are not as small as those in gridded tubes. It is possible to use klystrons equipped with the proper electrodynamic systems on higher-frequency bands, including centimeter and even millimeter ranges (bands S to K in Table PII.1).

In systems with continuous interaction that exploits the Cerenkov synchronism (Section 8.2.2), the quasistationarity of the working space is violated. These are traveling-wave tubes (TWTs), backward-wave oscillators (BWOs) of the O type, and their cross-field versions. They are used widely in the S to K bands. Note that magnetrons and magnetron amplifiers with electrons moving in crossed electric and magnetic static fields can also be referred to systems with noquasistationary interaction space or developed electrodynamic structures.

Finally, in the millimeter and submillimeter ranges (bands K to U), devices with curvilinear periodical electron beams are used where normal Doppler synchronism of electrons with the electromagnetic field takes place: classical electron masers and free electron lasers (FELs). These devices are usually relativistic or extreme relativistic with multimode electrodynamics systems. They provide the highest levels of output power in short wavelength bands in long-pulse and continuous-wave regimes.

The use of microwave vacuum electronic tubes is a huge sphere of study and embraces extensive fields of science, technology, transport, domestic equipment, consumer service, and the military. After 1960, when transistors penetrated radio equipment and microelectronics originated, solid-state devices practically excluded vacuum electron tubes from radio and television, in general, from the sphere of information technology. However, vacuum electronics was developed in an accelerated sense as a field of a high-power microwave (HPM). So vacuum and solid-state electronics today may be considered not as competitive areas but as a mutual supplement enriching other regions of electronics.

Concrete applications of vacuum microwave tubes are discussed in sections dedicated to consideration of corresponding electron devices. We should emphasis that this book cannot be used as a reference book on different microwave-constructive versions of electron devices (unlike, e.g., the comprehensive book of Gilmour, 1986). Here, the accent is on the detailed physical and analytical treatment of the mechanism of electron beams interacting with high-frequency fields. Considerable attention is given to detailed statements of problems, physics of operation, discussions of models and approximations, and the sequence of analysis. I have chosen material in a form acceptable for students and experts working not only in microwave electronics but also in adjacent fields of science and technology.

It must be admitted that the material in Part II is distributed unevenly. More attention is given to the physics and analysis of nonrelativistic and moderate relativistic microwave electronic tubes, such as klystrons, TWTs, BWOs, orotrons, magnetron amplifiers, magnetrons, gyromonotrons, and gyroklystrons. The theory of devices with high-current relativistic and especially extreme relativistic electron beams (e.g., CARMs and FELs), are given comparatively briefer treatment. The reader can find more detailed consideration of the physics and construction of corresponding devices in the books of Barker and Schamiloglu (2001), Barker et al. (2005), and Freund and Antonsen (1992).

CHAPTER SIX

Quasistationary Microwave Devices

6.1 INTRODUCTION

In this chapter we consider electron devices that retain the criteria of quasistationarity equations (PII.1) and (PII.2), but conditions (PII.4) and (PII.5), which exclude transit effects, are violated. Therefore, the electric field is potentially one that satisfies Eq. (PII.1), but the energy integral is not conserved. Furthermore, when the electric field acting on the particles is variable, the space-charge density is also variable and the continuity theorem must be written as

$$\operatorname{div} \mathbf{j} = -\frac{\partial \rho}{\partial t} \tag{6.1}$$

In addition, there may be cases where the outside field is constant but the space-charge density is changed for a time on the order of the transit time. Then the space-charge field is variable and we are again in the frame of microwave electronics.

6.2 CURRENTS IN ELECTRON GAPS. TOTAL CURRENT AND THE SHOCKLEY–RAMO THEOREM

First, we must revise the ordinary ("obvious") approach to currents flowing through the gaps when the transit time is long ($t_e \sim \widetilde{T}$).

Electron Beams and Microwave Vacuum Electronics. By Shulim E. Tsimring

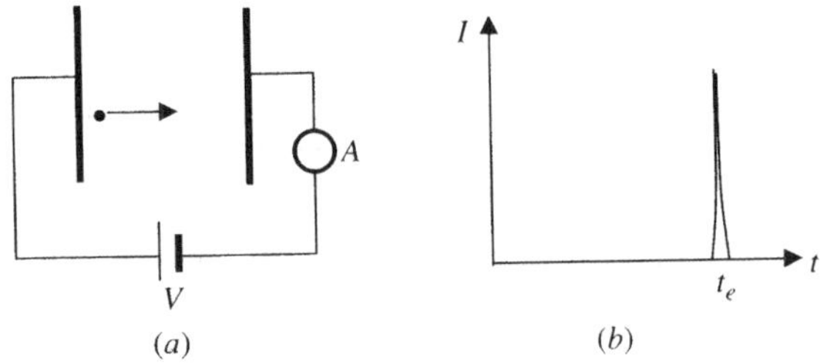

FIGURE 6.1 (*a*) Electron gap; (*b*) projected current curve.

Example 6.1 The point charge *e* starts at the time $t = 0$ from a cathode in a flat gap connected with a source of constant voltage *V* (Fig. 6.1*a*). What current would be measured by the ammeter in the gap circuit? It is expected that because $V = \text{const.}$ and the capacitive current is zero, the circuit current's shape will be spikelike (Fig. 6.1*b*).

Below we'll see that the form of a true current will be another shape.

6.2.1 Total Current. Continuity of Total Current

Definition of Total Current Let us write the Maxwell equation (I.5) as

$$\text{curl}\,\mathbf{H} = \mathbf{j} + \varepsilon_0 \frac{\partial \mathbf{E}}{\partial t} = \mathbf{j}_c + \mathbf{j}_d \tag{6.2}$$

where $\mathbf{j}_c$ and $\mathbf{j}_d$ are convection and displacement currents respectively. In this equation the variable electric field that determines the displacement current includes the space-charge field.

Let us define the total density current as a sum:

$$\mathbf{j}_t \equiv \mathbf{j}_c + \mathbf{j}_d \tag{6.3}$$

Correspondingly the total current is a sum:

$$I_t = I_c + I_d \tag{6.3a}$$

Note that the electric field in metal is zero; therefore, the total current in the metal is equal to the convection current:

$$I_{tm} = I_c \tag{6.4}$$

Continuity of Total Current According to the identity $\text{div}\,\text{curl}\,\mathbf{H} = 0$ of the vector analysis,

$$\text{div}\,\mathbf{j}_t = 0 \tag{6.5}$$

Let us choose a volume Ω in the electron gap (Fig. 6.2) confined by surfaces S_1, S_2, and S, where the completed surface S passes outside the electric field and convection current.

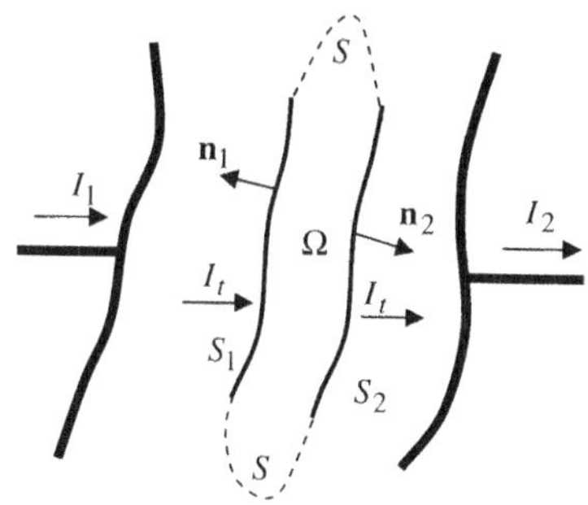

FIGURE 6.2 Continuity of total current I_t into a gap.

Application of Gauss's theorem to the integral of div $\mathbf{j}_t$ over the volume Ω gives

$$0 = \int_\Omega \text{div}\, \mathbf{j}_t\, d\tau = \int_{S_1} \mathbf{j}_{tn_1}\, d\sigma + \int_{S_2} \mathbf{j}_{tn_2}\, d\sigma = -I_{t1} + I_{t2} \tag{6.6}$$

where I_{t1} and I_{t2} are the total currents intersecting the surfaces S_1 and S_2. Thus, $I_{t1} = I_{t2} = I_t$, which indicates continuity of the total current. Because the displacement currents in wires connecting the sources to the gap are zero, the total current in the gap is equal to the convective currents in the leads: $I_t = I_1 = I_2$.

6.2.2 Total Current and the Shockley–Ramo Theorem

Assume that some source with voltage U is connected to the gap, and the electron current of the source is I_c. Let us write the power expenses in terms of the source: first, the work $\int_\Omega \mathbf{j}_c \mathbf{E}\, d\tau$ of the electric field over the electrons in the volume Ω; and second, a change in the electromagnetic field energy in the gap. Because of the condition of quasistationarity, the electromagnetic energy W_e is equal to the electrostatic energy $C_g U^2/2$. Then

$$\frac{dW_e}{dt} = \frac{d}{dt}\left(\frac{C_g U^2}{2}\right) = C_g U \frac{dU}{dt}$$

Thus, the source power is

$$IU = \int_\Omega \mathbf{j}_c \mathbf{E}\, d\tau + C_g U \frac{dU}{dt} \tag{6.7}$$

Here I is the current in the wires that, as shown, is equal to the total current. We obtain the total current theorem:

$$I_t = \frac{1}{U} \int_\Omega \mathbf{j}_c \mathbf{E}\, d\tau + C_g \frac{dU}{dt} \tag{6.8}$$

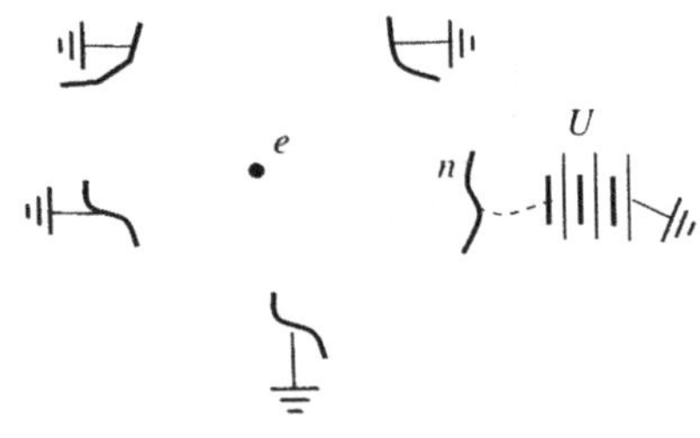

FIGURE 6.3 Shockley–Ramo theorem.

According to Eq. (6.8), the total current is equal to the sum of two components. The second component is the current through the capacitance formed by the gap electrodes. What is the first component? The answer to this question is given by the *Shockley–Ramo theorem* (Ramo, 1939b; Shockley, 1938).

Let us consider a system of grounded metal electrodes (Fig. 6.3) and put the point charge e into the interelectrode region. This charge induces charges on the electrodes: the electric field of charge e ejects charges of the same sign from the electrodes to Earth and retains the opposite (induced) charges on the electrode surfaces, maintaining the zero electric field inside the metal electrodes.

Values of the induced charges could be found from a Green's formula for a potential field (see, e.g., Smythe, 1950). So to find the charge induced on the nth electrode, a test voltage U between this electrode and the Earth is applied. The charge induced is expressed via the potential $U_e^{(n)}$, which originates at the position of the charge e:

$$Q_{\text{ind}}^{(n)} = -e\frac{U_e^{(n)}}{U} \tag{6.9}$$

Note that $Q_{\text{ind}}^{(n)}$ does not really depend on U because U and $U_e^{(n)}$ are proportional, so the ratio $U_n^{(n)}/U$ is a purely geometric factor that is determined by the position of the charge e and the configuration of the electrode system. The current induced on the nth electrode by moving charge is

$$I_{\text{ind}}^{(n)} = \frac{dQ_{\text{ind}}^{(n)}}{dt} = -e\frac{1}{U}\frac{dU_e^{(n)}}{dt} \tag{6.10}$$

Here the potential $U_e^{(n)}$ is the function of time through position $\mathbf{r}_e(t)$ of the charge e: $U_e^{(n)}(t) = U_e^{(n)}[\mathbf{r}_e(t)]$. Therefore, the derivative

$$\frac{dU_e^{(n)}}{dt} = \frac{\partial U_e^{(n)}}{\partial \mathbf{r}_e}\frac{d\mathbf{r}_e}{dt} = -\mathbf{E}_e^{(n)}\mathbf{v}_e$$

where $\mathbf{E}_e^{(n)}$ is the electric field that originates in the position of charge e. So the current induced on the nth electrode is

$$I_{\text{ind}}^{(n)} = e\frac{\mathbf{E}_e^{(n)}\mathbf{v}_e}{U} \tag{6.11}$$

Again, the ratio $\mathbf{E}_e^{(n)}/U$ is a purely geometric factor.

Induced Current Produced by a System of N Moving Charges The current is equal to the sum of the currents induced from each charge:

$$I_{\text{ind}}^{(n)}(t) = \frac{1}{U}\sum_{m=1}^{N} e_m \mathbf{E}_{ml}^{(n)}(t)\,\mathbf{v}_m(t) \tag{6.12}$$

It is necessary to remember that all values $\mathbf{E}_{ml}^{(n)}(t)$ and $\mathbf{v}_m(t)$ must to be taken at the same time, t. Note that the ratios $\mathbf{E}_{ml}^{(n)}/U$ are no longer purely geometric factors because they can be dependent on a space-charge field.

Induced Current Produced by a Continuous Distribution of Charges Assume that the distribution of moving charges in the gap is given a density $\rho(\mathbf{r}, t)$. Considering the elementary charge $\rho\, d\tau$ in the volume $d\tau$ as the point charge, we obtain the current induced:

$$I_{\text{ind}}^{(n)}(t) = \frac{1}{U}\int_{\Omega} \rho(\mathbf{r},t)\,\mathbf{v}(\mathbf{r},t)\,\mathbf{E}^{(n)}(\mathbf{r},t)\,d\tau = \frac{1}{U}\int_{\Omega} \mathbf{j}_c(\mathbf{r},t)\mathbf{E}^{(n)}(\mathbf{r},t)\,d\tau \tag{6.13}$$

Here the integration must be extended to the fixed time. Equations (6.11)–(6.13) express the Shockley–Ramo theorem. Previously, we determined the total current as the sum of the convection and displacement currents. The total current theorem represents the same current as another sum of induced and capacitive currents.

6.2.3 Particular Cases

Stationary Current in a Constant Electric Field In this case the derivative dU/dt and the displacement current are zero. According to Eqs. (6.3a) and (6.8), we obtain

$$I_t = I_c = I_{\text{ind}} \tag{6.14}$$

The total current is equal to the convection and induced currents together.

Moving Point Charge in a Flat Gap Let us reconsider Example 6.1. It is clear that in the circuit of Fig. 6.1 the total current circulates. According to Eq. (6.8), the total current is equal to the induced current because $U = \text{const}$. For this case, the induced current that was produced by motion of the point charge according to Eq. (6.11) is

$$I_{\text{ind}} = \frac{e}{U}Ev = \frac{ev}{d} \tag{6.15}$$

where d is the gap width. The velocity of a charge in the homogeneous electric field is $v = (\eta U/d)t$. Thus, the current in the circuit is shaped like a triangular pulse (Fig. 6.4):

$$I = I_{\text{ind}} = \frac{e\eta U}{d^2}t \tag{6.16}$$

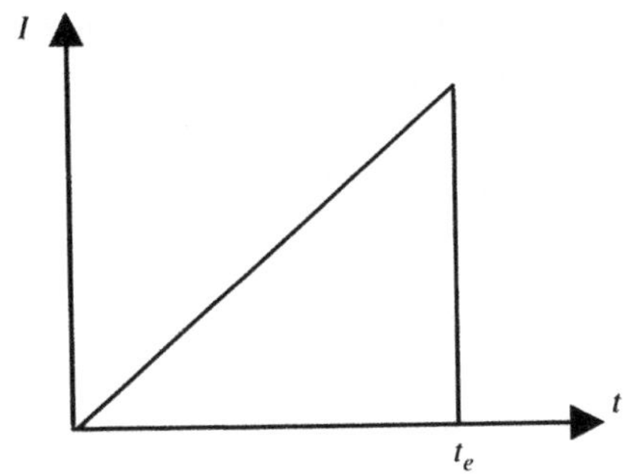

FIGURE 6.4 Current in the circuit of Fig. 6.1.

Total Current Theorem for a Flat Gap Assume that the electric field in the flat electron gap (Fig. 6.5) is directed along x. Then the volume integral in Eq. (6.13) can be written as

$$I_{\text{ind}}(t) = \frac{1}{U}\int_0^d dx \int_S j_c(\mathbf{r},t)E(\mathbf{r},t)\, d\sigma \tag{6.17}$$

If neglected by the space-charge field, the electric field would be written as $E = U/d$, and Eq. (6.17) takes the form

$$I_{\text{ind}}(t) = \frac{1}{d}\int_0^d dx \int_S j_c\, d\sigma = \frac{1}{d}\int_0^d I_c(x,t)\, dx \tag{6.18}$$

Thus, the total current theorem for the flat gap in the absence of the space charge is

$$I_t(t) = \frac{1}{d}\int_0^d I_c(x,t)\, dx + C_g \frac{dU}{dt} \tag{6.19}$$

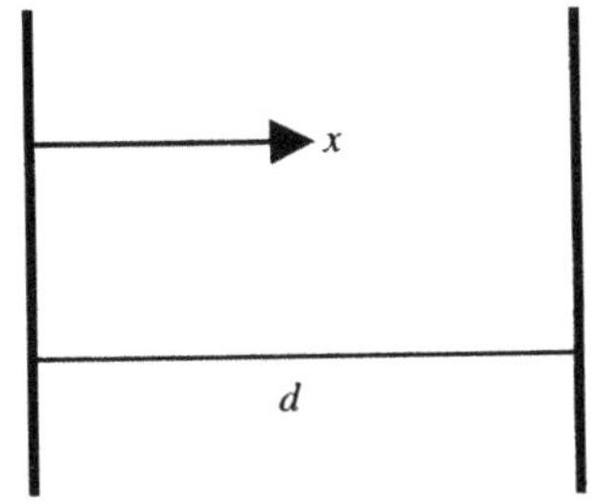

FIGURE 6.5 Flat electron gap.

6.3 ADMITTANCE OF A PLANAR ELECTRON GAP. ELECTRON GAP AS AN OSCILLATOR. MONOTRON

6.3.1 Formulation of the Problem. Scheme of the Solution

An electron gap is depicted in Fig. 6.6. The electrons are accelerated by the constant electric field between the cathode and the left grid of the gap, up to the velocity $v_0 = \sqrt{2\eta U_0}$. After entering the gap where a high-frequency voltage $\widetilde{U}\cos\omega t$ is applied, the problem consists of the calculation of the current in the circuit of the ac source. The solution contains the following idealizations: (1) nonrelativistic approximation, (2) one-dimensionality (all quantities depend only on x), and (3) neglect by the space-charge field. According to the assumed approximations, we will use Eq. (6.19), where, in principle, the convection current can be written as

$$I_c(x,t) = \rho_l(x,t)v(x,t) \tag{6.20}$$

Here $\rho_l(x,t)$ is the linear space-charge density (a space-charge density per unit axial length) and $v(x,t)$ is the electron velocity. The electric field in the gap is uniform:

$$E(x,t) = \frac{\widetilde{U}}{d}\cos\omega t \tag{6.21}$$

In the frame of a complex amplitude method,

$$E(x,t) = \frac{\widetilde{U}}{d}e^{i\omega t} \tag{6.22}$$

and the equation of the electron motion into the gap $(0, d)$ is

$$\frac{dv}{dt} = \frac{\eta U}{d}e^{i\omega t} \tag{6.23}$$

It is important to remember that the method of complex amplitudes is only applicable to linear relations between quantities. Then, on each step of an analysis we can replace a real equation by a complex form. But if the analysis is extended to a nonlinear stage, we must first transform the corresponding expressions or equations to their real form.

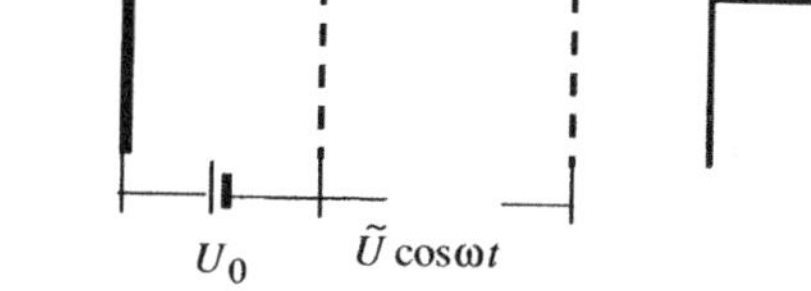

FIGURE 6.6 Electron gap.

Integration of Eq. (6.23) is simple, but before doing so, let us consider the problem of initial conditions. The choice of initial conditions in the constant field is simple because the velocity of the particles at each point do not depend on time. However, the motion of particles entering the gap with a variable field in different moments of time will also be different. Therefore, we must introduce into the description of motion an entry time τ of a particle as a separate variable. In the present case, the initial velocities of all particles are the same:

$$v(x,t)_{x=0} = v(\tau) = v_0 = \sqrt{2\eta U_0} \tag{6.24}$$

Integrating Eq. (6.23) with the initial condition (6.24), we obtain

$$v(\tau,t) = v_0 + \int_\tau^t \frac{\eta \tilde{U}}{d} e^{i\omega t}\, dt = v_0 \left[1 + \frac{i\eta\tilde{U}}{\omega v_0 d}(e^{i\omega\tau} - e^{i\omega t}) \right] \tag{6.25}$$

Integrating $v(\tau,t)$ gives the coordinate of the particle:

$$x(\tau,t) = \int_\tau^t v(\tau,t)\, dt = v_0(t-\tau) + \frac{i\eta\tilde{U}(t-\tau)}{\omega d} + \frac{\eta\tilde{U}}{\omega^2 d}(e^{i\omega\tau} - e^{i\omega t}) \tag{6.26}$$

where we assumed that $x_{t=\tau} = 0$.

Here we encounter the following difficulty: The velocity v in Eq. (6.25) was found as a function of τ and t, whereas in Eq. (6.20), $v = v(x,t)$. Note that τ and t are Lagrangian variables (see, e.g., Lamb, 1945). They describe electron flow via the motion of separate particles. So $v(\tau,t)$ is the velocity of a particle in a moment t that began to move in a moment τ, whereas x and t are Eulerian variables and $v(x,t)$ is the vector of the velocity field that describes the electron beam as a continuous medium. So $v(x,t)$ is the velocity of particles in the moment t in the plane x. It is possible to convert $v(\tau,t)$ in $v(x,t)$ using Eq. (6.26). To find $v(x,t)$ from $v(\tau,t)$ it is necessary to calculate a function $\tau(x,t)$. However, determining the function $\tau(x,t)$ from transcendental equation (6.26) is not a simple problem. In addition, the $v(x,t)$ obtained can be a multivalued function of x. At last, it remains a problem of obtaining, the function $\rho(x,t)$.

6.3.2 Law of Charge Conservation

Let us concentrate on the physical mining of the velocity multivaluedness of multistream states of beams with intersecting electron trajectories. In principle, it is possible to consider multistream beams; however, in the present case it is easier to convert Eq. (6.20) to Lagrangian variables for the current induced. This is reached by using the law of charge conservation.

Consider the motion of the electron in a planar gap (Fig. 6.7). Assume that the electron enters the gap through the grid $x = 0$ in the moment τ and finds itself in the plane x in a moment t where $x = x(\tau,t)$ according to Eq. (6.26). Now let us place an electron in the gap in the moment $\tau + d\tau$. It is clear that in the same moment t, the electron will be in the plane $x + dx$, where $dx = (\partial x/\partial\tau)_{t=\text{const}} < 0$. So the same charge $I_0\, d\tau$ that

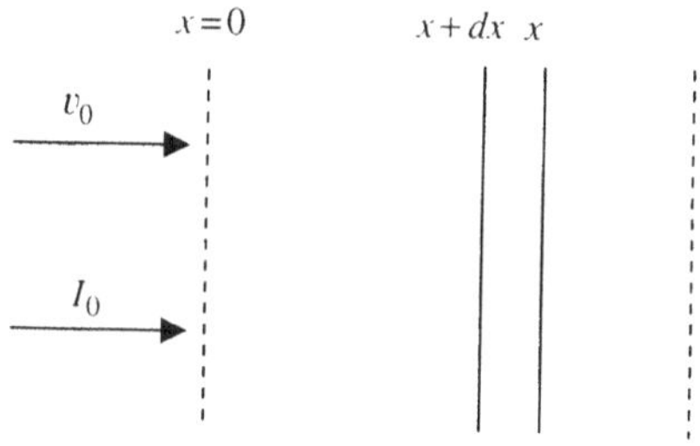

FIGURE 6.7 Law of charge conservation.

intersects the plane $x = 0$ for the time interval $d\tau$ will be disposed in the moment t on the line section dx. We obtain the *law of charge conservation*:

$$I_0 \, d\tau = -\rho_l \, dx \tag{6.27}$$

6.3.3 Calculation of Induced Current

Substituting Eq. (6.27) in Eq. (6.20), we determine the induced current as

$$I_{\text{ind}} = -\frac{1}{d}\int_0^d I_0 v(x, t) \, d\tau \tag{6.28}$$

In this integral, the independent variable is τ. Therefore, we can replace $v(x, t)$ by the function $v(\tau, t)$ from Eq. (6.25). We must also change the limits of the integration. The limit $\tau = t$ corresponds to $x = 0$. Because the integration in Eq. (6.28) must be carried out for the same time t [see the remark following Eq. (6.13)], the upper limit $x = d$ must correspond to $\tau = t - T$, where T is the transit time. Thus, Eq. (6.28) is replaced by

$$I_{\text{ind}}(t) = -\frac{I_0}{d}\int_t^{t-T} v(\tau, t) \, dt = \frac{I_0}{d}\int_{t-T}^{t} v_0\left[1 + \frac{i\eta\widetilde{U}}{\omega v_0 d}\left(e^{i\omega\tau} - e^{i\omega t}\right)\right] d\tau \tag{6.29}$$

The integration is developed easily. However, let us first introduce the following dimensionless variables that are typical for microwave electronics.

- $\psi = \omega t$ is a dimensionless time. Because $U = \widetilde{U}e^{i\omega t} = \widetilde{U}e^{i\psi}$, this variable is a phase of high-frequency voltage applied to the gap.
- $\varphi = \omega T$ is a transit angle. That is a change in the voltage phase for the transit time. Note that T and φ are not constants. The electrons that enter into the gap in the different moments of time move in different fields, and their transit time will be different. So $T = T(t)$ and $\varphi = \varphi(\psi)$.
- $\varphi_0 = \omega T_0$ is a nondisturbed transit angle; $T_0 = d/v_0$ is a nondisturbed transit time.
- $\alpha = \omega\tau$ is an input phase (i.e., the voltage phase in a moment of the electron's entry into the gap).

- $\xi = \widetilde{U}/U_0$ is a voltage factor.
- $\mu = \eta\widetilde{U}/\omega v_0 d = \xi/2\varphi_0$ is a velocity modulation factor.

Using these variables, we rewrite Eq. (6.25) as

$$v = v_0[1 + i\mu(e^{i\alpha} - e^{i\psi})] \tag{6.30}$$

After integration, the current induced in the dimensionless variables according to Eq. (6.29) would be

$$I_{\text{ind}} = \frac{I_0}{\varphi_0}\int_{\psi-\varphi}^{\psi} [1 + i\mu(e^{i\alpha} - e^{i\psi})]\, d\alpha = \frac{I_0}{\varphi_0}[\varphi + \mu e^{i\psi}(1 - e^{-i\varphi} - i\varphi)] \tag{6.31}$$

Outwardly, Eq. (6.31) is very simple. However, for calculating $I_{\text{ind}} = I_{\text{ind}}(\psi)$, we must know the relationship of the transit angle and the time [the function $\varphi = \varphi(\psi)$].

To get this function, let us use Eq. (6.26) for $x = d$. The electron in the plane $x = d$ at this moment t must enter the gap at the moment $\tau = t - T$. Then, setting $x = d$ and $\tau = t - T$, in Eq. (6.26) and transforming the equation to dimensionless variables, we obtain

$$\varphi = \varphi_0 + \mu e^{i\psi}(1 - e^{-i\varphi} - i\varphi e^{-i\varphi}) \tag{6.32}$$

This equation is transcendental for the desired function $\varphi(\psi)$. It does not have an analytical solution. According to Eq. (6.32), the function $\varphi(\psi)$ is periodical with period 2π (i.e., the transit time is periodical in t with a period $\widetilde{T}$ of the input signal). The current induced is also periodic in t with the period $\widetilde{T}$. But it is obvious that the current induced is not a harmonic function and includes a number of harmonics from the input signal. So, in general, the electron gap is a non-linear element. Note that because Eq. (6.32) is nonlinear, it is necessary to transform this equation to a real form before obtaining a numerical solution.

6.3.4 Linearization of Induced Current. Complex Admittance of an Electron Gap

It is interesting to explore Eq. (6.31) in a linear approximation when the amplitude of the input signal is small (i.e., the voltage factor $\xi = \widetilde{U}/U_0 \ll 1$, and the velocity modulation factor $\mu = \xi/2\varphi_0 \ll 1$). If we replace the term in parentheses in Eq. (6.32) by $\varphi \to \varphi_0$, we make a mistake on the order of ξ. Substituting φ from Eq. (6.32) into the first term φ in brackets in Eq. (6.31), we represent the current induced as

$$I_{\text{ind}} = I_0\left[1 + \frac{i\xi\varphi}{\varphi_0^2}e^{-i\varphi/2}\left(\frac{\sin(\varphi/2)}{\varphi/2} - \cos(\varphi/2)\right)e^{i\psi}\right] \tag{6.33}$$

We find that the periodic in the time term is proportional to ξ. Because φ differs from φ_0 by a value of order ξ, we will make an error on the order ξ^2 in I_{ind} if we replace φ in Eq. (6.33) by φ_0. So up to values of second order in ξ, we obtain

$$I_{\text{ind}} = I_0 + I_0 \frac{i\xi}{\varphi_0} e^{-i\varphi_0/2} \left[\frac{\sin(\varphi_0/2)}{\varphi_0/2} - \cos\frac{\varphi_0}{2} \right] e^{i\psi} \tag{6.34}$$

In this approximation, the current induced is the harmonic function of $\psi = \omega t$, and the electron gap is a linear element. Let us represent Eq. (6.34) as $I_{\text{ind}} = I_0 + I_e e^{i\psi}$, where the function

$$I_e = i\xi \frac{I_0}{\varphi_0} e^{-i\varphi_0/2} \left[\frac{\sin(\varphi_0/2)}{\varphi_0/2} - \cos\frac{\varphi_0}{2} \right] \tag{6.35}$$

is the complex amplitude of the current induced. Then the total current corresponding to Eq. (6.8) is equal to the sum of the induced and capacitive currents:

$$I_t = I_e e^{i\psi} + C_g \frac{dU}{dt} = (I_e + I_{\text{cap}}) e^{i\psi} = \tilde{I} e^{i\psi} \tag{6.36}$$

where $I_{\text{cap}} = i\omega C_g \tilde{U}$ is a complex amplitude of the capacitive current. A complex admittance of the electron gap as the linear element is

$$Y = \frac{\tilde{I}}{\tilde{U}} = \frac{I_e}{\tilde{U}} + \frac{I_{\text{cap}}}{\tilde{U}} = Y_e + i\omega C_g \tag{6.37}$$

where the complex electron admittance

$$Y_e = i\frac{Y_0}{\varphi_0} e^{-i\varphi_0/2} \left[\frac{\sin(\varphi_0/2)}{\varphi_0/2} - \cos\frac{\varphi_0}{2} \right] = Y_{er} + iY_{ei} \tag{6.38}$$

Here $Y_0 = I_0/U_0$ is a dc conductance of the electron gap. The real and imaginary parts of Y_e are correspondingly the conductance and susceptance of the gap:

$$Y_{er} = \text{Re}(Y_e) = \frac{Y_0}{2} \left[\left(\frac{\sin(\varphi_0/2)}{\varphi_0/2} \right)^2 - \frac{\sin\varphi_0}{\varphi_0} \right] \tag{6.39}$$

$$Y_{ei} = \text{Im}(Y_e) = Y_0 \left[\frac{\sin\varphi_0}{\varphi_0^2} - \frac{\cos^2(\varphi_0/2)}{\varphi_0} \right] \tag{6.40}$$

The quality plot of Y_{er} and Y_{ei} as functions of φ_0 is depicted in Fig. 6.8. Note that according to Eqs. (6.39) and (6.40), $Y_{er} = 0$ for $\varphi_0 = 0,\ 2n\pi$, and $Y_{ei} = 0$ for $\varphi_0 = 0, (2n+1)\pi \quad (n = 1, 2, \ldots)$.

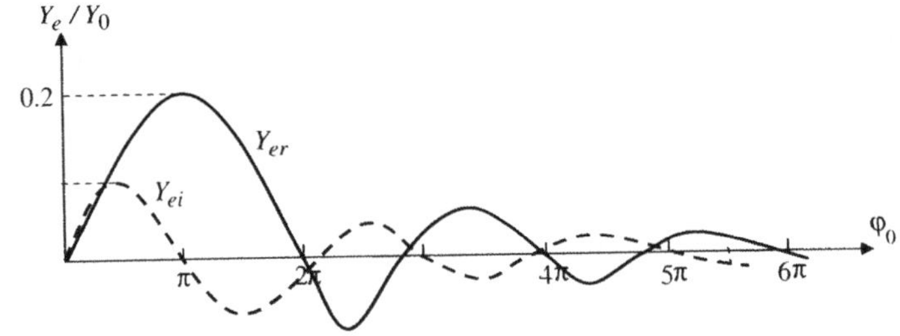

FIGURE 6.8 Conductance and susceptance of an electron gap as functions of a nondisturbed transit angle.

Discussion of Results The average oscillation period power expended by any source of a harmonic current is

$$\overline{IU} = \frac{1}{T}\int_0^T IU\,dt = Y_r\frac{\widetilde{U}^2}{2} \tag{6.41}$$

Because the capacitive admittance is a purely imaginary value, the conductance of the electron gap is the electron conductance (i.e., $Y_r = Y_{er}$). According to Fig. 6.8, the conductance of the gap in the range $0 < \varphi_0 < 2\pi$ is positive and the source accomplishes positive work (i.e., on the average over the period when the electrons are accelerated. In the range $2\pi < \varphi_0 \ll 3\pi$, the work of the source is negative, which means that on average over the period, the source retards the electrons.

Mechanism of Acceleration and Retardation Consider the following example: Assume that a nondisturbed transit angle is π (i.e., Y_{er} is positive). Assume that the voltage on the gap varies as $\widetilde{U}\sin\omega t$ (Fig. 6.9). Also assume that electron 1 enters the gap at the moment $t = 0$ and the nondisturbed transit time is equal to a half-period. It means that if the electron moves with a nondisturbed velocity v_0, it would leave the gap in the moment $t = \frac{1}{2}T_0$. But the electron moves in the accelerating field and its velocity would be $v > v_0$. Therefore, the electron leaves the gap $t_1 < \frac{1}{2}T_0$ at that moment. Now let us take electron 2, which enters the gap at the moment $t = -\frac{1}{2}T_0$. This electron begins the motion in the retarding

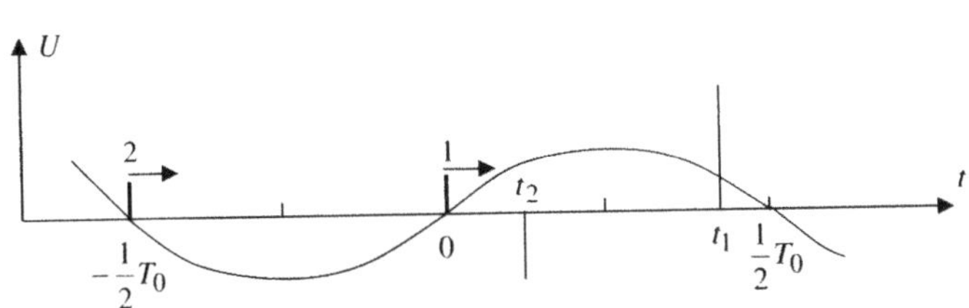

FIGURE 6.9 Electron bunching in an electron gap.

field (Fig. 6.9) with velocity $v < v_0$. Therefore, electron 2 leaves the gap at the moment $t_2 > 0$.

Following is a summary. The first electron spent all of its time moving in the accelerating field, while the second electron moved partly in the retarding and partly in the accelerating field. Therefore, *on average*, both particles are accelerated, and the source performs positive work. It is important to pay attention to the time interval changes between particles. In the beginning they were a half-period apart. In the end they have closed in. This means that *bunching* occurs. In the present case, bunching occurred in the accelerated phase of the field. Thus considered, the transit effect has led to the bunching of the particles and phasing of the bunches. Furthermore, we will see that bunching and phasing are typical for any microwave electron devices. Note that for nondisturbed transit angles in the range $2\pi < \varphi_0 < 3\pi$, the conductance is negative. This may be treated as bunching of electrons in the retarding phase of the field.

It is interesting to trace why $Y_r = 0$ for $\varphi_0 = 0$. In this case, when the particles intersect the gap, the field does not change and bunching is absent. Therefore, half of the particles move in constantly accelerating fields and the other half in retarding fields. The summary effect is equal to zero.

The imaginary part of the electron admittance (susceptance) arises as a result of a phase difference between the bunch and the field. Note that when the phase shift is equal to zero or to an odd multiple of π, the bunches are in phase or antiphase with the field, and the susceptanse is equal to zero. In general, susceptance has different signs. This means that for different transit angles, the electron gap acquires capacitive or inductive properties.

6.3.5 Equivalent Circuit of an Electron Gap. Electron Gap as an Oscillator. Monotron

The total admittance of an electron gap is

$$Y = Y_{er} + Y_{ei} + i\omega C_g \tag{6.42}$$

Summation of the conductance corresponds to their parallel connection in the equivalent circuit performance. The equivalent scheme is shown in Fig. 6.10.

Let us adopt the oscillator scheme shown in Fig. 6.11*a*. C_c, L_c, and R are parameters of the circuit; and R is a resistor that includes losses from L_c and C_c and from an output load. An equivalent circuit for the oscillator is shown in Fig. 6.11*b*. Here, the total capacity is $C = C_g + C_c$. The source $\tilde{U}e^{i\omega t}$ is connected to the circuit in

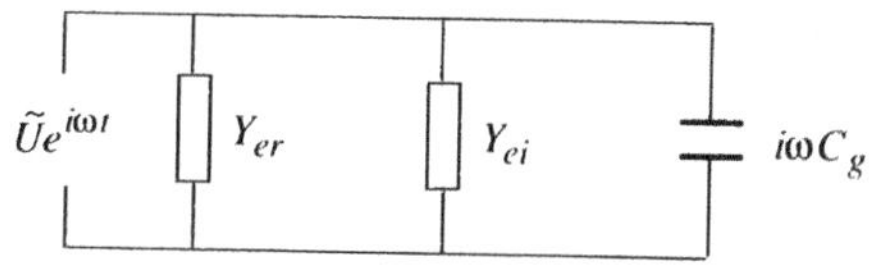

FIGURE 6.10 Equivalent circuit of an electron gap.

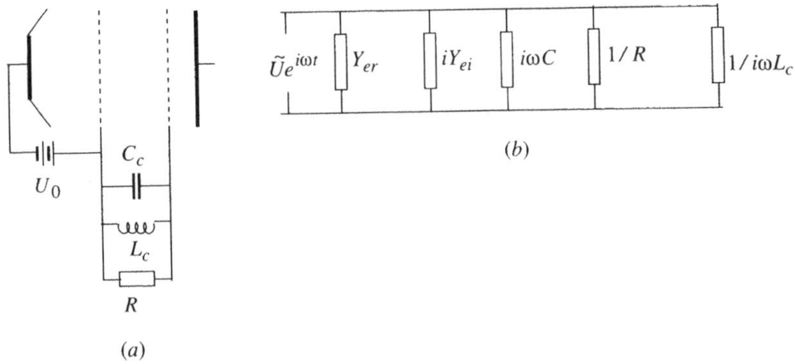

FIGURE 6.11 (*a*) Electron oscillator; (*b*) equivalent circuit.

Fig. 6.11*b*. But in the self-excitation mode, the source is absent. A condition of the self-excitation of the oscillations is equal to zero of the total complex admittance. This condition is

$$Y_{er} + \frac{1}{R} = 0 \qquad \text{real admittance} \tag{6.43}$$

$$iY_{ei} + i\omega C + \frac{1}{i\omega L_c} = 0 \qquad \text{imaginary admittance} \tag{6.44}$$

To satisfy Eq. (6.43), Y_{er} must be negative and sufficiently large in magnitude. A negative sign for Y_{er} is achieved by the choice of the transit angle, which is determined by the dc potential U_0, the distance d, and the frequency ω. The magnitude of Y_{er} is proportional to dc admittance $Y_0 = I_0/U_0$ (i.e., to the beam current), and does not depend on $\tilde{U}$. That is obviously a result of linear approximation. So the condition (6.43) gives a starting current of self-excited oscillations. A zero imaginary admittance [Eq. (6.44)] determines the frequency of oscillations (see below).

The oscillator in Fig. 6.11 is sometimes called a *monotron* (Birdsall and Bridges, 1966) because it works with a single oscillatory circuit. Experimental versions of monotrons have been developed successfully. However, the parameters of the current device have not been high because the processes of bunching and phasing in the monotron considered are not optimal.

6.4 EQUATION OF STATIONARY OSCILLATIONS OF A RESONANCE SELF-EXCITED CIRCUIT

Let us write the conditions of self-excitation of oscillations in explicit form. First, find the complex resonance frequency of the circuit (i.e., the eigenfrequency of the circuit without the electron beam). One is determined by the equation

$$Y_C = \frac{1}{R} + i\omega^{(0)}C + \frac{1}{i\omega^{(0)}L} = 0 \tag{6.45}$$

Assume that $\omega^{(0)}$ is close to the eigenfrequency of an ideal LC circuit, $\omega_0 = 1/\sqrt{LC}$: that is,

$$\frac{\left|\omega^{(0)}\right| - \omega_0}{\omega_0} \ll 1 \tag{6.46}$$

Then we obtain from Eq. (6.45),

$$\omega^{(0)} \approx \omega_0 + \frac{i}{2RC} = \omega_0 + i\omega_0' \tag{6.47}$$

The presence of the imaginary component $\omega_0' > 0$, of course, corresponds to an aperiodicity of the oscillations in the circuit without the electron beam. It is easy to show that

$$\omega_0' = \frac{1}{2RC} = \frac{\omega_0}{2Q} \tag{6.48}$$

where Q is the loaded quality factor:

$$Q = \omega_0 CR = \frac{\omega_0 W}{P_l} \tag{6.49}$$

W is the energy stored in the resonator and P_l is the power dissipated in the circuit, including losses of output power. Remember that Eq. (6.47) is true when $\omega_0' \ll \omega_0$ (i.e., $Q \gg 1$).

Now let us return to Eqs. (6.43) and (6.44). According to Eq. (6.48), $1/R = \omega_0 C/Q$; then, from Eq. (6.43),

$$Y_{er} + \frac{\omega_0 C}{Q} = 0 \tag{6.50}$$

Equation (6.44) can be written as

$$Y_{ei} + 2C(\omega - \omega_0) = 0 \tag{6.51}$$

To obtain Eq. (6.51), we took into account that the frequency of the self-excited oscillation ω is real and $|\omega - \omega_0| \ll \omega_0$. Equation (6.50) allows us to find a starting current if Y_{er} does not depend on the ac amplitude of the potential $\widetilde{U}$ (i.e., in the linear approximation of the admittance). In the same approximation, Eq. (6.51) gives the electron tuning of the frequency. Equations (6.50) and (6.51) can be rewritten when transferring from admittance to electron power. Express the capacity C through the energy stored in the circuit (resonator):

$$W = \frac{C\widetilde{U}^2}{2} \tag{6.52}$$

Here W is the sum of the energies stored in the capacity and the inductance. Let us multiply Eqs. (6.50) and (6.51) by $\widetilde{U}^2/2$ and take into account that $P_{er} = Y_{er}\widetilde{U}^2/2$ and $P_{ei} = Y_{ei}\widetilde{U}^2/2$ are the real and imaginary parts of the complex electron power. We then obtain

$$P_{er} + \frac{\omega_0 W}{Q} = P_{er} + P_l = 0 \tag{6.53}$$

$$\frac{\omega - \omega_0}{\omega_0} = -\frac{P_{ei}}{2\omega_0 W} = \frac{1}{2Q}\frac{P_{ei}}{P_{er}} \tag{6.54}$$

Remember that Eqs. (6.50) and (6.51) determine the starting current and frequency when the electron power P_e is proportional to $\widetilde{U}^2$ [i.e., when the electron admittance is linearized like Eqs. (6.39) and (6.40)]. However, Eqs. (6.53) and (6.54) exceed the limits of linear approximation. After multiplying Eqs. (6.50) and (6.51) by $\widetilde{U}^2/2$, we restore the nonlinear representation of the electron power, which is virtually contained in Eq. (6.7). As shown by Vainstein and Solntsev (1973) and by Vainstein (1988),[1] these equations describe, in general, the stationary oscillations of any single-mode self-excited oscillator when complex electron power at the input-to-oscillation circuit is known.

6.5 EFFECTS OF A SPACE-CHARGE FIELD. TOTAL CURRENT METHOD. HIGH-FREQUENCY DIODE IN THE ρ-MODE. LLEWELLYN–PETERSON EQUATIONS

The theory stated in Section 6.4 cannot describe the operation of devices in which an electron motion depends on the space-charge fields. When we wrote the equation of motion as $d^2x/dt^2 = \eta(U/d)$, we believed that $E = E(t) = -U/d$. However, when taking the space-charge field into account, the electric field must be written as $E = E(x, t)$, where the electric field in general is an unknown function of x. Then the equation of motion in the nonrelativistic approximation is

$$\frac{d^2x}{dt^2} = -\eta E(x, t) \tag{6.55}$$

Here, distribution of the space-charge field itself is determined by the motion of the particles, and we arrive at a typical self-matching problem.

6.5.1 Total Current Method

This method was first worked out by Lewellyn (1939). After Grinberg (1948) developed a method for arbitrary one-dimensional high-frequency electron gaps (i.e., cylindrical, spherical, etc.) and large transit angles. The theory of a planar electron gap with nonzero space charge is considered briefly below.

[1]Unfortunately, these two remarkable books have not been translated into English.

Let us take a time derivative of Eq. (6.55):

$$\frac{d^3x}{dt^3} = -\eta\frac{dE(x,t)}{dt} = -\eta\left(\frac{\partial E}{\partial t} + \frac{\partial E}{\partial x}\frac{dx}{dt}\right)$$
$$= -\frac{\eta}{\varepsilon_0}\left(\rho v + \varepsilon_0\frac{\partial E}{\partial t}\right) = -\frac{\eta}{\varepsilon_0}(j_c + j_d) = -\frac{\eta}{\varepsilon_0}j_t \tag{6.56}$$

(we used the relation $\partial E/\partial x = \text{div}\,\mathbf{E} = \rho/\varepsilon_0$). The density of the total current according to Eq. (6.5) depends only on t: $j_t = j_t(t)$.

In the theory of electron gaps with zero space charge, we set a voltage $U(t)$ and found a current. Now, we assume that the total current $I_t(t) = Sj_t(t)$ is given and we will search the voltage. Performing three successive integrations of Eq. (6.56) in the Lagrange space with initial acceleration $a(\tau,\tau)$, velocity $v(\tau,\tau)$, and coordinate $x(\tau,\tau)$, we obtain

$$\frac{d^2x}{dt^2} = a(\tau,t) = -\frac{\eta}{\varepsilon_0 S}\int_\tau^t I_t(t)\,dt + a(\tau,\tau) \tag{6.57}$$

$$\frac{dx}{dt} = v(\tau,t) = \int_\tau^t a(\tau,t)\,dt + v(\tau,\tau) \tag{6.58}$$

$$x(\tau,t) = \int_\tau^t v(\tau,t)\,dt + x(\tau,\tau) \tag{6.59}$$

On first glance the appearance of the initial acceleration is strange. It is known that motion of a particle in the field of forces is determined uniquely by its initial velocity and position. Why is it necessary to specify the initial acceleration? The reason is simple. In the present self-matching problem, the field is not given. The motion takes place in a field that must be found.

The method of finding a voltage is the following. From Eq. (6.59) we obtain the function $\tau = \tau(x,t)$. Then from Eq. (6.57) we find the electric field:

$$E(x,t) = -\frac{a(x,t)}{\eta} = -\frac{a[\tau(x,t),t]}{\eta} \tag{6.60}$$

and the voltage $U(t) = -\int_0^d E(x,t)\,dx$.

6.5.2 Analysis of a Diode for Current Limited by Space Charge

In Chapter 3 we analyzed the static diode in a ρ-mode (current limited by space charge). Let us consider briefly the results of the theory of high-frequency planar diode in the ρ-mode. The scheme of an ac source connected in series with a dc voltage source is depicted in Fig. 6.12. Let us use the basic equation of the total current method:

$$\frac{d^3x}{dt^3} = -\frac{I(t)}{\varepsilon_0 S} \tag{6.61}$$

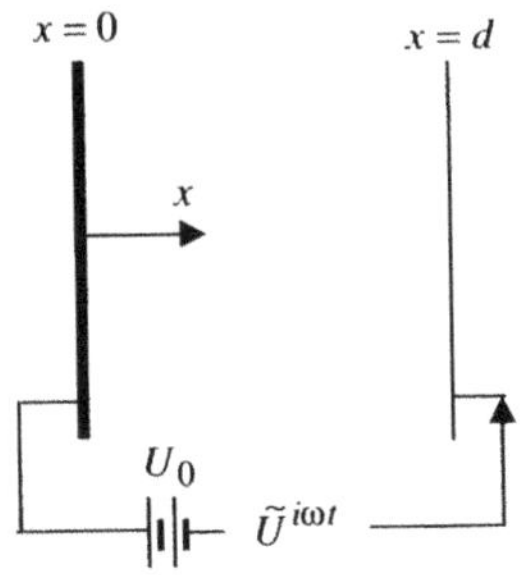

FIGURE 6.12 High-frequency flat diode.

When we take the total current as a harmonic function of time $I(t) = I_0 + \widetilde{I}e^{i\omega t}$, a consistent solution of the corresponding Eqs. (6.57)–(6.60) would obviously lead to a solution for the ac voltage as a nonharmonic function of time. However, if we use a linear approximation assuming the smallness of the quantities $\widetilde{I}/I_0$ and $\widetilde{U}/U_0$, the ac voltage would be a harmonic function of time as noted in Fig. 6.12.

Let us assume that both the initial position and the velocity of electrons found in the solution of Eq. (6.61) are equal to zero:

$$x(\tau, t)_{t=\tau} = 0 \tag{6.62}$$

$$v(\tau, t)_{t=\tau} = 0 \tag{6.63}$$

The cathode electric field in the ρ-mode with a zero initial velocity is zero. Therefore, the initial acceleration is also zero:

$$a(\tau, t)_{t=\tau} = 0 \tag{6.64}$$

The scheme for the solution corresponds to the chain of equations (6.57)–(6.60). Here we omit the calculations and give the results (see, e.g., Chodorow and Susskind, 1964). In contrast to the theory of the electron gap (Section 6.3.4), the solution is given in terms of the complex impedance $Z = \widetilde{U}/\widetilde{I} = R + iX$, where R is the resistance (a real part of the impedance) and X is the reactance (an imaginary part of the impedance). The resistance and the reactance as functions of the nondisturbed transit angle φ_0 are (Chodorow and Susskind, 1964)

$$R = R_0 \frac{12}{\varphi_0^4}[2(1 - \cos\varphi_0) - \varphi_0 \sin\varphi_0] \tag{6.65}$$

$$X = -R_0 \frac{12}{\varphi_0^4}\left[\frac{\varphi_0^3}{6} + \varphi_0(1 + \cos\varphi_0) - 2\sin\varphi_0\right] \tag{6.66}$$

Here R_0 is a dc admittance of the diode. Using the Child–Langmuir formula [Eq. (3.105)], $I_0 = \frac{4}{9}(S/d^2)\varepsilon_0\sqrt{2\eta}U_0^{3/2}$, and Eq. (3.103), $U(x) = U_0(x/d)^{4/3}$, for dc potential distribution, we obtain

$$R_0 = \frac{dU_0}{dI_0} = \frac{3d^2}{2\varphi\varepsilon_0 S\sqrt{2\eta U_0}},$$
$$\varphi_0 = \omega T_0 = \omega \int_0^d \frac{dx}{v} = \omega \int_0^d \frac{dx}{\sqrt{2\eta U(x)}} = \frac{3\omega d}{\sqrt{2\eta U_0}} \tag{6.67}$$

Resistance and reactance as functions of the nondisturbed transit angle are depicted in Fig. 6.13. According to Eqs. (6.65) and (6.66), $R = 0$ for $\varphi_0 = 2n\pi$ and $X = 0$ for $\varphi_0 = 0$. As we can see, the resistance is positive in the range $0 \leq \varphi_0 < 2\pi$ and negative in the range $2\pi < \varphi_0 < 2.7\pi$. The magnitude of the negative R is very small. So the maximum of $-R/R_0$ is ~ 0.022 ($\varphi_{0,\max} \approx 2.3\pi$). In contrast to the resistance, the reactance is a monotonic function of φ_0. The admittance of the diode is

$$Y = \frac{1}{Z} = \frac{R}{R^2 + X^2} - i\frac{X}{R^2 + X^2} \tag{6.68}$$

In the diode and in contrast to the electron gap, the number of electrons that enter the interelectrode space in the positive and negative half-periods of the potential are not the same because the current depends on the voltage. Therefore, the admittance for the zero transit angle is not zero either: $Y_{\varphi_0=0} = 1/R_{\varphi_0=0} = 1/R_i$.

For large transit angles ($\varphi_0 \sim \infty$), the resistance $|R_\infty| \ll |X_\infty|$, and according to Eq. (6.68), the reactance is

$$Y_\infty \approx -i\frac{1}{X_\infty} \approx i\frac{\varphi_0}{2R_i} \tag{6.69}$$

Substituting φ_0 and R_i from Eq. (6.67), we obtain $Y_\infty = \omega(\varepsilon_0 S/d) = \omega C$; that is, the admittance of the diode for large frequencies is purely reactive and is equal to the reactance of the cold capacity formed by electrodes of the diode.

The scheme of the diode oscillator is very simple, for it is only necessary to include an oscillatory circuit in the diode circuit (Fig. 6.14). For oscillation

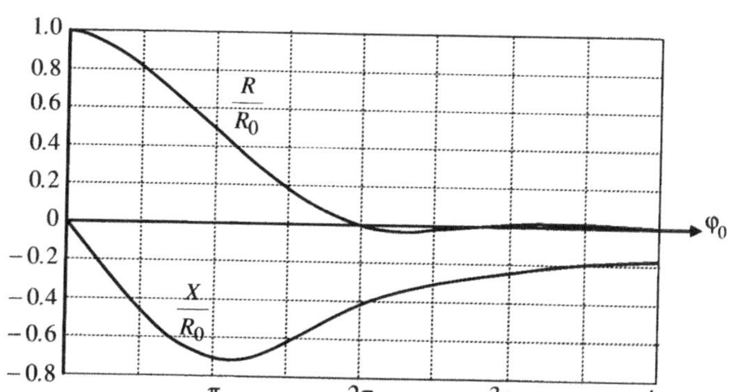

FIGURE 6.13 Diode impedance for current limited by a space charge.

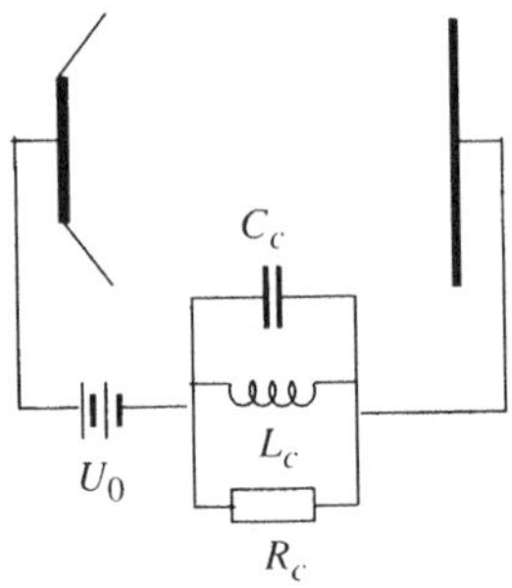

FIGURE 6.14 Diode oscillator.

excitation the real part of the admittance [according to Eq. (6.68), approximately equal to R/X^2 for $\varphi_0 > 2\pi$] must be taken to be negative and sufficiently large in magnitude. That is attained by the choice of transit angle φ_0. After choosing the oscillation frequency and according to Eq. (6.67), selection of the values d, S, and U_0 remains.

The oscillation frequency is determined by the capacity C_c and inductance L_c. If it remains within the range of transit angles where the conductivity is negative, it is possible to retune the frequency by a small variation of the voltage U_0. The efficiency of experimental diode oscillators is very low (less than 1%) (Llewellyn and Bowen, 1939; Muller, 1934), because of the noneffective processes of bunching and phasing.

6.5.3 Llewellyn–Peterson Equations

Llewellyn–Peterson (L-P) equations (Llewellyn, 1935; Llewellyn and Peterson, 1944) are the result of consistent application of the total current method (Section 6.5.1) to an analysis of planar electron gaps, taking the linear approximation and space-charge fields into account. So any quantity $\zeta_\Sigma(\tau, t)$ in the problem, including the transit time $T = t - \tau$, can be represented as $\zeta_\Sigma = \zeta_0 + \zeta = \zeta_0 + \tilde{\zeta} e^{i\omega t}$, where ζ_0 and ζ are dc and ac components of ζ_Σ and $\zeta \ll \zeta_0$, respectively. The equations relate input and output quantities in two planes of the gap (Fig. 6.15), where v_a and v_b are ac inputs and output velocities, respectively, and I_{ca} and I_{cb} are the corresponding ac convective currents. Finally, the I_t is the total ac current, which according to Section 6.2.1 is continued into the gap and equals the ac current in the leads. These equations follow:

$$\begin{aligned} U &= a_{11} I_t + a_{12} I_{ca} + a_{13} v_a \\ I_{cb} &= a_{21} I_t + a_{22} I_{ca} + a_{23} v_a \\ v_b &= a_{31} I_t + a_{32} I_{ca} + a_{33} v_a \end{aligned} \tag{6.70}$$

The coefficients a_{ij} are functions of the gap parameters (e.g., geometric parameters, an initial dc velocity, a dc potential U_0, an undisturbed transit angle φ_0 between

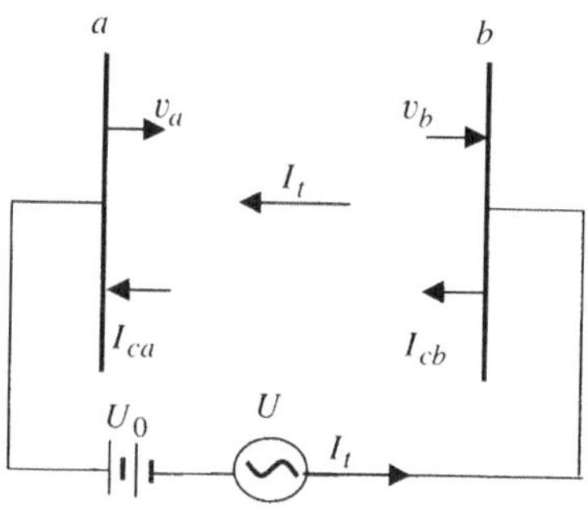

FIGURE 6.15 Variables in Llewellyn–Peterson equations.

planes a and b) and the frequency of the signal. The expressions for a_{ij} are very cumbersome and have been omitted. The reader can find them for example, in, Kleen (1958).

Let us write the L-P equations for a diode in the ρ-mode with zero initial velocity. Assume that both the cathode and anode are arranged correspondingly in planes a and b. Then $v_a = 0$ and $I_{ca} = I_t$ because the cathode electric field in the ρ-mode and the corresponding displacement current are zero. The L-P equations (6.70) are reduced to

$$\begin{aligned} U &= (a_{11} + a_{12})I_t \\ I_{cb} &= (a_{21} + a_{22})I_t \\ v_b &= (a_{31} + a_{32})I_t \end{aligned} \tag{6.71}$$

We see, in particular, that the diode impedance is $Z = U/I_t = a_{11} + a_{12}$.

The L-P equations have been used for analysis of multielement tubes (e.g., triodes, tetrodes) considering the regions between electrodes as successive electron gaps. The exit velocities v_{bi}, convention currents I_{cbi} total currents I_{ti} of any gap, and corresponding dc quantities, can be used as entrance conditions for a following gap, and so on. For example, the total current in the second gap can be written as

$$I_{t2} = I_{t1} - I_b \tag{6.72}$$

where I_b is the lead current in plane b.

CHAPTER SEVEN

Klystrons

7.1 INTRODUCTION

In Chapter 6 we have seen how transit effects radically revise traditional concepts of quasistationary devices. The novel essential properties were bunching and phasing of particles in the beam. That ensures electron-stimulated transition radiation in the gap and leads to such important properties of the electron gaps as negative or additional positive absorption of the ac electron beam energy. The negative admittance allows us, in particular, to realize high-frequency oscillators. However, as mentioned, these effects are weak and these devices have found few applications in microwave electronics.

The concept of a klystron, the first genuine microwave electronic device to take full advantage of the principle of bunching and phasing, was proposed by Russel and Sigurd Varian in 1935. During 1935–1939, Russel, an engineer, and his brother Sigurd, a pilot, developed their klystron idea. Together with William Hansen, Russel Varian worked out the theoretical basis of the klystron while Sigurd Varian built the device. The first klystron was tested in August 1937. The Varians named the new device the *klystron*, using the Greek word *clyso*, meaning "waves breaking on a shore."

Stanford University provided the Varian brothers with only $100 for all this work. However, the great value of the klystron was very quickly appreciated by contemporaries. The past 65 years have fully confirmed expectations. Until now, klystrons have found very extensive use among all microwave devices as power and superpower amplifiers in the UHF and SHF ranges. It is sufficient to mention such important areas as radar, radio navigation, space communication, television, radio repeaters, and charged particle accelerators. In 1948 the Varians established the company Varian Associates, now one of the largest corporations producing microwave electronic devices.

Electron Beams and Microwave Vacuum Electronics. By Shulim E. Tsimring

Let us quote a description of the klystron mechanism given by the Varians (Varian and Varian, 1939), which can be used as a good introduction to this chapter):

> A dc stream of cathode rays of constant current and speed is sent through a pair of grids between which is an oscillating electric field, parallel to the stream and of such strength as to change the speeds of the cathode rays by appreciable but not too large fractions of their initial speed. After passing these grids the electrons with increased speeds begin to overtake those with decreased speeds ahead of them. This motion groups the electrons into bunches separated by relatively empty spaces. At any points beyond the grids, therefore, the cathode ray current can be resolved into the original dc plus a nonsinusoidal ac. A considerable fraction of its power can then be converted into power of high frequency oscillations by running the stream through a second pair of grids between which is an ac electric field such as to take energy away from the electrons in bunches. These two ac fields are best obtained by making the grids form parts of the surfaces of resonators of type described in [this journal].

In Fig. 7.1 a diagram of the klystron that corresponds to this description is shown. Here a buncher cavity implements velocity modulation of the electron beam. After the buncher gap, electrons move with different velocities, depending on the phase of the ac field between buncher grids. In the drift tube between the buncher and the catcher, an electric field is absent, and the electrons move due to the force of inertia and are bunched. In the catcher cavity, which in general can be tuned to the nth harmonic of the input signal, the electron bunches deliver power to the ac field, and the output signal is formed.

The Cavities The buncher and catcher grids are installed in oscillatory circuits where the two pairs of grids play the role of capacitances. Usually, the klystron circuits are versions of the cavity resonators (sometimes, of toroidal cavities). The use of cavities allows us to accumulate the ac field energy and increase the field amplitude in order to intensify electron–field interaction. That property unifies

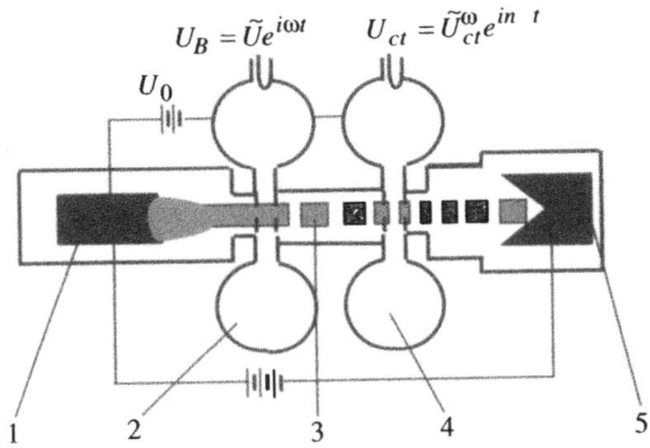

FIGURE 7.1 Double-cavity klystron. 1, Cathode; 2, buncher cavity; 3, bunched electron beam; 4, catcher cavity; 5, collector.

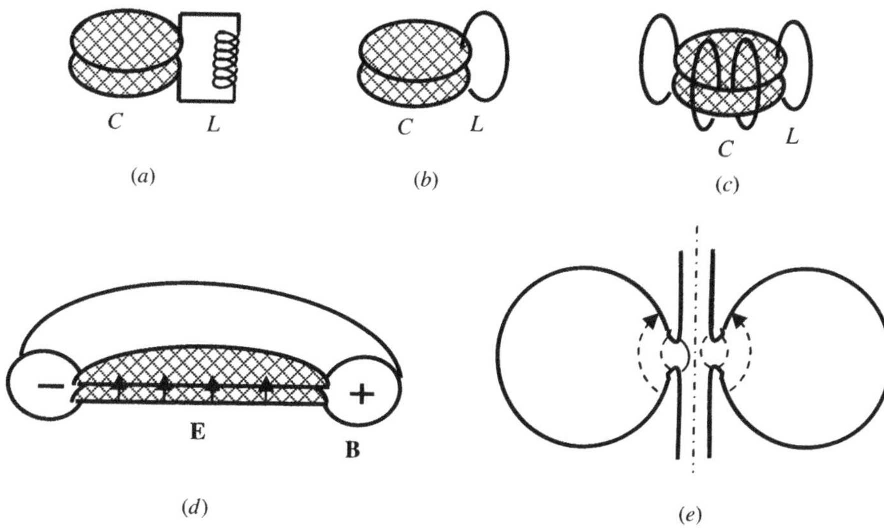

FIGURE 7.2 Transformation of a simple oscillation circuit in a toroidal cavity resonator.

the klystrons with other nonquasistationary microwave devices (e.g., TWTs, BWOs, magnetron amplifiers). The evolution of the ac circuits originating from the usual low-frequency oscillation circuit to the toroidal cavity is illustrated in Fig. 7.2. Replacement of the multiturn inductor (Fig. 7.2*a*) by a single half-turn wire (Fig. 7.2*b*) and further by a few inductors in parallel (Fig. 7.2*c*) decreases the inductive impedance of the circuit and allows its natural resonant frequency to increase. Finally, substitution of many half-turn wires by a wall of the hollow toroid transforms the system into a cavity resonator. In principle, many oscillation types (modes) are possible in this resonator. Usually, the lowest-frequency (fundamental) mode is used. In this mode the electric field is localized between grids, and the magnetic field is stored in a toroidal tube (Fig. 7.2*d*). It is important that radiation losses are nearly absent in the cavity resonator. Use of a cavity resonator in the klystron was a revolutionary idea of Hansen and the Varians. Hansen called these resonators *rumbatrons*. The first Varian klystron model underwent significant improvements in modern klystron systems. The largest modernization is related to application of multiple-cavity and multibeam klystrons (Section 7.7) and to abandonment of gridded cavities in high-power, high-current klystrons. The ac electric field that is localized in the central region of these gridless resonators (Fig. 7.2*e*) is quasiuniform and available for effective interaction with an electron beam.

7.2 VELOCITY MODULATION OF AN ELECTRON BEAM

A buncher can be considered as an electron gap with applied potential $U_B = \widetilde{U}e^{i\omega t}$. Let us examine the velocity modulation process, neglecting the space-charge field in the gap. This process was analyzed in Section 6.3. We have obtained Eq. (6.30) for the velocity in dimensionless Lagrange variables:

$$v = v_0[1 + i\mu(e^{i\alpha} - e^{i\psi})] \tag{7.1}$$

Let us recall that $v_0 = \sqrt{2\mu U_0}$ is the input electron velocity, $\alpha = \omega\tau$ is the input phase, $\mu = \xi/2\varphi_0$ is the velocity modulation factor, and $\xi = \widetilde{U}/U_0$ is the voltage factor. That expression gives an electron velocity at time t if the electron enters the gap at time τ.

We are interested in the output velocity of electrons from the gap; hence, if t is the time of electron departure, the input time and input phase will be, respectively,

$$\tau = t - T, \qquad \alpha = \psi - \varphi \tag{7.2}$$

where T is the transit time and $\varphi = \omega T$ is the transit angle. Substituting Eqs. (7.2) in Eq. (7.1), we obtain

$$v = v_0[1 + i\mu(e^{i(\psi-\varphi)} - e^{i\psi})] = v_0\left[1 + \frac{1}{2}\xi\frac{\sin(\varphi/2)}{\varphi_0/2}e^{i(\psi-\varphi/2)}\right] \tag{7.3}$$

In this equation the transit angle $\varphi = \varphi(\psi)$ is an unknown function of time that must be determined [see Eq. (6.32)]. However, it is necessary to remember that the klystron is an amplifier and the input signal (drive power) is weak. Therefore, the voltage factor $\xi \ll 1$. Because the small parameter ξ is in front of the temporal exponent in Eq. (7.3), replacing φ by φ_0 in this term leads to an error of second order in ξ [see also Eqs. (6.32) and (6.33)]. Thus, the approximate output velocity is a linear function of $\widetilde{U}$:

$$v = v_0\left[1 + \frac{1}{2}\xi\frac{\sin(\varphi_0/2)}{\varphi_0/2}e^{-i\varphi_0/2}\,e^{i\psi}\right] = v_0 + \tilde{v}\,e^{i\psi} \tag{7.4}$$

The velocity v is a sum of dc and ac components. The ac component is a harmonic function of time. The complex amplitude of the ac velocity component can be written as $\widetilde{v} = \frac{1}{2}v_0\xi M e^{-i\varphi_0/2}$, where $M = \sin(\varphi_0/2)/\varphi_0/2$ is the gap coupling coefficient. So the velocity is

$$v = v_0\left(1 + \tfrac{1}{2}\xi M e^{-i\varphi_0/2}\,e^{i\omega t}\right) \tag{7.5}$$

Discussion of Results The coupling coefficient M characterizes the influence of the transit angle on the velocity modulation. The phase factor $e^{-i\varphi_0/2}$ indicates that the ac velocity component is shifted backward in phase relative to the input signal by a half of the nondisturbed transit angle. It means that the ac output velocity component is in phase with the ac voltage taken at the time electrons pass the gap midplane.

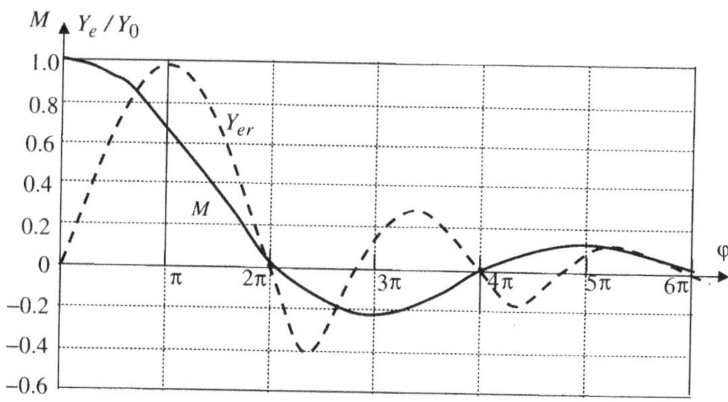

FIGURE 7.3 Coupling coefficient and conductance of an electron gap as functions of a nondisturbed transit angle.

The dependence of M on the nondisturbed transit angle is depicted in Fig. 7.3. If, for example, $\varphi_0 = 2\pi$, the electrons would be accelerated by the electric field within the first half of the gap and retarded during the second half. The resulting ac velocity would be zero. According to Fig. 7.3, the best value of the transit angle is $\varphi_0 \sim 0$. But this case experiences technological difficulties, especially for short wavelengths.

Let us take, for example, $U_0 = 3\,\text{kV}$, $\lambda = 4$ mm; then the transit angle is

$$\varphi_0 = \frac{\omega d}{v_0} = 2\pi \frac{c}{v_0} \frac{d}{\lambda}$$

For $U_0 = 3$ kV we obtain $c/v_0 \approx 16/\sqrt{U_{0\ \text{kV}}} \approx 10$, and the corresponding transit angle is $\varphi_0 \approx 20\pi(d/\lambda)$. Assume, for example, that $d = 0.2$ mm. That corresponds to $d/\lambda = 0.05$, and we obtain $\varphi_0 = \pi$. For $\varphi_0 = \pi/4$, the gap width d would be 0.05 mm, which is a very small value. Can we take $\varphi_0 = \pi$? In this case, according to Fig. 7.3, M is reduced to ~65%. This is not so bad. However, let us recall the gap conductance as a function of the transit angle (Section 6.3.4). According to Fig. 7.3, the gap conductance for $\varphi_0 = \pi$ is maximal. It leads to high power losses during the acceleration of electrons in the gap. The electrons "load" the cavity resonator, and its Q-factor is decreased. As a result, the accumulated circuit energy $[w = Q(P/\omega)]$, the corresponding ac electric field, and the modulation efficiency will drop. Note that for $\varphi_0 \sim 0$ the modulation is realized without energy consumption. But as we saw earlier, this process is not feasible technologically.

7.3 CINEMATIC (ELEMENTARY) THEORY OF BUNCHING

7.3.1 Qualitative Discussion

One may suppose that after leaving the buncher gap, electrons move in a channel where dc and ac fields are absent. It is the first approximation in klystron theory. The most

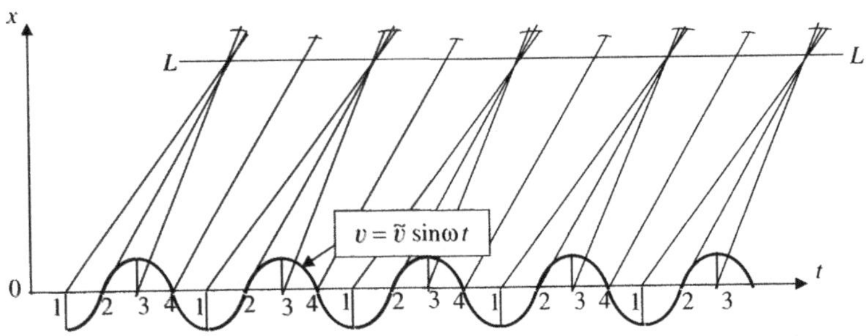

FIGURE 7.4 Distance–time diagram in a velocity-modulated beam.

important disadvantage of this approach that it ignores interaction forces between particles (i.e., space-charge fields). The theory will be refined correspondingly later.

So the particles move in a straight line with the velocity they acquired at the buncher. A distance–time diagram (Applegate diagram) is shown in Fig. 7.4. The slope of each straight line is determined by the velocity of the corresponding particle at the buncher output plane. According to Eq. (7.5), the velocities are sinusoidal functions of the exit time. In Fig. 7.4, the numbers on the t-axis represent different phases of the velocity. For points 2 and 4 the ac component of the velocity is zero, which means that the velocities of electrons leaving the buncher at these times are equal to the nondisturbed velocity. The electrons leaving the buncher at times 3 have maximal velocities. They will overtake the slower electrons (2) that left the buncher earlier. The intersection of straight lines 2 and 3 in the plane $x = L$ means that particles 3 surpass particles 2. The velocities of particles 1 are minimal. If these particles left the buncher first, they would be overtaken first by electrons 2 and then by electrons 1. Assume that linc 1 intersects lines 2 and 3 in the plane $x = L$. Then we arrive at an effect of the formation of ideal bunches in this plane. In general, this process can be treated as a mechanism of conversion from a velocity-modulated beam to a density-modulated beam.

According to Fig. 7.4, bunches in the plane $x = L$ alternate through time intervals equal to the period of the input signal. An observer in the plane $x = L$ detects a periodic current of frequency ω. Electrons 2 form centers of bunches. These electrons move with nondisturbed velocity and intersect in the midplane of the gap in moments when an ac electric field crosses zero from the retarding phase to the accelerating phase. Note that electrons 4 do not participate in the bunching. It can be shown that they will form second bunches with electrons 4 as centers. In principle, bunches of any high order can be obtained though the space-charge field effects that perturb this process.

7.3.2 Bunching of a Convection Current

Earlier we used the following formula for the convection current [Eq. (6.20)]:

$$I_c(x,t) = \rho_l(x,t)\, v(x,t)$$

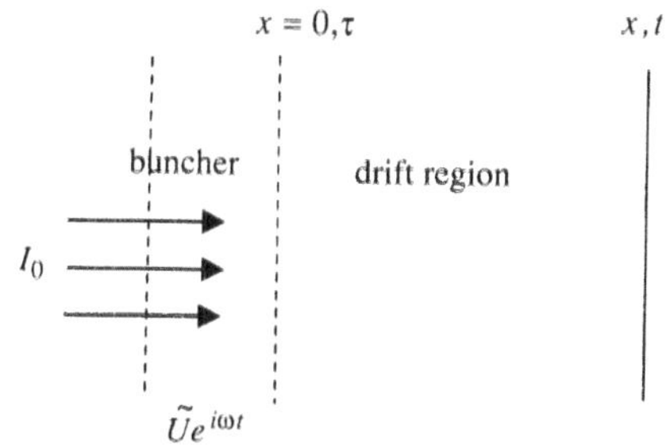

FIGURE 7.5 Klystron bunching space.

But the use of this equation meets the same difficulties as for a planar electron gap: absence of adequate relations for the linear space-charge density $\rho_l(x,t)$ and the velocity $v(x,t)$ in Euler variables. These difficulties are removed by using Lagrange variables in a manner similar to the theory of the flat electron gap, in particular, by using the law of charge conservation. However, it is convenient in this case to use another formulation of the charge conservation law.

In Fig. 7.5 a diagram of a bunching space is shown. The coordinate $x = 0$ is the beginning of the drift region. Assume that an electron enters the drift region at time τ and is located in plane x at time t. Now let us introduce another electron into the drift region, at time $\tau + d\tau$. The second electron will be in the same plane x at time $t + dt$. If the function $t = t(\tau)$ is known,

$$dt = \left(\frac{\partial t}{\partial \tau}\right)_{x=\text{const}} d\tau \tag{7.6}$$

According to the condition of charge conservation, all electrons that enter the drift region in the time interval $d\tau$ will intersect plane x in the time interval dt. The full charges that pass through the intervals indicated are $I_0\, d\tau$ and $I_c(x,t)\, dt$, where I_0 is the dc beam current. These are charges of the *same particles*. So we obtain the law of charge conservation:

$$I_0\, d\tau = I_c(x,t)\, dt \tag{7.7}$$

Using Eq. (7.6), we find that

$$I_c(x,t) = I_0 \Big/ \left(\frac{\partial t}{\partial \tau}\right)_{x=\text{const.}} \tag{7.8}$$

The relation between t and τ can be found if we take into account that the motion of the electrons in the region $x > 0$ is uniform:

$$t = \tau + T_{\text{tr}} = \tau + \frac{x}{v} \tag{7.9}$$

where T_{tr} is the transit time in the bunching region. Let us substitute in this relation velocity v from Eq. (7.5). But there t denoted the time departure from the gap. Therefore, we replace t by τ, where τ is the time of entrance into the drift region, and Eq. (7.9) becomes

$$t = \tau + \frac{x}{v_0\left[1 + \frac{1}{2}\xi M e^{i(\omega\tau - \varphi_0/2)}\right]} \tag{7.10}$$

Remember that $\xi \ll 1$ and $M < 1$. Then up to $O(\xi^2)$ we can write

$$t = \tau + \frac{x}{v_0}\left[1 - \frac{1}{2}\xi M e^{i(\omega\tau - \varphi_0/2)}\right] \tag{7.11}$$

Let us multiply Eq. (7.11) by ω and transform this equation to dimensionless variables [see the comments following Eq. (6.29)]. Denote the coefficient $\omega x/v_0 = \omega T_{tr} = \theta_0$. Here $T_{tr} = x/v_0$ is the nondisturbed transit time in the drift region and θ_0 is the corresponding transit angle. This gives rise to a new nondimensional parameter,

$$X = \frac{1}{2}\xi M\theta_0 = \frac{1}{2}\xi M\frac{\omega x}{v_0} \tag{7.12}$$

called the *bunching factor*. In essence, X is a dimensionless distance x. The dimensionless equation (7.11) becomes

$$\psi = \alpha + \theta_0 - X e^{i(\alpha - \varphi_0/2)} \tag{7.13}$$

where $\alpha = \omega\tau$. The dimensionless equation (7.8) is

$$I_c(x,t) = I_c(X,\psi) = I_0 \bigg/ \left(\frac{\partial\psi}{\partial\alpha}\right)_{X=\text{const.}} \tag{7.14}$$

Differentiating, we obtain the convection current,

$$I_c(X,\psi) = \frac{I_0}{1 - iXe^{i(\alpha-\varphi_0/2)}} \tag{7.15}$$

Discussion of Results The right-hand side of Eq. (7.15) is a function of (X,α); however, we are interested in the function of (X,ψ). To express α via ψ it is necessary to solve transcendental equation (7.10) or (7.11). This situation recalls the theory of the electron gap (Section 6.3.4). However, in contrast to ξ, the bunching factor X is not small because it contains the transit angle θ_0, which can be large for a balance of small velocity modulation.

Let us consider first some qualitative properties of the relations above. It may easily be shown that the convection current is a periodic function of time with period equal to the period of the input signal, $\widetilde{T} = 2\pi/\omega$. Indeed, according to Eq. (7.13), replacing ψ by $\psi + 2\pi$ changes α by 2π. That means that τ and I_c are periodic functions of t with period $\widetilde{T}$. Now let us examine qualitatively the temporal

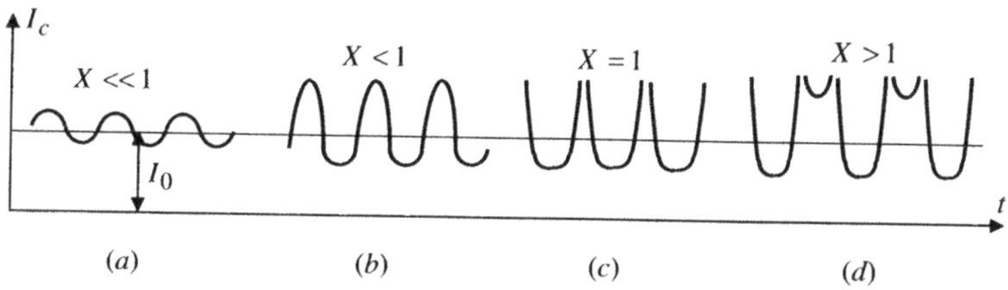

FIGURE 7.6 Convection current in a fixed plane for different values of the bunching factor.

dependence of the convection current on different values of the bunching factor X (Fig. 7.6). We consider the function $I_c(t)$ in the same plane. Then different values of X are distinguished by the voltage factor $\xi = \widetilde{U}/U_0$.

- $X \ll 1$. According to Eq. (7.13), $\alpha \approx \omega t - \theta_0$. We obtain from Eq. (7.15) the fact that the convection current is a harmonic function of t.
- *Large* $X < 1$. The current is a smooth periodic but nonharmonic function of time.
- $X = 1$. The denominator in Eq. (7.15) in some moments becomes zero and the convection current becomes, infinity. That occurs when

$$\alpha = \frac{\varphi_0}{2} + \frac{3}{2}\pi \tag{7.16}$$

  Because the potential was assumed to be $\widetilde{U}e^{i\omega t}$, Eq. (7.16) means that peaks of the beam pulse in Fig. 7.6*c* belong to the electrons that intersect the midplane of the buncher at moments when the ac field crosses zero from a retarding phase to an accelerating plane.
- $X > 1$. In this case the convection current becomes infinite twice during the period. The bunching process is reminiscent of the formation of a shock wave, which justifies the commonality of the terms *klystron* and *cliso*.

7.3.3 Fourier Expansion of a Convection Current

Equation (7.15) for a convection current has two defects. The first was mentioned earlier: The current in Eq. (7.15) is expressed as a function of (x, τ) but not (x, t), and a method for the transformation of variables is not obvious. The second disadvantage is the following: We are interested in the convection current as a sum of harmonics. Indeed, the periodic convection current excites the catcher resonator, which is a spectral device. It responds to specific frequencies. Both problems can be solved simultaneously and completely.

Let us write the Fourier expansion of the convection current:

$$I_c(x,t) = \sum_{-\infty}^{\infty} A_n(x)\, e^{in\omega t} \tag{7.17}$$

or in dimensionless form,

$$I_c(X,\psi) = \sum_{-\infty}^{\infty} A_n(X)\, e^{in\psi} \tag{7.18}$$

The series coefficients are determined by the formula

$$A_n(X) = \frac{1}{2\pi}\int_0^{2\pi} I_c(X,\psi)\, e^{-in\psi}\, d\psi \tag{7.19}$$

Let us apply the charge conservation law [dimensionless form of Eq. (7.7)] as

$$I_0\, d\alpha = I_c\, d\psi \tag{7.20}$$

Then we obtain

$$A_n(X) = \frac{I_0}{2\pi}\int_0^{2\pi} e^{-in\psi}\, d\alpha \tag{7.21}$$

Further, we can use Eq. (7.13), which gives ψ as a function of α. But because we will carry out a nonlinear operation, we first represent Eq. (7.13) in a real form:

$$\psi = \alpha + \theta_0 - X\cos\left(\frac{\alpha - \varphi_0}{2}\right) \tag{7.22}$$

Substitution of Eq. (7.22) in Eq. (7.21) gives

$$A_n(X) = \frac{I_0}{2\pi}\int_0^{2\pi} e^{-in[\alpha+\theta_0-X\cos(\alpha-\varphi_0/2)]}\, d\alpha \tag{7.23}$$

It may readily be shown (see, e.g., Abramowitz and Stegun, 1972) that the integral in Eq. (7.23) is an integral representation of a Bessel function:

$$A_n(X) = I_0 e^{-in(\varphi_0/2-\pi/2+\theta_0)} I_n(nX) \tag{7.24}$$

where I_n is a Bessel function of the first kind of order n. Substitution of Eq. (7.24) in Eq. (7.17) gives a Fourier expansion of the convection current. The corresponding relation can be simplified using the formula for Bessel functions:

$$I_n(nX) = -I_{-n}(-nX) \tag{7.25}$$

Combining the terms with positive and negative n and applying Eq. (7.25), we obtain the solution as a real series:

$$I_c(X,\psi) = I_0 + 2I_0 \sum_{1}^{\infty} I_n(nX)\cos n\left(\psi - \theta_0 - \frac{\varphi_0}{2} + \frac{\pi}{2}\right) \tag{7.26}$$

TABLE 7.1 Maximum Amplitudes and Bunching Factors of *n*th Harmonics

	n					
	1	2	3	5	10	20
$(nX)_{max}$	1.84	3.05	4.20	6.21	11.7	22.2
X_{max}	1.84	1.52	1.40	1.25	1.17	1.11
$2I_n(nX)_{max}$	1.16	0.96	0.86	0.74	0.60	0.48

The harmonics of the convection current are traveling waves with identical wavelengths $\lambda_c = 2\pi(\omega/v_0)$, and the amplitudes

$$I_c^{(n)}(X) = 2I_0 I_n(nX) \tag{7.27}$$

The maximum dimensionless amplitudes of some harmonics and corresponding values of X_{max} are given in Table 7.1. It can be seen from the table that the amplitudes of even very remote harmonics can be sufficiently high if one chooses adequate values of X. So the maximum amplitude of the tenth harmonic is 0.60 of the dc beam current. There is a tendency to approach $X \to 1$ for obtaining maximum amplitudes of remote harmonics. As we can see from a qualitative discussion of the bunching process, a form of the convection current for $X \to 1$ is close to a δ-function. That corresponds to Fourier series theory because Fourier expansion of the periodic sequence of δ functions has a uniform Fourier spectrum. The value of the bunching factor $X = \frac{1}{2}\xi M\theta_0 = \frac{1}{2}(\widetilde{U}/U)M\theta_0 \sim \widetilde{U}x$ is proportional to the product of the amplitude of the modulating signal and the bunching length. That is easy to treat qualitatively: The smaller the signal amplitude, the smaller the velocity modulation. But the optimal bunching is then attained at correspondingly greater length.

7.4 INTERACTION OF A BUNCHED CURRENT WITH A CATCHER FIELD. OUTPUT POWER OF A TWO-CAVITY KLYSTRON

7.4.1 Interaction of a Bunched Current with a Catcher Field

A diagram of an interaction region and its notation are given in Fig. 7.7. We assume that the resonator of the catcher is tuned to the nth harmonic of the input signal. Let us calculate the power expense of the output source when the bunching current intersects the catcher. If the power is negative, it would correspond to positive output power. Let us write the potential of the catcher corresponding to the nth harmonic of the input signal as

$$U_{ct} = \widetilde{U}_{ct}^{(n)} e^{i(n\omega t+\beta_n)} \tag{7.28}$$

The average over a period power expended by the catcher source is

$$P_{ct}^{(n)} = \frac{1}{T_{ct}} \int_0^{T_{ct}} I_{ct}^{(n)}(t) U_{ct}^{(n)}(t)\, dt = \frac{1}{2}\ \mathrm{Re}[\widetilde{U}_{ct}^{(n)} \widetilde{I}_{ct}^{(n)*}] e^{i\beta_n} \tag{7.29}$$

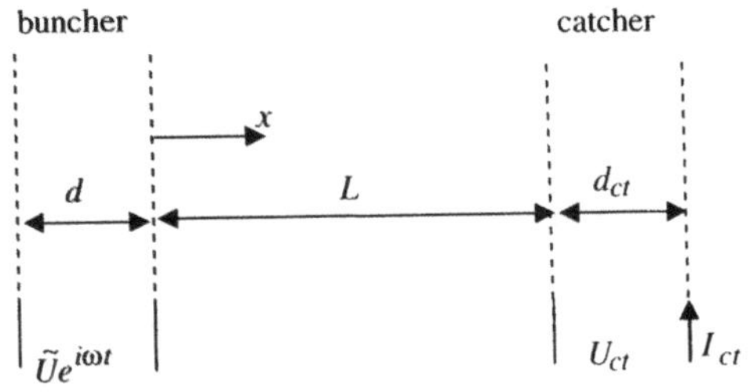

FIGURE 7.7 Interaction region of a two-cavity klystron.

where $\tilde{U}_{ct}^{(n)}$ and $\tilde{I}_{ct}^{(n)}$ are complex amplitudes on the nth harmonic of the voltage and current in the catcher circuit respectively. I_{ct} is, as we know, equal to the total current in the catcher area:

$$I_{ct} = (I_{ct})_{\text{ind}} + C\frac{dU_{ct}}{dt} \tag{7.30}$$

The capacitive current can be ignored in the calculation of power because the phase shift of this current with respect to the catcher potential is $\pi/2$. According to Eq. (6.18), for the current induced in the planar gap we find that

$$(I_{ct})_{\text{ind}} = \frac{1}{d_{ct}}\int_L^{L+d_{ct}} I_c(x,t)\, dx \tag{7.31}$$

Substitution of $I_c(x,t)$ in Eq. (7.26) gives in dimensionless variables

$$\begin{aligned}(I_{ct})_{\text{ind}} &= \frac{1}{\varphi_{ct}}\int_{\theta_L}^{\theta_L+\varphi_{ct}} I_c(X,\psi)\, d\theta \\ &= \frac{1}{\varphi_{ct}}\int_{\theta_L}^{\theta_L+\varphi_{ct}} 2I_0\sum_1^{\infty} I_n(nX)\cos\left[n\left(\psi-\theta-\frac{\varphi_0}{2}+\frac{\pi}{2}\right)\right] d\theta\end{aligned} \tag{7.32}$$

where $\varphi_0 = \omega d/v_0$, $\varphi_{ct} = \omega d_{ct}/v_0$, $\theta_L = \omega L/v_0$, $\psi = \omega t$, $X = \frac{1}{2}\xi M\theta_0$, and $\theta_0 = \omega x/v_0$. We can assume a small variation in the bunching factor in the catcher gap; therefore, the average of the Bessel function over the catcher gap is

$$\overline{I_n(nX)} = [I_n(nX)]_{x=L} = I_n(nX_L) \tag{7.33}$$

Now we can extract the Bessel function in Eq. (7.32) outside the integral. Then the integration is reduced to

$$(I_{ct})_{\text{ind}} = 2I_0\sum_1^{\infty} I_n(nX_L)M_{ct}^{(n)}\cos\left[n\left(\psi-\theta_{\Sigma}+\frac{\pi}{2}\right)\right] \tag{7.34}$$

where $M_{ct}^{(n)} = \sin(n\varphi_{ct}/2)/n\varphi_{ct}/2$ is the coupling coefficient of the catcher gap at the nth harmonic, and $\theta_{\Sigma} = \theta_L + \varphi_0/2 + \varphi_{ct}/2$ is the nondisturbed transit angle between the buncher and catcher midplanes.

7.4.2 Output Power and Electron Efficiency in a Kinematic Approximation

Let us find the complex amplitude of the nth harmonic of the current induced. The nth term from the series in Eq. (7.34) is read

$$(I_{ct})_{\text{ind}}^{(n)} = 2I_0 I_n(nX_L)M_{ct}^{(n)} \cos\left[n\left(\psi - \theta_\Sigma + \frac{\pi}{2}\right)\right] \tag{7.35}$$

Replacing the cos factor here by the complex exponent and separating the temporal factor $e^{in\psi}$, we obtain the desired complex amplitude of the nth harmonic as

$$(\widetilde{I}_{ct})_{\text{ind}}^{(n)} = 2I_0 I_n(nX_L)M_{ct}^{(n)} e^{in[(\pi/2)-\theta_\Sigma]} \tag{7.36}$$

The value $R_g^{(n)} \equiv U_{ct}^{(n)}/\widetilde{I}_{ct}^{(n)}$ is called the *gap resistance*. According to Eq. (7.36),

$$R_g^{(n)} = \frac{1}{2}\frac{U_0}{I_0}\frac{\xi^{(n)} e^{-in[(\pi/2)-\theta_\Sigma]}}{I_n(nX_L)M_{ct}^{(n)}} \tag{7.36a}$$

and $\xi^{(n)} = U_{ct}^{(n)}/U_0$.

According to Eq. (7.29), the average power that the source gives away to electrons is

$$\begin{aligned} P_{ct}^{(n)} &= \frac{1}{2}\text{Re}[\widetilde{U}_{ct}^{(n)} 2I_0 I_n(nX_L)M_{ct}^{(n)} e^{i[n\theta_\Sigma - n(\pi/2)+\beta_n]}] \\ &= I_0 \widetilde{U}_{ct}^{(n)} I_n(nX_l)M_{ct}^{(n)} \cos\left(n\theta_\Sigma - n\frac{\pi}{2} + \beta_n\right) \end{aligned} \tag{7.37}$$

The output power is obviously the power that the source gets from electrons (i.e., $P_{\text{out}}^{(n)} = -P_{ct}^{(n)}$. Evidently, the condition of maximum output power is

$$\cos\left(n\theta_\Sigma - n\frac{\pi}{2} + \beta_n\right) = -1 \tag{7.38}$$

which yields the maximum output power

$$(P_{\text{out}}^{(n)})_{\text{ext}} = I_0 \widetilde{U}_{ct}^{(n)} M_{ct}^{(n)} I_n(nX_L) \tag{7.39}$$

This power can be called the *electron power*: It is not equal to the output power, which depends on dissipation of the energy in the catcher circuit and in a load resistance. We can define maximum electron efficiency as

$$\begin{aligned} (\eta_{\text{el}}^{(n)})_{\text{ext}} &= \frac{(P_{\text{out}}^{(n)})_{\text{ext}}}{I_0 U_0} = \frac{\widetilde{U}_{ct}^{(n)}}{U_0} M_{ct}^{(n)} I_n(nX_L) \\ &= \xi_{ct} M_{ct}^{(n)} I_n(nX_L) \end{aligned} \tag{7.40}$$

Attainable Maximum Electron Efficiency The greatest possible electron efficiency can be obtained for two rather unrealistic conditions: (1) $M_{ct}^{(n)} = 1$, and (2) $\xi_{ct} = 1$. Condition 1 requires a zero length of the catcher gap. Realization of the approximate condition $M_{ct}^{(n)} \sim 1$ is especially difficult for $n > 1$. The alternative to condition 2, $\xi_{ct} > 1$ [i.e., $\widetilde{U}_{ct}^{(n)}(n) > U_0$], results from the reflection back from the catcher gap of some part of the electrons, with a decrease in efficiency. Thus, the absolute maximum of electron efficiency is

$$\eta_{\max}^{(n)} = I_n(nX_{\max}) \tag{7.41}$$

where the quantities $X_{\max}$ maximize the function $I_n(nX_L)$. According to Table 7.1, these quantities are

$$\begin{aligned} n &= 1: \quad X_{\max} = 1.84, \ \eta_{\max} = 58\% \\ n &= 2: \quad X_{\max} = 1.52, \ \eta_{\max} = 48\% \\ n &= 5: \quad X_{\max} = 1.25, \ \eta_{\max} = 37\% \end{aligned} \tag{7.42}$$

The case $n = 1$ corresponds to an amplifier klystron, and $n > 2$ corresponds to multiplier klystrons.

Let us dwell on the physical meaning of the phase condition for maximal efficiency [Eq. (7.38)], which leads to the equality

$$n\theta_\Sigma - n\frac{\pi}{2} + \beta_n = (2m + 1)\pi \tag{7.43}$$

Let us take for simplicity $n = 1$ (an amplifier). Assume also that the phases of the ac potentials in the buncher and the catcher are identical ($\beta_1 = 0$). Then condition (7.42) gives up the form

$$\theta_\Sigma = 2m\pi + \tfrac{3}{2}\pi \tag{7.44}$$

We see that the nondisturbed transit time between the buncher and catcher midplanes must be equal $(\frac{3}{4} + m)\widetilde{T}$, where $\widetilde{T}$ is the oscillation period. Now let us recall the Applegate diagram (Fig. 7.4). We saw that the centers of the bunches are electrons 2, which move with undisturbed velocity and intersect the midplane of the buncher gap in the moments when the ac electric field crosses zero from the retarding phase to the accelerating phase. Then, according to Eq. (7.43), the center of the bunch will cross the midplane of the catcher at a moment when the electric field in the buncher will be at the maximum retarding phase. But because the fields in the buncher and in catcher were supposed co-phased, we come to the following physically transparent result: *For maximum efficiency, the centers of bunches must arrive at the catcher midplane in the maximum retarding phase.*

7.5 EXPERIMENTAL CHARACTERISTICS OF A TWO-RESONATOR AMPLIFIER AND FREQUENCY-MULTIPLIER KLYSTRONS

There are two types of operations: voltage amplifier mode and power amplifier mode. In the *voltage amplifier mode* (amplification of a small drive power), efficiency is not essential. Maximum amplification is reached when the drive power is minimum and determined by space-charge debunching (see below), by losses in the catcher circuit, and by noise (Chodorow and Susskind, 1964).

Parameters of the *power amplifier mode* are determined by the maximum efficiency attainable. They are much worse than the idealized values given in Eq. (7.42). For example, the following efficiencies of experimental klystrons have been obtained (Hamilton et al., 1966): for $n = 1$, $\eta \sim 10\%$; for $n = 11$, $\eta \sim 0.4\%$. Three main factors are responsible for this disadvantage. First, it is space-charge debunching that is a result of the action of repulsive forces in bunches and correspondingly, space-charge waves in velocity-modulated electron beams. (This effect is considered in the next section.) The second factor is related primarily to the impossibility of reaching optimal bunching in two-cavity klystrons. As we have seen, not all electrons of the beam participate in the formation of bunches. Also, the spread of electron velocities in the bunch is too great. These factors can be overcome successfully in multicavity klystrons (Section 7.7). The third factor is connected with the gap coupling coefficients $M_{ct}^{(n)} = \sin(n\varphi_{ct}/2)/(n\varphi_{ct}/2)$. The reduction in efficiency due to this factor is especially notable in a multiplier klystron on high harmonics. One way to enhance efficiency is to increase the drive power (the coefficient ξ). Therefore, in power-amplifier mode, gain is diminished.

7.6 SPACE-CHARGE WAVES IN VELOCITY-MODULATED BEAMS

7.6.1 Formulation of the Problem

Space-charge fields describe repulsive forces in dense beams of charged particles. These forces induce oscillations of particles with plasma frequency $\omega_p = |\rho|\eta/\varepsilon_0$, which in a moving medium have the form of a propagating wave (i.e., the space-charge wave). In general, space-charge waves are connected with electromagnetic waves: for example, in traveling-wave tubes. However, in klystrons, the electromagnetic waves can be ignored. The theory of "pure" space-charge waves was first developed by Hahn (1939).

The basic assumptions that are used in a one-dimensional model of space-charge waves in velocity-modulated beams are as follows:

1. All values depend only on a single space variable x.
2. The electric field has only an x-component, $E = -\partial U/\partial x$.
3. There are no transverse velocities of electrons.
4. Ac values are small compared with dc values.
5. Electrons have a constant dc velocity.
6. The electron velocities $v \ll c$.
7. Electron beams are nondense:

$$\omega_p \ll \omega \tag{7.45}$$

8. Dc quantities ρ_0 and v_0 are constant.

9. There is wave variation in all values:

$$E = \tilde{E}e^{i(\omega t-\beta x)}, \qquad v_{\approx} = \tilde{v}e^{i(\omega t-\beta x)}, \qquad j_{\approx} = \tilde{j}e^{i(\omega t-\beta x)},$$
$$\rho_{\approx} = \tilde{\rho}e^{i(\omega t-\beta x)} \tag{7.46}$$

where β is a propagation constant (to be determined).

This formulation is a base of the simplified version of the space-charge wave theory, which is applicable only in the approximation of infinitely wide beams and for a strong magnetic field when any transverse motion can be excluded. The comprehensive theory was developed by many authors, including Chodorow and Susskind (1964), Hahn (1939), Ramo (1939a), and Trotman (1966). The assumptions above do not strongly contradict klystron model theory, except for assumption 4, which is, however, inherent in other versions of the theory. To summarize, assumptions 1 to 8 correspond to the one-dimensional small-signal low-velocity model of a beam neutralized by ions (Trotman, 1966).

7.6.2 Equations of Space-Charge Waves

Let us divide the derivation of space-charge wave equations into two particular subproblems.

Expression of an Electric Field Through a Current The basic equations are

$$\frac{dv}{dt} = -\eta E, \qquad j = \rho v, \qquad \frac{\partial j}{\partial x} = -\frac{\partial \rho}{\partial t},$$
$$v = v_0 + v_{\approx}, \qquad j = j_0 + j_{\approx}, \qquad \rho = \rho_0 + \rho_{\approx} \tag{7.47}$$

where quantities with the subscript $\approx$ denote an ac component of a beam. It is convenient to transfer from a current density to a current. Write $I = \rho_l v$, where ρ_l is a line charge density. Then taking into account that all values are functions of t and x and supposing further that ρ is the line charge density, we can rewrite Eqs. (7.47) as

$$\frac{dv_{\approx}}{dt} = \frac{\partial v_{\approx}}{\partial t} + \frac{\partial v_{\approx}}{\partial x}(v_0 + v_{\approx}) = -\eta E$$
$$I = I_0 + I_{\approx} = \rho_0 v_0 + \rho_{\approx} v_0 + \rho_0 v_{\approx} + \rho_{\approx} v_{\approx} \quad \text{or}$$
$$I_{\approx} = \rho_{\approx} v_0 + \rho_0 v_{\approx} + \rho_{\approx} v_{\approx} \tag{7.48}$$
$$\frac{\partial I_{\approx}}{\partial x} = -\frac{\partial \rho_{\approx}}{\partial t}$$

Now let us linearize these equations:

$$\left(\frac{\partial}{\partial t} + v_0 \frac{\partial}{\partial x}\right) v_{\approx} = -\eta E$$

$$I_{\approx} = \rho_0 v_{\approx} + \rho_{\approx} v_0 \qquad (7.49)$$

$$\frac{\partial I_{\approx}}{\partial x} = -\frac{\partial \rho_{\approx}}{\partial t}$$

Eliminating $\rho_{\approx}$ from Eqs. (7.49), we obtain

$$\left(\frac{\partial}{\partial t} + v_0 \frac{\partial}{\partial x}\right) I_{\approx} = \rho_0 \frac{\partial v_{\approx}}{\partial t} \qquad (7.50)$$

Differentiation of the first equation of Eqs. (7.49) with respect t and taking Eq. (7.50) into account gives us an equation that connects $I_{\approx}$ and E:

$$\rho_0 \eta \frac{\partial E}{\partial t} = -\left(\frac{\partial}{\partial t} + v_0 \frac{\partial}{\partial x}\right)^2 I_{\approx} \qquad (7.51)$$

Expression of a Current Through a Field The Poisson equation is[1]

$$\frac{\partial E}{\partial x} = \frac{\rho_{\approx}}{\varepsilon_0 S} \qquad (7.52)$$

The continuity equation is

$$\frac{\partial I_{\approx}}{\partial x} = -\frac{\partial \rho_{\approx}}{\partial t} \qquad (7.53)$$

Eliminating $\rho_{\approx}$ from Eqs. (7.52) and (7.53), we obtain

$$\frac{\partial I_{\approx}}{\partial x} = -\varepsilon_0 S \frac{\partial^2 E}{\partial x\, \partial t} \qquad (7.54)$$

7.6.3 Fast and Slow Space-Charge Waves

Let us substitute the wave representation [Eq. (7.46)] in Eqs. (7.51) and (7.54):

$$(\omega - v_0 \beta)^2 I_{\approx} = i\omega \rho_0 \eta E \qquad (7.55)$$

$$I_{\approx} = -i\omega \varepsilon_0 S E \qquad (7.56)$$

[1]The equation div $\mathbf{E} = \rho_{\approx}/\varepsilon_0 S$ is reduced to Eq. (7.52) only for a single component of the electric field. In this case the electric field has a potential, and Eq. (7.52) is the Poisson equation.

The condition for the existence of nonzero solutions of Eqs. (7.55) and (7.56) gives the dispersion equation

$$(\omega - v_0\beta)^2 = \omega_p^2 \tag{7.57}$$

or

$$\beta = \frac{\omega \pm \omega_p}{v_0} \tag{7.58}$$

where $\omega_p^2 = -\rho_0\eta/\varepsilon_0 S$ is the positive electron plasma frequency (remember that $\rho_0 < 0$ is the nondisturbed line electron beam density). According to Eq. (7.58), β is real. So there are two nonattenuated waves with phase velocities

$$v_p^{(s)} = \frac{v_0}{1 + \omega_p/\omega} \tag{7.59}$$

$$v_p^{(f)} = \frac{v_0}{1 - \omega_p/\omega} \tag{7.60}$$

Since $\omega_p \ll \omega$, the "slow" wave [Eq. (7.59)] has a phase velocity slightly lower than the nondisturbed electron velocity, and the "fast" wave [Eq. (7.60)] has a slightly higher phase velocity. Let us write Eq. (7.58) as

$$\begin{aligned} \beta^{(s)} &= \beta_e + \beta_p \\ \beta^{(f)} &= \beta_e - \beta_p \end{aligned} \tag{7.61}$$

where $\beta_e = \omega/v_0$ and $\beta_p = \omega_p/v_0$ are propagation constants of the electron and space-charge waves, correspondingly . In general, any value connected with a space-charge wave in the electron beam is a linear superposition of both waves: for example, for a velocity

$$v_{\approx} = A_1 e^{i[\omega t - \beta^{(s)} x]} + A_2 e^{i[\omega t - \beta^{(f)} x]} \tag{7.62}$$

Discussion of Results We have seen that ac disturbances can propagate along the beam as space-charge waves. Let us discuss briefly the properties of these waves. The following relation between the ac velocity and ac density in the beam may be found readily from the first of Eq. (7.49), and Eqs. (7.46), and (7.52):

$$(\omega - v_0\beta)v_{\approx} = \frac{\omega_p^2}{\beta}\frac{\rho_{\approx}}{\rho_0} \tag{7.63}$$

or according to Eqs. (7.59) and (7.60), and assuming that $\beta \approx \beta_e = \omega/v_0$ on the right-hand side of Eq. (7.63),

$$\frac{v_{\approx}}{v_0} \approx \pm \frac{\omega_p}{\omega}\frac{\rho_{\approx}}{\rho_0} \tag{7.64}$$

where the + and − signs correspond to fast and slow waves, respectively. We find that in a fast wave, the ac velocity and density are in phase, whereas in a slow wave, these quantities have opposite phases. Thus, in a fast wave the number of particles traveling with velocity greater than the dc velocity is more than the number of particles traveling with velocity less than the average velocity, and vice versa. So the beam kinetic energy increases when a fast wave is excited: that is, if we want to excite the fast space-charge wave, we must add the energy to the beam to "heat" it. For a slow wave, the number of particles with decreased velocities prevail. Hence, the beam energy is decreased when a slow wave is excited. For excitation of a slow wave we must take away the beam's kinetic energy—"cool" the beam. Usually, it is said *a fast space-charge wave carries positive energy, and a slow wave carries negative energy.*

7.6.4 Plasma Frequency Reduction Factor

So far we have considered only infinite electron beams. In finite beams, in general, one should consider transverse particle motion. This motion can be excluded, however, if one uses a strong magnetic field. But independent of that, longitudinal interaction forces between particles can be suppressed due to the fringing effect caused by the presence of conducting walls. In this case, a part of the force lines is closed by the walls (Fig. 7.8).

The theory of finite electron beams was developed by Hahn (1939) and Ramo (1939a) and involves the solution of Maxwell's equations with corresponding boundary conditions. It was shown that finite electron beams with excluded transverse motion of the electrons can be described with reduced plasma frequency ω_q, which depends on the frequency, electron velocity, and geometric parameters of the drift tube. One can show that if ω_q has been calculated, one can replace ω_p by ω_q in all beam relations. The ratio $R = \omega_q/\omega_p < 1$ is called the *plasma reduction factor*. Some curves $R = R(\omega a/v_0)$ with various a/b values are shown in Fig. 7.9 (Branch and Mihran, 1955; Collin, 2001).

7.6.5 Debunching of a Velocity-Modulated Beam by a Space Charge

The space-charge waves that are generated in the velocity-modulated beam obviously interfere with bunching. This process diminishes ac current and electron power. Because of the linearity of the beam equations, ac beam current as well as ac velocity are determined by a linear combination of two particular solutions

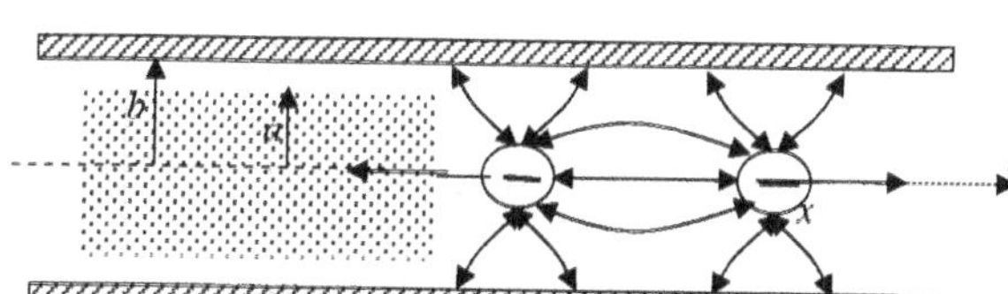

FIGURE 7.8 Reduction in the space-charge field pattern in a finite beam.

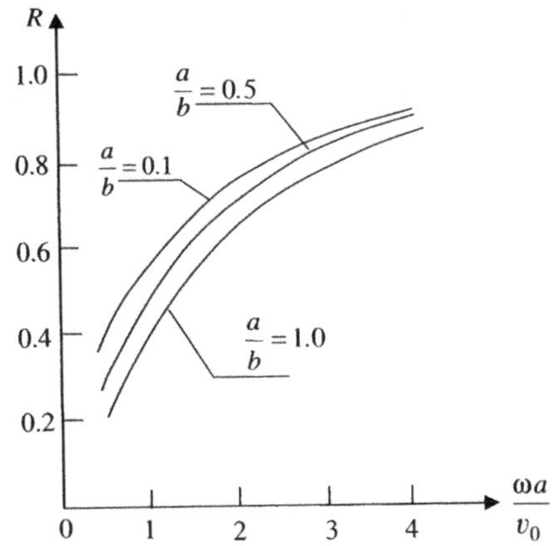

FIGURE 7.9 Space-charge reduction factor for a cylindrical electron beam inside a circular tube.

corresponding to two space-charge waves. Therefore, we can write, similarly to Eq. (7.62),

$$I_{\approx} = B_1 e^{i[\omega t - \beta^{(s)} x]} + B_2 e^{i[\omega t - \beta^{(f)} x]} \tag{7.65}$$

The constants A_i and B_i can be found from Eq. (7.50), which links the ac components of the current and velocity, and from the following initial conditions (plane $x = 0$ is the buncher output border):

$$v_{\approx x=0} = \tfrac{1}{2} v_0 \xi M e^{i\omega t} \tag{7.66}$$

$$I_{\approx} = 0 \tag{7.67}$$

Equation (7.66) is taken from Eq. (7.5) without dc velocity and with the phase multiplier $e^{-i\phi_0/2} \approx 1$. Equation (7.67) assumes that bunches do not have time to form in the short buncher gap and that the ac convection current into the buncher gap is zero.

Using Eqs. (7.50), (7.57), (7.62), and (7.65), we obtain the following relations between amplitudes of the current and velocity:

$$B_1 = -\frac{\omega}{\omega_p} \rho_0 A_1, \qquad B_2 = \frac{\omega}{\omega_p} \rho_0 A_2 \tag{7.68}$$

Substitution of Eq. (7.68) into Eq. (7.65) and application of Eqs. (7.62), (7.66), and (7.67) leads to a system of two equations determining the coefficients A_1 and A_2. As a result, we obtain

$$A_1 = A_2 = \tfrac{1}{4} v_0 \xi M \tag{7.69}$$

Substitution of Eqs. (7.69) and (7.68) in Eqs. (7.62) and (7.65) gives the following relations for the ac velocity and current in the bunched beam:

$$I_{\approx} = -\frac{1}{4}\frac{\omega}{\omega_p}\rho_0 v_0 \xi M e^{i\omega t}[e^{-i\beta^{(s)}x} - e^{-i\beta^{(f)}x}] \tag{7.70}$$

$$v_{\approx} = \frac{1}{4} v_0 \xi M e^{i\omega t}[e^{-i\beta^{(s)}x} + e^{-i\beta^{(f)}x}] \tag{7.71}$$

Using Eq. (7.58) and $I_0 = \rho_0 v_0$, $\theta = \omega x / v_0$, and $X = \frac{1}{2}\xi M \theta$ [from Eq. (7.12)], we obtain

$$I_{\approx} = iI_0 e^{i\omega(t - x/v_0)} X \frac{\sin \beta_p x}{\beta_p x} \tag{7.72}$$

$$v_{\approx} = \tfrac{1}{2} v_0 \xi M e^{i\omega(t - x/v_0)} \cos \beta_p x \tag{7.73}$$

where $X = \frac{1}{2}\xi M \theta_0$ is the bunching factor that was defined in Eq. (7.12).

Let us compare Eq. (7.72) with Eq. (7.15) for the convection current that was found in the kinematic bunching theory. If we assume that $X \ll 1$ in this formula, we find the ac current (with the multiplier $e^{-i\varphi_0/2} = 1$) as

$$I_{c\approx} = iI_0 X e^{i\alpha} \tag{7.74}$$

Because $\alpha = \omega\tau = \omega(t - x/v_0)$, we see that in the linear approximation ($X \ll 1$), space-charge debunching is described by the term

$$\frac{\sin \beta_p x}{\beta_p x} = \frac{\sin(\omega_p x / v_0)}{\omega_p x / v_0}$$

in the ac current. For small ω_p and a small length of the bunching region ($\beta_p x \ll 1$), this coefficient is close to 1. The reduction in the bunched current with increased β_p can be considerable. It is interesting to note that when the factor $\beta_p x$ approaches $\pi/2$, the bunched current and the output power cease to depend on the length of the bunching region because X/x is approximately constant. It is obviously a result of the linear approximation. To diminish the debunching effect without changing ω_p it is desirable to shorten the length of the bunching region while increasing the drive power (coefficient ξ). That way of increasing the efficiency and power decreases the gain at small signals (voltage amplifier mode). Webster (1939) recommended the following formula for bunched current, which takes into account space-charge debunching in a nonlinear approximation:

$$I_{\approx} = 2I_0 J_1(X) \frac{\sin \beta_p x}{\beta_p x} \tag{7.75}$$

The origin of this formula is not clear. It is probably a distinctive interpolation from Eqs. (7.27) and (7.72) (i.e., it is a space-charge correction of the nonlinear kinematic bunching theory).

It is necessary to note that the peculiarity of frequency harmonics is not displayed in the Eq. (7.72). In particular, there is a drop in efficiency with n harmonics even if one does not take into account the decrease in coupling coefficient $M_{ct}^{(n)}$ [see Eqs. (7.34), (7.37)], which is not seen. The reason for this disadvantage is obviously connected with the linearity of space-charge bunching theory. According to the kinematic nonlinear approximation, the appearance of noticeable amplitudes of high harmonics is possible for $X \geq 1$ when the form of the bunched current is essentially nonsinusoidal (Section 7.3.3). Meanwhile, space-charge debunching that smoothes the form of bunches must significantly reduce the amplitudes of high harmonics. This effect can be traced using numerical analysis.

7.7 MULTICAVITY AND MULTIBEAM KLYSTRON AMPLIFIERS

As we have seen in Section 7.5, two-cavity klystron amplifiers do not display high efficiency and amplification. During World War II and in the first after-war years, the magnetron oscillators were used widely in the radars. They were practically the only sources of powerful pulse microwave radiation in those years. Meanwhile, new problems that have appeared in radars, charged particle accelerators, and other powerful radio-electronic systems could not be solved without frequency synchronization of many high-frequency (HF) sources. For magnetrons (as for oscillators) it was a very difficult problem. In addition, the structural features of magnetrons do not allow the realization of powerful CW and long-pulse devices. The best solution was the use of powerful amplifiers. Therefore, the problem of effective high-power klystron amplifiers was very urgent. The solution has been found with the multicavity klystrons in the L and partially in the S to X ranges (see the introduction to Part II). The earliest powerful pulse three-cavity klystron was developed at Stanford in 1948 for the Mark III electron accelerator (Chodorow et al., 1953).

A multicavity klystron is depicted in Fig. 7.10. A velocity-modulated electron beam in the first cavity (the buncher) enters the second cavity. This cavity is unloaded and has a high Q-factor; therefore, even in the case of weak bunching, the high ac power is stored in the resonator. The amplified ac field in the second cavity gap causes additional velocity modulation of the beam. As result, under the action of both the first and second electron gaps, a complicated velocity modulation is created. Effective bunching is achieved for favorable conditions in the drift region between the second and third cavities. Further, the process is repeated, and a high gain and/or high efficiency can be realized. A multicavity klystron is a very complicated device with many independent parameters. There are three important modes that must be considered in choosing these parameters: the voltage amplifier, power amplifier, and bandwidth amplifier modes.

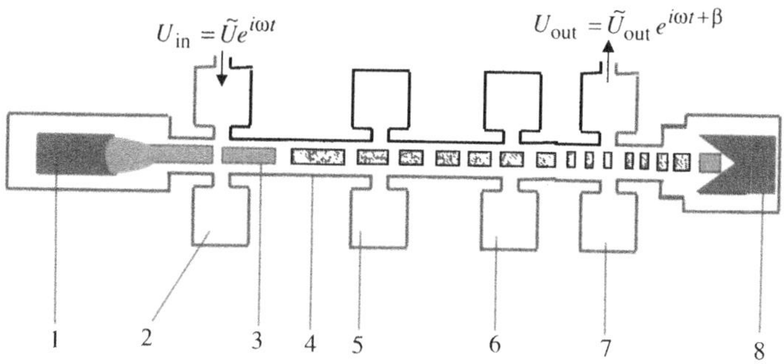

FIGURE 7.10 Four-cavity klystron. 1, Electron gun; 2, input cavity; 3, bunched electron beam; 4, drift tube; 5, intermediate cavity; 6, penultimate cavity; 7, output cavity; 8, collector.

7.7.1 Voltage Amplifier Mode (Maximum Gain)

In this mode, the resonance frequencies of all cavities are chosen to be identical, equal to the input (operating) frequency (the *synchronous tuning regime*). At resonance, when the natural frequencies of cavities are equal to the frequency of the input signal, the excited ac fields in the cavities are maximal. Upon arrival of the bunch centers in the gap, the ac field phase induced aligns itself at maximum retarding. If this condition is satisfied in all cavities, it leads to high amplification for a sufficiently small beam current (essentially, that in this case the space-charge debunching is weak). In principle, an arbitrarily large gain can be obtained in the voltage amplifier mode for a sufficient number of cavities. There is the following rough empirical formula for the dependence of the maximum power gain on the number N of cavities in the klystron:

$$K_P = 15 + 20(N - 2)\,\text{dB} \tag{7.76}$$

Really attainable maximum amplification values are on the order of 50 to 70 dB. The main limiting factors are noise and self-excitation of the klystron because of parasitic feedback between cavities.

The voltage amplifier mode is realized with a small drive power. Increased drive power leads to a gain saturation level long before the optimal efficiency is attained, because the velocity modulation in this case is determined by a nonoptimal ac field.

7.7.2 Power Amplifier Mode

The main principle in the realization of high efficiency is obtaining compact bunches entering the last (output) resonator. The term *compact* means a short length of the bunch, minimum velocity spread, and effective filling of the bunch by beam electrons. Compact bunching, combined with an appropriate value for a retarding ac field in the last gap, allows one to extract from an output resonator an overwhelming share of the electron kinetic energy with high efficiency.

Let us consider in more detail a realization of the power amplifier mode in a four-cavity klystron (Fig. 7.10). The first cavity is tuned synchronously to obtain a maximum velocity modulation. The second (intermediate) unloaded cavity must be tuned slightly upward in frequency (i.e., the resonance frequency ω_2 must be slightly greater than the operating frequency ω_0). Then the susceptance of this resonator reduced to the electron gap is inductive. As a result, the ac field in the gap advances the phase of the ac convention current. According to an analysis by Webber (1958), optimal bunching is achieved when the ac potential at the second cavity outruns the current by $\pi/2$. The bunch center enters the gap not at the maximum retarded phase of the ac field but at the moment when the field crosses zero from the retarded phase to the accelerated phase (see the Applegate diagram in Fig. 7.4). In this case the fast electrons in front of the bunch are decelerated, and the slow electrons behind the bunch are accelerated. In addition, former "neutral" particles participate in the bunching process. Thus, although the detuning of the intermediate cavity reduces the ac potential induced, the bunching after this cavity is improved significantly.

The detuning of the third (penultimate) cavity is also positive ($\omega_3 > \omega_0$). However, the difference $\omega_3 - \omega_0$ must be smaller for excitation of a higher ac output potential. Simultaneously, the ac potential in the penultimate gap must be increased significantly (up to 0.3 to 0.4 of the dc potential). Then it is possible to shorten the length of the last drift tube. That leads to the weakening of space-charge debunching and diminishing the velocity spread in bunches. The small velocity spread allows all electrons of a bunch to overcome the strong retarding ac electric field in the output electron gap without reflections. The features indicated provide a high efficiency. Note that an optimum is reached with an increased drive power, which in general reduces the gain.

Multicavity klystrons usually employ a larger number of resonators: five or six, sometimes up to eight. In some cases, one or more resonators are tuned on the second harmonic of the drive frequency to improve the bunch compactness. As an example, the following data are from a superpowerful CW klystron (VKP-7952, developed by Communications & Product Industries in Palo Alto, California) (Lenci and Bohlen, 2002):

- *Frequency:* 700 MHz
- *Cathode voltage:* 92 kV
- *Beam current:* 17 A
- *Output power:* 1.02 MW
- *Efficiency:* 66%
- *Gain:* 43 dB
- *Drive power:* 51 W

This klystron has been delivered to the Los Alamos National Laboratory for accelerator-assisted production of tritium.

7.7.3 Bandwidth Amplifier Mode

The klystron bandwidth is determined by two main factors: the bandwidth of the effective velocity modulation and the bandwidth of the output resonator. Optimization of the velocity modulation is reached by the detuning of intermediate resonators and reduction in their Q-factors. In Fig. 7.11 a scheme for tuning in five-cavity klystrons is presented. As we see, the first and output (fifth) resonators are tuned synchronously with the drive signal. This strengthens the velocity modulation and allows us to extract the maximum beam energy through the output resonator. The intermediate cavities (2,3) are detuned downward and upward. Although this operation worsens the bunching and finally, the efficiency, it broadens the bandwidth $\Delta\omega$. Detuning upward of the penultimate (fourth) cavity corrects bunching following the intermediate cavities. The typical bandwidth of multicavity klystrons is on the order on 1%.

The parameters of the output resonator are very significant. The most important is the choice of the loading Q-factor and gap resistance $R_g^{(1)}$ [Eq. (7.36*a*)] (Caryotakis, 1998). In particular, the bandwidth is increased with reduction of the loading Q-factor and gap resistance. Very effective for increasing bandwidth is the use of coupled cavities (Gilmour, 1986). To support high output power, it is desirable to use high perveance of the beam. However, that increases space-charge debunching and diminishes the efficiency. Thus, the development meets the contradictory requirements.

7.7.4 Multibeam Klystrons

One promising solution of these problems is attained in a *multibeam klystron* (MBK). A typical MBK is depicted in Fig. 7.12. After formation in the gun with a common cathode (1), an anode (2), and a common focusing magnetic field, the electron beams (4) are transported through separate thin tunnels that reduce the plasma frequency. The beams are coupled with radio-frequency cavity fields via common electron gaps. Therefore, a large total perveance is combined with the small perveances and currents of each beam. That weakening of the debunching space-charge effect simplifies the problem of beam formation and transport. In particular, the focusing magnetic field can be reduced. The possibility of achieving a high total level of perveance allows us to

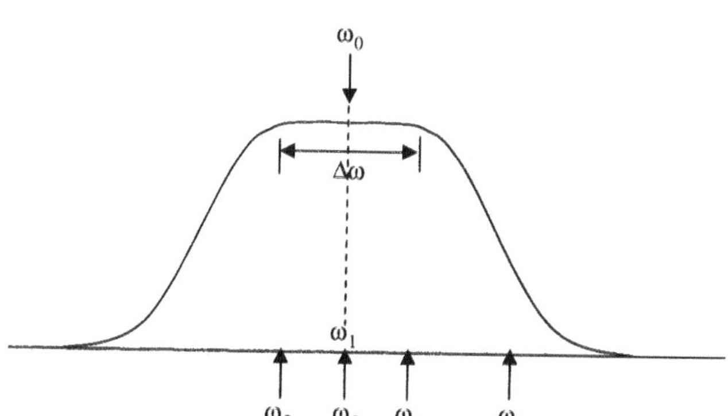

FIGURE 7.11 Optimal tuning of the resonators in a five-cavity klystron.

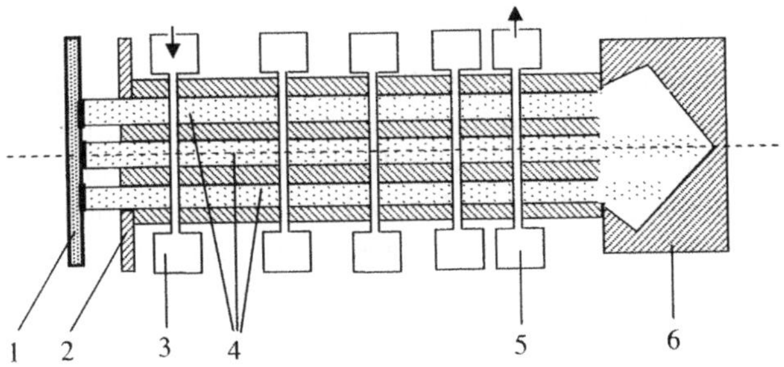

FIGURE 7.12 Five-cavity three-beam klystron. 1, Cathode; 2, anode; 3, input resonator; 4, beams; 5, output resonator; 6, collector.

work with reduced dc potential and to broaden the MBK bandwidth significantly compared with single-beam multicavity klystrons (Korolyov et al., 2004).

A pulse seven-beam six-cavity klystron for a Tesla linear collider was developed by Thomson Tubes Electroniques in France. The operating frequency, output power, gain, RF pulse duration, and efficiency of this device were, respectively, 1300 MHz, 10 MW, 48 dB, 1.5 ms, and 65% (Beunas and Fallon, 1998). The number of the beams in an advanced MBK can be increased up to 18 (Nguen et al., 2005).

7.8 RELATIVISTIC KLYSTRONS

Development of superpowerful klystrons amplifiers with corresponding enhancement of the voltage up to the relativistic electron energy is dictated primarily by needs in charged particle accelerator physics and in particular, RF linear accelerators (linacs) for TeV electron colliders. In general, three essentially different types of relativistic klystrons are known (see, e.g., Fazio et al., 1994).

The first type is low-perveance klystrons with a solid beam, such as the pulse three-cavity relativistic klystron with voltage 400 kV developed at Stanford University in 1949 by the team of Chodorow and Ginzton (Chodorow et al., 1953). The operating frequency of this klystron and the output power were correspondingly 3 GHz and 10 MW. Advanced X-band SLAC klystrons of this type operate at voltage 1.3 MV, a current of 600 A, and produce 300 MW of power.

The second type is relativistic multibeam klystrons of a gigawatt power level (GMBK). The Stanford University designed a 10-beam L-band GMBK (Caryotakis et al., 1996, 1998). This GMBK should produce 1 GW with an efficiency of 40%. The beam voltage of this klystron is 500 kV, and the total current is 5.0 kA, so the current of the each beam is 500 A. The peculiarity of the GMBK is the use of the periodic permanent magnetic focusing (see Section 5.8.2). In the proposed klystron, the spacing between beams is large and the considered focusing is very effective.

The third type of relativistic klystrons is a high-current, high-perveance, mildly relativistic tube with an annular beam, generated by an explosive emission magnetically insulated diode (see Section 4.7.3), and coaxial cavities. These klystrons were investigated originally at the Naval Research Laboratory (NRL) (Friedman et al., 1990). The thin annular electron beam, propagating near a conducting wall of a large-diameter tube, can carry much higher currents than solid beams can. The typical value of the accelerating voltage is 500 kV, and the current of 5 to 20 kA is approximately an order of magnitude greater than the current in SLAC klystrons. The output power of the 1.3-GHz klystron was about 3 GW, the efficiency 50%, and the pulse duration up to 100 ns, with 37 dB gain (Lau et al., 1990).

Physical processes in the relativistic klystrons have a number of important peculiarities, so the velocity modulation in a klystron of the first type is very low compared with the nonrelativistic version. The correct relativistic formula for the ac velocity at the output buncher is (Chodorow et al., 1953)

$$\tilde{v} \approx \frac{2v_0}{\gamma(\gamma+1)} \xi M e^{-i\varphi_0/2} \tag{7.77}$$

where γ is a relativistic factor. Let us compare this relation with Eq. (7.5). For $v = 600$ kV, we obtain $2/(\gamma(\gamma+1)) \approx 0.29$, more than a threefold reduction of the ac velocity. The gap coupling coefficient M is not changed compared with the nonrelativistic version. The small velocity modulation is balanced by a significant lengthening of the drift tunnels, although according to the results of the analysis, the velocity modulation for the attainment of optimal efficiency should not be very large. So in the last drift tube of a 500-kV klystron, the maximum deviation of the velocity from the average value does not exceed 0.05. The energy extraction in relativistic klystrons also has interesting features that do not require a strong deceleration of electrons in the output gap because the energy draws out from the particle mass. So for $U_0 \sim 1$ MV, the minimum electron velocities for efficiency exceeding 80% can comprise 0.5 to 0.8 of the average value.

In relativistic klystrons of the third type, space-charge waves on tenuous electron beams in conventional klystrons are modified dramatically by the strong self-electric field of beams. The bunching mechanism in the modified beams is unique; unlike the case of klystrons of the first type, a longer drift region is unnecessary. The second important property of the modified space charge in the third type of klystron is the strong electrostatic insulation against a vacuum breakdown at high power levels.

The analysis of optimal regimes in relativistic multicavity klystrons was developed by a team working with A. A. Kuraev (Aksenchik et al., 1982). A klystron has been developed in which both efficiency and gain grow with the number of cavities. For eight-cavity klystrons, for example, with a voltage of 1500 kV, the maximum designed efficiency and the amplification coefficient were respectively 84% and 70 dB. For a current of 450 A, the maximum output power was 0.56 GW. The optimal ac electric field in the optimized regimes increases with the dc potential. For $U_0 = 1.5$ MV, $E_{\text{opt}} = 75$ kV/mm, which is close to the electrical breakdown in a vacuum. An effective method of reduction of the optimal electric field is the use of output resonators as coupled cavity systems of the type used in sections of a slow-wave structure (Pohl, 1965). We thus arrive at a scheme for a twistron (see, e.g., Gilmour, 1986). In general, the design of optimal multicavity klystrons is a difficult task. It is an optimization control problem involving the nonlinear interaction of the electron beams and ac electromagnetic field, taking into account many parameters (a space charge, eigenfrequencies and quality factors of resonators, lengths of the drift tunnels, parameters of electron gaps, etc.). Experimental examination involves adjusting many units and is also very complicated because of the high amplification rate.

The basic use for relativistic klystrons is as high-energy drivers for electron and ion accelerators and colliders. An important use is in compact RF electron accelerators to drive free electron lasers and possible even high-frequency drivers for direct energy weapons (see, e.g., Barker and Schamiloglu, 2001).

Increasing the radio frequency and pulse duration is a very important way to create multi-TeV linear colliders. Also, perspective radars and communication systems should operate at the W-band, where there is an atmospheric window. High-power efficiency relativistic klystrons do not exist at these frequencies because the dimensions of basic component became too small; frequencies of order 10 GHz are practically the upper boundary for these klystrons. Relativistic gyroamplifiers have a number of advantages (see Chapter 10).

7.9 REFLEX KLYSTRONS

Reflex klystrons (RKs) are velocity-modulation single-cavity tubes that for long were widespread microwave low-power oscillators applied in diverse fields of microwave engineering. The frequency range of the RK family is striking: from the L up to the K_a band. The highest frequency of the RK is near 150 GHz ($\lambda = 2$ mm). The positions of RKs have lately been weakened by the advent of solid-state high-frequency oscillators; however, some advantages of RKs, especially its radiation steadiness, justify its use especially in the short-wave millimeter range. In any case, the physics of the RK as an oscillator with intrinsic electron feedback is of general interest. RKs were invented independently in 1939 by R. Varian and W. Hansen at Stanford University, W. C. Hahn and G. Metcalf at Schenectady, and A. Arsenjeva-Heil with O. Heil in Germany and the USSR (see Ginzton and Harrison, 1946; Pierce, 1945). A scheme of RK is shown in Fig. 7.13.

The only cavity of the tube accomplishes the functions of a buncher and a catcher simultaneously. The reflector potential is negative around the cathode, and the

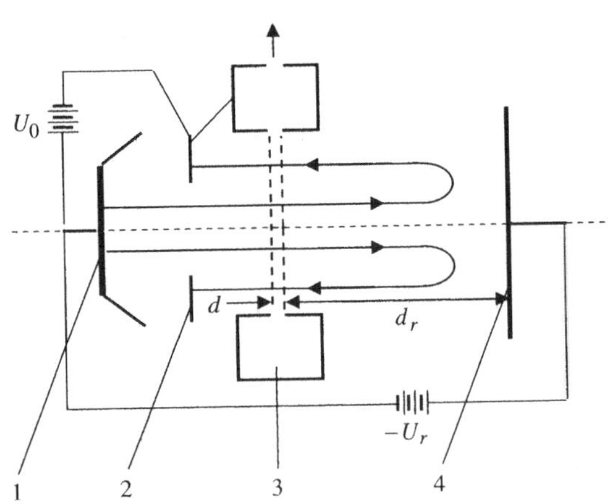

FIGURE 7.13 Reflex klystron. 1, Cathode; 2, anode; 3, cavity; 4, reflector.

electrons are reflected and move backward through the cavity gap at the anode. In the gap between the reflector and the cavity (the reflection space), the electron beam is bunched. As a result, a periodic convection current is formed that intersects the cavity gap in the back motion. Self-excitation oscillations are attained for favorable phasing between an ac potential and an ac current.

Given below is a detailed description of the reflex klystron mechanism that is both elegant and instructive.

7.9.1 Bunching of an Electron Beam in a Retarded Field

The velocity modulation of the electron beam in the cavity gap is the same as in a simple electron gap (Section 7.2). Applying formula (7.5) for the electron velocity, we obtain

$$v = v_0 + \tilde{v}e^{i(\omega\tau - \varphi_0/2)} = v_0[1 + \tfrac{1}{2}\xi M e^{i(\omega\tau - \varphi_0/2)}] \tag{7.78}$$

where $v_0\sqrt{2U_0}$, $\xi = \widetilde{U}/U_0$, $M = \sin(\varphi_0/2)/\varphi_0/2$, and $\varphi_0 = \omega d/v_0$. The time τ denotes the moment of the departure of electrons from the gap cavity into the reflection space. Assume that the space-charge effects in the reflection space can be neglected and the electric field is uniform. Then the motion of electrons in this region can be considered as decelerated uniformly with the acceleration and velocity, respectively:

$$a = \frac{\eta(U_0 + U_r)}{d_r}, \qquad v = v_0 - at \tag{7.79}$$

An electron in a retarding dc field moves to the reflector until it stops. The total nondisturbed transit time in the reflection space is

$$T_0 = \frac{2v_0}{a} = \frac{2v_0 d_r}{\eta(U_0 + U_r)} = \frac{2d_r\sqrt{2\eta U_0}}{\eta(U_0 + U_r)} \tag{7.80}$$

The ac transit time determined by the velocity modulation is

$$T = \frac{2v}{a} = \frac{2v_0}{a}\frac{v}{v_0} = T_0[1 + \tfrac{1}{2}\xi M e^{i(\omega\tau - \varphi_0/2)}] \tag{7.81}$$

where we used Eq. (7.78). The corresponding transit angle is

$$\theta = \theta_0 + Xe^{i(\alpha - \varphi_0/2)} \tag{7.82}$$

where $\theta = \omega T$, $\theta_0 = \omega T_0$, $\alpha = \omega\tau$, and the bunching factor

$$X = \frac{1}{2}\xi M\theta_0 = \frac{1}{2}\frac{\widetilde{U}}{U_0}M\theta_0 \tag{7.83}$$

According to Eqs. (7.81) and (7.82), the transit time and angle oscillate with frequency ω of the drive signal. In Fig. 7.14 a version of the Applegate diagram for motion of electrons in the reflection space is presented. According to Eq. (7.81), the transit time is proportional to the initial velocity. That is analogous to the motion of a ball thrown up in the air: The greater the initial velocity of the ball, the higher it raises and the later it returns. The parabolic trajectories in Fig. 7.14 are depicted in such a way (which is the most favorable case) that for a given reflector potential, they return

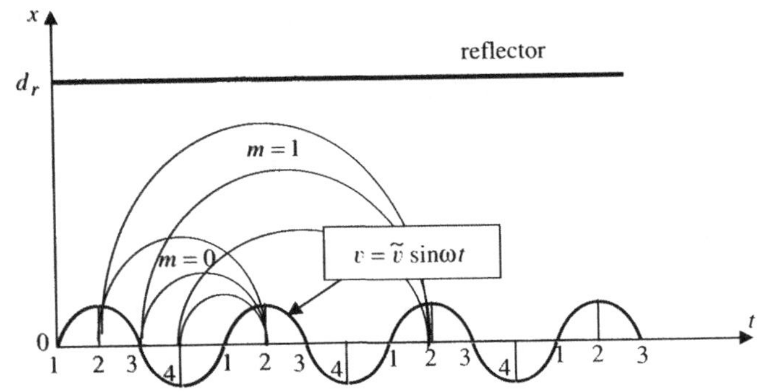

FIGURE 7.14 Distance–time Applegate diagram for a velocity-modulated beam for two values of the reflector potential.

in the buncher simultaneously. Electron 2 enters first. At that moment, the ac gap potential is maximal. Therefore, electron 2 is "raised" to the greatest height and its transit time is also the greatest. Then electron 3 begins; its velocity is slower. Electron 4 has the slowest velocity. It moves the shortest time and overtakes electrons 2 and 3. As a result, the bunch is formed and it returns to the input gap in the maximum retarding phase of the field (for backward motion). The curves marked in Fig. 7.14 as $m = 0$ correspond to the maximum retarding reflector potential. Other trajectories ($m = 1$) correspond to a smaller reflector potential that also form the bunch.

Electron 3 is a bunch center. It moves with a nondisturbed velocity and intersects the midplane of the buncher gap at the moment when the ac electric field crosses zero from an acceleration phase to a retarding one. As can be seen from comparison between Figs. 7.14 and 7.4, the bunch center is phase-shifted by π relative to the bunch centers in two- or multiresonator klystrons. According to Fig. 7.14, the optimal nondisturbed transit angle is $\frac{3}{2}\pi$ or the transit time is $\frac{3}{4}\widetilde{T}$. The other optimal transit angles are $\theta_0 = \frac{3}{2}\pi + 2m\pi$ or $T_0 = \frac{3}{4}\widetilde{T} + m\widetilde{T}$. Note that the optimal transit angle is the same as in a drift tube klystron. That is explained by the additional shift by π in the returning phase compared with an ordinary klystron (the ac electric field must be accelerating for forward motion).

7.9.2 Calculation of Electron Power and Efficiency

The calculation method is analogous to the method used in Sections 7.3.3 and 7.4.2 for ordinary klystrons. The convection current $I_c(x,t)$ in dimensionless variables is $I_c(X,\psi)$, where $\psi = \omega t$. The bunching factor X from Eq. (7.83) is equivalent to the factor X in drift tube klystrons [Eq. (7.12)]. Let us write the expansion of the convection current into a Fourier series in dimensionless form:

$$I_c(X,\psi) = \sum_{-\infty}^{\infty} A_n(X)e^{in\omega t} \qquad (7.84)$$

where

$$A_n(X) = \frac{1}{2\pi}\int_0^{2\pi} I_c(X,\psi)e^{-in\psi}\,d\psi \tag{7.85}$$

Applying the charge conservation law (7.20), we obtain

$$A_n(X) = \frac{I_0}{2\pi}\int_0^{2\pi} e^{-in\psi}\,d\psi \tag{7.86}$$

For calculation of the integral in Eq. (7.86), we need the function $\psi = \psi(\alpha)$. In this case, $t = \tau + T$, or in dimensionless variables, $\psi = \alpha + \theta$. Using Eq. (7.82), we find that

$$\psi = \alpha + \theta_0 + Xe^{i(\alpha - \varphi_0/2)} \tag{7.87}$$

Before substituting the function [Eq. (7.87)] into the integral [Eq. (7.86)], we must take its real part (see Section 7.3.3). Then we obtain

$$A_n(X) = \frac{I_0}{2\pi}\int_0^{2\pi} e^{-in[\alpha + \theta_0 + X\cos(\alpha - \varphi_0/2)]}\,d\alpha \tag{7.88}$$

This integral is again reduced to a Bessel function. We can omit the corresponding calculation. It is sufficient to note that this integral differs from the integral in Eq. (7.23) only by the sign of X. Therefore, we can take Eq. (7.26) for the convection current and replace X by $-X$. Furthermore, we can use Eq. (7.37) for the nth harmonic of a complex electron power:

$$P_e^{(n)} = I_0\widetilde{U}_{ct}^{(n)} I_n(nX_L)M_{ct}^{(n)} e^{i[n\theta_\Sigma - n(\pi/2) + \beta_n]} \tag{7.89}$$

and transform this formula, taking into account the following obvious features of a reflex klystron:

$$n = 1, \qquad \beta_n = 0, \qquad \widetilde{U}_{ct}^{(n)} = -\widetilde{U}, \quad X \to -X,$$

$$M_{ct}^{(n)} \to M = \frac{\sin\varphi_0/2}{\varphi_0/2}, \qquad \theta_\Sigma \to \theta_0$$

As a result, we obtain the electron power:

$$P_e = I_0 M(-\widetilde{U})I_1(-X)e^{i(\theta_0 - \pi/2)} \tag{7.90}$$

So the real and imaginary parts of the electron powers are

$$P_{er} = I_0 M\widetilde{U} I_1(X)\sin\theta_0 = 2\frac{I_0 U_0}{\theta_0} X I_1(X)\sin\theta_0 \tag{7.91}$$

$$P_{ei} = -I_0 M\widetilde{U} I_1(X)\cos\theta_0 = -2\frac{I_0 U_0}{\theta_0} X I_1(X)\cos\theta_0 \tag{7.92}$$

The optimal electron power that is delivered to the cavity according to Eq. (7.91) is achieved for $\sin\theta_0 = -1$. That gives us

$$\theta_0^{(m)} = \tfrac{3}{2}\pi + 2m\pi \tag{7.93}$$

Equation (7.93) agrees with the qualitative discussion of the Applegate diagram in Fig. 7.14 and represents the condition of return of the bunch in the phase of the maximum retarding field. As a result, optimal electron output power and optimal electron efficiency are

$$P_e^{(\text{opt})} = I_0 M \widetilde{U} I_1(X), \qquad \eta_e^{(\text{opt})} = \frac{P_e}{I_0 U_0} = M\xi I_1(X) \tag{7.94}$$

Using Eqs. (7.83) and (7.94), we obtain

$$\eta_e^{(\text{opt})} = \frac{2XI_1(X)}{\frac{3}{2}\pi + 2m\pi} \tag{7.95}$$

A value of the bunching factor X giving the maximum efficiency can be found from the equation

$$\frac{d}{dX}[XI_1(X)] \equiv I_0(X) = 0 \tag{7.96}$$

The first zero of the Bessel function I_0 is $X \approx 2.4$. Then

$$\eta_e^{(\max)} = \frac{(2)(2.4)I_1(2.4)}{\frac{3}{2}\pi + 2m\pi} \approx \frac{2.5}{\frac{3}{2}\pi + 2m\pi} \tag{7.97}$$

The greatest electron efficiency $\eta_e^{(\max)} = 0.53$ is obtained for $m = 0$. In this case the transit angle is minimal and is $(\theta_0)_{\text{opt}} = \frac{3}{2}\pi$.

The reflector potentials corresponding to the optimal transit angles [Eq. (7.93)] can be found from Eq. (7.80):

$$\theta_0^{(m)} = \frac{3}{2}\pi + 2m\pi = \frac{2\omega d_r\sqrt{2\eta U_0}}{\eta[U_0 + U_r^{(m)}]} \tag{7.98}$$

According to this equation, $U_r^{(m)}$ decreases with m. The minimum of U_r is zero, because for a positive reflector potential the electrons do not return to the cavity. As a result, a number m is restricted. $m_{\max}$ can be found if we set $U_r = 0$ in Eq. (7.100).

The efficiency quickly drops with increased m. Let us return to the Applegate diagram (Fig. 7.14). We see that the bunch can pass the retarding maximum at various distances. But for any m, the optimal bunching factor $X = \frac{1}{2}\xi M\theta_0$ found from Eq. (7.96) is the same. Therefore, with increasing θ_0 the corresponding value $\xi = \widetilde{U}/U_0$ is decreased (i.e., with increased m the optimal amplitude of the ac potential, and also the power and the efficiency, are decreased).

7.9.3 Equation of Stationary Oscillations. Starting Current. Electronic Frequency Tuning. Oscillation Zones

The equation of stationary oscillations can be obtained by substitution of electron power equations (7.91) into the general equations (6.53) and (6.54) of the oscillator:

$$2\frac{I_0 U_0}{\theta_0} X I_1(X) \sin\theta_0 = -P_l \tag{7.99}$$

$$\frac{\omega - \omega_0}{\omega_0} = \frac{1}{2Q}\frac{P_{ei}}{P_{er}} = -\frac{\cot\theta_0}{2Q} \tag{7.100}$$

The power dissipated in the cavity $P_l = \frac{1}{2} Y_l \widetilde{U}^2$, where Y_l is the loading cavity conductance. Using Eq. (7.83), we obtain

$$\frac{I_1(X)}{X} = -\frac{U_0}{I_0}\frac{Y_l}{M^2\theta_0 \sin\theta_0} \tag{7.101}$$

This equation defines the amplitude of stationary oscillations of the loading reflex klystron.

The starting current corresponds to oscillations with an amplitude close to zero. Assuming that $X \to 0$ and taking into account that $I_1(X)/X|_{X\to 0} = \frac{1}{2}$, we obtain from Eq. (7.101) the starting current

$$I_{st} = -\frac{2Y_l U_0}{M^2\theta_0 \sin\theta_0} \tag{7.102}$$

As we see, I_{st} is minimal for the optimal transit angles $\theta_0^{(m)} = \frac{3}{2}\pi + 2m\pi$ when $\sin\theta_0^{(m)} = -1$. A deviation of θ_0 from the optimal $\theta_0^{(m)}$ increases the starting current and decreases the power. Also, the starting current decreases with m.

The limiting deviations $\theta_{min}^{(m)}$ and $\theta_{max}^{(m)}$ of $\theta_0^{(m)}$ correspond to a range of transit angles where $I_1(X)/X < \frac{1}{2}$. So the dependence of the output power of the transit angle acquires the form of the separated zones (oscillation zones). Obviously, the starting current in each zone is decreased with the zone number m. That can be explained by an increase of θ_0 in the bunching length. Therefore, the amplitude of the ac potential $\widetilde{U}$ required for proper velocity modulation is reduced. That decreases cavity losses and correspondingly, the starting current.

How can we control the transit angle? According to Eq. (7.98), it is possible by changing U_0 or U_r. It is more beneficial to change U_r because the current of the reflector at the negative electrode is zero. The corresponding control θ is realized without energy consumption.

In Fig. 7.15, a qualitative view of the power and oscillation frequency as functions of the reflector potential is shown for the first three oscillation zones. The reflector potentials at the centers of the zones correspond to the optimal transit angles [Eq. (7.98)]. According to Eqs. (7.97) and (7.98), the output power is maximal in the zone centers and decreases with m. The frequencies in the centers

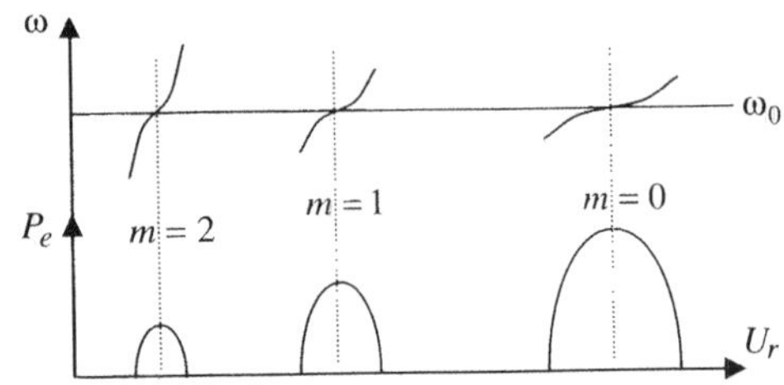

FIGURE 7.15 Output power and frequency for the first three oscillation zones as functions of a reflector potential.

of the zones according to Eq. (7.100) are the same and equal to the nondisturbed cavity frequency. This is a result of the reactive electron admittance being zero at the centers of each zone. The ability to control the radiation frequency by the reflector potential (*electronic tuning*) is a very valuable property of the reflex klystron, which does not require energy consumption and is practically without inertia. Usually, the electronic tuning is about 1%, but it can be enhanced to 10% by using a cavity with a small Q-factor [Eq. (7.100)] and increasing the beam current.

7.9.4 Efficiency and Applications of Reflex Klystrons

Above we estimated the maximum efficiency for a zero oscillation zone as ~53%. In the experimental klystrons the efficiency is significantly smaller. The main reason is connected with the influence of the space charge, which has not been taken into account in the previous analysis. The disturbance in the electron motion, especially space-charge debunching in the reflex klystron, is stronger than in the amplified klystrons because of electron deceleration near the reflector and the corresponding increase in the space-charge density. As result, even in the zeroth zone, the experimental efficiency is less than 10%. The space-charge effects are reinforced with m because of expansion of the bunching region. The efficiency in zones with $m \geq 1$ is only of order 1%, but one is balanced by small starting currents in these zones and an increased electronic tuning band (see Fig. 7.15). Therefore, the reflex klystrons are designed for large values of θ with correspondingly low collector potentials and used as oscillators of small power.

CHAPTER EIGHT

Traveling-Wave Tubes and Backward-Wave Oscillators (O-Type Tubes)

8.1 INTRODUCTION

Traveling-wave tubes (TWTs) and backward-wave oscillators (BWOs) are microwave amplifiers and oscillator electron tubes, respectively, in which a rectilinear electron beam interacts with a traveling electromagnetic field. In the operational space of both TWTs and BWOs, dc electromagnetic fields are absent, except for a focusing magnetic field that runs parallel to the beam. Such tubes (including klystrons) are called *O-type devices*. These contrast with M-type devices, in which electron beams interact with electromagnetic waves in crossed dc electric and magnetic fields.

The following principal disadvantages of klystron amplifiers stimulated the development of O-type TWTs.

1. *Narrow bandwidth.* Ac electric fields in klystrons are localized in very short gaps. Therefore, for effective control of electron velocities, strong ac electric fields are required, along with a correspondingly large stored energy, W, in the klystron resonators. Recall that the quality factor of a resonance circuit, Q, is equal to a ratio of frequency to resonance width:

$$Q = \frac{\omega W}{P} = \frac{\omega}{\Delta\omega} \tag{8.1}$$

Therefore, a device with high Q will necessarily possess a correspondingly narrow bandwidth $\Delta\omega$.

Electron Beams and Microwave Vacuum Electronics. By Shulim E. Tsimring

2. *High-frequency limitations*. The effect indicated above becomes more and more significant as the wavelength is shortened. Therefore, expansion of the klystrons into short centimeter and especially into millimeter bands leads to a drop in output power.

The traveling-wave tube was invented in 1942 by Rudolf Kompfner in England, although Lindenblad was perhaps the first to propose the idea of a helix traveling-wave amplifier in a patent (Lindenblad, 1942) (filed May 1940). Kompfner (1964) described the history of this invention. It is interesting that microwave electronics began as a hobby for Kompfner; he was trained as an architect in Vienna.

The design of traveling-wave tubes addresses the following key issues: Is it possible to extend the interaction length if the field is variable in the time? Can the electrons "see" the same ac field phase over an extended period? Both are achieved through the interaction of particles with a traveling wave whose phase velocity is close to the particle velocity. Since the electron velocity in nonrelativistic beams is much less than the light velocity, it is necessary to apply some waveguide structure that would significantly reduce a velocity of the electromagnetic wave. Kompfner used a helix as a slow-wave structure for this purpose. The brilliant service of Kompfner was the embodiment of this idea in a real working tube. The first comprehensive theory of O-type TWTs (TWTOs) was formulated by Pierce (1947).

8.2 QUALITATIVE MECHANISM OF BUNCHING AND ENERGY OUTPUT IN A TWTO

8.2.1 Scheme of a TWTO

In Fig. 8.1 an O-type traveling-wave tube is shown. As we see, the basic TWTO components are an electron gun, a slow-wave structure (here the helix structure), a collector, a power supply source, device input–output, and a focusing magnet. Occasionally, solenoids are used, but periodic magnet systems are usually employed (see Section 5.8).

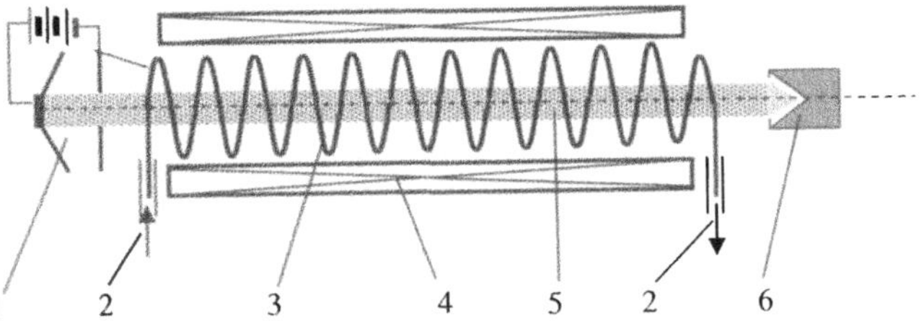

FIGURE 8.1 Traveling-wave tube. 1, Electron gun; 2, input–output; 3, helix slow-wave structure; 4, focusing magnet; 5, electron beam; 6, collector.

8.2.2 Qualitative Mechanism of Bunching and Energy Radiation in a TWTO

So there is a traveling electromagnet wave synchronized with the moving electrons. Let's pass to the frame of reference where both the wave field and the electrons are immovable. A representation of the sinusoidal electric field together with particles is shown in Fig. 8.2. At first, the electrons fill all wave phases uniformly (Fig. 8.2*a*). But in a moving frame of reference, electrons move to the right in the accelerating phase of the wave and to the left in the retarding phase. As a result, they fill the leading wave edge (Fig. 8.2*b*). At some time at a sufficiently large distance from the entry point, the electrons form a bunch at the point where the field crosses zero from the accelerating phase to the retarding phase (Fig. 8.2*c*). Thus, ideal bunching is achieved through exact synchronism and small space-charge repulsion.

But the phasing of the bunch is not favorable: The bunch arises at a zero field and cannot exchange energy with the wave. It is necessary to shift the bunch in the retarding phase of the field. Let us create some desynchronism, making the velocity of the electrons little more than the wave phase velocity $v > v_{ph}$. The bunch will move into the retarding phase and take energy out of the beam into the field. However, after some time the electrons turn into an accelerating phase and begin to take back wave energy. To avoid that, the interaction must be terminated. So the bunch must not pass a distance greater than half the wavelength with respect to the wave field during the transit time: that is,

$$(v - v_{ph})T \leq \frac{\lambda}{2} \tag{8.2}$$

According to Eq. (8.2), the maximum relative desynchronism must be approximately equal to

$$\delta v = \frac{\Delta v}{v_{ph}} = \frac{\lambda}{2Tv_{ph}} = \frac{\widetilde{T}}{2T} \ll 1 \tag{8.3}$$

The inequality in Eq. (8.3) assumes that the transit time T is large in TWTOs.

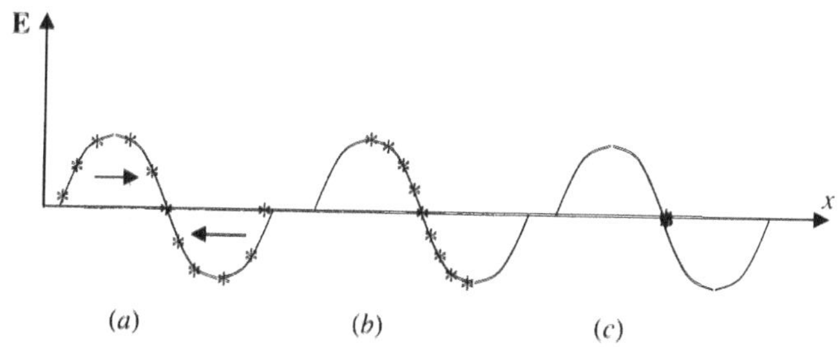

FIGURE 8.2 Electron bunching in a traveling wave.

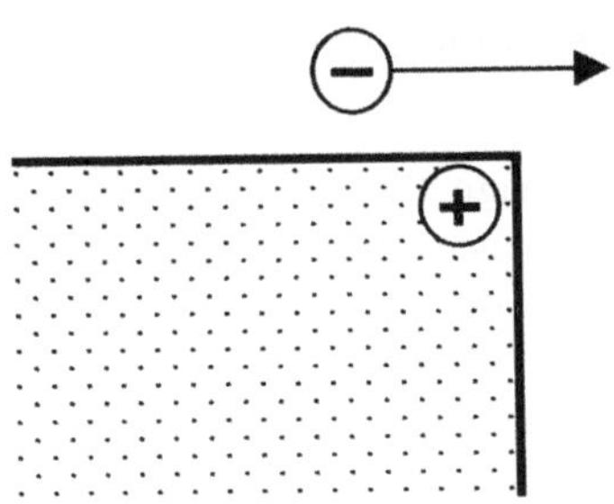

FIGURE 8.3 Transitive radiation on an edge nonhomogeneity.

Another interpretation of the TWTO mechanism comes from electromagnetic radiation theory. As has been shown (see, e.g., Barker et al., 2005; Jackson, 1999; Landau and Lifshitz, 1995), a charged particle that executes nonuniform or curvilinear motion always radiates an electromagnetic field. But a uniformly rectilinearly moving particle radiates in only two cases.

Cerenkov (Cherenkov) radiation arises when a charged particle moves with a speed that is greater than the speed of an electromagnetic wave in the same medium, $v > v_{\text{ph}}$. In general, each particle radiates independently. Therefore, the radiation is noncoherent and not effective for radio electronics. For the radiation to become coherent, it is necessary to form bunches. This is possible when the difference $v - v_{\text{ph}}$ is not large (i.e., in the presence of synchronism). We come to the same condition. Note that the bunching is forced, and the radiation is not spontaneous but is induced by the ac field. So in TWTOs, stimulated (induced) Cerenkov radiation takes place.

Transition radiation is observed when a particle moves in a nonhomogeneous dielectric medium, or more generally, in a medium containing nonhomogeneities. For example, consider an electron passing near an angular nonhomogeneity (Fig. 8.3). As we see, the electron induces positive charge in metal or dielectric, and together with this charge forms a dipole that has a variable dipole moment and radiates. The stimulated transition radiation is realized in narrow gaps of klystrons also, owing to the bunching of electrons and the phasing of bunches.

8.3 SLOW-WAVE STRUCTURES

As we have seen, the TWTO interaction is possible in the presence of electromagnetic structures that support propagation of "slow" electromagnetic waves with a phase velocity less than the speed of light. The first such structure, used by Kompfner, was a helix (Fig. 8.4). The theory of the helix as an electrodynamic slow-wave structure was developed by A. Sommerfeld long before the invention of TWTOs. In the simplest model of this structure, the wave propagates along a wire

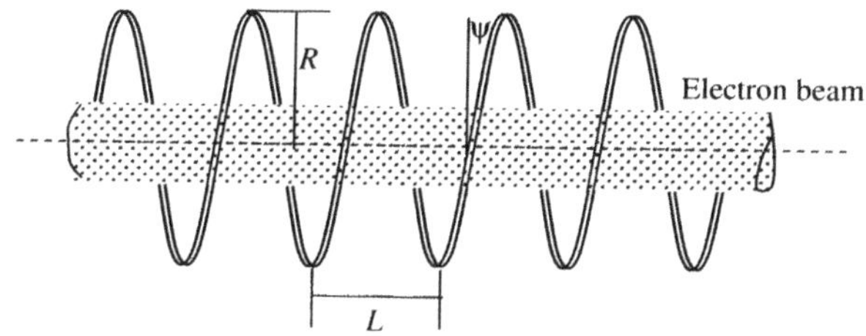

FIGURE 8.4 Helical slow-wave structure.

at the speed of light and therefore moves in an axial direction with the phase velocity

$$v_{\text{ph}} = c \sin \psi \tag{8.4}$$

where the helix pitch angle

$$\psi = a \tan \frac{L}{2\pi R} \tag{8.5}$$

According to Eq. (8.4), the phase velocity does not depend on the frequency (zero dispersion). This is a very important property, because the synchronism of the beam and the wave would be supported at all frequencies, which implies the broad bandwidth of TWTOs. In reality, the dispersion of the helix is not zero. The phase velocity increases at low frequencies, and at a high frequency where $\lambda \to 2L$ (Fig. 8.4), and the wave field in general is strongly perturbed. Nevertheless, the bandwidth of the helix is striking: So far there are no other known slow-wave structures whose dispersion comes close to the helix. Disadvantages of the helix include a significant limitation in the propagating high-frequency power, losses, and mechanics problems. Huge versions of slow-wave structures have been developed that satisfy many specific requirements, depending on the required frequency, power, and other properties of the device.

The examples in Fig. 8.5 show two typical slow-wave structures: a comb structure and an interdigital structure. The structures shown in Fig. 8.5 are chosen from a broad class of periodic structures, which are formed by electromagnetic coupling

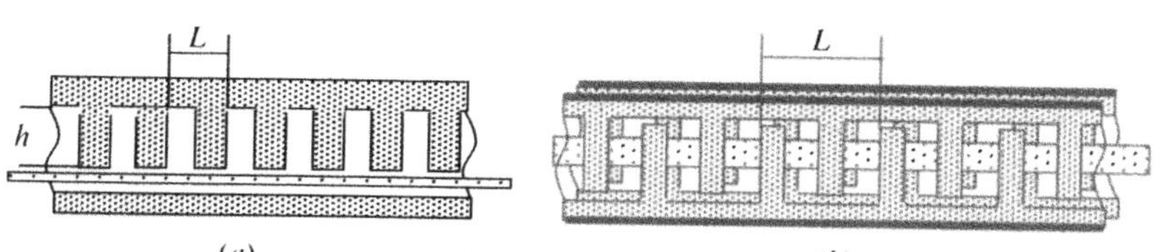

FIGURE 8.5 Slow-wave structures with electron beams: (*a*) comb slow-wave structure; (*b*) interdigital structure.

of the cavities through specially designed holes. The frequency dispersion of the phase velocity in these structures is considerably stronger than in the helix, and the bandwidth is correspondingly narrower (Appendix 4). Coupled cavity structures, however, are mostly metallic, and they have significantly better mechanical and thermal properties than those of helix structures. As a result, coupled-cavity TWTOs allow much greater average power, especially in the short-wave bands of the microwave range (see Section 8.7).

In principle, uniform dielectric waveguides can also be used as slow-wave structures in TWTOs. In this case, the electron beam propagates in a hollow tube inside the dielectric. The phase velocity equals approximately $v_{\text{ph}} = c/\sqrt{\varepsilon}$. Since an attainable permittivity is roughly $\varepsilon_{\max} \sim 10$, these structures can be employed in high-voltage TWTOs. One serious disadvantage of using dielectric waveguides is the charging of dielectric surfaces by the electron beam. Accumulated negative charge finally closes the beam. Therefore, only metal slow-wave structures are used in modern TWTOs. We turn next to some general properties of these periodic structures.

8.4 ELEMENTS OF SWS THEORY (see, e.g., Brillouin, 1953)

8.4.1 Floquet's Theorem

Assume that the electromagnetic field is a harmonic function of time:

$$\mathbf{E}(\mathbf{r},t) = \mathbf{E}(x,y,z)e^{i\omega t} \tag{8.6}$$

Consider a structure that is periodic along z with period L. Then a modulus of the field will also be a periodic function with the same period: $|\mathbf{E}(x,y,z+L)| = |\mathbf{E}(x,y,z)|$. Hence, the complex amplitude after a shift by one period along z gets the phase multiplier $e^{-i\varphi}$, where φ is a real constant:

$$\mathbf{E}(x,y,z+L) = \mathbf{E}(x,y,z)e^{-i\varphi} \tag{8.7}$$

Now consider the function of coordinates $\mathbf{F}(x,y,z) = \mathbf{E}(x,y,x)e^{i\beta_0 z}$, where $\beta_0 = \varphi/L$. It is readily verified that this function is periodic:

$$\mathbf{F}(x,y,z+L) = \mathbf{F}(x,y,z) \tag{8.8}$$

Thus, a field in a periodic system with period L is a periodic function of z with the same period L up to the factor $e^{-i\beta_0 z}$:

$$\mathbf{E}(x,y,z) = \mathbf{F}(x,y,z)e^{-i\beta_0 z} \tag{8.9}$$

Equations (8.8) and (8.9) express *Floquet's theorem*. According to Eq. (8.6), we can write the Floquet theorem as

$$\mathbf{E}(x,y,z,t) = \mathbf{F}(x,y,z)e^{i(\omega t-\beta_0 z)} \tag{8.10}$$

8.4.2 Spatial Harmonics

Using the periodicity of the function $\mathbf{F}(x, y, z)$, let us write it as a Fourier series:

$$\mathbf{F}(x, y, z) = \sum_{-\infty}^{\infty} \mathbf{F}_n(x, y) e^{-i(2\pi/L)nz} \tag{8.11}$$

Substituting Eq. (8.11) in Eq. (8.10), we obtain

$$\mathbf{E}(x, y, z, t) = \sum_{-\infty}^{\infty} \mathbf{F}_n(x, y) e^{i(\omega t - \beta_n z)} \tag{8.12}$$

where

$$\beta_n = \beta_0 + \frac{2\pi}{L} n, \qquad n = -\infty, \ldots, -1, 0, 1, \ldots, \infty \tag{8.13}$$

So the field in the periodical structure is a superposition of wave type: $\mathbf{F}_n(x, y) e^{i(\omega t - \beta_n z)}$. These waves have equal frequencies but different spatial structures. In particular, they have different wavelengths $\lambda_n = 2\pi/\beta_n$. They also have different transversal distributions. The function $\mathbf{F}_n(x, y)$ is obtained by solving Maxwell's equations with the corresponding boundary conditions. These waves are called *spatial harmonics*. Below we describe informally the key properties of spatial harmonics.

1. Wavelength:

$$\lambda_n = \frac{2\pi}{\beta_n} = \frac{2\pi}{\beta_0 + 2\pi n/L}, \qquad n = \ldots, -1, 0, 1, \ldots \tag{8.14}$$

2. Phase velocity:

$$v_{\mathrm{ph},n} = \frac{\omega}{\beta_n} = \frac{\omega}{\beta_0 + 2\pi n/L}, \qquad \beta_0 = \frac{\omega}{v_{\varphi 0}} \tag{8.15}$$

3. Group velocity:

$$v_g = \frac{\partial \omega}{\partial \beta_n} = \frac{\partial \omega}{\partial \beta_0} \tag{8.15a}$$

As we see, the group velocity is the same for all harmonics. It can be shown that in systems without losses, $v_g = v_{\mathrm{en}}$, where v_{en} is the velocity of an energy motion: $v_{\mathrm{en}} = P/\overline{W}$, $P = \int_S [\mathrm{E} \times \mathrm{H}]_z \, d\sigma$, and $\overline{W} = (1/2L) \int_{\Omega_L} (\varepsilon_0 E^2 + \mu_0 H^2) \, d\tau$ are the energy flux through a cross section of the structure and the average energy stored per unit length, respectively, Ω_L is a volume of one cell of the structure.

4. The field of each harmonic satisfies the Maxwell equations but not the boundary conditions. The period of harmonic λ_n does not equal a multiple of the structure period L. Only the total field [Eq. (8.12)] satisfies the boundary condition. Thus, if we excite any single harmonic (e.g., by interaction with a synchronous electron beam), all other harmonics would emerge with amplitude and phase relationships that are determined by the geometry of the structure and frequency. A specific set of harmonics determines a certain mode. In general, an infinite number of modes are possible for any periodic structure.

5. For sufficiently large n, harmonics are slow waves. This follows from Eq. (8.15).

6. All slow harmonics ($v_{\mathrm{ph},n} < c$) have properties of surface waves. It may be shown that the field of the nth harmonic decreases with distance x from the boundary of the periodical structure as

$$e^{-\sqrt{\beta_n^2 - k^2}x} = e^{-(2\pi x/\lambda_n)\sqrt{1-(v_{\mathrm{ph},n}/c)^2}}$$

So the field of the slow harmonic diminishes by a factor of $e^{2\pi\sqrt{1-(v_{\mathrm{ph},n}/c)^2}}$ along a distance of λ_n. Thus, the energy of the slow harmonic propagates in the *skin layer*,

$$\Delta_{\mathrm{sk}} = \frac{\lambda_n}{2\pi\sqrt{1-(v_{\mathrm{ph},n}/c)^2}} = \frac{\lambda}{2\pi}\frac{v_{\varphi n}}{c}\frac{1}{\sqrt{1-(v_{\mathrm{ph},n}/c)^2}} \tag{8.16}$$

For a large delay ($v_{\mathrm{ph},n} \ll c$),

$$\Delta_{\mathrm{sk}} = \frac{\lambda_n}{2\pi} = \frac{\lambda}{2\pi}\frac{v_{\mathrm{ph},n}}{c}$$

7. The function $\omega = \omega(\beta_0)$ is periodic with period $2\pi/L$. If we increase β_0 by $2\pi/L$, β_n takes the place of β_{n+1}. As a result, the label of each β_n is changed, but the full set of harmonics does not change. For the same reason, we can conclude that ω is an even function of β_0: Inverting the sign for β_0 together with the rest of the harmonics β_{-n}, gives the same wave traveling in the $-z$ direction. In Fig. 8.6 a qualitative diagram of the dispersion function $\omega(\beta_0)$ is provided. The lower curve corresponds to the fundamental mode $N = 1$. Propagation of the wave in this mode is possible in the frequency bandwidth $\Delta\omega_1$ (Fig. 8.6). In the next mode the bandwidth is $\Delta\omega_2$. In principle, the frequency domain of any periodic system consists of separate transmission bands divided by attenuation bands (i.e., any periodic structure similar to some bandpass filter).

8. Taking into account the properties of the dispersion function $\omega = \omega(\beta_0)$ indicated above, we can deduce that the propagation constant of the zero harmonic β_0 is enclosed in the range

$$0 \le \beta_0 L \le \pi \tag{8.17}$$

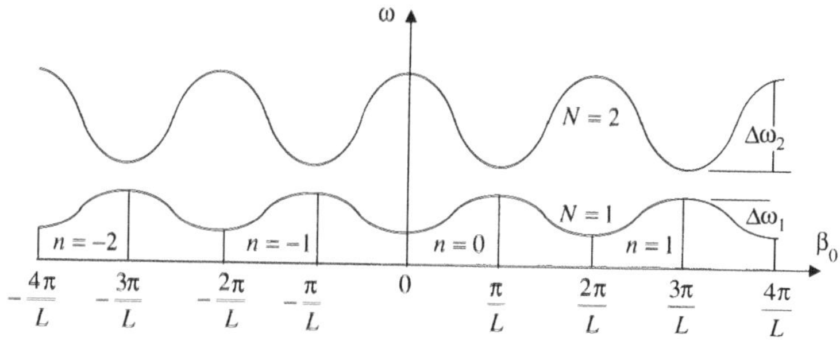

FIGURE 8.6 Dispersion diagram for a periodic structure.

So the maximum phase shift of the zero harmonic per period is equal to π. The group velocity v_g for the range boundaries $\beta_0 = 0, \pi$ is obviously zero.

9. According to Eqs. (8.13) and (8.17), we find that the module of the propagation constant (-1)th harmonic is within the range $\pi/L \leq |\beta_{-1}| \leq 2\pi/L$. As such, the modules of the harmonic propagation constants can be arranged as

$$|\beta_0| \leq |\beta_{-1}| \leq |\beta_1| \leq |\beta_{-2}| \leq |\beta_2| \leq \cdots \tag{8.18}$$

It can be shown that the fields of the harmonics decrease with $|\beta_n|$. Therefore, Eq. (8.18) gives the law of harmonic amplitude decrease. The zero harmonic has the greatest amplitude. Recall that the field of the harmonic decreases with distance x from the boundary as $e^{-\beta_n x}$. So at a sufficient distance from the boundary, we see only zero harmonics (i.e., the field has the structure of a regular waveguide field).

10. All harmonics with $n \geq 0$ have a positive phase velocity. On the contrary, the harmonics with $n < 0$ have a negative phase velocity. Assume that the curve $\omega = \omega(\beta_0)$ has a positive slope similar to the curve shown in Fig. 8.6 for the first mode in the range of Eq. (8.17). We find that all positive harmonics have coinciding directions of phase and group velocities, so they are forward waves. All negative harmonics having phase and group velocities in the opposite direction are backward waves. For mode 2 in Fig. 8.6, all positive harmonics are backward waves, and the negative harmonics are forward waves. When the phase and group velocities of a fundamental harmonic of a structure have the same directions the structure is said to have positive dispersion. In a structure with negative dispersion the zero harmonic and all positive harmonics in the fundamental mode, the phase and group velocities have opposite directions. The comb and interdigital structures in Fig. 8.5 are examples of structures with positive and negative dispersion, respectively.

Note that the group velocity of nth spatial harmonic can be represented as

$$\frac{1}{v_{g,n}} = \frac{d\beta_n}{d\omega} = -\frac{2\pi}{\lambda_n^2}\frac{d\lambda_n}{d\omega}$$

Therefore, for structures with positive dispersion, the wavelengths of the zero and all positive spatial harmonics are decreased with frequency.

The theory of the slow-wave structures reduces to solving the Maxwell equations with periodic boundary conditions. The problem is sufficiently complicated due to complex boundary conditions. A number of computational methods (e.g., integral equations, variational methods, direct simulation) have been proposed (see, e.g., Bevensee, 1964; Lines et al., 1950; Tsimring, 1957). The final results [dispersion $\omega(\beta)$ and distribution of electromagnetic fields] are usually developed numerically. An equivalent circuit method occupies a particular place. The method gives a detailed qualitative description of dispersion characteristics for very complex structures: in particular, systems of coupled-cavity TWTOs (see, e.g., Gilmour, 1994).

8.5 LINEAR THEORY OF A NONRELATIVISTIC TWTO. DISPERSION EQUATION, GAIN, EFFECTS OF NONSYNCHRONISM, SPACE CHARGE, AND LOSS IN A SLOW-WAVE STRUCTURE

8.5.1 Statement of the Problem

The exchange of energy between the bunched electron beam and the electromagnetic field in a TWTO is accompanied by the formation of a variable wave amplitude, which is not given beforehand. On the other hand, bunching of the electron beam is determined by an unknown distribution of the electromagnetic field. Thus, we are presented with a typical self-matched problem. The usual method for solving such problems is to split them into subproblems and then match the solutions to the subproblems. In the case of TWTOs, we come to the following subproblems: calculation of a bunched electron beam in the wave field, and excitation of the wave in the slow-wave structure by the convection current obtained. Naturally, the field and current in both subproblems include some arbitrary parameters that are defined in the process of matching. In the linear theory, the ac values (current $I_{\approx}$, velocity $v_{\approx}$, and space-charge density $\rho_{\approx}$) are assumed to be small compared with the dc values. Therefore, in the frame of the linear theory variation, the dc electron energy is ignored.

8.5.2 Bunching of a Convection Current in a Traveling-Wave Field

This problem is similar to the one considered in Section 7.6 on space-charged waves in velocity-modulated beams, with the exception of a condition of potentiality in the electric field. We accept the following:

1. One-dimensionality (all values are functions of a single spatial variable, z)
2. A strong focusing magnetic field B_z, which excludes transverse motion of the electrons
3. Smallness of ac values as compared with dc values

4. Constant dc quantities v_0 and ρ_0
5. A nonrelativistic approximation of $v \ll c$
6. Consideration of only a fundamental (zero) spatial harmonic of the field
7. Wave behavior for all values
8.
$$E = \widetilde{E}e^{i(\omega t - \beta z)}, \qquad v_{\approx} = \tilde{v}e^{i(\omega t - \beta z)},$$
$$I_{\approx} = \widetilde{I}e^{i(\omega t - \beta z)}, \qquad \rho_{\approx} = \tilde{\rho}e^{i(\omega t - \beta z)} \tag{8.19}$$

where $I_{\approx}$ and $\rho_{\approx}$ are the ac beam current and ac line charge density correspondingly.

We can see that these idealizations do not include condition 2 from Section 7.6.1, according to which the ac electric field has only a z-component. One leads to the Poisson equation [Eq. (7.52)] and indicates the potentiality of the ac electric field. This equation is added to conditions 1 to 7 only for an analysis of the space charge in TWTOs.

Unlike in Eq. (7.46), here the propagation constant is assumed to be complex: $\beta = \beta_1 + i\beta_2$, where β_1 and β_2 are not given and should be found via matching. According to Eq. (8.19), if $\beta_2 > 0$, the wave is amplified along the z-coordinate. The real component, β_1, determines a phase velocity $v_\varphi = \omega/\beta_1$.

Temporarily we will neglect losses in the slow-wave structure without the electron beam. So the propagation constant β_0 in the equation of the undisturbed wave

$$E^{(0)} = \widetilde{E}^{(0)}e^{i(\omega t - \beta_0 z)} \tag{8.20}$$

is real. In general, $\beta_1 \neq \beta_0$. If $\beta_1 > \beta_0$, the phase velocity of a "hot" wave is less than $v_{\varphi 0}$. This wave is called a *slow wave*. Waves where $\beta_1 < \beta_0$ are *fast waves*.

As we see, the equations describing the bunching process coincide with Eqs. (7.47)–(7.49), which were used for an analysis of space-charge waves. This is because conditions 1 to 7, and the corresponding conditions from Section 7.6.1 are identical. Therefore, we can apply Eq. (7.51) directly. (Here we replaced x by z):

$$\left(\frac{\partial}{\partial t} + v_0\frac{\partial}{\partial z}\right)^2 I_{\approx} = -\rho_0\eta\frac{\partial E}{\partial t} \tag{8.21}$$

where v_0 is the dc electron velocity and E is the z-component of the electric field. Substituting Eq. (8.19) in Eq. (8.21), we obtain

$$(\beta - \beta_e)^2\,\widetilde{I} = \frac{i\rho_0\eta\beta_e}{v_0}\widetilde{E} \tag{8.22}$$

where $\beta_e = \omega/v_0$. Equation (8.22) determines the amplitude of the bunched convection current in a given field. According to Eq. (8.22), the bunched current is maximal for the *synchronism condition*:

$$\beta \approx \beta_e \quad \text{or} \quad v_{\text{ph}} \approx v_0 \tag{8.23}$$

Separation of the Space-Charge Field The field E is the longitudinal component of an ac electric field. It can be represented as a sum of the electric field of the slow-wave structure and the ac space-charge field:

$$E = E_w + E_p$$

E_w can be considered a z-component of the purely vortical field $\mathbf{E}_w = \text{curl}\,\mathbf{B}_w$, where $\mathbf{B}_w$ is a magnetic field in the slow-wave structure. Therefore, $\text{div}\,\mathbf{E}_w = 0$. In the first approximation, the field E_w is not coupled locally with the electron beam. The space-charge field E_p, on the contrary, is a purely longitudinal wave field with a z-component only. It is expressed through the ac space-charge line density $\rho_{\approx}$ via the Poisson equation,

$$\frac{\partial E_p}{\partial z} = \frac{\rho_{\approx}}{\varepsilon_0 S} \tag{8.24}$$

This is a key equation in the theory of space-charge waves. According to Eqs. (8.24) and (8.19),

$$\rho_{\approx} = -i\varepsilon_0 S\beta E_p \tag{8.25}$$

On the other hand, using the continuity equation $\partial I_{\approx}/\partial z = -\partial\rho_{\approx}/\partial t$, we obtain

$$\beta I_{\approx} = \omega\rho_{\approx} \tag{8.26}$$

Equations (8.25) and (8.26) allow us to express the space-charge field through the current

$$E_p = i\frac{I}{\omega\varepsilon_0 S} \tag{8.27}$$

Substitute $E = E_w + E_p$ and E_p from Eq. (8.27) in Eq. (8.22). We find that

$$(\beta - \beta_e)^2 I = \frac{i\rho_0\eta\beta_e}{v_0}E_w + \frac{\rho_0\eta}{S\varepsilon_0 v_0^2}I \tag{8.28}$$

Finally, taking into account that

$$v_0^2 = 2\eta U_0, \qquad \rho_0 v_0 = -I_0, \qquad \omega_p^2 = -\frac{\rho_0\eta}{\varepsilon_0 S}, \qquad \beta_p = \frac{\omega_p}{v_0} \tag{8.29}$$

we obtain the desired connection between I and E_w:

$$[(\beta - \beta_e)^2 - \beta_p^2]I = -i\frac{I_0\beta_e}{2U_0}E_w \tag{8.30}$$

8.5.3 Excitation of the Field E_w in a Slow-Wave Structure by a Given Convection Current

Let us represent the slow-wave structure as a generalized transmission line (Fig. 8.7). It can be considered as a first approximation in the theory of waveguide

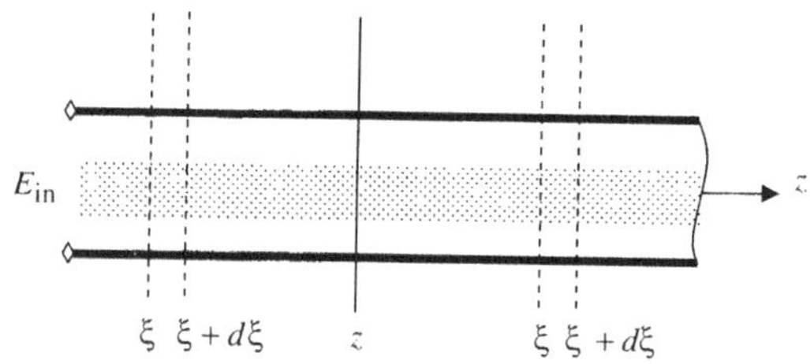

FIGURE 8.7 Transmission line excited by an electron beam.

excitation. We postpone discussion of this issue, however, to the next section. The bunched electrons exchange energy with the electric field in each plane of the structure. Let us consider a layer $\xi \Leftrightarrow \xi + d\xi$ as an elementary source of a radiation. The field in the z plane is formed as a sum of the input wave $E_{in}(z,t) = \widetilde{E}_{in}e^{i(\omega t-\beta_0 z)}$ and waves of the secondary sources $E_{sec}(\xi, t)$. So the field in this plane is

$$E(z,t) = \widetilde{E}_{in}e^{i(\omega t-\beta_0 z)} + \sum_{\xi<z} E_{sec}(\xi,t) + \sum_{\xi>z} E_{sec}(\xi,t) \tag{8.31}$$

Determination of Secondary Fields Assume that $P(\xi)$ is the energy flux of the wave $E_w(\xi)$ in plane ξ. This field forms in the gap $\xi \Leftrightarrow \xi + d\xi$ with a voltage of $dU(\xi) = -E_w(\xi)d\xi$. The source accomplishes the work $\frac{1}{2}\text{Re}[\widetilde{I}(\xi)d\widetilde{U}^*(\xi)]$. That work is equal to the increment in the energy flow:

$$dP(\xi) = -\frac{1}{2}\text{Re}[\widetilde{I}(\xi)\widetilde{E}_w^*(\xi)]\,d\xi \tag{8.32}$$

On the other hand, the power flow is proportional to the square of the field amplitude modulus:

$$P(\xi) = K|\widetilde{E}_w(\xi)|^2 \tag{8.33}$$

The constant K is usually expressed via another constant, $Z = 1/2\beta_0^2 K$. We obtain

$$Z = \frac{|\widetilde{E}_w(\xi)|^2}{2\beta_0^2 P(\xi)} \tag{8.34}$$

It is readily seen that the quantity Z has dimensions of the impedance. It is called a coupling impedance. Let us find from Eq. (8.33) the power increment in the element $\xi \Leftrightarrow \xi + d\xi$:

$$dP(\xi) = \frac{1}{2\beta_0^2 Z}d|\widetilde{E}_w(\xi|^2 = \frac{1}{\beta_0^2 Z}\text{Re}[d\widetilde{E}_w(\xi)\widetilde{E}_w^*(\xi)] \tag{8.35}$$

Equating $dP(\xi)$ of Eq. (8.35) to $dP(\xi)$ of Eq. (8.32), we come to the relation between the field increment in the plane ξ and the beam current:

$$d\widetilde{E}_w(\xi) = -\frac{\beta_0^2 Z}{2}\widetilde{I}(\xi)\,d\xi \tag{8.36}$$

This perturbation propagates to the z plane as a wave and forms in this plane the field

$$d\widetilde{E}_w(\xi, z) = d\widetilde{E}_w(\xi)e^{-i\beta_0(z-\xi)} = -\frac{\beta_0^2 Z}{2}\widetilde{I}(\xi)e^{-i\beta_0(z-\xi)}\,d\xi \tag{8.37}$$

Similarly, a perturbation propagating to the z plane from the right source ($\xi > z$) is

$$d\widetilde{E}_w(\xi, z) = -\frac{\beta_0^2 Z}{2}\widetilde{I}(\xi)e^{i\beta_0(z-\xi)}\,d\xi \tag{8.38}$$

Thus, according to Eq. (8.31), the total electric field excited in plane z is equal to

$$\widetilde{E}_w(z) = \widetilde{E}_{\text{in}}\,e^{-i\beta_0 z} - \frac{\beta_0^2 Z}{2}\int_0^z \widetilde{I}(\xi)e^{-i\beta_0(z-\xi)}\,d\xi - \frac{\beta_0^2 Z}{2}\int_z^L \widetilde{I}(\xi)e^{i\beta_0(z-\xi)}\,d\xi \tag{8.39}$$

To obtain $E_w(z,t)$, this expression must be multiplied by $e^{i\omega t}$. The integral representation of the field [Eq. (8.39)] is not convenient for analysis. Let us transform it instead into a differential equation. Taking a double differential $\widetilde{E}_w(z)$ from Eq. (8.39) and adding the resultant with $\beta_0^2\widetilde{E}_w(z)$, we arrive at the linear second-order differential equation

$$\frac{d^2\widetilde{E}_w}{dz^2} + \beta_0^2\widetilde{E}_w = i\beta_0^3 Z\widetilde{I} \tag{8.40}$$

Again representing the field as $\widetilde{E}_w(z) \sim e^{-i\beta z}$, we obtain

$$(\beta_0^2 - \beta^2)\widetilde{E}_w = i\beta_0^3 Z\widetilde{I} \tag{8.41}$$

8.5.4 Dispersion Equation of the TWTO

The linear equations (8.41) and (8.30) must form a self-consistent system. As a result, we obtain the dispersion equation

$$(\beta^2 - \beta_0^2)[(\beta - \beta_e)^2 - \beta_p^2] = -\frac{\beta_0^3\beta_e I_0 Z}{2U_0} \tag{8.42}$$

which determines the propagation constant β as a function of the frequency and other parameters of the system. Equation (8.42) is a fourth-degree algebraic equation and has four complex roots: β_1, β_2, β_3, and β_4. The resulting field and other TWTO quantities ($\tilde{I}$, $\tilde{\rho}$, and $\tilde{v}$) are superpositions of the waves that are determined by these roots. Solution of this equation needs four boundary conditions. They can be taken as voltages at the two ends of the circuit and input values of the ac velocity and the current.

Our discussion below of the linear theory of TWTO is based on an analysis of the dispersion equation (8.42). We note that the foregoing reduction of the excitation problem to the excitation of an idealized transmission line is an essential simplification. In reality, the electric field E_w is the superposition of many waves (modes). Among these waves there is one (synchronous) wave whose phase velocity is close to the undisturbed electron velocity v_0. This is a *resonant wave*. The remainder of the field is a sum of asynchronous and nonpropagation modes. Nonresonant modes can be considered as modes locally dependent on the ac current. In sum, they supplement the quasistatic field up to the space-charge field of the the beam in a specific slow-wave structure. As shown in relevant works (see, e.g., Pierce, 1950; Solntsev, 1968; Vainstein, 1956), an account of these modes for a sufficiently small gain is equivalent to replacement of the plasma frequency ω_p by the reduced plasma frequency ω_q. The space-charge parameter β_p^2 in Eqs. (8.30) and (8.42) now means that $(\omega_q/v_0)^2 = F^2(\omega_p/\omega)^2$, where $F = \omega_q/\omega_p$ is the reduction factor. Then the space-charge field $\tilde{E}_p$ in Eq. (8.27) should be replaced by a reduced electric field:

$$\tilde{E}_p^* = i\frac{F^2\tilde{I}}{\omega\varepsilon_0 S} \tag{8.43}$$

The electric field $\tilde{E}_w$ in Eqs. (8.23) and (8.32)–(8.41) should now imply a resonant (synchronous) wave. Let us consider some particular cases.

1. $Z = 0$. Coupling between the electron beam and the electromagnetic field is absent. We obtain two independent coupled waves: $\beta = \pm\beta_0$ and $\beta = \beta_e \pm \beta_p$. These are two waveguide waves and [according to Eq. (7.61)] two space-charge waves in the electron beam.

2. $I_0 = 0$ ("cold" field). The right side of Eq. (8.42) and β_p are equal to zero. Then $\beta = \pm\beta_0$. These are two waves (direct and reflected) in the unperturbed transmission line.

3. $I_0 \neq 0$, but I_0 is a small quantity. J. R. Pierce, the pioneer of TWTO theory, introduced the quantity

$$C^3 = \frac{I_0 Z}{4U_0} \tag{8.44}$$

which, as will be shown, is a small parameter in ordinary TWTOs. This quantity appears on the right side of Eq. (8.42). C is frequently called the *Pierce parameter*. We will see that C determines TWTO gain, and as such is also called the *gain parameter*.

Equation (8.42) becomes

$$(\beta^2 - \beta_0^2)[(\beta - \beta_e)^2 - \beta_p^2] = -2\beta_0^3\beta_e C^3 \tag{8.45}$$

Let us estimate C for a TWTO of medium power. Take the current $I_0 = 0.1$ A and the voltage $U_0 = 1000$ V. The coupling parameter Z is by an order of magnitude equal to the characteristic impedance of a transmission line $\sim 100\ \Omega$. Then $C^3 \approx 1/400$ and $C \sim 0.1$.

Assume that β is close to β_0. Then $\beta^2 - \beta_0^2 \approx 2\beta_0(\beta - \beta_0)$, and Eq. (8.44) becomes

$$(\beta - \beta_0)[(\beta - \beta_e)^2 - \beta_p^2] = -\beta_0^2\beta_e C^3 \tag{8.46}$$

So Eq. (8.42) has been reduced to a third-degree equation. In this approximation, we in fact ignore a reflected wave.

If, on the contrary, β is close to $-\beta_0$ (which corresponds to taking into account only the backward wave), $\beta^2 - \beta_0^2 \approx -2\beta_0(\beta + \beta_0)$ and $(\beta - \beta_e)^2 - \beta_p^2 \approx 4\beta_0^2$. We obtain from Eq. (8.45),

$$\beta_4 = -\beta_0\left(1 - \frac{C^3}{4}\right) \tag{8.47}$$

We see that the backward wave propagates as a reflected waveguide wave without amplification or attenuation.

8.5.5 Dimensionless Parameters, Initial Conditions, and Gain of a Nonrelativistic TWTO

Further, we will use the following dimensionless parameters that were introduced by Pierce.

1. Parameter of nonsynchronism:

$$b = \frac{v_e - v_{\varphi 0}}{v_{\varphi 0} C} = \frac{\beta_0 - \beta_e}{\beta_e C} \tag{8.48}$$

2. Space-charge parameter:

$$q = 4QC = \frac{\beta_p^2}{\beta_e^2 C^2} = \frac{\omega_q^2}{\omega^2 C^2} = \frac{F^2\omega_p^2}{\omega^2 C^2} \tag{8.49}$$

This relation takes into account a reduction in the plasma frequency ω_p. The Pierce notation $q = 4QC$ (Pierce, 1950) is often used in the literature.

3. Incremental propagation constant:

$$\delta = -i\frac{\beta - \beta_e}{\beta_e C} \tag{8.50}$$

Substituting $\beta = \beta_1 + i\beta_2$ into Eq. (8.50), we obtain

$$\delta = \frac{\beta_2}{\beta_e C} - i\frac{\beta_1 - \beta_e}{\beta_e C} \tag{8.51}$$

Hence, the wave grows when $\operatorname{Re}\delta > 0$, and vice versa. The imaginary part of δ is proportional to the difference between the wave phase velocity and the initial electron velocity. If $\operatorname{Im}\delta < 0$, the wave is slow (i.e., $v_\varphi < v_0$), and in the reverse situation, where $\operatorname{Im}\delta > 0$, the wave is fast.

Next, we transform the dispersion equation (8.46) into new parameters. We divide Eq. (8.46) by $\beta_e^3 C^3$, taking into account that

$$\frac{\beta - \beta_0}{\beta_e C} = \frac{\beta - \beta_e}{\beta_e C} - \frac{\beta_0 - \beta_e}{\beta_e C} = i\delta - b$$

The constant on the right side of the transformed equation is $\beta_0^2/\beta_e^2 = (1 + Cb)^2$. The dispersion equation takes the form

$$i(\delta + ib)(\delta^2 + q) = (1 + Cb)^2 \tag{8.52}$$

This equation has three complex roots: δ_1, δ_2, and δ_3. The partial solutions to Eqs. (8.21) and (8.40) corresponding to these roots are

$$\widetilde{E}_k(z) = e^{-i\beta_k z} = e^{-i\beta_e z} e^{\beta_e C \delta_k z}, \qquad k = 1, 2, 3 \tag{8.53}$$

Next, we find partial solutions for the current and the velocity. After reducing this equation to the dimensionless parameters [Eqs. (8.48)–(8.50)], we obtain from Eq. (8.30)

$$\widetilde{I}_k(z) = i\frac{I_0}{2U_0\beta_e C^2}\frac{1}{\delta_k^2 + q}\widetilde{E}_k(z), \qquad k = 1, 2, 3 \tag{8.54}$$

Finally, according to Eqs. (7.50) and (8.19), we find that

$$\tilde{v}_k(z) = \frac{1}{\omega\rho_0}(\omega - v_0\beta_k)\widetilde{I}_k(z) = \frac{\eta}{v_0\beta_e C}\frac{\delta_k}{\delta_k^2 + q}\widetilde{E}_k(z), \qquad k = 1, 2, 3 \tag{8.55}$$

A general solution of the equations is

$$\begin{aligned}
\widetilde{E}(z) &= \sum_{k=1}^{3} A_k \widetilde{E}_k(z) \\
\widetilde{I}(z) &= \sum_{k=1}^{3} A_k \widetilde{I}_k(z) = i\frac{I_0}{2U_0\beta_e C^2}\sum_{k=1}^{3}\frac{1}{\delta_k^2 + q}A_k\widetilde{E}_k(z) \\
\tilde{v}(z) &= \sum_{k=1}^{3} A_k \tilde{v}_k(z) = \frac{\eta}{v_0\beta_0 C}\sum_{k=1}^{3}\frac{\delta_k}{\delta_k^2 + q}A_k\widetilde{E}_k(z)
\end{aligned} \tag{8.56}$$

where A_1, A_2, and A_3 are arbitrary constants which can be found from boundary conditions. Assume, for example, that the boundary conditions are set at the input plane:

$$\widetilde{E}(0) = \widetilde{E}_0, \qquad \widetilde{I}(0) = \widetilde{I}_0, \qquad \tilde{v}(0) = \tilde{v}_0 \tag{8.57}$$

Then we obtain from Eq. (8.56) a system of three linear algebraic equations, which determine A_k, where, according to Eq. (8.53), $\widetilde{E}_k(0) = 1$. Using the Cramer rule, we find the following complex amplitudes of the electric field:

$$A_k = \frac{\delta_k^2 + q}{(\delta_k - \delta_l)(\delta_k - \delta_m)} \left\{ \widetilde{E}_0 + 2\beta_e C U_0 \left[\frac{\tilde{v}_0}{v_0}(\delta_l + \delta_m) - iC\frac{\widetilde{I}_0}{I_0}(\delta_l\delta_m - q) \right] \right\} \tag{8.58}$$

If the velocity and current modulation at the tube input are absent ($\tilde{v}_0 = \widetilde{I}_0 = 0$),

$$A_k = \frac{\delta_k^2 + q}{(\delta_k - \delta_l)(\delta_k - \delta_m)} \widetilde{E}_0, \qquad k, l, m = 1, 2, 3 \tag{8.59}$$

According to Eqs. (8.56), (8.53), and (8.59), the voltage gain in this case,

$$G_E = \frac{\widetilde{E}(L)}{\widetilde{E}(0)} = \frac{1}{\widetilde{E}(0)} \sum_{k=1}^{3} A_k \widetilde{E}_k(L)$$

$$= e^{-i\beta_e L} \sum_{k=1}^{3} \frac{\delta_k^2 + q}{(\delta_k - \delta_l)(\delta_k - \delta_m)} e^{\delta_k \beta_e C L}, \qquad l, m = 2, 3 \tag{8.60}$$

where L is the space interaction length. The power gain of the tube is

$$G = 10 \log |G_E|^2 = 20 \log \left| \sum_{k=1}^{3} \frac{\delta_k^2 + q}{(\delta_k - \delta_l)(\delta_k - \delta_m)} e^{\delta_k \beta_e C L} \right| \text{ dB} \tag{8.61}$$

8.5.6 Particular Cases

Small Nonsynchronism, $Cb = (\beta_0 - \beta_e)/\beta_e \ll 1$ The dispersion equation (8.52) assumes the form

$$i(\delta + ib)(\delta^2 + q) = 1 \tag{8.62}$$

Taking $\delta = x + iy$, we obtain the following system of equations with respect to the real x and y:

$$\begin{aligned} (x^2 - y^2)(y + b) + 2x^2 y + q(y + b) + 1 &= 0 \\ x[(x^2 - y^2) - 2y(y + b) + q] &= 0 \end{aligned} \tag{8.62a}$$

These equations can be solved numerically. Recall that $x > 0$ corresponds to a growing wave; $y < 0$ corresponds to a wave with $v_\varphi < v_0$ (a slow wave) [see Eq. (8.51)].

Equation (8.62) has two complex-conjugate roots, δ_1 and δ_2, and one imaginary root, $x_3 = 0$. First, we obtain from Eqs. (8.62*a*) for $x_3 = 0$,

$$(y_3 + b)(y_3^2 - q) = 1 \tag{8.62b}$$

It can be shown (see, e.g., Vainstein, 1956) that complex roots exist within a limited range of the nonsynchronism parameter: $b_{\min} \leq b \leq b_{\max}$. Inside this range the imaginary parts of the roots are $y_1(b,q) = y_2(b,q)$, and the real parts are $x_1(b,q) = -x_2(b,q)$. Outside this range δ_1 and δ_2 are purely imaginary. The minimum value $b_{\min} = -3/\sqrt[3]{4} \approx -1.9$ is attained for $q = 0$.

Synchronous Mode with a Small Space Charge Define the synchronous mode as the equality $\beta_e = \beta_0$ [i.e., the initial electron velocity and the phase velocity of the input (cold) wave are equal, and $b = 0$]. Assume also a small space charge $q/\delta^2 \ll 1$; then Eq. (8.62) reduces to

$$\delta^3 = -i \tag{8.63}$$

We obtain the following three roots:

$$\delta_1 = \frac{\sqrt{3}}{2} - i\frac{1}{2}, \qquad \delta_2 = -\frac{\sqrt{3}}{2} - i\frac{1}{2}, \qquad \delta_3 = i \tag{8.63a}$$

These roots correspond to three waves. According to Eq. (8.53), we can write the amplitudes $\widetilde{E}_k$ as

$$\widetilde{E}_k = e^{-i\beta_k z} = e^{-i\beta_e z} e^{\beta_e C \delta_k z}, \qquad n = 1, 2, 3 \tag{8.64}$$

We see according to Eq. (8.51) that wave 3 has a constant amplitude. Waves 1 and 2 are slow waves because $\operatorname{Im}\delta_1 = \operatorname{Im}\delta_2 < 0$. Wave 1 is exponentially growing, since $\operatorname{Re}\delta > 0$ and wave 2 is an attenuating wave.

This result can be interpreted in the following way: After neglecting the backward waveguide wave, three waves remain: a direct waveguide wave and two space-charge waves. Although we ignored β_p^2 in Eq. (8.45), these waves do exist. They are almost synchronous with the electrons, because according to Eq. (7.61),

$$\frac{|\beta - \beta_e|}{\beta_e} \sim \frac{\beta_p}{\beta_e} = \frac{\omega_p}{\omega} \ll 1$$

During the interaction the slow space-charge wave absorbs the beam energy and grows as a result. The fast wave gives energy to the beam and therefore attenuates. The waveguide wave changes only in phase velocity.

The gain can be found from Eq. (8.59), and for the case in question ($b = q = 0$) all three amplitudes are the same (i.e., $A_1 = A_2 = A_3$). Because $(\widetilde{E}_k)_{z=0} = 1$, we

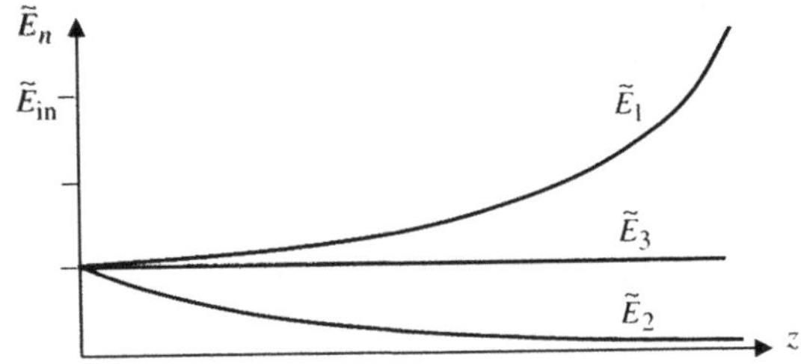

FIGURE 8.8 Amplitude distribution of partial waves for a synchronous mode.

obtain from Eqs. (8.56) that $\tilde{E}_0 = A_1 + A_2 + A_3$ and $A_k = \tilde{E}_0/3$. The gain in that case, according to Eq. (8.61), is equal to

$$G = 20\log\frac{1}{3}\sum_{k=1}^{3} e^{\delta_k \beta_e CL}$$

A qualitative diagram of the amplitude distribution of three partial waves is shown in Fig. 8.8.

Consider the gain of a TWTO for a large length of the system. The output signal is represented only by the growing wave ($\tilde{E}_1$). Hence, the gain is

$$20\log\left(\frac{1}{3}e^{(\sqrt{3}/2)\beta_e CL}\right) = -20\log 3 + 20\frac{\sqrt{3}}{2}\beta_e CL\log e$$

$$= -20\log 3 + 10\sqrt{3}\frac{2\pi L}{\lambda}C\log e = -9.54 + 47.3CN$$

where $N = L/\lambda$ is the number of slow wavelengths in the system. The constant -9.54 is called the *coefficient of initial losses*. It appears because the initial field of the growing wave is one-third of the input field. Take as an example, $C = 0.1$, $U_0 = 1000$ V, and $L = \lambda_0$, where λ_0 is the wavelength in the free space. The coefficient of deceleration [see Eq. (1.10)] is

$$\frac{c}{v_0} \approx \frac{16}{\sqrt{U_{0\,\mathrm{kV}}}} = \frac{\lambda_0}{\lambda} = 16$$

Meanwhile, the gain is $K_{\mathrm{dB}} = -9.54 + 47.3 \times 0.1 \times 16 \approx 66$ dB. However, due to a variety of factors (some of which are discussed below), this value overestimates the real gain in the TWTO.

Effect of a Space Charge In Fig. 8.9, solutions of Eq. (8.62), as functions of nonsynchronous parameter q ($4QC$) are shown. We can see that the zone boundaries

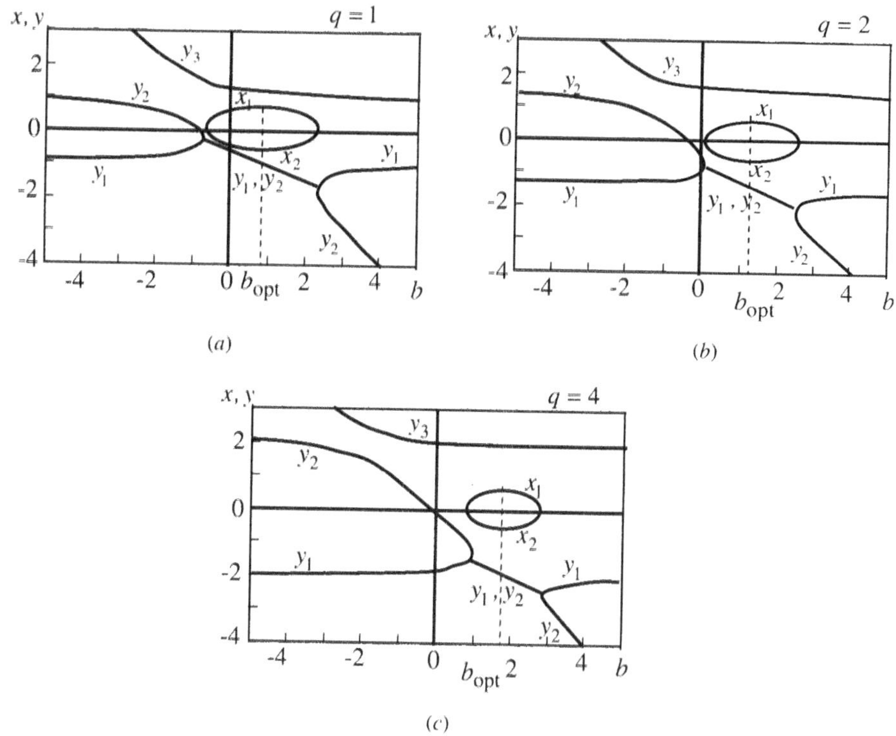

FIGURE 8.9 Real and imaginary parts of the incremental propagation constant δ as functions of the nonsynchronism parameter b for three values of space-charge parameter $q = 4QC$. (From Pierce, 1950.)

$b_{\min}$ and $b_{\max}$ draw together and shift to more b with increasing q. Also, the value b_{opt}, corresponding to maximum x_1, shifts to more b with increasing q (Fig. 8.9). One can be shown that $b_{\text{opt}} \approx \sqrt{q}$ for large q. According to Eqs. (8.48) and (8.49), that gives

$$\frac{\beta_{0,\text{opt}} - \beta_e}{\beta_e C} \approx \frac{\beta_p}{\beta_e C}, \quad \text{or} \quad \beta_{0,\text{opt}} = \beta_e + \beta_p$$

Compare this equation with Eq. (7.61). We conclude that for a large space charge, the electromagnetic wave in the line corresponding to maximum gain is synchronous with the slow space-charge wave. In the case of large **b** and small **C**, we find from Eq. (8.52) that $\delta^2 + q \approx -1/b \approx 0$. Then $\delta \approx \pm i\sqrt{q}$ and according to Eqs. (8.49) and (8.50), $\beta = \beta_e \pm \beta_p$.

We have arrived at the equation for space-charge waves. Thus, for a large nonsynchronism parameter and small current, the field in the line essentially does not interact with the electrons, and pure space-charge waves of constant amplitude propagate into the electron beam. Of course, this is true only in the framework of the linear theory.

Next we estimate the space-charge parameter for the following conditions: $F = 1$, $j = 10\,\text{A/cm}^2$, $U_0 = 1\,\text{kV}$, $f = 10\,\text{GHz}$, and $C = 0.1$. We obtain

$$q = \frac{\omega_p^2}{\omega^2 C^2} = \frac{j\eta}{\sqrt{2\eta U_0 \varepsilon_0 4\pi^2 f^2 C^2}} \approx 2.7$$

Judging by these values, quantity $q > 1$ is a rather high value for a space-charge parameter.

Effects of Attenuation (Loss) on a Slow-Wave Structure In this case the amplitude of an unperturbed wave attenuates with distance as $e^{-\beta_d z}$, where β_d is the loss coefficient. Then the amplitude of an unperturbed wave in a slow-wave structure can be written as $e^{-i\beta_0^* z}$, where $\beta_0^* = -i\beta_d + \beta_0$. Let us define the loss parameter as follows: $d = \beta_d/\beta_e C$. Then

$$\beta_d = \beta_e dC \quad \text{and} \quad \beta_0^* = \beta_0 - i\beta_e dC \tag{8.65}$$

Next, we substitute β_0^* for β_0 in Eq. (8.45). Repeating the calculation with $d = 0$, we obtain the dispersion equation

$$i(\delta + d + ib)(\delta^2 + q) = (1 + Cb)^2 \tag{8.65a}$$

As expected, numerical solution of this equation shows (Gilmour, 1994; Pierce, 1950) a decrease in gain associated with the losses. For small losses, reduction of the gain grows with an increase in space charge, and for a small space charge, the reduction of the gain grows with losses into the line.

8.6 NONLINEAR EFFECTS IN A NONRELATIVISTIC TWTO. ENHANCEMENT OF TWTO EFFICIENCY (VELOCITY TAPERING, DEPRESSED COLLECTORS)

8.6.1 Introduction

Predicted by linear theory, unlimited exponential growth in the field amplitude with the length of the interaction space and the absence of harmonic content in the amplified signal has not yet been affirmed for comparatively low efficiency and output power. In reality the amplifier TWTO characteristics have a form similar to that shown in Fig. 8.10.

In principle, the limitation of output power with the length of the tube and the lowering of the saturation threshold with the level of the input power are both simple consequences of the energy conservation law. However, these effects can

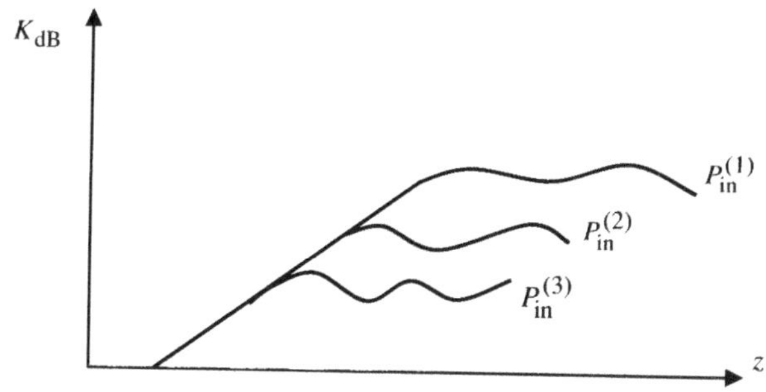

FIGURE 8.10 Gain coefficient as a function of interaction length for various power values for an input signal ($P_{in}^{(3)} > P_{in}^{(2)} > P_{in}^{(1)}$).

result in significantly different TWTO efficiencies. Naturally, only nonlinear analysis can provide a quantitative analysis of these "anomalies" and, most important, indicate ways to increase the real gain, efficiency, and nondistortion of broadband signals.

In the nonlinear theory of TWTOs, a number of important effects that are ignored in the framework of linear theory should be taken into account; for example:

- Change in the average electron velocity
- Electrons overtaking one another
- Deformation of bunches and their trapping in the frame of reference of the resonant wave
- Emergence of harmonics of the drive signal in the output signal
- Stopping and reversal of electrons

The first four of these effects are the most important, because they can be seen to influence TWTO characteristics significantly even at medium output power. The nonlinear theory is naturally more complicated than the linear theory. As a rule, the results can be found through numerical solution of approximated nonlinear equations.

Initially, the nonlinear theory of TWTO was developed by Nordsiek (1953). The principal results were later obtained by Rowe (1965), Tien (1956), and Vainstein (1957a,b).

8.6.2 Derivation of Nonlinear Equations of a TWTO (Vainstein, 1957a,b)

The scheme for nonlinear analysis is similar to the self-matched scheme for the linear theory. Both involve the formation of an ac current on the ac field and excitation of the field by the ac current.

Calculation of Current The velocity $v(z,t)$ in general is not a single-valued function of z,t, due to overtaking effects which occur when a number of particles with different velocities can turn out in the same place or when some particles stop and move in the opposite direction. Even ignoring the reverse motion, the need to account for the overtaking effects significantly complicates the Eulerian description of an electron beam as a continuous medium. It should be replaced by the Lagrangian description. We will use as Lagrangian variables the input time t_0 and the coordinate z (in Section 7.3.2 we used the notation τ, x). The time of arrival of the electron in the z plane is determined by the function $t = t(z, t_0)$. Then the difficulties raised by overtaking disappear: In fact, we follow the particle that enters the interaction region at moment t_0.

The velocity and acceleration in Lagrangian variables are equal to

$$
\begin{aligned}
v(z, t_0) &= \frac{1}{(dt/dz)_{t_0}} = \frac{1}{\partial t/\partial z} \\
\frac{dv}{dt} &= \left(\frac{\partial v}{\partial z}\right)\left(\frac{1}{\partial t/\partial z}\right) = -\frac{\partial^2 t/\partial z^2}{(\partial t/\partial z)^3}
\end{aligned}
\tag{8.66}
$$

so the equation of motion is

$$
\frac{\partial^2 t}{\partial z^2} = -\left(\frac{\partial t}{\partial z}\right)^3 \eta E \tag{8.67}
$$

where $E = E_w + E_p$ is the sum of z-components of the resonant wave and the space-charge field.

We define the function $t = t(z, t_0)$ as

$$
t = t_0 + \frac{z}{v_0} + \Delta(z, t_0) \tag{8.68}
$$

where Δ is a perturbation of the arrival time in the z plane determined by the ac field. Multiplying Eq. (8.68) by ω, we obtain the phase relation

$$
\omega t = \omega t_0 + \beta_e z + \theta(\xi, t_0) \tag{8.69}
$$

where $\theta(\xi, t_0) = \omega\Delta(z, t_0)$. Here we replace z on a slow variable, $\xi = C\beta_e z$. The meaning of this replacement follows from the linear theory. According to Eq. (8.50), all ac quantities change with z as $e^{\beta z} \approx e^{\beta_e(1+kC)z}$, where $k \sim 1$ and $C \ll 1$. Therefore, the perturbation of the electron phase is $k\beta_0 Cz$.

Let us represent Eq. (8.69) as

$$
\omega t = \xi/C + u \qquad \text{where} \quad u(\xi, t_0) = \omega t_0 + \theta \tag{8.70}
$$

Quantity u represents the electron phase in the reference frame that moves with the initial electron velocity. Without perturbation, the phase would be equal to the initial phase ωt_0.

We obtain from Eqs. (8.66) and (8.70),

$$\frac{\partial t}{\partial z} = \frac{\partial t}{\partial \xi}\frac{\partial \xi}{\partial z} = \frac{1}{v_0}\left(1 + C\frac{\partial u}{\partial \xi}\right) \tag{8.71}$$

$$\frac{\partial^2 t}{\partial z^2} = \frac{1}{v_0}\frac{\partial}{\partial \xi}\left(1 + C\frac{\partial u}{\partial \xi}\right)\frac{\partial \xi}{\partial z} = \frac{\omega}{v_0^2}C^2\frac{\partial^2 u}{\partial \xi^2} \tag{8.72}$$

The equation of motion (8.67) takes the form

$$\frac{\partial^2 u}{\partial \xi^2} = -\frac{\eta}{\omega v_0 C^2}\left(1 + C\frac{\partial u}{\partial \xi}\right)^3 E \tag{8.73}$$

Let us express the current as the Fourier series

$$I(z,t) = I_0 + \mathrm{Re}\sum_{n=1}^{\infty}\widetilde{I}_n(z)e^{in\omega t} \tag{8.74}$$

The complex Fourier coefficients are

$$\widetilde{I}_n(z) = \frac{1}{\pi}\int_0^{2\pi} I(z,t)e^{-in\omega t}\, d(\omega t) \tag{8.75}$$

Application of the law of charge conservation (Section 7.3) gives

$$I(z,t)\, dt = I_0\, dt_0 \tag{8.76}$$

Then using Eq. (8.70), we obtain

$$\widetilde{I}_n(z) = \frac{I_0}{\pi}\int_0^{2\pi} e^{-in\omega t}\, d(\omega t_0) = I_0 e^{-in\beta_e z}\widetilde{\Phi}_n(\xi) \tag{8.77}$$

where

$$\widetilde{\Phi}_n(\xi) = \frac{1}{\pi}\int_0^{2\pi} e^{-inu(\xi,\, u_0)}\, du_0 \tag{8.78}$$

Equations (8.74), (8.77), and (8.78) give the convection current through a given electric field that determines the function $u(\xi, t_0)$ from the equation of motion (8.73). Initial conditions for the equation of motion (8.73) are

$$u_{\xi=0} = u_0 = \omega t_0, \qquad \left(\frac{\partial u}{\partial \xi}\right)_{\xi=0} = 0 \tag{8.79}$$

The first condition determines the input phase; the second ensures that the initial electron velocity is equal to v_0. That follows from Eq. (8.71).

Excitation of the Field E_w of a Resonant Wave by an AC Convection Current Above we obtained Eq. (8.40) as the excitation equation of a synchronous wave based on the integral representation (8.39) and taking into account the forward and backward waves propagating from their sources. If we consider only forward waves, however, Eq. (8.40) is replaced by a simpler first-order equation. Retaining only forward waves in Eq. (8.39), we arrive at the integral representation

$$\widetilde{E}_w(z) = \widetilde{E}_{\text{in}} e^{-i\beta_0 z} - \frac{\beta_0^2 Z}{2} \int_0^z \widetilde{I}_1(\xi) e^{-i\beta_0(z-\xi)}\, d\xi \tag{8.80}$$

Differentiating $\widetilde{E}_w(z)$ with respect to z and summing with $i\beta_0 E_w$, we obtain the equation

$$\frac{d\widetilde{E}_w}{dz} + i\beta_0 \widetilde{E}_w = -\frac{\beta_0^2 Z}{2} \widetilde{I}_1(z) \tag{8.81}$$

If $\widetilde{E}_w(z)$ is not monochromatic, we expand it into a Fourier series:

$$E_w(z,t) = \text{Re} \sum_{n=1}^{\infty} \widetilde{E}_{wn}(z) e^{in\omega t} \tag{8.82}$$

Now we should write equations similar to Eq. (8.81) separately for each complex component of the field spectrum. Note that the benefit of this representation is related primarily to the account of the space-charge field, which is essential for examination of the configuration of bunches at the end of the interaction space. We omit this analysis here and use only the first harmonic [i.e., Eq. (8.81)].

Let us introduce a dimensionless complex amplitude of the synchronous wave:

$$\widetilde{F}_w(z) = \frac{\widetilde{E}_w(z) e^{i\beta_e z}}{2\beta_e U_0 C^2} \tag{8.83}$$

Note that according to definition (8.83), $\widetilde{F}_w(z)$ is in fact a complex amplitude of the synchronous field in the moving reference frame. According to the equality

$$\frac{d}{dz} = \frac{d}{d\xi}\frac{d\xi}{dz} = \beta_e C \frac{d}{d\xi}$$

we obtain

$$\frac{d\widetilde{E}_w(z)}{dz} + i\beta_0 \widetilde{E}_w(z) = 2\beta_e^2 C^3 U_0 \left[\frac{d\widetilde{F}_w(\xi)}{d\xi} + \frac{i(\beta_0 - \beta_e)}{\beta_e C} \widetilde{F}_w \right] = -\frac{\beta_0^2 Z}{2} \widetilde{I}_1(z) e^{i\beta_e z}$$

Recall that $(\beta_0 - \beta_e)/\beta_e C = b$ [see Eq. (8.48)] and

$$2\frac{\beta_e^2 C^3 U_0}{\beta_0^2 Z} = I_0 \frac{\beta_e^2}{2\beta_0^2} = \frac{I_0}{2} \frac{1}{(1 + Cb)^2}$$

[see Eq. (8.44)]. Then we find that

$$\frac{d\widetilde{F}_{w}(\xi)}{d\xi} + ib\widetilde{F}_{w}(\xi) = -(1 + Cb)^2 \frac{\widetilde{I}_1(z)}{I_0} e^{i\beta_e z} \tag{8.84}$$

Substituting $\widetilde{I}_1(z)$ from Eq. (8.77) and using Eq. (8.78), we come to the excitation equation of the resonant wave field by the convection current:

$$\frac{d\widetilde{F}_{w}(\xi)}{d\xi} + ib\widetilde{F}_{w}(\xi) = -(1 + Cb)^2 \tilde{\Phi}_1(\xi) \tag{8.85}$$

where

$$\tilde{\Phi}_1(\xi) = \frac{1}{\pi} \int_0^{2\pi} e^{-iu(\xi, u_0)}\, du_0 \tag{8.86}$$

The initial condition for this equation, according to Eqs. (8.80) and (8.83), is

$$[\widetilde{F}_{w}(\xi)]_{\xi=0} = \frac{\widetilde{E}_{w}(0)}{2\beta_e U_0 C^2} = \frac{\widetilde{E}_{\text{in}}}{2\beta_e U_0 C^2} \tag{8.87}$$

We express through $\widetilde{F}(\xi)$, the right side of the equation of motion (8.73):

$$\frac{\eta E}{\omega v C^2} = \frac{\eta}{\omega v C^2} \operatorname{Re}(\widetilde{E} e^{i\omega t}) = \frac{\eta}{\omega v C^2} \operatorname{Re}[2\beta_e U_0 C^2 \widetilde{F}(z) e^{i(\omega t - \beta_e z)}] = \operatorname{Re}(\widetilde{F}(z) e^{iu})$$

Here $\widetilde{F}(\xi) = \widetilde{F}_{w}(\xi) + \widetilde{F}_p(\xi)$, where $\widetilde{F}_p$ is the dimensionless complex amplitude of the space-charge field. Finally, the equation of motion is

$$\frac{\partial^2 u}{\partial \xi^2} + \left(1 + C\frac{\partial u}{\partial \xi}\right)^3 \operatorname{Re}\{[\widetilde{F}_{w}(\xi) + \widetilde{F}_p(\xi)] e^{iu}\} = 0 \tag{8.88}$$

We have obtained two equations [(8.88) and (8.85), with Eq. (8.86)] for two functions, $\widetilde{F}_{w}(\xi)$ and $u(\xi)$ (slowly varying amplitude and phase), with initial conditions (8.79) and (8.87). These are nonlinear integrodifferential equations. The solution of this system requires computation of the space-charge field $\widetilde{F}_p(\xi)$, which is not a trivial problem (see, e.g., Rowe, 1965; Tien, 1956; Vainstein, 1957a,b). If $\widetilde{F}_p(\xi)$ is neglected, this system fully determines the nonlinear characteristics of TWTOs in the monochromatic approximation. We limit further discussion by this approximation.

Linearization of Equations These equations can be transformed into linear equations if we assume a small phase perturbation $\theta = |u - u_0| \ll 1$, $C|\partial\theta/\partial\xi| \ll 1$. The linearized equation describes characteristics of a TWTO only in the initial part of the tube. As a result, a system of two equations of the linear theory of TWTOs can be deduced. Matching this system leads to the dispersion equation (8.65*a*) with $d = q = 0$ if we do not take the line losses into account and if we neglect the space-charge field ($\widetilde{F}_p = 0$). Note that these solutions can be used as initial conditions for solving the nonlinear equations. In the initial section of the TWTO, three waves will be excited according to the linear theory. If one of these waves is growing, it will dominate, acting as if there are no other waves. This wave can be considered an initial condition. However, this approach is not applicable to the analysis of TWTOs with finite or large input signals.

Conservation of Energy in an Electron Beam Interacting with a Traveling Wave To begin, we average the equation of motion (8.88) over the initial phase u_0. Take as a simple example $C \ll 1$ and $\widetilde{F}_p = 0$. We obtain the law of conservation of energy in the following form:

$$\frac{I_0}{e_0}\left(\frac{mv_0^2}{2} - \frac{1}{2\pi}\int_0^{2\pi}\frac{mv^2}{2}\,du_0\right) = \frac{1}{2}\int_0^z \mathrm{Re}(\widetilde{E}_w\widetilde{I}^*)\,dz \tag{8.89}$$

Here $v = v(z, u_0)$ is the electron velocity in the z plane as a function of the initial phase. The left-hand side of Eq. (8.89) is averaged over the period of loss of the electrons' kinetic energy. The right-hand side is equal to the (retarding) work over a period produced by the synchronous wave field. Such a relationship could not be obtained in the linear theory, because within that theory the average energy of electrons is constant.

8.6.3 Phasing of Electrons and Trapping of Bunches in a Traveling Wave

For simplicity, let us consider the mechanism of nonlinear electron phasing, supposing that $C \ll 1$ and neglecting the space-charge field $\widetilde{F}_p = 0$. The equation of motion (8.88) in this case can be written as

$$\frac{\partial^2 u}{\partial \xi^2} = -\mathrm{Re}[\widetilde{F}_w(\xi)e^{iu}] \tag{8.90}$$

Let us represent

$$\widetilde{F}_w(\xi) = Fe^{-i\varphi} \tag{8.91}$$

where $F(\xi) = |\widetilde{F}_w(\xi)|$ and $\varphi(\xi)$ are the slowly varying modulus and phase of the resonant wave field in the moving reference frame.

The factor $\mathrm{Re}[Fe^{i(u-\varphi)}] = F\cos(u - \varphi)$ on the right-hand side of Eq. (8.90) can be considered as a derivative of the function $V = F\sin(u - \varphi)$ with respect to u. So we can write the equation of motion in the form

$$\frac{\partial^2 u}{\partial \xi^2} = -\frac{\partial V}{\partial u} \tag{8.92}$$

It can be shown that the variable $-u$ is the dimensionless coordinate in the moving reference frame. As such, according to Eq. (8.70),

$$-u = \beta_e z - \omega t = \beta_e(z - v_0 t) \tag{8.93}$$

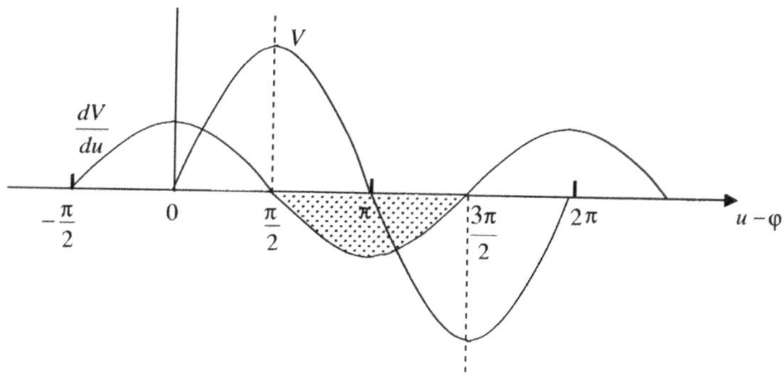

FIGURE 8.11 Dimensionless potential V and force $\partial V/\partial d$ as functions of phase $u-\varphi$.

Therefore, the function $V(\xi, u)$ represents a dimensionless potential in the moving reference frame, and $\partial V/\partial u$ is the dimensionless force. ξ is a very slowly varying variable since $C \ll 1$ and $\xi = C\beta_e z$. Therefore, the potential in the moving frame depends only on u. Functions V and dV/du are shown in Fig. 8.11. The shaded zone in the figure is the zone of the retarding (negative) force dV/du.

Consider the equation of excitation (8.85) for $C \ll 1$, where we substitute Eq. (8.91) and represent the function

$$\widetilde{\Phi}_1(\xi) = \Phi(\xi)e^{-u_b(\xi)} \tag{8.94}$$

We obtain

$$\frac{dF}{d\xi} - i\frac{d\varphi}{d\xi}F + ibF = -\Phi e^{i(\varphi - u_b)} \tag{8.95}$$

Take the real part of Eq. (8.95):

$$\frac{dF}{d\xi} = -\Phi\cos(u_b - \varphi) \tag{8.96}$$

where $u_b - \varphi$ is the phase shift between the current and the field of the resonant wave. According to Eq. (8.96), the condition of the wave growing along the tube $(dF/d\xi > 0)$ is $\cos(u_b - \varphi) < 0$; that is,

$$\frac{\pi}{2} < u_b - \varphi < \frac{3\pi}{2}$$

It is interesting to discuss the physical meaning of the current phase u_b. We can show that u_b is the coordinate of the bunch center in a moving frame. Indeed, we can write the beam current, using Eqs. (8.74) and (8.77) for $n = 1$ and Eqs. (8.70) and (8.94), as

$$\begin{aligned} I(z,t) &= I_0 + I_0\mathrm{Re}[e^{i(\omega t - \beta_e z)}\widetilde{I}_1(\xi)] = I_0 + I_0\Phi\,\mathrm{Re}[e^{i(\omega t - \beta_e z)}e^{-u_b}] \\ &= I_0[1 + \Phi\cos(u - u_b)] \end{aligned} \tag{8.97}$$

We see that the maximum for the current is attained when $u = u_b$. When the bunch center moves in the shaded zone in Fig. 8.11, the amplitude of the wave grows. Thus, the wave grows when the bunch center is retarded by the wave field. Note that the retarding region is at the right slope of the potential $V(u - \varphi)$.

After reaching the potential well, the bunch finds itself in the accelerating field. The wave amplitude decreases and velocity v increases. Then, according to Eq. (8.71), the dimensionless velocity $\partial u/\partial \xi$ in the moving reference frame is decreased and the bunch can return to the retarding phase—if its velocity is not too large. As a result, the bunch begins to oscillate near the potential well. The net effect is to trap the bunch with the resonant traveling-wave field.

The final results are determined by the choice of the nonsynchronism parameter b. Assume that $b = b_{opt}$, where b_{opt} corresponds to the maximal gain in the linear theory. Then the bunch is formed in the maximum of the retarded field ($u - \varphi = \pi$ in Fig. 8.11). Continuing to move in the retarding phase, the bunch gives the energy to the field until it reaches the minimum potential ($u - \varphi = 3\pi/2$). At this point, the power has reached its maximum. Passing through the minimum of the potential energy, the bunch is accelerated and takes the field energy. So the power is decreased, and at some distance it reaches its minimum. When the bunch returns to the potential minimum, the second maximum of the power is attained. The magnitude of the second maximum of the electromagnetic field is a little different from the first. The process is repeated further, and the movement of the bunch is similar to oscillations of a pendulum.

Thus, the electron movement breaks into two stages: (1) bunching, as determined by the linear theory, and (2) oscillations of the bunches about the minima of the potential. The amplitude of oscillations determines the level of the saturated power and the difference between maximum and minimum of the electron power. So if the bunch was formed near the minimum of the potential $u - \varphi = 3\pi/2$ (minimum in excess of the electron velocity), the difference in power would be near zero, and the level of the saturation of the power would be low. Obtaining more power requires the choice of $b = b_{max}$, when bunches are formed in $u - \varphi = \pi/2$ (i.e., in the beginning of the retarding phase and when the bunches spend energy along the fully retarded zone). Note that the linear gain for $b = b_{max}$ is decreased because the strong nonsynchronism lowers the growth rate of the exponentially growing wave. Thus, the conditions of reaching maximum gain and maximum efficiency are essentially distinct, and the maximum excessive velocity (value of b_{max}) is determined by the permissible minimum of the amplification.

In the framework above, treatment two mechanisms of limitations of the efficiency and output power can be indicated: (1) retarding of the electrons, and as a result, attainment of the saturation level when retarded electrons are transferred in the accelerated phase of the traveling-wave field, and (2) instability of the bunches. Because the formation of bunches is a result of the difference electron velocities, continuation of the same factor causes a decay in bunches.

Recall that these results are valid only for the essentially simplified model of TWTO (low space-charge field, low magnitude of the Pierce parameter C, low input power, and finally, low losses in the slow-wave structure). Numerical analysis of the idealized scheme estimates the maximum efficiency TWTO at about 50% for $C \geq 0.1$ and $b \approx 1.8$.

Taking the space charge into account changes these optimistic estimates significantly. The space-charge field accelerates the decay of bunches. Note that the space-charge interaction takes place not only inside a bunch but also between neighboring bunches. For a large space charge, a second bunch is formed during decay of the main bunch, and power reaches the maximum. There remain a large number of irregular-phase electrons. After completion of the initial process of bunching and the decay of bunches, the maximum output power is expressed unclearly (Vainstein, 1957a,b).

An interesting illustration to these processes in a typical TWTO is shown in Fig. 8.12. We see that the initial bunching is performed strictly in the retarding phase of the field. But after formation, the bunches slowly lag from the retarded phase of the wave field, and power reaches a maximum. Further, the splitting of bunches begins. As can be seen, at this phase in each period two bunches arise within each period. So the frequency of the current doubles and the first current harmonic is close to zero.

8.6.4 Enhancement of TWTO Efficiency. Velocity Tapering. Depressed Collectors

Velocity Tapering As we saw above, the saturation of power and the corresponding limitation in efficiency is due in large part to the reduction in electron velocity

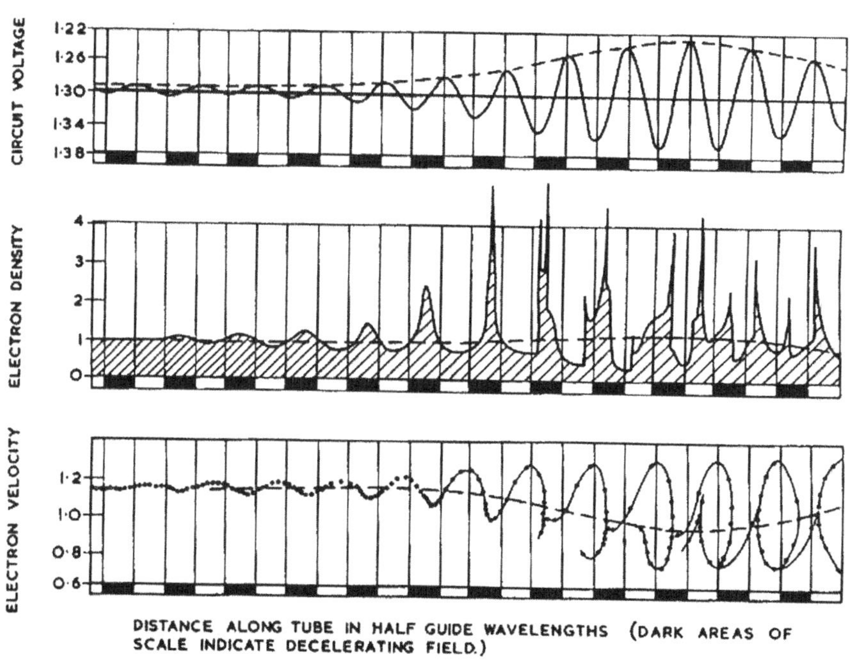

FIGURE 8.12 Interaction mechanism in a TWTO. (From Hess, 1960.)

caused by transfer of the beam kinetic energy to the electromagnetic field. Understanding this effect stimulated the idea of modifying slow-wave structures to reduce the phase velocity of the slow wave toward the end of the interaction space and extend the existence of bunches in the retarded phase of the circuit field. This idea of *velocity tapering* was suggested in several early works (see, e.g., Meeker and Rowe, 1962; Pierce, 1950; Rowe, 1965). Early attempts to use velocity tapering, however, were not successful and achieved little or no improvement in efficiency. Development of computer analyses of TWTO nonlinear theory resulted in a better understanding of the velocity-tapering possibilities and eventually led to the design of tapers that produced appreciable improvements in efficiency and gain.

In principle, a number of versions of tapering are possible: maintaining a fixed nonsynchronism of both disturbed electrons and the wave during interaction; maintaining fixed detuning between the averaged electron velocity and the phase velocity of the nonperturbed circuit wave; maintaining optimal phasing between the first current harmonic and the circuit wave; and so on. Let us consider as an example a condition of fixed detuning between the perturbed averaged velocity of the electrons and the wave phase velocity. This can be written as a condition of constancy of the ratio of the average electron velocity over a period of the initial phase of the circuit-wave phase velocity:

$$\frac{1}{2\pi v_{\text{ph}}^2}\int_0^{2\pi} v^2\, du_0 = \left(\frac{v_0}{v_{\text{ph},0}}\right)^2 \tag{8.98}$$

where v and v_{ph} are the electron and phase velocities that are variable along z as a result of interaction, and v_0 and $v_{\text{ph},0}$ are the initial (nonperturbed) velocities. According to the definition of the nonsynchronism parameter [Eq. (8.48)], their ratio is equal to

$$\frac{v_0}{v_{\text{ph},0}} = 1 + b_0 C \tag{8.99}$$

A perturbed electron velocity is [Eq. (8.71)]

$$v = \frac{1}{\partial t/\partial z} = \frac{v_0}{1 + C(\partial u/\partial \xi)} \tag{8.100}$$

To find the perturbed phase velocity of the circuit wave, we use Eq. (8.82) (for $n = 1$), and Eqs. (8.83) and (8.91). Then the electric field of the resonant wave can be represented as

$$E_w(z,t) = \text{Re}\,\widetilde{E}_w(z)e^{i\omega t} = MF\,\text{Re}[e^{i(\omega t - \beta_e z - \varphi)}] \tag{8.101}$$

where M is a constant. The phase velocity is found from the following condition of the constant phase:

$$\frac{d}{dt}(\omega t - \beta_e z - \varphi) = \omega - \left[\frac{\partial}{\partial z}(\beta_e z + \varphi)\frac{dz}{dt}\right] = \omega - \frac{dz}{dt}\left(\beta_e + \frac{\partial \varphi}{\partial z}\right) = 0 \tag{8.102}$$

Taking into account that $\xi = C\beta_e z$, we obtain the phase velocity:

$$v_{\text{ph}} = \frac{dz}{dt} = \frac{\omega}{\beta_e + \partial\varphi/\partial z} = \frac{v_0}{1 + C(\partial\varphi/\partial\xi)} \tag{8.103}$$

Substituting Eqs. (8.99), (8.100), and (8.103) into Eq. (8.98) gives the relation

$$\frac{1}{2\pi}\int_0^{2\pi} \frac{du_0}{[1 + C(\partial u/\partial\xi)]^2} = \frac{(1 + b_0 C)^2}{[1 + C(\partial\varphi/\partial\xi)]^2} \tag{8.104}$$

Equation (8.104) must be added to the fundamental system equations (8.85), (8.86), and (8.88). The system of equations that results includes the function $b(\xi)$, along with the unknown functions $\widetilde{F}_w(\xi)$ and $u(\xi)$. Having $b(\xi)$ and using Eqs. (8.99), we can obtain the desired distribution of the phase velocity:

$$v_{\text{ph}}(z) = v_{\text{ph},0}\frac{1 + b_0 C}{1 + Cb(\xi)_{\xi = C\beta_e z}} \tag{8.105}$$

The resulting efficiency and gain depend on a number of parameters $(b_0, C, P_{\text{in}}, q, L)$, where P_{in} is the input power, q is the space-charge parameter, and L is the length of the interaction space. According to the results of the analysis, velocity tapering in general ensures the essential enhancement of efficiency. For a large space charge ($q > 1$), however, the tapering does not increase and even diminishes the efficiency. It can be explained by the increase in velocity spread in the bunches (Hess, 1960). For large q, two bunches are formed (Fig. 8.12) which move with essentially different velocities. As a result, velocity tapering does not make sense in this context.

Note that the increase in efficiency takes place for comparatively small values of parameter b which correspond to the maximum linear gain. In this case the efficiency reaches maximum values of nontapering TWTOs with large values of b. For large b, tapering has no effect on efficiency. As a result, we obtain the remarkable feature of tapering TWTO: combining in one tube high gain with high efficiency. In addition, velocity tapering in general expands the region of linearity of amplification. This is important for communication applications of TWTOs, where low distortion is required.

In work of Gulyaev et al. (2004), a method of enhancing of the efficiency of TWTOs has been developed in a general statement: search of the optimal distribution of the cold phase velocity in arbitrary irregular slow-wave structures of TWTOs. The problem is solved in the paper by Gulyaeu et al. using a variation-iteration optimization method. The results of computations confirm possibilities of increased efficiency.

Depressed Collectors Depressed collectors convert some of the kinetic energy remaining in beams after interaction (*spent electron beams*) into potential electric energy, thus reducing the power consumed by the tube. This reduction is accomplished by depression of the potentials of collecting electrodes located outside the tube body. As a result, the overall efficiency

$$\eta_{ov} = \frac{P_e}{I_0 U_0 - P_{rec}} \tag{8.106}$$

can significantly exceed the electron efficiency $\eta_e = P_e/I_0 U_0$ (P_e is the electron power and P_{rec} is the power recovered by the collector from the spent beam). In addition to saving power, the use of depressed collectors diminishes the problems of cooling the collector. This feature is very important in high-power tubes. In medium-power tubes, depressed collectors allow use air cooling of collectors instead of water cooling. Finally, depressed collectors reduce the X-ray radiation in high-voltage tubes.

Consider a simple scheme of a depressed collector with a single electrode (a *single-stage depressed collector*; Fig. 8.13). A retarded voltage V_{coll} is applied between the body of the tube and the collector. When the collector voltage increases, electrons of the spent beam can be reflected and the collector beam decreases. Figure 8.14 shows the curve $I_{coll} = I_{coll}(V_{coll})$ of the collector cutoff current. We ignore the current of secondary electrons. So the area under the cutoff curve $P_s = \int_0^{V_{max}} I_{coll}\, dV_{coll}$ is the power of the spent beam. The tail between U_0 and V_{max} belongs to the small current of the electrons that are accelerated during interaction with the ac field. The shape of the curve depends on the nature of the interaction between the beam and the ac field. The more the efficiency, the less the spent beam power, and the curve is lowered. For example, the cutoff curves of highly effective klystrons are arranged essentially lower than TWTO curves. The dashed curve in Fig. 8.14 is the approximate cutoff current curve. Here V_{knee} corresponds to the lowest energy of electrons in the spent beam. So if the potential of the collector with respect to the cathode is equal to $U_{ck} = U_0 - V_{knee}$, the collector current does not change but the power recovered by the collector will be

$$P_{rec} = I_0 V_{knee} = I_0(U_0 - U_{ck}) \tag{8.107}$$

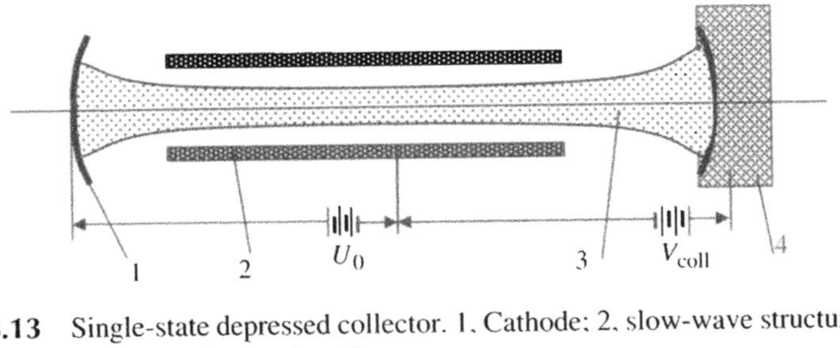

FIGURE 8.13 Single-state depressed collector. 1, Cathode; 2, slow-wave structure (body of the tube); 3, spent electron beam; 4, collector.

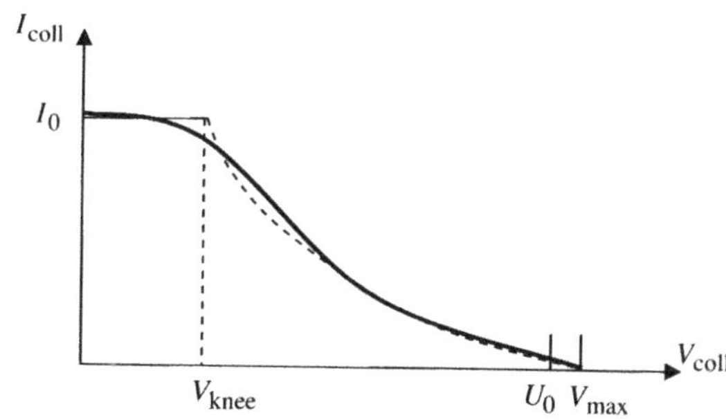

FIGURE 8.14 Collector cutoff current.

Substituting Eq. (8.107) into Eq. (8.106), we obtain the overall efficiency of a single-state depressed collector:

$$\eta_{ov} = \frac{P_e}{I_0 U_{ck}} = \eta_e \frac{U_0}{U_{ck}} \tag{8.108}$$

If a potential of the collector negative to the tube body exceeds V_{knee}, the reflected current and correspondingly, the body current increase, and the current efficiency drops. Note that Eq. (8.108) overestimates the efficiency because the knee of the curve is normally not defined as it is shown on the approximate curve in Fig. 8.14.

Disadvantages of single-state depressed collectors include low collector efficiency $\eta_{coll} = P_{rec}/P_{sp}$ and a high current of secondary electrons. The secondary currents flow into the tube body. These diminish the collector efficiency, causing excessive noise and heating of the slow-wave structure.

Modern *multistage depressed collectors* (MDCs) use several electrodes of special shape with a variety of potentials. Electrons of the spent beam are sorted by the collector into different energy classes and collected by corresponding electrodes. The MDCs can recover a large fraction of the spent beam power (up to 90% and more). According to Sterzer (1958), the maximum overall efficiency that can be obtained with an n-stage collector is given by

$$\eta_{ov} = \frac{\eta_e}{\sum_{m=1}^{n} K_m(\alpha_m - \alpha_{m-1}) + 1 - \alpha_n} \tag{8.109}$$

where $K_m = U_m/U_0$, $U_1 < U_2 < \cdots < U_n < U_0$, α_m is the fraction of electrons that have been slowed down to less than U_m, and $\alpha_0 = 0$. The positive electrode potentials are adjusted so that the denominator in Eq. (8.109) is minimal.

The design for a four-stage depressed collector is shown in Fig. 8.15. A spent beam in the figure corresponds to a TWTO operating at a saturation output power. The numbers on the left side of the plot are potentials of electrodes with respect to the tube body. The numbers on the right side are energies of particles. As we can see, the electrons hit the back sides of the electrodes, and the electric field between electrodes retards secondary electrons. Therefore, the bulk of the secondary electrons return to the emission zones.

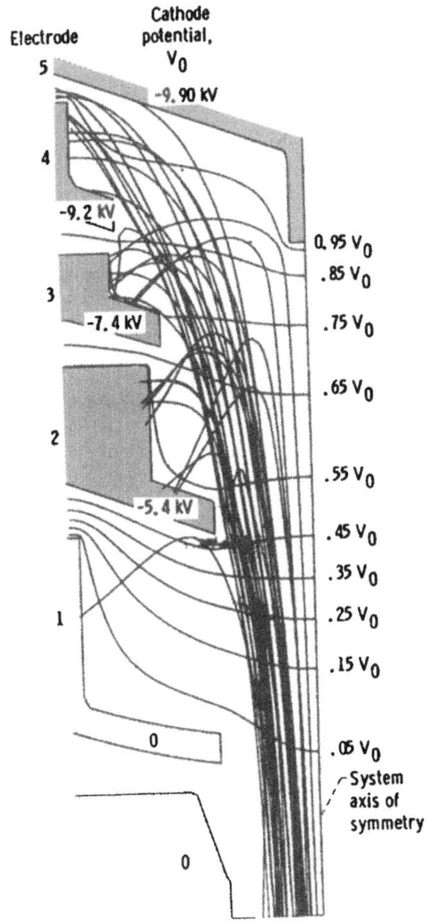

FIGURE 8.15 Design for a four-stage depressed collector. (From Kosmel, 1982.)

8.7 BASIC CHARACTERISTICS AND APPLICATIONS OF NONRELATIVISTIC TWTOs

8.7.1 Introduction

The sphere of applications of TWTOs, beginning with their invention in 1942, expanded continuously correspondingly to progress in the theory (including computer simulation), experiment, and technology, which in turn was stimulated by the applications. TWTO efficiency, power, and gain in all primary microwave bands (including the K_a-band) progressed rapidly. The lifetime of space TWTOs is about 15 years. Applications of TWTOs have penetrated science, communications, radar, electronic countermeasure systems, and other branches.

Satellite TWTOs helped to open the era of space communications (in 1962). After launch of a geostationary satellite by the United States in 1965, progress in

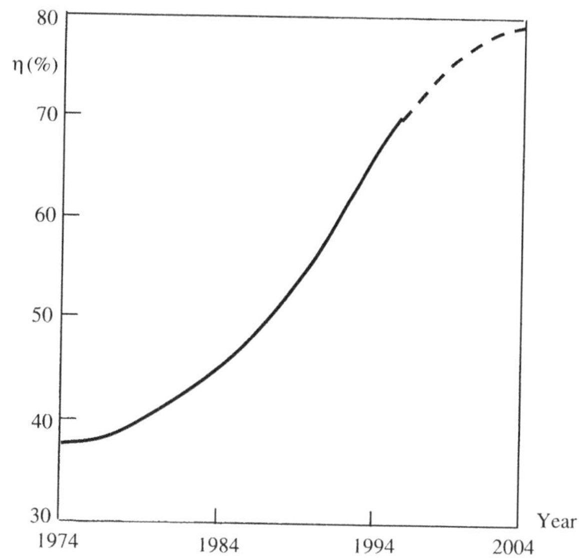

FIGURE 8.16 Efficiency of TWTOs in the K_u-band vs. time. Dashed part of the curve, projected up to year 2004. (From Mahie et al., 1998.)

space communication TWTOs (including commercial and military ground applications) acquired in the character of a revolution. According to the data of Mahie et al. (1998), the worldwide TWTO market already in 1998 was on the order $500 million ($250 million in the United States). The total TWTO space market has been estimated then at $250 million.

In Fig. 8.16 an informative graph of the Thompson (AERG) Corp. is shown where improvement of TWTO efficiency over time and projections up to 2004 are given. DJM Electronics Corp. has developed a TWTO satellite amplifier in the K_u-band with output power 500 W and gain 70 dB (Gulyaev et al., 2004). Such powerful TWTOs are necessary for satellite TV broadcasting in both analog and digital systems. Radar TWTOs are characterized by high-power pulsed transmission of relatively low duty cycle. Their progress is equally impressive. According to published information (Anon., 2003), powerful radar TWTOs developed in the frequency range 9.4 to 10 GHZ have reached an output power of 50 kW and a saturated gain of 55 dB (maximum duty 1%).

Communication applications require the development of both sensitive low-power TWTOs and powerful high-efficiency TWTOs with linear gain. TWTO properties that determine the above-mentioned characteristics are discussed briefly below.

8.7.2 TWTO Noises

Noise determines the *sensitivity* of an amplifier: the minimum drive power that can be distinguishable after amplification. The principal issue in noise theory is the problem of noise measurement, which depends on the choice of a noise standard source and a noise unit. The best source is a thermal (Johnson) noise that is generated by the thermal agitation of electrons in a conductor. Thermal noise is a naturally reproductive source; its important property is uniform energy distribution in the frequency spectrum. The power (W) of the thermal noise generated in a conductor with temperature T is

$$P = 4kT\,\Delta f \tag{8.110}$$

where k is the Boltzmann constant and $\Delta f = f_2 - f_1$ is the frequency bandwidth in hertz. The effective voltage produced by the thermal noise on the resistance R is given by

$$e^2 = 4kT \int_{f_1}^{f_2} R(f)\,df \tag{8.111}$$

If R does not depend on the frequency,

$$e^2 = 4kTR\,\Delta f \tag{8.112}$$

When the resistor R is connected with a matched conductor, the noise voltage is divided equally between R and the load conductor. Then the thermal power delivered to the load is

$$P_L = \frac{(e/2)^2}{R} = kT\Delta f \tag{8.113}$$

P_L is the theoretical minimum for the noise level. At room temperature (290 K) and 1 MHz bandwidth, $P_L \approx 4 \times 10^{-15}$ W.

Definition of Noise Figure If noise of a receiver (amplifier) with a gain G is absent, the output noise power according to Eq. (8.113) is $P_{\text{out}}^{(0)} = kTG\,\Delta f$. The noise figure of the real amplifier with the noise is defined as

$$F = \frac{(P_{\text{out}})_{T=290}}{(P_{\text{out}}^{(0)})_{T=290}} = \frac{(P_{\text{out}})_{T=290}}{290kG\,\Delta f} \tag{8.114}$$

A noise figure in decibels is

$$F_{\text{dB}} = 10\log_{10} F \tag{8.115}$$

For a noise-free amplifier, $F_{\text{dB}} = 0$. By definition, the equivalent noise temperature T_e corresponds to a power of

$$P_a = kT_e G\,\Delta f \tag{8.116}$$

where P_a is the noise power that is added by the amplifier. It follows that

$$P_a = (P_{\text{out}})_{T=290} - 290kG\,\Delta f = (F-1)290kG\,\Delta f \tag{8.117}$$

Then the equivalent noise temperature according to Eq. (8.116) is

$$T_e = (F - 1)290 \tag{8.118}$$

This factor is generally used for amplifiers with small noise figures. For example, $T_e = 75\,\text{K}$ for $F = 1\,\text{dB}$.

Noise in a TWTO Electron Gun The main sources of noise in a TWTO gun are shot noise and velocity noise. *Shot noise* is the result of the random nature of the emission of electrons as discrete particles. *Velocity noise* arises as a consequence of the velocity distribution of electrons emitted by a hot cathode.

Certain effects in the gun promote the reduction of noise. The important factor for shot noise reduction is related to the existence of a potential minimum near the thermionic cathode (Section 3.4.5). Slow electrons are reflected from the potential barrier. This equalizes the velocity distribution and the current fluctuations beyond the barrier. One method for a reduction in gun noise is control of the amplitude of standing space-charge wave patterns which are generated by shot and velocity noises and the location of the helix entrance at a position that minimizes the noise induced by waves. In particular, a transformer of the beam impedance is used. The transformer comprises several electrodes within the electron gun which control the rate of beam acceleration.

Partition Noise Partition noise is generated by transverse components of velocities that cause fluctuations in the current intercepted by electrodes in a gun and by the slow-wave structure. Analog noises are produced by the transverse motion of electrons in a slow-wave structure, even when random currents are not intercepted by the structure. This effect is connected with the essentially nonuniform transverse distribution of the ac field in a surface slow wave.

There are also many (more than 10) other known noise sources, but these sources fall outside the scope of this discussion.

The *minimum noise figure* of TWTOs was estimated by Haus and Robinson (1955). Assuming the absence of any correlation between the shot noise and the velocity noise at the potential minimum leads to the relation

$$F_{\text{min}} = 1 + \frac{T_c}{290} \tag{8.119}$$

where T_c is the cathode temperature. For $T_c = 1000\,\text{K}$, $F_{\text{min}} \approx 5\,\text{dB}$.

Haus and Robinson also predicted the possibility of F_{min} approaching zero in low-noise TWTO through the use of very large magnetic fields. Experimental confirmation of this effect was obtained by Hammer and Wen (1964).

8.7.3 Influence of Wave Reflections on Gain. Self-Excitation of TWTOs. Attenuation and Severing

In the analysis of TWTO amplification, it is assumed that input and output transitions are matched perfectly. However, real transitions are mismatched significantly, especially if one takes into account the broad bandwidth of TWTO. As a result, feedback arises, which leads to the frequency variations of amplification and even to the self-excitation of TWTO.

Let us denote the input and output ac electric fields (without taking reflections into account) as E_i and E_o, respectively. The amplification coefficient then is $K = E_o/E_i$. Following reflection of the signal from both the output and input terminals, an additional input field E_F appears:

$$E_F = E_o \rho_o K_L \rho_i e^{i\varphi} = E_i K K_L \rho_o \rho_i e^{i\varphi} \tag{8.120}$$

where ρ_o and ρ_i are the magnitudes of the output and input reflection coefficients (in voltage), K_L is the loss coefficient of the cold structure (from output to input), and φ is the phase difference, including the phases of the reflection coefficients and the phase shift of wave propagation. Thus, the total input field is the sum $E_i + E_F$, and an effective amplification coefficient is

$$K_E = \frac{E_o}{E_i + E_F} = \frac{K}{1 + K K_L \rho_o \rho_i e^{i\varphi}} \tag{8.121}$$

We see that the effective amplification changes with the frequency. For $\varphi = \pi$ the maximum K_E will be

$$K_{E,\max} = \frac{K}{1 - K K_L \rho_o \rho_i} \tag{8.122}$$

According to Eq. (8.122), self-excitation of the RWT may occur when $K > 1/K_L \rho_o \rho_i$ or when

$$G > G_{\max} = L - 20 \log (\rho_o \rho_i)$$

where G and L are the gain and the loss in decibels. Taking $L = 6\,\text{dB}$ and $\rho_o \rho_i = 0.1$ for real estimates we obtain $G_{\max} = 26\,\text{dB}$. To limit the variation in gain with frequency, the actual gain must be no more then 20 dB. Note that the maximum gain can be reduced some more as a result of reflected wave amplification via elastic reflected secondary electrons (Peter and Ruetr, 1953).

There are two known methods of increasing the stable gain.

1. *Local attenuators*. Some sections of a slow-wave structure, for example, the dielectric support rods in a helical TWTO (Fig. 8.17), are coated with a lossy material that forms an attenuator. The attenuation of the backward wave weakens the feedback significantly. On the other hand, the forward wave is also attenuated. After the attenuator, the amplitude of the forward wave can be close to zero. But the bunching of the beam before the attenuator leads to intensification of the amplification after the attenuator.

One disadvantage of this method is the need for a lengthy attenuation section for matching the attenuator impedance with the rest of the slow-wave structure. As a result, there is a loss of control of the bunching process that increases the velocity

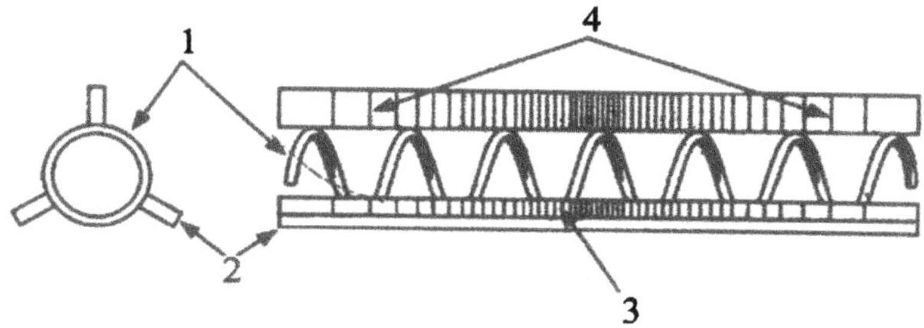

FIGURE 8.17 Lossy film attenuator in a helical TWTO. 1, Helix; 2, support rods; 3, lossy film; 4, thickness of the film is gradually reduced at ends of the helix.

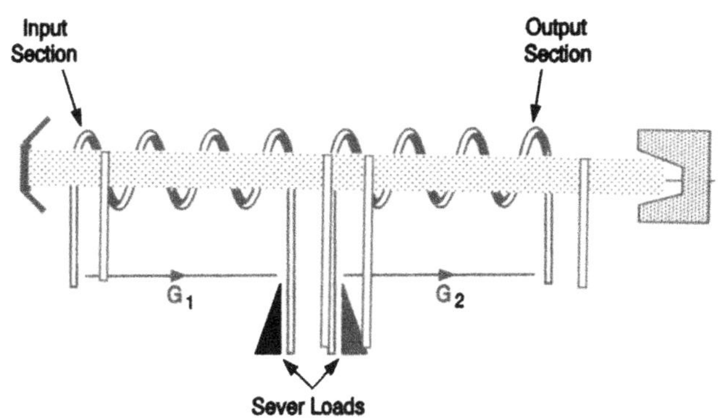

FIGURE 8.18 TWTO with two-section severed helix.

spread in bunches and reduces the efficiency. Finally, the lossy film technique for producing attenuation is not attractive for high-power TWTO.

2. *Severing slow-wave structures.* This is a more adequate technique for suppressing the reflected wave in high-power TWTOs. The forward wave is lost at the break, but the bunching of the beam produced in the input section remains and leads to further amplification in the output section (Fig. 8.18). The severed region should be as short as possible to preserve bunching. Because the reflection coefficients at the severed terminals are not greater than those in the input and output terminals, the limiting gain of each section G_1 and G_2 can be estimated similar to the gain of the TWTO without severing (i.e., about 20 dB). Then the limiting gain in two-section TWTOs will be about 40 dB. Greater gains are possible with additional severing.

8.7.4 Intermodulation Distortion

Intermodulation distortion results from the nonlinearity of an amplifier's amplitude characteristic. If one carrier is being amplified, nonlinear distortion produces

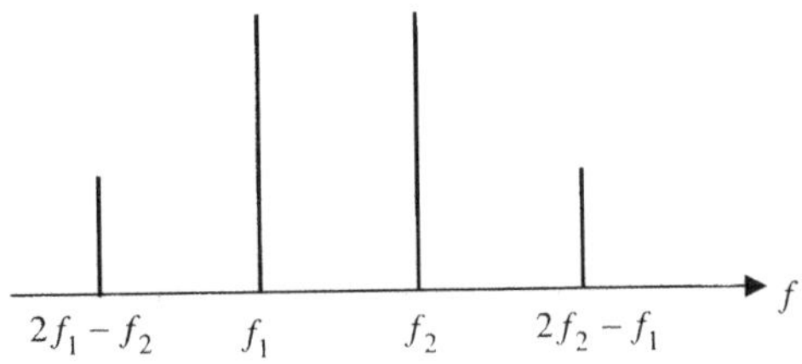

FIGURE 8.19 Intermodulation products of third order.

harmonics of the signal. For most communication TWTOs (with bandwidth less than an octave), these harmonics can be eliminated through filtering. However, if two or more carriers are present, beat products are generated in the vicinity of the input signal. These new signals are known as *intermodulation distortion* (IMD) *products*. They are located at frequency intervals equal to the separation of the input carriers. Filtering cannot eliminate these products because they are located at the same frequency or nearby. Most IMD products are of third order, which is closest to the signal frequency and largest in amplitude (Fig. 8.19). The frequencies of the fifth-order product are $3f_1 - 2f_2$ and $3f_2 - 2f_1$. If TWTOs are driven to saturation, the distortion is significant. According to Gilmour (1994), third-order products are below the input signal by 8 to 10 dB for equal amplitudes of two carriers. IMD significantly influences communication in both analog and digital signals.

The linearity of TWTOs can be improved by optimization of the taper in a helical TWTO and also by operation of the tube at a level significantly lower than saturation. However, backed-off operation requires higher power capabilities than does operation at saturation. According to Goebel et al. (2000), an L-band TWTO operated at 8 dB, reduced from 1600 W at saturation, produced over 250 W of continuous output power with level IMD products at −28 dB and an efficiency of over 25%. Another method of suppressing intermodulation products is based on harmonic injection. Wirth et al. (2002) obtained suppression of third-order IMD products at 25 dB without back-off from saturation.

8.7.5 Characteristics of Helical and Coupled-Cavity TWTOs

Helix and coupled-cavity TWTOs, which use helix and coupled-cavity slow-wave structures, respectively, each have specific properties and application problems.

Helical TWTOs The most important property of helix tubes is their great bandwidth, which follows from almost near-zero dispersion of the helix circuit in a broad frequency band (see Section 8.3). Some helical TWTOs have a bandwidth of over two octaves. Most other TWTOs have a bandwidth no greater than 10%.

The helix structure in general is not an adequate system for generating large average power in high-frequency microwave bands. So the diameter of the helix in the K_a-band ($\lambda \sim 1$ cm) must be about 0.5 mm, with an electron beam diameter of ~0.4 mm. The obvious difficulties in designing such powerful TWTOs lie in

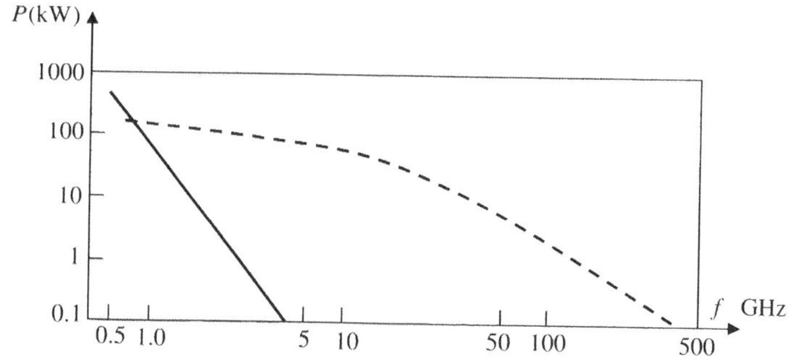

FIGURE 8.20 Maximum average power of TWTOs as a function of frequency. Solid curve, helical TWTO; dashed curve, coupled-cavity TWTO. (From Whitaker, 1999.)

formation of the electron beam, losses in slow-wave systems, and their cooling. Nevertheless, the development of powerful and sophisticated software that performs three-dimensional electromagnetic simulations, together with overall progress in the technology, have allowed the development of helical TWTOs with incredible parameters. According to Fig. 8.20, the maximum average power of a helical TWTO in the S-band can reach 10 kW. In general, helical TWTOs are the best amplifiers for communication and particularly for space communication applications. In the L and S bands, helical TWTO demonstrates high efficiency (more than 30%) and high gain (more than 50 dB).

Wide bandwidth, together with highly nonlinear bunching processes, gives rise to some problems that are not typical for other TWTOs.[1] For example:

1. *Harmonic generation.* If the second harmonic is generated near the low part of the frequency band, the corresponding signals are amplified when they are within the bandwidth, for example, of the octave TWTO.

2. *Backward-wave oscillations.* These oscillations are the result of beam interaction with the first negative spatial harmonic of the structure (see Sections 8.4.2 and 8.8). It is a major problem that determines the upper frequency of the bandwidth and the peak power capability of the tubes. One method of suppressing BWO oscillation is the introduction into the helix of frequency-sensitive attenuation. This problem can also be solved by applying other helix-like structures (Fig. 8.21) but with some narrowing of the bandwidth.

3. *Matching of transitions.* The transitions at the TWTO input and output terminals must provide reasonable impedance matching. This process is difficult when we take the great bandwidth into account. As we saw in Section 8.3, the presence of even small reflections from transitions leads to a frequency dependence of the gain and to self-excitation of the tube.

[1] A detailed analysis of these problems is provided by Gilmour (1994).

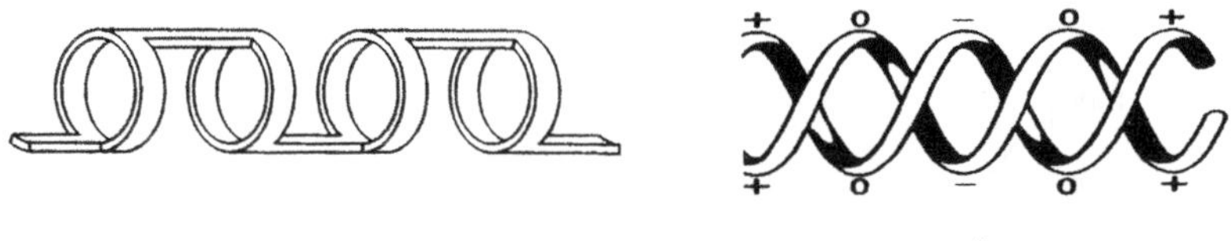

(a) (b)

FIGURE 8.21 Helixlike structures with suppressed BWO oscillations: (*a*) ring-bar circuit; (*b*) bifilar helix.

Coupled-Cavity TWTOs (CCTs) The slow-wave structure of a CCT is a linear chain of several tens of cavities provided by coupling holes or slots and a beam tunnel. The slow-wave structures in Fig. 8.5 represent particular cases from among the huge family of CCT structures. Some schemes of CCT structures are also given as examples in Fig. 8.22.

The cloverleaf structure (Fig. 8.22*a*) (Chodorow and Craig, 1957) is commonly used in high-power CCT and as an output section of a twistron (a combination of a high-power klystron as an input section and a TWTO output section). The high-power twistron amplifier has a bandwidth of up to 15% and output power on the order of 4 MW. Twistrons with a peak power output of up to 10 MW have been built. The average power exceeds 30 kW and can be increased by an order of magnitude (Gilmour, 1994).

The stagger slot CCT (Fig. 8.22*b*) has an output power on the order of 10 kW in the range 10 to 10.5 GHz with an overall efficiency of 20 to 30% (Staprans et al., 1973). Aligned slot CCTs (Fig. 8.22*c*) are characterized by a smaller bandwidth. However, for a given frequency this structure has larger dimensions, and higher powers are possible. Ladder core structures (Fig. 8.22*d*) have a comparatively simple geometry which allows the use of microfabrication in their manufacture.

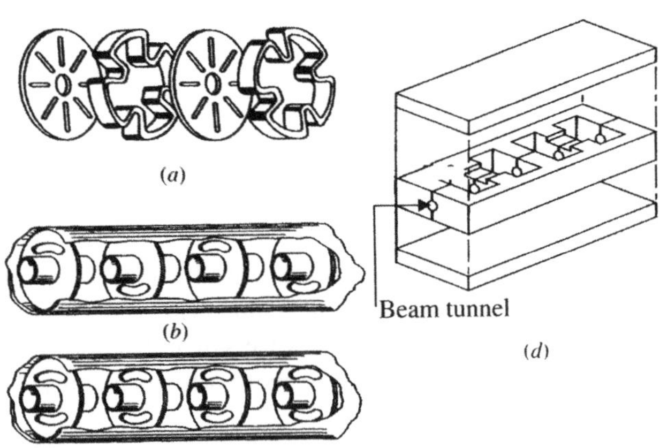

FIGURE 8.22 Slow-wave structures of coupled-cavity TWTOs: (*a*) Cloverleaf; (*b*) staggered slot; (*c*) aligned slot; (*d*) ladder-core structure with staggered coupled slots.

These structures are adequate for the millimeter range. CCTs with a ladder structure have demonstrated 20% bandwidth in the range 100 GHz with a power output of 100 W in continuous-wave operation (Gilmour, 1994).

Note the following general properties of CCTs:

- Larger dispersion and correspondingly narrower bandwidth than for helical TWTOs
- Large coupling impedance
- High wave phase velocity and corresponding electron velocity
- Mostly metallic construction
- Low thermal resistance

We see that coupled-cavity TWTOs are significantly more powerful as amplifiers (both average and peak power) than helical TWTOs, especially in high-frequency bands (Fig. 8.20), but with considerably reduced bandwidths. The high-power properties of CCTs allowed their use as ground-based and airborne radars. According to L3 Communications (2003), radar CCTs have been produced in San Carlos, California, with power levels of up to 150 kW peak, 5 kW of average power, and 10% bandwidth in the X and K_u bands.

8.8 BACKWARD-WAVE OSCILLATORS

8.8.1 Traveling-Wave Tube as an Oscillator

In Section 8.7.2. we saw how reflection waves excite autooscillations in TWTOs. However, this property of TWTOs cannot be designed in advance. The regular way to create a TWTO oscillator is setting a special feedback line in the TWTO (Fig. 8.23). The feedback signal must be in phase with an input signal, which is not provided in all frequencies. It can be shown that for optimal positive feedback, the total phase shift in the circuit, including forward and backward waves propagation, must be a multiple of 2π. In principle, the slow wave–electron synchronism makes it possible to vary the oscillation frequency continuously by tuning the beam potential. However, the excited frequencies must also satisfy the phase condition mentioned above. As a result, we obtain a frequency characteristic (Fig. 8.24) reminscent of the frequency zones of the reflex klystron. Within each zone, the

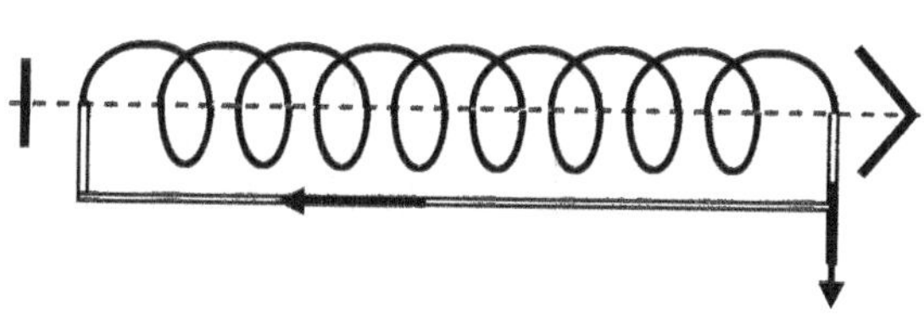

FIGURE 8.23 TWTO oscillator.

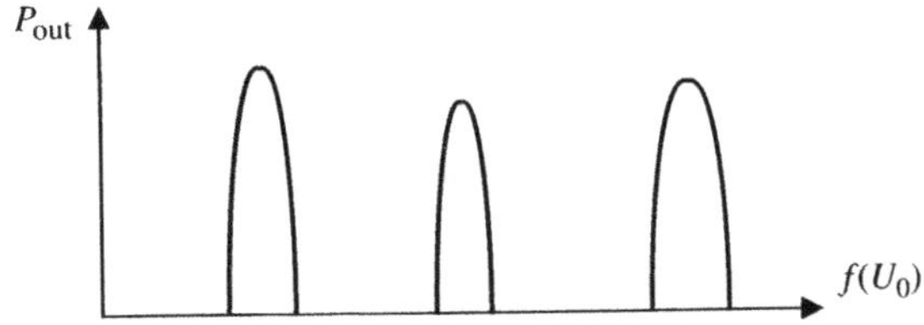

FIGURE 8.24 Frequency characteristic of a TWTO with positive feedback.

frequency variations correspond to the dispersion characteristic of a slow-wave structure. At the maxima of the frequency zones, synchronism and the phase condition are satisfied simultaneously.

8.8.2 Backward-Wave Amplifier Tubes: Principles of Operation

BWT performance for amplifiers and oscillators is based on an interaction between the electron beam and the negative spatial harmonics of the slow-wave structure. As we saw in Section 8.4.2, the energy velocity in the slow-wave structure moves opposite to the phase velocity of the negative harmonics. If the electron beam is synchronized with some negative harmonic (i.e., if the *synchronous condition* is fulfilled [see Eq. (8.15)],

$$|\beta_{-n}| \approx \beta_e \quad \text{or} \quad \frac{2\pi n}{L} - \frac{\omega}{v_{\text{ph},0}} \approx \frac{\omega}{v_0}, \qquad n > 0 \tag{8.123}$$

the energy that the electron beam transfers to the field is transported back to the cathode. So the output terminal must be positioned near the gun end of the tube (Fig. 8.25).

The interaction of an electron beam with a backward wave essentially changes the character of the field and the current distribution in comparison with forward TWTOs (Fig. 8.26). In the TWTO the current and field are increased exponentially in the direction of electron movement. In BWTs, because the energy is transferred to the left, the field amplitude increases to the left and its maximum is attained near the cathode. Consequently, electron bunching is accomplished in the lowering field. In the region near to collector where the field is small, the current does not change

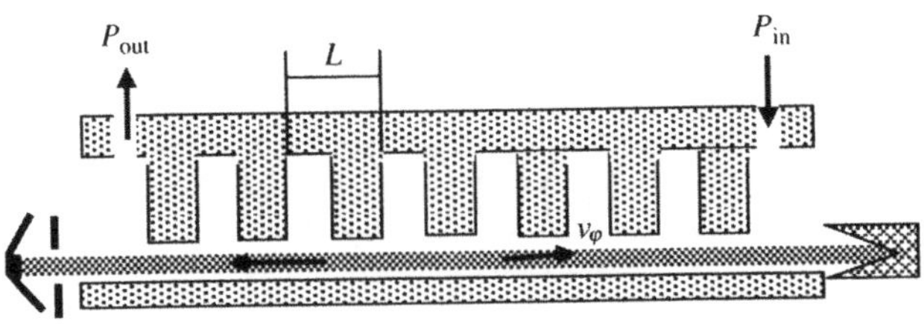

FIGURE 8.25 Backward-wave amplifier.

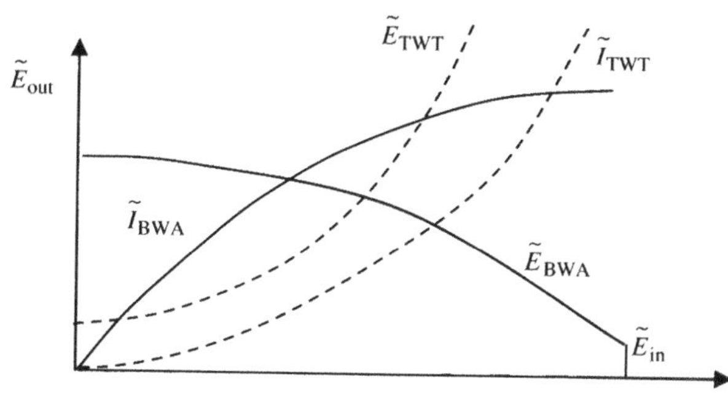

FIGURE 8.26 Amplitude distribution of convection current and electric field in a BWA amplifier (solid curves) and a TWTO amplifier (dashed curves).

practically. The maximum rate of bunching will be near the gun where the field is maximum. Thus, the distribution of the ac convection current has the form of the curve with saturation (the current distribution is similar to the first quarter of a sinusoid for a large space charge) (Heffner, 1954). On the other hand, the field near the cathode changes very little because a small convection current gives a small contribution to the energy. So the field distribution has the form of the lowering curve with distribution similar to that of the first quarter of the cosinusoid (Heffner, 1954). So the BWTs are typical regenerative amplifiers. Note that in these systems the backward wave is alone, and the phasing of the feedback signal is set automatically.

The magnitude of the output field depends on the dc beam current. Now let us imagine that we decrease the input signal and increase the beam current simultaneously. For some current magnitude, a finite output signal will be formed when the input signal is zero. Then the amplifier is converted in the oscillator [the backward-wave oscillator (BWO)]. The BWO was invented by R. Kompfner in 1952. He reported on this invention with N. Williams at the conference in Ottawa (Kompfner and Williams, 1952). At the same conference, Epsztein (1952) reported on the independent invention in France of a backward-wave oscillator of M-type, the *carcinotron*.

8.8.3 Gain in Backward Amplifiers and Starting Conditions of BWOs

According to the discussion above, the difference between bunching processes in BWAs and TWTOs is determined solely by the quantitative distribution of the ac field. Therefore, we can use Eq. (8.30) to describe the relation between the current and the field. Using dimensionless variables [Eqs. (8.48)–(8.50)], this relation has the form

$$(\delta^2 + q)\widetilde{I} = i\frac{I_0}{2U_0\beta_e C^2}\widetilde{E}_w \tag{8.124}$$

Let us turn to the problem of excitation of the field with a negative spatial harmonic. We can use Eq. (8.39) for an excited field, where we will retain the integral corresponding to the propagation of the back wave:

$$\widetilde{E}_w = \widetilde{E}_0 e^{-i\beta_0 z} - \frac{\beta_0^2 Z}{2} \int_z^L \widetilde{I}(\xi) e^{i\beta_0(z-\xi)}\, d\xi \tag{8.125}$$

Taking into account the opposite directions of phase and group velocities, we must reverse the sign of β_0 in the integral. Finally, the amplitude of the input signal is $\widetilde{E}_{\text{in}} = \widetilde{E}_0 e^{-\beta_0 L}$ at the plane $z = L$. As a result, we obtain

$$\widetilde{E}_w = \widetilde{E}_{\text{in}} e^{-i\beta_0(z-L)} - \frac{\beta_0^2 Z}{2} \int_z^L \widetilde{I}(\xi) e^{-i\beta_0(z-\xi)}\, d\xi \tag{8.126}$$

Following the same logic as that used in deriving Eq. (8.81), we pass from Eq. (8.126) to the differential equation

$$\frac{d\widetilde{E}_w}{dz} + i\beta_0 \widetilde{E}_w(z) = \frac{\beta_0^2 Z}{2} \widetilde{I}(z) \tag{8.127}$$

This equation differs from Eq. (8.81) only in the sign on the right-hand side. Note that we also omit the subscript 1 in the first harmonic of the ac convection current $\widetilde{I}_1(z)$. Substituting $\widetilde{E}_w, \widetilde{I} \sim \widetilde{A} e^{-i\beta z}$ into Eq. (8.127), we obtain an equation with dimensionless variables [Eqs. (8.48)–(8.50)]:

$$(\delta + ib)\widetilde{E}_w = \frac{\beta_0^2 Z}{2\beta_e C} \widetilde{I} \tag{8.128}$$

Matching Eqs. (8.124) and (8.128), and taking Eq. (8.44) into account, we obtain the dispersion equation

$$i(\delta + ib)(\delta^2 + q) = -(1 + Cb)^2 \tag{8.129}$$

The difference between the dispersion equations for TWTOs [Eq. (8.52)] and BWTs is in the sign on the right-hand side.

Solution of the remainder of the problem follows the solution for TWTOs [Eqs. (8.53)–(8.60)]. Assume that the three complex roots of Eq. (8.129) are δ_1, δ_2, and δ_3. The partial solutions of Eqs. (8.124) and (8.127) corresponding to these roots are

$$\widetilde{E}_k(z) = e^{-i\beta_e z} e^{\beta_e C \delta_k z} \tag{8.130}$$

The partial solutions for the current and the electron velocity coincide with Eqs. (8.54) and (8.55).

A general solution of the equation repeats the derivation of Eqs. (8.56). The arbitrary constants A_1, A_2, and A_3 are found from the following boundary conditions at the plane $z = 0$:

$$\tilde{E}(0) = \tilde{E}_0, \qquad \tilde{I}(0) = \tilde{v}(0) = 0 \tag{8.131}$$

The boundary condition follows Eq. (8.57) except that $\tilde{E}_0$ is now the complex amplitude of the output but not the input ac electric field. It is obvious that constants A_k are determined by the same Eq. (8.59) as that used above, and we do not repeat the procedure here.

The input (collector) complex amplitude $\tilde{E}(L)$ is determined by the first of Eqs. (8.56), where amplitudes of the partial solutions are $\tilde{E}_k(L) = e^{-i\beta_e z} e^{\beta_e C \delta_k L}$. As a result, the complex gain in BWA is

$$G_E = \frac{\tilde{E}(0)}{\tilde{E}(L)} = e^{i\beta_e L} \frac{1}{\displaystyle\sum_{k=1}^{3} \frac{\delta_k^2 + q}{(\delta_k - \delta_l)(\delta_k - \delta_m)} e^{\delta_k \beta_e CL}} \tag{8.132}$$

The starting conditions for BWOs (starting current and frequency) correspond to the infinite complex gain or to the zero of the denominator in Eq. (8.132). Let us write this condition as

$$\frac{(\delta_1^2 + q)e^{2\pi CN\delta_1}}{(\delta_1 - \delta_2)(\delta_1 - \delta_3)} + \frac{(\delta_2^2 + q)e^{2\pi CN\delta_2}}{(\delta_2 - \delta_1)(\delta_2 - \delta_3)} + \frac{(\delta_3^2 + q)e^{2\pi CN\delta_3}}{(\delta_3 - \delta_1)(\delta_3 - \delta_2)} = 0 \tag{8.133}$$

where N is the length of the interaction space in electronic wavelength, or

$$\beta_e L = 2\pi N \tag{8.134}$$

Taking into account loss in the slow-wave structure [see Eqs. (8.65) and (8.65a)], the dispersion equation (8.129) for $Cb \ll 1$ assumes the form

$$i(\delta - d + ib)(\delta^2 + q) = -1 \tag{8.135}$$

Note that besides the sign on the right-hand side, Eq. (8.135) also differs from Eq. (8.65a) for TWTOs in the sign of d. This definition is chosen so that $d > 0$ corresponds to the attenuation of a backward wave in a slow-wave structure.

Simple Lossless Case d = 0 We introduce the incremental propagation constant

$$\delta = i\chi \tag{8.136}$$

Then Eq. (8.135) becomes a cubic equation with real coefficients:

$$(\chi + b)(q - \chi^2) = 1 \tag{8.137}$$

In general, this equation can have one real root and two complex roots, or three real roots. Assume that the real root is χ_1. Then the two complex roots will be

$$\chi_{2,3} = -\frac{\chi_1 + b}{2} \pm \frac{i}{2}\sqrt{3\chi_1^2 + 2\chi_1 b - b^2 - q} \tag{8.138}$$

We obtain one wave of constant amplitude and two waves with increasing and decreasing amplitudes, respectively. However, in the case of BWO, the growing wave does not become dominant at the output end of the tube, because the initial amplitudes of the three waves at the input end ($z = L$) must satisfy Eq. (8.133) and can in principle have any magnitude. Finally, depending on the magnitudes of the coefficients b and q, Eq. (8.137) can have three real roots (i.e., δ_1, δ_2, and δ_3 will be imaginary). In this case, the output signal is the superposition of three waves with comparable amplitudes. The gain mechanism, is therefore, the result of an interference effect. The same holds for $d > 0$ as well.

The numerical solutions to Eq. (8.135) or (8.137), together with the real and imaginary parts of Eq. (8.133), provide starting values of δ_1, δ_2, and δ_3 as well as values for CN_{st} and b_{st} which determine the starting current and the frequency, respectively. Taking into account the definitions of C [Eq. (8.44)] and b [Eq. (8.48)], we obtain

$$I_{\text{st}} = 4U_0 \frac{(CN_{\text{st}})^3}{ZN^3} \tag{8.139a}$$

$$\frac{v_0}{v_{\text{ph},0}} = \frac{b_{\text{st}}}{N}(CN_{\text{st}}) + 1 \tag{8.139b}$$

Equation (8.139b) determines the oscillation frequency, ω, when the dispersion of the slow-wave structure $v_{\text{ph},0} = v_{\text{ph},0}(\omega)$ is known.

The reader can find the results of computations of BWO starting conditions in papers by Heffner (1954) and Johnson (1955). We note that for $q = 0$, the following relation is valid with good precision:

$$(\beta_0 - \beta_e)L = (2n + 1)\pi \tag{8.139c}$$

The onset of oscillations in the tube follows from a noise level of electromagnetic waves that are excited by a noisy electron beam. The frequency spectrum of these waves is extremely wide. The noisy waves propagate with different phase velocities that are determined by dispersion of the slow-wave structure. The condition of effective interaction [Eqs. (139b,c)] selects only the wave that is synchronous with the electron beam. This backward wave then modulates the electron beam, creating a mechanism for maintaining stationary oscillations.

We saw in Section 8.2.2 [Eq. (8.2)] that the physical meaning of relation (8.139c) for $n = 0$ is that the bunch must pass a distance of one half-wavelength relative to the wave during transit time (i.e., it must move permanently in the retarding phase of the field). For $n > 0$, condition (8.139c) requires that the bunch pass a distance of $(\lambda/2)/(2n + 1)$ in the retarding phase. According to Eq. (8.139c), the electrons

TABLE 8.1 Starting Condition BWO for Different Orders of Oscillations

n	$(\beta_0 - \beta_e)L$	CN_{st}
0	3.003	0.314
1	9.860	0.588
2	16.388	0.762
3	21.403	1.046

Source: Heffner (1954).

pass n half-wavelengths in the accelerated phase and $(n + 1)$ in the retarding phase, which is equivalent to shortening the structure $2n + 1$ times. Obviously, the starting current in this case should increase significantly.

Some computed values for $(\beta_0 - \beta_e)L$ and CN_{st} for $q = 0$ are given in Table 8.1. According to the table and Eq. (8.139*a*), the starting current for $n = 1$ is greater than the starting current for $n = 0$: approximately 6.6 times. It is important to note that the conditions in Eq. (8.139*c*) can be satisfied simultaneously for different frequencies. Thus, if a current exceeds the starting current for $n \geq 1$, self-excitation of the tube is possible at very closely spaced frequencies, and the output signal becomes nonmonochromatic. Therefore, for stable operation of BWOs, the current should remain in the range of $(I_{st})_{n=0} < I < (I_{st})_{n=1}$.

The values $(\beta_0 - \beta_e)L$ and CN_{st} (frequency detuning and the starting current, respectively) grow with an increase in loss d and space charge q. It can be shown (see, e.g., Chodorow and Susskind, 1964) that for large q, and in the absence of loss, the parameter of nonsynchronism $b \approx \sqrt{q}$. This b is equal to b_{opt}, which corresponds to the maximum gain in TWTO for a large space charge. In that case, as shown in Section 8.5.6, the circuit wave is synchronous with the slow space-charge wave. The corresponding CN_{st} is equal to

$$(CN)_{st} = \frac{(2n+1)^{2/3}}{2}\left(\frac{\beta_p L}{2\pi}\right)^{1/3} \tag{8.140}$$

where β_p is determined by Eq. (8.49).

Gain in Backward Tubes for Current Below the Starting Current Applying a Taylor expansion to the denominator in Eq. (8.132) around the $G = 0$ point, we obtain a first-order approximation of the voltage gain:

$$G = \frac{k}{I_s - I} \tag{8.141}$$

An analogous expansion can be developed around the starting value $b = b_{st}$. Thus, the gain in BWTs is highly nonuniform near the starting frequency. As such, BWAs are considered narrowband amplifiers in contrast with TWTOs. The central frequency of the amplifier band can be shifted by varying the voltage. In this mode, BWAs can be used as voltage-tuning selective microwave filters.

8.8.4 Frequency Tuning in BWOs

According to relation (8.139*c*), the starting difference $(\beta_0 - \beta_e)$ should not depend on $\beta_e = \omega/\sqrt{2\eta U_0}$. Therefore, when U_0 is changed, the self-excitation of BWOs is accompanied by a corresponding change $\beta_0 = \omega/v_{\text{ph},0}(\omega)$ (i.e., by frequency tuning). Voltage frequency tuning is an extremely valuable property of BWOs. Proper choice of the slow-wave structure and construction of the tube (electron gun, magnetic focusing system, impedance matching of the output terminal over the entire band) allows us to attain up to three-octave band frequency tuning. In particular, good results are obtained with different modifications of the interdigital structure (Fig. 8.5*b*), where the first negative spatial harmonic is the fundamental one.

In wideband BWOs it is essential to match not only the output terminal but also the collector end of the slow-wave structure. The reflection of the energy from the collector end creates a wave with the phase velocity directed along the electron velocity. As a result, the parasitic reflected wave is amplified, and it interferes with the basic wave. This effect distorts the frequency characteristics of the oscillator. Matching is achieved through the application of a special absorber at the collector end of the slow-wave structure.

8.8.5 Properties of Nonrelativistic BWOs for a Current Greater Than the Starting Value

When the beam current exceeds the starting value, the behavior of the BWOs should be described by nonlinear theory. Grow and Watkins (1955) calculated the efficiency of lossless BWOs based on the assumption that the oscillation level of the tube is limited by the saturation of the ac beam current. According to Grow and Watkins, the efficiency for a zero space charge changes approximately as

$$\eta \cong C \tag{8.142a}$$

The analysis in Grow and Watkins (1955) does not, however, yield frequency detuning as a function of the beam current.

A comprehensive nonlinear theory of stationary oscillations in nonrelativistic BWOs was developed by Rappoport (1964) and Rowe (1965). The details are beyond the scope of this work, but we consider briefly the results of the numerical computation of efficiency and the parameter b (frequency detuning). The curves $\eta = \eta(I/I_{\text{st}})$ and $b = b(I/I_{\text{st}})$ for different values C and q are shown in Fig. 8.27. As shown, the efficiency and nonsynchronous parameter decrease with the space-charge parameter q. The relation between the efficiency and C (Fig. 8.27*a*) is comparable to relation (8.142) of Grow and Watkins (1955).

In general, the maximum efficiency for this type of wideband BWOs is significantly lower than the corresponding value for TWTOs. This is the natural result of unfavorable distribution of the ac electric field in BWOs (Fig. 8.26): the ac field is minimal where the ac current is maximal, and vice versa.

Resonant BWO is a version of BWOs in which instead of a matched absorber, a reflector on the collector end is employed. Furthermore, the gun terminal is supplied

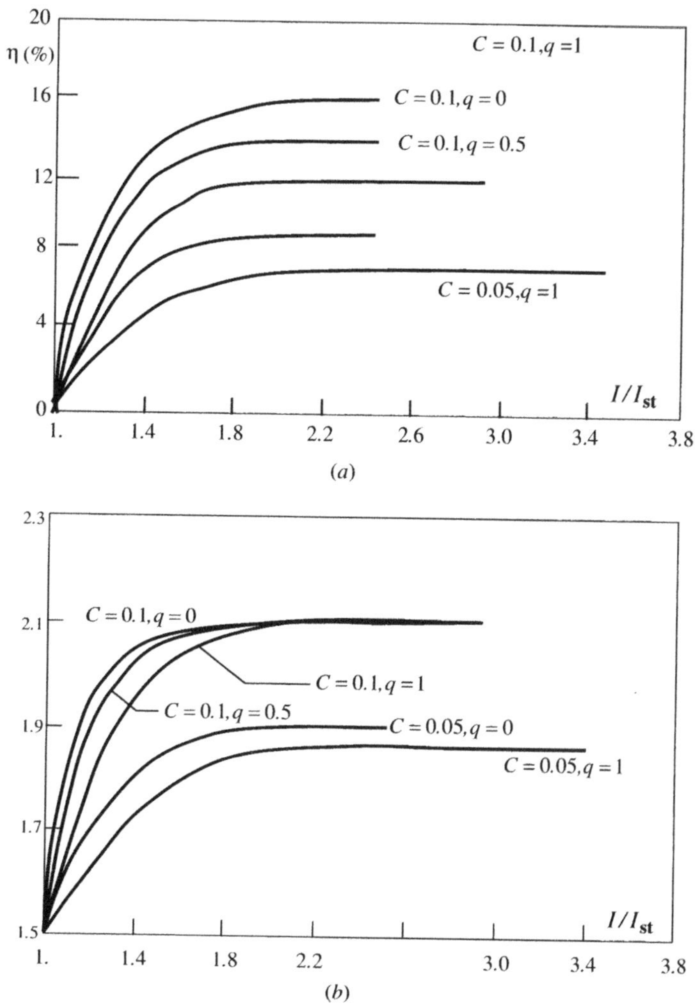

FIGURE 8.27 Efficiency (*a*) and nonsynchronism parameter (*b*) as functions of I/I_{st}. (From Rowe, 1965.)

with a special ac transformer that provides the required matching of the tube with the output line. In such a tube there are multiple reflections of the ac electric field that make its distribution along the interaction space almost uniform. As a result, a significant increase in the efficiency and a reduction in the starting current in the selective values of the phasing $\beta_e L$ can be achieved. The disadvantage of resonant BWOs is the discontinuity of the frequency characteristics. The tube works in a narrow frequency band that is determined by the resonance condition in the feedback circuit. Therefore, the generating frequency is changed discontinuously with voltage.

The low efficiency of BWOs makes their applications advisable as the oscillators of a small output power but with a very wideband voltage detuning and monochromatic spectrum. Structurally, BWOs in the centimeter frequency range are usually carried out as compact devices with built-in periodical permanent magnetic systems (see Section 5.8.2). But in short millimeter and submillimeter BWOs, the solenoidal magnetic systems with a uniform magnetic dc field are used (see Section 8.9).

8.9 MILLIMETER NONRELATIVISTIC TWTOs, BWOs, AND OROTRONS

8.9.1 Structural Features of Millimeter Nonrelativistic TWTOs and BWOs

Consider as an example the construction characteristics of both TWTOs and BWOs at a wavelength $\lambda = 2\,\text{mm}$. Assume an accelerating voltage $U_0 = 3\,\text{kV}$. The nonrelativistic deceleration of the electron then is (see Section 1.3)

$$\xi_n = \frac{c}{v_0} \approx \frac{16}{\sqrt{U_{0\,\text{kV}}}} = \frac{16}{\sqrt{3}} \approx 10$$

This is also the deceleration of the electromagnetic wave that interacts with the electron beam. Therefore, the delayed wavelength is equal to $\lambda_s = \lambda/10 = 0.2\,\text{mm}$. Recall (Section 8.4.2) that slow waves are surface waves and that their fields decrease with a distance x from the structure surface as $e^{-(2\pi/\lambda_s)x}$. This means that the ac field in the slow-wave structure is localized in the characteristic layer (the skin layer) at a depth of $\Delta x = \lambda_s/2\pi \approx \frac{1}{6}\lambda_s$ adjoining the structure surface. In our example $\Delta x \approx 0.03\,\text{mm} = 30\,\mu\text{m}$. So the thickness of the electron beam must be less than $30\,\mu\text{m}$. Note that the beam must be very smooth. The radius of a helical trajectory in the magnetic field is equal to $r_\perp = v_\perp/\eta B$. Assume that the transverse electron velocity at the entrance to the interaction space is $v_\perp = 0.05 v_0$. Then, if one takes the maximum admissible ripple amplitude $r_\perp = \frac{1}{3}\Delta x = 10\,\mu\text{m}$, the magnetic field required will be

$$B > \frac{0.05v}{\eta r_\perp} = \frac{0.05c}{\eta \xi_n r_\perp} \approx 0.8\,\text{T}$$

Thus, attainable output power for both TWTOs and BWOs in the millimeter-wave range diminishes sharply with increased frequency. The common construction elements of millimeter TWTOs and BWOs are tiny slow-wave structures, very thin electron beams, high current density, and strong focusing magnetic fields. It is important to emphasize that the features indicated for nonrelativistic TWTOs and BWOs are the result of Cerenkov radiation of uniformly moving electrons, since this requires the use of slow electromagnetic waves with a phase velocity much less than the light velocity. Here we discuss only nonrelativistic electron beams. Some properties of relativistic TWTOs and BWOs are considered in Section 8.10.

Let us estimate typical dimensions of for the slow-wave structure. Take, for example, the comb structure (Fig. 8.5*a*). The height of the tooth for this structure is $h \approx \frac{1}{4}\lambda = 0.5\,\text{mm}$, the pitch $L \approx \frac{1}{2}\lambda = 1\,\text{mm}$. Losses in the structure grow with

TABLE 8.2 CW Output Power of Nonrelativistic Millimeter BWOs

λ (mm)	8	4	1	0.2
P_{out}	100 W	10 W	50 W	0.5 mW

the frequency, approximately as $\sqrt{\omega}$. More strict limitations are typical for helical TWTOs. The diameter of the helix in a 4-mm TWTO must be on the order of 0.5 mm, and the diameter of the electron beam, ~0.4 mm. The rods support the helix. Therefore, wideband helix TWTO amplifiers are not used on wavelengths shorter than 4 mm. Wideband BWOs are produced and used for frequencies up to 500 GHz.

Table 8.2 lists some typical CW output powers for nonrelativistic millimeter BWOs with $U_0 = 10\,\text{kV}$. As shown, output power decreases faster than λ^2 (which would correspond to the cross section of the electron beam area). Here it is necessary to take into account the growth of losses and the nonideal focusing of the electron beams.

Below we consider two more advanced versions of millimeter oscillators.

8.9.2 Clinotrons

The clinotron has been proposed as a modification of convention BWOs to eliminate some disadvantages of the latter. The basic principle of electron–wave interaction in the clinotrion (see, e.g., Levin et al., 1992; Schuneman and Variv, 1999) is similar to that in ordinary BWOs. The distinguishing features of this oscillator consist of the following. First, the electron beam is inclined to the surface of a grating as illustrated in Fig. 8.28. By varying the small angle α, it is easy to adjust the effective interaction length L without changing the grating geometry. Second, the beam thickness is large compared with that in conventional BWOs and more than the thickness Δ of the skin layer. Third, requirements as to the smoothness of the electron beam and consequently, the value of the focusing magnetic field and the precision of the beam formation weaken essentially. As a result, the level of the clinotron output power is almost an order of magnitude larger than for ordinary millimeter BWOs while small physical dimensions, the possibility of using a low operating voltage and other advantages of BWOs are is retained. Some characteristics of clinotrons currently in production are listed in Table 8.3.

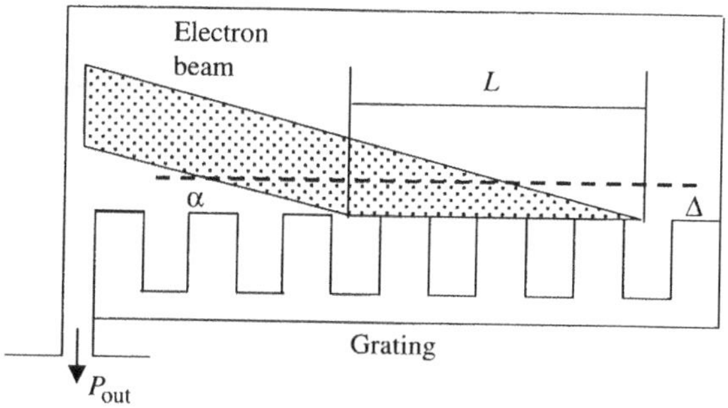

FIGURE 8.28 Clinotron.

TABLE 8.3 Parameters of Advanced Clinotrons

Range of Frequency Tuning (GHz)	Maximum Output Power (W)	Anode Voltage (kV)	Weight[a] (kg)
53–63	11.0	4.0	1.2
113–122	3.0	4.3	1.2
137–151	2.0	4.5	1.2
345–390	0.1	5.0	12
442–510	0.05–0.1	5.5	12

[a]Together with a focusing magnet.

8.9.3 Orotrons

The orotron is a generator of coherent electromagnetic radiation in the millimeter and submillimeter range. A diagram of the oscillator is shown in Fig. 8.29. The electron beam (1) moves over a grating (2) which is placed on the plane mirror of a quasioptic resonator. The second mirror (3) of the resonator is usually formed as a spherical reflector. As a result, caustic surfaces (4) are formed that provide a high value of the quality factor Q for the resonator.

The device in Fig. 8.29 was developed by Rusin and Bogomolov (1969) and called the *orotron*. It also is known by other names, such as *laddertron* (Mizuno et al., 1973) and *generator of diffraction radiation* (Balaklitskii et al., 1973). The electron beam in the orotron is synchronous with the mth spatial harmonic of the periodic structure (usually, $m = 1$). It is similar to the synchronism in TWTOs and BWOs (depending on a sign of m), and the scheme of the orotron is reminiscent of resonant TWTOs or BWOs. However, the important distinction of the orotron is that the zero and perhaps a few other lowest spatial harmonics are fast waves. Therefore, the electromagnetic field generated is not propagated as a pure surface wave but is basically radiated through fast harmonics into the volume of the quasioptic resonator. The fundamental wave, corresponding to zero spatial harmonic, is an eigenmode of an open resonator. The phase velocity of this harmonic, $v_{0,\mathrm{ph}} \gg c$; that is,

$$\beta_0 = \frac{\omega}{v_{0,\mathrm{ph}}} \ll \frac{2\pi}{\lambda} \tag{8.142b}$$

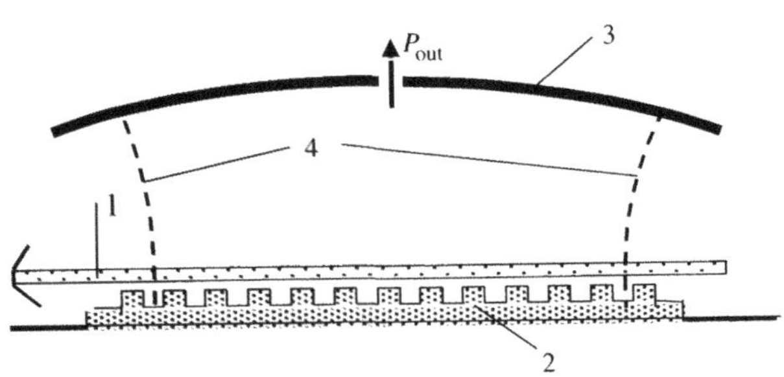

FIGURE 8.29 Orotron.

The radiation of electrons passing over a grating was discovered by Smith and Purcell (1953). But they did not use any resonators that in general could provide feedback and bunching of the electron beam. Therefore, their radiation was spontaneous, incoherent, and had very little output power. Note that, in principle, the Smith–Purcell effect can be considered a particular case of transition radiation (Section 8.2.2) over periodical nonhomogeneities.

If a proper condition of synchronism between the electron and phase velocities of a slow spatial harmonic is satisfied and if the electron beam current exceeds the starting current, the orotron will radiate coherently at a frequency close to one of the resonant frequencies of an open resonator. In this sense the orotron is a device with stimulated (induced) Smith–Purcell radiation.

The primary difference between orotron theory and classic TWTO and BWO theory is reduced to taking into account the specific electromagnetic structure of the open resonator. However, that does not change the mechanism of the interaction between an electron beam and a high-frequency field compared with closed tubes. The starting current and frequency are determined from the conditions of stationary oscillation of the orotron, which can be written similar to Eqs. (6.53) and (6.54):

$$\begin{aligned} \frac{\omega_0 W}{Q} &= -P_{er} \\ \frac{\omega - \omega_0}{\omega_0} &= \frac{1}{2Q}\frac{P_{ei}}{P_{er}} \end{aligned} \tag{8.143}$$

where for the starting mode, P_{er} and P_{ei} are the linearized real and imaginary electron power, respectively. W is the energy stored in the open resonator. (We omit various other details of this analysis.) It can be shown (Vainstein et al., 1983) that the starting current has a magnitude on the order of

$$I_{st} = K\frac{\omega_0 U_0}{\beta_e r^3}\frac{\hat{V}}{Q} \sim \frac{U_0^{-3/2}}{Q}\frac{\hat{V}}{r^3} \tag{8.143a}$$

where $\hat{V}$ is the effective volume of the resonator and r is the radius of a caustic zone that significantly influences the starting current.

The nonlinear theory of the orotron is analogous to the nonlinear theory of the TWTO (Vainstein et al., 1983). Approximate analytic nonlinear theory is developed by Myasin et al. (1990).

The orotron has the following potential advantages as an oscillator in the millimeter and especially in the submillimeter range.

1. Fewer losses, because the energy is translated and radiated through the fast spatial harmonics.
2. A high quality factor Q value of the resonator, leading to high-frequency stability and small starting currents.

3. A large volume of space occupied by the ac electromagnetic field. [This property of the orotron makes it usable, for example, in high-sensitivity molecular spectrometers (Surin et al., 2001). The gas being examined is placed into the orotron resonator and the absorption signal is detected as a function of the tuning frequency.]
4. Technological advantages in manufacturing slow-wave structures, electron-optical focusing systems, and output circuits.
5. In relativistic orotrons (Bratman et al., 1987) with $v_0 \approx v_{\mathrm{ph},1} \leq c$, a period L of the SWS close to the radiation wavelength λ, as follows from Eq. (8.142).

As a whole, orotrons are more adequate than BWOs as power oscillators in the submillimeter range, especially at relativistic energies. The main disadvantage of an orotron is the limitation on electron frequency tuning. Variation of the oscillation frequency is possible only via simultaneous adjustment of both the voltage and natural frequency of the resonator.

8.10 RELATIVISTIC TWTOs AND BWOs

8.10.1 Introduction

An increase in electron velocities up to relativistic velocities will significantly weaken limitations connected with the surface nature of slow electromagnetic waves in O-type devices. In particular, taking into account the synchronism condition $v_{\varphi n} = v_0$, we obtain from Eq. (8.16) the thickness of the skin layer for relativistic beams:

$$\Delta_{\mathrm{sk}} = \frac{\lambda}{2\pi}\frac{v_{\varphi n}}{c}\frac{1}{\sqrt{1-(v_{\varphi n}/c)^2}} = \frac{\lambda}{2\pi}\gamma\frac{v_{\varphi n}}{c} \approx \frac{\lambda}{2\pi}\gamma$$

For example, $\Delta_{\mathrm{sk}} \approx \lambda$ for $U_0 = 2.5$ MV. A very important aspect of relativistic beam use is determined by weakening of the space-charge limitation (see Sections 3.1, 4.8, and 5.1 to 5.3). Thus, transfer to relativistic TWTOs and BWOs allows us essentially to increase both the frequency and the energetic limits of amplifiers and oscillators being considered.

Creation of these tubes is possible to a considerable extent by the effect of explosive electron emission that has ensured a generation of superintense relativistic electron beams (see Section 4.8). Only two years later, Nation (1970) investigated the interaction between a relativistic beam and a backward wave. Nation obtained an output power of about 10 MW in the X-band with a beam current of 30 kA and an energy of 500 keV. Although the efficiency was not high (~0.05%), it was a promising result. Kovalev et al. (1973) obtained radiation with an efficiency of 12 to 15% at a 3-cm wavelength, using a design of relativistic BWOs with a corrugated slow-wave structure (Fig. 8.30) and electron energy 700 keV. Relativistic TWTOs appeared in about 1985–1991 (Korovin et al., 1985; Shiffler et al., 1991). Below some principal features of powerful BWOs and TWTOs are described briefly.

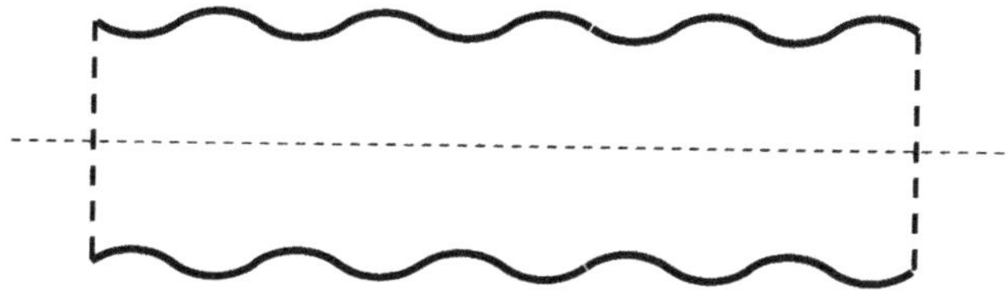

FIGURE 8.30 Corrugated slow-wave relativistic BWO–TWTO.

8.10.2 Equations of Relativistic BWOs and TWTOs. Relations of Similarity

Considered below are equations (Petelin, 1970) based on a simplest approximation: one-dimensional motion of electrons in a given high-frequency electric field:

$$E = E(z)e^{i\omega t} = \widetilde{E}(z)e^{i\omega(t-z/v_\varphi)}$$

The relativistic equation of motion

$$\frac{d(mv)}{dt} = \text{Re}[F(z)e^{i\omega t}] \tag{8.144}$$

[where $F(z) = -e_0E(z) = \widetilde{F}(z)e^{-i\omega(z/v_\varphi)}$] can be written in the form of two first-order equations:

$$c^2\frac{dm}{dz} = m_0c^2\frac{d\gamma}{dz} = \text{Re}\,[F(z)e^{i\omega t}] \tag{8.145}$$

$$\frac{dt}{dz} = \frac{1}{v} = \frac{\gamma}{c\sqrt{\gamma^2 - 1}} \tag{8.146}$$

$\gamma = m/m_0 = 1/\sqrt{1-(v/c)^2}$ is the relativistic factor. The solution of Eqs. (8.145) and (8.146)—$\gamma = \gamma(z, t_0)$ with initial conditions

$$(\gamma)_{z=0} = \gamma_0, \qquad (t)_{z=0} = t_0 \tag{8.147}$$

determines the motion of an electron that enters the interaction space at the moment $t = t_0$. The electron efficiency is an average of the relative energy that the electrons give to the ac field in the interaction length L:

$$\eta = \frac{1}{\gamma_0 - 1}\left[\gamma_0 - \frac{1}{T}\int_0^T \gamma(L, t_0)\,dt_0\right] \tag{8.148}$$

Let us consider a case of ultrarelativistic electronic devices: a beam with extreme relativistic energy of electrons $\gamma \gg 1$. Equations (8.146) and (8.148) in this case are transformed to the forms

$$\frac{dt}{dz} = \frac{1}{c}\left(1 + \frac{1}{2}\gamma^{-2}\right) \tag{8.149}$$

$$\eta = 1 - \frac{1}{T}\int_0^T \frac{\gamma(L, t_0)}{\gamma_0}\, dt_0 \tag{8.150}$$

The system of equations (8.145), (8.149), and (8.150) can be reduced to a form that does not contain the initial energy γ_0. Here we introduce the dimensionless quantities

$$\mu = \frac{\gamma}{\gamma_0}, \qquad \Phi = \omega\left[t - \left(1 + \frac{1}{2\gamma_0^2}\right)\frac{z}{c}\right] \tag{8.151}$$

We obtain from Eq. (8.149) the equation

$$\frac{1}{\omega}\frac{d\Phi}{dz} = \frac{1}{2c\gamma_0^2}\frac{1}{\mu^2} - \frac{1}{2c\gamma_0^2} \tag{8.152}$$

Defining the dimensionless length

$$z = \frac{2c\gamma_0^2}{\omega}\zeta = \frac{\lambda\gamma_0^2}{\pi}\zeta \tag{8.153}$$

we reduce the system of equations (8.145), (8.152), and (8.150) to the form

$$\frac{d\mu}{d\zeta} = \mathrm{Re}[A(\zeta)e^{i\Phi}] \tag{8.154}$$

$$\frac{d\Phi}{d\zeta} = \frac{1}{\mu^2} - 1 \tag{8.155}$$

$$\eta = 1 - \frac{1}{2\pi}\int_0^{2\pi} \mu(\Lambda, \Phi_0)\, d\Phi_0 \tag{8.156}$$

where

$$\Lambda = \frac{\pi}{\gamma_0^2}\frac{L}{\lambda} \tag{8.157}$$

$$\begin{aligned} A(\zeta) &= \frac{2\gamma_0}{m_0\omega c}F(z)e^{i(\omega t - \Phi)} = \frac{\lambda\gamma_0}{\pi m_0 c^2}\widetilde{F}(z)e^{i[(1+1/2\gamma_0^2)(\omega z/c) - \omega z/v_\varphi]} \\ &= \frac{\lambda\gamma_0}{\pi m_0 c^2}\widetilde{F}\left(\frac{\lambda\gamma_0^2}{\pi}\zeta\right)e^{i\delta\zeta} \end{aligned} \tag{8.158}$$

$\delta = 1 - 2\gamma_0^2[(c/v_\varphi) - 1]$ is the ultrarelativistic parameter of nonsynchronism.

The system (8.154)–(8.156) allows a set of problem to search the function $A(\zeta)$, which ensures maximum efficiency [Eq. (8.156)]. Such a function in principle may be physically unrealizable. But taking a number of limitations into account (e.g., the equations for the electromagnetic field, configuration waveguide, losses) it is possible to reduce the problem to searching for optimum efficiency. If the corresponding optimal distribution $A(\zeta)$ is found, important scaling rules arise which follow from relations (8.157) and (8.158). In particular, the optimal length L/λ of the tube should be chosen proportional to γ_0^2 (the square of the initial beam energy). It is interesting that the amplitude of the optimal ac electric field is inversely proportional to γ_0, while at low voltages the amplitude increases with V. According to Eq. (8.158), the optimal field amplitude grows with shortening wavelength.

The theory described above does not take into account the dependence of $F(z)$ on beam bunching, space charge, and so on. So the theory can be applied to devices with a fixed spatial structure of the ac field (e.g., resonant TWTOs, BWOs, and orotrons). Self-consistent equations for the electron beam and electromagnetic field have been obtained by Kovalev and Smorgonsky (1975) similar to equations in the theory of nonrelativistic TWTOs and BWOs (Sections 8.5, 8.6, and 8.8). In this case the system of equations (8.154) and (8.155) is supplemented by the equation of ac electric field excitation for the case of interaction of the electron beam with a single mode of the traveling wave. Corresponding equation in is the ultrarelativistic version of the nonrelativistic equations (8.85) and (8.86). This equation has the form

$$\frac{dA}{d\zeta} - i\delta A = \mp I \int_0^{2\pi} e^{-i\Phi}\, d\tau_0 \qquad (8.159)$$

The term on the right side of Eq. (8.159) is proportional to a first harmonic of the ac convection current. The sign on the right side of Eq. (8.159) should be chosen negative or positive for beam interaction with the forward and backward spatial harmonics of the periodic slow-wave structure, respectively.

8.10.3 Relativistic BWOs

The diagram of a relativistic BWO with a tubular electron beam is shown in Fig. 8.31. The electron beam in the tube interacts with the synchronous

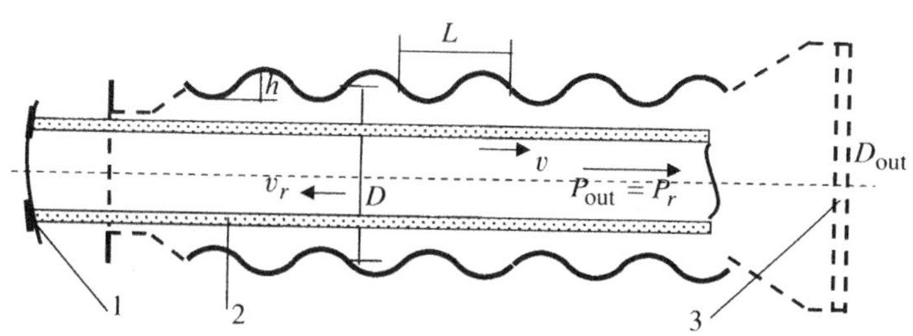

FIGURE 8.31 Relativistic single-stage BWO. 1, Cathode; 2, electron beam; 3, output window.

-1 spatial harmonic: usually, the TM_{01} mode in a circular corrugated waveguide or the HE_{01} mode of a rectangular waveguide. The depth of the corrugations is small, $h/\lambda \ll 1$, because of the small deceleration of relativistic electrons. The beam tunnel from the regular slow-wave structure to the gun is tapered up to the cutoff waveguide in order to exclude ac power leakage into the gun. The synchronous (backward) harmonic of the wave reflected from the input waveguide does not interact with the electron beam, and the wave radiates through the waveguide in the collector end. The widening of the output waveguide ($D_{out}/\lambda \gg 1$) excludes high-frequency breakdown in a dielectric output window.

After obtaining the first results (Kovalev et al., 1973; Nation, 1970), the development of BWOs was determined by requirements of enhancement output power, efficiency, and coherence of radiation. The common way of increasing the output power is using space-developed (overmoded) slow-wave structures (SWSs). On the one hand, there is an increase in the volume of an active medium that is formed by electrons and the electromagnetic field, and on the other hand, it prevents RF breakdown and development of processes that lead to output pulse shortening (Barker and Schamiloglu, 2001; Benford and Benford, 1997; Kovalev et al., 1998) (the RF pulse becomes shorter than the applied electric pulse). According to a number of works (see, e.g., Kovalev et al., 1998), the main factor in output pulse shortening is a multifactor effect in the strong HF electric field. Bombardment by secondary electrons of SWS walls stimulates desorption of gas and finally, formation of plasma that stops microwave generation. The characteristic time of this process at the gigawatt power level according to data of Kovalev et al. (1998) is much shorter than the time of processes determined by the cathode and collector plasma expansion.

Bugaev et al. (1990) developed an ultrarelativistic oscillator with a periodic two-section diaphragmatic waveguide and the ratio of the diameter to the wavelength (D/λ) up to 13. The output power and efficiency reached 15 GW and 50%, respectively, in the X-band. The energy of electrons in the beam generated by the magnetically insulated diode and the beam current were ~2 MeV and 15 kA with a pulse duration of ~70 ns. The electron beam interacted with a set of harmonics of different modes near the short-wave boundary of the fundamental mode passband. As such, this tube can be considered simultaneously a resonant BWO and a TWTO oscillator. The authors called their tube a *multiwave Cerenkov generator*. Although the oscillator had a high degree of time coherence, the spatial coherence could not in practice be ensured by the multimode indefinite output structure of a signal. Thus, effective application of space-developed SWS required solution of a mode selection problem.

A number of electrodynamic and electron methods of mode selection are known (see, e.g., Bratman et al., 1987). In particular, Abe et al. (1998) have studied mode selection in relativistic BWO with over-moded corrugated SWS up to a D/λ ratio of ~3. Selection of the fundamental mode TM_{01} has been achieved through variation of the corrugated structure parameters L, h, and D (Fig. 8.31) in order to merge the interaction range of TM_{01} with the short-wave edge of the passband, thus increasing frequency separation from the competing mode TM_{02} (electrodynamic selection).

The beam was also placed closer to the corrugated wall in order to increase the coupling impedance of the chosen harmonic (the electron selection). With these modifications, BWOs have generated an output power of 200 MW in the frequency range 5.2 to 5.7 GHz with an efficiency of 4%. The time coherence ($\Delta\omega/\omega < 0.5\%$) and spatial coherence according to the experimental data of Abe et al. (1998) were both sufficiently high.

Effective cyclotron-resonance mode selection was proposed and developed by Abubakirov et al. (1983). Let us choose the focusing uniform magnetic field B according to the equality

$$\omega_c = \frac{\eta B}{\gamma} = \frac{2\pi}{L} v_0 \tag{8.160}$$

Then, taking the Cerenkov condition of synchronism for the -1 harmonic [Eq. (8.123)] into account, $2\pi/L - \omega/v_{\text{ph},0} = \omega/v_0$, we arrive at the equation

$$\omega_c = \omega\left(1 + \frac{v_0}{v_{\text{ph},0}}\right) \tag{8.161}$$

This equation corresponds to the Doppler condition of cyclotron resonance for a beam with zero spatial harmonic which propagates to the cathode (i.e., against the electrons). The electron beam in the BWO is practically rectilinear, and the electrons are similar to linear oscillators. In this case, the electron beam forms an absorbing medium (see Section 10.1), and all backward modes of the slow-wave structure whose zero harmonics have a phase velocity close to $v_{\varphi 0}$ will be damped. However, if we choose the radius r of the electron beam according to the condition of zero absorption at the cyclotron resonance

$$I_{m\pm 1}(\kappa r) = 0 \tag{8.162}$$

where I_m is a Bessel function of first kind, m order, and κ is the transverse wave number of the operative mode, this mode does not decay. This method was realized successfully in X-band BWOs with operative mode TM_{02}.

The theoretical efficiency of an idealized model of ultrarelativistic BWOs with uniform SWS is 15% (Kovalev et al., 1979). Possibilities of efficiency enhancement are connected with using nonuniform SWSs to compensate for the ordinary BWO unfavorable axial structure of the ac electric field. Using a jump in SWS coupling impedance allows increase in the theoretical efficiency up to 45% (Kovalev et al., 1979). It is attained by changing the amplitude and period of the ripples in corrugated SWSs (Fig. 8.31). In particular, the elevation of the coupling impedance near the end of an SWS has allowed increased experimental efficiency: up to 35% in the X-band (Korivin et al., 1992a). Another possibility lies in changing the phase velocity at the end of the interaction space (Korovin et al., 1992b; Moreland et al., 1994). In this case the mechanism for efficiency improvement is similar to velocity tapering in nonrelativistic TWTOs (see Section 8.6.3).

Powerful relativistic BWOs can operate in both the single-shot and repetitive regimes. Maximum output power is attained in single-shot operation. So far, these sources have been used in some plasma experiments and in the exploration of super-powerful relativistic electron beams. Usually, the energy pulse parameter $E = W\tau$, where τ is a pulse duration, is used for comparing ultrarelativistic electron guns. This parameter is limited by the pulse-shortening effect. For ordinary BWOs, E is significantly less than 1 kJ. The record $E_{\max} \approx 1$ kJ belongs to the aforementioned multiwave BWO with $D/\lambda \sim 13$.

BWOs in the repetitive operation regime are characterized by a comparatively high average power if one takes the nanosecond pulse duration into account. So the X-band BWO described by Bykov et al. (1989) reached an average radiation power of 400 W with a pulse repetition frequency of 100 Hz and a pulse duration of 5 ns. In single-shot operation this tube produced 0.8 to 1 GW. Use of repetitive operation significantly increases the precision of measurements of frequency and output power.

Extension of this line of research has led to the creation of nanosecond radars, for which a series of compact and reliable high-voltage modulators (SINUS generator of Clunie et al., 1997) for repetitive operation are employed. A BWO in the X-band as a radar microwave source had an output pulse power of 500 MW, a pulse duration of 5 ns, and a repetition frequency of 150 Hz. Use of a short-duration pulse and high peak microwave power have ensured unique resolution for the detection of fast-moving targets traveling above the sea surface. The slowly changing clutter returns were suppressed by the over-period subtraction. Some characteristics of a radar built in Russia under contract to the United Kingdom, as well as the results of experimental trials of a radar system on the southern UK coast, are described by Clunie et al. (1997).

8.10.4 Relativistic TWTOs

Relativistic TWTO amplifiers combine high power, efficiency, and the possibility of phase synchronization, which is very important for such uses as high-energy accelerators (colliders), phase array radars, and communication systems. The following key problems specify trends in the investigation and construction of these tubes.

1. Enhancement of the pulse output power and pulse duration. This goal requires enlargement of the interaction space and cross section of the output waveguide to avoid microwave breakdown and pulse shortening. This inevitably leads to the problem of mode selection. Note that this problem is more vital for TWTOs than for BWOs because of the need to ensure phase synchronization.

2. The necessity of getting high amplification, which in turn is dictated by the big gap between the gigawatt output power of ultrarelativistic TWTOs and the power of traditional microwave sources as master oscillators. For example, the maximum output of magnetrons in the X-band is on the order of 100 kW. Therefore, to reach 1 GW of output power in a TWTO, the necessary magnitude of gain should be no less than 40 dB. Another important problem for high-gain TWTOs is the

TABLE 8.4 Output Parameters of Single- and Two-Stage 8.8-GHz TWTOs

Type of Tube	Single Stage	Two Stage
Gain (dB)	13–35	28–35
Peak output power (mw)	100	400
Efficiency (%)	11	45
3-dB bandwidth (MHz)	20	150

Source: Shiffler et al. (1997).

self-noise of electron beams generated by explosive emission guns. The high level of this noise limits the low level of the input signal.

Note that if the electron beam in a TWTO is chosen to be synchronous with the $+n$th spatial harmonic and the phase velocity of the zero harmonic is equal to c, the SBS spatial period L is very large for relativistic energies. Indeed, assuming that $\beta_0 = \omega/c$, we find from Eq. (8.15) that

$$L = \frac{2\pi n v_{\text{ph},n}}{\omega(1 - v_{\text{ph},n}/c)} = \frac{n v_{\text{ph},n}}{(1 - v_{\text{ph},n}/c)}\lambda \tag{8.163}$$

Taking into account that the electron velocity is $v_0 \approx v_{\text{ph},n}$, we find that for relativistic velocities the denominator in Eq. (8.163) is $\sim 1/2\gamma^2$, so $L \approx 2n\gamma^2\,\lambda$.

The first relativistic TWTOs had a comparatively high output power, ~400 MW, but they did not use oversized SWSs. For these TWTOs the most important problem was suppressing backward self-excitation. Recall (see Section 8.7.2 and Gittins, 1965) that this is also the key problem for nonrelativistic TWTOs. This problem is usually solved by using severed structures. A comprehensive exploration of single- and two-stage severed amplifiers was carried out by Shiffler et al. (1991). In that work, a tube with frequency 8.8 GHz was operated in the TM_{01} mode, and the electron energy was 850 keV. For a single-stage amplifier, a maximum gain

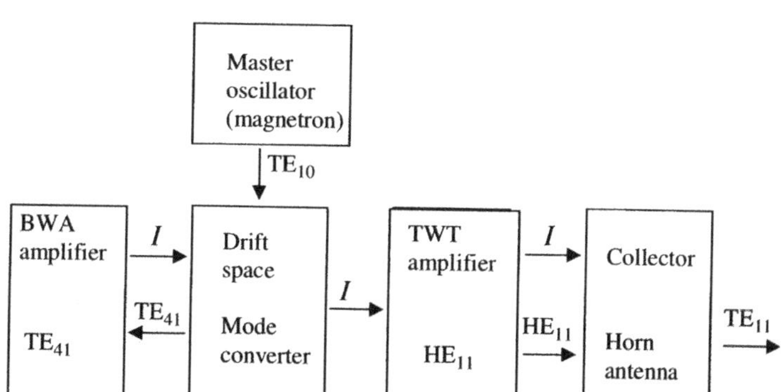

FIGURE 8.32 Two-section TWTO amplifier with mode transformation.

of 33 dB and an output power of 110 MW in the 0.2% band were achieved with a beam current of 1.6 kA. Attempts to operate with higher currents a have led to oscillation at the input frequency. A two-stage TWTO with stages separated by a *sever* (a graphite section with an attenuation of ~30 dB) shows enhanced output power, efficiency, gain, and bandwidth (Table 8.4). The primary disadvantage of this device is the presence of intense oscillations in sidebands. These are always present in severed amplifiers, but in single-stage devices, sidebands appear at power levels in excess of 70 MW. The microwave power in the sidebands can reach up to 50% of the total output power. The results of simulations did not give a definite explanation for this phenomenon. Note, however, that the use of severed structures does not resolve the problem of mode selection in oversized SWSs when synchronous spatial harmonics of different modes have a close-coupling impedance.

A different and very effective method of sectioning ultrarelativistic TWTOs was proposed and realized by Abubakirov et al. (2002). Two sections of a TWTO were formed by an SWS with different azimuthal symmetries and types of interaction (backward and forward). As a result, only electron coupling between sections was present. A block diagram of the amplifier is shown in Fig. 8.32. The first stage is a regenerative backward-wave amplifier (BWA). It operates at the TE_{41} mode of a helical corrugated waveguide. The second stage is an amplifier at a forward synchronous harmonic of the mode HE_{11} in the azimuthally symmetrical corrugated waveguide. The external signal from a 100-kV magnetron enters the quasioptical mode converter, which transforms the TE_{10} mode of a rectangular waveguide into a hybrid whispering gallery mode TE_{41} with a synchronous -1 harmonic. A backward interaction in that section provides transportation of the amplified energy in the direction of the gun. This effect, along with the essentially different spatial structure of TE_{41} and HE_{11} modes, ensured very weak electrodynamic coupling between sections. The specifications for the BWA narrowband of the amplification (see Section 8.8.3) favors mode selection and reducing the noise in a bunched electron beam. Also, the choice of operating in a high whispering gallery mode TE_{41} reinforced the mode selection. The surface structure of operating harmonic determines the sharp dependence of the gain on the initial radius of the electron beam. The electron bunching, coupling of the sections, and gain essentially depend on drift section length.

The maximum for the electron beam current and the energy were 6 kA and 9.8 MeV, respectively. The radiated power of the X-band tube in the optimal operation exceeded 1 GW with a gain of 47 dB and efficiency at 23%. The output power growth was accompanied by pulse shortening. Thus, the increase in output power from 0.3 GW to 1.1 GW reduced the duration of the pulse from 250 ns to 90 ns. Use of the HE_{11} mode as an output wave and application of an adiabatically widening horn antenna obtains an output signal close to a Gaussian wave beam, which can also be considered a result of the high mode selection.

CHAPTER NINE

Crossed-Field Amplifiers and Oscillators (M-Type Tubes)

9.1 INTRODUCTION

Classical high-frequency crossed-field oscillators (magnetrons) have a much longer history than that of O-type traveling wave tubes (TWTOs), which appeared in 1942. However, the fundamental idea of synchronism between traveling waves and electrons in magnetron theory was adopted from TWTOs. The earliest magnetrons (1921) used smooth cylindrical anodes and demonstrated poor output power and efficiency. Later, in 1935, magnetrons with multisegment anodes (which, in fact, appear as closed-loop slow-wave structures) were proposed by Posthumus (1935). These magnetrons radiated 500 MHz with an efficiency of up to 50%. Then in 1938, Alekseev and Malairov developed a prototype of the modern cavity magnetron with a copper anode and hole-slot resonators (Alekseev and Malairov, 1944). The cavity magnetron was invented independently by Boot and Randall in 1940 (Boot and Randall, 1946). By May 1940, experimental radar containing a pulsed-power 3-GHz cavity magnetron was in operation. It would not be an exaggeration to say that magnetron radars played a central role in the defense of Great Britain against German missiles and submarines. By the end of World War II, the magnetron had become the most widely used powerful microwave tube.

Other devices of M type (e.g., injected-beam amplifiers and oscillators, magnetron amplifiers) were proposed later (in the early to mid-1950s) (Brown, 1984; Epsztein, 1952; Warnecke et al., 1950, 1955). Their appearance resulted in significant progress in power microwave electronics and its application, especially in radar technology.

Alongside the development of construction and technology of magnetrons and other crossed-field tubes, great attention has been paid to the theory of these

Electron Beams and Microwave Vacuum Electronics. By Shulim E. Tsimring

devices. Note that despite intense theoretical investigation and the application of large-scale computer simulations, magnetron theory met with significant difficulties, connected primarily with the inherent instability of electron clouds. A possible cause of these difficulties is related to the unique property of the magnetron as a device that is stable only in the large-signal regime (i.e., large ac amplitudes), where the motion of electrons cannot be described by simple equations.

Let us first take up two classes of M-type devices: planar M-type traveling-wave tubes and backward-wave oscillators with an injected beam: MTWTs and MBWOs (carcinotrons). Below we refer to both MTWT and MBWO devices as MTWT unless otherwise noted.

9.2 ELEMENTARY THEORY OF A PLANAR MTWT

9.2.1 Scheme of a Planar MTWT. Electron Beam in DC Crossed Fields

The anode of the tube is implemented as a slow-wave structure (SWS) (see Fig. 9.1, where the comb structure is shown). The electron beam is injected by an M-type electron gun (Appendix 2) into a space with dc electric and magnetic crossed fields. Assume that the undisturbed electron beam is rectilinear. Then the sum of electrical and magnetic forces $e_0E_0 - e_0v_{z0}B$ should be zero. We obtain the undisturbed velocity $v_{z0} = E_0/B$ (the direction of B, shown in Fig. 9.1, is chosen for $v_z > 0$). We assume that nonhomogeneities of the dc electric field formed by the slow-wave structure decay quickly with Δ (Fig. 9.1), so the dc electric field acting on the beam $E_0 = \text{const}$.

Another approach is based on drift equations (Section 1.5.3). Because $v_\perp = 0$, we obtain from Eq. (1.106) the fact that $\mathbf{v} = d\mathbf{R}_d/dt$. Uniformity of the dc magnetic field and the absence of velocity along the magnetic field $v_B = 0$ allow us to reduce Eq. (1.110) to

$$\mathbf{v} = \frac{\mathbf{E}_0 \times \mathbf{B}}{B^2} \quad \text{or} \quad v_{z0} = \frac{E_0}{B} \tag{9.1}$$

Here v_{z0} is the velocity of the guiding center. Thus, Eq. (9.1) is true in the drift approximation, and it allows us to ignore small disturbances in the dc electric field created by the periodic slow-wave structure (SWS).

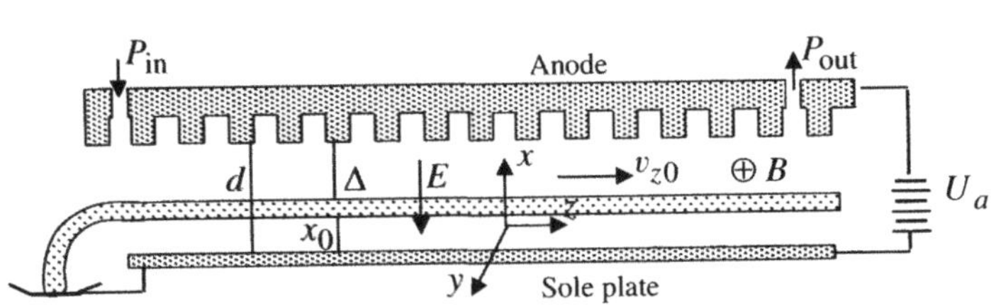

FIGURE 9.1 Planar M-type traveling-wave tube.

Suppose that the beam is thin and we can neglect the dc space-charge field as well. This gives us a simple relation for the beam position. Because v_{z0} is the full velocity, this value is determined by the energy integral $v_{z0} = \sqrt{2\eta U_{x_0}}$, where $U_{x_0} = (U_a/d)\, x_0 = E_0 x_0$ is the potential of the beam relative to the cathode (the sole plate). From Eq. (9.1) we obtain $E_0/B = \sqrt{2\eta E_0 x_0}$. Then

$$x_0 = \frac{E_0}{2\eta B^2} = \frac{U_a}{2\eta B^2 d} \tag{9.2}$$

9.2.2 Motion and Bunching of Electrons in a High-Frequency Field (Adiabatic Approximation)

Assume that some spatial harmonic of the SWS is synchronous with the electron beam. For example, for a comb structure, a zero (direct) spatial harmonic is the best. Let us describe the electron motion in a frame of reference that moves with respect to a laboratory coordinate system with undisturbed velocity v_{z0}. In this frame the undisturbed electrons are not moving. Let us draw qualitatively electrical field lines (Fig. 9.2). We used the following properties of the spatial harmonics (Section 8.4).

1. The field of the nth harmonic is periodic, with the spatial period equal to its wavelength:

$$\lambda_n = \frac{2\pi}{\beta_0 + 2\pi n/L} \tag{9.3}$$

where L is the period of an SWS.

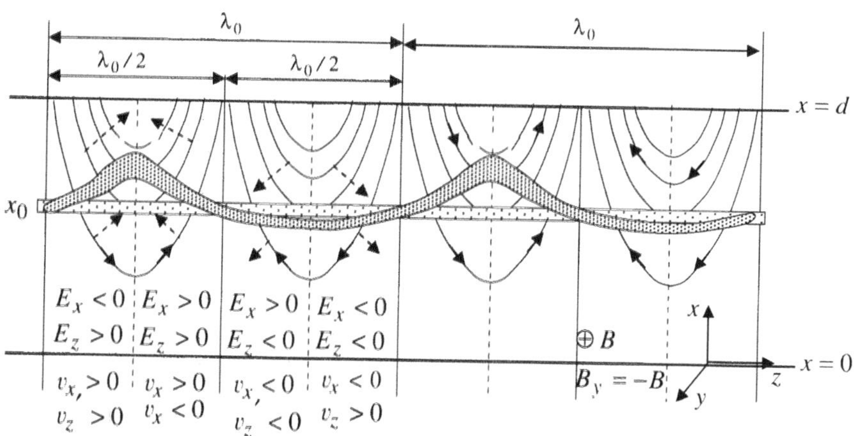

FIGURE 9.2 Electron bunching in crossed fields. Solid arrows, electric field lines; dashed arrows, velocities of the particles; light shading, undisturbed beam; dark shading, disturbed beam.

2. The field of each spatial harmonic satisfies Maxwell's equations. In particular, field lines are either closed curves, or they begin and finish at infinity. In the present example, the field lines are closed around sources outside the interaction space.

3. The synchronous spatial harmonic is the slow wave. Therefore, the latter is a surface wave and its field lines become more rarefied away from the SWS surface. In Fig. 9.2, dashed arrows denote the velocities in different quadrants of the wavelength. Long arrows indicate velocities at the upper boundary of the beam, and short arrows indicate velocities at the lower boundary, where the ac electric field is weaker because of the surface character of the slow wave.

Disturbed Electron Motion Suppose that the high-frequency fields are weakly nonhomogeneous: that is, the scale of nonhomogeneity of the electric field is significantly greater than the characteristic spatial scale of the electron trajectory in the magnetic field. For this case it is the Larmor radius, $r_\perp$. Then we can represent the velocity as a sum of the drift and gyrating velocities:

$$\mathbf{v} = \mathbf{v}_d + \mathbf{v}_\perp \tag{9.4}$$

Taking into account the shortness of the Larmor radius compared with the wavelength, and ignoring the resonance interaction of the high-frequency (HF) field with gyrated electrons,[1] we can ignore the Larmor gyration, as it does not significantly affect the position of particles on the scale of the wavelength. Thus, the influence of electron rotation on bunching and energy transfer is negligibly small. This type of approximation in MTWT theory is called *adiabatic approximation*. So the electron velocity is a pure drift velocity:

$$\mathbf{v} = \frac{\mathbf{E}_0 \times \mathbf{B}}{B^2} + \frac{\mathbf{E} \times \mathbf{B}}{B^2} \tag{9.5}$$

where $\mathbf{E}_0$ is the dc electric field. The first term in this sum is the undisturbed drift velocity, which is equal to zero in the moving frame of reference. In this frame the electron velocity is equal to $\mathbf{v}' = (\mathbf{E} \times \mathbf{B})/B^2$. The electric field has components E_x and E_y. The magnetic field is directed along $-y$ (i.e., $B_y = -B$; see Fig. 9.1). As a result, we obtain

$$v_x = \frac{E_z}{B}, \quad v_z' = -\frac{E_x}{B} \tag{9.6}$$

The z' coordinate in the moving reference frame is

$$z' = z - v_{z0}t \tag{9.7}$$

where $v_{z0} = -E_{x0}/B$.

Disturbed Form of a Beam Let us define the retarded and accelerated phases of the electric field with respect to the longitudinal electron velocity: In the retarded

[1] The interaction of electron oscillators with ac fields is taken up in Chapter 10.

phase, $E_z > 0$, and in the accelerated phase, $E_z < 0$. Based on the analysis of Fig. 9.2 and Eq. (9.6), we reach the following conclusions:

- *Retarded Phase*. The electron beam is expanded and lifted. The potential energy of the electrons decreases in this phase, and the beam is then fed by electrons from the accelerated phase.
- *Accelerated Phase*. The electron beam descends and the potential energy of the electrons increases. The electrons transfer to the retarded phase.

Let us express the components of the electron velocity as functions of the ac field [Eqs. (9.6) and (A5.10) from Appendix 5]:

$$v_x = -\frac{1}{B}\frac{\partial U}{\partial z} = \frac{1}{B}E_z = i\frac{\beta U_d}{B}\frac{sh\beta x}{sh\beta d}e^{i(\omega t-\beta z)}$$
$$v_z' = \frac{1}{B}\frac{\partial U}{\partial x} = -\frac{1}{B}E_x = \frac{\beta_0 U_d}{B}\frac{ch\beta x}{sh\beta d}e^{i(\omega t-\beta z)} \tag{9.8}$$

where $U_d = \int_0^d |E_{0x}(x)|dx$.

Taking Eq. (9.8) into account, it can readily be shown, that electrons in the moving frame move along equipotential lines of the potential $U(x,z)$. Assume for simplicity that static fields and the position of the beam are chosen to ensure exact synchronism between the undisturbed electron beam and the high-frequency field (i.e., $v_{z0} = v_{\text{ph}} = \omega/\beta$, where v_{ph} is the wave phase velocity). Then taking Eq. (9.7) into account, we obtain the fact that in the moving frame the real part of the potential can be written as [see also Eq. (A5.9)]

$$U(x,z') = \frac{U_d}{sh\beta_0 d}\cos\beta z' sh\beta x \tag{9.9}$$

The equipotential lines corresponding to Eq. (9.9) are shown in Fig. 9.3 for $x > 0$. According to Fig. 9.2, the centers of the bunches are located in the planes $\beta z' = (\pi/2) + 2n\pi$. The electrons (guiding centers) in these planes move to the anode. The direction of the electron velocities at these trajectories (the arrows

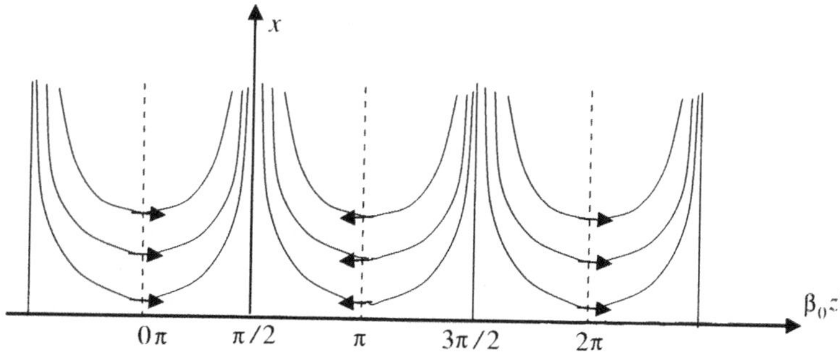

FIGURE 9.3 Equipotentials and trajectories.

in Fig. 9.3) is determined from Eqs. (9.8) and (9.9). Also, in these planes the component E_z is positive (retarded) and maximal. Thus, the bunching and phasing of the electrons takes place in the retarded phase of the synchronous harmonic, and on average, the potential energy of electrons in the beam decreases. It is important that the synchronism is preserved so that the electron kinetic energy does not change.

Note in Fig. 9.3 that only *possible* trajectories are shown. The actual form of the beam is determined by real filling of the trajectories by the guiding centers. The actual positioning of the guiding centers is determined in turn by the lifetime of electrons in the interaction space and by the manner of electron input into this space. In the framework of the linear theory, the electron beam can be considered thin. Qualitatively, its form corresponds to Fig. 9.2.

According to Eq. (9.8) div $\mathbf{v}' = \partial v_x/\partial x + \partial v_z'/\partial z' = 0$. Then we obtain from the continuity theorem, div $\mathbf{j}$ = div $(\rho\mathbf{v}') = \rho$ div $\mathbf{v}' + \mathbf{v}'\nabla\rho = -\partial\rho/\delta t$, that

$$\frac{\partial\rho}{\partial t} + \mathbf{v}'\nabla\rho \equiv \frac{d\rho}{dt} = 0 \tag{9.10}$$

so the space-charge density in crossed-field beams is invariant, and the guiding centers move as particles of incompressible fluid. Assume that the initial space-charge density is constant, $(\rho)_{t=0} = \rho_0$. Then

$$\tilde{\rho} = (\tilde{\rho})_{t=0} = 0 \tag{9.11}$$

in the moving frame.

9.2.3 Comparison of Basic Features of Bunching and Energy Transfer in MTWTs and TWTOs

1. The energy transfer in both TWTOs and MTWTs is controlled by the longitudinal component of the ac electric field. However, TWTO beams exploit kinetic energy, whereas MTWT beams exploit potential energy.

2. In TWTOs, bunching is determined by the longitudinal component of the ac electric field. Bunching in MTWTs is also connected with the longitudinal motion of particles, but here it is controlled by the transverse field component.

3. In TWTOs the transfer of energy from the beam occurs when $v_0 > v_\varphi$. Furthermore, in O-type tubes, efficiency is determined by the deceleration of electrons within a range of excess velocity. In M-type tubes, however, energy is transferred when the electron velocity $v_e \cong v_\varphi$, and it does not change in the process of interaction. Thus, the nature of electron radiation in crossed-field devices is essentially different from that of Cerenkov radiation, and the motion of guiding centers is different from the electron motion seen in the rectilinear beams of O-type devices. However, the synchronism required between an electron beam and a slow wave, as well as the surface character of the wave, brings us very close to the frequency limitations of both TWTMs and TWTOs (see Section 9.4.3).

4. In O-type tubes, bunching is accompanied by an increase in space-charge density, which leads to a decrease in electronic efficiency. The space-charge density in MTWT is not affected by bunching.

5. The active area of the beam is variable in MTWTs but not in TWTOs.

6. Ac fields in TWTOs determine the acceleration rate of particles. Therefore, if particles acquire different velocities (as a result of different accelerations), the velocities remain constant following the release of force, and bunching continues under inertia. This inertial bunching takes place in klystrons, TWTOs, BWOs, and other devices. The motion of particles in MTWTs (and in other M-type devices) is basically a drift motion. In this case, the force determines velocity directly. Therefore, if the force is released, the drift motion stops, as does the bunching. The latter is called *forced bunching* or *M-type bunching*, as opposed to the inertial O-type bunching in O-type devices.

It is interesting to compare the motion of an electron toward the anode in an MTWT with a falling body in the atmosphere, where a gravitational force is balanced by a frictional force. The body moves with constant velocity and loses only potential energy, which is turned into heat. The electrons in MTWTs fall on the anode, and the potential energy is converted into electromagnetic energy. As such, this process can be called *electromagnetic friction*.

9.3 MTWT AMPLIFICATION

The linear theory of amplification can be reduced to the derivation and analysis of dispersion equations. This problem in TWTO theory has been split into two subproblems: calculating the bunching of the electron beam in a wave field, and calculating the excitation of the wave in the slow-wave structure by the ac convection current. The general approach for the analysis in MTWT theory is similar. However, the two-dimensional character of motion and the interaction in MTWTs, and the considerable transversal heterogeneity of the electromagnetic field complicate the analysis, requiring revision of the equations used in Chapter 8. Derivation of the dispersion equation based on the equation of excitation of equivalent transmission lines with lumped parameters is given by Gittins (1965), Kleen (1958), and Mourier (1961). A consistent electrodynamic analysis has been developed by Gaiduk (1971) and Hutter (1965).

9.3.1 Equation of Excitation

Let us use the formula for excitation of the waveguide (Vainstein, 1956), ignoring the spacecharge and using only one waveguide eigenmode (one fundamental spatial harmonic), $\mathbf{E}_1(x,z)$:

$$E_x(x,z,t) = CE_{1x}(x,z,t) \tag{9.12}$$

$$E_z(x,z,t) = CE_{1z}(x,z,t) \tag{9.13}$$

where

$$C = \frac{1}{4P} \int_{z_1}^{z} dz \int_{S_p} [j_z(x,z,t) E_{1z}^*(x,z,t) + j_x(x,z,t) E_{1x}^*(x,z,t)] d\sigma \tag{9.14}$$

$$P = \frac{1}{2} \mathrm{Re} \int_{S_p} (\widetilde{\mathbf{E}}_1 \times \widetilde{\mathbf{H}}_0^*)_z \, dS \tag{9.15}$$

is the real power transmitted by the undisturbed wave in the direction of the z-axis. S_p is the beam cross-sectional area. Further, we use the notation

$$\mathbf{j}(x,z,t) = \tilde{\mathbf{j}}(x) e^{i(\omega t - \beta z)}, \quad \mathbf{E}_1(x,z,t) = \widetilde{\mathbf{E}}_1(x) e^{i(\omega t - \beta_0 z)} \tag{9.16}$$

where variables with a tilde ($\sim$) denote complex amplitudes. Substituting Eqs. (9.16) into Eq. (9.14), we obtain

$$\begin{aligned} C &= \frac{e^{i\omega t}}{4P} \int_{z_1}^{z} e^{i(\beta_0 - \beta)z} dz \int_{S_p} [\tilde{j}_z(x) \widetilde{E}_{1z}^*(x) + \tilde{j}_x(x) \widetilde{E}_{1x}^*(x)] \, d\sigma \\ &= \frac{e^{i\omega t}}{4iP(\beta_0 - \beta)} e^{i(\beta_0 - \beta)z} \int_{S_p} [\tilde{j}_z(x) \widetilde{E}_{1z}^*(x) + \tilde{j}_x(x) \widetilde{E}_{1x}^*(x)] \, d\sigma \end{aligned} \tag{9.17}$$

Next we use the factor

$$C_1 = \frac{1}{4iP(\beta_0 - \beta)} \int_{S_p} [\tilde{j}_z(x) \widetilde{E}_{1z}(x) + \tilde{j}_x(x) \widetilde{E}_{1x}(x)] \, d\sigma \tag{9.18}$$

Then substituting Eq. (9.17) into Eqs. (9.12) and (9.13) and using Eq. (9.18), we obtain

$$\widetilde{E}_x(x) = C_1 \widetilde{E}_{1x}(x) \tag{9.19}$$

$$\widetilde{E}_z(x) = C_1 \widetilde{E}_{1z}(x) \tag{9.20}$$

Let us introduce perturbations $\tilde{\mathbf{r}}$ of the coordinates $\mathbf{r} = \mathbf{r}_0 + \tilde{\mathbf{r}}(x) e^{i(\omega t - \beta z)}$. In the framework of the linear theory, these perturbations are considered small. We find $\tilde{\mathbf{r}}$ according to Eq. (9.6):

$$v_x = \frac{E_z}{B} = \frac{dx}{dt} = \frac{\partial x}{\partial t} + v_0 \frac{\partial x}{\partial z} = i(\omega - v_0 \beta) x = i v_0 (\beta_e - \beta) x \tag{9.21}$$

Then, using Eqs. (9.8), we obtain the amplitude of the perturbations:

$$\tilde{x}(x) = \frac{1}{iv_0 B(\beta_e - \beta)} \widetilde{E}_{1z}(x) = \frac{U_d}{U_a} \frac{\beta d}{\beta_e - \beta} \frac{sh\beta x}{sh\beta d} \tag{9.22}$$

Similarly,

$$\tilde{z}(x) = -\frac{1}{iv_0 B(\beta_e - \beta)} \widetilde{E}_{1x}(x) = -\frac{U_d}{U_a} \frac{i\beta d}{\beta_e - \beta} \frac{ch\beta x}{sh\beta d} \tag{9.23}$$

It has been shown by Gaiduk (1971) that the constant C_1 can be expressed through amplitudes $\tilde{x}$ and $\tilde{y}$:

$$C_1 = -\frac{\rho_0 \omega}{4i\beta_0 P(\beta - \beta_0)} \int_{S_p} \left[-i\beta_0 \tilde{z}(x) \widetilde{E}^*_{1z} + \tilde{x}(x) \frac{\partial \widetilde{E}^*_{1z}}{\partial x} \right] d\sigma \tag{9.24}$$

It is important that the integral in Eq. (9.24) be taken over the undisturbed beam cross section. Using Eqs. (A.5.7) and (A.5.8) and equality $\widetilde{E}^*_{1z}(x) = -\widetilde{E}_{1z}(x)$, we obtain

$$C_1 = -\frac{\rho_0 \omega}{4P(\beta - \beta_0)} \int_{S_p} (\tilde{z}(x) \widetilde{E}_{1z} - \tilde{x} E_{1x}(x))\, d\sigma \tag{9.25}$$

9.3.2 Dispersion Equation

Let us rewrite the integrand in Eq. (9.25) using Eqs. (9.19), (9.20), (9.22), and (9.23):

$$\tilde{z}\widetilde{E}_{1z} - \tilde{x}\widetilde{E}_{1x} = -C_1 \frac{2\widetilde{E}_{1x}(x)\widetilde{E}_{1z}(x)}{iv_0 B(\beta_e - \beta)} \tag{9.26}$$

Substituting Eq. (9.26) into the integral in Eq. (9.25), we find that

$$C_1 = -\frac{\rho_0 \omega i d}{2P(\beta - \beta_0)(\beta_e - \beta) U_a} \int_{S_p} C_1 \widetilde{E}_{1x}(x) \widetilde{E}_{1z}(x)\, d\sigma \tag{9.27}$$

Canceling out C_1 and noting that $\widetilde{E}_{1x} = i\widetilde{E}^*_{1z} \coth \beta_0 x_0$ [Eqs. (9.22) and (9.23)], we obtain the dispersion equation

$$(\beta - \beta_0)(\beta_e - \beta) = \frac{\beta_0^2 I_0 \beta_e d}{U_a} Z_M \tag{9.28}$$

where $Z_M = (|\widetilde{E}^2_{1z}|/2\beta_0^2 P) \coth \beta_0 x_0$ is the coupling impedance. This value differs from the coupling impedance used in TWTO theory by a factor of $\coth \beta_0 x_0$.

The dispersion equation accepts the form

$$(\beta - \beta_0)(\beta_e - \beta) = \beta_0^2 D^2 \tag{9.29}$$

where

$$D^2 = \frac{I_0 \beta_e d}{U_a} Z_M \tag{9.30}$$

9.3.3 Gain of an MTWT

First we introduce dimensionless parameters similar to those used in TWTO theory (Section 8.5.5):

- Nonsynchronism parameter

$$b = \frac{\beta_0 - \beta_e}{\beta_e D} \tag{9.31}$$

- Incremental propagation constant

$$\delta = -i \frac{\beta - \beta_e}{\beta_e D} \tag{9.32}$$

The dispersion equation (9.29) accepts the form $\delta(\delta + ib) = (1 + bD)^2$ or

$$\delta(\delta + ib) = 1 \tag{9.33}$$

for $b \ll 1$ (i.e., when the tube operates near synchronism). Equation (9.33) has two roots, corresponding to two waves that are the result of coupling of the beam with a slow-wave structure wave. In general, roots of Eq. (9.33) are complex-conjugate roots. As such, one wave will be growing exponentially and another will be decaying exponentially. The general solution is determined by the sum

$$\widetilde{\mathbf{E}} = \widetilde{\mathbf{E}}_1 e^{-i\beta_1 z} + \widetilde{\mathbf{E}}_2 e^{-i\beta_2 z} \tag{9.34}$$

where $\beta_{1,2}$ corresponds to the two roots of Eq. (9.33). The amplitudes $\widetilde{\mathbf{E}}_{1,2}$ can be found from boundary conditions. Assume, for example, that the beam is unmodulated at the input plane ($z = 0$) and that the input signal is given as

$$\tilde{x}(0) = 0 \quad \text{and} \quad \widetilde{E}_z(0) = E_0 \tag{9.35}$$

According to Eqs. (9.22) and (9.32),

$$iv_0 \, B\tilde{x}(0) = \frac{\widetilde{E}_z(0)}{\beta_e - \beta} = \frac{i\widetilde{E}_z(0)}{\beta_e D\delta} \tag{9.36}$$

Taking Eq. (9.35) into account, we obtain

$$\begin{aligned} \widetilde{E}_{z1}(0) + \widetilde{E}_{z2}(0) &= E_0 \\ \frac{\widetilde{E}_{z1}(0)}{\delta_1} + \frac{\widetilde{E}_{z2}(0)}{\delta_2} &= 0 \end{aligned} \tag{9.37}$$

so

$$\widetilde{E}_{z1}(0) = \frac{E_0}{1 - \delta_2/\delta_1}, \qquad \widetilde{E}_{z2}(0) = \frac{E_0}{1 - \delta_1/\delta_2} \tag{9.38}$$

Consider, for example, the case of rigorous synchronism $b = 0$ ($\beta_e = \beta_0$). The roots of Eq. (9.33) are $\delta_{1,2} = \pm 1$. According to Eq. (9.38),

$$\widetilde{E}_{z1}(0) = \widetilde{E}_{z2}(0) = \frac{E_0}{2} \tag{9.39}$$

The amplification coefficient K is determined by the growing wave with $\delta = \delta_1 = +1$ and according to Eqs. (9.32) and (9.34) is

$$K = \tfrac{1}{2} e^{-i\beta_1 x} = \tfrac{1}{2} e^{-i\beta_e x} e^{\beta_e D x} \tag{9.40}$$

The gain is

$$G = 20 \log|K| \approx -6 + 55DN \quad dB \tag{9.41}$$

where N is the number of wavelengths along the interaction length. The initial 6-DB loss corresponds to the splitting of the signal into two waves. Note that the gain is proportional to $I_0^{1/2}$ [see Eq. (9.30)], in contrast with TWTO, where $G \sim I_0^{1/3}$.

The general case of arbitrary nonsynchronism as represented in Eq. (9.33) is illustrated in Fig. 9.4*a*, where real and imaginary parts of the incremental propagation constant $\delta = x + iy$ are shown as functions of b. As seen in the figure, the maximum gain becomes $\beta_e = \beta_0$. This value corresponds to Eq. (9.41). Amplification takes place for $-2 \leq b \leq 2$. The propagation constants for $|b| < 2$ (in the zone of amplification) are the same for both waves. Since in this zone $y_{1,2} = -b/2$, we find that the two waves travel with equal velocities. This velocity is intermediate between the undisturbed phase velocity and the beam velocity.

9.3.4 Effects of Space Charge and Losses

The space charge increases the order of the dispersion equation. This is the result of oscillations of electrons under the space-charge effect exciting space-charge waves which are coupled with circuit waves. According to Gould (1957), taking the space charge and losses into account, the dispersion equation is

$$(\delta + ib + d)(\delta^2 + 2igS\delta - S^2) = \delta \tag{9.42}$$

where the space-charge parameter is

$$S = \frac{\omega_p^2}{\omega\omega_g D} \tag{9.43}$$

and the geometric parameter is

$$g = \frac{\tanh \beta_e(d - x_0) - \tanh \beta_e x_0}{\tanh \beta_e(d - x_0) + \tanh \beta_e x_0} \tag{9.44}$$

The loss parameter is

$$d = \frac{\beta_d}{\beta e D} \tag{9.45}$$

where β_d is an imaginary attenuation correction factor of the real propagation constant. For the beam in the middle, between the sole and the anode (the slow-wave structure), $x_0 = d/2$ and $g = 0$. In this case the dispersion equation is

$$(\delta + ib + d)(\delta^2 - S^2) = \delta \tag{9.46}$$

The real and imaginary parts of the incremental propagation constant δ as functions of b, following from Eq. (9.46) for $S = 1$ (moderate space-charge parameter), are shown in Fig. 9.4*b*. Comparing Fig. 9.4*a* and *b* leads us to conclude that *space charge increases gain*, independent of the sign of b, even for big values of $|b|$. This property of MTWT is different from that of TWTOs and corresponds to the difference in signs of the space-charge terms in Eqs. (9.46) and (8.62), According to Fig. 8.9, the growth of the space charge for TWTOs leads to a narrowing of the amplification zone. Note that according to Fig. 9.4*b*, the imaginary parts of $\delta_{1,2}$ are equal and small. So the phase velocities of both waves (growing and attenuated) are close to the undisturbed electron velocity. We see that the presence of the positive factor d in Eq. (9.46) always decreases the magnitude of the real part of δ and as a result, the gain.

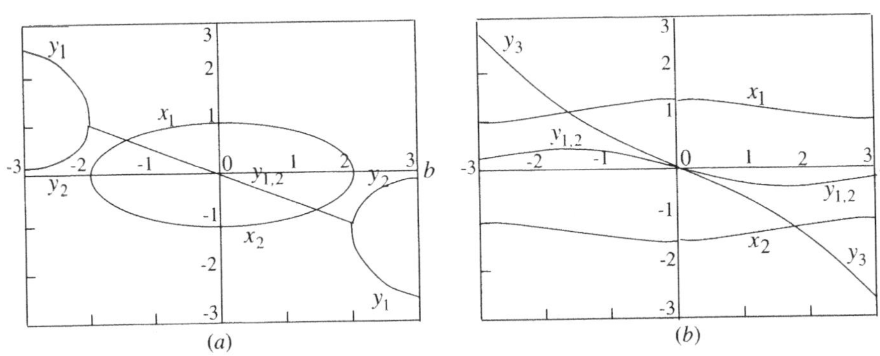

FIGURE 9.4 Real and imaginary parts of an incremental propagation constant as functions of nonsynchronous parameter b for the beam in the middle between the anode and the cathode: (*a*) space-charge parameter $S = 0$; (*b*) $S = 1$. (From Gould, 1957.)

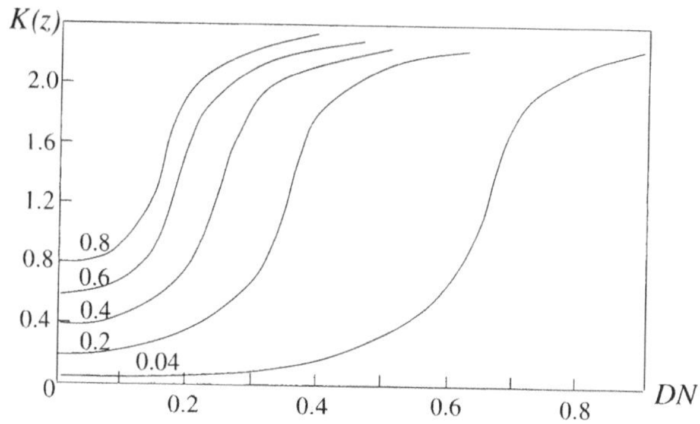

FIGURE 9.5 Amplitude of a normalized ac field as a function of the normalized distance along the MTWT for various values of the input field.

9.3.5 Nonlinear Gain of an MTWT

The nonlinear theory of MTWT was developed in a number of works (see, e.g., Feinstein and Kino, 1961; Gandhi and Rowe, 1961; Sedin, 1961. The theory takes into account the effects of space charge and circuit loss. However, complicated nonlinear equations admit only numerical solutions.

Here we consider briefly the results of a numerical nonlinear analysis of amplification in MTWT, ignoring space-charge effects. This analysis was carried out by Feinstein and Kino (1961). The growth of the ac field $K(z) = \widetilde{E}(z)/E_0 D$ as a function of DN for various values of $\widetilde{E}_z(0)$ is plotted in Fig. 9.5 ($\widetilde{E}$ is the output amplitude of the ac field, E_0 is the dc electric field in the interaction space, and N is the number of wavelengths along the tube). We see that the rate of signal growth along the tube is initially small and increases when the beam approaches the anode and enters a region with a strong ac field. Further, as the beam is gradually collected at the anode, the rate of growth decreases until the output becomes constant. The theory indicates one essential difference in the operation of MTWTs and TWTOs: Below saturation level the gain in MTWTs can increase with the input signal. For example, according to Fig. 9.5, when the input field $\widetilde{E}_1(0) = 0.04E_0D$, the power gain for distance $DN = 0.35$ is ~ 10. But when $\widetilde{E}_1(0) = 0.2E_0D$, the power gain at the same distance is ~ 30. The reason for this behavior is that as the input level increases, a large part of the beam is raised nearer the anode, where the ac field is stronger, so the *effective impedance* seen by the beam is increased.

9.3.6 Efficiency of an MTWT

According to Figs. 9.2 and 9.3, electrons in the center plane of the bunch do not change their longitudinal velocities in the process of interaction with an RF field and move to the anode. Here we assume that all electrons of the beam have bunched in the retarded phase. The electrons begin moving from level x_0, where they have energy $e_0 U_{x_0} = e_0 (U_a/d) x_0$. If the electron bunches in the process of interaction approach the anode, the kinetic energy of all electrons at the anode is the same. But if the electrons arrive at the anode moving in the dc electric field, their kinetic energy is e_0U_a. The energy difference

$$\Delta W = e_0U_a - e_0U_{x0} = eU_a\left(1 - \frac{x_0}{d}\right)$$

transfers to the electromagnetic field. Thus, the electronic efficiency is

$$\eta_e = \frac{\Delta W}{e_0U_a} = 1 - \frac{x_0}{d} \tag{9.47}$$

This relation ensures high efficiency, in principle. For example, $\eta_e = 90\%$ when $x_0/d = 0.1$. The real electronic efficiency is significantly less. First, not all electrons turn out in the retarded phase; some of the electrons appear in the accelerated phase and are collected at the sole. Second, the real undisturbed beam is not strictly rectilinear. The initial beam can be rippled, with the dimension of ripples approximately equal to the Larmor radius ρ. Because the drift to the anode does not change ρ, the electrons are collected by the anode when they are at the top of the cycloid, where their velocity is $v = v_0 + \omega_g\rho$. Then Eq. (9.47) should be replaced by

$$\eta_e = 1 - \frac{x_0}{d}\left(1 + \frac{\omega_g\rho}{v_0}\right)^2 \tag{9.48}$$

The relation (9.48) can be represented as [see Eq. (9.2)]

$$\eta_e = 1 - \frac{1}{4}\left(1 + \frac{\omega_g\rho}{v_0}\right)^2\left(\frac{B_c}{B}\right)^2 \tag{9.49}$$

where $B_c = \sqrt{2U_a/\eta}/d$ is the minimal magnetic field that provides the cutoff for the anode current and forms a magnetic isolation in the planar diode. The efficiency η^e is called the *saturated efficiency* because it is assumed that the tube is sufficiently long to collect all the electrons at the anode.

Saturation efficiency in TWTOs is determined by the maximum permissible excess velocity (Section 8.6.2) and the space-charge effect. As a result, the efficiency of MTWT is significantly higher. Typical parameters for both types of tube are given in Table 9.1.

The high efficiency, output power, broad bandwidth, and compactness of M-type amplifiers (especially in pulse operations) make them ideal for a variety of applications. A very important application of MTWTs, in particular, is related with radar-phased antenna arrays.

TABLE 9.1 Comparative Parameters of TWTM and TWTO

Amplifier	Gain (dB)	Bandwidth (%)	Efficiency (%)
TWTM	15–25	25–35	40–60
TWTO	20–60	10–100	5–35

Sources: Mourier and Okress (1961).

9.4 M-TYPE INJECTED BEAM BACKWARD-WAVE OSCILLATORS (MWO, M-CARCINOTRON)

9.4.1 Introduction

The two types of backward-wave oscillators (BWOs and MBWOs) were invented almost simultaneously in 1952: the BWO by R. Kompfner and the MBWO by the R. Warnecke team. Despite differences in bunching processes and energy transfer from the beam to the electromagnetic field, the two oscillators are based on the same fundamental idea: synchronous interaction of an electron beam with a negative spatial harmonic of a slow-wave structure. As a result, the two carcinotrons have similar operating principles. In particular, as in the theory of BWOs [Eqs. (8.65*a*) and (8.135)], the dispersion equation for MBWOs can be derived from the dispersion equation of MTWTs [Eq. (9.42)] simply by replacing the signs on the right side of the equation and the loss parameter d, due to the reverse propagation of energy. Here we consider the midway position of the beam between the sole and the anode, as shown in Fig. 9.6. According to Eq. (9.46), the dispersion equation of the MBWO is given as

$$(\delta + ib - d)(\delta^2 - S^2) = -\delta \tag{9.50}$$

where $\delta = -i[(\beta - \beta_e)/\beta_e D]$, and the parameter of nonsynchronism $b = (\beta_0 - \beta_e)/\beta_e D$.

9.4.2 Starting Conditions of an MBWO

With no space charge or loss, Eq. (9.50) is reduced to a second-degree equation:

$$(\delta + ib)\delta = -1 \tag{9.51}$$

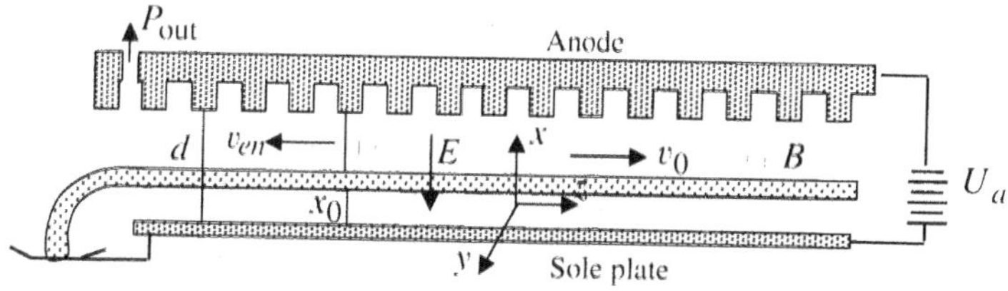

FIGURE 9.6 Planar M-type backward oscillator.

We obtain two waves with purely imaginary δ:

$$\delta_{1,2} = -i\left(\frac{b}{2} \pm \sqrt{\frac{1+b^2}{4}}\right) \tag{9.52}$$

As a result, exponentially growing waves are absent. The general wave solution is thus

$$\widetilde{\mathbf{E}} = \widetilde{\mathbf{E}}_1 e^{-i\beta_1 z} + \widetilde{\mathbf{E}}_2 e^{-i\beta_2 z} \tag{9.53}$$

The amplitudes $\widetilde{\mathbf{E}}_{1,2}$ can be found from boundary conditions similar to Eq. (9.35) for the TWTM:

$$\widetilde{E}_z(0) = E_0, \qquad \tilde{x}(0) = 0 \tag{9.54}$$

This gives us

$$\widetilde{E}_{1z}(0) = \frac{\widetilde{E}_0}{1 - \delta_2/\delta_1}, \qquad \widetilde{E}_{2z}(0) = \frac{\widetilde{E}_0}{1 - \delta_1/\delta_2} \tag{9.55}$$

Let us assume that the tube is matched at both ends, so that reflections are absent. Then the field at the collector end ($z = L$) should be equal to zero under conditions of self-excitation. Substituting Eq. (9.55) in Eq. (9.53), we obtain for the z-component of the field the condition

$$\widetilde{E}_z(L) = \frac{\widetilde{E}_0}{\delta_1 - \delta_2}\left(\delta_1 e^{-i\beta_1 L} - \delta_2 e^{-i\beta_2 L}\right) = 0 \tag{9.56}$$

Hence,

$$\frac{\delta_1}{\delta_2} = e^{i(\beta_1 - \beta_2)L} = e^{\beta_e DL(\delta_2 - \delta_1)} = e^{i2\beta_e DL\sqrt{1+b^2/4}} \tag{9.57}$$

Separating real and imaginary parts of Eq. (9.56) or (9.57) and using Eq. (9.52), we obtain

$$b = 0, \qquad \delta_1 = -\delta_2 = -i, \qquad \cos 2\beta_e D_{st} L = -1 \quad \text{or}$$

$$D_{st} N = \frac{2n-1}{4}, \quad n = 1, 2, \ldots \tag{9.58}$$

Here $\beta_e L = 2\pi N$ and N is the length of the slow-wave structure in wavelengths. We see that the undisturbed phase velocity in the starting regime equals the beam velocity. This property provides electronic tuning of the carcinotron. Note that disturbed phase velocities of both waves, which are determined by $\delta_{1,2}$, are not equal to the electron velocity.

Using the definition of D [Eq. (9.30)], we obtain the starting current

$$I_{st} = \frac{(2n-1)^2 U_a}{16 N^2 \beta_e dZ_M} = \frac{(2n-1)^2 U_a L}{32\pi N^3 dZ_M}, \qquad n = 1, 2, \ldots \tag{9.59}$$

According to this relation, there are several orders of oscillation at the same frequency, and the starting current increases rapidly with n. For example, $I_{st}^{(2)} = 9I_{st}^{(1)}$.

Taking the space charge into account leads to the dispersion equation (9.50) with three waves. Starting conditions in this case are determined by an equation of Gould (1957):

$$\sum_{\substack{i=1 \\ j,k=1-3}}^{3} \frac{\delta_i^2 - S^2}{(\delta_i - \delta_j)(\delta_i - \delta_k)} e^{2\pi D N \delta_i} = 0 \tag{9.60}$$

This equation follows from the analogous equation for BWOs [Eq. (8.133)] if one replaces q by $-S^2$ and C by D. The change of sign in the space-charge term indicates an opposite effect of the space charge on the starting current, which decreases with S. The starting current and the factor $(\beta_0 - \beta_e)L = 2\pi NDb$ are shown as functions of S in Fig. 9.7.

As shown, the starting current decreases considerably with S. Starting currents of oscillations with $n > 1$ decrease much more. So for zero space charge, the ratio $I_{st2}/I_{st1} = 9$. For large S, $I_{st2}/I_{st1} \sim 3$ to 5. Note that for $S = 0$, oscillations with any n have the same value as the parameters $b_n = 0$, but with an increase in S, values of b_n become nonzero and different. As a result, frequencies of oscillations with different n are not the same ($\Delta f \sim 1\%$), and for the beam current $I > I_{st2}$, the radiation spectrum of the M-carcinotron becomes worse.

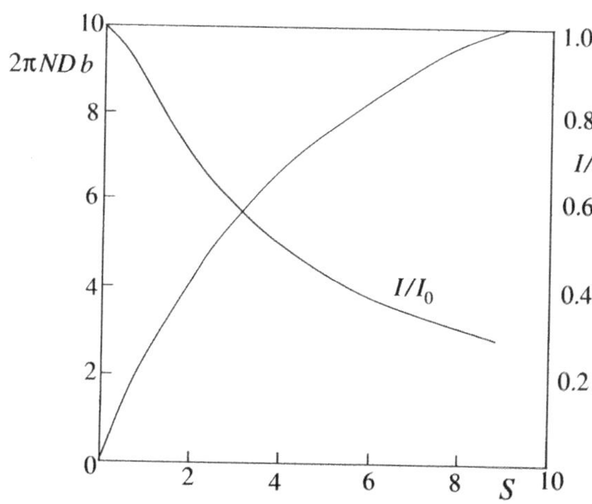

FIGURE 9.7 Starting current of an MBWO and a nonsynchronism parameter as functions of a space-charge parameter.

9.4.3 Large-Signal Effects

Despite the smallness of the ac field near the collector end seen in all carcinotrons, the choice of proper tube length and magnitude of the beam current in M-carcinotrons permit collection of beam electrons at the anode with high saturation efficiency, according to Eq. (9.49). Thus, in contrast to BWOs, where the saturation is fundamentally different (see Section 8.6.2) and the negative effect of the space charge is very strong, MBWOs are highly effective devices and can be designed as powerful oscillators.

The analysis of MBWOs for currents greater than the starting current is a topic of interest within the nonlinear theory. Feinstein and Kino (1961) considered models for MTWTs and MBWOs with a thin beam, ignoring the space charge. It is interesting that space-charge effects for M-carcinotrons do not play as important a role in the large-signal mode as they do in small-signal operation. This is because ac fields are stronger, and they suppress space-charge fields. The amplitude of the normalized ac field as a function of the normalized distance along the tube is shown in Fig. 9.8. If we take into account the same full lengths of the SWS, different curves in Fig. 9.8 correspond to different beam currents. The output power grows quickly with the current ($\sim E_{\text{out}}^2$). One interesting feature predicted by these curves is that in large-signal mode the same output power should be obtained with a tube length shorter than the starting value $D_{\text{st}}N = 0.25$. This follows from small-signal theory [see Eq. (9.58)]. This effect is similar to increasing gain in MTWT by increasing the signal level (Fig. 9.5). Obviously, under large-signal conditions, the beam is raised toward the anode and the effective impedance of the slow-wave structure increases. One might expect that a tube with $DN < 0.25$ would not be self-running, but the application of external power would start oscillations, which would then continue after the external power is removed. This effect is typical for generators with hard excitation.

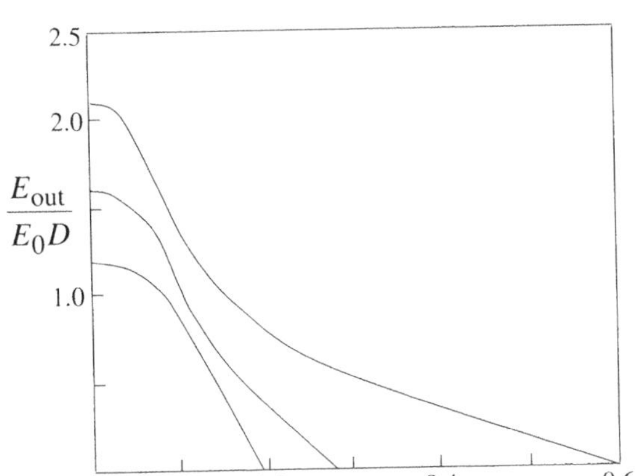

FIGURE 9.8 Normalized amplitude of an ac output field of an M-carcinotron as a function of normalized distance along a tube for various beam currents.

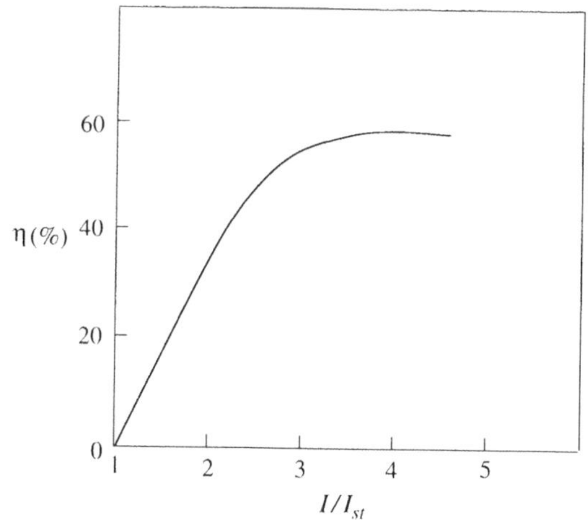

FIGURE 9.9 Efficiency of an M-carcinotron as a function of the ratio of the operation current to the starting current. (Experimental data from Doehler, 1961.)

According to the theory and confirmed by experimental data, the efficiency of MBWOs grows almost linearly with beam current and becomes constant for large currents (saturation effect). The experimental efficiency as a function of current for an L-band M-carcinotron is shown in Fig. 9.9.

9.4.4 Parameters and Construction of an MBWO

Typical values of parameters for M-carcinotrons include the following: Output power is a few hundred watts between the X and S bands in continuous-wave (CW) operations and a few megawatts in pulsed operations; the voltage frequency tuning is over 20 to 50%; and the efficiency approaches 60%, depending on the frequency band and the operating voltage. These tubes operate on frequencies from 300 MHz to 35 GHz.

The output power and efficiency of MBWOs both decrease sharply in the millimeter range. The cause is the same as that seen in BWOs: the interaction of electrons with slow waves that are surface waves. This factor is even more important for MBWOs because electrons in the skin layer, of width $\Delta = (1/2\pi)(v_\varphi/c)\lambda$ (see Section 8.4.2), must move in the transverse direction and deliver the potential energy. For $\lambda = 1$ cm and $U_a = 5\,\text{kV}, \Delta \approx 0.2\,\text{mm}$. The maximum CW power for a 35-GHz MBWO is ~10 W (Feinstein and Kino, 1961). Thus, the frequency range of powerful M-carcinotrons is limited by frequency bands L to X.

Three versions of M-carcinotrons and MTWTs are used: planar (Figs. 9.1 and 9.6), cylindrical, and coaxial. The cylindrical version (Fig. 9.10*a*) is more compact and is used more often, especially with M-carcinotrons. In cylindrical MTWTs, maximum gain is limited by self-excitation because of stronger parasitic feedback between the input and output of the tube. A photograph of a pulse M-carcinotron in the S-band is shown in Fig. 9.11. In coaxial designs, a magnetic

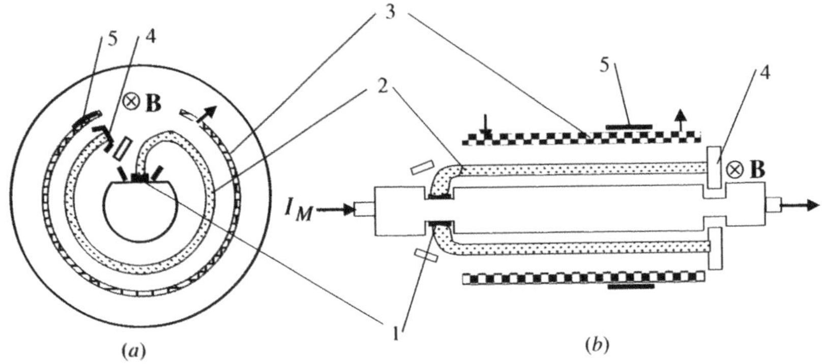

FIGURE 9.10 M-carcinotron: (*a*) cylindrical construction; (*b*) coaxial M-traveling-wave tube. 1, Cathode; 2, electron beam; 3, slow-wave structure; 4, collector; 5, absorber; I_M, current of magnetization.

field is usually created by a strong axial current I_M fed by a special pulse source through the central rod (Fig. 9.10*b*). The current of magnetization is on the order of kiloamperes. The coaxial M-type devices are rarely applied.

M-type backward devices are also used as M-type backward amplifiers. Naturally, in these cases the electrons are synchronous with some backward spatial harmonic, which is fundamental for SWSs with negative dispersion (e.g., interdigital SWSs) or -1 for SWSs with positive dispersion (e.g., with comb structure). Sometimes, MBWOs are used as synchronizing oscillators. In

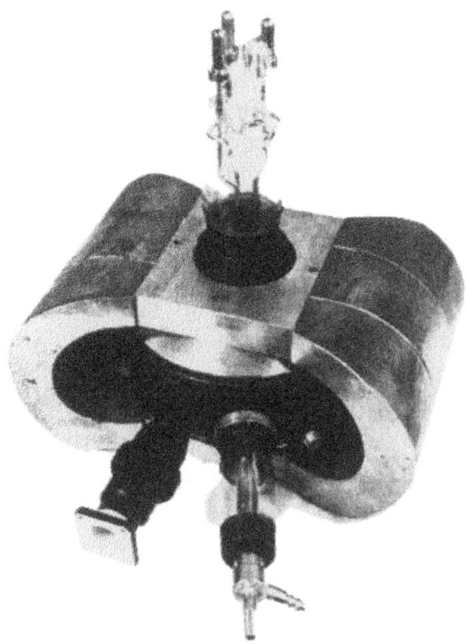

FIGURE 9.11 S-band pulse cylindrical M-carcinotron with an output power of 150 kW.

these cases, the current is set higher than the starting current, but a strong input signal is applied to the collector end of the tube. A synchronized MBWO ensures narrowband amplification with controlled frequency.

The high power, efficiency, and very broad voltage tuning (sometimes up to an octave) of MBWOs ensure their effective application in radar countermeasure systems.

9.5 MAGNETRONS

The scheme of a magnetron is shown in Fig. 9.12. The following characteristics distinguish the magnetron from M-type crossed-field tubes with injected electron beams considered above.

- Cylindrical construction only.
- The reentrant electron beam is emitted from the negative electrode (as opposed to the sole electrode in injected beam devices) into the interaction space. The electron beam, without ac electromagnetic fields and in a stable state, forms an axisymmetrical cloud.
- The reentrant anode periodic structure forms a resonator.

9.5.1 Electromagnetic Field in a Magnetron Resonator

If we mentally cut and straighten the structure in Fig. 9.12, we obtain a regular slow-wave structure (if extended to infinity). The dispersion curve of the fundamental spatial harmonic as dependent on the delay factor of the wavelength for a structure with positive dispersion (see Section 8.4.2 and Appendix 4) is depicted in Fig. 9.13.

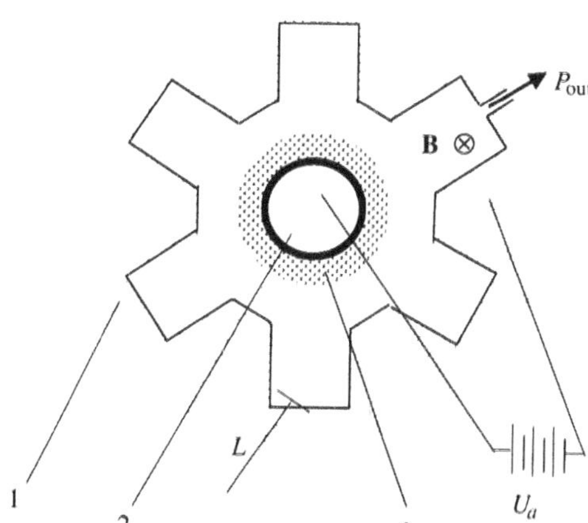

FIGURE 9.12 Six-resonator magnetron. 1, Anode with resonators; 2, cathode; 3, electron beam.

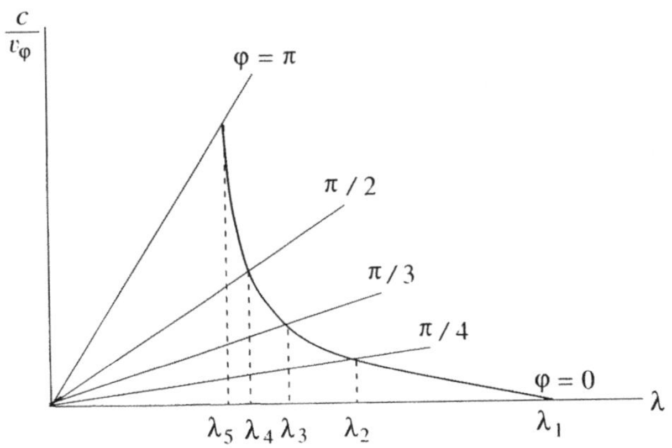

FIGURE 9.13 Dispersion curve of a structure with positive dispersion and the spectrum of an eight-cavity resonator.

Recall that since $\beta_{0,\max} = \pi/L$ [Eq. (8.17)], the boundary of the bandwidth is determined by the relation

$$\lambda_{g,\min} = \frac{2\pi}{\beta_{0,\max}} = 2 \qquad (9.61)$$

where λ_g is the wavelength in the waveguide and L is a period of the structure. Now let us restore the initial structure by bending and closing the ends. Because the phases of the ends must coincide, phase progression along the revolution length must be a multiple of 2π. Then the full length of the structure $\Lambda = NL$ (N is a number of the cavities in the magnetron resonator) must be equal to an integer multiple of λ_g:

$$NL = n\lambda_g, \qquad n = 0, 1, 2, \ldots \qquad (9.62)$$

According to Eq. (9.61), $\lambda_{g,\min} = 2L$. Hence, the number of eigenmodes in the resonator will be equal to $n_{\max} \sim N/2 + 1$. If N is even, $n_{\max}$ is precisely equal to $N/2 + 1$ (i.e., $n = 0, 1, 2, \ldots, N/2$). The dispersion curve in Fig. 9.13 corresponds to an eight-cavity magnetron. The phase shift per period for the shortest wavelength is $\varphi = \pi$. This is referred to as the π-*mode* of operation. Fields in neighboring cavities for the π-*mode* oscillate in the antiphase without energy exchange with their neighbors. Even in an open (i.e., nonclosed) transmission line, forward and reverse traveling waves coalesce and are indistinguishable in the π-mode. The dispersion curve passes through the stationary point corresponding to the zero group velocity. At this point, the characteristic impedance for a loss-free line is infinite. For mode $n = 0$, $\varphi = 0$ and the fields in all cavities are co-phased.

It can readily be shown that the spectrum of the resonator on each spatial harmonic belonging to the same mode is the same. Assume that the field of the mth harmonic satisfies the closing condition $NL = n\lambda_{g,m}$. Because according to Eq. (8.14), $\lambda_{g,m} = 2\pi/(\beta_0 + 2\pi m/L)$ where $\beta_0 = 2\pi/\lambda_g$, we obtain $NL(\beta_0 + 2\pi m/L) = 2\pi n$

or $NL = (n - mN)\lambda_g$. We arrive at the same formula: Eq. (9.62). Thus, the entire spectrum of spatial harmonics is represented within each oscillation mode.

In reality there are $(N/2) + 1$ other oscillations in the resonator, because the same set of oscillations can be obtained from waves that propagate opposite to the transmission line. Each pair of waves, traveling in opposite directions, would have identical frequencies after closure to the origin line. Hence, oscillations in the resonator have two-fold degeneracy, excluding the π- and 0-modes, because these modes are not traveling waves. So there is a total of N oscillations: $(N/2) - 1$ twofold degenerate oscillations and two nondegenerate oscillations. This is important when a perturbation is introduced into the resonator. An example of such a perturbation is an anode output power hole (Fig. 9.12). It is known that any disturbance with lesser symmetry destroys the degeneracy. Then each of the degenerate oscillations would split into two oscillations (a doublet) with close frequencies. Only the π- and 0-modes do not form such doublets. This property of the π-mode, along with its high coupling impedance and high delay factor, make the π-mode preferable as the operative mode for a magnetron.

Separation of the π-Mode Despite the absence of a frequency split in the π-mode, the frequency separation between the π-mode and the nearest $[(N/2) - 1]$-mode is too small. For example, in an eight-cavity resonator, the frequency separation is only ~1 to 2%. This sometimes leads to a loss of operative stability and, in particular, to frequency jumps. An increase in mode separation can be attained by different methods. Let us consider here briefly two frequently used techniques: straps and rising-sun structure.

In the strapped structure (Fig. 9.14*a*), alternate anode segments are connected by wires (straps). For the π-mode of operation, the alternate segments and corresponding strap ends have identical ac potentials. Thus, no additional inductance will be introduced. Straps will, however, add capacitance to the circuit and alter the π-mode frequency. The voltage drops between alternate segments are nonzero for modes other than the π-mode, and therefore the straps introduce inductance, resulting in a different frequency shift with respect to the π-mode. According to measurements, this mode

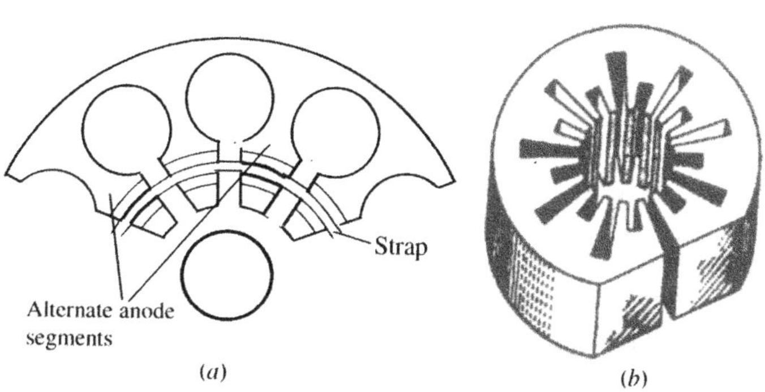

FIGURE 9.14 (*a*) Hole-and-slot block of a strapped magnetron resonator; (*b*) rising sun resonator.

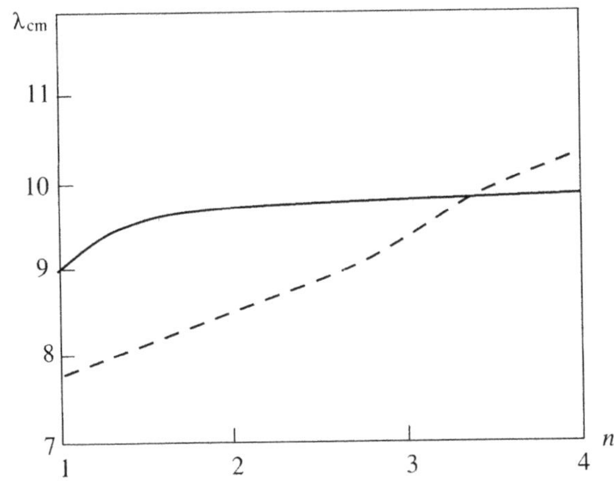

FIGURE 9.15 Frequency separation of frequencies by mode strapping. Solid curve, unstrapped structure; dashed curve, strapped structure. Mode $n = 4$ is an π-mode.

separation reaches 10 to 35%, depending on the tube power. The effect of strapping in an eight-cavity magnetron is shown in Fig. 9.15.

The mechanism for mode separation in a rising sun resonator is based on the different influence of short and long slots on the frequencies of low and high modes. Rising sun resonators are generally used in the millimeter wavelength range, where dimensions are small and fabrication of strapped structures is difficult.

Rotational Waves in a Magnetron Resonator The ac electric field in a resonator can be represented as

$$\mathbf{E} = \mathbf{E}(\mathbf{r}, t) = \vec{E}(r, z, \psi)e^{i\omega t} \tag{9.63}$$

For the n-mode of the resonator, the angular period of the field is $\psi_T = 2\pi/n$. Let us expand $\vec{E}$ in a real Fourier series:

$$\vec{E}(r, z, \psi) = \sum_{m=1}^{\infty} \vec{E}_m(r, z) \cos \frac{2\pi m}{\psi_T} \psi = \sum_{m=1}^{\infty} \vec{E}_m(r, z) \cos mn\psi \tag{9.64}$$

The first (main) term of the series is $\vec{E}_1(r, z)\ \cos n\psi$. Using Euler's formula, we obtain

$$\mathbf{E}(\mathbf{r}, t) = \tfrac{1}{2}\vec{E}_1(r, z)e^{i(\omega t - n\psi)} + \tfrac{1}{2}\vec{E}_1(r, z)e^{i(\omega t + n\psi)} \tag{9.65}$$

Thus, the full field can be represented as the sum of two waves rotating in opposite directions. The angular phase velocity of these waves is $\Omega = \omega/n$. For the π-mode $n = N/2$, and the angular phase velocity is

$$\Omega_\pi = \frac{2\omega}{N} \tag{9.66}$$

9.5.2 Electron Beam in a Magnetron

State of a Magnetron Electron Beam The state of the electron cloud in a magnetron is one of the oldest problems in vacuum high-frequency electronics. It has been discussed since the 1920s, before the multicavity magnetron was even invented. Basically, smooth magnetron diodes were proposed (i.e., cylindrical diodes with magnetic field B directed along the axis). The magnetic field distorts the electron trajectories. Depending on the B magnitude, the following three types of trajectories are possible (Fig. 9.16):

- *Cutoff regime*, $B = B_c$: electron trajectories touch the anode surface.
- *Before-cutoff regime*, $B < B_c$: electron trajectories intersect the anode surface.
- *After-cutoff regime*, $B > B_c$: electrons do not reach the anode but instead, return to the cathode (the magnetically isolated anode).

Cutoff Relation. Hull Parabola The relation describing the cutoff regime was first obtained by Hull (1921). We assume axial symmetry of the electron beam and the existence of a static regime. In this case, two quantities are conserved: the energy integral and the generalized angular momentum (Busch's theorem). For the cutoff regime, the electron velocity at the anode has only an angular component. Using the energy integral and assuming zero velocity at the cathode, we can write in the nonrelativistic approximation:

$$(v)_{r_a} = r_a \dot{\varphi}_{r_a} = \sqrt{2U_a} \tag{9.67}$$

Busch's theorem [Eq. (1.127)] yields

$$\dot{\varphi}_{r_a} = \frac{\eta}{2\pi r_a^2}(\Psi_{r_a} - \psi_{r_c}) = \frac{\eta B_c}{2}\left(1 - \frac{r_c^2}{r_a^2}\right) \tag{9.68}$$

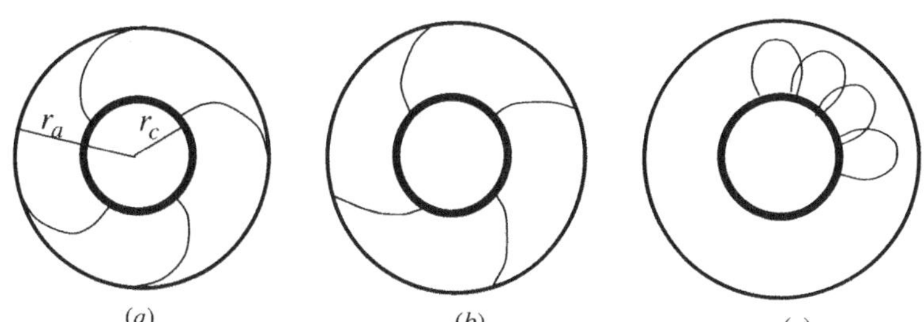

FIGURE 9.16 Three types of electron trajectories in cylindrical magnetron diodes: (*a*) cutoff regime (trajectories touch the anode surface); (*b*) before-cutoff regime; (*c*) after-cutoff regime.

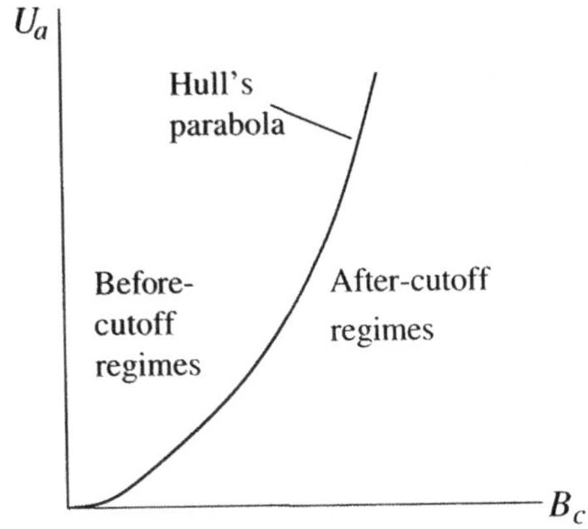

FIGURE 9.17 Hull's parabola (cutoff regime).

where B_c is the value of the magnetic field corresponding to the cutoff regime. Combining Eqs. (9.67) and (9.78), we obtain the cutoff relation

$$U_a \equiv U_c = \frac{\eta B_c^2 r_a^2}{8}\left(1 - \frac{r_c^2}{r_a^2}\right)^2 \tag{9.69}$$

The curve $U_a = U_a(B_c)$ is a parabola called *Hull's parabola* (Fig. 9.17). The relation (9.69) is exact for an arbitrary space-charge distribution if the latter is stationary and axisymmetric.

Failure of the Cutoff Relation From the cutoff relation we can conclude that the electron current in the magnetron diode, as a function of the anode potential U_a for the constant magnetic field B_c, must be zero until $U_a = U_c$. Thereafter, it would jump to $I = I_0$, with subsequent increase in U_a. The corresponding volt–ampere characteristic is shown by the dashed line in Fig. 9.18. However, in reality, volt–ampere characteristics follow the solid curve in Fig. 9.18.

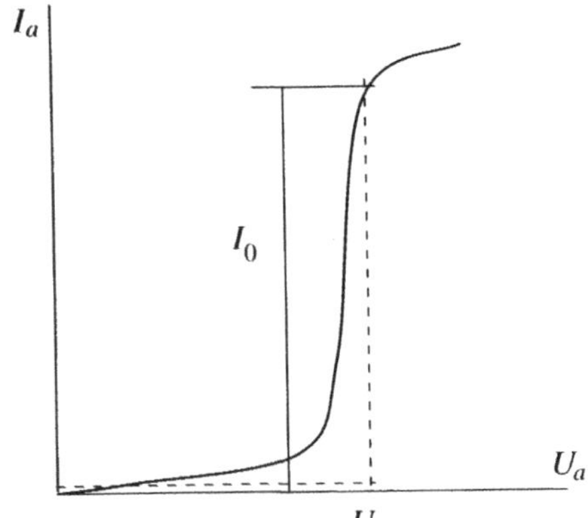

FIGURE 9.18 Idealized and real volt–ampere characteristics of a magnetron diode.

As we see, the anode current appears in the essentially *after-cutoff regimes*. A number of investigations have attempted to find an explanation of this effect within the framework of static approximation. In particular, the effects of electron emission velocities, the tilt of the magnetic field, the cathode eccentricity, and fringe fields have been examined (Harvey, 1942). However, none of these effects could agree with the magnitude of deviation from the cutoff state. Furthermore, essentially nonstationary effects were found, among them the existence of electrons with excess energy up to 10^5 °C (Linder, 1938) and back-bombardment of the cathode (Jepsen and Miller, 1951). Thus, the nature of the cutoff curve anomalies is related to oscillations in the electron cloud.

Steady States of an Electron Cloud Let us dwell on the states of the electron cloud in the after-cutoff regime, because in this regime only, the electrons spend a long time in the interaction space. The question is: If the above-mentioned oscillations are small perturbations of some steady states of the electron cloud, what are these steady states? The actual form of the steady states has been the subject of much investigation, and two stationary solutions that satisfy the boundary conditions have been proposed:

1. The *double-stream state*, in which electrons follow the orbits that intersect the cathode surface. The configuration of the stream is close to Fig. 9.16*c*.
2. The *single-stream state*, in which electrons move in concentric circles around the cathode. This state was first pointed out by Brillouin (1941) and is known as the *Brillouin stream*.

Examination of the double stream indicates that this state is unstable (see, e.g., Kadomtsev, 1960). The nature of this instability is close to the classical two-stream instability of two rectilinear streams moving in opposite directions (e.g., Lawson, 1988). It turns out that the instability of the double-stream state in a magnetron diode develops on a time scale on the order of the time of electron motion along one loop. As a result, the double-stream state is not realized at all (Kadomtsev, 1960).

Let us consider the Brillouin stream in more detail (Fig. 9.19). The electrons in the stream have only an angular velocity. In this case, the full velocity in the nonrelativistic approximation is

$$v(r) = r\dot{\varphi} = \sqrt{2\eta U(r)} \tag{9.70}$$

According to Busch's theorem,

$$\dot{\varphi} = \frac{\eta(\Psi - \Psi_c)}{2\pi r^2} = \frac{\eta(B\pi r^2 - B\pi r_c^2)}{2\pi r^2} = \frac{\eta B}{2}\left(1 - \frac{r_c^2}{r^2}\right) \tag{9.71}$$

We obtain from Eqs. (9.70) and (9.71) the potential distribution

$$U(r) = \frac{r^2 \eta B^2}{8}\left(1 - \frac{r_c^2}{r^2}\right)^2 \tag{9.72}$$

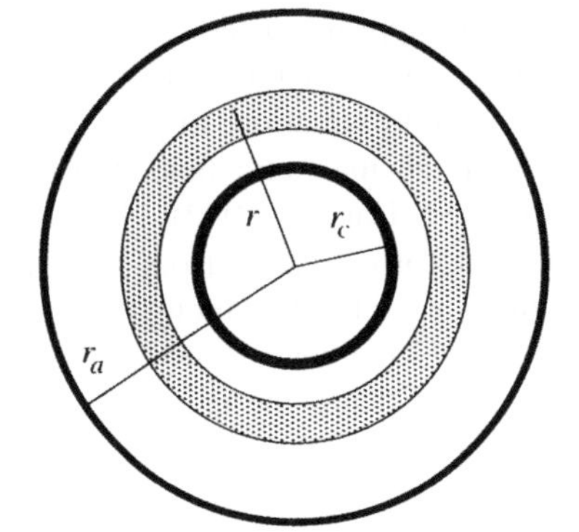

FIGURE 9.19 Brillouin flow.

The self-matched solution must also satisfy Poisson's equation,

$$\Delta U = \frac{1}{r}\frac{d}{dr}\left[r\frac{dU(r)}{dr}\right] = -\frac{\rho}{\varepsilon_0}$$

Because the distribution $U = U(r)$ is known, we obtain for the space-charge density

$$|\rho| = \varepsilon_0 \frac{1}{r}\frac{d}{dr}\left[r\frac{dU(r)}{dr}\right] = \frac{\varepsilon_0 \eta B^2}{2}\left(1 + \frac{r_c^4}{r^4}\right) \tag{9.73}$$

Thus, all characteristics of the stream have been found: trajectories, velocities, potential, and space-charge density. Note that according to Eqs. (9.72) and (9.69), the trajectories at each radius of the Brillouin stream satisfy the cutoff condition.[2]

The stability of the Brillouin solution has been studied by many authors. In particular, Harris (1953) has shown that the electron cloud is stable for an infinitely thin cathode. This is to be expected, in general, because in this case the cloud is analogous to a solid rotator ($\dot{\varphi} = \eta B/2 = \text{const.}$). Therefore, according to the Larmor theorem, in a coordinate system that rotates with angular velocity $\eta B/2$, the effect of the magnetic field on the electron motion is eliminated, and the electron cloud is comparable to a quasineutral plasma. The perturbations of the electron density are reduced to purely harmonic oscillations and do not lead to instabilities. For $r_c \neq 0$, slipping of the electron layers occurs. In this case, the electron cloud is unstable with respect to nonsymmetrical disturbances. However, this instability is significantly weaker than the double-stream instability. Two considerations make the existence of the Brillouin stream questionable: (1) it seems extraordinary as a process for launching electrons from the cathode into equilibrium ($r = \text{const.}$) orbits; and (2) the back-bombardment of the cathode needs an explanation.

[2]Note that although the electron velocity at the cathode is equal to zero, the space-charge density $|\rho_c| = \varepsilon_0 \eta B^2 \neq \infty$ since the current density $j = \rho v_r = 0$. It is readily verified that the plasma frequency ω_p at the cathode is equal to the gyrofrequency $\omega_g = \eta B$. It is a specific space-charge limitation for Brillouin flow.

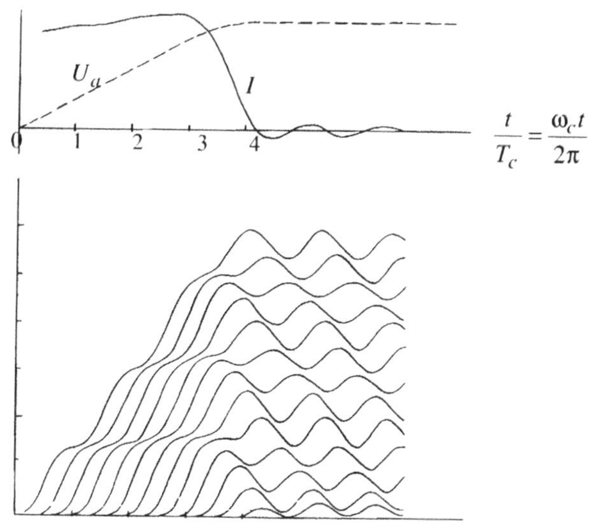

FIGURE 9.20 Formation of a single Brillouin state.

In principle, the nonlinear character of the processes of formation of an electron cloud provides a number of quasistable states. It is essential to retrace the transient processes. Corresponding "computational experiments" were carried out by Hartree and Nicolson (1943) (see also Buneman, 1961). They integrated the equations of motion numerically for a large number of electron sheets, starting with an empty diode at time $t = 0$. In Fig. 9.20, formation of the Brillouin stream is shown in the planar model when anode potential rises over 4 gyroperiods. As we see, the stream remains single at all times. After the establishment of the anode potential, the emission of new electrons practically ceases, and the electron sheets perform steady oscillations about their equilibrium (Brillouin) positions. This result is instructive: we see that instabilities are developed in such a way as to diminish the electron velocities in a direction perpendicular to the cathode.

9.5.3 Dynamic Regime of a Magnetron. Threshold Voltage

In the dynamic regime, in addition to the static electric field there is a rotating high-frequency wave field in the interaction space. According to Section 9.5.1, the angular velocity of this wave is

$$\Omega_n = \frac{\omega}{n}, \qquad n = 1, 2, \ldots, N/2 \tag{9.74}$$

where N is the (even) number of cavities in the resonator. Below, the after-cutoff regime ($B > B_c$) is considered. Interaction of the electron beam with a synchronous ac field is similar to the bunching and energy transfer processes in planar MTWTs discussed in Section 9.2: Spokes are formed; the centers of the spokes are located in the maximum of the retarded phase of the longitudinal ac electric field; and electrons interacting with the ac field drift to the anode along equipotentials of the ac electric field in the rotating frame. Unlike in MTWTs, in magnetrons the transfer of electron energy to an electromagnetic field with zero anode current is not possible, because a dc source would not perform any work without an anode current.

Below, the condition of attainment for a nonzero anode current is formulated for a small ac field energy that corresponds to a starting regime of a magnetron. We consider the frame of reference that rotates with angular frequency Ω_n. In this frame the electron angular velocity at the anode is

$$\dot{\theta}_a = \dot{\varphi}_a - \Omega_n \tag{9.75}$$

where according to Eq. (9.71),

$$\dot{\varphi}_a = \frac{\eta B}{2}\left(1 - \frac{r_c^2}{r_a^2}\right) \tag{9.76}$$

Using Eq. (9.75), we find that the electron kinetic energy at the anode in the rotating frame is

$$W_a = \frac{m}{2}\left(\dot{r}_a^2 + r_a^2\dot{\theta}_a^2\right) = \frac{m}{2}(\dot{r}_a^2 + r_a^2\dot{\varphi}_a^2 - 2r_a^2\dot{\varphi}_a\Omega_n + r_a^2\Omega_n^2) \tag{9.77}$$

Assuming that the RF field energy is small, we can write

$$\frac{m}{2}(\dot{r}_a^2 + r_a^2\dot{\varphi}_a^2) = e_0 U_a \tag{9.78}$$

where U_a is the dc anode voltage. Thus, the kinetic energy of electrons at the anode is

$$W_a = e_0(U_a - U_{\text{BH}}) \tag{9.79}$$

where

$$U_{\text{BH}} = \frac{1}{\eta} r_a^2 \dot{\varphi}_a \Omega_n - \frac{r_a^2}{2\eta}\Omega_n^2 \tag{9.80}$$

Taking Eq. (9.76) into account, we obtain

$$U_{\text{BH}} = \frac{B}{2}(r_a^2 - r_c^2)\Omega_n - \frac{r_a^2}{2\eta}\Omega_n^2 \tag{9.81}$$

The value U_{BH} is called the *threshold voltage* or *Buneman–Hartree threshold voltage*. When the anode voltage is increased, oscillations in the magnetron can begin from $U_a \geq U_{\text{BH}}$. In this case, according to Eq. (9.79), the kinetic energy of electrons at the anode in the rotating frame is $W_a \geq 0$. The function

$U_{BH} = U_{BH}(B)$ is a straight line. Let us show that the threshold line is located below the cutoff parabola. According to Eqs. (9.69) and (9.81), the difference $U_c - U_{BH}$ is equal to

$$U_c - U_{BH} = \frac{r_a^2}{2\eta}\left[\frac{\eta B}{2}\left(1 - \frac{r_c^2}{r_a^2}\right) - \Omega_n\right]^2 = \frac{r_a^2}{2\eta}(\dot{\varphi}_a - \Omega_n)^2 = \frac{r_a^2 \dot{\theta}_a^2}{2\eta} \geq 0 \qquad (9.82)$$

Thus, the threshold line does not intersect the cutoff parabola. It can only be a tangent to the parabola. According to Eq. (9.82), the magnetic field in the tangency point is

$$B_0 = \frac{2\Omega_n}{\eta(1 - r_c^2/r_a^2)} \qquad (9.83)$$

The corresponding value of the anode voltage is minimal and can be obtained by substituting Eq. (9.83) in the cutoff relation (9.69):

$$U_{a,\min} = \frac{r_a^2 \Omega_n^2}{2\eta} \qquad (9.84)$$

The Hull parabola and threshold lines corresponding to different $\Omega_n = \omega/n$ ($n = 1, 2, 3, 4$) for an eight-cavity magnetron are shown in Fig. 9.21. As we see, the threshold voltage is minimal for the π-mode (i.e., the latter is excited in the first place). In region 1 the value of the voltage is insufficient for electrons to reach the anode. In

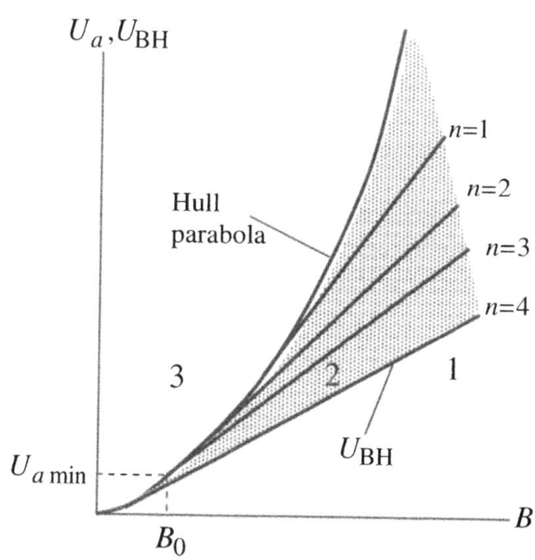

FIGURE 9.21 Threshold voltage tangents to a Hull parabola as functions of the magnetic field for an eight-cavity resonator. The line $n = 4$ corresponds to the π-mode.

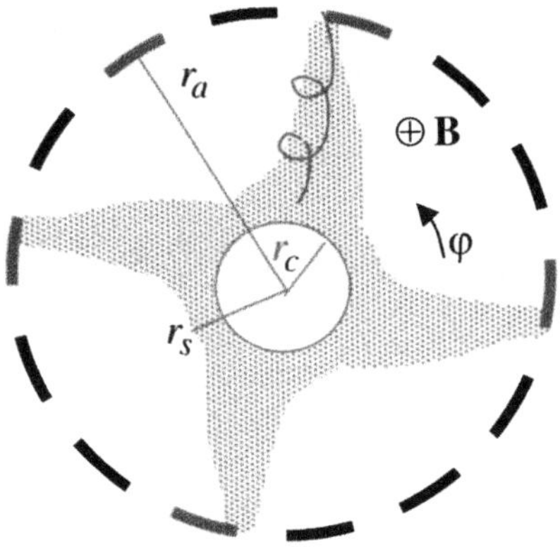

FIGURE 9.22 Electron beam in an oscillating magnetron.

region 2 the electrons synchronous with the rotating wave are bunched in rotating spokes that are attached to the anode. A typical configuration for this electron "rotor" formed by the rotating spokes is shown in Fig. 9.22. The illustration corresponds to the π-mode in an eight-cavity magnetron. The base of the spokes $r = r_s$ (synchronous radius) is determined by the condition $\dot{\varphi} = \Omega_n$ in the Brillouin stream. In the threshold regime, the value of r_s is adjusted by the magnetic field selection: According to Eqs. (9.70), (9.71), and (9.81),

$$\begin{aligned} r_s^2\Omega_n^2 &= 2\eta U_{\mathrm{BH}} = \eta B(r_a^2 - r_c^2)\Omega_n - r_a^2\Omega_n^2 \\ \Omega_n &= \frac{\eta B}{2}\left(1 - \frac{r_c^2}{r_s^2}\right) \end{aligned} \tag{9.85}$$

Note that the configuration of the rotating electron cloud shown in Fig. 9.22 is true as a result of an averaging of the real electron motion. In fact, the electrons besides the drift also perform an orbital motion. One such trajectory is shown in Fig. 9.22.

In region 3 in Fig. 9.21 an excitation of the magnetron is impossible because it is the region of the before-cutoff regime (Fig. 9.17), and the electrons travel directly to the anode without interacting with the RF field.

9.5.4 Magnetron Efficiency

Magnetrons are high-efficiency microwave devices that share a common property of crossed-field tubes. The relation for the magnetron electronic efficiency can be obtained in the same way that we obtained the efficiency for MTWT (Section 9.3.6). Let us assume that the undisturbed state of the electron cloud has the form of a Brillouin stream. An electron interacts with the high-frequency field beginning from the synchronous radius r_s with kinetic energy $m(r_s\Omega_n)^2/2$. We assume that the magnetic field is chosen to provide synchronism between electrons and the rotating wave in the threshold regime $[(\dot{\varphi})_{U_a=U_T} = \Omega_n]$, and the kinetic energy stays the same when the electron arrives at the anode. The energy that the electron transfers to the HF field is then $e_0U_a - e_0U_T$. According to Eq. (9.85), the electronic efficiency is

$$\eta_e = 1 - \frac{U_{\mathrm{BH}}}{U_a} = 1 - \frac{B(r_a^2 - r_c^2)\Omega_n - r_a^2\Omega_n^2/\eta}{2U_a} \tag{9.86}$$

Formula (9.86) is an estimate. It does not take into account the orbital motion of electrons in the spokes or the presence of electrons in the accelerated phase of the HF field, which bombard the cathode. The latter process significantly affects electron emission, especially in powerful tubes. Both factors lead to a decrease in efficiency η_e compared to Eq. (9.86).

The conversion efficiency of the magnetron is equal to $\eta = \eta_e\eta_c$, where η_c is a circuit efficiency that is determined by the loss in the output circuit. Nevertheless, the conversion efficiency of the experimental magnetrons in the S-band can exceed 80%. Note that according to Eq. (9.86) and Fig. 9.21, the efficiency is maximal for the π-mode.

9.5.5 Magnetron Performance

The voltage–current characteristics of magnetrons depend strongly on the magnetic field. As such, these characteristics are usually displayed on performance charts like the one shown in Fig. 9.23, where the anode voltage, efficiency, and output power of a 50-GHz pulse magnetron are shown as functions of the current for various values of the magnetic field.

The current depends on the anode voltage, the magnetic field, and the cathode temperature, although secondary electrons in the oscillating magnetron make a significant contribution to the current as a result of back-bombardment. Usually, the current influences the frequency as a result of electronic impedance variation. The change in frequency with current is called *frequency pushing*.

Average and CW power levels are limited by the ability to dissipate heat by the anode and cathode bombarded by electrons. Therefore, power magnetrons are generally used in pulse regimes. Only these magnetrons are suitable for radars and electronic countermeasure means. Overall power characteristics for CW and pulsed magnetrons are shown in Fig. 9.24. As can be seen, the peak power exceeds 1 MW. The pulse duration for these magnetrons is ~1 μs. The peak power of relativistic magnetrons (see below) exceeds 1 GW with a pulse duration of ~10 ns. One notable use is that of magnetrons in microwave ovens. These tubes typically operate at a power level around 0.8 to 1 kW and a standard frequency of 2.45 GHz. Mass production of these tubes (~10 million in a year) forced the design of very effective, reliable, lightweight, and inexpensive devices.

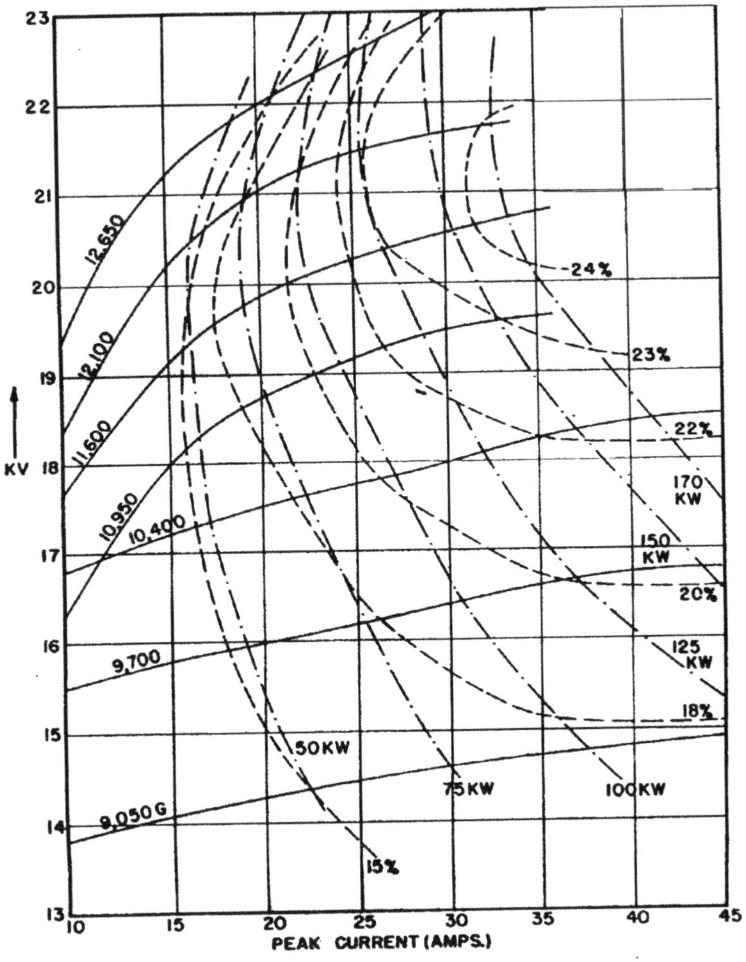

FIGURE 9.23 Operation characteristics of a pulse 50-GHz magnetron. (From Bernstein and Kroll, 1961.)

Magnetrons as a whole have high noise levels. This is the result of the common instability of electron clouds in crossed fields. This instability and the competition of the π-mode with other $(N/2) - 1$ modes are the cause of an additional low-frequency instability. Frequency fluctuations can lead to serious difficulties in some radar applications: for example, in Doppler radars. In this respect, amplifiers are preferable to oscillators.

9.5.6 Magnetron Frequency Tuning

The ability to change the radiation frequency (tuning) is an important operation characteristic of magnetrons. Below we consider two methods of magnetron tuning.

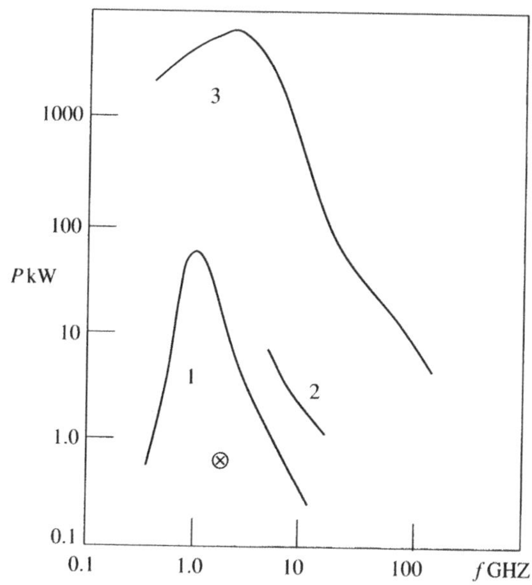

FIGURE 9.24 Magnetron power capability. 1, CW magnetrons; 2, beacon magnetrons; 3, pulsed magnetrons; ⊗, magnetrons for microwave ovens.

Mechanically Tuned Magnetrons Two basic methods of mechanical tuning are known. The more straightforward approach employs a change in capacitance or inductance of the anode cavities. For example, the capacitance of the cavity can be varied by a movable element in the vicinity of the slot portion of the cavity (Fig. 9.25). This method makes broadband tuning possible. However, the tuning mechanism often negatively affects other characteristics of the magnetron.

An alternative method is based on changing the reactance of single or multiple external circuits coupled with anode cavities. This method is convenient for powerful magnetrons because it does not impede anode cooling and does not use small gaps with possible breakdown. The maximum range of tuning is close to the frequency separation between π- and $(N/2) - 1$ modes. According to Vaccaro (1961), using four external resonators gives the best results. Typical characteristics for an X-band magnetron are as follows:

Range of tuning	8500–9600 MHz
Output peak power	220 kW
Full efficiency in tuning range	37%
Anode voltage	22 kV
Pulse duty factor	0.001

Voltage-Tunable Magnetrons Two important properties distinguish voltage-tunable magnetrons (VTMs) from conventional narrowband magnetrons: injection

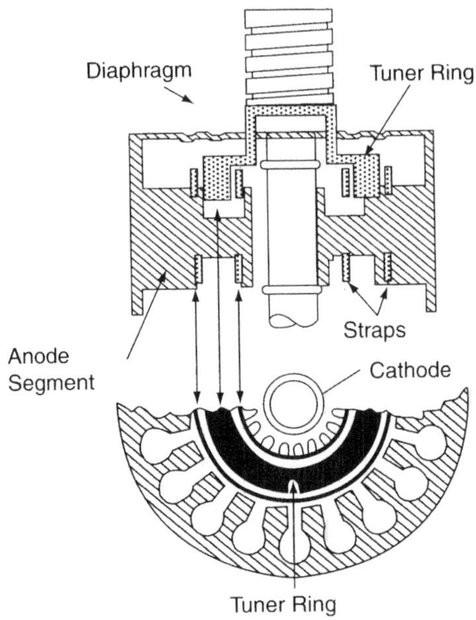

FIGURE 9.25 Configuration of capacitive tuning elements of a mechanically tuned magnetron. (From Coleman, 1982.)

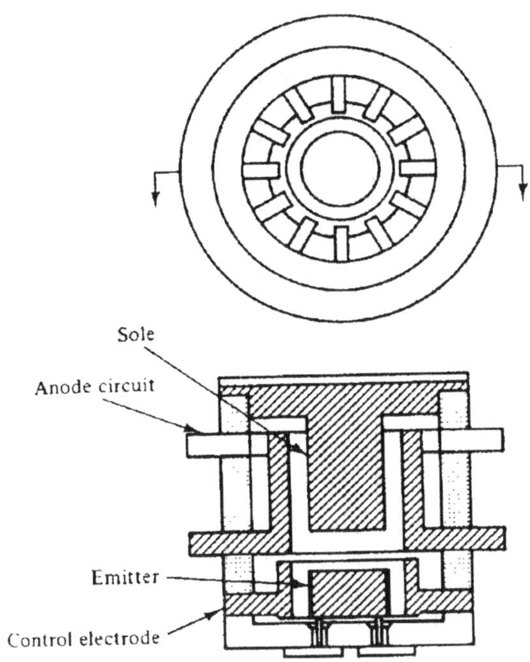

FIGURE 9.26 Voltage tunable magnetron. (From Varian Associates, *Microwave Tube Manual*, Air Force Publication T.O. 00-25-251, October 1979.)

of an electron beam into the interaction space from an external cathode and the very low Q-factor of the resonator. One possible scheme for a VTM is shown in Fig. 9.26. The low Q-factor is attained by the choice of anode structure. The interdigital anode is an example of a heavily and uniformly loading structure (Wilbur and Peters, 1961).

Electrons are accelerated between the emitter and the control electrode in the vertical magnetic field, and the hollow beam that is formed is injected into the appropriate magnetron region. This device is similar to the magnetron injected gun (Section 4.7). The structure of the beam rotating between the sole and the anode circuit is similar to a Brillouin stream in the before-threshold voltage regime. Let us estimate the threshold voltage. Because the anode angular velocity is $\dot{\varphi}_a > \Omega_n/2$, we can neglect the second term in Eq. (9.81) and the threshold voltage will be equal to

$$U_{\mathrm{BH}} \approx \frac{B}{2} r_a^2 \left(1 - \frac{r_c^2}{r_a^2}\right) \Omega_n \tag{9.87}$$

According to Eq. (9.66) for the π-mode $\Omega_n = 2\omega/N = 4\pi f/N$, so

$$U_{\mathrm{BH}} = \frac{2\pi B f}{N} r_a^2 \left(1 - \frac{r_c^2}{r_a^2}\right) \tag{9.88}$$

For VTMs the operation voltage $\Delta U = U_{\mathrm{op}} - U_{\mathrm{BH}} \approx 1.1 U_{\mathrm{BH}}$ (Wilbur and Peters, 1961). Therefore, we can assume that $U_{\mathrm{BH}} \approx U_{\mathrm{op}}$, and relation (9.88) thus ensures the linear tuning characteristic. The beam reactance for conventional magnetrons with a high Q-factor consists of a small portion of the resonator circuit reactance; therefore, the frequency tuning is very small. The increase in ΔU raises the beam current and the output power, but not the frequency.

The low Q-factor of VTM resonators and beam current limitations diminish the ac circuit voltage. The reactance of the circuit turns out to be close to the beam reactance. As a result, the beam reactance essentially determines the resonator frequency, and the condition of constant current determines the output power. Thus important properties of VTMs include the broadband linear frequency characteristic and a weak dependence of the output power on the anode voltage. For low-power low-frequency VTMs, the tuning range reaches 75% and the efficiency ~30%. When these values are increased, the tuning range shrinks rapidly. Usually, VTMs operate in the CW regime, so their output power and efficiency are not high. Bandwidth capabilities of VTM are illustrated by Fig. 9.27.

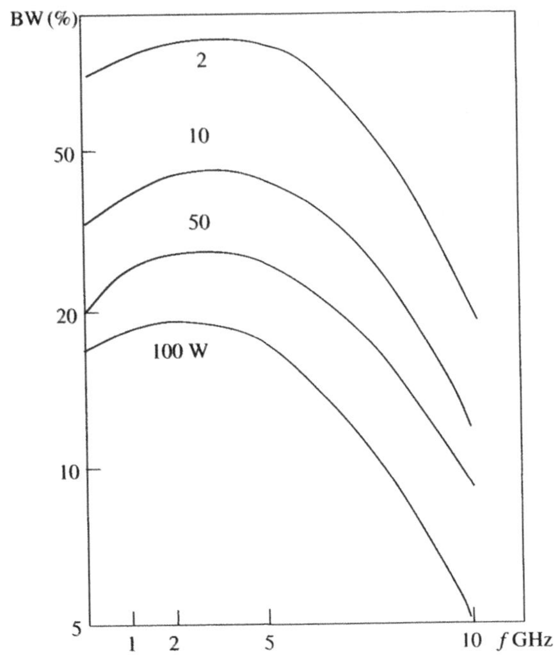

FIGURE 9.27 Tuning bandwidth BW as a function of frequency for different values of the output power in a VTM. (From Gilmour, 1986.)

9.6 RELATIVISTIC MAGNETRONS

9.6.1 Introduction

A fundamental distinction between relativistic and conventional magnetrons is the application of much higher voltages and explosive emission cathodes. That allows one to overcome current limitations connected with cathode emission and space charge (the Child–Langmuir law) and to raise the current and output power by approximately two orders of magnitude.

Bekefi and Orzechowski (1976) at the Massachusetts Institute of Technology were the first to develop a relativistic magnetron. Theirs was an S-band magnetron equipped with a six-sector cavity resonator, a graphite cathode, and a tapered output waveguide (Fig. 9.28). The magnetron operated at a 10-cm wavelength with a potential of 360 kV, a magnetic field 8 kG, and a field emission current of 35 kA (without the magnetic field). Taking into account that this was the first relativistic magnetron, the results were sensational: output power 1.7 GW, conversion efficiency 35%, and pulse duration 30 ns. This large pulse duration was essentially determined by the highly favorable diode orientation of the magnetic field in the magnetron for explosive emission, similar to magnetically insulated diodes (see Section 4.8.3). Average comparative parameters of conventional and relativistic magnetrons are given in Table 9.2. Characteristic dimensions (in centimeters) of several relativistic

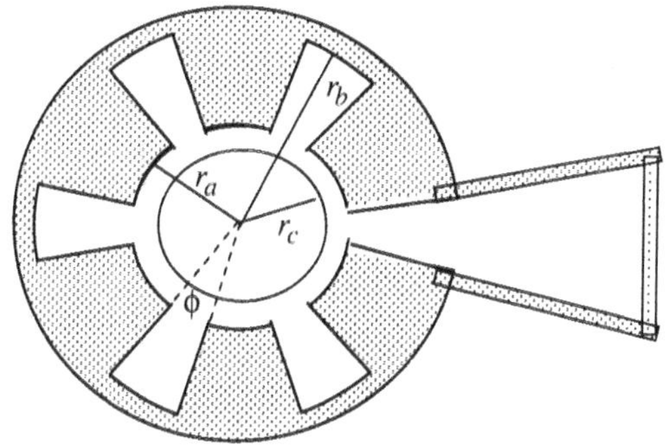

FIGURE 9.28 Bekefi relativistic magnetron. (From Palevski and Bekefi, 1979.)

TABLE 9.2 Comparative Parameters of Conventional and Relativistic Magnetrons

Parameter	Conventional	Relativistic
Maximum voltage	~10 kV	~1 MV
Maximum current	~100 A	~10 kA
Emission	Thermoionic and secondary	Field
Maximum power	~10 MW	~1 GW
Maximum efficiency	~85%	30%
Pulse duration	>1 μs	<100 ns
Maximum frequency	100 GHz	10 GHz

TABLE 9.3 Characteristic Dimensions (in centimeters) of Several MIT Relativistic Magnetrons

Magnetron	r_c	r_a	r_b	L	φ (deg)	f GHz	N
A6	1.58	2.11	4.11	7.2	20	2.34	6
B6	1.58	2.11	4.11	3.81	20	2.34	6
D6	1.83	2.46	4.83	8.42	20	1.98	6
J6	1.69	2.22	3.82	7.2	18	2.81	6

magnetrons (Fig. 9.28) are given in Table 9.3. The frequency of the π-mode is computed for an infinitely long anode block.

9.6.2 Hull and Buneman–Hartree Conditions

According to experimental and simulation data, the fundamental properties of relativistic magnetrons can be considered as a relativistic extension of the corresponding properties of conventional magnetrons. In particular, it was proven that the operation space for the relativistic magnetron is limited by the Hull and Buneman–Hartree curves (see, e.g., Benford, 1987; Benford and Swegle, 1992).

Similar to Eqs. (9.67) and (9.68), the relativistic version of the Hull cutoff condition follows from the energy integral and Busch's theorem [see Eq. (1.126) for $\dot{\theta}_c = 0$ and uniform magnetic field]:

$$v_a = r_a \dot{\phi}_{r_a} = \frac{c}{\gamma_a}\sqrt{\gamma_a^2 - 1} \tag{9.89}$$

$$\dot{\phi}_{r_a} = \frac{\eta B_c}{2\gamma_a}\left(1 - \frac{r_c^2}{r_a^2}\right) \tag{9.90}$$

where B_c is the cutoff value of the magnetic field and

$$\gamma_a = \frac{1}{\sqrt{1 - v_a^2/c^2}} = 1 + \frac{\eta U_a}{c^2}$$

is the relativistic factor, corresponding to velocity v_a as a full velocity. Combining Eqs. (9.89) and (9.90), we obtain the cutoff relation:

$$B_c = \frac{2}{\sqrt{\eta}\, r_a(1 - r_c^2/r_a^2)}\sqrt{2U_a + \frac{\eta U_a^2}{c^2}} \tag{9.91}$$

A comparison of Eq. (9.91) with Eq. (9.69) shows that the nonrelativistic approximation (the Hull parabola) can be used for $U_a < 300$ kV. On the contrary, the extreme relativistic approximation when the voltage U_a is a linear function of B_c is correct for $U_a > 5$ MV.

Derivation of the relativistic Buneman–Hartee condition (the threshold voltage) can be found in Collins (1948, Chap. 6). That relation in our notation is

$$U_{\mathrm{BH}} = \frac{B}{2}(r_a^2 - r_c^2)\Omega_n - \frac{c^2}{\eta}\left[1 - \sqrt{1 - \left(\frac{r_a \Omega_n}{c}\right)^2}\right] \tag{9.92}$$

Note that the first term on the right-hand side of Eq. (9.92) coincides with the corresponding term in the nonrelativistic Eq. (9.81). According to the experimental data (see, e.g., Benford, 1987; Benford and Swegle, 1992), operational values of U and B, which give maximum efficiency, are near the Buneman–Hartree straight line.

9.6.3 Performance of Relativistic Magnetrons

Efficiency The theoretical estimation of electronic efficiency according to Benford and Swegle (1992) is

$$\eta_e = 1 - \frac{2m_0 U}{e_0\{B^2[(r_a^2 - r_c^2)^2/r_a^2] - U^2/c^2\}} \tag{9.93}$$

The conversion efficiency η should take the circuit efficiency into account. The experimental efficiency (Benford and Swegle, 1992) turns out to be lower than η_e

from Eq. (9.93) typically by a factor of 2 or more. There are several causes, including emission from the cathode tip and dissipation of energy in the electron plasma. Note that according to recent simulations (Barker and Schamiloglu, 2001, Sec. 3.2), spokes are developed under only one vane and their energy is extracted in a short transit time.

Power Extraction and Limitations of Output Power Two methods of power extraction are known for relativistic magnetrons: radial and axial (Fig. 9.29). Radial extraction (Fig. 9.29*a*) is more common and involves radial transfer of ac power through apertures in the anode. Apertures and the corresponding extraction waveguides can be installed in one or many vanes. TE_{10} modes are excited in each rectangular waveguide. Extraction power as a function of waveguide number was studied by Sze et al. (1987). Maximum power for the A6 magnetron was achieved with three open vanes, reaching ~3.5 GW.

The axial extraction method was applied for X-band magnetrons by Kovalev et al. (1980). A tapered coaxial waveguide (Fig. 9.29*b*) produces TM_{np} modes, where p is the radial index. These modes have a central null and propagate further in the tapered hollow waveguide. The design is similar to the diffraction power output in gyrotrons. Effective mode transformers have been developed for the transformation of these modes into Gaussian beams (see Section 10.9.3). An output power of about 4 GW has been obtained (Kovalev et al., 1989).

Next we turn to some fundamental effects in relativistic magnetrons which limit output power.

1. *Maximum Axial Current.* The angular component of the magnetic field produced by the axial current, together with a radial electric field, creates an axial drift of electrons that leads to their departure from a resonator before they produce high power. According to estimates of Benford and Swegle (1992), the maximum axial current is equal to

$$I_{\max}(\text{kA}) \sim 0.8B(\text{kG})L(\text{cm}) \tag{9.94}$$

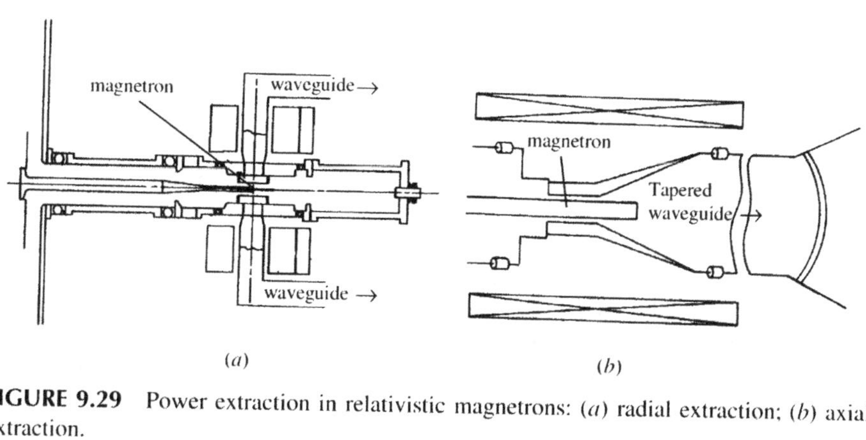

(*a*) (*b*)

FIGURE 9.29 Power extraction in relativistic magnetrons: (*a*) radial extraction; (*b*) axial extraction.

where L is the resonator length. For example, Eq. (9.94) gives $I_{max} = 24$ kA at $B = 4.6$ kG in an A6 magnetron (Table 9.3). The threshold current observed was 15 to 20 kA.

2. *Upper Limit of ac Power.* HF electric and magnetic fields at high output power can cause a loss of electrons across the gap. In particular, because the HF magnetic field $\widetilde{B}$ is axial, the difference $|B - \widetilde{B}|$ can lead to violation of the Hull cutoff with a corresponding loss of electrons, at least within a part of the ac period.

9.6.4 Pulse Duration, Repetitive Operations, and Phase Locking in Relativistic Magnetrons

Pulse Duration Because the total energy in the pulse for a constant output power is proportional to the pulse duration, increased pulse duration is problematic for many applications of relativistic magnetrons, in particular for radar. The time scale of the relativistic magnetron operation, typically tens of nanoseconds, is substantially less than the duration of an electrical pulse. Three main factors determine the pulse duration.

1. *Pulse Rise Time* τ_r. This is basically determined by the relation $\tau_r \approx Q/\omega$ for a Q-factor of ~50 to 100, and $f = 3$ GHz is $\tau_r = 5$ to 10 ns. This quantity is negligibly small for conventional magnetrons but represents a substantial part of an electrical pulse for relativistic magnetrons.

2. *Voltage Variations during the Operation Part of a Pulse.* There is only a narrow range of voltages for which the magnetron can oscillate with a given magnetic field. The small input impedance of the magnetron circuit compared to conventional magnetrons impedes reaching the necessary pulse flatness.

3. *Closure of the Anode–Cathode Gap by the Cathode Plasma.* The gap closure effect is significantly weakened in coaxial magnetically insulated diodes with a strong axial magnetic field (see Section 4.8.2). There the pulse duration can be greater than 1 μs. According to some authors, however, the high HF electromagnetic field in relativistic magnetrons significantly accelerates transport of the gap plasma and truncates pulses. It is necessary to take into account as well that the cathode plasma motion leads to an alteration of the resonance conditions essentially prior to real gap shorting, even if the voltage pulse is constant. So retarding the plasma motion is very important for increasing effective pulse duration.

Two effects determine the rate of plasma expansion: the mass of the plasma ions and the plasma temperature, because $v = \sqrt{kT/m}$. In this respect the application of carbon fiber cathodes coated by heavy elements (Barker and Schamiloglu, 2001) is very promising (Section 9.3). Application of carbon velvet cathodes with fibers coated by cesium iodide (CsI) (Garate et al., 1995; Shiffler et al., 2002) is particularly interesting, because besides the large mass of cesium and iodine ions, plasma formation in CsI is accompanied by photoemission. As a result, the number of emission centers is increased, the local current density is lowered, ohmic heating is reduced, and gap closure by the cathode plasma is delayed. The application of CsI in HPM for many years was limited by the electric field and emission current density. Today, however, adequate fabrication technology allows an increase in the maximum electric field up to 300 kV/cm with a speed of closure $\sim 0.7 \times 10^6$ cm/s (Benford, 1998).

Repetitive Operations Repetitive operations allow the average power of pulse devices to increase and in general to broaden their application domains. Two primary factors limit the rate and the operational lifespan of relativistic magnetrons: diode recovery and anode erosion. Electron bombardment releases plasma and neutral particles, and the system must be cleared before voltage can again be applied. To ensure long-term operation, it is necessary to provide sufficiently high pumping speed. Anode erosion is a result of two effects: fracture of the surface with detachment of metallic particles and melting of the surface layer. In the latter case it is necessary to maintain the temperature of the surface below its melting value. One drastic measure for doing so is raising magnetron efficiency. However, rate repetition, peak power, pulse duration, and operation frequency must, in reality, be chosen carefully.

According to Smith et al. (1991), there was no detected erosion in an L-band magnetron at a repetition rate of 100 Hz after several thousands of shots, despite high power (up to 3 GW), because of the larger resonator volume. A repetition rate of 50 Hz in an S-band magnetron with a peak power of 360 MW was obtained (Vasil'ev et al., 1987).

Phase Locking Many applications for microwave sources require high radiated energy. The upper limit of the high-frequency power of a single magnetron raises the problem of phase locking for a group of magnetrons. The physics of phase locking of high-power microwave oscillators has been discussed at length (see, e.g., Benford and Swegle, 1992; Woo et al., 1989). Phase locking of two oscillators is considered for cases in which the coupled oscillators drive one another or one oscillator is driven without feedback. The phase locking of two coupled oscillators can occur only when the connector contributes zero or π phase delay.

For relativistic magnetrons, locking time is essential. According to Benford and Swegle (1992), locking time is inversely proportional to the ratio ρ of the incoming and outgoing fields in the output of coupled magnetrons. This time turns out to be sufficiently small for $\rho < \sim 1$. In experiments with $\rho = 0.6$ (Benford et al., 1989), phase locking occurred at an output power of 1.5 GW with a locking time of ~ 10 ns. The phase discriminator signal for this case is shown in Fig. 9.30. A fourfold increase in power density was reported relative to a single magnetron.

9.7 MAGNETICALLY INSULATED LINE OSCILLATORS

A magnetically insulated line oscillator (MILO) is a cross-field oscillator that differs from the magnetron primarily in that the insulating dc magnetic field, which together with the orthogonal dc electric field determines the electron drift, is generated by its own electron beam instead of the external magnetic field. This possibility requires such high current density that it can only be provided by an explosive electron

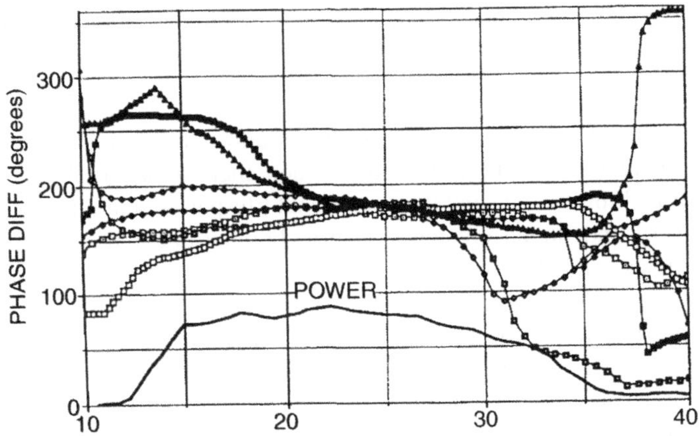

FIGURE 9.30 Phase difference Δ° for seven consecutive shots of two magnetrons as a function of time (τ nanoseconds).

emission. Thus, MILO is an astonishing combination of magnetic insulation, explosive emission, and interaction with retarding electromagnetic waves. The first MILO was implemented by Clark et al. (1988).

9.7.1 MILO Principles of Operation. Choke-Equipped MILOs

A MILO scheme is shown in Fig. 9.31. The absence of a heavy supply of dc magnetic field simplifies the construction of MILO significantly compared with relativistic magnetrons and especially with conventional magnetrons. The explosive emission current in MILO is provided by the emitter, which is made out of a material with a low emission threshold (e.g., velvet). The emitter is set on the cathode in the region between the upstream side of the first vane and the end of the cathode. The rest of the cathode surface is aluminum, which has a large emission threshold. The coaxial resonator in Fig. 9.31 is the section of an SWS with mode TM_{00}. The lengths

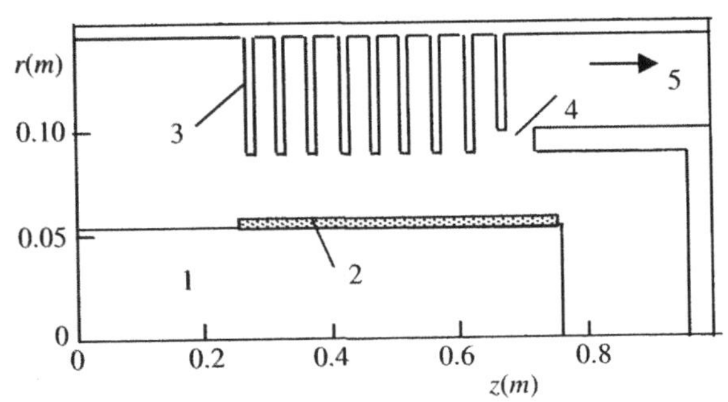

FIGURE 9.31 MILO. 1, Cathode; 2, emitter; 3, vane of anode (SWS); 4, extractor gap; 5, microwave output. The dimensions indicated correspond to those of a 1.2-GHz MILO. (From Lemke et al., 1997.)

of vanes are chosen to be approximately one-fourth of the wavelength, to provide operation on the π-mode.

After turning on the high voltage, the state of the MILO passes through three distinct phases:

1. *Initial Phase.* The electron beam is uninsulated (Fig. 9.32*a*).
2. *Short Intermediate Phase.* The electron beam is insulated, but the ac field is insufficient to form spokes.
3. *Final, Steady Phase.* High-density spokes drift in the crossed dc electric and magnetic fields and interact with the HF electromagnetic field (Fig. 9.32*b*).

The original MILO had $N \geq 6$ cavities, and radial power was extracted through apertures in the cavity walls. The level of output power was ~50 MW. The considerable increase in power (~300 MW) was obtained as a result of the transition to an axial extraction scheme, even though the group velocity is near zero for the π-mode.

Experiments (Lemke et al., 1997) with MILOs of the type shown in Figs. 9.31 and 9.32, with a number of cavities varying from 4 to 15, revealed significant leakage of microwave power toward the pulse-power source. The leakage power was reflected at some location and returned to the SWS out of the oscillation phase, thereby degrading tube performance.

Using results from both simulations and experiments, Lemke et al. (1997) have modified the configuration Fig. 9.31 by applying chokes (cavities with a long vane) and decreasing the total number of vanes (Fig. 9.33). The cathode is covered with velvet from the upstream side of the third choke vane to the end. The choke structure is not an active element of the MILO. It probably operates as a rejection filter. It turns out that implementation of the choke, besides reducing leakage significantly, also increases output power (by a factor 2 or more).

An experimental MILO with a peak power of 1.3 GW, peak efficiency of ~8%, and frequency of 1.8 GHz has been developed. The input voltage and the current were 510 kV and 34 kA, respectively. The pulse duration in these MILOs reached ~150 ns with an output power of ~1 GW, and the microwave energy per pulse reached more than 200 J (Barker and Schamiloglu, 2001). A delay between the

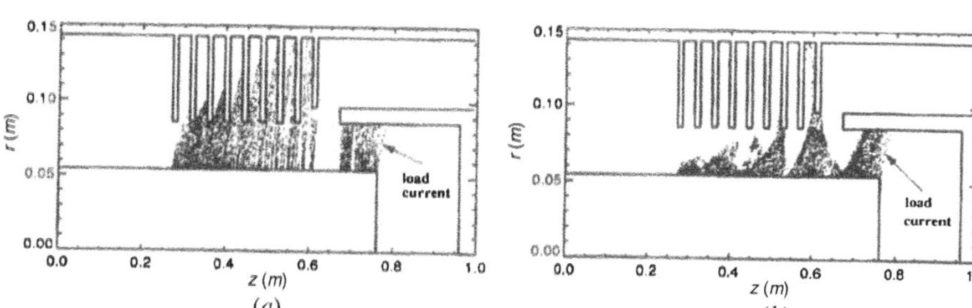

FIGURE 9.32 Phases of an electron cloud state: (*a*) uninsulated electron cloud; (*b*) steady phase with drifting electron spokes. (From Lemke et al., 1997.)

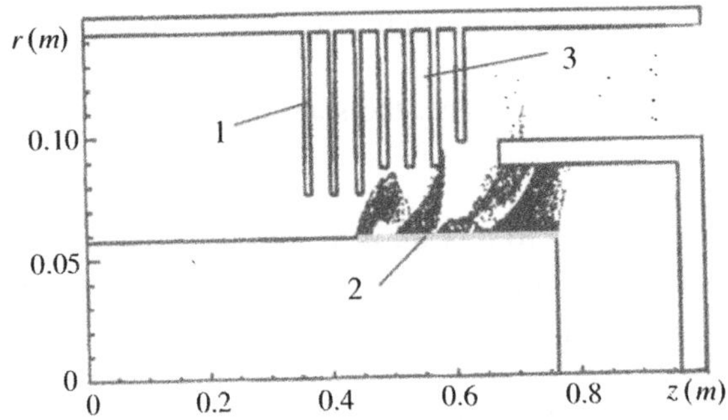

FIGURE 9.33 Optimized configuration of a MILO equipped with chokes. 1, Choke; 2, velvet emitter; 3, primary SWS.

onset of the beam current and microwaves was usually on the order of 50 ns. This delay corresponds to the time for establishing self-magnetic insulation.

9.7.2 Tapered MILOs

A tapered MILO, developed by Eastwood et al. (1998), is distinguished by combining the cavity depth tapering section and the diode configuration into a slow-wave interaction space (Fig. 9.34). The driver section ensures interaction at the π-mode and defines the oscillation frequency. The group velocity in this section is nearly zero. The group velocity at the next gently tapered section increases gradually at the same frequency, but the small variation in the phase velocity supports synchronous interaction of the spokes with the wave. Two factors determine the choice of taper: power flow and amplification. Lengthening of the taper improves the power extraction but worsens phase focusing because an increasing fraction of electrons would be below the resonance threshold. Eastwood et al. (1998) concluded that phase focusing in a tapered MILO allows increased gain by using M-type interaction but avoids the self-oscillation problem for different frequencies.

The sharp tapered section provides a transition to the coaxial line and additionally, diode extraction of power from the electron spokes that flow from the diode

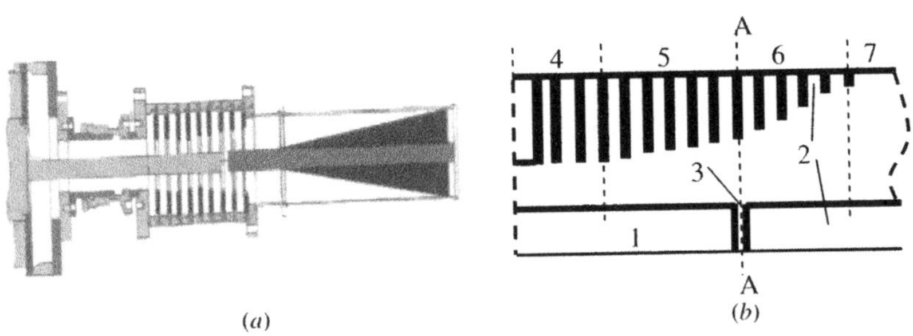

FIGURE 9.34 (*a*) Tapered MILO; (*b*) MILO interaction space. 1, Cathode; 2, anodes; 3, diode gap; 4, driver section; 5, gentle section; 6, sharp section; 7, coxial output.

end of the cathode. This jet, according to a simulation (Eastwood et al., 1998), feeds the wave mainly by giving up its kinetic energy, while the M-type interaction provides the transfer of potential beam energy into the driver and tapered sections.

The current in a MILO can be divided into two components: the first flows to the outer anode (SDS structure), and the second flows to the inner anode (to the right of the diode gap). The latter current component is largely responsible for the insulating magnetic field, while the former feeds the microwave energy. This component is an essential part of the dc current I_9 in the plane (A–A) crossing the last (ninth) vane before the diode. Figure 9.35 shows I_9 as a function of the input voltage U. The Hull cutoff and the Buneman–Hartree curves (Eastwood et al., 1998) are also shown. These functions can be represented in the following forms. Hull cutoff condition:

$$\frac{eR_c I_H}{c} = \frac{1}{\sqrt{\eta}}\sqrt{2U_H + \frac{\eta U_H^2}{c^2}} \tag{9.95}$$

Buneman–Hartree threshold voltage:

$$U_{\rm BH} = Z_c I_{\rm BH}\beta - \frac{c^2}{\eta}\left(1 - \sqrt{1-\beta^2}\right) \tag{9.96}$$

where $\beta = L/h$, $Z_c = (Z_0/2\pi)\ln(r_A/r_K)$, Z_0 is the input resistance of the system, L is the period of the SWS, h is the cavity depth, and r_A and r_K are the outer and inner radii of the coaxial line. Note that these equations are exact analogs of the corresponding relativistic magnetron conditions (9.91) and (9.92) if we replace the magnetic field B with the dc magnetic insulation current I. As we see, the operation current curve I_9 is located between the Hull and B-H threshold curves. Comparison of these curves shows that at low voltage the magnetic insulation is lost and the potential energy contribution to the microwave power is reduced. At the high

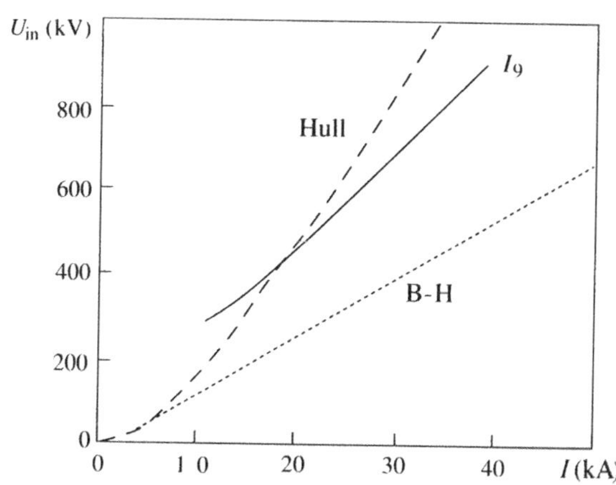

FIGURE 9.35 Dc operation current MILO in the A–A plane (Fig. 9.34). Hull and B–H currents as functions of input voltage.

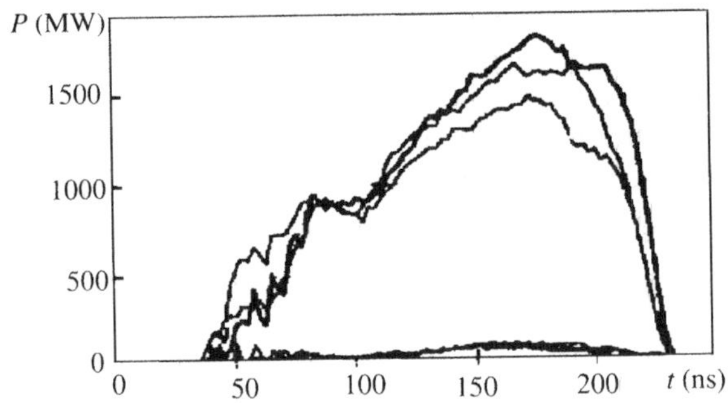

FIGURE 9.36 Measured output ac power of a 1-GHz MILO for three successive shots.

voltage, electrons remain close to the cathode and the ac power is reduced. According to Eastwood et al. (1998), peak performance is reached when the current I_9 is close to that of the Hull curve (~550 kV).

Computer modeling of a prototype 1-GHz MILO with input impedance 15 Ω predicts 25% efficiency (Eastwood et al., 2000). Experimental shots at 475 kV yielded 250 J per shot and a mean power of ~1 GW, with a pulse length of 200 ns (the latter was limited by the power supply). Profiles of the measured output power for three shots at 475 kV are shown in Fig. 9.36.

9.8 CROSSED-FIELD AMPLIFIERS

Crossed-field amplifiers can be classified as either distributed-emission or injected-beam devices. The latter [M-type traveling-wave tubes (MTWTs)] were considered in Sections 9.2 and 9.3. Distributed crossed-field amplifiers (CFAs) comprise a family of devices that differ from the magnetron in that their slow-wave circuits are non-reentrant and nonresonant. The reentrant electron beam in CFAs is similar to the magnetron beam. CFAs were invented in 1953 by Brown (1961, 1965, 1984) under the name *platinotron*.

9.8.1 Basic Types of CFAs

Diagrams of two primary CFAs are presented in Fig. 9.37. The *amplitron* (Fig. 9.37*a*) has broad applications as a high-power amplifier in L and S bands. It combines a number of positive features, including efficiency in excess of 70% and output power of several hundred kilowatts in the CW regime and tens of megawatts in the pulse regime. The amplitron employs a cold secondary emission cathode and has a broad bandwidth, together with both good phase linearity and low phase pushing. The tube has a simple structure and is mechanically rugged.

As can be seen in Fig. 9.37, the structural differences between the amplitron and the ultron include the type of interaction of the beam with the slow wave and the corresponding type of SWS employed. The amplitron is based on a backward interaction (i.e., the angular velocity of the spokes Ω_s and the corresponding angular phase velocity of the wave v_{ph} are opposite to the group velocity v_g). In the *ultron*

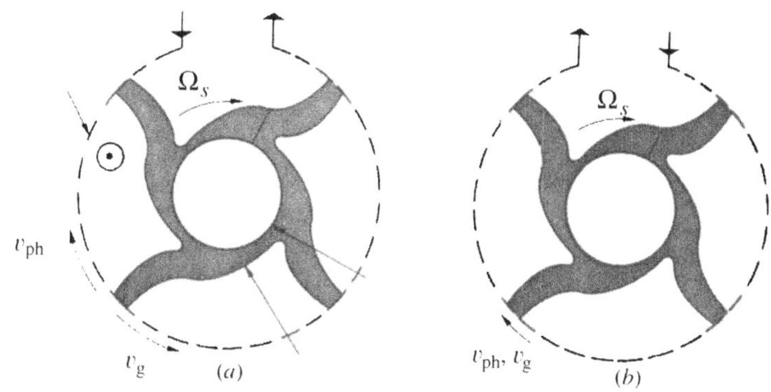

FIGURE 9.37 Platinotrons: (*a*) amplitron; (*b*) ultron. C, cathode; EC, electron cloud; Ω_s, angular velocity of spokes; v_{ph} and v_g, phase and group velocities of waves in SWS.

(Fig. 9.37*b*), forward interaction takes place (i.e., the above-mentioned velocities have coincident directions). This difference significantly influences the structure of the electron cloud in the dynamic regime. The crucial point is that the high-frequency field is absent in the gap between input and output, and therefore bunches entering this gap partly defocus as a result of the forced nature of bunching in M-type systems (see Section 9.2.3). But bunches in the amplitron after the gap are located in the strong output field, and as such are restored very fast, (i.e., the bunching process in an amplitron is essentially similar to bunching in the magnetron). This leads to the formation of identical rotating spokes (Section 9.5.3) in which the electrons move synchronously with the corresponding spatial harmonic of the ac field and drift slowly (in the rotating frame of reference) to the anode along equipotential lines (Section 9.2.2). The number of spokes corresponds to the periodicity of the synchronous component of the ac field (Fig. 9.37). (Each spoke is located in the retarded phase of the ac field).

Because the bunching processes are very similar in the amplitron and magnetron, we can apply the Buneman–Hartree threshold relation [Eq. (9.81)] for the magnetron directly to the amplitron. First we rewrite the phase of the rotating wave from Eq. (9.65) as

$$\omega t - \theta \frac{r_a \phi}{L} = \omega t - \frac{\theta N}{2\pi} \varphi \qquad (9.97)$$

where φ is an angle and θ is a phase shift in the SWS period. As a result, the angular phase velocity is

$$\Omega = \frac{2\pi\omega}{N\theta} = \frac{4\pi^2 c}{N\lambda\theta} \qquad (9.98)$$

So the factor $N\theta/2\pi$ replaces the coefficient n in the formula for the magnetron angular velocity $\Omega_M = \omega/n$ [see Eq. (9.66)]. Substitution of Ω in

Eq. (9.81) gives the Buneman–Hartree threshold voltage for the amplitron:

$$U_{\mathrm{BH}} = \frac{B}{2}(r_a^2 - r_c^2)\Omega - \frac{r_a^2}{2\eta}\Omega^2 = U_0\left(\frac{2B}{B_0} - 1\right) \tag{9.99}$$

where

$$U_0 = \frac{8\pi^4 c^2 r_a^2}{\eta\theta^2\lambda^2 N^2}, \qquad B_0 = \frac{8\pi^2 c}{\theta\lambda N(1 - r_c^2/r_a^2)} \tag{9.100}$$

There are also two other important versions of the CFA: the stabilotron and the reflex amplitron. The *stabilotron* (Fig. 9.38*a*) is a derivative of the amplitron. Part of the output power in a stabilotron is reflected back through the tube to cavity resonator 1 at the input. The amplitron is characterized by high directivity: the backward signal will be neither increased nor decreased (for the most part). After reflection from the cavity at its resonance frequency, the signal is amplified and directed into the output. Any nonresonance signal is absorbed in the matched load at the input. So the stabilotron is characterized by high power, high stability, and a tunable capability.

The stabilotron is tuned and stabilized in frequency simultaneously by resonator 1. The tuned band reaches 6 to 10%. However, for effective operation of the stabilotron within a frequency band, it is necessary to draw both the frequency of the tunable cavity and the natural frequency of the amplitron system. The latter is attained through phase shifter tuning. Without a phase shifter the frequency band is only about 0.5%. The frequency stability of the stabilotron is greater than that of the magnetron by approximately one order of magnitude (Brown, 1961). Due to this high stability, the efficiency of the stabilotron in the X-band can be close

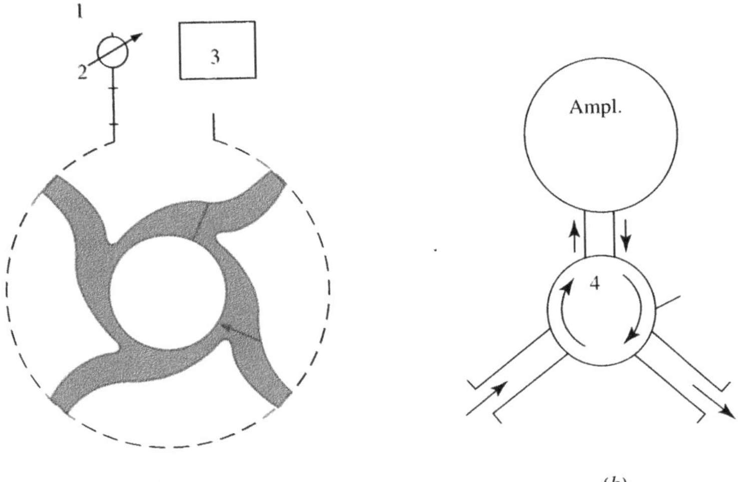

FIGURE 9.38 (*a*) Stabilitron; (*b*) reflex amplitron. 1, High-*Q* tunable cavity; 2, phase shifter; 3, output load; 4, circulator.

to the efficiency in the S-band. Another important property of the stabilotron is the possibility of obtaining a long-pulse duration (up to 20 to 40 μs). That is a perquisite for radar in the L-band. The stabilotron was first used in the Hawk missile system in 1954 and is still used today, more than 50 years later.

The *reflex amplitron* (Fig. 9.38*b*) is a single-port device. It is transformed due to circulator 4 into a two-port amplifier. The signal is introduced into the tube through one of two circulator terminals. It propagates further without gain or appreciable attenuation to the input of the amplitron, where it is amplified and reflected entirely. The amplified signal is delivered to the output through another port of the circulator. Use of a reflex amplitron greatly simplifies construction of the tube. These amplifiers have sufficiently high efficiency but comparatively narrow bandwidth. For example, in the X-band a gain of 20 dB has been obtained with a pulse power of 10 kW, an efficiency of 40%, and a bandwidth of 5%.

9.8.2 Slow-Wave Structures of CFAs

As seen in Fig. 9.37, CFAs can be backward- or forward-wave tubes. Taking into account the high power of CFAs, and setting aside issues of dispersion, coupling impedance, and bandwidth, the choice of SWS is determined essentially by the thermal capability of the structure. We point out that SWS in the CFA, as in the magnetron, operates as a collector for the electron beam.

Among the broad class of SWS configurations, helical SWSs are used more often in ultrons. These structures have the best dispersion properties and can be configured as heavily loaded systems (see, e.g., Gilmour, 1986). In principle, any waveguide-loaded forward-wave circuits can be used. The interdigital structures (Fig. 8.5) are employed in the amplitron as backward SWSs. But most applied SWSs are strapped bar lines. According to Fig. 9.39, the strapped structure consists of a system of vanes and two circular straps, one of which is attached to all even vanes, another to all odd vanes. The ends of the straps are connected with input and output devices.

Note that, in general, the dispersion of the phase velocity in backward structures is stronger than in forward structures. Therefore, the bandwidth of an amplitron, all else being equal, is narrower than the bandwidth of an ultron.

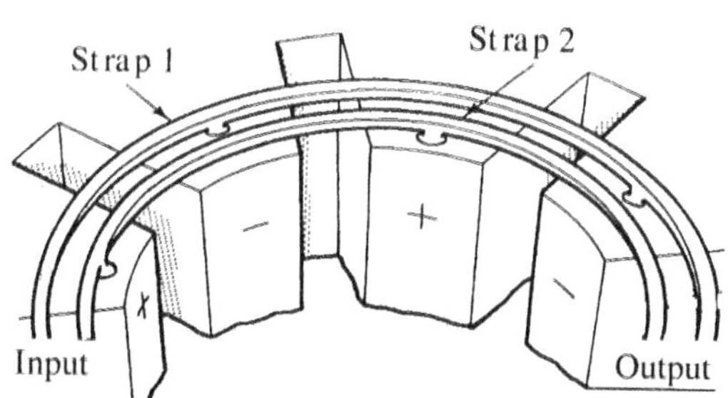

FIGURE 9.39 Strapped bar structure of an amplitron.

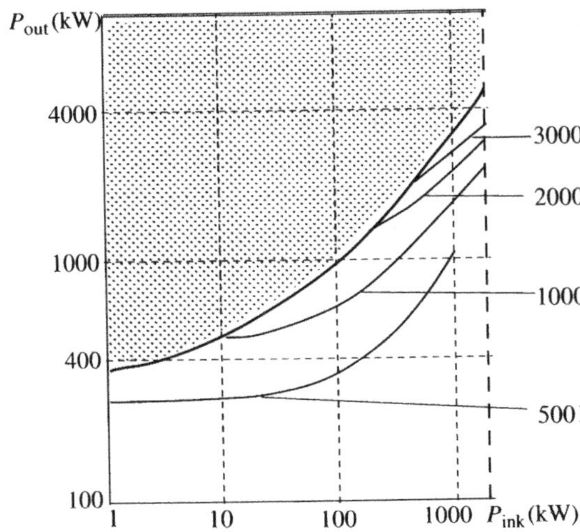

FIGURE 9.40 Peak HF power as a function of peak drive power for different input dc power.

9.8.3 Amplification and Efficiency of Amplitrons

Typical characteristics of the amplitron as a circuit element are represented in Fig. 9.40, where the dependence of the peak ac output power is shown as a function of the peak ac input power for different values of the input dc power.[3] As we see, the amplification $G = P_{out}/P_{in}$ essentially diminishes with the growth in input power, so the amplitron behaves as a saturated amplifier. For each value of the dc input power, there is a minimum input power level that provides stable amplification. In Fig. 9.40 a nonadmissible area is shown where the output signal is not controlled by the input power. In this zone the output signal is noise, poorly defined with a frequency different from a driving signal. The output power is maximum at the boundary of this zone and, according to Brown (1961), is determined by the empirical relation

$$P_{out} = \frac{K^2}{4} + K\sqrt{P_{in}} + P_{in} \tag{9.101}$$

where K is a constant. P_{in} and P_{out} should be expressed in kilowatts. Thus, the maximum gain as a function of ac input power is equal to

$$G_{max} = 1 + \frac{K^2}{4P_{in}} + \frac{K}{\sqrt{P_{in}}} \tag{9.102}$$

[3]These characteristics belong to the classical amplitron QK434, the first CFA, which replaced the magnetron 5J26, the backone of many L-band radars during World War II, with exceptional success (Brown, 1965).

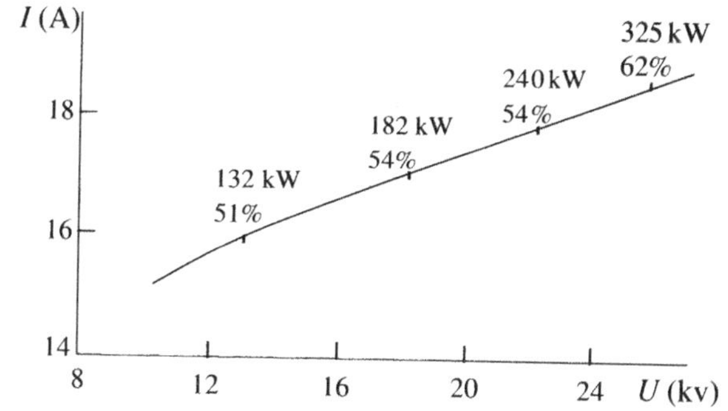

FIGURE 9.41 Voltage–current characteristic of an amplitron. Frequency = 440 MHz, $P_{in} = 25$ kW.

The parameter $K \approx 43$ for the amplitron QK-434. So for $P_{in} = 100$ kW, $G_{max} = 9.9$ ($\sim$10 dB); for $P_{in} = 1000$ kW, $G_{max} = 2.8$ (4.5 dB). It should be noted that although the gain for a large input power is itself not large, the efficiency is very high, increasing with both dc and output power. This can be seen in Fig. 9.40. So for the dc power 2000 kW, $P_{max} \gg 1000$ kW, which corresponds to $\eta \approx 60\%$. Recall that input power is not lost but appears as a part of the output power. Figure 9.41 shows the typical voltage–current characteristic of the amplitron along with values of output power and efficiency in isolated points of the characteristic. Note that powerful CFAs are characterized by relatively low voltage. For example, the superpower amplitron GR1224, developed by Raytheon Co., delivered a CW output power of 450 kW for an anode voltage of 20 kV. The amplitron had a gain of $\sim$10 dB and an efficiency of $\sim$70%. The same amplitron has delivered a pulse output power of 35 MW with 60% efficiency for an anode voltage of 120 kV.

We can trace the location of effective regimes along the Buneman–Hartree threshold line [Eqs. (9.99.) and (9.100)]. In Fig. 9.42 the optimal wavelength λ and the corresponding phase shift $\theta = \theta(\lambda)$ were used (Brown, 1961).

9.8.4 Amplitron Bandwidth

Figure 9.43 shows efficiency as a function of frequency for a typical amplitron at a fixed input. As we see, efficiency remains almost constant over the 10% frequency band. This property, along with exceptionally low phase pushing (see below), is a key characteristics of the amplitron as a powerful microwave amplifier.

Bandwidth shrinks with increase in gain. The operation range at a low-frequency border is limited by the current shift in the breakdown area (Tseytlin et al., 1978). The output signal in this case is not controlled by the input power (see Fig. 9.40). The limitation of the bandwidth at the high-frequency-side end is determined by the sharp decay of amplification as well as by mode interference and competition. The gain–frequency characteristics shown below were calculated by Dombrowski for a single-frequency input signal with a dc voltage that is adjusted by the Buneman–Hartree condition. The envelope of a family of such characteristics

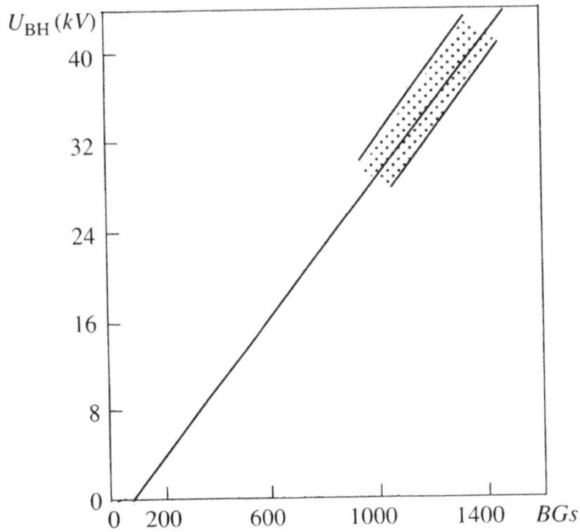

FIGURE 9.42 Buneman–Hartree line and the experimental zone of effective regimes for a QK434 amplitron. (From Brown, 1961.)

with different input powers is shown in Fig. 9.44. It can be assumed that the dB gain–bandwidth product is more or less independent of the relative drive power level.

The overall power capabilities of the main CFA types are shown in Fig. 9.45. As shown, amplitrons have a smaller bandwidth than ultrons. This can be explained by the lower dispersion of forward-wave SWSs. The output power of an ultron exceeds the megawatt level at a frequency 10 GHz. However, pulse amplitrons have significantly higher output power.

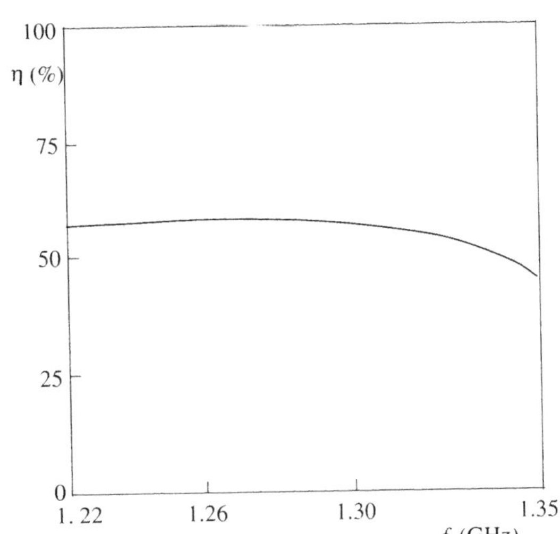

FIGURE 9.43 Efficiency of a typical amplitron as a function of frequency. (From Brown, 1961.)

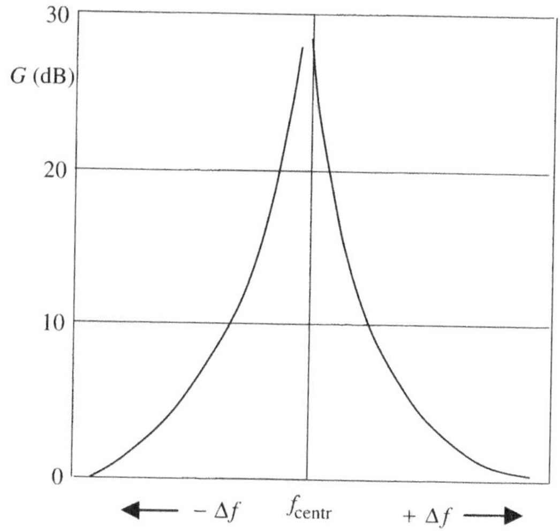

FIGURE 9.44 Envelope of the gain–frequency characteristics of a typical amplitron. (From Dombrowski, 1959.)

9.8.5 Amplitron Phase Characteristics

A phase shift between output and driver signals in any amplifier device depends on all of the operating parameters. Among them, phase sensitivity to voltage and current (*phase pushing*) has special practical importance. In TWTOs the phase changes are about 30° for a 1% change in beam voltage. In CFAs the angular velocity of the spokes is locked to the phase velocity of waves in SWSs. In addition, the electric length of the ac circuit is not large. As a result, the phase sensitivity to anode voltage changes in CFAs is significantly less than that seen in TWTOs.

Phase characteristics are determined by the electric length Φ of a device, which can be assumed to be

$$\Phi = \Phi_0 + \Phi_e \tag{9.103}$$

where Φ_0 and Φ_e are the lengths of slow-wave structures without an electron beam and in a dynamic regime, respectively. The qualitative dependence of Φ_e on the anode current in the amplitron, according to Tseytlin et al. (1978), can be represented by Fig. 9.46. Because the cold phase shift Φ_0 as a function of I is practically constant, the curve in Fig. 9.46 can be assumed to be a full phase shift. A near-zero slope for the curve can be seen for some current value. Within that region the phase pushing changes sign and is very small. In general, phase pushing in the amplitron is negligibly small compared with the injected beam high-power microwave amplifier. Experimentally measured averaged phase-pushing values of an L-band amplitron (degree/ampere) are represented in Table 9.4. According to the table data, the phase pushing changes sign in the approximate current range 15 to 25 A.

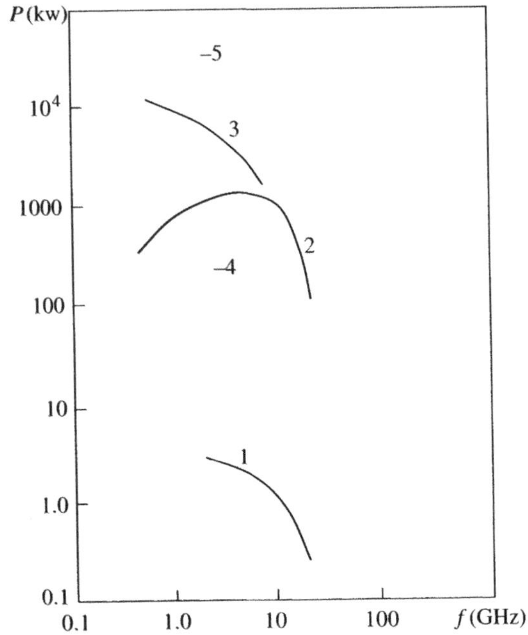

FIGURE 9.45 Power capabilities of various CFAs. 1, Injected beam CW CFAs (CW MTWT); 2, Pulsed ultrons; 3, pulsed amplitrons; 4, superpower CW amplitron; 5, superpower pulsed amplitron. (From Gilmour, 1986.)

TABLE 9.4 Phase-Pushing Values (degree/ampere) of L-Band Amplitron

	ΔI		
U_{kV}	10–20 A	20–30 A	30–35 A
36	−0.8	+0.34	+0.85
35	−0.45	+0.22	−0.8
33	−0.44	+0.22	+0.9
30	−0.6	+0.40	+0.8
26	−0.8	+0.60	+0.9

Source: Brown (1961).

The possibility of obtaining small amounts of phase pushing in an amplifier has considerable importance in the design of many radar systems. For example, phased-array antennas provide an advance in radar system performance far beyond the capabilities of the rotating antenna radars of their past.

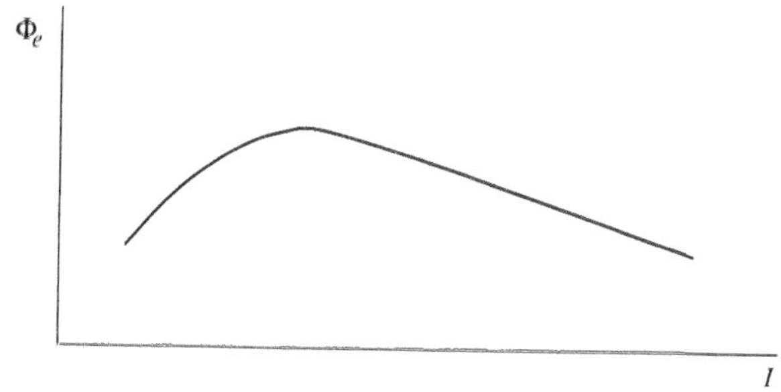

FIGURE 9.46 Electron phase shift of an amplitron as a function of beam current.

9.8.6 Electron Emission in CFAs

The effects of cathode back-bombardment by out-of-phase electrons are seen in all microwave M-type tubes where the cathode is positioned in the interaction space. For CFAs these effects are particularly important because the combination of large ac output power and small amplification results in the need for a powerful drive signal. In this case, a relatively small density of free electrons near the cathode surface is sufficient for the initiation of secondary emission. These electrons appear from various emission sources, including even negligible thermionic emission at room temperature. Further, a process of secondary emission-back-bombardment multiplication is developed that is finished by the built-up emission level. This process can be performed until the output power builds up to a sufficient dc voltage and input power. The resulting emission buildup time is very short ($\sim$10 ns) with a small jitter.

In magnetrons where the drive power is absent, secondary electron emission is possible once oscillations have begun. Therefore, the use of thermionic cathodes or other initial electron sources is necessary. In water-cooled CFAs, purely secondary emission cathodes can be used. Obvious advantages of these cathodes include a potentially very high lifetime, simplicity of construction and technology, short starting time, and of course, a very high emission current density.

Because of the emission dependence on the cathode heat regime, the theory of emission effects that are stipulated by bombardment by out-of-phase electrons is very complicated, especially with regard to the space-charge field. These effects are also essential for steady-state and transient processes, and in particular, for setting the Brillouin flow near the cathode.

It is known that a stable secondary-emission current in operating CFAs is maintained automatically. However, this effect has not been investigated thoroughly.

CHAPTER TEN

Classical Electron Masers and Free Electron Lasers

10.1 INTRODUCTION

Until the mid-1960s, almost all vacuum microwave amplifiers and oscillators exploited radiation of electrons with rectilinear and uniform motion (i.e., Cerenkov and transition radiation). Even when trajectories of particles were not strictly rectilinear, this difference had no real effect on the interaction mechanism. Despite efforts, output power and efficiency of devices inevitably declined with frequency. So the shortest nonrelativistic BWOs have a maximum output power of ~ 1 mW in the range of a fraction of a millimeter. As we saw in Chapters 7 to 9, this tendency, is not the result of design or technological miscalculation, but arises from the nature of Cerenkov and transition radiations. In both cases, effective interaction of the electron beam with the electromagnetic field is possible at a distance of a fraction of a wavelength from the beam to the circuit surface. This also concerns crossed-field devices in which synchronism of the electrons with slow electromagnetic waves should also be provided, which predetermines in the MSM range (0.1 to 4 mm) a small volume of the active medium and exceedingly tense specific loading of the circuits. Meanwhile, practical applications of microwave devices have required considerably more power. Among applications requiring it are long-range radars, countermeasure means, space communication, thermonuclear fusion experiments, and in general studies of the effect of strong radiation on matter, and in many other technological and medical applications. However, serious demand for such powerful devices in the MSM range has been damped, because demand usually follows supply, which was absent at the time. The possibility of creating powerful relativistic Cerenkov devices [i.e., O-type traveling-wave tubes (TWTOs), backward-wave oscillators (BWOs), orotrons] into the MSM range was difficult

Electron Beams and Microwave Vacuum Electronics. By Shulim E. Tsimring

to realize in practice for long-pulse or CW regimes because the advantages of these devices occur only for the relativistic factor $\gamma \gg 1$ and short-pulse explosion emission beams (Kovalev et al., 1979).

A break into the range of short electromagnetic waves occurred in 1960 and was caused by lasers, which opened the era of quantum electronics. The most effective and powerful lasers work in visible and near-infrared ranges, where the energy of quanta and the efficiency of pumping are significant. However, both decrease quickly with increased wavelength. As a result, there exists a long-wavelength threshold for the operation of powerful lasers, and this threshold coincides with the short-wave boundary of the MSM range. Thus, the MSM range turns out to be a peculiar gap in the spectrum of powerful coherent electromagnetic oscillators, where both classic microwave devices (e.g., klystrons, TWTs, BWOs, magnetrons) and lasers are not effective.

New ideas emerged toward the end of the 1950s and the beginning of the 1960s. At that time, attention turned to beams with curvilinear trajectories of electrons, because a charged particle whose trajectory has a nonzero curvature radiates for any relationship between its velocity and the wave phase velocity in the given medium. It is clear that if the trajectory of the particle is periodic in a uniform medium, its radiation should also be periodic in time. So in this case, a synchronous interaction of electrons with fast electromagnetic waves can be realized. However, this does not solve the problem. First, the electron radiation should be coherent; only in this case it does have a value for radio electronics. Second, because in general the spatial period of a trajectory (at least for subrelativistic energies) is small compared with the wavelength of radiation, it is not so simple to form electron-optical systems with elements that are not small on the wavelength scale. Otherwise, the difficulties arise for transfer from electrodynamics to electron optics, which would, as before, limit the volume of the active medium.

Two ideas, proposed simultaneously and independently by Gaponov (1959), Schneider (1959), Twiss (1958), and Zheleznyakov (1960), turned out to be decisive: (1) discovery of a mehanism of stimulated radiation via curvilinear relativistic electron beams as an ensemble of nonisochronous (anharmonic) classical electron oscillators, and (2) the formation of periodical electron trajectories in a uniform magnetic field. The reader can find a detailed account of the beginning phase of this field of research in historical surveys by Hirshfield and Granatstein (1977) and Petelin (1999).

As is well known, the electron in an atom is an oscillator with an essentially nonequidistant energy spectrum. The radiation of electrons induced in an electromagnetic field with frequency ω can be performed through a transition between levels with an energy difference $\Delta W = \hbar\omega$. Therefore, to obtain stimulated radiation, it is sufficient to place the corresponding matter into a resonator with a high Q-factor and to create an inversion population in this two-level system. The considerable energy of a quantum in the optical range and the high density of the active medium together form a fundamental basis for high-power lasers.

Electrons moving in a magnetic field from a quantum physics standpoint are oscillators with a discrete energy spectrum but with a very small difference between the nearby levels in comparison with the oscillation energy of particles (see, e.g., Landau and Lifshitz, 1977). Their frequency in the relativistic approximation depends on the energy (see Section 1.5). Therefore, the energy spectrum of these anharmonic oscillators is weakly nonequidistant. As has been shown (see, e.g., Gaponov, 1960), it is possible to provide radiation transitions between multiple pairs of levels. As a result, each electron can perform many successive transitions downward and transfer a considerable share of its energy to the field.

These properties of relativistic electrons in the magnetic field provide conditions for stimulated radiation (bunching and energy transfer) of electrons in quasiuniform electromagnetic fields for any relation of electron and wave phase velocities. Application of a uniform magnetic field considerably simplifies the electron-optical problem of the formation of a periodic beam in large volumes (on the wavelength scale).

Electrons moving in spatial-periodical magnetic fields possess analogous properties. Stimulated radiation generated by ultrarelativistic electrons in a spatially periodic magnetic field, as shown first by Madey (1971), constitutes the basis for free electron lasers (FELs).

Devices based on the stimulated radiation of classical electron oscillators (normal Doppler synchronism; see Section 10.2) are called *classical electron masers* (CEMs). The acronym ECM (*electron cyclotron maser*), introduced by Hirshfield and Wachtel (1964), is used for devices using stimulated radiation of electrons in a dc magnetic field. According to their operational principle, ECMs occupy an intermediate place between conventional microwave devices and optical lasers, and fortunately, ECMs do not encounter the difficulties considered above in the MSM range. A unique place among ECMs belongs to *gyrotrons* (Flyagin et al., 1977), which are distinguished by helical electron beams and weakly irregular open cavities as resonators. Today, gyromonotrons and their varieties (i.e., gyroklystrons, gyroTWTs, etc.) are the most powerful long-pulse and CW oscillators and amplifiers in the MSM range.

Devices based on the stimulated radiation of ultrarelativistic electrons in spatially periodic magnetic fields are called *free electron lasers* (FELs). These startling systems cover not only infrared and optical, but even ultraviolet and soft x-ray, ranges.

The mechanism of spontaneous and stimulated electron radiation is considered in Sections 10.2 and 10.3. Examples of electron CEMs are given in Section 10.4. In Sections 10.5 to 10.9, various gyrotron issues are treated. Some properties of gyroklystrons, gyroTWTs, masers on the cyclotron autoresonance (MCARs), FELs, and applications of gyrotrons are discussed in Sections 10.10 to 10.14.

10.2 SPONTANEOUS RADIATION OF CLASSICAL ELECTRON OSCILLATORS

10.2.1 Introduction

Assume that an electron performs periodic motion with velocity $\mathbf{v}_1$ and transfers energy to a plane wave with a wave vector $\mathbf{k}$, where $k = \omega/v_{\text{ph}}$. Following Zheleznyakov (1960), let us consider the result of radiation on the basis of the quantum laws of energy and momentum conservation and in the end, proceeding to the classical limit by approaching the Planck constant $\hbar \to 0$.

When an electron in a laboratory frame of reference emits a photon with energy $\hbar\omega$ and momentum $\hbar\mathbf{k}$, it changes its momentum from $\mathbf{p}_1$ to $\mathbf{p}_2$. The electron in the guiding center frame (the *rest frame*) can be considered as an oscillator with nonzero internal energy and a rest mass m_1, which is changed to m_2 as a result of the emission of the quantum $\hbar\omega_0$. Therefore, the condition of the energy conservation in this reference frame is

$$m_1c^2 - m_2c^2 = \hbar\omega_0 \tag{10.1}$$

The conservation laws in the laboratory frame are

$$\sqrt{m_1^2c^4 + p_1^2c^2} - \sqrt{m_2^2c^4 + p_2^2c^2} = \hbar\omega \tag{10.2}$$

$$\mathbf{p}_1 - \mathbf{p}_2 = \hbar\,\mathbf{k} \tag{10.3}$$

Eliminating terms with $\mathbf{p}_2$ and m_2 from these equations, we obtain

$$\omega_0c^2m_1 + \mathbf{k}\mathbf{p}_1c^2 - \frac{\omega m_1c^2}{\sqrt{1-\beta_1^2}} = \frac{\hbar}{2}(\omega_0^2 + \omega^2 - c^2k^2) \tag{10.4}$$

where $\beta_1 = v_1/c$. Transforming Eq. (10.4) in the classical limit ($\hbar = 0$), we find that

$$\omega = \frac{\omega_0}{\gamma\left(1 - \dfrac{v_1}{v_{ph}}\cos\vartheta\right)} \tag{10.5}$$

where $v_1\cos\vartheta$ is the projection of $\mathbf{v}_1$ on $\mathbf{k}$, $\gamma = 1/\sqrt{1-\beta^2}$ is the relativistic factor, $\beta = v_0/c$, and v_0 is the full velocity. Equation (10.5) expresses the Doppler effect: The radiation of a source with frequency ω_0 in a moving frame is perceived in the laboratory frame as a wave with frequency ω if the source moves with the velocity v at angle ϑ to the wave vector. It can be seen that Eq. (10.5) expresses a condition of resonance of the oscillator and the high-frequency field with a Doppler effect correction. Note that an ensemble of uncoupled electrons can produce only incoherent, spontaneous radiation.

Let us consider some particular cases.

1. $\omega_0 = 0$. The internal energy of the particle does not change (i.e., it behaves as a charged particle performing uniform rectilinear motion). According to Eq. (10.5), $\omega \neq 0$ only for $v_{ph}/v = \cos\vartheta < 1$. This relation expresses the necessary condition of the Cerenkov synchronism (i.e., radiation of rectilinear and uniformly moving particles is possible only when they are synchronous with slow electromagnetic waves).

2. $\omega_0 > 0$. The particle loses oscillatory energy with radiation. In this case there should be $v/v_{ph} < 1/\cos\vartheta$. This inequality can be fulfilled for any relation between v and v_{ph}. This is a case of *normal Doppler synchronism.*

3. $\omega_0 < 0$. The oscillatory energy of the particle increases with radiation. According to Eq. (10.5), radiation is possible only for slow waves where $v/v_{ph} > 1/\cos\vartheta$. This is a case of an *anomalous Doppler effect.* Thus, the particle radiates and simultaneously increases its oscillatory energy. Clearly, the source of both energies is the energy of the drift motion. The necessity of operating with slow waves makes the anomalous Doppler effect less attractive in the MSM range. Note, however, that the corresponding CEM can in principle be very effective, because it is possible to introduce a nonoscillating beam into a system while maintaining the possibility of subsequent emergence of oscillations.

10.2.2 Magnetic Bremsstrahlung. Doppler Frequency Up-Conversion

Radiation of an electron rotating in a *uniform magnetic field* is a specific case of magnetic bremsstrahlung (magnetic braking radiation). An oscillation frequency ω_0 in the Doppler formula [Eq. (10.5)] is the gyrofrequency e_0B/m_0 in the rest frame. Then Eq. (10.5) can be rewritten as

$$\omega = \frac{\omega_g}{1 - \frac{v}{v_{ph}}\cos\vartheta} \tag{10.6}$$

where $\omega_g = \omega_0/\gamma = e_0B/m_0\gamma$ is the gyrofrequency in the laboratory frame.

Another variant of magnetic bremsstrahlung is the radiation of an electron in a *spatially periodic magnetic field.* The simplest device used to produce oscillatory motion in a spatially periodic magnetic field (called an *undulator* or *wiggler*) is depicted in Fig. 10.1. It can be shown that the fundamental frequency of the electron oscillations in the laboratory frame is

$$\omega_u = \frac{2\pi v}{L} \tag{10.7}$$

where v is the velocity of the guiding center along the z-axis. The frequency in the rest frame will then be $\omega_0 = 2\pi v/L'$, where L' is the period of the structure for the observer in the rest frame. Let us next take into account Lorentz contraction,

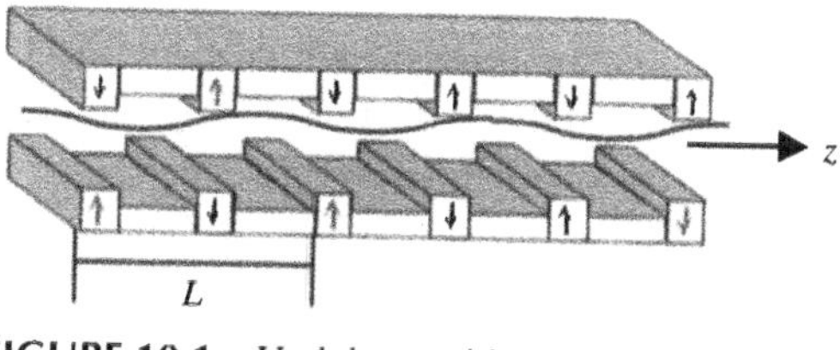

FIGURE 10.1 Undulator with permanent magnets.

$L' = L\sqrt{1-\beta^2}$. Then $\omega_0 = (2\pi v/L)\gamma = \omega_u \gamma$. Substituting ω_0 in Eq. (10.5), we obtain the Doppler formula for the bremsstrahlung with ω_u in the laboratory frame,

$$\omega = \frac{\omega_u}{1 - (v/v_{\text{ph}})\cos\vartheta} \tag{10.8}$$

Doppler frequency up-conversion (frequency gain) consists of an increase in the radiation frequency ω for a given ωg or ω_u by means of decreasing the factor $[1-(v/v_{\text{ph}})\cos\vartheta]$. The application of frequency up-conversion is especially effective for the radiation of relativistic electrons into a wave that propagates with velocity $v_{\text{ph}} = c$ in the direction of the translation motion of the particle (i.e., for $\vartheta = 0$). In this case the Doppler formula [Eq. (10.8)] takes the form

$$\omega = \frac{\omega_u}{1 - v/c} \tag{10.9}$$

Let us substitute the relativistic factor γ into Eq. (10.9), taking into account the particle oscillatory velocity $v_\perp$:

$$\gamma^{-2} = 1 - \frac{v_0^2}{c^2} = 1 - \frac{v^2 + v_\perp^2}{c^2} \quad \text{or} \quad 1 - \frac{v}{c} = \frac{\gamma^{-2} + v_\perp^2/c^2}{1 + v/c} \tag{10.10}$$

Then Eq. (10.9) can be written as

$$\omega = \frac{1 + v/c}{1 + (v_\perp^2/c^2)\gamma^2}\gamma^2\omega_u \tag{10.11}$$

Equation (10.11) for an ultrarelativistic beam with the condition

$$\frac{v_\perp}{c}\gamma \ll 1 \tag{10.12}$$

is reduced to

$$\omega = 2\gamma^2\omega_u \tag{10.13}$$

As we see, the frequency ω does not depend on $v_\perp$ for the condition (10.12). According to Eq. (10.13), it is possible to attain a very large frequency gain for a sufficiently large γ. This relation expresses the fundamental property of FELs.

A similar relation can be derived from Eq. (10.6) for the bremsstrahlung in a uniform magnetic field:

$$\omega = 2\gamma^2\omega_g \tag{10.14}$$

However, the real frequency gain in this case is a linear function of γ only because the gyrofrequency ω_g changes as $1/\gamma$.

10.3 STIMULATED RADIATION OF EXCITED CLASSICAL ELECTRON OSCILLATORS

10.3.1 Admittance of an Ensemble of Classical Oscillators

The conditions for coherent stimulated radiation by classical oscillators can be analyzed in a highly general form (Gaponov et al., 1967). We will follow a simplified model (presented by Gaponov et al., 1967). Let us consider an ensemble of identical one-dimensional oscillators formed by electrons that oscillate in a potential well $U_0(x)$ [i.e., under the action of the force $F_0(x) = -dU_0(x)/dx$]. The oscillators are also under the action of a high-frequency field $F(t) = -e_0E_0e^{i\omega t}$, which for the sake of simplicity is assumed uniform and sufficiently weak for the linear approximation to be applicable. The equation of motion in these fields reads

$$m\frac{d^2x}{dt^2} - F_0(x) = F(t) \tag{10.15}$$

The undisturbed periodical motion is described by the equation

$$m\frac{d^2x^{(0)}}{dt^2} - F_0(x^{(0)}) = 0 \tag{10.16}$$

The solution of this equation can be represented by the Fourier series:

$$x^{(0)} = \sum_{k=-\infty}^{k=\infty} x_k\, e^{ik(\omega_0 t + \phi_0)} \tag{10.17}$$

where $x^{(0)}(t)$ is the real function. Therefore,

$$x_{-k} = x_k^* \tag{10.18}$$

Taking into account the assumed weakness of the perturbation created by the high-frequency field and also the limitations of the oscillator lifetime (see below), we can represent the solution of Eq. (10.15) as $x(t) = x^{(0)}(t) + x^{(1)}(t)$, where $x^{(1)} \ll x^{(0)}$. Therefore, this equation in the first-order approximation is

$$m\frac{d^2x^{(1)}}{dt^2} - \frac{dF_0(x^{(0)})}{dx^{(0)}}x^{(1)} = F(t) \tag{10.19}$$

Unlike Eq. (10.15), Eq. (10.19) is a linear equation because $dF_0(x^{(0)})/dx^{(0)}$ is a known function of t. Its general solution by the Lagrange method of variations of coefficients yields

$$x^{(1)} = \frac{1}{mW}\int_{t_0}^{t} [\xi_2(t)\xi_1(t') - \xi_1(t)\xi_2(t')]F(t')\,dt' \tag{10.20}$$

Here $\xi_1(t)$ and $\xi_2(t)$ are linearly independent solutions of the homogeneous equation

$$m\frac{d^2x^{(1)}}{dt^2} - \frac{dF_0(x^{(0)})}{dx^{(0)}}x^{(1)} = 0 \tag{10.21}$$

$W = \xi_1(t)\dot{\xi}_2(t) - \xi_2(t)\dot{\xi}_1(t)$ is the Wronskian of solutions. It can readily be shown that for the capacity of ξ_1 and ξ_2, the partial derivatives $\partial x^{(0)}(t, C_{1,2})/\partial C$ can be used, where $C_{1,2}$ are parameters specified by initial conditions. Let us, for example, choose $C_{1,2}$ as the initial phase ϕ_0 and energy E; that is,

$$\xi_1 = \frac{\partial x^{(0)}}{\partial \phi_0}, \qquad \xi_2 = \frac{\partial x^{(0)}}{\partial E} \tag{10.22}$$

Using Eq. (10.17), it is readily verified that in this case the Wronskian is equal to

$$W = \frac{1}{m\omega_0} \neq 0 \tag{10.23}$$

and hence the ξ_1 and ξ_2 chosen are linearly independent.

The next stage of the analysis of the radiation properties of the ensemble of oscillators consists of calculation of the current produced by the ac field and then of the system admittance. At this point, it is necessary to take two principal factors into account:

1. The particles oscillate in different phases. Therefore, the total effect can be found by phase averaging of the current obtained.

2. It is necessary to express explicitly the condition of restricting oscillator lifetime τ in the ac field. Particles with infinite τ would continuously exchange energy with the ac field, and the average effect would be zero.

Let us introduce a lifetime distribution function $g(\tau)$. Then the density of particles with a lifetime in the interval $[\tau, \tau + d\tau]$ is equal to

$$dN = N_0 g(\tau)\,d\tau \tag{10.24}$$

where N_0 is the total density of particles. The current density will be

$$j = -e_0 \int \dot{x}^{(1)} \, dN = -e_0 N_0 \int_0^{\infty} \dot{x}^{(1)} g(\tau) \, d\tau \tag{10.25}$$

Measured current should be averaged over initial phases ϕ_0. The latter are usually distributed uniformly and included explicitly in ξ_1 and ξ_2 (see below):

$$\begin{aligned} \bar{j} &= \frac{1}{2\pi} \int_0^{2\pi} j \, d\phi_0 = -e_0 N_0 \int_0^{\infty} g(\tau) \left[\frac{1}{2\pi} \int_0^{2\pi} \dot{x}^{(1)} \, d\phi_0 \right] d\tau \\ &= -e_0 N_0 \int_0^{\infty} g(\tau) \bar{\dot{x}}^{(1)} \, d\tau \end{aligned} \tag{10.26}$$

The average value $\bar{\dot{x}}$ is determined by the integral in Eq. (10.20), taking Eq. (10.23) into account:

$$\bar{\dot{x}} = \frac{\omega_0}{2\pi} \int_{t_0}^{t} \int_0^{2\pi} [\dot{\xi}_2(t)\xi_1(t') - \dot{\xi}_1(t)\xi_2(t')] F(t) \, dt' \, d\phi_0 \tag{10.27}$$

Let us rewrite expressions for $\xi_{1,2}$ using the Fourier series [Eqs. (10.17) and (10.22)]:

$$\begin{aligned} \xi_1 &= \frac{\partial x^{(0)}}{\partial \phi_0} = \sum_{k=-\infty}^{\infty} ikx_k e^{ik(\omega_0 t + \phi_0)} \\ \xi_1 &= \frac{\partial x^{(0)}}{\partial E} = \sum_{k=-\infty}^{\infty} \frac{\partial}{\partial E} [x_k e^{ik(\omega_0 t + \phi_0)}] \end{aligned} \tag{10.28}$$

The derivatives $\dot{\xi}_1$ and $\dot{\xi}_2$ can be found by differentiating expressions (10.28) with respect to t. Substitution of these series into Eq. (10.27) leads to integrals such as

$$\frac{1}{2\pi} \int_0^{2\pi} \left(\sum_{k=-\infty}^{\infty} a_k e^{ik\phi_0} \sum_{m=-\infty}^{\infty} b_m e^{im\phi_0} \right) d\phi_0$$

Obviously, only terms with $m = -k$ in the double series are nonzero. These integrals are reduced to the form $\sum_{k=-\infty}^{\infty} a_k b_{-k} = \sum_{k=-\infty}^{\infty} a_k b_k^*$ [taking Eq. (10.17) into account]. Then the integral in Eq. (10.27) of ϕ_0 is equal to

$$\frac{1}{2\pi} \int_0^{2\pi} [\dot{\xi}_2(t)\xi_1(t') - \dot{\xi}_1(t)\xi_2(t') \, d\phi_0 = \sum_{k=-\infty}^{\infty} k^2 \frac{\partial}{\partial E} \left[\omega_0 |x_k|^2 e^{ik\omega_0(t-t')} \right] \tag{10.29}$$

Assuming that the HF field has the form $F(t) = -e_0 E_0 e^{i\omega t}$, we find, using Eqs. (10.27) and (10.29), that

$$\bar{\dot{x}} = -\omega_0 e_0 E_0 \sum_{k=-\infty}^{\infty} k^2 \frac{\partial}{\partial E}\left[\omega_0 |x_k|^2 e^{ik\omega_0 t} \int_{t_0}^{t} e^{i\Delta\omega_k t'}\, dt'\right] \tag{10.30}$$

where $\Delta\omega_k = \omega - k\omega_0$. Finally, taking into account that $t - t_0 = \tau$, we obtain

$$\bar{\dot{x}} = i\omega_0 e_0 E_0 \sum_{k=-\infty}^{\infty} k^2 \frac{d}{dE}\left(\omega_0 |x_k|^2 \frac{1 - e^{-\Delta\omega_k \tau}}{\Delta\omega_k}\right) \tag{10.31}$$

Let us assume the following lifetime distribution: $g(\tau) = \nu e^{-\nu\tau}$. It is readily verified that the average lifetime for this distribution is $\bar{\tau} = 1/\nu$. Substituting

$$\frac{1}{2\pi}\int_0^{2\pi}\left(\sum_{k=-\infty}^{\infty} a_k e^{ik\phi_0} \sum_{m=-\infty}^{\infty} b_m e^{im\phi_0}\right) d\phi_0$$

into Eqs. (10.27) and (10.25) and integrating, we obtain the current density and the complex admittance of the medium formed by the ensemble oscillators:

$$Y = \frac{j}{E} = e_0^2 N_0 \omega_0 \sum_{k=-\infty}^{\infty} k^2 \frac{d}{dE}\left(\frac{\omega_0 |x_k|^2}{i\Delta\omega_k + \nu}\right), \qquad \nu = \frac{1}{\bar{\tau}} \tag{10.32}$$

Limiting cases:

1. $\nu = 0$ (infinite lifetime). According to Eq. (10.32), the admittance of the system oscillators–HF field for $\Delta\omega_k \neq 0$ is reduced to pure susceptance. The subsystems of this system exchange energy, but the average energy of each subsystem is constant.
2. $\nu = \infty$ (zero lifetime). In this case, $Y = 0$; interaction is absent.

10.3.2 Linear Oscillators Near a Resonance

Oscillators that move in a parabolic potential well $[F(x^{(0)}) = -a^2 x^{(0)}]$ are harmonic oscillators, which means that only terms $k = \pm 1$ should be retained in the series (10.32). Also, the natural frequency of oscillators ω_0 does not depend on their energy: $d\omega_0/dE = 0$. Taking into account that $E = 2m\omega_0^2 |x_1|^2$, we obtain $d|x_1|^2/dE = 1/2m\omega_0^2$, and according to Eq. (10.32),

$$Y = Y' + iY'' = \frac{e_0^2 N_0}{2m}\left[\frac{1}{i(\omega - \omega_0) + \nu} + \frac{1}{i(\omega + \omega_0) + \nu}\right] \tag{10.33}$$

Let us examine Y near the resonance $\omega - \omega_0 \ll \omega_0$ with a lifetime significantly exceeding the period (i.e., $\bar{\tau} = 1/\nu \gg 2\pi/\omega_0$. The first (resonant) term in Eq. (10.33) for these conditions is much larger than the second term. Then conductance Y' and susceptance Y'' are

$$Y' = \frac{e_0^2 N_0 \nu}{2m[(\Delta\omega)^2 + \nu^2]}, \qquad Y'' = -\frac{e_0^2 N_0 \Delta\omega}{2m[(\Delta\omega)^2 + \nu^2]} \tag{10.34}$$

A plot of Y' and Y'' as functions of $\Delta\omega$ is shown in Fig. 10.2. We see that Y' is positive for any frequency (i.e., harmonic oscillators form a purely absorbing medium). The absorption maximum is related to the exact resonance $\omega = \omega_0$ and grows with an increase in lifetime. A half-width of the resonance curve is equal to $2\nu = 2/\tau$: The longer the lifetime, the more exact the resonance should be. The susceptance is zero at $\omega = \omega_0$ and changes sign together with the sign of $\Delta\omega$, which is typical for any resonance circuit.

A quantum interpretation of the behavior of this system is as follows. The spectrum of the energy levels of the harmonic oscillators is equidistant. In this case (see, e.g., Petelin, 1999) the probability of transitions between neighboring levels is proportional to the energy of the output level. Then it is obvious that the probability of the transition to the upper level is greater than to the lower level, and the oscillators absorb energy, on average. This effect can be explained from the classical standpoint by the isochronism of linear oscillators. Interaction of particles with the HF field proceeds in this case independent of the energy. But oscillators with zero energy can absorb energy only.

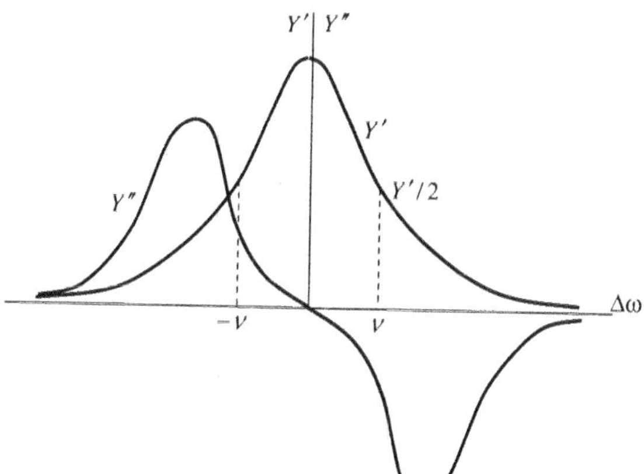

FIGURE 10.2 Conductance and susceptance of an ensemble of oscillators as functions of resonance mismatch, $\Delta\omega = \omega - \omega_0$.

10.3.3 Stimulated Radiation of Nonlinear Oscillators Near Resonance on the *n*th Harmonic ($|\Delta\omega_n| = |\omega - n\omega_0| \ll \omega_0$)

Only the nth term should be retained in the sum Eq. (10.32) for the admittance:

$$Y = e_0^2 N_0 \omega_0 n^2 \frac{d}{dE}\left(\frac{\omega_0 |x_n|^2}{i\Delta\omega_n + \nu}\right) \equiv Y_1 + Y_2 \qquad (10.35)$$

where

$$Y_1 = \frac{e_0^2 N_0 \omega_0 n^2}{i\Delta\omega_n + \nu}\frac{d}{dE}(\omega_0 |x_n|^2), \qquad Y_2 = \frac{i e_0^2 N_0 n^3 \omega_0^2 |x_n|^2}{(i\Delta\omega_n + \nu)^2}\frac{d\omega_0}{dE} \qquad (10.36)$$

The denominators in Y_1 and Y_2 are linear and quadratic functions, respectively, of the resonance factor $i\Delta\omega_n + \nu$. Let us consider the real parts of Y_1 and Y_2:

$$Y_1' = \frac{e_0^2 N_0 n^2 \nu \omega_0}{(\Delta\omega_n)^2 + \nu^2}\frac{d}{dE}(\omega_0 |x_n|^2), \qquad Y_2' = \frac{2 e_0^2 N_0 n^3 \omega_0^2 |x_n|^2 \nu \Delta\omega_n}{[(\Delta\omega_n)^2 + \nu^2]^2}\frac{d\omega_0}{dE} \qquad (10.37)$$

Y_1' and Y_2' are even and odd functions, respectively, of the mismatch $\Delta\omega_n$. A qualitative plot of the functions $Y_1'(\Delta\omega_n)$ and $Y_2''(\Delta\omega_n)$ for values $d(\omega_0|x_n|^2)/dE > 0$ and $d\omega_0/dE > 0$ is depicted in Fig. 10.3. According to Eq. (10.37), Y_1' has an extreme at the exact resonance. The sign of Y_1' coincides with the sign of $d(\omega_0|x_n|^2)/dE$. In contrast, the sign of Y_2' can be arbitrary for a given nonsynchronism $d\omega_0/dE$. This is achieved by the choice of $\Delta\omega_n$. The maximum of Y_2' is achieved for a mismatch from the resonance of approximately $|\Delta\omega_n| \sim \nu$.

What is notable in the behavior of the nonlinear oscillator ensemble? Essentially, it depends on the lifetime. According to Eqs. (10.37), the modulus of Y_1' is proportional to $\bar{\tau}$, while Y_2' is a quadratic function of $\bar{\tau}$. Thus, for large values of the lifetime, the conductance of the system is determined by Y_2' (i.e., by the

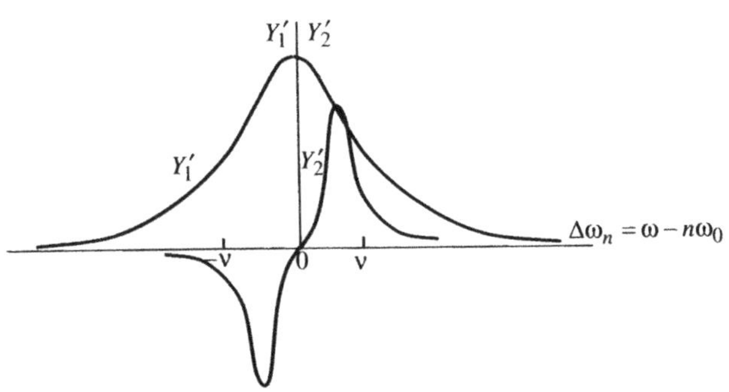

FIGURE 10.3 Frequency dependence of linear and quadratic conductances in the vicinity of resonance on the nth harmonic.

nonsynchronism of oscillators). For an arbitrary sign of $d\omega_0/dE$, the system of oscillators can be made radiating by choice of the sign of $\Delta\omega_n$.

Quantum Interpretation of the Behavior of Nonisochronous Oscillators In this case (Gaponov, 1960; Gaponov et al., 1967) the energy spectrum of the classic oscillator is nonequidistant. Therefore, the frequencies of the transitions $\omega_{p,p+n}$ (from level p to level $p+n$) and $\omega_{p,p-n}$ are different. One can show (Gaponov, 1960) that the conductance Y_2', which is an even function of $\Delta\omega$ [Eq. (10.37)], is connected with the difference in transition probabilities $p \to p+n$ and $p \to p-n$. The negative sign of Y_2' for the oscillators with a large negative derivative $d\omega_0/dE$ is a result of $\omega_{p,p+n} - \omega_{p,p-n} < 0$ (i.e., of the excess of quantum radiated over quantum absorbed.

Example 10.1: Relativistic Electrons in a Magnetic Field The natural frequency of electron oscillations is $\omega_0 = \omega_g = e_0 B/m_0\gamma = (E_0/E)\omega_g^{(0)}$, where $E_0 = m_0c^2$, $E = mc^2$, and $\omega_g^{(0)} = \eta B$ is the nonrelativistic gyrofrequency. The derivative $d\omega_0/d\mathrm{E} = -E_0\omega_g^{(0)}/E^2 = -\omega_g/E$. According to Eq. (10.37), the stimulated magnetic bremsstrahlung takes place for $\Delta\omega_n > 0$ (i.e., for $n\omega_g < \omega$).

Linear and Quadratic Bunching One question arises: Why does the ensemble of oscillators distributed randomly over initial phases possess nonzero admittance; that is, why does it give a specific response to high-frequency perturbation? The reason for this behavior is that phase bunching of particles occurs with respect to the phase of the HF field. Let us calculate the average perturbation of the electron coordinate over the initial phase and the lifetime. Applying Eq. (10.20), we obtain

$$\bar{x}_1^{(1)} = \frac{\omega}{2\pi}\int_0^{2\pi}\int_0^{\infty}\int_{t_0}^{t} [\xi_2(t)\xi_1(t') - \xi_1(t)\xi_2(t')]F(t')g(\tau)\,dt'\,d\tau\,d\phi_0 \tag{10.38}$$

Substitution ξ_1 and ξ_2 from Eq. (10.28), $g(\tau) = \nu e^{-\nu\tau}$, $F(t') = -e_0E_0e^{i\omega t'}$, and integration give

$$\bar{x}_1^{(1)} = ie_0E\omega_0 \sum_{k=-\infty}^{\infty} k\frac{d}{dE}\left(\frac{|x_k|^2}{i\,\Delta\omega_k + \nu}\right) \tag{10.39}$$

Separating the resonance term $k = n$, we obtain

$$\frac{\bar{x}_1^{(1)}}{E} = e_0\omega_0\left[\frac{in}{i\,\Delta\omega_n + \nu}\frac{d|x_n|^2}{dE} - \frac{n^2}{(i\,\Delta\omega_n + \nu)^2}\frac{d\omega_0}{dE}\right] \tag{10.40}$$

This formula shows that the average shift of the particles is nonzero and oscillates with the field (i.e., phase bunching takes place). An imaginary part of $|\bar{x}_1^{(1)}|/E$ determines the phase shift of the bunch with respect to the field. The first term in

Eq. (10.40) has the value $\sim 1/\nu$ (i.e., it grows linearly with the average lifetime and describes *linear bunching*. The second term, connected with nonsynchronism of the oscillators, varies as $\bar{\tau}^2$ and describes *quadratic bunching*.

Nonisochronous Parameter An effectiveness of quadratic bunching that ensures negative absorption of the medium (i.e., radiation), can be estimated through the special nonisochronous parameter

$$\mu = \left|\frac{Y_2'}{Y_1'}\right|_{\Delta\omega_n=\nu} = \frac{\omega_0 n|d\omega_0/dE||x_n|^2}{\nu|d(\omega_0|x_n|^2)/dE} \tag{10.41}$$

Example 10.2: Nonisochronous Parameter μ of Electron Oscillators in a Magnetic Field In this case, $\omega_0 = \omega_g$. Using $d\omega_0/dE$ from Example 10.1 and taking into account that fact that $m\omega_g$ does not depend on E, we obtain from Eq. (10.41),

$$\mu = \frac{nm\omega_g^3|x_n|^2}{\nu Ed(m\omega_g^2|x_n|^2)/dE} \tag{10.42}$$

Consider further a *subrelativistic approximation* $E \approx m_0c^2 + m_0v_\perp^2/2$, $m \approx m_0$, and $\omega_g \approx \omega_g^{(sr)}$. We find that $m\omega_g^2|x_n|^2 \approx m_0v_\perp^2$ and $d(m_0v_\perp^2)/dE \approx 2$. Then taking Eq. (10.32) into account, we obtain

$$\mu = \frac{nm_0\omega_g^{(0)}v_\perp^2\bar{\tau}}{2m_0c^2} = \frac{n\omega_g^{(0)}\beta_\perp^2\bar{\tau}}{2} = \pi\beta_\perp^2 N = \frac{\pi\beta_\perp^2 L}{\beta_z\lambda} \tag{10.43}$$

where L is the interaction length and $N = \bar{\tau}/T$ is the number of revolutions of an electron during the lifetime, $\beta_\perp = v_\perp/c$ and $\beta_z = v_z/c$. Note that relativistic properties of the electron oscillator validate this approximation since $d\omega_0/dE \neq 0$. The dimensionless parameter μ in the gyrotron theory is sometimes called the *normalized interaction length* (see, e.g., Kartikeyan et al., 2004).

Radiation via Linear Bunching. Phase Selection Let us consider the case of a lifetime that is insufficient for developing quadratic bunching, and the main part of the conductance is determined by the term Y_1'. The condition of radiation in this case, according to Eq. (10.37), is

$$\frac{d}{dE}(\omega_0|x_n|^2) < 0 \quad \text{or} \quad \omega_0\frac{d|x_n|^2}{dE} < -|x_n|^2\frac{d\omega_0}{dE}$$

If the derivative $d\omega_0/dE > 0$, the derivative $d|x_n|^2/dE$ must be negative, which has a small probability. It remains to assume that $d\omega_0/dE < 0$, and the value $|d\omega_0/dE|$ must be sufficiently large. This case can be realized when a potential well has a "shelf" (Fig. 10.4). If the shelf is long (dashed line), this effect becomes more pronounced.

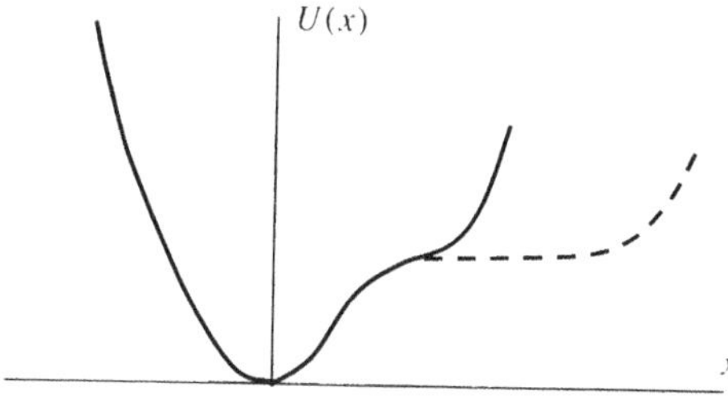

FIGURE 10.4 Potential well with a shelf.

A particle at a long shelf practically stops oscillating and interacting with the field. This type of phasing is called *phase selection*. The phase selection effect leads to radiation when electrons in the accelerated phase of the HF field are excluded from the interaction.

10.3.4 Spatial Bunching of Classical Oscillators

The spatial uniformity of an HF field and the one-dimensional motion of particles assumed above in general represent significant idealizations. In more general situations, other mechanisms of phasing and stimulated radiation are also possible. These mechanisms were considered in detail by (Gaponov et al., 1967). In particular, even linear oscillators can form a radiative medium if the HF field has alternate signs on the particle trajectory.

As an example, we consider particles that perform nondisturbed oscillations along the x-axis, drift in the Y-direction, and are placed in the HF field with components E_x and E_y. Further, we assume that the HF field is nonuniform in the x-direction with a nonuniformity scale on the order of the amplitude of oscillations. In this case, the oscillators drift along the y-axis with different velocities that depend on the HF field phase. As a result, bunches emerge and radiate in the retarding phase of the E_x field. It turns out that this is quadratic bunching (Gaponov et al., 1967). This spatial bunching is not connected with the nonisochronism of oscillators. Note that for "pure" phase bunching, oscillators are stirred and spatial compressions in the medium are absent.

10.4 EXAMPLES OF ELECTRON CYCLOTRON MASERS

10.4.1 Oscillator of Barkhausen–Kurtz

The scheme of the oscillator of Barkhausen–Kurtz (1920) is shown in Fig. 10.5. Structurally, this is a triode with a positive charged grid. Electrons of an undisturbed beam perform oscillations, passing through the grid. The LC circuit that is tuned to the frequency near the frequency of free electron oscillations is introduced between the cathode and the anode. Electrons in the retarded phase of an HF field perform

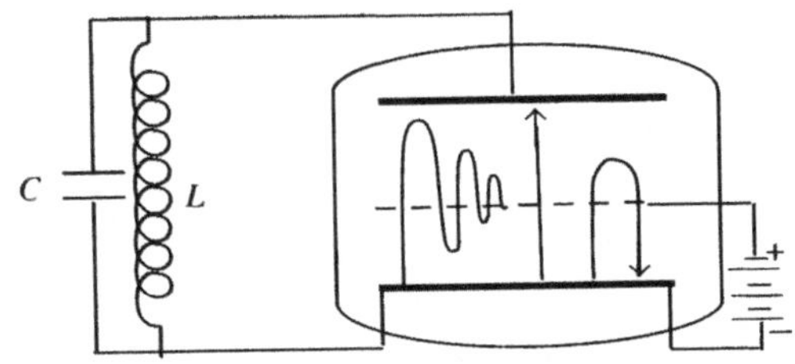

FIGURE 10.5 Barkhausen–Kurtz oscillator.

damping oscillations near the grid and transfer their energy to the circuit. Electrons in the accelerated phase of the HF field depart to the anode or the cathode with small energy. The oscillator is a typical system with phase selection.

10.4.2 Strophotron

A three-electrode structure is placed into a uniform dc magnetic field (Alfven and Rommel, 1954), as shown in Fig. 10.6. Electrons emitted from the cathode perform oscillations in the potential well along the magnetic field. The presence of a y-component of the dc electric field, together with the dc magnetic field **B**, provides a drift of electrons in the crossed fields along the plates, so that the total particle trajectories have a form of flat helices.

A strophotron can operate in two modes: phase selection and phase bunching. If a potential well has a parabolic structure, the oscillations are harmonic. Then stimulated radiation is determined by phase selection: Accelerated particles are deposited on the cathode plates. Retarded particles oscillate with reduced amplitudes and transfer energy to the *LC* circuit. The second mode is a result of a nonparabolic potential well: The electrons become nonisochronous oscillators and phase bunching arises. This mode was investigated by Agdur (1957), a graduate student of H. Alfven. Note that Agdur was probably the first person to formulate the principle

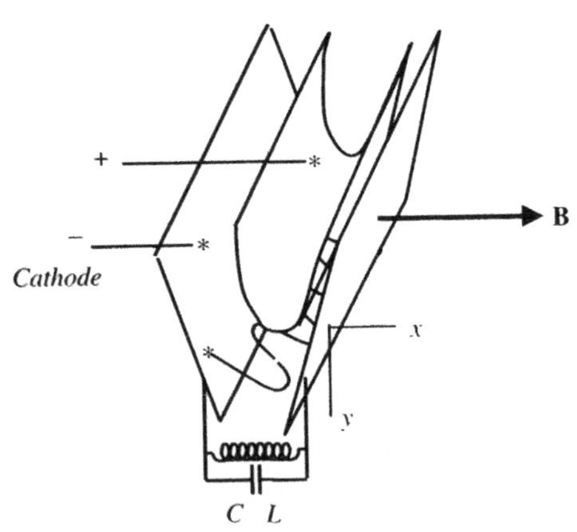

FIGURE 10.6 Strophotron.

of phase bunching of nonisochronous oscillators. In reality, the operational mode of the strophotron is a combination of the two modes indicated.

10.4.3 Ubitron

Spontaneous radiation of electrons in a spatial-periodic magnetic field discovered by Motz (1951) used high-energy electron beams (up to 3 MeV) and produced millimeter and submilimeter radiation. Phillips (1960) developed a microwave amplifier based on this principle which he called the *ubitron* (undulating beam interaction). Stimulated radiation of an electron beam propagating in a periodic magnet system (*undulator*) has been implemented on the basis of synchronous interaction of electrons with a TE_{01} mode of a circular waveguide. The scheme of the undulator with permanent magnets is shown in Fig. 10.1 in connection with magnetic bremsstrahlung. The synchronism condition is

$$\omega = \frac{\omega_u}{1 - v/v_{ph}} \tag{10.44}$$

where the undulator frequency $\omega_u = 2\pi v/L$. v and v_{ph} are the electron and the wave phase velocities, respectively, and L is the period of the magnetic structure. The ratio $v/v_{ph} \ll 1$ for subrelativistic ubitrons, and $\omega \approx \omega_u$.

Development of ubitrons by the Phillips team continued until 1966, when the project was terminated for a lack of financial support. The final 70-kV ubitron produced 150 kW of peak power at 54 GHz with a maximum efficiency of 6% and a maximum gain of 30 dB (Phillips, 1988). Intensive investigations of the ubitron on the basis of relativistic Doppler frequency up-conversion (see Section 10.2.1) did not begin until 1971. This development transformed the ubitron into a free electron laser (FEL) (see Section 10.14).

10.4.4 Peniotron

The electrodynamic system of the first peniotron (Yamanouchi et al., 1964) was a double-pair ridged rectangular waveguide (Fig. 10.7). The electric field of the TE wave in the waveguide is concentrated mainly between ridges (regions I and III) and is practically absent in region II. A field frequency is chosen equal to one of the even harmonics of the gyrofrequency (usually, the second or fourth). The axis

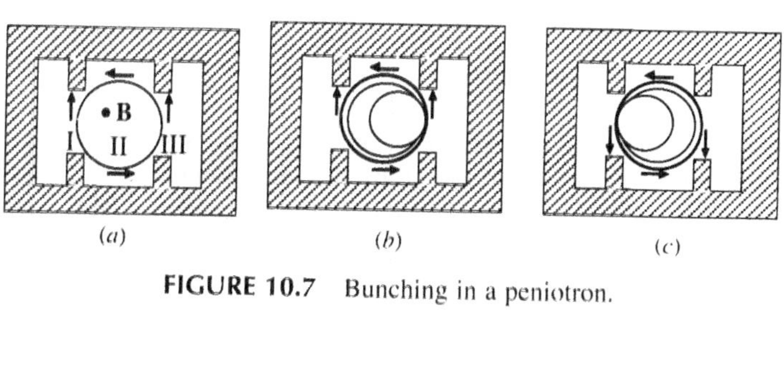

(*a*) (*b*) (*c*)

FIGURE 10.7 Bunching in a peniotron.

of the undisturbed thin helical electron beam coincides with the axis of symmetry of the waveguide cross section. The dimensions of the cross section and the Larmor radius are chosen so that an undisturbed electron beam intersects all three regions.

The vertical arrows in Fig. 10.7*a* and *b* correspond to the phase of the ac electric field in which the electrons in region I are accelerated. When the particles pass half of the circle and get to region III, the field phase is changed by 2π and the electrons are retarded. In the beginning, the two actions indicated are equal. However, it is necessary to take into account the drift of the guiding center in the cross fields (static magnetic field and ac electric field). This drift for the phase of the ac electric field shown in Fig. 10.7*b* is directed to the right. Therefore, rotating particles are subjected mainly to retardation. The radii of the orbits are decreased, and the beam kinetic energy is transferred to an electromagnetic field. For the inverse phase of the electromagnetic field, the "left" electrons are retarded, but the direction of the drift is also reversed. Therefore, the retardation also prevails, and the orbits tighten to the left ridge (Fig. 10.7*c*). We see that nonsynchronism of oscillators is not important in this case, and bunching can be considered spatial.

The peniotron has a number of attractive properties (e.g., high efficiency, possibility of operation on harmonics of the gyrofrequency). Its main disadvantage is a small volume of active medium in the wavelength scale and very tough requirements for electron optics. Certain improvements have been made in the modified peniotron (Razeghi et al., 1985), in which a ridged waveguide is replaced by a rectangular waveguide with a TE_{11} mode. This version produces certain increases in efficiency and output power in the centimeter range.

10.4.5 Magnicon

Magnicon is a scanning-beam microwave amplifier that is an advanced version of the *girocon*, invented by G.I. Budker in 1967 (Budker et al., 1974). The input cavity of a girocon with a circular polarized ac field works as a deflector, deflecting electron trajectories of a relativistic beam with frequency ω. The scanned beam can interact synchronously with the rotating electromagnetic field of the output resonator without bunching. This creates the preconditions for a significant increase in efficiency compared with other types of microwave amplifiers. In contrast to the girocon, in the magnicon (see, e.g., Nezhevenko, 1994) the output resonator is placed in a dc magnetic field, which is conducive to the formation of helical trajectories. The gyrofrequency of electrons ω_g is chosen to be equal to natural frequencies of both input and output resonators ω. As a result, a stimulated synchronous bremsstrahlung arises. Thus, a magnicon can be referred to as one of the class of electron cyclotron masers (ECMs).

The scheme of the simplest version of the magnicon is shown in Fig. 10.8. The electron trajectories of the beam emitted by the high-voltage gun (1), are deflected by the rotating ac magnetic field of TM_{110} mode in a cavity (2) at angle α_0. The radially deflected electrons travel along straight lines, forming a conic surface. After the electrons enter the magnetic field formed by the magnet (3), and the

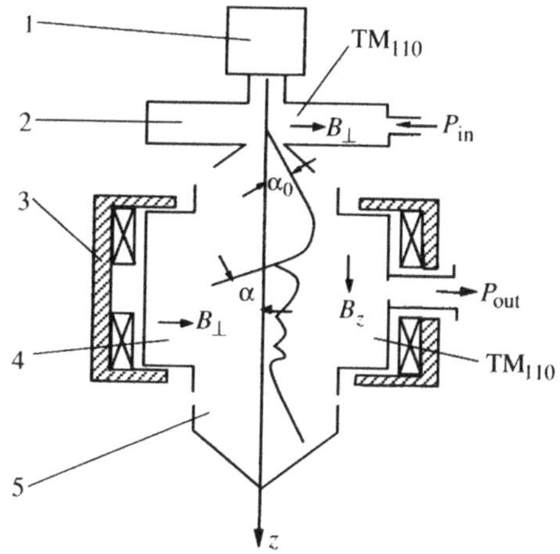

FIGURE 10.8 Magnicon.

electron trajectories are transformed into helixes with a pith angle $\alpha > \alpha_0$. Then the helical trajectories enter the output cavity (4) with a rotating entry point at the drive frequency ω, and the mode TM_{110} is excited at the same frequency ω of the input signal. If the gyrofrequency ω_g is equal to ω, the interaction is supported during many periods of RF oscillations. Particle energy is transformed into electromagnetic field energy due to the decrease in transverse velocity. It can be shown (Nezhevenko, 1994) that axial electron velocity is conserved. Therefore, to achieve a high efficiency, the maximum values of angle α should be provided. The large time and long length of the interaction make it possible to diminish the RF field and ohmic losses in the output cavity. In addition, the presence of a strong dc magnetic field decreases beam interception. As a result, the magnicon can produce significantly higher power at shorter wavelengths than can the gyrocon. The early magnicon (1985) had the following parameters (Nezhevenko, 1994): frequency, 915 MHz; beam voltage and current, 300 kV and 12 A; output power, 2.6 MW; efficiency, 73%, RF pulse duration, 30 μs; and gain, 30 dB.

Later versions of the magnicon included two interim (passive) cavities, which made it possible to increase the deflection angle at the entry to the output resontor. The latter is tuned to the second harmonic (mode TM_{210}) of the drive signal. Parameters of this magnicon were as follows: frequency, 7 GHz; output power, 50 MW; beam voltage and current, 420 kV and 240 A; efficiency, 50%; RF pulse duration, 5 μs; and gain, 50 dB. Note that the required small thickness of the electron beam was provided by the electron gun with the extraordinary electrostatic compression of 1000 : 1 in area.

An advanced version of the magnicon worked in the millimeter range (34 GHz) (Nezhevenko et al., 2003). The RF system consisted of seven cavities. The last output cavity was tuned to the third harmonic (mode TM_{311}) of the drive frequency 11.2 GHz. The beam voltage and current were 500 kV and 215 A. The output power measured was about 4 MW, and the pulse duration was 1.5 μs. The beam diameter was less than 1 mm, which that a cathode current density of 12 A/cm^2 corresponded to an electrostatic compression of $\sim$2000 : 1 in area.

As we have seen, the CEM structures considered above for creating periodic motion of electrons have typical transversal dimensions much larger than the wavelength. But transversal dimensions of the electron beam are on the order of the spatial oscillator scale (i.e., considerably less than wavelengths). So the difficulties of short-wavelength TWTs and BWOs connected with the surface nature of slow electromagnetic waves are transferred to the electron optics. Note that transversal distributions of both fields (static fields in electron masers and slow-wave electromagnetic fields in conventional TWTs and BWOs) are described by the same Laplace equation. The difficulties indicated can be partially overcome by the use of relativistic and ultrarelativistic electron energies. However, these devices in general are not suitable as sources of powerful CW or long-pulse radiation in the MSM range.

In light of the foregoing discussion, free electron masers in which periodic trajectories are formed by dc uniform (homogeneous) fields currently attract the greatest attention. Of course, the dc magnetic field is an obligatory element. Thus, CEMs with uniform dc fields are always electron cyclotron masers (ECMs). Only two types of periodic trajectories are possible in uniform static fields (Section 1.5): helical trajectories (e.g., uniform magnetic field) and trochoidal trajectories (e.g., perpendicular electric and magnetic fields, nonrelativistic approximation). Trajectories of relativistic electrons in perpendicular E and B fields are periodic for $E/B < c$. However, these trajectories are not trochoids and are quite complex (see Landau and Lifshitz, 1987, and Section 1.5.4).

10.4.6 Trochotron (ECM with Trochoidal Electron Beam)

In Fig. 10.9 the interaction zone of an ECM with a trochoidal beam (a *trochotron*) (Antakov et. al., 1960) is shown. In a weakly relativistic approximation (see Section 1.5.4), electron trajectories in uniform crossed fields are formed by superposition of the rotation with the gyrofrequency in the plane perpendicular to B_0 and the transversal drift of the guiding center with velocity

$$\mathbf{v}_t = \frac{\mathbf{E} \times \mathbf{B}_0}{B_0^2} \qquad (10.45)$$

For the proper length of the interaction zone, the quadratic bunching caused by the relativistic dependence of a gyrofrequency on an energy prevailed, and stimulated bremsstrahlung follows. These ECMs can be used as amplifiers and oscillators,

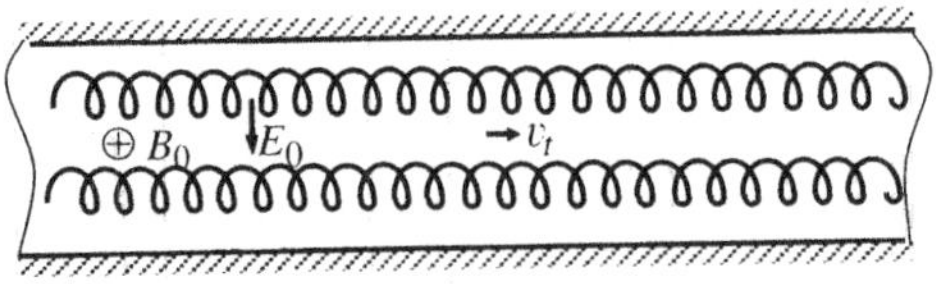

FIGURE 10.9 Interaction zone in a trochotron.

depending on the type of electrodynamics system. The CW output power of a 35-GHz trochotron amplifier reached a record value of 1 kW in 1966.

The drift velocity v_t diminishes with frequency because the magnetic field $B_0 = \omega_g/\eta \approx \omega/\eta$. Therefore, it is necessary to increase the value of the electric field E_0 for maintaining the drift velocity. However, the electric field is limited by the effect of the electric breakdown of the interelectrode gap ($E_{0,\max} \sim 100$ kV/cm). As a result, the drift velocity is diminished when the wavelength is shortened, and the space-charge density increases for a given beam current. For example, $B_0 \approx 10$ T for a wavelength of 1 mm, and according to Eq. (10.45), a drift velocity with $E = E_{\max}$ is equal to $v_t = 10^6$ m/s, which corresponds to the energy of the drift motion 3 eV. The latter leads to a strong potential depression in an electron beam and inhomogeneity of the electric field. These effects significantly limit the output power of trochoidal ECMs in the short-wavelength band of the MSM range.

10.4.7 Gyromonotron: Oscillator with a Helical Electron Beam in a Quasiuniform Waveguide Near Its Cutoff Frequency

The fundamental flaw of the trochoidal ECM is absent in another ECM, the *gyrotron*, in which electrons move in a purely magnetic uniform field B_0 and the electron trajectories are helix lines. The scheme of the gyromonotron (a gyrotron oscillator with a single resonator) (Flyagin et al., 1977; Gaponov et al., 1976; Antakov et al., 1966) is shown in Fig. 10.10.

Stimulated radiation arises as a result of quadratic bunching. Taking the Doppler effect into account, the frequency of radiation is

$$\omega = \frac{n\omega_g}{1 - v_d/v_{ph}}, \qquad n = 1, 2, \ldots \tag{10.46}$$

where v_{ph} is the phase velocity and v_d is the drift velocity of the guiding center. In contrast to v_t, in Eq. (10.45) the velocity v_d depends on the conditions of beam

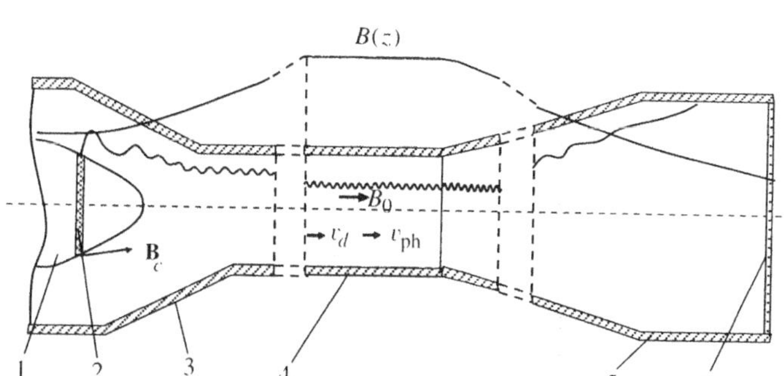

FIGURE 10.10 Gyromonotron. 1, Cathode; 2, emitter; 3, anode; 4, resonator; 5, collector; 6, output window.

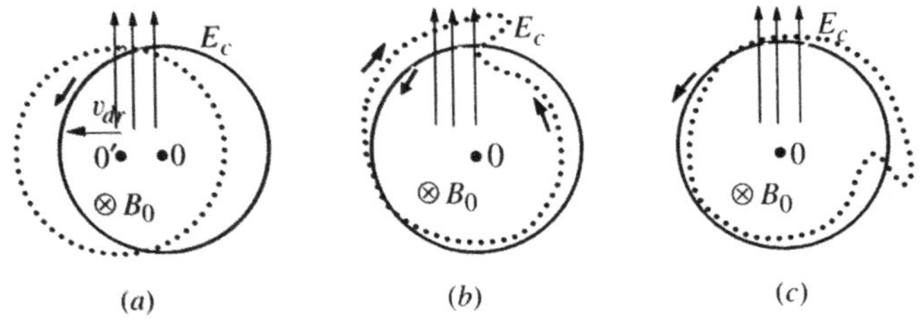

FIGURE 10.11 Phasing of electrons in a gyrotron.

formation (see, e.g., Section 4.3), and in principle it is not the same for different electrons of the beam. However, these distinctions are not significant for $v_{ph} \gg v_d$. Therefore, the electrodynamic system of a gyrotron is chosen as sections of a weakly irregular waveguide with the cross section close to cutoff at an operation frequency. So $v_{ph} \gg c$, and the frequency is practically independent of v_d ($\omega \approx \omega_g$) if the distribution of the dc magnetic field at the resonator is uniform.

Let us illustrate the mechanism of the phasing of electron oscillators based on their nonisochronism. Assume that rotating electrons interact with a wave in which the electric field is linearly polarized in a plane perpendicular to the dc magnetic field. In this plane, undisturbed electron trajectories have a circular shape. One such trajectory of an electron emitted from a certain point on a cathode is shown in Fig. 10.11 by solid circles. Electrons emitted from the same point on a cathode at equal time intervals will fill that circle uniformly.

Let us decompose the linearly polarized HF electric field into the sum of two rotating circular polarized fields. One component, E_c, rotates synchronously with the electrons and will affect their motion significantly. In the frame of reference that rotates with the field frequency about the laboratory frame, this component looks like a static uniform field, and together with the dc magnetic field, it produces a drift of electrons and their shift to the left with the velocity v_{dr} (without taking Coriolis acceleration into account). Therefore, over time the electrons will come to occupy the dashed circle with center O′ (Fig. 10.11*a*). In the laboratory frame they rotate around the center O but with different radii: The radii of "left" electrons increase and those of "right" electrons decrease. The increments of the radii of the left and right electrons in the first approximation are equal. However, the change of energy is proportional to the square of the radius; therefore, the change in left electrons' energy prevails. Thus, an ensemble of isochronous oscillators on average absorbs HF field energy. This effect has been established in a general form in Section 10.3.2.

Let us take into account nonisochronism of the electron rotation. An increase in electron energy at an accelerated HF field is accompanied by an increase in mass and a decrease in gyrofrequencies. Therefore, particles in the rotating frame not only drift to the left but turn clockwise. Electrons in the retarded phase lose mass and turn counterclockwise. As a result, a bunch is formed in the neutral phase of the HF field (the upper part of the Larmor circle) (Fig. 10.11*b*). This system could begin to radiate if the bunch is displaced in the retarded phase of the field (Fig. 10.11*c*). This is achieved by choosing a positive frequency detuning

$\Delta\omega = \omega - \omega_g > 0$. Subsequent passage of the bunch further into an accelerated phase can be eliminated by limiting the transit time (lifetime) $\bar{\tau}$: $\Delta\omega\bar{\tau} < \pi$.

It may readily be shown that the orbital bunching considered belongs to a type of inertial bunching (see Section 9.2.3). Indeed, let us imagine that electrons have interacted with an HF field and after that entered a region free from the HF field. Then the difference of masses and the corresponding difference of angular velocities that were accumulated during interaction would be conserved and would lead to continuation of the bunching. This bunching has much in common with the bunching seen in rectilinear electron beams, which is being used in conventional O-type electron beam devices (e.g., klystrons, TWTOs, BWOs). This analogy suggests a correspondence between O-type devices and the various types of gyrotrons. Table 10.1 presents sketches of devices of both classes. The longitudinal distributions of the HF field shown in the table are the simplest approximations that were assumed in estimation of the efficiency in a gyrotron theory (Flyagin et al., 1977). In particular, the uniform approximation and the narrow gap approximation were used in gyromonotron and the gyroklyston theory, respectively. The form of electrodynamic systems in gyroTWTs and gyroBWOs was assumed to be uniform waveguides. In all cases, the magnetic field was assumed uniform. The orbital efficiency shown in Table 10.1 is by definition equal to the ratio of the electron power to the power of the oscillatory motion in the electron beam. Experimental data in general are in accord with the theory. It is important to note that optimization of the longitudinal field distribution may lead to a considerable rise in efficiency. The principal advantages of the gyrotrons described above have been realized in an extensive set of theoretical and experimental work. These studies have demonstrated that gyrotrons do not have competitors as sources of CW and long-pulse radiation in the MSM range. Below, the physics of these devices is considered in more detail. Primary attention is paid to the gyromonotron (note that this oscillator is often called simply a gyrotron

TABLE 10.1 Similar Features of Gyrotron Devices and Electron Beam Devices of O-Type

O-Type Devices	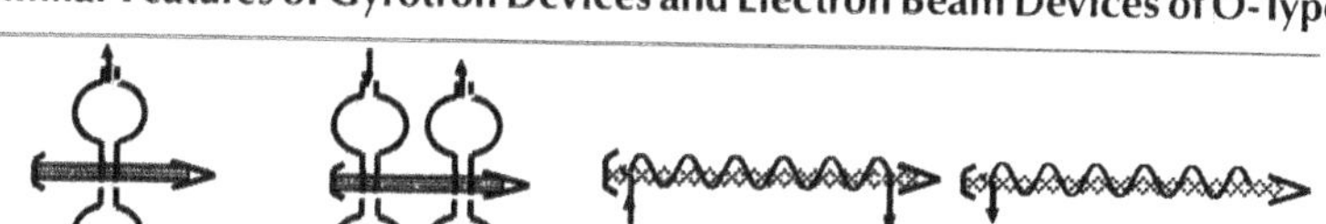			
	Monotron	Klystron	TWT	BWO
Gyrotron Devices				
	Gyromonotron ("gyrotron")	Gyroklystron	Gyro-TWT	Gyro-BWO
HF Field Structure				
Orbital Efficiency	42%	70%	34%	20%

in the literature). The most significant theoretical and experimental results in theory and experiments with cyclotron masers have been obtained with gyromonotrons. Finally, a large part of gyromonotron theory can be extended directly to other types of gyrotrons. Note that the main differences between magnicon and gyrotron designs (but not mechanisms!) consist of the method used to create electron oscillatory motion.

10.5 RESONATORS OF GYROMONOTRONS (FREE AND FORCED OSCILLATIONS)

10.5.1 Construction and Basic Parameters of Resonators

As mentioned earlier, we will focus on the gyromonotron: that is, a gyrotron-type oscillator with a single cavity. The configuration of the resonator at first glance is very simple (Fig. 10.12). In reality, one should take into account a number of factors that arise from the necessity of obtaining high power in the MSM range, peculiarities of the bunching and radiation mechanism, design considerations, and so on. Let us discuss some of the most important requirements.

1. Axial symmetry has significant technological and constructive advantages. In particular, it is optimal for a magnetic system (usually, solenoids), and for just this reason, it is preferable for electron-optical systems.

2. Condition $v_{ph} \gg c$ leads to design of the resonator as a section of a quasi-regular waveguide that is partially closed at the ends and has a cross section close to cutoff. Assume that κ is the transversal wave number for a chosen wave type. Then the condition of quasicriticality is evidently

$$\beta = \sqrt{k^2 - \kappa^2} \ll k \tag{10.47}$$

3. Transverse dimensions of the resonator must be large in the wavelength scale: that is,

$$\frac{R}{\lambda} \gg 1 \tag{10.48}$$

This condition is not only the result of a desire to have a large volume of the active medium but also of necessity to dissipate high Joule losses on resonator walls.

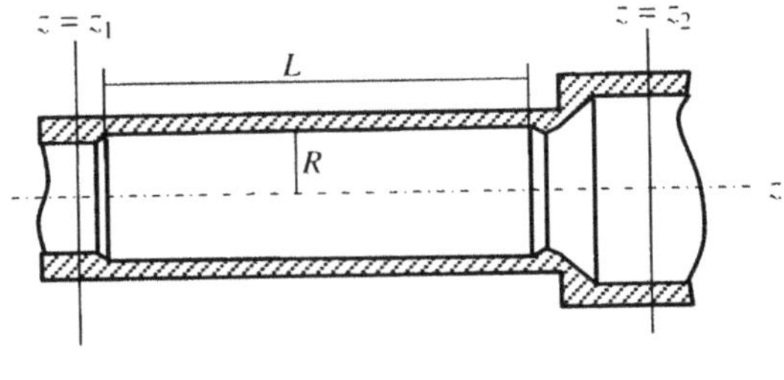

FIGURE 10.12 Gyromonotron resonator.

4. The length L of the resonator should be sufficient for the development of quadratic bunching. From the results of gyrotron theory (Nusinovich, 2004; Schneider, 1959; Yulpatov, 1981), an acceptable efficiency for the gyromonotron can be obtained for the value of the nonisochronous parameter $\mu \gg 1$. Thus, the length of the resonator according to Eq. (10.43) equals

$$\frac{L}{\lambda} = \mu \frac{\beta_z}{\pi \beta_\perp^2} = \mu \frac{1}{\pi g \beta_\perp} \gg \frac{1}{\pi g \beta_\perp} \tag{10.49}$$

where $g = v_\perp / v_z$. Let us estimate L/λ, assuming that $\mu = 5$, $g = 2$, and $\beta_\perp = 0.3$ ($U_0 \approx 30$ kV). We obtain $L/\lambda \approx 4$. Note that assumed values of μ and g agree with the parameters of real gyromonotrons.

5. In regular hollow waveguides, propagation of both TE and TM modes is possible. According to the theory (see, e.g., Bratman et al., 1977), interaction of electrons in the helical beam with TM modes, having a longitudinal component of the HF electric field, is significantly less effective for subrelativistic electron energy than with TE modes, as a result of the disbunching effect of the HF longitudinal electric field. A balance of TE and TM modes is restored for relativistic energy. Below we use the mode notation, TE_{mnq}, where m is an azimuthal (angular) index, n is a radial index, and q is a longitudinal index. The index q will often be omitted.

6. Slight waveguide irregularity [condition (10.47)] leads promptly to the need to apply high modes in the circular waveguides, for which one or both transversal indexes are significantly greater than 1. For example, for the TE_{11} mode, the ratio R/λ is only about 0.3. But the high modes are not stable: Sharp variation of the radius results in the transformation of a chosen mode in other modes, which radically changes the efficiency of the interaction and the characteristics of the radiation. Note that a partial reflection of the operation field mode takes place for a small length of the narrowed section and leads to a leak of the radiation power into an output waveguide. A choice of the narrow spot geometry then allows us to use this element as a power output (*diffraction power output*) and also as a corrector of the resonator's Q-factor. The idea of the diffraction power output belonging to Gaponov and Petelin in, 1965 (Gaponov et al., 1976) formed an important landmark in the history of the gyrotron.

Taking into account all of the foregoing, we can summarize that cylindrical, long, and slightly irregular waveguides should be used as gyromonotron resonators that are supplied by the diffraction power output and operate on the TE_{mnq} modes with a frequency close to the cutoff.

10.5.2 Free Oscillations of Gyrotron Resonators

Below, we consider axially symmetric resonators and neglect ohmic losses. In this case, damping of free oscillations is caused only by the energy radiation through the open end cross sections $z = z_1$ and $z = z_2$ (Fig. 10.12).

It can be shown (Katselenbaum, 1998) that a transverse distribution of the electromagnetic field in any cross section of a sufficiently weakly nonuniform waveguide coincides with the distribution of the field in a regular waveguide of the same crosssection. (The latter is called a *comparison waveguide*.) For TE_{mn} modes the axial component of the magnetic field $B_z(\mathbf{r}_\perp, z) \neq 0$ and can be represented as (Appendix 6)

$$B_z(\mathbf{r}_\perp, z) = C(z) B_{z\perp}(\mathbf{r}_\perp) \tag{10.50}$$

where $B_{z\perp}$ is a membrane function that satisfies the equation

$$\nabla_\perp^2 B_{z\perp} + \kappa^2 B_{z\perp} = 0 \tag{10.51}$$

with the differential operator

$$\nabla_\perp = \frac{\partial}{\partial r}\mathbf{i}_r + \frac{1}{r}\frac{\partial}{\partial \varphi}\mathbf{i}_\varphi = \frac{\partial}{\partial x}\mathbf{i}_x + \frac{\partial}{\partial y}\mathbf{i}_y \tag{10.52}$$

The boundary condition at the cross section of the comparison waveguide should be

$$\frac{\partial B_{z\perp}(\mathbf{r}_\perp)}{\partial n} = 0 \tag{10.53}$$

where $\kappa \equiv \kappa_{mn}$ is an eigenvalue of the membrane equation (10.51). The eigenfunction in the axially symmetrical resonator is

$$B_z(r, \varphi, z) = C(z) \mathrm{I}_m(\kappa r) e^{-im\varphi} = A f(z) \mathrm{I}_m(\kappa r) e^{-im\varphi} \tag{10.54}$$

where I_m is the mth-order Bessel function of the first kind. The boundary condition in this case is

$$\left.\frac{d\,\mathrm{I}_m(\kappa r)}{dr}\right|_{r=R} = 0 \tag{10.55}$$

The expression for B_z as a function of Cartesian coordinates is given in Appendix 6.

The boundary value problem (10.51)–(10.53) determines the spectrum of eigenvalues κ. Because the contour of the transverse cross section (in our case, a circle) is different for each cross section, the radius of the circle in Eq. (10.55) is a function of z, $R = R(z)$. Therefore, the eigenvalues of Eq. (10.51), $\kappa = \kappa(z)$. The transverse field components are expressed through B_z (see, e.g., Jackson, 1999):

$$\mathbf{B}_\perp = -\frac{i\beta}{\kappa^2}\nabla_\perp B_z, \qquad \mathbf{E}_\perp = \frac{i\omega}{\kappa^2}(\nabla_\perp \times \mathbf{i}_z) B_z \tag{10.56}$$

Next we find the amplitude $C(z)$. It can be shown (Appendix 6) that

$$\frac{d^2C}{dz^2} + \beta^2 C = 0 \tag{10.57}$$

Equation (10.61) is distinguished from the analogous equation for a regular waveguide by the value of $\beta^2(z) = k^2 - \kappa^2(z)$, which is a function of z. For the free oscillations considered, the field at the end cross sections $z = z_1$ and $z = z_2$ (Fig. 10.12) is determined by the radiation conditions:

$$\left.\frac{dC(z)}{dz}\right|_{z=z_1} = i\beta C(z_1), \qquad \left.\frac{dC(z)}{dz}\right|_{z=z_2} = -i\beta C(z_2) \tag{10.58}$$

Nontrivial solutions of Eq. (10.57) with conditions (10.58) exist for some discrete complex values $\omega = \omega_0 = \omega_{01} + i\omega_{02}$ that determine the natural frequency ω_{01} and the diffraction Q-factor of the resonator

$$Q_{\text{dif}} = \frac{\omega_{01}}{2\omega_{02}} \tag{10.59}$$

Obviously, $\omega_{02} > 0$ for damping oscillations.

10.5.3 Energy Balance for Free and Forced Stationary Oscillations

Let us turn to Maxwell's equations for complex amplitudes [Eqs. (A6.3) in Appendix 6]. Multiplication of the first equation by the factor $\mathbf{E}^*\,d\tau$ and integration over the resonator volume between cross sections $z = z_1$ and $z = z_2$ (Fig. 10.12) gives

$$\int_V \mathbf{E}^* \operatorname{curl} \mathbf{B}\, d\tau = \frac{i\omega}{c^2}\int_V |\mathbf{E}|^2 d\tau + \mu_0 \int_V \mathbf{j}\mathbf{E}^*\, d\tau \tag{10.60}$$

Integration of the left side by its parts, application of Gauss's theorem, and substitution of the second of Eqs. (A6.3) leads to the following equation:

$$\frac{\omega}{c^2}\int_V |\mathbf{E}|^2\, d\tau - \omega^* \int_V |\mathbf{B}|^2\, d\tau = i\oint_S (\mathbf{E}^* \times \mathbf{B})_n\, d\sigma + i\mu_0 \int_V \mathbf{j}\mathbf{E}^*\, d\tau \tag{10.61}$$

Equating separately the real and imaginary parts of Eq. (10.61) and taking $\omega = \omega_1 + i\omega_2$ into account gives

$$\omega_1 \int_V \left[\varepsilon_0 |\mathbf{E}|^2 - \frac{1}{\mu_0}|\mathbf{B}|^2\right] d\tau = -\frac{1}{\mu_0}\operatorname{Im}\int_S (\mathbf{E}^* \times \mathbf{B})_n\, d\sigma - \operatorname{Im}\int_V \mathbf{j}\mathbf{E}^*\, d\tau \tag{10.62}$$

$$2\omega_2 \int_V \left[\varepsilon_0 |\mathbf{E}|^2 + \frac{1}{\mu_0}|\mathbf{B}|^2\right] d\tau = \frac{1}{\mu_0}\operatorname{Re}\int_{S_{1,2}} (\mathbf{E}^* \times \mathbf{B})_n\, d\sigma + \operatorname{Re}\int_V \mathbf{j}\mathbf{E}^*\, d\tau \tag{10.63}$$

In Eqs. (10.62) and (10.63), we have replaced the integration along a closed surface of the resonator by the integration along the end cross sections $S_{1,2}$ due to zero ohmic losses. Consider two particular cases.

Free Oscillations (j = 0) All values in this case will be denoted by the index 0. Note that the integral $\int_S (\mathbf{E}_0^* \times \mathbf{B}_0)_n \, d\sigma$ can be considered as a real function of a real frequency ω because the waves at the end cross sections are purely traveling, for which **E** and **B** are in phase. Then we obtain from Eq. (10.62)

$$W_0 = \int_V \varepsilon_0 |\mathbf{E}_0|^2 \, d\tau = \frac{1}{\mu_0} \int_V |\mathbf{B}_0|^2 \, d\tau \tag{10.64}$$

where W_0 is the total energy (sum of electric and magnetic energies) stored in the resonator. The equality of the time-averaged electrical energy stored in the resonator to the magnetic energy is a well-known property of closed resonators without ohmic losses. If the natural frequency is complex, Eq. (10.64) is satisfied approximately for open resonators with a high Q-factor.[1]

Another common property of free oscillations follows from Eq. (10.63). Taking $\mathbf{j} = 0$ and using Eq. (10.64), we obtain

$$2\omega_{02} W_0 = P_0 \tag{10.65}$$

where $P_0 = 1/2\mu_0 \operatorname{Re} \int_S (\mathbf{E}_0^* \times \mathbf{B}_0)_n \, d\sigma$ is the average power of diffraction losses. Using the definition of the Q-factor, $Q = \omega_{01} W_0 / P_0$, we arrive at formula (10.59).

Forced Stationary Oscillations with a Fixed Transverse Field Structure Assume that $\mathbf{j} \neq 0$ and oscillations are stationary (i.e., the balance of energies radiated by the beam and radiated from the resonator has reached a steady state). For these oscillations, $\omega_2 = 0$ and ω is real. The important assumption that transverse distributions of the HF field in question coincide with the structure of the free oscillations field is not evident. Nevertheless, the detailed theory, based on numerical results (Bratman et al., 1973, 1981a), shows that structures of both fields are close for a sufficiently high Q-factor of the resonator (i.e., the validity of the fixed field structure approximation).

Two relations that determine the amplitude and frequency of the stationary oscillations can be obtained (Appendix 7):

$$2\omega_{02} W = \operatorname{Re} P \tag{10.66}$$

$$2(\omega - \omega_{01}) W = \operatorname{Im} P \tag{10.67}$$

[1]The value of the Q-factor of the gryotron resonator can be considered high when $Q/Q_{\min} \gg 1$, where $Q_{\min} = 4\pi(L/\lambda)^2$ is the Q-factor of the resonator that is formed by a section of a uniform waveguide with length L that is matched with a load (Bratman et al., 1973).

where the complex power extracted by the resonator from the electron beam is

$$P = -\frac{1}{2}\int_V \mathbf{j}\mathbf{E}^*\,d\tau = -\frac{1}{2\pi}\int_V\int_0^{2\pi} \mathbf{j}(\mathbf{r},t)\,\mathbf{E}^* e^{-i\omega t}\,d\omega t\,d\tau \tag{10.68}$$

Note that Eqs. (10.66) and (10.67) coincide with the respective equations of the resonant self-excited circuit: Eqs. (6.53) and (6.54). The latter is also true only when $Q \gg 1$.

If the density of the exciting current is given, and the distribution of the HF field is determined by an approximation of the fixed field structure, the right-hand sides of Eqs. (10.66) and (10.67) can be considered known. Then Eq. (10.66) determines the amplitudes of the fields that are contained in W, and Eq. (10.67) yields the mismatch between the frequency of stationary oscillations and the natural frequency of the resonator.

However, in reality the HF current density is not given. The relations described by Eqs. (10.66) and (10.67), can then be considered solutions of the first subproblem of the general self-consistent problem. The second subproblem consists of obtaining $\mathbf{j}(\mathbf{r}, t)$, which is based on the analysis of particle motion in an HF field, the amplitude of which should be found. This analysis for the gyrotron is simplified significantly by averaging of the motion equations.

10.6 THEORY OF A GYROMONOTRON[2]

10.6.1 Equation of Electron Motion in a TE Wave at Quasi-Cutoff Frequency

The equation of the electron motion in a static magnetic field $\mathbf{B}_0$ and HF fields $\mathbf{E}$ and $\mathbf{B}$ reads

$$\frac{d\mathbf{p}}{dt} = -e_0\mathbf{E} - e_0\mathbf{v}\times(\mathbf{B}_0 + \mathbf{B}) \tag{10.69}$$

The magnetic field $\mathbf{B}_0$ is assumed uniform. According to Eq. (10.56), $\mathbf{B}_\perp = 0$, because the propagation constant $\beta = \sqrt{k^2 - \kappa^2}$ is close to zero at the cutoff frequency. So the HF magnetic field is directed along the z-axis (i.e., parallel to $\mathbf{B}_0$). The HF electric field $\mathbf{E}$ in the TE wave has only a transverse component. Thus, the equations of motion [Eq. (10.69)] accept the form

$$\frac{d\mathbf{p}_\perp}{dt} = -e_0\mathbf{E}_\perp - e_0(B_0 + B_z)(\mathbf{v}_\perp \times \mathbf{i}_z) \tag{10.70}$$

$$\frac{dp_z}{dt} = 0 \tag{10.71}$$

[2]The content of Sections 10.6 and 10.7 essentially follows a paper of Yulpatov (1981).

Obviously, the conservation of longitudinal momentum is a direct consequence of using quasi-cutoff waves for which the phase velocity $v_{ph} \gg c$. In this case the wave propagates almost perpendicular to the waveguide axis.

Let us rewrite Eqs. (10.70) and (10.71) by taking $\mathbf{p}(t) = \mathbf{p}[z(t)]$ and replacing the independent variable t by z $[d/dt = v_z d/dz]$. We obtain the equations

$$\frac{d\mathbf{p}_\perp}{dz} + \beta_g \mathbf{p}_\perp \times \mathbf{i}_z = -\frac{e_0 m}{p_z}\mathbf{G} \tag{10.72}$$

$$\frac{d\mathbf{r}_\perp}{dz} = \frac{\mathbf{p}_\perp}{p_z} \tag{10.73}$$

$$\frac{dt}{dz} = \frac{m}{p_z} \tag{10.74}$$

$$m = m_0\gamma = \frac{m_0}{\sqrt{1 - v^2/c^2}} \tag{10.75}$$

$$p_z = \text{const.} \tag{10.76}$$

where $\beta_g = e_0 B_0/p_z = \omega_g/v_z = \text{const.}$ is the electron propagation constant of the wave whose frequency is the gyrofrequency, and the phase velocity is the electron velocity v_z. This constant has the meaning of the turning angle of the electron at a Larmor circle when a particle shifts by a unit of length along z. The factor $\mathbf{G} = \mathbf{E}_\perp + B_z \mathbf{v}_\perp \times \mathbf{i}_z$ is an HF Lorentz force acting on a unit positive charge. Using Eqs. (10.56), we obtain $\mathbf{G} = (i\omega/\kappa^2)(\nabla_\perp \times \mathbf{i}_z)B_z + \mathbf{v}_\perp \times \mathbf{i}_z B_z$. For the cutoff frequency, $\kappa \approx k = \omega/c$. Then

$$\mathbf{G} = \frac{ic}{k}(\nabla_\perp \times \mathbf{i}_z)B_z + \mathbf{v}_\perp \times \mathbf{i}_z B_z \tag{10.77}$$

A direct solution to the system (10.72)–(10.76) presents significant difficulties if one takes into account the fact that the Lorentz force $\mathbf{G}$ is a complicated function of time and particle coordinates. However, the motion is quasiperiodic, and it is possible to simplify the equations considerably by averaging over rapidly oscillating variables (this is a standard method in the theory of nonlinear oscillations).

10.6.2 Reducing Eqs. (10.72) and (10.73) to Slow Variables

Let us introduce the following change of variables:

$$\begin{aligned} x &= X + r_\perp \cos\theta, & y &= Y + r_\perp \sin\theta \\ p_x &= -p_\perp \sin\theta, & p_y &= p_\perp \cos\theta \\ r_\perp &= \frac{p_\perp}{\beta_g p_z}, & \theta &= \beta_g z + \Psi \end{aligned} \tag{10.78}$$

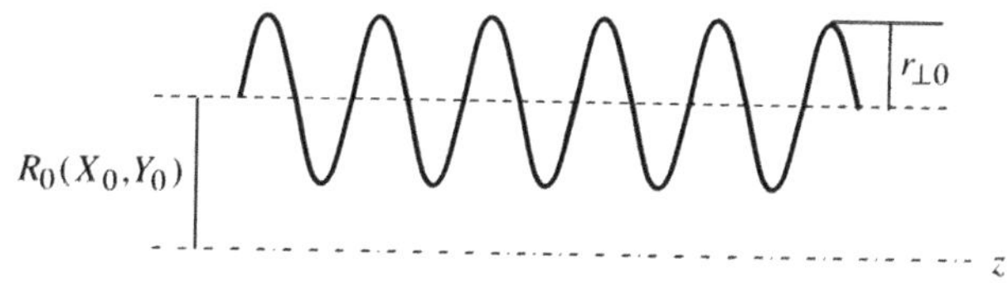

FIGURE 10.13 Parameters of an undisturbed electron trajectory.

Substituting Eqs. (10.78) into Eqs. (10.72) and (10.73), after simple manipulations we arrive at the following equations with respect to the new variables X, Y, $p_\perp$, and Ψ.

$$\begin{aligned} p'_\perp &= \frac{e_0}{v_z} G_\theta, & p_\perp \Psi' &= -\frac{e_0}{v_z} G_r \\ X' &= \frac{1}{B_0 v_z} G_y, & Y' &= -\frac{1}{B_0 v_z} G_x \end{aligned} \tag{10.79}$$

where

$$G_\theta = G_x \sin\theta - G_y \cos\theta, \qquad G_r = G_x \cos\theta + G_y \sin\theta \tag{10.80}$$

The primed symbols mean d/dz.

Let us discuss the meaning of the new variables. Assume that the HF force $\mathbf{G} = 0$. Then Eqs. (10.72) and (10.73) describe the helical motion of the electron (Fig. 10.13). Here X_0 and Y_0 are transverse coordinates of the guiding center (the axis of a helix) and $r_{\perp 0}$ is the Larmor radius. The coordinates and momenta of particles at the helical trajectory are equal to

$$\begin{aligned} x &= X_0 + r_{\perp 0}\cos\theta_0, & y &= Y_0 + r_{\perp 0}\sin\theta_0 \\ p_{x0} &= m\dot{x} = -p_{\perp 0}\sin\theta_0, & p_{y0} &= m\dot{y} = p_{\perp 0}\cos\theta_0 \end{aligned} \tag{10.81}$$

where $\theta_0 = \omega_g t + \Psi_0 = \beta_g z + \Psi_0$, $r_{\perp 0} = v_{\perp 0}/\omega_g = p_{\perp 0}/e_0 B_0$, and $p_{\perp 0}$ is the undisturbed oscillatory momentum. It is clear that X_0, Y_0, $p_{\perp 0}$, $r_{\perp 0}$, and Ψ_0 are constants. Comparison of Eqs. (10.81) with Eqs. (10.78) shows that X, Y, $p_\perp$, $r_\perp$, and Ψ can be considered coordinates of the guiding center, the transverse electron momentum, the Larmor radius, and the phase, respectively. However, although the new values are variable, they should change only *slightly* during a gyroperiod. Otherwise, the synchronism conditions would be violated, and resonance would be impossible.

10.6.3 Averaging of Equations

The slowness of new variables does not exclude their small pulsations. Such "tremors" are induced by components of the HF force $\mathbf{G}$ on the right-hand side of Eqs. (10.79) explicitly and indirectly (through coordinates of rotating electrons), depending on the fast variable θ. However, the presence of fast oscillations does not influence the characteristics of the radiation in the frame of the resonance mechanism. Therefore, averaging of the right-hand sides in the equations indicated is an adequate procedure. In this case, the slowly changing components are isolated, leading to significantly simpler, "reduced" equations.

The HF force averaged over the rotation phase θ is equal to

$$\overline{\mathbf{G}(\mathbf{r},t)} = \frac{1}{2\pi}\int_0^{2\pi} \mathbf{G}(\mathbf{r},t)\,d\theta = \frac{1}{2\pi}\operatorname{Re}\int_0^{2\pi} \mathbf{G}(x,y,z)e^{i\omega t}\,d\theta \tag{10.82}$$

Since x and y are periodic functions of θ [Eqs. (10.78)], the complex amplitude $\mathbf{G}(x,y,z)$ can be represented by the Fourier series

$$\mathbf{G}(x,y,z) = \sum_{m=-\infty}^{\infty} \mathbf{G}_m e^{-im\theta} \tag{10.83}$$

where amplitudes of harmonics are equal to

$$\mathbf{G}_m \equiv \mathbf{G}_m(X,Y,r_\perp,z) = \frac{1}{2\pi}\int_0^{2\pi} \mathbf{G}(x,y,z)e^{im\theta'}\,d\theta' \tag{10.84}$$

Substitution of Eq. (10.83) into Eq. (10.82) yields

$$\overline{\mathbf{G}(\mathbf{r},t)} = \operatorname{Re}\frac{1}{2\pi}\sum_{m=-\infty}^{\infty}\int_0^{2\pi} \mathbf{G}_m e^{i(\omega t - m\theta)}\,d\theta \tag{10.85}$$

Because $\mathbf{G}_m$ does not depend on θ, the averaging reduces Eq. (10.85) to the calculation of integrals:

$$\frac{1}{2\pi}\int_0^{2\pi} e^{i(\omega t - m\theta)}\,d\theta \tag{10.86}$$

The phases $\omega t - m\theta$ in general change quickly in the interval $0 \div 2\pi$, so these integrals are small. But let us assume that for some $m = n$, the phase $\omega t - n\theta$ is approximately constant. Then the corresponding integral yields the main contribution to the sum in Eq. (10.85). Using the last equation from Eqs. (10.78), we can obtain the condition of the phase constancy

$$\frac{d}{dt}(\omega t - n\theta) = \omega - n\beta_g v_z - n\Psi' = 0$$

Because Ψ' is small, we obtain the *condition of the gyroresonance* $\omega - n\omega_g \approx 0$.

Thus, we can retain in the series equation (10.85) only the nth term that contains the integral

$$\frac{1}{2\pi}\int_0^{2\pi} e^{i(\omega t - n\theta)}\, d\theta = e^{i(\omega t - n\theta)} \tag{10.87}$$

As a result, we obtain

$$\overline{\mathbf{G}(\mathbf{r},t)} = \mathrm{Re}(\mathbf{G}_n e^{-i\vartheta}) \tag{10.88}$$

where

$$\vartheta = -\omega t + n\theta = -\omega t + n\beta_g z + n\Psi \tag{10.89}$$

The value of ϑ changes slowly. As such, this is sometimes called a *slow phase*.

Note that if a drift of electrons (with the velocity v_z) is present, Eq. (10.87) is valid in the guiding center frame only. In the laboratory frame, instead of Eq. (10.82) the average value of **G** should be written as

$$\overline{\mathbf{G}(\mathbf{r},t)} = \frac{1}{2\pi}\mathrm{Re}\int_0^{2\pi} \mathbf{G}e^{i(\omega t - \beta z)}\, d\theta \tag{10.90}$$

Therefore, the exponential factor in the integral of Eq. (10.87) should be replaced by $\omega t - \beta z - n\theta$. Then the gyroresonance condition becomes $\omega - \beta v_z \approx n\omega_g$ (i.e., the Doppler factor appears).

Let us replace components G_θ, G_r, G_x, and G_y in Eqs. (10.79) by their averaged values $G_{\theta n}$, G_{rn}, G_{xn}, and G_{yn} [Eqs. (10.84) and (10.88)], and also take $n\Psi' = \vartheta' + \omega/v_z - n\beta_g$. Then we obtain averaged ("reduced") gyrotron equations:

$$p'_\perp = \frac{e_0}{v_z}\mathrm{Re}(G_{\theta n}e^{-i\vartheta}) \tag{10.91}$$

$$p_\perp\left(\vartheta' + \frac{\omega}{v_z} - n\beta_g\right) = \frac{e_0 n}{v_z}\mathrm{Re}(G_{rn}e^{-i\vartheta}) \tag{10.92}$$

$$X' = \frac{1}{v_z B_0}\mathrm{Re}(G_{yn}e^{-i\vartheta}) \tag{10.93}$$

$$Y' = -\frac{1}{v_z B_0}\mathrm{Re}(G_{xn}e^{-i\vartheta}) \tag{10.94}$$

Calculation of the Fourier coefficients $G_{\theta n}$, G_{rn}, G_{xn}, and G_{yn} is described in Appendix 8.

10.6.4 Hamiltonian Form of Averaged Equations

The averaged equations can be rewritten in a form in which the relativistic electron energy

$$E = mc^2 = \sqrt{m_0^2 c^4 + (p_\perp^2 + p_z^2)c^2} \tag{10.95}$$

and the slow phase ϑ forms a pair of variables similar to a canonical conjugated pair for Hamiltonian equations (I.32).

Because $p_z = \text{const.}$, it is easy to see from Eq. (10.95) that $E' = v_\perp p'_\perp$. Multiplying Eq. (10.91) by $v_\perp$ and taking the first of Eqs. (A8.9) into account leads to

$$E' = e_0 \frac{p_\perp}{p_z} \text{Re}\left(G_{\theta n} e^{-i\vartheta}\right) = e_0 \frac{p_\perp}{p_z} \text{Re}\left(-ic \frac{dI_n}{d\xi} L_n e^{-i\vartheta}\right) = e_0 \text{Re}\left(\frac{\partial H}{\partial \vartheta}\right) \tag{10.96}$$

where we introduce the Hamiltonian

$$H = c \frac{p_\perp}{p_z} \frac{dI_n}{d\xi} L_n e^{-i\vartheta} \tag{10.97}$$

and

$$L_n = D^n B_z(X, Y, z) = \frac{1}{k^n}\left(\frac{\partial}{\partial X} + i \frac{\partial}{\partial Y}\right)^n B_z(X, Y, z) \tag{10.97a}$$

The second equation for the Hamiltonian pair is obtained from Eq. (10.92) and the second equation in Eq. (A8.9):

$$\vartheta' + \frac{\omega}{v_z} - n\beta_g = \frac{e_0 n}{v_z p_\perp} \text{Re}\left[-\frac{c}{n} \frac{d}{d\xi}\left(\xi \frac{dI_n}{d\xi}\right) L_n e^{-i\vartheta}\right] \tag{10.98}$$

where

$$\xi = kr_\perp = k \frac{v_\perp}{\omega_g} = k \frac{p_\perp}{e_0 B_0} \tag{10.99}$$

Then substitution of Eqs. (10.97) and (10.99) into Eq. (10.98) and using $p_z = \text{const.}$ yield

$$\begin{aligned} \vartheta' + \frac{\omega}{v_z} - n\beta_g &= -\text{Re}\left(\frac{km}{p_\perp B_0} \frac{\partial H}{\partial \xi}\right) = -\text{Re}\left(\frac{km}{p_\perp B_0} \frac{\partial H}{\partial z} \Big/ \frac{\partial \xi}{\partial z}\right) \\ &= -\text{Re}\left(\frac{e_0}{v_\perp p'_\perp} \frac{\partial H}{\partial z}\right) = -e_0 \text{Re}\left(\frac{\partial H}{\partial E}\right) \end{aligned} \tag{10.100}$$

Similarly, it is straightforward to convert Eqs. (10.93) and (10.94) using Eqs. (A8.9). Finally, the Hamiltonian version of the gyrotron equations takes the form

$$E' = e_0 \mathrm{Re}\left(\frac{\partial H}{\partial \vartheta}\right) \tag{10.101}$$

$$\vartheta' + \frac{\omega}{v_z} - n\beta_g = -e_0 \mathrm{Re}\left(\frac{\partial H}{\partial E}\right) \tag{10.102}$$

$$X' = -\frac{1}{\omega B_0}\mathrm{Re}\left(\frac{\partial H}{\partial Y}\right) \tag{10.103}$$

$$Y' = \frac{1}{\omega B_0}\mathrm{Re}\left(\frac{\partial H}{\partial X}\right) \tag{10.104}$$

10.6.5 Energy Integral of Averaged Equations

The current density **j** and the electric field **E** in Eq. (10.68) for the complex power P have not been expressed through slow variables. Therefore, P contains small pulsations and should be averaged. Let us first convert P using the approximation of a thin beam. In this case, electron velocities and the HF field are considered uniform over the beam cross section. For TE modes, the factor $\mathbf{jE}^* = \mathbf{j}_\perp \mathbf{E}_\perp^* = j_z \mathbf{p}_\perp / p_z \mathbf{E}_\perp^*$. Then Eq. (10.68) can be written as

$$P = -\frac{1}{2\pi}\int_0^{2\pi}\int_V j_z \frac{\mathbf{p}_\perp \mathbf{E}_\perp^*}{p_z} e^{-i\omega t}\, d\omega t\, d\tau \tag{10.105}$$

The volume integral of $j_z\ d\tau$ in the thin beam approximation is reduced to $I(t, z)\, dz$. Next, let us apply the law of charge conservation (Section 7.3.2), $I\, dt = -I_0\, dt_0$, where t_0 and I_0, respectively, are the input time and the positive current. Using the equality $\mathbf{p}_\perp \mathbf{E}^* = p_\perp E_\perp^* = p_\perp G_\theta^*$, we obtain

$$P = \frac{I_0}{2\pi p_z}\int_0^{2\pi}\int_0^L p_\perp G_\theta^* e^{-i\omega t}\, d\omega t_0\, dz$$

The orbital average complex power would then be equal to

$$\overline{P} = \frac{I_0}{2\pi p_z}\int_0^{2\pi}\int_0^L d\omega t_0\ dz\ \frac{1}{2\pi}\int_0^{2\pi} p_\perp G_\theta^*\, d\theta \tag{10.106}$$

The calculation of the inner integral coincides with the procedure for averaging the right-hand sides in Eqs. (10.79): expansion of G_θ^* in a Fourier series and separation of a resonance term. Using Eqs. (10.87), (10.89), (A8.9) and (10.97), we obtain the averaged complex power

$$\overline{P} = \frac{I_0}{2\pi p_z}\int_0^{2\pi}\int_0^L p_\perp G_{\theta n}^* e^{i\vartheta} dz\, d\omega t_0 = -\frac{I_0}{2\pi}\int_0^{2\pi}\int_0^L \frac{\partial H^*}{\partial \vartheta}\, dz\, d\vartheta_0 \tag{10.107}$$

where $d\omega\ t_0 = -d\vartheta_0$ because $\vartheta_0 = (\vartheta)_{z=0} = n\Psi_0 - \omega t_0$.

The energy integral is equal to the real part of the averaged complex power:

$$\begin{aligned}\mathrm{Re}(\overline{P}) &= -\frac{I_0}{2\pi}\int_0^{2\pi}\int_0^L \mathrm{Re}\left(\frac{\partial H}{\partial \vartheta}\right) dz\, d\vartheta_0 = -\frac{I_0}{2\pi e_0}\int_0^{2\pi}\int_0^L \frac{dE}{dz}\, dz\, d\vartheta_0 \\ &= \frac{I_0}{e_0}\frac{1}{2\pi}\int_0^{2\pi}[E(0) - E(L)]\, d\vartheta_0\end{aligned} \tag{10.108}$$

Here the first of Eqs. (10.101) was used. The physical meaning of Eq. (10.108) is clear: Averaging over the period of the real power that an electron beam delivers to the HF field is equal to averaging the electron power loss over the initial phase.

10.6.6 Averaged Equations in Polar Coordinates

Let us turn to polar coordinates of the guiding center:

$$X = R\cos\Psi, \qquad Y = R\sin\Psi \tag{10.109}$$

In this case, averaged equations (10.101) and (10.102) do not change. Equations (10.103) and (10.104) are converted to

$$RR' = -\frac{1}{\omega B_0}\mathrm{Re}\left(\frac{\partial H}{\partial \Psi}\right) \tag{10.110}$$

$$R\Psi' = \frac{1}{\omega B_0}\mathrm{Re}\left(\frac{\partial H}{\partial R}\right) \tag{10.111}$$

Let us express function H through the coordinates R and Ψ using Eqs. (10.97) and (10.54). Taking into account Eqs. (A8.8), (10.54), and (10.109), we obtain

$$\begin{aligned}L_1 &= \frac{1}{\kappa}\left(\frac{\partial}{\partial X} + i\frac{\partial}{\partial Y}\right)C(z)I_m(\kappa R)e^{-im\Psi} \\ &= \frac{e^{i\Psi}}{\kappa}\left(\frac{\partial}{\partial R} + \frac{i}{R}\frac{\partial}{\partial \Psi}\right)C(z)I_m(\kappa R)e^{-im\Psi} \\ &= C(z)I_{m-1}(\kappa R)e^{-i(m-1)\Psi} = Af(z)I_{m-1}(\kappa R)e^{-i(m-1)\Psi} \\ L_n &= Af(z)I_{m-n}(\kappa R)e^{-i(m-n)\Psi}\end{aligned} \tag{10.112}$$

Then

$$H = c\frac{p_\perp}{p_z}\frac{dI_n}{d\xi}L_n e^{-i\vartheta} = Af(z)c\frac{p_\perp}{p_z}\frac{dI_n(\xi)}{d\xi}I_{m-n}(kR)e^{-i(m-n)\Psi}e^{-i\vartheta} \tag{10.113}$$

Equations (10.110), (10.111), and (10.113) together with Eqs. (10.101) and (10.102) and the averaged complex power equation (10.107) form a full system of averaged equations for the axially symmetric gyrotron. These equations have, besides the energy integral, an additional integral of motion. According to Eq. (10.113),

$$\frac{\partial H}{\partial \Psi} = -i(m-n)H = (m-n)\frac{\partial H}{\partial \vartheta}$$

Then, using Eqs. (10.110) and (10.111), we obtain

$$RR' = \frac{1}{2}(R^2)' = -\frac{m-n}{\omega B_0}\mathrm{Re}\left(\frac{\partial H}{\partial \vartheta}\right) = -\frac{m-n}{e_0\omega B_0}E'$$

Integration of this equation gives

$$E - E(0) = \frac{e_0\omega B_0}{2(n-m)}[R^2 - R^2(0)] \qquad (10.114)$$

The change in energy turns out to be proportional to the radial shift of the guiding center. If the order of the synchronous harmonic n is greater than the azimuthal index m of the mode, the transfer of energy by electrons is accompanied by a decrease in R. For example, an electron interacting with the mode TE_{01} always, on average, approaches the resonator axis; the radius of the guiding center for interaction with the mode TE_{1p} at the first harmonic (fundamental cyclotron resonance), does not change; and so on.

The changes indicated in R are negligible for subrelativistic energies, however. It can be readily estimated from Eq. (10.114) that

$$\frac{R - R(0)}{R(0)} \approx \frac{1}{4\pi^2}\frac{E - E(0)}{E_0}\left[\frac{\lambda}{R(0)}\right]^2 n(n-m) \qquad (10.115)$$

For example, for $m = 0$, $R(0) = \lambda$, $E - E(0) = 100$ keV, and $E_0 = 500$ keV (the rest energy of the electron), we obtain $\Delta R/R(0) \approx 0.05$ even for $n = 3$.

Applying Eq. (10.114) allows us to reduce the system of four equations [Eqs. (10.101) and (10.102) and Eqs. (10.110) and (10.111)] to a system of two equations (Yulpatov, 1981):

$$\begin{aligned} \frac{dE}{dz} &= e_0\mathrm{Re}\left(\frac{\partial H}{\partial \bar{\vartheta}}\right) \\ \frac{d\bar{\vartheta}}{dz} + \frac{\omega}{v_z} - n\beta_g &= -e_0\mathrm{Re}\left(\frac{dH}{dE}\right) \end{aligned} \qquad (10.116)$$

where the new phase is $\bar{\vartheta} = \vartheta + (n-m)\Psi$, and the total derivative is

$$\frac{dH}{dE} = \frac{\partial H}{\partial E} + \frac{\partial H}{\partial R^2}\frac{dR^2}{dE}$$

These equations form a complete system: They do not contain Ψ because according to Eq. (10.113) $H = H(R, \bar{\vartheta})$, and the radius R included in H is expressed simply through E [Eq. (10.114)].

10.7 SUBRELATIVISTIC GYROTRONS

10.7.1 Introduction

In the traditional scheme of a gyrotron (i.e., in the synchronism $\omega = n\omega_g$), high efficiency can be realized when a reduction in the electron energy during interaction does not change substantially the particle mass and hence the gyrofrequency. Therefore, the electron energy must satisfy the condition $(E - E_0)/E_0 \ll 1$, where E_0 is the electron rest energy. This inequality is the condition of the subrelativism $v/c \ll 1$ and allows us to consider the gyrotrons as classical *subrelativistic devices*. More careful analysis (Bratman et al., 1981b) shows, however, that even in an approximation of the fixed field structure, the efficiency of relativistic gyromonotrons may remain on a sufficiently high level in the moderate energy range (200 to 500 keV). In this case the attainment of 20 to 30% efficiency is provided by reducing the interaction space length, thus expanding the cyclotron resonance bandwidth. Bratman et al. used an approximate model based on the reduction of equations of a relativistic gyromonotron to a subrelativistic version, which reinforces the universality of the subrelativistic equations considered below.

These results have been confirmed in experiments with relativistic high-power gyromonotrons Black et al. (1990). An output power of 100 MW at 35 GHz in the TE_{62} mode with an efficiency of approximately 8% of a 800-kV gyromonotron has been achieved. A helical electron beam in the oscillator was formed by the magnetically insolated explosive emission diode (see Section 4.8) with a pulse duration of ~50 ns and a short pump magnet, arranged after the diode, with a nonadiabatic change in the axial magnetic field. As a result, electrons passing the region of the pump magnet acquire an initial transversal magnetic moment. The latter is increased further in a routine adiabatically increased magnetic field.

We can also refer to the important result of recent theoretical and experimental optimization of a relativistic gyromonotron (Zaitsev et al., 2002). The numerical simulation was based on the solution of relativistic equations of motion combined with the nonuniform string equation for the stationary regime (see, e.g., Nusinovich, 2004):

$$\frac{d^2C}{dz^2} + \beta(z)^2 C = \bar{I} \int_0^{2\pi} w^{n/2} e^{i\vartheta}\, d\vartheta_0 \qquad (10.117)$$

where $\bar{I}$ is a dimensionless current and w is determined below in Eq. (10.123). Unlike the analogous equation (10.57) for free oscillations, this equation determines a self-consistent field profile. The results have shown that the reduction in efficiency after passing over to the relativistic range and with an optimization of the field profile is considerably lower than predicted by theory (Bratman et al., 1981b). The parameters achieved for an X-band 280-keV experimental gyromonotron were: efficiency, 50% at a power of 7 MW; and efficiency, 30% at a power of 11 MW. A gyromonotron gun was provided by a thermoionic cathode, and the microwave pulse duration was about 6 μs.

10.7.2 Derivation of Subrelativistic Averaged Equations

In the derivation below, we assume Eqs. (10.101) and (10.102), the Hamiltonian (10.110), and the expression for active power (10.105). Equation (10.101), which determine the motion of guiding centers, are unnecessary, because the guiding center displacement is negligible in subrelativistic approximation [Eq. (10.115)]. Thus, H can be calculated using the input coordinates of these centers. Let us introduce the following dimensionless variables.

1. Relative oscillatory and drift velocities at the input to the interaction space:

$$\beta_{\perp 0} = \frac{v_{\perp 0}}{c}, \qquad \beta_z = \frac{v_{z0}}{c} \tag{10.118}$$

It is clear that these values are small in a subrelativistic approximation.

2. Longitudinal coordinate:

$$\zeta = \frac{\pi\beta_{\perp 0}^2}{\beta_z}\frac{z}{\lambda} \tag{10.119}$$

3. Parameter $\xi = kr_\perp$ [argument of the Bessel function $I_n(\xi)$] is small in the case of synchronism and when n is not very large:

$$\xi = kr_\perp \approx \frac{n\omega_g r_\perp}{c} = n\frac{v_{\perp 0}}{c}\frac{p_\perp}{p_{\perp 0}} = n\beta_{\perp 0}\sqrt{w} \ll 1 \tag{10.120}$$

Therefore, the derivative $dI_n(\xi)/d\xi$ that enters the Hamiltonian H [Eq. (10.97)] can be replaced by the first term of the power series $dI_n(\xi)/d\xi \approx n\xi^{n-1}/2^n n!$. Thus, H adopts the form

$$H = \frac{m_0 c\omega\beta_{\perp 0}^4}{4e_0\beta_z} w^{n/2} e^{-i\vartheta} F \tag{10.121}$$

with the field parameter

$$F = \frac{e_0}{m_0\omega}\frac{n^n}{2^{n-2}n!}\beta_{\perp 0}^{n-4} L_n = Af(z)\frac{e_0}{m_0\omega}\frac{n^n\beta_{\perp 0}^{n-4}}{2^{n-2}n!} I_{m-n}(\kappa R)e^{-i(n-m)\Psi} \tag{10.122}$$

L_n was defined in Eqs. (10.112).

4. Subrelativistic oscillatory energy:

$$w = \frac{E_\perp}{E_{\perp 0}} = \frac{p_\perp^2}{p_{\perp 0}^2} \leq 1 \tag{10.123}$$

This ratio allows us to rewrite Eq. (10.101) in the following way. Since $p_z^2 = \text{const.}$, we can see from Eq. (10.95) that $E^2 - E_0^2 = p_{\perp 0}^2 c^2 (w - 1)$. The derivative of this relation gives

$$\frac{dw}{dz} = \frac{2EE'}{p_{\perp 0}^2 c^2} \tag{10.124}$$

Then using Eqs. (10.116), (10.121), and (10.119) and assuming that $E = mc^2$, we obtain

$$\frac{dw}{d\zeta} = \frac{\partial}{\partial \vartheta} \text{Re}(w^{n/2} F e^{-i\vartheta}) \tag{10.125}$$

The term $(\omega / v_z) - n\beta_g$ in Eqs. (10.116) is equal to

$$\frac{\omega}{v_z} - n\beta_g = \frac{\omega}{p_z}[m - m(0)] + \frac{\omega}{v_z(0)} - n\beta_g$$

where the mass difference is approximately equal to

$$m - m(0) = \frac{E - E(0)}{c^2} \approx \frac{p_{\perp 0}^2}{2 m_0 c^2}(w - 1)$$

Then we obtain

$$\frac{\omega}{v_z} - n\beta_g = \frac{\mu}{L}(\Delta + w - 1) \tag{10.126}$$

where

$$\Delta = \frac{2[\omega - n\omega_g(0)]}{\omega \beta_{\perp 0}^2} \tag{10.127}$$

is the dimensionless frequency mismatch. The parameter

$$\mu = \frac{\pi \beta_{\perp 0}^2}{\beta_{z0}} \frac{L}{\lambda} \tag{10.128}$$

has the meaning of the normalized length of the interaction space. We saw before [Eq. (10.43)] that this "nonisochronous parameter" characterizes the efficiency of the quadratic bunching.

Using Eq. (10.124) and Eqs. (10.119), (10.121), (10.126), and (10.128), we can reduce Eq. (10.102) to the form

$$\frac{d\vartheta}{\partial \zeta} + \Delta + w - 1 = -\frac{\partial}{\partial w} \text{Re}[(w^{n/2} F e^{-i\vartheta})] \tag{10.129}$$

Equations (10.125) and (10.129) are the averaged equations of the subrelativistic gyrotron. The energy w on the left-hand side of Eq. (10.129) is directly related to the relativistic nonisochronism of oscillators. It determines the inertial bunching that can develop in a drift section where the HF field is absent, as in the case of klystrons. On the other hand, the right-hand side of Eq. (10.129) is proportional to the field amplitude and determines the force bunching. Note the bunching in the input section of the interaction space, where $w \approx 1$ is determined mainly by the mismatch term Δ.

According to Eqs. (10.107) and (10.121), the subrelativistic complex power is equal to

$$\overline{P} = -i \frac{I_0}{2\pi} \frac{m_0 c \omega \beta_{\perp 0}^4}{4 e_0 \beta_z} F \int_0^{2\pi} \int_0^L w^{n/2} e^{i\vartheta} \, dz \, d\vartheta_0$$

or, replacing z by ζ, according to Eq. (10.119), we obtain

$$\overline{P} = -i \frac{I_0 m_0 c^2 \beta_{\perp 0}^2}{4\pi e_0} F \int_0^{2\pi} \int_0^{\mu} w^{n/2} e^{i\vartheta} \, d\zeta \, d\vartheta_0 \qquad (10.130)$$

The active power in a subrelativistic approximation can be obtained by substitution of the difference $E(0) - E(L) \sim (p_\perp^2 - p_{\perp 0}^2)/2m_0$ into Eq. (10.108). As a result, $\text{Re}(\overline{P}) = P_{\perp 0}(1/2\pi) \int_0^{2\pi} (1 - w) \, d\vartheta_0$, where $P_{\perp 0} = (I_0/e_0)(p_{\perp 0}^2/2m_0)$ is the input oscillatory power of the electron beam. The value

$$\eta_\perp = \frac{\text{Re}(\overline{P})}{P_{\perp 0}} = \frac{1}{2\pi} \int_0^{2\pi} (1 - w) \, d\vartheta_0 \qquad (10.131)$$

is called the *electronic orbital efficiency* (the ratio of the power that the electron beam transfers to the electromagnetic field in a resonator to the input oscillatory power).

10.7.3 Results of Integration of Subrelativistic Gyrotron Equations

Equations (10.125) and (10.129) contain parameters Δ and F, which depend on the distribution and amplitude of the HF field [$B_z = Af(z)I_m(\kappa r)e^{-im\varphi}$], frequency ω, and the parameters of the electron beam, $\beta_\perp$ and ω_g. The solutions also depend on the normalized interaction length μ, (i.e., on the size of the integration domain). Even if the function $f(z)$ can be considered as given in a fixed field approximation, the amplitude A and oscillation frequency ω remain unknown. As already mentioned, that uncertainty can by removed using Eqs. (10.66) and (10.67) for the amplitude and frequency of stationary oscillations, together with Eqs. (10.125) and (10.129).

There is, however, an alternative, which sees the reduction of this task as a problem of finding the optimal efficiency. Let us consider the values Δ, F, and μ as varying parameters [if the field distribution $f(\zeta)$ is given, only the amplitude A should vary]. Then for a certain choice of indicated parameters, Eqs. (10.125) and (10.129) are integrated numerically, and the orbital efficiency [Eq. (10.130)] is calculated. This process is repeated until the maximum $\eta_\perp$ is obtained. If the optimal parameters are found, we can use the balance equations (10.66) and

TABLE 10.2 Optimal Parameters of a Gyrotron with Uniform Distribution $F(\zeta)$ for the First Five Harmonics of the Gyrofrequency

n	$\eta_\perp(\%)$	μ	Δ	$\lvert F_0\rvert^2 \times 10^2$
1	42	7.5	0.60	8
2	30	8.0	0.55	6
3	22	10.0	0.40	2
4	17	12.5	0.35	1
5	14	17.5	0.25	4

(10.67) along with formulas (10.122), (10.127) and (10.128), which determine dimensionless parameters for the design of the resonator and electrical regime.

In Tables 10.2 and 10.3 are listed optimal parameters of gyrotrons that were calculated by Yulpatov (1981) for two of the simplest longitudinal distributions of the HF field in a resonator (Fig. 10.14): $F = F_0$ and $F = F_0\zeta/\mu$. As we see, the efficiency of the uniform distribution is substantially less than that of the triangular distribution. The principal cause is the following: The electrons in the input section of the interaction space where linear bunching prevails absorb in a strong field a large fraction of the field energy (see Section 10.3). This process is especially intense on harmonics of the gyrofrequency. According to Table 10.3, the triangular distribution is devoid of this defect: The efficiency is considerably higher and decays slowly with the harmonic number.

Note that the triangular distribution is a significant idealization, but calculations with other distributions yield similar results, provided that these distributions have a weak field at the initial section and a strong field in the region of the energy interchange, where dense bunches are formed as a result of the quadratic bunching. The values given in Table 10.3 correspond to the Gaussian distribution $F = F_0 e^{-3(2\zeta/\mu-1)^2}$, which is close to real field distributions in gyrotron resonators, similar to the one shown in Fig. 10.12. A comprehensive analysis of the efficiency and starting currents of subrelativistic gyrotrons operating on the first (fundamental) through fifth harmonics of the gyrofrequency was developed by Danly and Temkin (1986). These authors used the longitudinal field distribution close to Gaussian. Optimized maximum orbital efficiencies for harmonics 2, 3, 4, and 5 were 0.72, 0.57, 0.45, and 0.36, respectively.

After the optimal orbital efficiency has been found, further constructive and electric characteristics of the device should be determined. These characteristics are, in

TABLE 10.3 Optimal Parameters of a Gyrotron with Triangular (Up) and Gaussian (Down) Distributions for the First Three Harmonics of the Gyrofrequency

n	$\eta_\perp(\%)$	μ	Δ
1	71/72	14/17	0.55/50
2	64/71	14/16	0.55/55
3	56/55	20/22	0.40/40

Source: Nusinovich and Erm (1972).

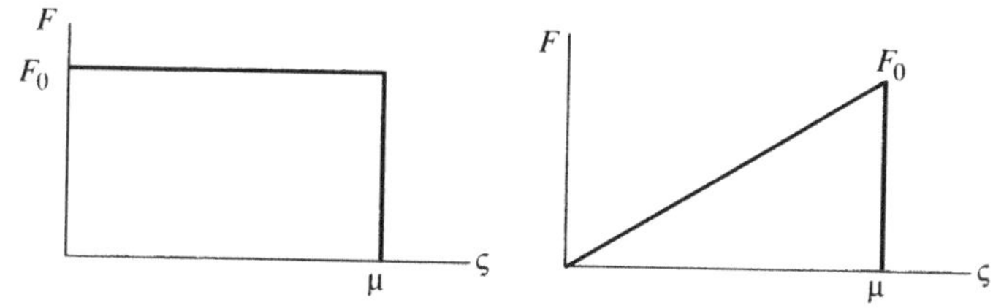

FIGURE 10.14 Uniform and triangular distributions of a field parameter.

particular, the parameters of the electron beam R_0, $v_\perp$, v_z, and ω_g, the axial distributions of the HF field and dc magnetic field, the Q-factor of the resonator, and its profile, including the configuration of the diffraction power output. Finally, of an electron collector and an output window, important components of the design, are constructed. In general, obtaining all these characteristics is a problem that does not have a unique solution. Certain difficulties arise from consideration of a number of important additional factors: the character of excitation of the operation mode (hard or soft self-excitation), the relation to competing modes (Kartikeyan et al., 2004; Nusinovich, 1981, 2004), the stability of the electron beam (Tsimring, 2001), heat effects, and others. These factors play a different role in different frequency bands and for different energetic characteristics of devices (e.g., output power, overall efficiency, pulse duration). Some characteristics are considered briefly later.

10.7.4 Linearization of Subrelativistic Gyrotron Equations

The numerical solution mentioned above is insufficiently transparent. It is more interesting to obtain a solution in analytical form. This can be done via linearization of the averaged equations. This will also allow us to calculate the starting current in explicit form.

If the field amplitude is small, the field parameter F, increments of the energy w, and the phase ϑ are also small parameters. Let us represent w and ϑ [taking into account Eqs. (10.89), (10.120), and (10.126)–(10.128)] in the form

$$w = w(0) + w_1 = 1 + w_1, \quad \vartheta = -\omega t + n\beta_g z + n\Psi = z\left(-\frac{\omega}{v_z} + n\beta_g\right) + n\Psi$$
$$= -\Delta\zeta + \vartheta_0 + \vartheta_1 \tag{10.132}$$

where w_1 and ϑ_1 are small increments of the energy and phase arising from the HF interaction. Let us linearize Eqs. (10.125) and (10.129), retaining terms of first order with respect to w_1 and ϑ_1. We obtain the linear equations

$$\frac{dw_1}{d\zeta} = -\mathrm{Re}[iFe^{i(\Delta\zeta-\vartheta_0)}] \tag{10.133}$$

$$\frac{d\vartheta_1}{d\zeta} = -w_1 - \frac{n}{2}\mathrm{Re}[Fe^{i(\Delta\zeta-\vartheta_0)}] \tag{10.134}$$

Here we used the following linearized representation of the the factor $w^{n/2}e^{i\vartheta}$ in expression (10.130) for the complex power:

$$w^{n/2}e^{i\vartheta} \approx \left(1+\frac{n}{2}w_1\right)(1+i\vartheta_1)e^{i(\vartheta_0-\Delta\zeta)} \approx \left(1+\frac{n}{2}w_1+i\vartheta_1\right)e^{i(\vartheta_0-\Delta\zeta)} \quad (10.135)$$

The unity in the parentheses can be canceled because the integral $\int_0^{2\pi} e^{i\vartheta_0}d\vartheta_0 = 0$. This means physically that the bunching is absent, and the average power is equal to zero when the phase does not change. The averaged complex power according to Eq. (10.130) equals

$$\overline{P} = P_{\perp 0}\frac{1}{2\pi}\int_0^{2\pi}\int_0^{\mu}\left(\vartheta_1 - \frac{in}{2}w_1\right)e^{i(\vartheta_0-\Delta\zeta)}F^*\,d\vartheta_0\,d\zeta \quad (10.136)$$

Integration of Linearized Equations The integral of Eq. (10.133) can be written as

$$w_1(\sigma) = -\mu\int_0^{\sigma}\mathrm{Re}[iF_0e^{i(\varphi\sigma'-\vartheta_0)}f(\sigma')\,d\sigma' \quad (10.137)$$

Here we introduce new dimensionless variables and a normalized longitudinal field distribution:

$$\sigma = \frac{\zeta}{\mu} = \frac{z}{L}, \quad \varphi = \Delta\mu = \frac{\omega - n\omega_g(0)}{v_z}L = [\omega - n\omega_g(0)]T, \quad F = F_0f(\sigma) \quad (10.138)$$

where σ and φ are the dimensionless interaction length and the cyclotron detuning, respectively.

The integration of Eq. (10.134) in the new variables gives

$$\vartheta_1(\sigma) = -\mu\int_0^{\sigma}w_1(\sigma')\,d\sigma' - \frac{n}{2}\int_0^{\sigma}\mathrm{Re}[e^{i(\varphi\sigma'-\vartheta_0)}]F_0f(\sigma')\,d\sigma' \quad (10.139)$$

Substituting w_1 from Eq. (10.137) in the first integral of Eq. (10.139) leads to a double integral. If we form the difference $\theta_1 - (in/2)w_1$ and substitute it in the integral of Eq. (10.136) for the complex averaged power, we obtain a sum of four- and threefold multiple integrals that in general cannot be taken. However, as shown by Yulpatov, the result of integration for the active averaged power is expressed through a relatively simple one-dimensional integral:

$$\mathrm{Re}(\overline{P}) = -\frac{P_{\perp 0}\mu^2|F_0|^2}{4}\left(n+\mu\frac{\partial}{\partial\varphi}\right)\left|\int_0^1 e^{i\varphi\sigma'}f(\sigma')\,d\sigma'\right|^2 \quad (10.140)$$

where $|F_0|$ should be taken from Eq. (10.122), replacing $f(z)$ by $\overline{f} = 1/L\int_0^L f(z)\,df$. As we see, the active power $\mathrm{Re}(\overline{P})$ can be positive only due to the second term in Eq. (10.140) and for sufficiently large μ. Note that this term arises from the term $-w_1$ in Eq. (10.134), which [see the comment following Eq. (10.129)] is connected directly to the relativistic nonsynchronism of oscillators and the inertial bunching.

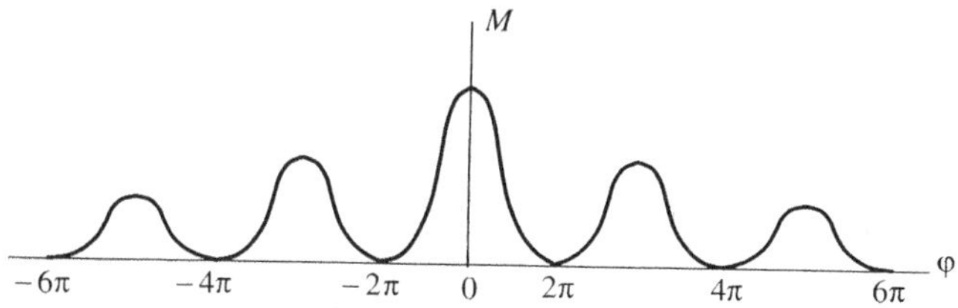

FIGURE 10.15 Linearized active power of a subrelativistic gyrotron for uniform field distribution as a function of the mismatch angle: $\varphi = [\omega - n\omega_g(0)]T$.

The integral in Eq. (10.140) can be calculated in closed form with some simple distributions $f(\sigma)$. For example, the uniform distribution $f(\sigma) = 1$, $0 \le z \le L$, leads to the integral

$$M = \left| \int_0^1 e^{i\varphi\sigma'} d\sigma' \right|^2 = \left[\frac{\sin \varphi/2}{\varphi/2} \right]^2$$

This function is depicted in Fig. 10.15 (we met the analogous function in the theory of the electron gap, see Section 6.3.4). The intervals in φ with negative slope of the curve M correspond to the positive contribution of the term $\mu(\partial/\partial\varphi)$ in $\mathrm{Re}(\overline{P})$. Clearly, the best interval is $0 \le \varphi \le 2\pi$. Note that according to Eqs. (10.138), the mismatch angle φ is the progressive phase of the rotating electron relative to the field phase for the transit time T. The corresponding frequency band is equal to

$$\frac{\Delta\omega}{\omega} = \frac{(\Delta\varphi)_{\max}}{\omega T} = \frac{2\pi}{\omega T} = \frac{T_g}{nT}$$

and coincides with the bandwidth of the cyclotron absorption for linear bunching when the parameter μ is small and $(\overline{P})$ in Eq. (10.140) is determined by the first term.

10.7.5 Starting Regime of a Gyromonotron

The starting condition of an oscillator can be formulated as an equation of the energy balance with infinitesimally small amplitude (i.e., in the regime that is described by linearized equations). Let us begin with Eq. (10.66) for the active power balance. This equation is valid for arbitrary amplitudes. If we use the linearized active power from Eq. (10.140), then because both the stored energy W and $\mathrm{Re}(\overline{P})$ are proportional to the square of the field amplitude, the latter is canceled and the equality (10.66) gives us the starting current.

Combining Eqs. (10.66) and (10.140) leads to

$$2\omega_{02} W \approx \frac{\omega W}{Q} = \mathrm{Re}(\overline{P}) = -P_{\perp 0} \frac{\mu^2 |F_0|^2}{4} q(\varphi) \qquad (10.141)$$

where

$$P_{\perp 0} = \frac{I_0}{e_0} \frac{p_{\perp 0}^2}{2\, m_0}, \qquad q(\varphi) = (n+\mu)\frac{\partial}{\partial \varphi}\left|\int_0^1 e^{i\varphi\sigma'} f(\sigma')\, d\sigma'\right|^2$$

Let us represent F_0 and W in polar coordinates. For this case according to Eqs. (10.122) and (10.140),

$$|F_0| = |A| \bar{f} \frac{e_0}{m_0 \omega} \frac{n^n}{2^{n-2} n!} \beta_{\perp 0}^{n-4} I_{m-n}(\kappa R_0) \tag{10.142}$$

where $\bar{f} = 1/L \int_0^L f(z)\, dz$. The stored energy is calculated from the integral

$$\begin{aligned} W &= \frac{1}{2\mu 0} \int_V |B_z|^2\, d\tau 0 \\ &= \frac{|A|^2}{2\mu_0} \int_0^L f(z) \int_0^{R_r} I_m{}^2(\kappa r) 2\pi r\, dr\, dz \\ &= \frac{|A|^2 \pi R_r^2 L \bar{f}}{2\mu_0} [I_m^2(\kappa R_r) - I_{m-1}(\kappa R_r) I_{m+1}(\kappa R_r)] \end{aligned} \tag{10.143}$$

Substituting these values in Eq. (10.141) and canceling $|A|^2$, we obtain the starting current:

$$I_{st} = \frac{\pi m_0 \omega R_r^2 (n!)^2 \cdot 4^n \beta_z^2 [I_m^2(\kappa R_r) - I_{m-1}(\kappa R_r) I_{m+1}(\kappa R_r)]}{Q \mu_0 e_0 \beta_{\perp 0}^{2n-2} L n^{2n} q(\varphi) I_{m-n}^2(\kappa R_0) \bar{f}} \tag{10.144}$$

Let us take $q(\varphi)$, corresponding to a Gaussian distribution of the field. Then it can be shown that the minimal starting current (in amperes) for the optimum mismatch angle φ is equal to

$$I_{st,\min} = 17.5 \frac{4^n (n!)^2 \mu^{2(2-n)}}{(Q \times 10^{-3}) n^{2n} G(L/\lambda)^{5-2n} (\pi g)^{2(3-n)}} \tag{10.145}$$

where the pitch factor of the helical beam and the structure factor are equal to

$$g = \frac{\beta_{\perp 0}}{\beta_{z0}}, \qquad G = \frac{I_{m-n}^2(\kappa R_0)}{[(\kappa R_r)^2 - m^2] I_m^2(\kappa R_r)} \tag{10.146}$$

respectively. It is important that the values $\varphi_{st} \sim [\omega - n\omega_g(0)_{st}]$, corresponding to the minimum of the starting current, do not coincide with the values $\varphi_\eta \sim [\omega - n\omega_g(0)_\eta]$, providing the maximum of the orbital efficiency. In the latter case the values of the starting current can be greater than the operation beam current, which corresponds to the hard self-excitation regime.

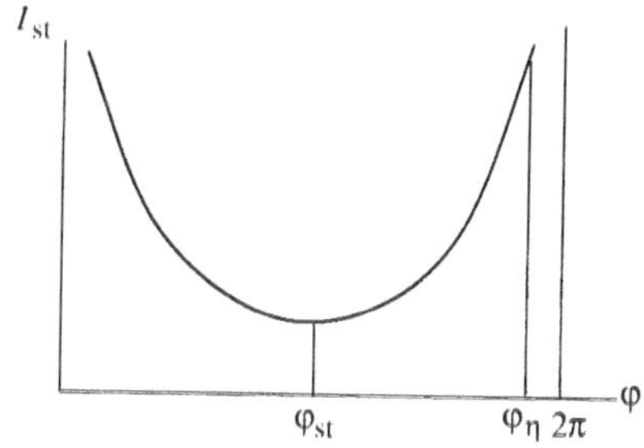

FIGURE 10.16 Starting current as a function of normalized mismatch: $\varphi = [\omega - n\omega_g(0)]T$.

The qualitative dependence of the starting current of the mismatch φ is shown in Fig. 10.16. According to the figure, the current I_{st} is maximal near $\varphi = 0,\ 2\pi$, where the slope of the curve $M(\varphi)$ is close to zero (Fig. 10.15). Note, however, that the optimal efficiency is realized in regimes with a large initial mismatch φ_η because during the interaction, the electron energy diminishes, the gyrofrequency grows, and φ decreases.[3] If the starting current exceeds the operation current, the condition of the energy balance excludes self-excitation of the device, beginning from the noise level of the HF amplitude. The energy balance is possible for finite amplitudes of the HF field because of the nonlinear nature of the electron power $\overline{P}$. In this case, oscillations begin when the initial amplitude exceeds a certain bifurcation threshold. This is a case of hard self-excitation. The hard self-excitation regime can be achieved by other methods as well. In particular, in CW operation, an initial magnetic field can be chosen that is greater than optimal (Kisel et al., 1974). Then φ is less, and a regime of soft self-excitation is realized. Next, the magnetic field is decreased and the optimal efficiency is restored. However, this method cannot be used in pulse regimes. In the latter case a startup scenario is realized in which the regime of the soft self-excitation is achieved through a pulse voltage switch process (see Nusinovich, 1974; Nusinovich et al., 2004; Whaley et al., 1994; and Section 10.9.2.). Note that the region of soft self-excitation at gyroharmonics is significantly narrower than that for fundamental gyroresonance ($n = 1$).

According to Eq. (10.144), the starting current (or the product $I_{st}Q$) quickly increases with n (due mostly to $\beta_{\perp 0}$ in the denominator). The presence of this factor and of the factor $I_{m-n}(\kappa R_0)$ is determined by nonuniformity of the HF field at the Larmor circle. For radiuses of the latter on the order of the wavelength, the nth component of the effective field is equal to $(r_\perp/\lambda)^{n-1}(\overline{\mathbf{E}^2})^{1/2}$. Therefore, the starting current increases with n and decreases with the particle energy. Bunching of electron oscillators in harmonic fields has significant differences compared with bunching at the fundamental harmonic. In particular, the scheme of phasing in Fig. 10.11 is changed: The number of electron bunches at the Larmor circle

[3]This effect is analogous to the influence of the nonsynchronism parameter b in TWTOs (see Section 8.6.3).

increases and becomes equal to the harmonic number. Note that the starting current at the fundamental harmonic increases with energy due to β_z in the numerator of Eq. (10.144), which is related to the decrease in the interaction duration.

10.8 ELEMENTS OF GYROTRON ELECTRON OPTICS

10.8.1 Parameters of Helical Electron Beams

Let us assume an axially symmetric configuration for electron-optical systems and corresponding fields in gyrotrons. Any particle trajectory in a helical electron beam (HEB) can be characterized by four parameters: the gyrofrequency ω_g, the axis radius R_0 of the helical trajectory (Fig. 10.17), and the oscillatory $v_\perp$ and longitudinal (drift) v_z velocity components. Instead of $v_\perp$ and v_z, the full velocity v and the parameter $g = v_\perp / v_z$ can be utilized. Then

$$v_\perp = v\frac{g}{\sqrt{1+g^2}}, \qquad v_z = v\frac{1}{\sqrt{1+g^2}} \tag{10.147}$$

The parameter g characterizes the angle of the helical winding of the trajectory and is called the *pitch-factor*. Because the radiation energy in gyrotrons is extracted from oscillatory electron motion, the magnitude of the pitch factor should be $g > 1$. This parameter is determined by an entire complex of factors participating in the formation of the helical trajectory. Varying these factors along different trajectories signifies a velocity spread and a corresponding spread in the values g.

All practical methods of helical beam formation use static electric and magnetic fields. Therefore, if we neglect initial velocities and dynamic instabilities (these issues are considered later), the spread of full electron velocities is equal to zero:

$$\Delta v^2 = \Delta(v_\perp^2 + v_z^2) = 0 \tag{10.148}$$

Then the values of the relative oscillatory and drift velocity spread are coupled by the following approximate relations:

$$\delta v_z \equiv \frac{\Delta v_z}{\bar{v}_z} \approx g^2 \delta v_\perp = g^2 \frac{\Delta v_\perp}{\bar{v}_\perp} \tag{10.149}$$

If $g^2 \gg 1$, $\delta v_z \gg \delta v_\perp$. For example, for $g = 2$ and $\delta v_\perp = 0.1$ (the latter value is rather moderate for an HEB (see, e.g., Avdoshin et al., 1973), we get $\delta v_z = 0.4$.

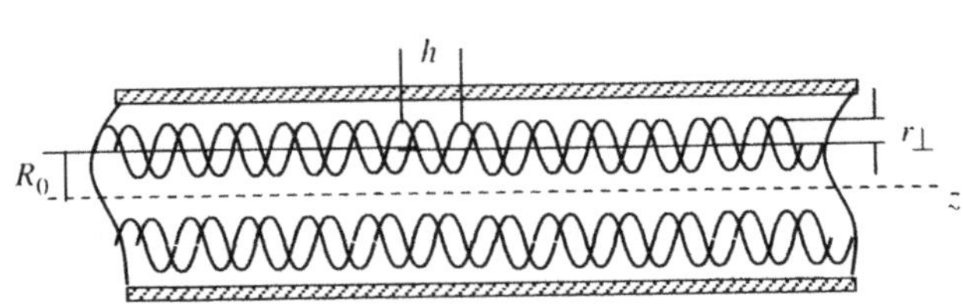

FIGURE 10.17 Helical electron beam in a uniform magnetic field.

The large values of δv_z are responsible for two undesirable effects: (1) the spread of the transit time of electrons in the interaction space, and (2) the reflection back to the cathode of electrons with small v_z (maximum g). The first effect diminishes the interaction efficiency. Taking into account the cutoff structure of the electromagnetic field in gyrotron resonators, we can conclude that the Doppler suppression of resonance is not significant. However, the velocity spread has other important consequences. In particular, bunching of the electrons at the initial section of the resonator where the HF field is weak (e.g., for Gaussian or triangular field distributions) is determined mostly by the mismatch $\Delta = 2(\omega - n\omega_g)/\omega\beta_{\perp 0}^2$ (see Section 10.7.1) with the dispersion of order of $2\delta v_\perp$. This effect can significantly reduce efficiency.

An analysis of the velocity spread effects (Ergakov et al., 1980) has shown that this decrease in efficiency as a function of the velocity spread can be weakened by an optimization of gyrotron construction and electric regimes. For example, the optimized orbital efficiency, even for $\delta v_\perp = 0.3$, can exceed $0.6\eta_0$, where η_0 is the optimal orbital efficiency for $\delta v_\perp = 0$. The change of optimal input parameters of the gyrotron (i.e., μ_0, I_0, g_0, Δ_0) can be considerable. So, according to Ergakov et al. (1980), optimum values of g_0 in the range $0.1 < \delta v_\perp < 0.8$ are determined by the relation $g_{\text{opt}}^2 \delta v_\perp = \text{const}$. Naturally, a decrease in g_{opt} with $\delta v_\perp$ can lead to a considerable loss of efficiency. We should point out an interesting result in a paper by Ergakov et al. (1980): an extension of a zone of soft self-excitation regimes with velocity spread.

The second effect, reflection of some electrons with large g back to the cathode, taking into account a very small probability of their capture by the cathode, leads to an accumulation of reflected particles in the region between the cathode and the resonator and to an excitation of instabilities in the electron beam. This effect is considered later.

Every known system of formation of HEBs contains an extended section with a weakly inhomogeneous distribution of electric and magnetic fields in which a transversal (orbital) invariant is conserved (see Section 1.6):

$$J = \frac{p_\perp^2}{B} = \text{const.} \tag{10.150}$$

where $p_\perp = mv_\perp$ is the electron oscillatory momentum. It can readily be shown that in this case relative spreads of oscillatory velocities or momenta can also be considered as adiabatic invariants. Indeed, according to Eq. (10.150), oscillatory momenta or the velocities of all particles in any two cross sections of the beam are in the same relation, which is determined by the ratio of the magnetic fields and potentials in these cross sections. This means that a distribution of the oscillatory momenta is identical in all cross sections of the beam in weakly inhomogeneous fields, including the relative spread of oscillatory momenta or velocities. (In the last case, corresponding planes are assumed equipotential). The spread indicated can be taken as the fundamental parameter of the beam transverse momentum (velocity) distribution. According to Eq. (10.150), oscillatory momentum and velocity increase when the beam moves in the direction of a growing magnetic field, which allows one to pump the oscillatory energy in the beam. Note that the relative spread of longitudinal velocities in this case increases continuously.

Other characteristic parameters of HEBs are the Larmor radius $r_\perp$ and the pitch h of the helical trajectory (Fig. 10.17). Taking into account the condition of the gyroresonance $\omega = n\omega_g$ and neglecting the Doppler effect, it can readily be shown that

$$r_\perp = \frac{\sqrt{\gamma^2-1}}{\gamma}\frac{g}{\sqrt{1+g^2}}\frac{n\lambda}{2\pi} \approx 0.01\left[\frac{g\sqrt{U_0}}{(1+0.0015U_0)\sqrt{1+g^2}}\right]n\lambda \qquad (10.151)$$

$$h = 2\pi\frac{r_\perp}{g}$$

An error of the approximate formula (10.151) (U_0 is expressed in kilovolts) is on the order of $(\eta U/4c^2)^2$, which is $\sim 1\%$ for $U_0 = 200\,\text{kV}$.

10.8.2 Systems of HEB Formation

The criterion of weakly inhomogeneous fields required for the validity of Eq. (10.150) is (see Section 1.6)

$$\varepsilon = \Lambda\frac{1}{|\mathbf{F}|}\left|\frac{d\mathbf{F}}{d\mathbf{r}}\right| \ll 1 \qquad (10.152)$$

where Λ is the Larmor radius or the pitch (the largest of the two) and $|\mathbf{F}|$ is the modulus of electric or magnetic fields. This criterion is a basis for the classification of HEB guns, defining their essential features.

Adiabatic guns are, by definition, systems for which criterion (10.152) is satisfied in an entire zone of formation. Therefore, the electrons must acquire oscillatory energy just after emission from the cathode. That is possible if the latter is immersed in crossed electrical and magnetic fields. Otherwise, if $v_\perp = 0$ at the cathode, according to Eq. (10.150), oscillatory motion will be absent. A typical adiabatic system is the *magnetron injection gun* (MIG) (see, e.g., Goldenberg and Petelin, 1973), whose scheme is shown in Fig. 10.18. Electrons that are emitted move along quasi-throchoidal trajectories in weakly inhomogeneous crossed fields. A component of the electric field that is parallel to the magnetic field provides an injection of the beam in the transitional area with a purely magnetic field and transforms the beam to a helical beam. The configuration of electrodes in a gyrotron's MIG is analogous to the configuration in a Kino–Tailor gun (see Section 4.7.2), which is used to obtain rectilinear electron beams. However, typical gyrotron MIGs have important distinctions. In particular, the T-mode (see Section 3.4.4) is generally used with the electric field at the cathode weakly disturbed by the space charge.[4]

[4]Note that in advanced MIGs developed by a numerical optimization, the current can be increased up to a significant fraction of the space-charge current (Section 10.8.4). Furthermore, even MIGs working in the ρ-mode can be designed (Kuftin et al., 1993; Manuilov et al., 1995), although strictly speaking, the guns cannot be called completely adiabatic.

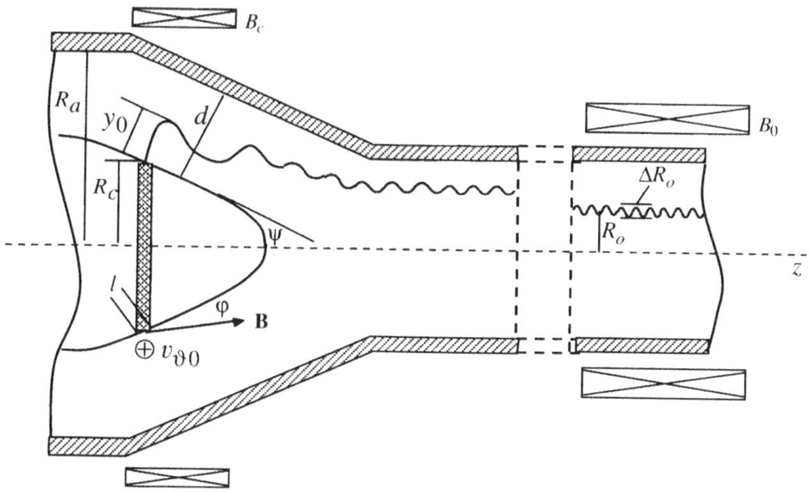

FIGURE 10.18 Magnetron injection gun of a gyrotron.

Below, an elementary theory of the gyrotron MIG neglecting the space-charge field is represented.

Nonadiabatic guns form helical beams when an initially rectilinear beam intersects a sufficiently short section with transverse magnetic or electric fields. Note that a number of diverse nonadiabatic guns can be proposed, whereas the fundamental features of adiabatic guns follow immediately from the assumed requirement of adiabaticity.

Probably, the first nonadiabatic gun applied in an ECM was the *Chow–Pantell gun* for a gyroresonance BWO oscillator (Chow and Pantell, 1960) (Fig. 10.19). The construction of the gun resembles that of a Harris gun (Section 5.5), in which electrons of the rectilinear beam after an annular slot with a radial magnetic field are introduced into a coaxial tube with a radial electric field and move along helical trajectories that encircle the z-axis. The electrons in a Chow–Pantell gun after the annular slot find themselves in a uniform magnetic field and form a hollow helical beam with electron trajectories that in general do not encircle the z-axis. Note that the gun can operate in the ρ-regime. It is readily shown that the system is efficient when the gap of an annular slot is on the order of the Larmor radius or the pitch of the helical trajectory.

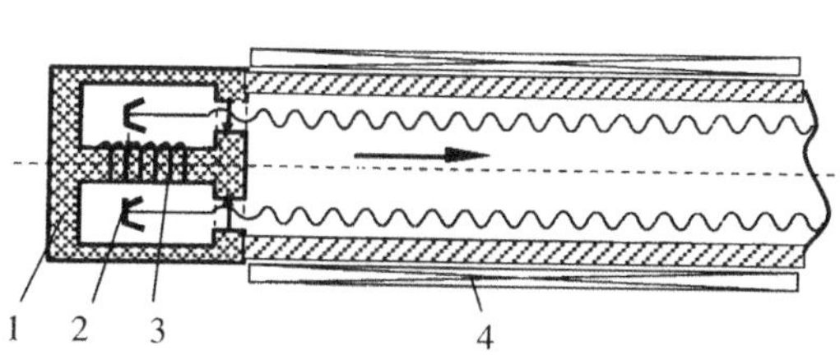

FIGURE 10.19 Chow–Pantell gun. 1, Ferromagnetic yoke; 2, cathode; 3, drive winding; 4, solenoid.

A wide class of nonadiabatic guns is formed by systems with a jump in the magnetic field (Section 1.7.2). In particular, thin helical electron beams encircling the axis of an axially symmetric system are produced by cusp-injected guns (Section 1.7.2). The latter are utilized in large orbit relativistic gyrotrons, providing selective generation of short-wave radiation at the nth cyclotron harmonic (Idehara et al., 2004; Lawson et al., 1985; McDermott et al., 1983) when the annular helical electron beam interacts with the TE_{np} mode of an axially symmetric waveguide.

Nonadiabatic systems producing an annular helical beam in large-orbit relativistic gyrotrons are based on the usage of kickers (Bratman et al., 1999) in the form of rectangular current frames sloping to the z-axis and producing a practically homogeneous transverse magnetic field close to the initial rectilinear beam. Bratman et al. have implemented a large-orbit gyrotron with an electron energy of ~300 kV, generating gyroharmonics up to fifth order. To this type of electron guns, relativistic guns formatted by an explosive emission diode and a nonadiabatic pump magnet can be referred (see Sections 4.8.2 and 10.7.1).

It is necessary to keep in mind that in all enumerated gyrotron electron-optical systems, the magnets that create a quasiuniform magnetic field in the cavity and a nonuniform field in the formation region and near the collector are obligatory components. The value of the magnetic field in a cavity B_0 must satisfy Doppler conditions [Eq. (10.46)] and determine the output frequency of the gyrotron. Thus, possibilities for the attainment of the greatest output frequency of the gyrotron are determined to a decisive degree by limitation of the magnetic field that can be created by the magnet. Some available magnetic systems are considered briefly in Appendix 9.

10.8.3 Theory of an Adiabatic Magnetron-Injected Gun in a Nonrelativistic Approximation

According to Eq. (10.152), fields in a near-cathode area of adiabatic guns can be assumed uniform. Below, all values in this area are marked by the subscript c. The electron velocity can be represented as a sum of oscillatory and drift velocities [Eq. (1.65)]:

$$\mathbf{v} = \mathbf{v}_\perp + \mathbf{v}_d \tag{10.153}$$

where the drift velocity [see Eq. (1.72)]

$$\mathbf{v}_d = v_B \frac{\mathbf{B}}{B} + \frac{1}{B^2} \mathbf{E}_\perp \times \mathbf{B} \tag{10.154}$$

$\mathbf{E}_\perp$ is the component of the electric field perpendicular to the magnetic field $\mathbf{B}$. The components of Eqs. (10.53) and (10.54) at the plane perpendicular to $\mathbf{B}$ near the cathode are

$$v_{\perp c} = \frac{E_{\perp c}}{B_c} \pm v_\vartheta = \frac{E_c \cos\varphi}{B_c} \pm v_\vartheta \tag{10.155}$$

where v_ϑ is the modulus of the initial azimuthal velocity in the ϑ-direction (the direction perpendicular to the magnetic field $\mathbf{B}_c$ on the cathode surface; see Fig.10.18). The condition of the adiabatic invariant conservation [Eq. (10.150)] and the energy integral in the nonrelativistic approximation determine the oscillatory and drift velocities in the operation space of the gyrotron:

$$v_{\perp o} = \left(\frac{E_c \cos\varphi}{B_c} \pm v_\vartheta\right)\sqrt{\alpha}, \qquad v_{zo} = \sqrt{2\eta U_o - v_\perp^2} \tag{10.156}$$

where $\alpha = B_o/B_c$ is the magnetic compression coefficient, φ is the slope of the magnetic field to the cathode surface (Fig. 10.18), and B_o and U_o are the magnetic field and the potential in the operation space, respectively. Further, the theory is built with the following conditions that characterize a simplified model of an adiabatic MIG (Goldenberg and Petelin, 1973):

1. Axial symmetry
2. Small space-charge field
3. Quasiplanar gun construction typical of short-wave gyrotrons. This leads to the cathode electric field

$$E_c = \frac{U_a}{d} \tag{10.157}$$

   where U_a is the anode voltage and d is the distance anode–cathode (Fig. 10.18).
4. Paraxial approximation for the magnetic field:

$$B_z(r, z) \approx B_z(0, z) \tag{10.158}$$

Taking Busch's theorem and conditions 1 and 4 into account, it can be shown that guiding centers of electron orbits move in the operation space along the magnetic lines of force that pass through cathode entry points of the corresponding particles. Then the radius R_o of the guiding center in the operation space according to Eq. (10.158) is

$$R_o = \frac{R_c}{\sqrt{\alpha}} \tag{10.159}$$

where R_c is the cathode radial position of a given particle.

Let us introduce the following dimensionless parameters (see Fig. 10.18):

1. The ratio of the beam current to the Child–Langmuir current in the quasiplanar nonmagnetic diode formed by the gun electrodes:

$$t_j = I/\sqrt{2.33 \times 10^{-6}(2\pi R_c l/d^2)U_a^{3/2}} \tag{10.160}$$

2. The nonrelativistic oscillatory electron energy in operation space:

$$t_\perp = \frac{v_{\perp o}^2}{v^2} = \frac{g^2}{1+g^2} \tag{10.161}$$

3. The width of the emitter:

$$t_l = \frac{l}{d} \tag{10.162}$$

4. The height of the electron trajectory at the cathode:

$$t_y = y\sqrt{d[1+(\pi^2/4)\tan^2\varphi]} \tag{10.163}$$

This formula is based on the relation for the first maximum of the electron trajectory in the planar magnetron with an inclined magnetic field (see Section 3.5.3):

$$y = \frac{2E_c\cos^2\varphi}{\eta B_c^2}\left(1+\frac{\pi^2}{4}\tan^2\varphi\right) \tag{10.164}$$

For $\varphi \ll 1$, Eqs. (10.156)–(10.163) lead to the following relations for the beam current, the cathode beam density, the anode voltage, and the magnetic compression:

$$I = 0.93\, t_j t_l t_y^{-1/2} t_\perp^{1/2} U_o^{1/2} E_c R_o \tag{10.165}$$

$$j_c = \frac{I}{2\pi R_c l} = 4.52\, t_j t_y^{1/2} t_\perp^{-1/3} U_o^{-1/3} E_c^{5/3} B_0^{1/3} \tag{10.166}$$

$$U_a = 2.63\, t_y^{-1} t_\perp^{2/3} U_o^{2/3} E_c^{2/3} B_o^{-2/3} \tag{10.167}$$

$$\alpha = 1.52\, t_\perp^{1/3} B_o^{2/3} E_c^{-2/3} U_o^{1/3} \tag{10.168}$$

Here the following respective units were used: A, A/cm^2, kV, kV/mm, mm, and kG. e_0U_0 is the electron energy in the working space (cavity).

The right-hand sides of Eqs. (10.165)–(10.168) contain only the parameters specified. In particular, the magnetic field is determined by wavelength. According to the conditions of gyroresonance and the energy integral,

$$B_o = \frac{2\pi c}{n\lambda\eta}\gamma = \frac{2\pi c}{n\lambda\eta}\left(1+\frac{\eta U_o}{c^2}\right) \approx \frac{107}{n\lambda}(1+0.002U_o)$$

It can readily be shown that the beam thickness in the cavity (Fig. 10.18)

$$\Delta R_o \approx \frac{l\sin\varphi}{\sqrt{\alpha}} \tag{10.169}$$

The average beam radius $\overline{R}_o$ is determined based on the operation mode chosen from the condition of the effective interaction of the wave with the electron beam. The parameter t_j can be considered an empirical constant: Its maximum value corresponds to a current in which the space-charge field determines the velocity spread (see Section 10.8.4). The parameter t_y characterizes the proximity of the regime to the cutoff regime for the planar magnetron formed by the cathode–anode gap in the magnetic field. This parameter is usually equal to $t_y = 0.2$ to 0.7. The parameter $t_l \sim 0.5$ to 1.0. For $t_l > 1$ it is difficult to get an electrode configuration that excludes the *position velocity spread* determined by the difference in the fields at different points of the emitter. In this case, the electrode geometry can only be found using numerical optimization. Finally, the value of $t_\perp$ corresponds to the pitch factor chosen [Eq. (10.161)].

According to Eq. (10.165), the beam current increases proportionally to the cathode electric field E_c. The magnetic compression and cathode radius in this case are determined by Eqs. (10.168) and (10.159). The limiting current is determined by vacuum breakdown of the anode–cathode gap. The corresponding electric field is usually on the order of 5 to 10 kV/mm. It is necessary, however, to control the cathode current density [Eq. (10.166)]. If the latter exceeds the maximum current for the type of cathode chosen, the limiting current is estimated with j_c but not with E_c (i.e., $t_j = I/2\pi R_c l j_c$).

Example 10.3: Calculations for an Adiabatic Gun Let us assume the following parameters for the gyrotron: frequency $f = 140$ GHz; fundamental harmonic $n = 1$; mode $\mathrm{TE}_{26.6}$; electron energy $e_0 U_o = 80$ keV; $E_c = 5$ kV/mm; $g = 1.5$; $t_j = 0.3$; $t_y = 0.2$; $t_l = 0.1$. The following values are calculated:

- Wavelength λ: $c/f = 2.14$ mm.
- Beam radius R_o: found from the conditions of the minimal starting current, which according to Eq. (10.146), corresponds to the maximum of the function $I_{m-1}(\kappa R_o)$ or the null of the function $I'_{m-1}(\kappa R_o) = I'_{25}(\kappa R_o)$. Usually, the first (largest) maximum is chosen. In this case, $\kappa R_o \approx (2\pi/\lambda)R_o = 27.4$. Then $R_o = 27.4(\lambda/2\pi) = 9.3$ mm.
- Magnetic field B_0: $(2\pi f/\eta)\gamma \approx (2\pi f/\eta)(1 + 0.002U_{0\mathrm{kV}}) \approx 58$ kG.
- Parameter $t_\perp$: $g^2/(1+g^2) = 0.69$.
- Magnetic compression α: According to Eq. (10.168), α is 29.7.
- Emitter radius R_c: $R_o\sqrt{\alpha} = 9.3\sqrt{29.7} = 51$ mm.
- Anode voltage [Eq. (10.167)] U_a: 37 kV.
- Beam current [Eq. (10.165)] is I: 43 A.
- Cathode current density [Eq. (10.166)] j_c:

$$\frac{I}{2\pi R_c l} = \frac{I}{2\pi R_c^2 t_l} = 2.7\ \mathrm{A/cm^2}$$

10.8.4 Advanced Design of an MIG

The task of MIG optimization is solved through numerical analysis in three parts: the self-consistent solution to beam equations, which are reduced to relativistic equations of motion for a chosen number of trajectories (or large particles), the continuity equation, and the boundary problem for the Poisson equation, taking into account the self-magnetic field of the beam (see, e.g., Borie et al., 1995; Lygin et al., 1987; Rayski and Tsimring, 1996). The numerical methods are distinguished by their approach to solving the Poisson equation [e.g., method of integral equations, finite differential method (net method), PIC method]. Initial data for the design are usually taken using Eqs. (10.165)–(10.169). The first-order data taking relativistic effects into account can be chosen correspondingly to Edgcombe (1988b).

It is necessary to mark the class of gyrotron MIGs working in a space-charge-limited current regime as an advanced version of gyrotron guns. The latter use laminar helical electron beams (see Section 10.8.5) and possess the following potential advantages:

- Absence of the growth emission current effect as a result of heating of the cathode by trapped electrons (see Section 10.8.5)
- Weakening of the influence of factors leading to the velocity spread in HEBs (see Section 10.8.5): roughness of an emitting surface and nonuniform distribution of cathode emission

Some results of theoretical and experimental investigation of these guns are given in Edgcombe (1988b), Kuftin et al. (1993), and Manuilov et al. (1995). Note that the initial state of a laminar beam can be calculated by the synthesis method based on an application of the Dryden flow (see Section 3.5 and Dryden, 1962; Edgcombe, 1988a; Manuilov and Tsimring, 1978; Tsimring, 1977). A detailed account of these topics goes beyond the scope of this book.

10.8.5 Velocity, Energy Spread, and Instabilities of HEBs

The major factor responsible for gyrotron characteristics is the quality of a helical electron beam (HEB). What is the quality of a HEB? It includes the homogeneity of current, velocity, and energy distributions, and beam stability: the absence of parasitic oscillations and the spontaneous rise of the emission current. Finally, high-quality HEBs must have a sufficient share of electron oscillatory energy determined by the pitch factor g. According to theory and experimental data, the most important factor providing high-quality HEB is the minimization of velocity and energy spread (VESP).

The influence of velocity spread was discussed briefly in Section 10.8.1. The energy spread is a result of dynamic effects in the beam. According to experiments, the energy spread in real gyrotron beams in general is on the order of a few percentage points (Dumbrais and Koponen, 1999; Glyavin et al., 1999). However, due to the relativistic dependence of electron gyrofrequency on energy, the efficiency of gyrotrons in which particles usually perform some large number N of cyclotron rotations is sensitive to the spread of the initial energy. The theory has shown that even for the energy spread $\Delta E/E << 1/N$, the effect can be considerable (Dumbrais and Koponen, 1999).

VESP in helical electron beams has been studied in a number of papers (see, e.g., Glyavin et al., 1999; Kuftin et al., 1992b; Tsimring, 1972, 2001; Tsimring and Zapevalov, 1996). The following factors leading to VESP in a magnetron-injection gun (MIG) are considered:

1. Spread of initial electron velocities
2. Roughness of an emitting surface
3. Nonuniform distribution of electric and magnetic fields determined by the geometry of electrodes and magnets
4. Nonuniform distribution of an emission current on the cathode
5. A space-charge field in the beam
6. Convective instabilities in the beam
7. Global instability in the beam

Factors 1 to 4 and 5 (in part) are static. Note that these factors mostly determine the dynamic factors 5 to 7 and also depend on them.

Factor 1: Spread of Initial Electron Velocities The nature of this spread is contained in Eq. (10.155), according to which, unlike the total velocity, the oscillatory velocity is determined by the sum of velocities rather than energies. This causes a substantially greater dependence of helical beam parameters than straight beams on small initial velocity perturbations. Let us denote the average initial electron energy in the ϑ-direction by $e_0\overline{U_{\vartheta 0}}$. Then extreme values of the initial velocity are $\pm\sqrt{2\eta\overline{U_{\vartheta 0}}}$ and the quantity

$$2v_{\vartheta 0} = 2\sqrt{2\eta\overline{U_{\vartheta 0}}} \tag{10.170}$$

is the spread of the initial velocities. The relative spread of oscillatory velocities according to Eq. (10.155) is then given by

$$\delta v_{\perp} = \frac{2v_{\vartheta 0}}{v_{\perp c}^{(0)}} = \frac{2B_c\sqrt{2\eta\overline{U_{\vartheta 0}}}}{E_c\cos\varphi} \tag{10.171}$$

Let us express $\delta v_{\perp}$ using Eq. (10.164) in terms more convenient for characterization of the gun operation mode:

$$\delta v_{\perp} = 4\sqrt{\frac{\overline{U_{\vartheta 0}}}{U_a}}\sqrt{\frac{d}{y_0}\left(1 + \frac{\pi^2}{4}\tan^2\varphi\right)} \tag{10.172}$$

where U_a is the anode voltage and y_0 is the height of the first vertex of the electron trajectory above the cathode (Fig. 10.18) for zero initial velocity. The coefficient 4 in Eq. (10.172) should be replaced by 3.6 for the normal (Gaussian) distribution of initial velocities (Tsimring, 1972).

The spread is very small for thermal velocities ($e_0\overline{U_{\vartheta 0}} \sim \text{kT} \sim 0.1$ eV). So for typical values $U_a = 50\,\text{kV}$, $d/y_0 = 10$, and $\varphi = 25^\circ$; we obtain $\delta v_\perp = 2\%$. However, the magnitude of $\overline{U_{\vartheta 0}}$ for real emitters can be substantially larger (up to 1 eV). The principal cause is the nonequipotentiality of active emitter centers. This may also be the result of emission inhomogeneity (see, e.g., Ilyin et al., 1998) associated with the spread of the work function and the cathode temperature. In this case, $\delta v_\perp$ may be more than 10%. Another cause of the increase in initial velocities is secondary emission as a result of cathode bombardment by trapped electrons and subsequent development of instability of the electron beam (see the discussion of factor 7).

Factor 2: Roughness of the Emitter Surface Roughness affects the spread in a manner similar to the nonequipotentiality of the emitter. Velocities that electrons acquire in local fields of nonuniformities can be treated as initial velocities on the surface of a smooth emitter.

Let us estimate the velocity component v_{x0} of electrons emerging from a nonuniformity, which we will consider a hemisphere with radius $r_0 \ll d$ located on the cathode surface (Fig. 10.20) in a uniform magnetic field. Integration of the equations of motion using the general expression for the electric field of a hemisphere over the conducting plane (Smythe, 1950) shows that the maximal value of v_{x0} is reached in the vicinity of the hemisphere where $d^2x/dt^2 = O(r_0^3/y^3)$. The typical value v_{x0} according to Tsimring (1972) is equal to

$$v_{x0} = \left(\frac{dx}{dt}\right)_{x=r_0, y=1.5r_0} = 0.4\sqrt{2\eta U_a \frac{r_0}{d}} \tag{10.173}$$

Using values v_{x0} as the initial velocities $v_{\vartheta 0}$ in Eq. (10.171) allows us to use Eq. (10.172) and find the spread of the oscillatory velocities:

$$\delta v_\perp = 1.6\sqrt{\frac{r_0}{y_0}\left(1 + \frac{\pi^2}{4}\tan^2\varphi\right)} \tag{10.174}$$

One can show that the effect of the magnetic field on the motion of electrons near the hemisphere is very small. In light of the fact that the ratio r_0/y_0 enters Eq. (10.174) in the power of $\frac{1}{2}$, even relatively little roughness may cause considerable spread. So

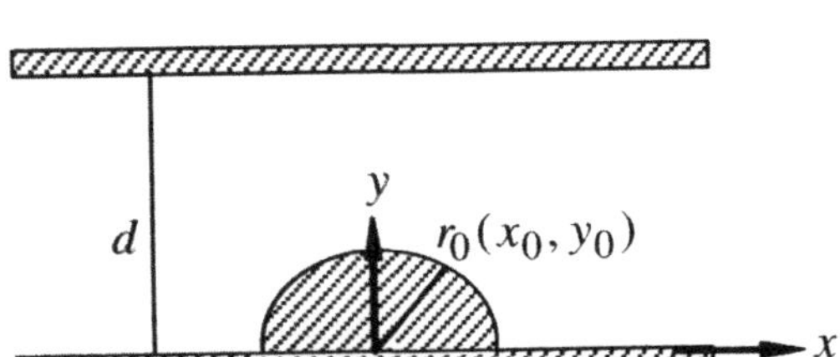

FIGURE 10.20 Roughness of an emitter surface modeled by a hemisphere in a uniform magnetic field.

for $r_0 = 2\,\mu\text{m}$, $y_0 = 2\,\text{mm}$, and $\varphi = 20^\circ$, we have $\delta v_\perp \approx 6\%$. This example proves that the requirements for the smoothness of the emitter must be very stringent.

Factor 3: Nonuniform Distribution of Electric and Magnetic Fields Near the Cathode Electron trajectories in adiabatic fields are determined by the distribution of static fields on the cathode. For an axially symmetric system there is software available for optimizing the gun geometry up to small values of the velocity spread. However, thermal deformations and inaccuracy of the installation of electrodes or magnets can induce asymmetry in the gun assembly. In these cases, radial displacement of the cathode is most dangerous. The corresponding velocity spread is determined by the approximate formula (Avdoshin et al., 1975)

$$\delta v_\perp \approx \frac{2a\cos\theta}{d} \tag{10.175}$$

where a is the radial displacement of the cathode and θ is the inclination of the cathode generatrix to the system axis.

Factor 4: Nonuniform Distribution of the Emission Current This factor is connected partially with factor 2, because the roughness of the cathode surface and subsequent increase in the velocity spread may induce nonuniformity of the emission current distribution. The strong emission inhomogeneity of gyrotron emitters has sometimes been blamed on the spread of experimental characteristics of identically constructed gyrotrons and from the temporal instability of their characteristics during long-term operation.

Raster electron microscopy has been applied to obtain an emission image of the surface of large cathodes (Andronov et al., 1995). An electron probe of a specific diameter was used with a resolution better than 0.1 mm. According to Andronov et al. (1995), angular variation of the density of the emission current can achieve a factor of 4.

Two mechanisms can be suggested to explain the influence of the emission current nonuniformity on the velocity spread. In the first, the oscillatory spread is the consequence of nonuniform space-charge fields (especially in the case of angular variation of the emission; see the discussion of factor 6). The second (indirect) effect of spatial fluctuations of the emission current lead to the instability of HEBs and therefore to further increase in the velocity and energy spread. We discuss this topic in more detail below.

Factor 5: Space-Charge Fields in HEBs The influence of the space-charge field in HEBs on the velocity spread is often the most important factor. The effect of the space charge depends on the position of electrons in the beam and on the beam topology. The latter is determined by the angle φ of inclination of the magnetic field to the conical cathode (Fig. 10.18). Two types of beams are represented in Fig. 10.21: a beam with regularly intersecting electron trajectories (Fig. 10.21a) and a laminar beam (Fig. 10.21b). Typically, regularly intersecting beams are

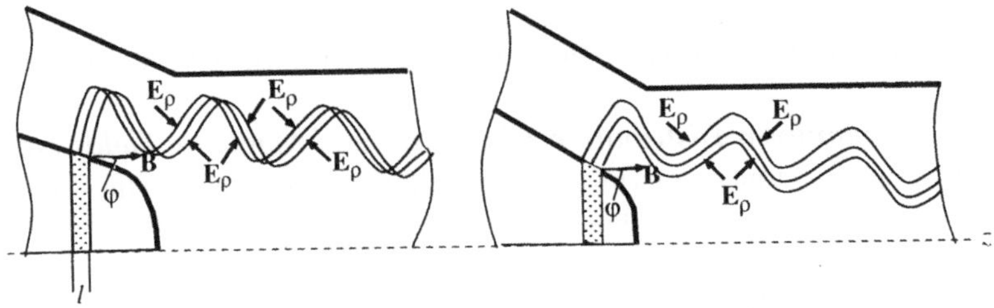

FIGURE 10.21 Topologies of a helical beam: (*a*) beam with regular intersections of trajectories; (*b*) laminar beam.

obtained for $\varphi \approx 10$ to 20°, whereas laminar beams are obtained for $\varphi > 24^\circ$ in the T-mode. The boundary of the laminar topology onset shifts to smaller angles with an increase in the beam current (see Fig. 3.10). These results are applicable for a planar model of the gun. A more complicated type of border between laminar beams and beams with regularly intersecting trajectories is obtained for the cylindrical geometry of the gun (see Manuilov and Tsimring, 1978, Fig. 2).

Let us consider a regularly intersecting beam (Fig. 10.21*a*). As we see, the space-charge force has a specific spatial resonant character: electrons emitted from the left part of the emitter are continuously decelerated in the direction of the magnetic field by the space-charge force $\mathbf{E}_\rho$, whereas the "right" electrons are accelerated continuously. This effect creates a velocity spread. According to Kuftin et al. (1992b),

$$\delta v_\perp = \frac{I}{I_\rho} \tag{10.176}$$

where I is the beam current and $I_\rho = A(\varphi, \Psi) R_c E_c^2 \alpha / B_0$. The coefficient A depends on φ and the electron transit angle Ψ through a region of regularly intersection of trajectories. Using Eq. (10.176), we can estimate the velocity spread as ~10% for $I = 20$ A. Numerical simulations typically give higher values (Kuftin et al., 1992b) (Table 10.4). According to Table 10.4, the velocity spread in the laminar beam ($\varphi = 27^\circ$) is significantly less for a large beam current than in a regularly intersecting beams. This effect is explained by the absence of spatial resonance: Space-charge forces acting on electrons in the frame moving with the particles change sign (Fig. 10.21*b*).

TABLE 10.4 Velocity Spread in Laminar and Regularly Intersecting Beams

	Beam current, I					
	0		20 A		30 A	
φ (deg)	$t_\perp$	$\delta v_\perp(\%)$	$t_\perp$	$\delta v_\perp(\%)$	$t_\perp$	$\delta v_\perp(\%)$
12	0.64	0.6	0.54	21	0.51	30
17	0.62	5.0	0.57	16	0.55	25
27	0.61	6.0	0.56	8	0.54	10

Experimental investigations confirm the results of simulations with the exception that the experimental spread for laminar beams usually exceeds the velocity spread calculated. The cause of this discrepancy is probably related to the stronger influence of reflected electrons, because these electrons in laminar HEBs produce regular intersections with trajectories of the primary beam (Manuilov et al., 1990; Tsimring and Zapevalov, 1996).

Factor 6: Convective Instabilities in HEBs Instabilities in HEBs are strongly connected with velocity and energy spreads: As a rule, they are determined by the velocity spread and themselves lead to a spread increase. Intense HEBs are similar to a dense plasma and so can be prone to many types of instabilities. Below we consider two types that are better investigated and presumably have the most significance.

Electrostatic Cyclotron Instability (ESCI) ESCI is also known as *a negative-mass instability* (NMI) or *Berstein mode instability* (see, e.g., Bratman and Savilov, 2000; Li and Antonsen, 1994). It takes place in any ensemble of electrons moving in a magnetic field and interacting via a space-charge field. The mechanism of NMI is similar to the gyrotron mechanism (Sections 10.6 and 10.7) with the exception that the high-frequency field of the cavity is replaced by the space-charge field of the beam. The simplest model of an HEB, taking into account space-charge interaction, is a flat thin layer of particles rotating in a uniform magnetic field (Fig. 10.22). The leading centers of the Larmor orbits are located in the plane $x = 0$. The Coulomb forces acting on the electrons in the beam can be represented as a superposition of fields of thin layers at planes $x =$ const. Each layer is formed by particles rotating in the phase interval $[\psi,\ \psi + d\psi]$. In the theory of NMI (Bratman and Savilov, 2000), a method of averaged equations has been used. High-frequency initial perturbations are amplified in the drift direction along the magnetic field. The distribution of the energy spread δE computed is shown in Fig. 10.23. The saturation ($\delta E \approx 7\%$) is reached in a region after the plateau of the function $B(z)$ (beyond the cavity). The measured energy spread (ES) in experimental gyrotrons was relatively low, ~1 to 2% (Glyavin et al., 1999). However, the ES at the linear stage depends on initial perturbations that are determined by emitter quality. Besides this, an enlargement of the drift region length (the distance between the cathode and the cavity) can increase the ES considerably: for example, in quasioptical gyrotrons (Bratman and Savilov, 2000).

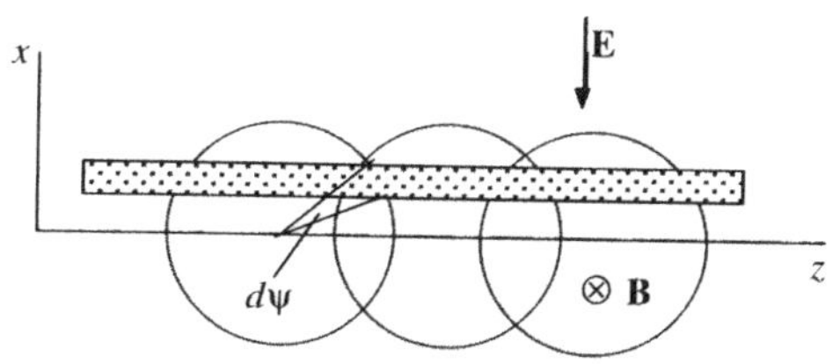

FIGURE 10.22 Flat thin layer formed by electrons rotating in a magnetic field.

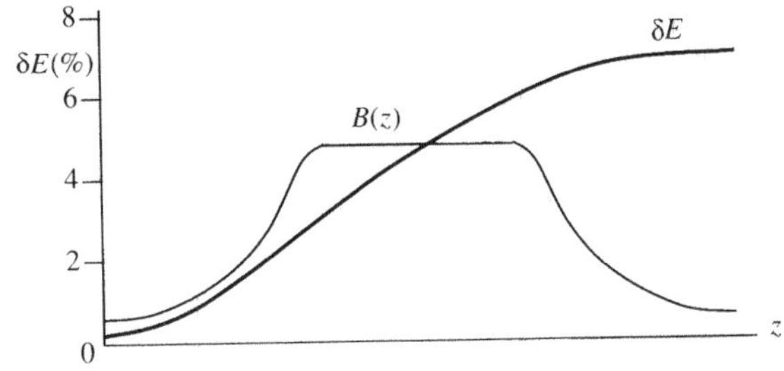

FIGURE 10.23 Distribution of a magnetic field and the electron energy spread in an electron beam for the nominal regime of a gyrotron. (From Glyavin et al., 1999.)

Diochotron Instability (DI) So far, the influence of the space-charge field on the velocity spread has been considered for an idealized axially symmetrical distribution of the charge density. However, according to experimental results (see the discussion of factor 4; see also Andronov et al., 1995), the angular distribution of the emisson current for real cathodes can be highly irregular. Hollow electron beams transported in a magnetic field are unstable with respect to angular perturbations of emission (Cutler, 1956; Kyhl and Webster, 1958). The latter produce an unbalanced space-charge field in a direction that increases perturbations further. As a result, initial weakly angular nonuniformity breaks up (at a sufficient distance from the cathode) into a series of specific nebulas. This effect was first observed in straight electron beams by Kyhl and Webster (1958). According to the linear theory (e.g., Davidson, 1990), the growth rate of this spatial DI is on the order of $\Delta \sim \omega_p^2/\omega_g v_z$, where v_z is the electron velocity along the magnetic field. The motion of particles leading to the DI has the character of drift in the space-charge and magnetic crossed fields. Therefore, DI is also developed in HEBs with the difference that the drift of electrons in straight beams is replaced by drift of the leading centers of Larmor orbits. A peculiarity of these thin, hollow gyrotron beams is the strong transverse nonuniformity of their space-charge distribution (see Fig. 10.22). The linear theory of the DI in HEBs was given by Lygin et al. (1991), where the spatial growth rate of the diochotron perturbations for different radial distributions of the space-charge density was recorded. However, real estimations of the DI effect on the velocity spread require a study of the beams in saturated regimes.

Direct simulation of HEBs with azimuthal nonuniformity was carried out by Kuftin et al. (1993). They considered an MIG with irregular (periodic) angular current distribution but within axially symmetric electrodes and magnets. The Fourier expansion of the current density has the form

$$j(\varphi) = j_0 + \mathrm{Re}\sum_{n=1}^{\infty} j_n e^{iMn\varphi}, \qquad M = 1, 2, \ldots \tag{10.177}$$

where M is a given value for the initial azimuthal current variations and n is the number of an azimuthal harmonic. Application of the expansion (10.177) to all

TABLE 10.5 Velocity Spread for Two Guns with Periodic Irregular Distribution of an Emission

Gun	f (GHz)	M	t_j	$\|\Delta I\|/I$	$\delta v_\perp$
1	20	6	0.1	0	0.24
				0.3	0.35
				0.5	0.44
				1.0	0.62
2	140	6	0.1	0	0.16
				0.3	0.44
				0.5	0.22
				1.0	0.28

variables and using the equation of motion gives relations between complex amplitudes of the azimuthal modes. The main difficulty for further analysis is related to solution of the Poisson equation with periodic boundary conditions, for which the special numerical code EPOSR was used (Kuftin et al., 1993). The final stage of the procedure is calculation of the velocity spread according to the trajectories found. In Table 10.5, values of the velocity spread obtained for two MIGs are shown as functions of $|\Delta I|/I$, where I is the beam current and ΔI is the current deflection from the average value; M is the number of current variations on the full angle. According to the table, an increase in $\delta v_\perp$ reaches 150% and 80% for guns 1 and 2, respectively. The theory above was limited to considerations of spatial DI only. Temporal DI obviously leads to an energy spread of HEBs.

Factor 7: Global Instability As a rule, analysis and synthesis of gyrotron HEBs use models with a static electric field. The theory allows us to determine the electrode configuration required and distribution of the magnetic field and electric regimes. Experimental values of the gun parameters generally agree with the static model (Tsimring and Zapevalov, 1996).

However, in the framework of the foregoing theory, it is difficult to explain some important observations:

1. The experimental velocity spread (VS) as a function of the beam current sometimes exceeds the calculated spread significantly.

2. There are limiting values of the pitch factor g for each gun construction. Larger values of g are unattainable by any regime change, including an increase in the magnetic compression α.

3. Intense quasiperiodic oscillations of the current and an electromagnetic radiation in the frequency range 10 to 1000 MHz (low-frequency oscillations) are observed for large currents (Goldenberg, 1976; Gorelov et al., 1999; Luksha and Sominski, 1998).

These effects are accompanied by the growth of the anode current, although the cathode–anode gap in guns of powerful short-wave gyrotrons is an order of magnitude greater than the height of the vertex of static electron trajectories above the cathode. Growth of the emission current is also observed, which can lead to a failure of the corresponding operation regimes for a large pulse duration, especially in a CW mode.

Clearly, as a result of VS, the distribution of electron oscillatory velocities does not have a sharp boundary on the large oscillatory velocity (small drift velocity) side. Therefore, reflection of some fraction of electrons from the magnetic mirror is inevitable. These particles in static fields cannot return to the cathode because a coincidence of direct and return trajectories near the cathode has a probability of practically zero. The reflected electrons in static fields also cannot reach the anode because the magnetron diode is in the over-cutoff regime. As a result, reflected electrons are locked in the trap between the cathode and the magnetic mirror preceding the cavity. The lifetime of trapped electrons (TEs) in a typical gyrotron adiabatic distribution of electric and magnetic fields can be very large because of the adiabatic invariant conservation. Therefore, an accumulation of TEs leads to the onset of instability. The latter can be called a *global instability* because the disturbances arise at every point of the confined space between the cathode and the magnetic mirror.

The following issues are of the most interest:

1. The origin and evolution of the instability
2. Possibility of the existence of stationary states of a beam with TEs
3. Nature of low-frequency electromagnetic oscillations
4. Possibilities of suppression of space-charge oscillations and electromagnetic low-frequency radiation

Origin of TE Instability. Stationary States The results of numerical simulations of the dynamic processes in an HEB with trapped electrons are presented briefly below. More detailed consideration of this problem has been given by Rayski and Tsimring (1996) and Tsimring (2001).

The algorithm of the simulation was implemented as a succession of temporal steps. At each step, large particles move in a given magnetic field and in the electric field found from the distribution of the space charge calculated at the preceding step. Simultaneously, particles with charges $e_m = I_m\tau$, $m = 1, 2, \ldots, N$, are injected into the beam from the cathode. Here τ is the temporal step and I_m is the emission current emitted from the mth cathode section. The distribution of initial oscillatory velocities was taken close to a normal distribution. The value of the magnetic field compression α was taken such that the share of reflected particles with oscillatory velocities $v_\perp > v$ was about 10%. A quasilaminar undisturbed beam has been assumed.

Some computed characteristics of the beam evolution are given below.

- *0–2 ns:* First, reflected electrons appear in the beam. The electric field on the cathode equals approximately 0.9 of a "cold" field.

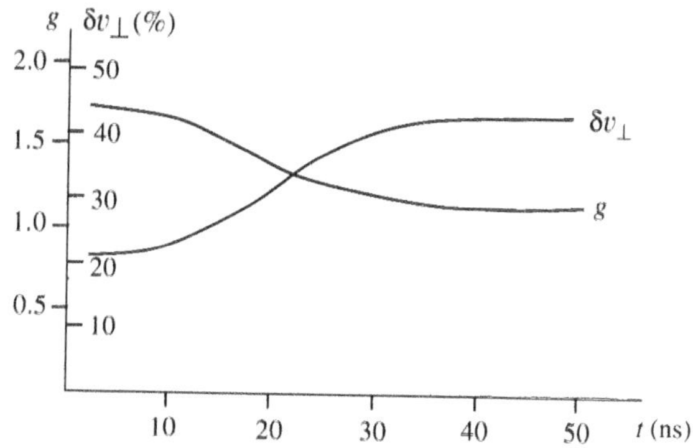

FIGURE 10.24 Values computed for pitch factor and velocity spread as functions of time.

- *2–4 ns:* Next, reflected electrons reach the cathode. The trajectories of reflected particles are phased with direct trajectories. This effect increases the velocity spread in the direct beam (Manuilov et al., 1990).
- *4–20 ns:* The beam is gradually filled by electrons locked in the trap. The phases of the trajectories are mixed, and TEs form a halo around direct trajectories. Accumulation of TEs near the cathode leads to further depression of the electric field. This field is nonstationary and oscillates. After several oscillations, some electrons lose part of their oscillation energy and can leave the trap. The number of new particles in the direct beam that can be reflected from the magnetic mirror also decreases, because these electrons begin to move in the weakened electric field.
- *>20 ns:* The beam acquires a quasistationary state in which the balance of reflected and departed particles is reached. Electrons can leave the trap by three different routes: into the operation space (through the magnetic mirror), toward the cathode, and toward the anode. Temporal variations for averaged beam parameters are illustrated in Fig. 10.24. According to Fig. 10.24, the asymptotic values of g and $\delta v_{\perp}$ are worse than values in the absence of TEs, (i.e., for $t < 10$ ns).

We can conclude that even a relatively small share of reflected electrons will worsen the beam quality considerably. Furhtermore, we can now understand the above-mentioned existence of the limiting value of g. For example, an attempt to enhance g by raising the magnetic compression α increases the number of reflected electrons. This leads to growth of the electric field depression near the cathode, and the pitch factor returns approximately to its old value. It is clear that the limiting value of g increases with a decrease in the initial spread $\delta v_{\perp}$.

Space-Charge Oscillations The establishment of a quasistationary state with TEs does not exclude the existence of space-charge oscillations. The origin of these oscillations is obviously connected with axial oscillations of TEs in the adiabatic trap (see an estimation of the frequency of these oscillations a paper by Tsimring, 2001). But even in the presence of TEs, the excitation of the space-charge oscillations is not a simple problem: In the absence of particle bunching, temporal oscillations of space-charge density are impossible. At least two mechanisms of high-frequency instability of an ensemble of electron oscillators may be proposed (see Section 10.3).

1. *Nonisochronous Oscillators* $d\omega_l/dE \neq 0$, where ω_l is the frequency of axial oscillations of the electrons in the trap and E is the oscillator energy. The sign and value of $d\omega_l/dE$ depend on the static magnetic field distribution. The latter can be controlled by appropriate design of the magnetic system.

2. *Phase Selection* This takes place when some trapped electrons depart the beam. Departure to the cathode is important because particles can heat up the cathode. In this case, trapped electrons that are in the accelerated phase of the space-charge field oscillate around their leading centers along loops of larger size and depart to the cathode, leaving oscillating retarded particles in the beam. This mechanism is similar to the Bakhausen–Kurz bunching of oscillators (Section 10.4.1). Note that particles can also reach the anode due to the diochotron effect.

Parasitic Electromagnetic Radiation In general, electromagnetic and space-charge oscillations are linked in a unified process. A very general overview is provided below. First, note that the transformation of low-frequency space-charge oscillations in electromagnetic radiation is possible only in the gun area where TEM oscillations are permitted. But the gap cathode–anode is short-circuited by the power supply circuit. Thus, an electron gun with a power supply circuit may be considered an open *LC* circuit. The power supply can be considered as an element of the circuit with distributed admittance, and the gap anode–cathode, as a nonlinear capacitance. We can interpret the electromagnetic radiation as an emission of an open *LC* circuit excited by the oscillating space charge. It is very important that oscillations of the anode voltage can amplify oscillations of the space-charge field. The usual recommendations for weakening oscillations are to increase dissipation in the power supply circuit and to detune the natural frequency of the circuit from the frequency of the space-charge oscillations.

The spectrum of parasitic electromagnetic oscillations in a powerful 110-GHz gyrotron was observed by Gorelov et al. (1999). The gyrotron was used in electron cyclotron heating and noninductive electron cyclotron current drive (see Section 10.12.2) in a DIII Tokamak (General Atomics, San Diego, California). The maximum output power was about 2 MW with a pulse duration of 2 s. The frequency of the parasitic emission was about 100 MHz near the frequency of the axial electron oscillation in the trap (Tsimring, 2001). According to Gorelov et al. (1999), the threshold electron beam current of the parasitic emission was about 4 A. At low currents, sporadic oscillations were observed over the range ± 5 MHz centered at 100 MHz. For currents $4\ \text{A} < I < 9\ \text{A}$, the averaged spectrum is filled by harmonics. For $I > 9$ A, the spectrum was narrowed drastically and was not changed up to a maximum operating current of 33 A. In that range, the fully developed instability had a rather monochromatic spectrum.

10.8.6 Potential Depression and Limiting Current of HEBs

The potential depression (PD) is one of the limiting factors of powerful gyrotrons. This factor leads to a limitation of a beam current similar to that in rectilinear beams (Bogdankevich and Rukhadze, 1971). However, in HEBs this effect is expressed more strongly because the drift velocity that determines the beam space-charge density is substantially less than the full velocity. Besides, PD in HEBs perturbs the gyrofrequency and leads to beam broadening.

PD and limiting current (LC) in HEBs were first analyzed by Drobot and Kim (1981) and Ganguly and Chu (1984). Influence of the velocity spread on PD and LC was studied by Antonsen and Levush (1986) and Tsimring (1993) for uniform distribution of oscillatory velocities, which in fact assumes the absence of electron reflection. Rayleigh distribution, which takes into account the reflection of electrons, was also treated by Tsimring (1993). It was shown in those papers that the velocity spread leads to further considerable reduction in the LC. Below we consider briefly the PD and LC in monovelocity HEBs and in helical beams with velocity spread, including the effect of the electron reflections. A nonrelativistic approximation is used by Tsimring (1993). Corresponding results are available up to a voltage of 100 kV, which is typical for subrelativistic gyrotrons.

The ion compensation of the PD is neglected. According to Varentsov and Tsimring (1983), the ion background may be ignored inside a MIG with a gaseous pressure below 10^{-6} torr. The effect of the ion compensation can be considerable for a sufficiently large pulse duration in the cavity where the static electric field is small and the ion departure time is large.

Next, we consider the problem of limiting current for various velocity distributions.

Monovelocity HEB The beam inside a circular tube with radius R_a (Fig. 10.25) is assumed thin, so the radial potential variation on the beam cross section may be neglected. The PD of the beam relative to the tube well with potential is

$$\Delta U = \frac{I}{2\pi\varepsilon_0 v_z} \ln \frac{R_a}{R_0} \tag{10.178}$$

Equation (10.178) determines ΔU indirectly, because this value also enters the longitudinal (drift) velocity v_z, which in the nonrelativisic approximation is

$$v_z = \sqrt{2\eta(U_a - \Delta U) - v_\perp^2} \tag{10.179}$$

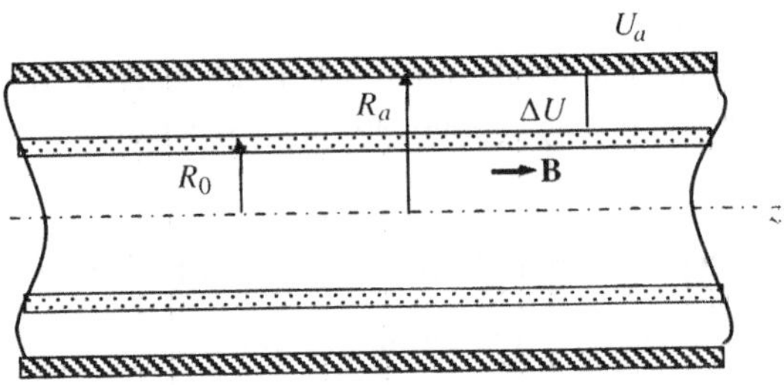

FIGURE 10.25 Hollow electron beam in a gyrotron cavity.

Here $v_\perp$ is the oscillatory velocity, which does not depend on ΔU (the result of conservation of the adiabatic invariant). We obtain from Eq. (10.179)

$$v_z = \sqrt{2\eta U_a}\sqrt{\frac{1}{\sqrt{(1+g_0^2)} - x}} \tag{10.180}$$

where $x = \Delta U/U_a$ is the relative PD, $g_0 = v_\perp/v_{z0}$ is the nonperturbed pitch factor (v_{z0} is the drift velocity without PD). Substituting Eq. (10.180) into Eq. (10.178), we find that the current is

$$I = \frac{2\pi\varepsilon_0\sqrt{2\eta}\,U_a^{3/2}x\sqrt{1/(1+g_0^2)-x}}{\ln(R_a/R_0)} \tag{10.181}$$

According to this relation, the current initially grows with x, but after reaching the maximum, it drops. It is readily verified that the maximum I corresponds to

$$x_{\max} = \frac{2}{3(1+g_0^2)} \tag{10.182}$$

Substitution of Eq. (10.182) into Eq. (10.181) gives the maximum current,

$$I_{\max} \cong \frac{0.4U_a^{3/2}}{(1+g_0^2)^{3/2}\ln(R_a/R_0)} \tag{10.183}$$

Here I and U are expressed in amperes and kilovolts, respectively. The beam is unstable for $x > x_{\max}$. Therefore, the current $I_{\max}$ is the *limiting current* (LC). The limiting disturbed pitch factor

$$g_{\max} = \left[\frac{g_0}{\sqrt{1 - x(1+g_0^2)}}\right]_{x=x_{\max}} = \sqrt{3}\,g_0 \tag{10.184}$$

The change in g influences the interaction efficiency.

Homogeneity of the Velocity Distribution of Electron Beams The presence of an electron portion with large oscillatory and accordingly, small drift velocities in multivelocity beams increases the space-charge density and diminishes the LC. In general, if a velocity distribution is similar to a normal distribution with typical "tails," the reflection of some part of the electrons is inevitable, and the situation is complicated. Let us first assume a uniform gyrovelocity distribution without reflections. The flow distribution function is used; that is, the beam current related to the gyrovelocity interval $v_\perp \div v_\perp + dv_\perp$ equals $dI = f(v_\perp)\,dv_\perp$, where

$$f = \begin{cases} 0, & v_\perp \le v_{\perp\,\min} \\ N, & v_{\perp\,\min} < v_\perp < v_{\perp\,\max} \\ 0, & v_\perp \ge v_{\perp\,\max} \end{cases} \tag{10.185}$$

Here $N = I/(v_{\perp\max} - v_{\perp\min})$. Let us define the velocity spread as $\delta v_\perp = (v_{\perp\max} - v_{\perp\min})/\bar{v}_\perp$, where

$$\bar{v}_\perp = \frac{v_{\perp\,\max} + v_{\perp\,\min}}{2} = \frac{g_0}{\sqrt{1+g_0^2}}\sqrt{2\eta U_a}$$

Then it can be shown (Tsimring, 1993) that the current

$$I = 1.04\frac{U_a^{3/2}}{\ln(R_a/R_0)}F(x) \tag{10.186}$$

where

$$F(x) = \frac{x g_0^2 \delta v_\perp}{\sqrt{1+g_0^2}\left[a\sin\dfrac{g_0(1+\delta_\perp/2)}{\sqrt{(1-x)(1+g_0^2)}} - a\sin\dfrac{g_0(1-\delta v_\perp/2)}{\sqrt{(1-x)(1+g_0^2)}}\right]}$$

The limiting current can be found numerically from the condition of the maximum value of the function $F(x)$. The ratio of the LC of multivelocity and monovelocity beams as a function of the velocity spread for different values of g_0 is shown in Fig. 10.26. Note that the geometric factor $\ln(R_a/R_0)$ in Eq. (10.186) and in other appropriate equations of this section must be replaced by $\ln(R_a/R_0)\ln(R_0/R_b)/\ln(R_a/R_b)$ for a coaxial structure with the inner rod radius R_b and the same radii of the beam R_0 and the anode R_a.

Helical Electron Beams with Electron Reflection Let us turn to a situation where the primary velocity distribution does not have a sharp boundary at the side of large oscillatory velocities, and the reflection of some electrons from the magnetic mirror is inevitable. What are the general features of the velocity distribution in the cavity (i.e., in the uniform magnetic field beyond the magnetic mirror)? First, there is a limiting oscillatory velocity in this distribution,

$$v_{\perp\,\text{lim}} = \sqrt{2\eta(U - \Delta U)} \tag{10.187}$$

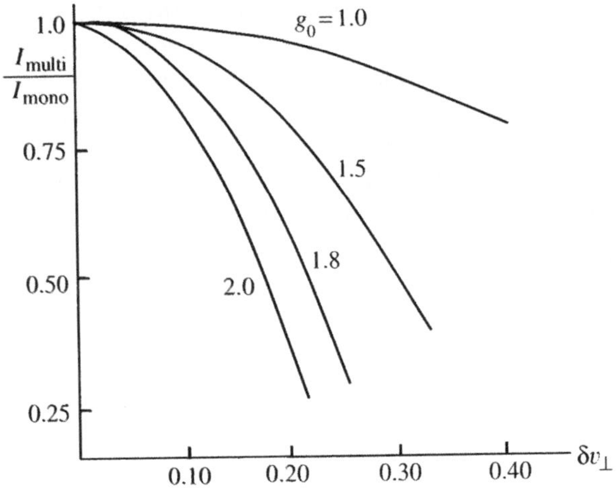

FIGURE 10.26 Ratio of limiting currents of multivelocity to monovelocity beams as a function of velocity spread for various nondisturbed pitch factors.

were ΔU is the potential depression. It is clear that the drift velocities of the electron become zero for $v_\perp = v_{\perp \lim}$. The distribution $f(v_\perp)$ with lesser $v_\perp$ maintains the features of the primary distribution [(i.e., there is a velocity $v_{\perp m}$ corresponding to the maximum of the distribution, and $f(v_\perp)$ falls down behind $v_{\perp m}$]. The corresponding distribution is depicted in Fig. 10.27. In a paper by Tsimring (1993), a Rayleigh distribution in the interval $0 \div v_{\perp \lim}$,

$$f(v_\perp) = Av_\perp \exp\left[-\frac{(v_\perp - v_{\perp 0})^2}{d^2}\right] \tag{10.188}$$

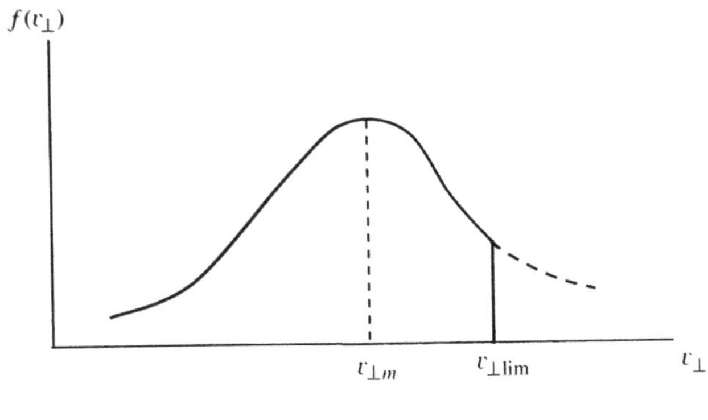

FIGURE 10.27 Velocity distribution in a cavity with the reflection of electrons in the region of a magnetic compression.

was used for simplicity. This distribution differs from the Gaussian distribution by the presence of the multiplier $v_\perp$. Thus, the normalized velocity distribution, corresponding to Fig. 10.27, can be written as

$$dI = \begin{cases} Ad^2 \exp[-(y - y_0)^2]d, & y \le y_{\lim} \\ 0 & y > y_{\lim} \end{cases} \tag{10.189}$$

where $y = v_\perp/d$, $y_0 = v_{\perp 0}/d$, and $y_{\lim} = v_{\perp\lim}/d$. The velocity spread can be expressed through the dispersion D and the mean oscillatory velocity of the distribution [Eq. (10.189)]:

$$\delta v_\perp = \frac{2\sqrt{D}}{\bar{v}_\perp} \tag{10.190}$$

Calculation of LC and PD has been carried out by Tsimring (1993) according to the scheme of calculation of the LC for a beam with zero velocity spread: The current I is expressed as a function of ΔU, and then LC and the limiting PD are found from the condition $dI/d\Delta U = 0$.

A normalized limiting current (the ratio of the HEB limiting current to the LC of a thin rectilinear beam) as a function of the unperturbed velocity spread $\delta v_{\perp 0}$ (neglecting the potential depression) with different unperturbed pitch factors g_0 is depicted in Fig. 10.28. The qualitative peculiarities of the curves in the figure may be treated in the following way. According to Eq. (10.181), the beam current can be written as $I = S\,\Delta U\,\bar{v}_z$, where $\bar{v}_z$ is some mean drift velocity. First the current grows with ΔU because v_z decreases weakly, but further, the current falls

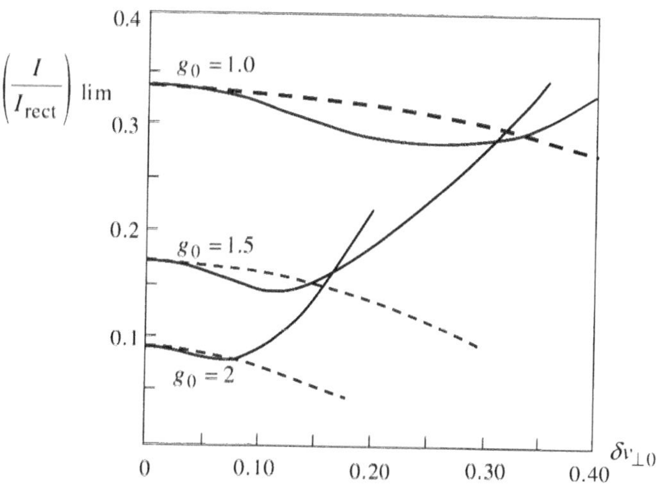

FIGURE 10.28 Normalized limiting current as a function of unperturbed velocity spread for different unperturbed pitch factors. Solid curves, Rayleigh velocity distribution; dashed curves, uniform velocity distribution.

due to the decrease in v_z. Consider first the case of uniform gyrovelocity distribution while setting aside the electron reflection. Let us take the mean oscillatory velocity equal to the velocity in the monovelocity beam. Then that portion of the beam that has a gyrovelocity greater than the mean velocity decreases the drift velocity more strongly than the symmetric portion increases the drift velocity. As a result, $\bar{v}_z$ is lower than in monovelocity beams (the dashed curves in Fig. 10.28).

The situation is different for the distribution (10.189). Here the growth of the velocity spread reinforces electron reflection. As a result, the upper boundary of the gyrovelocities is lowered. For small $\delta v_{\perp 0}$, this effect is not significant. Therefore, changes in the LC and PD are similar to the case of uniform velocity distribution. However, with the growth of $\delta v_{\perp 0}$ and g_0, the effect of the upper boundary limitation changes the velocity distribution considerably. That weakens the decrease of v_z with an increase of ΔU. Therefore, the maximum of the current is reached for a greater potential depression, and LC increases (solid curves in Fig. 10.28).

Influence of Potential Depression on Gyrofrequency The increase in the gyrofrequency ω_g is a direct consequence of depressing particle energy due to PD. It is readily verified that normalized perturbation of the gyrofrequency

$$\frac{\Delta\omega_g}{\omega} = \frac{x}{1 + m_0c^2/e_0U_o} \tag{10.191}$$

where $x = \Delta U_o/U_o$ and U_o is the unperturbed potential of the interaction space.

Beam Broadening The combined influence of the space-charge and magnetic fields excites an azimuthal rotation of beam layers. The angular drift velocity of the external layer with radius R_e is

$$\Omega = \frac{E_p}{R_eB} \tag{10.192}$$

where $E_p = Q/2\pi R_e\varepsilon_0$ is the space-charge field at radius R_e. Using Eq. (10.178), we obtain

$$\Omega = \frac{U_o x}{R_e^2 B \ln(R_c/R_e)} \tag{10.193}$$

According to the Busch theorem [Section 1.7, Eq. (1.128)]

$$\Omega = \Delta\dot{\theta} = \frac{\eta}{2\pi R_e^2}\Delta\Psi = \frac{\eta B}{2\pi R_e^2} 2\pi R_e\,\Delta R_e$$

Then

$$\frac{\Delta R_e}{R_e} = \frac{\Omega}{\omega_g} \tag{10.194}$$

and we obtain

$$\frac{\Delta R_e}{R_e} = \frac{\eta U_o x}{R_e^2 \omega_g^2 \ln (R_c/R_e)} \tag{10.195}$$

It can be estimated that this value is significant only for long-wavelength gyrotrons.

10.9 MODE INTERACTION AND MODE SELECTION IN GYROTRONS. OUTPUT POWER SYSTEMS

A distinctive feature of the gyromonotron as an oscillator in the submillimeter and short-millimeter ranges is its high power in long-pulse and continuous regimes. We have seen that the bremsstrahlung nature of the electron radiation in gyrotrons, together with peculiarities of the magnetron injection guns, lift restrictions on the volume of an active medium. However, at least two fundamental difficulties remain. The first is to ensure spatial coherence of radiation while retaining sufficient efficiency under conditions in which many modes in the spatially developed medium can be excited simultaneously. The second is the canalization of large fluxes of energy with a simple spatial transverse radiation structure. In this section we discuss briefly methods for solving these problems in experimental gyromonotrons.

10.9.1 Mode Interaction

Any researcher or designer of powerful oscillators sooner or later encounters the problem of mode competition or general mode interaction. This problem marks the juncture where the straight path of development turns into a labyrinth littered with many dead ends and blind alleys. The gyromonotron is an example of a system that moved along a straight path farther than many others. This is a result of peculiarities of the interaction mechanism which allow a designer to apply open resonators effectively with their extraordinarily selective properties. Nevertheless, even in gyrotrons, after achieving a certain level of power, the problem of mode interaction becomes acute.

The spectral density N of free oscillations of an arbitrary cavity is determined by the well-known Rayleigh–Jeans relation $N \sim V/\lambda^3$, where V is the resonator volume. In light of the foregoing [Eqs. (10.48) and (10.49)], it is clear that resonators of powerful gyrotrons must have a dense spectrum. If two or more modes get into a cyclotron resonance band, their interaction can negatively affect the gyrotron efficiency, monochromaticity, and spatial coherence of radiation (see, e.g., Kreischer et al., 1984). The mode interaction in gyrotrons is characterized by the following two basic effects (Nusinovich 1999, 2004):

1. *Mode Suppression and Coexistence*. Mode suppression takes place when excitation of one mode worsens the starting conditions for other modes. Usually, the first excited mode suppresses other modes. Analysis shows that the effect of

mode suppression requires strong coupling among modes. In the gyrotron this situation arises when frequencies of interacting modes are inside the gyroresonance band, and the mode competition is determined by the extraction of energy by the HF field of each mode from the same active medium (helical beam) (Nusinovich, 2004). Mode coexistence (multifrequency oscillations) is possible for weak mode coupling. So in HEBs with a large velocity spread, modes can interact with separate velocity fractions of the beam.

2. *Nonlinear Mode Excitation*. Frequency separation of competing modes is on the order of the gyroresonance band. According to Nusinovich (2004), one mode can improve the condition of excitation of another mode for a proper phase of the bunch. The effect of nonlinear excitation is pronounced in gyrotrons operating at high gyroharmonics (Zapevalov et al., 1977, 1993; Zarnitsina and Nusinovich, 1977) in the presence of an interaction with the beam at the fundamental resonance (first harmonic). So the quadrupole nature of bunching of electrons in the field of the second harmonic leads to fast saturation of the oscillation amplitude. But the bunched current of the second harmonic mode reduces the starting current of the fundamental harmonic when the latter is in the zone of hard excitation. The growth of the fundamental harmonic field can thus suppress the beginning of the operational mode. This effect has been observed in experiments by Idehara and Shimizu (1994) and Zapevalov and Tsimring (1978).

10.9.2 Mode Selection

The mechanism of mode suppression is the decisive factor that determines the possibility of single-mode operation under the conditions of a dense mode spectrum. The minimal requirement for the realization of this effect is selection of the desired mode, so that the starting current of the latter is smaller than for other modes. There are two common methods of mode selection: electrodynamic and electronic.

Electrodynamic Selection Methods of selection in this case are differentiated depending on the type of mode index TE_{mnq} selected.

Axial selection (of the longitudinal indexes q) is very important if one takes into account large values L/λ, because many modes with different values of q would be competitive. For a given q, the length of the resonator is equal to $L \approx q(\lambda_w/2)$, where λ_w is the wavelength in a waveguide. Therefore, the wavelength in free space corresponding to the mode TE_{mnq} is equal to

$$\lambda = \frac{\lambda_c}{\sqrt{1 + (\lambda_c/\lambda_w)^2}} = \frac{\lambda_c}{\sqrt{1 + (q\lambda_c/2L)^2}} \qquad (10.196)$$

where λ_c is the cutoff wavelength of the mode TE_{mn} in a regular waveguide. We see that λ is maximal for $q = 1$ and decreases quickly with q. For modes with small q, where $q(\lambda_c/2L) \ll 1$, the wavelength $\lambda \approx \lambda_c[1 - (q^2/8)(\lambda_c/L)^2]$, so the

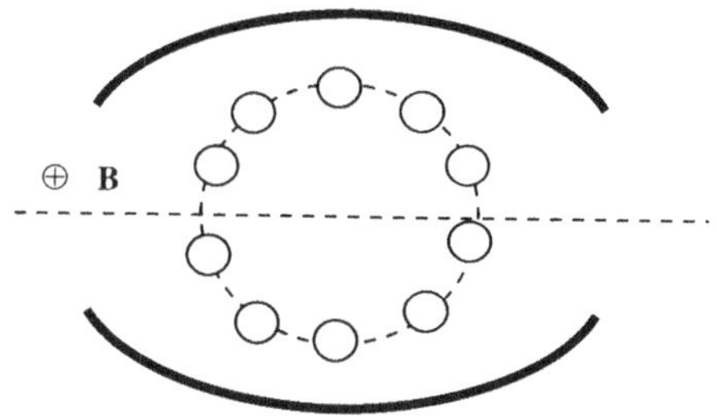

FIGURE 10.29 Cross section of a resonator in a quasioptical gyrotron.

wavelengths of the first oscillations are close to cutoff. The modes TE_{mn1} are the most interesting. By forming a little narrowing of the cross section at the output end of the waveguide (on the right side of Fig. 10.12), it is possible to choose the dimension of the corresponding cross section that will correspond to the super cutoff for the mode TE_{mn1} and to after the cutoff for modes with $q > 1$. As a result, the mode TE_{mn1} is reflected, whereas modes with $q > 1$ radiate freely, and the corresponding Q-factors of these modes are small. For example, the Q-factor for the mode TE_{mn2} is two to three times less than that of the TE_{mn1} mode. Q-factors of modes with $q \geq 3$ are negligibly small. Thus, the method of selection described rarefies the spectrum approximately $2L/\lambda$ times.

Selection over azimuthal indexes m necessitates disturbing the axial symmetry of the resonator (Gaponov et al., 1981). The cross section of such a resonator with an electron beam is depicted in Fig. 10.29. The axial slots ensure suppression of whispering modes (modes with $m \gg n$), so the TE_{1p} mode may have the highest value of Q. Resonators of the type shown in Fig. 10.29 are similar to the open resonators of lasers and masers, and the corresponding gyrotrons are called *quasioptical gyrotrons* (see, e.g., Fliflet et al., 1990). An attractive feature of such gyrotrons is the possibility of mechanical tuning of their eigenfrequency by varying the distance between mirrors. The disadvantage of quasioptical gyrotrons is the absence of correspondence between the transverse structures of both the HF field and the annular electron beam. Therefore, the efficiency of quasioptical gyrotrons is significantly lower than that of axially symmetric gyrotrons.

Electrodynamic selection on radial indexes n may be attained in a number ways. Let us consider the two most important cases.

1. *Coaxial resonators* (Gaponov et al., 1981; Piosczyk et al., 1999; Vlasov et al., 1976). A diagram of the coaxial cavity is shown in Fig. 10.30. For the most part, whispering gallery modes ($m \gg n$) do not "feel" the presence of the inner rod, and their transverse structures are basically the same as in empty cavities. Modes with greater n, having a radius of caustic surfaces close to the radius R_d of the inner tapered conductor, are radiated, and the cavity Q-factor for these modes diminishes considerably. Thus, the mode spectrum of the radial indexes in the coaxial cavity is significantly rarefied compared with an empty cavity of the same

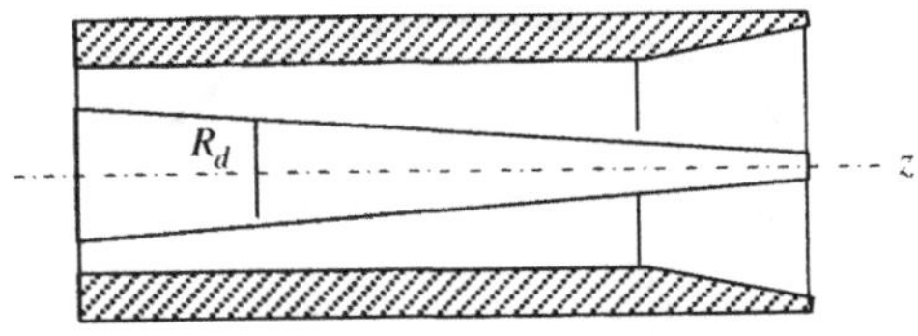

FIGURE 10.30 Cavity in a coaxial gyromonotron.

cross-sectional size. It naturally opens possibilities for single-mode operation with high modes. For example, a coaxial gyrotron with a $TE_{31,17}$ mode has been developed by Piosczyk et al. (1999) (frequency, 165 GHz; output power, 1.7 MW; efficiency without depressed collector, ~30%). Another positive feature of coaxial cavities is a reduced potential depression.

2. *Complex cavities (coupled resonators with mode conversion)* The system contains circular open resonators (Fig. 10.31*a*) 1 and 2 with equal eigenfrequencies of modes $TE_{mn_1 1}$ and $TE_{mn_2 1}$, which are characterized by coinciding azimuthal and different radial indexes. In the irregular waveguide 3, mutual mode conversion occurs, providing coupling between the resonators. Under certain conditions, amplitudes of the fields of normal oscillations in both resonators are comparable (Tsimring and Pavelyev, 1982; Malygin et al., 1983). This circumstance, as well as the coincidence of radii corresponding to maxima of structure factors that do not depend on the radial indexes of the modes [Eq. (10.146)], favors the efficient interaction of HEBs with an electromagnetic field over the entire length of the coupled system. Some other pairs of modes TE_{m_2,p_3} and TE_{m_2,p_4} can also form coupled oscillations. However, the number of these pairs is small, and the frequency separation of these

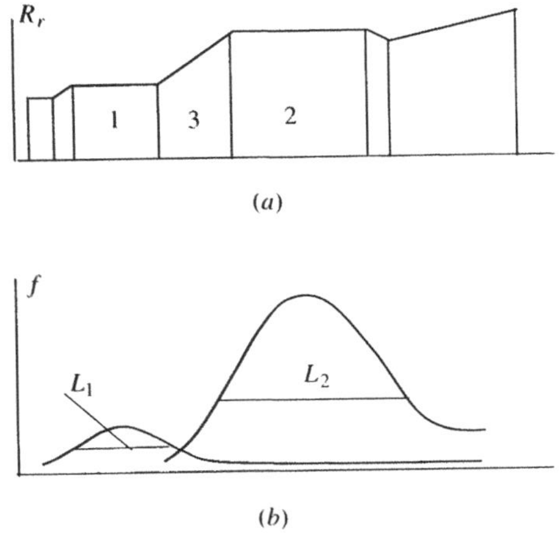

FIGURE 10.31 (*a*) Longitudinal cross section of a complex resonator; (*b*) distribution of field amplitudes of coupled modes.

pairs from the frequency of the fundamental pair is enlarged. Uncoupled modes can be excited only separately; their interaction lengths (L_1 or L_2) are shorter than $L_1 + L_2$ (Fig. 10.31*b*), and the starting currents are significantly higher. Thus, the spectrum of effective (strongly coupled) modes is rarefied compared with that of conventional cavities. At the same time, using a high mode in output cavity 2 provides an adequate working space and reduces the wall losses needed for power operation.

Complex cavities were originally proposed by Pavelyev and Tsimring (1979). The first results of gyrotron application of complex resonators were published by Gaponov et al. (1981) and Carmel et al. (1983). Operations with complex cavities in gyrotrons on the second (Pavelyev et al., 1987) and third harmonic (Malygin, 1986) of the gyrofrequency have confirmed the effectiveness of the method of mode selection considered and allowed for an increase in output power and efficiency. One disadvantage of complex resonator structures, however, is their more stringent requirements for fabrication.

Electronic Selection The choice of an optimal radius R_0 for the electron beam in the resonator is the most obvious method. The radius of the electron beam must correspond to the maximum of the structure factor G as a function of R_0 [Eq. (10.146)]. So the beam on the fundamental harmonic should be located at the inner peak of the electric field. This ensures a minimum starting current for the mode chosen.

A startup scenario (Nusinovich, 2004; Nusinovich et al., 2004) is the specific protocol of temporal varying gyrotron parameters (i.e., magnetic field, beam energy, anode and cavity voltages, and beam current) in the process of turning on a gyrotron, which leads to the onset of the single-mode stable oscillation desired. The choice of a proper startup scenario is determined by working in the hard self-excitation regime, which often is the only possible way to achieve high efficiency and suppressing of competitive modes. The problem of an adequate startup scenario was addressed in early investigations of gyrotrons (see, e.g., Nusinovich, 1974). In a CW regime, the simplest startup scenario is reduction of the dc cavity magnetic field B_o after attainment of starting oscillations. In this case, the power increases from the minimum starting value to a value ensuring stable maximum of the output power, usually with maximum admissible mismatch (Kisel et al., 1974). In general, the startup scenario can also be realized by variation of the anode voltages (for triode guns), the dc magnetic field of the cathode coil B_c (see Fig. 10.18), and the beam current (Nusinovich et al., 2004).

In a pulse regime, some of the methods described above are not applicable because of the inertia of gyrotron magnetic systems and the thermal inertia of a cathode heater. Instead, a scenario connected with the choice of the voltage feed in the gyrotron power supply can be used. Note, however, that the scenario with variable beam current can be realized in the MIG in the ρ-mode (space-charge-limited current) (see, e.g., Kuftin et al., 1993).

In triode-type guns in a pulse gyrotron with a separate anode (Fig. 10.18), two methods of feeding are possible: either proportional variation of the resonator and anode voltages (V_0 and V_a, respectively) due to a voltage divider, or keeping up a constant difference ($V_0 - V_a$) between the voltages during the voltage rise. Each scenario gives a different character for the pitch factor and the electron energy variation during the voltage rise. For a dense mode spectrum, these scenarios determine the sequence of excitation of the operation and parasitic modes. It has been shown that the second scenario provides better conditions for single-mode excitation of the desired mode (see, e.g., Nusinovich, 2004). For diode-type guns, only the first scenario with $V_a = V_0$ can be realized.

For multi-beam gyromonotrons (Wang Hui et al., 1985; Zapevalov and Tsimring, 1990; Zapevalov et al., 1989, 1991) the main beam is supplemented with one or several hollow HEBs coaxial with it. Depending on the parameters of the supplementary beams (mainly the oscillatory and drift velocities), the latter can execute functions of radiative or absorbing beams. This creates rich possibilities for mode selection. There are a number of constructive versions of multibeam MIGs (Zapevalov et al., 1991). Two possible schemes with supplementary radiative and absorbing beams are depicted in Fig. 10.32. According to Zapevalov and Tsimring (1990), the introduction of supplementary radiative beams (SRB) (Fig. 10 32*a*) increases the margin of oscillation stability for zero coupling of the SRB to the field of a parasite mode even though the coupling of the SRB to the field of the operating mode generally differs from optimal. The main factor here is that the enhanced oscillation stability makes it possible to draw closer to the high-efficiency regime of the main beam. The spatial separation of partial beams weakens the effect of the space charge on the velocity spread and reduces the probability of two-beam instability, which in principle can occur because of the multihump distribution function of longitudinal velocities.

Supplementary absorbing beams (SABs) should have parameters that ensure a maximum of cyclotron absorption of the parasitic mode at minimum interaction with the field of the operation mode. It can be shown that power absorbing by the helical beam is proportional to $\beta_\perp^{2(n-1)}$ (Zapevalov et al., 1991), hence the absorbing intensity does not depend on $\beta_\perp$ at the fundamental resonance and is zero for interaction of the rectilinear beam with the field at the harmonic of the gyrofrequency. Two particular cases are of greatest interest.

Operational and parasitic modes are synchronous with a beam at the fundamental resonance. A SAB should be introduced at the zero value of the operational mode

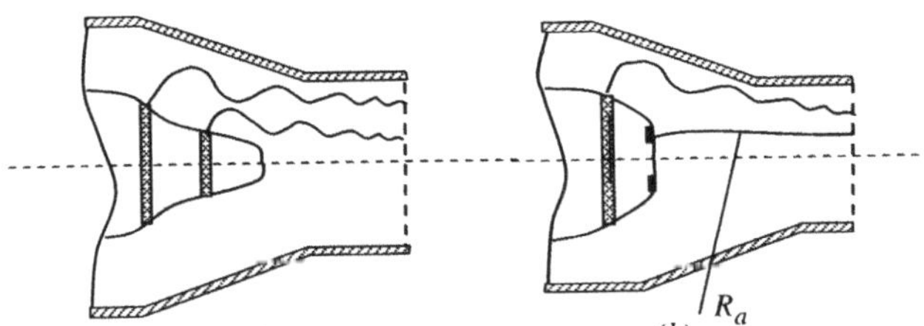

FIGURE 10.32 Two-beam MIGs: (*a*) gun with both radiative beams; (*b*) gun with one radiative and one absorbing beam (beam R_a).

field so that coupling of the beam with the field of the parasitic mode are maximal. It can be shown that the absorption power $P_a \sim U_a^2$, where $e_0 U_a$ is the energy of electrons in the SAB, so beams with minimum energy are best. However, in practice the energy should not be very low because of the detuning of the cyclotron resonance. Note that SABs are monovelocity beams even for a high degree of mode selection and can be recuperated effectively.

The operational mode is synchronous at the second harmonic of the gyrofrequency, and the parasitic mode is synchronous at the first harmonic. This is the most efficient case. The injection radius of an SAB should be chosen only under conditions of the best interaction with the parasitic mode field. Introduction of a nonoscillating beam ensures absorption of the first harmonic modes and has no effect on the field of the operating second harmonic.

Experiments by Zapevalov et al. (1993) with a gyrotron at the second harmonic of the gyrofrequency ($\lambda = 1.2$ cm, mode TE_{03}) confirmed the effectiveness of mode selection using absorbing beams. The maximum stable output power reached with an SAB was 600 kW, whereas without an SAB it was only 360 kW.

10.9.3 Output Power Systems

The purpose of an output system is to extract the RF power from the cavity through the RF window into the output waveguide with small losses. Simultaneously, an important function of the output system is to transform a resonator mode having a complicated transverse structure into a simple wave beam (e.g., a Gaussian beam) appropriate for channeling the RF power without distortions along an oversized transmission line into a load. Two possible output systems are depicted in Fig. 10.33.

A *straight output system* (Fig. 10.33*a*) unites producing transport of an electron beam to the collector with transmission of RF power to the window. However, use of this outwardly simple system to transmit high power meets significant difficulties. Because the output waveguide serves as an electron collector, the ratio R_{col}/R_a should be very large for gyrotrons with a high level of average power. In this case, elimination of mode transformation in the taper, even with optimization of its profile (Mobius and Thumm, 1993) is very difficult and requires using exceedingly long tapers. Besides, the problem of conversion of the operational mode into a Gaussian wave beam remains unsolved.

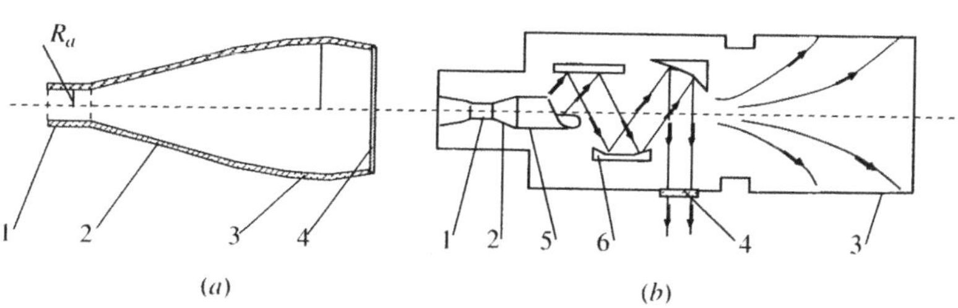

FIGURE 10.33 Output power systems: (*a*) straight output system; (*b*) quasioptical converter. 1, Resonator; 2, taper; 3, collector; 4, HF window; 5, launcher; 6, reflector.

A quasioptical converter (Fig. 10.33b), proposed and developed by Vlasov and Orlova (1974) and Vlasov et al. (1975), provides a pivotal solution to these problems. The mechanism of the converter is based on the Brillouin concept, representing the waveguide mode as a beam of rays. Straight Brillouin rays in a circular waveguide with the mode TE_{mn} are inclined to the waveguide axis with the angle $\beta = \sin^{-1}(\kappa_{mn}/kR_a)$. Each ray crosses a distance in the axial direction between two subsequent reflections from the waveguide wall equal to

$$L_B = 2R_a\left[1 - \left(\frac{m}{\kappa_{mn}}\right)^2\right]^{1/2} \cot\beta \tag{10.197}$$

and shifts by the angle

$$\Delta\varphi = 2\cos^{-1}\frac{m}{\kappa_{mn}} \tag{10.198}$$

Thus, the reflection points of each ray make a full revolution at the following distance along the waveguide axis:

$$H = L_B\frac{2\pi}{\Delta\varphi} = \frac{2\pi R_a}{\cos^{-1}(m/\kappa_{mn})}\cot\beta \tag{10.199}$$

It can be shown (Denisov et al., 1992) that all rays corresponding to the mode considered in the regular waveguide belong to the region G, having the form of a parallelogram of length H and width $R_a\,\Delta\varphi$. Thus, by cuting out segment G from the waveguide wall, we can obtain the flux of all rays (Fig. 10.34) through this waveguide cut with angular divergence in the azimuthal plane. The flux can be transformed into a beam of quasiparallel rays, due to a specially designed (Vlasov and Likin, 1980) quasiparabolic mirror (Fig. 10.34). The beam after the first mirror transmits across two correcting reflectors and propagates radially through the output window (Fig. 10.33*b*), shaped as a Gaussian wave beam. The best excitation coefficient (in terms of power) of this beam in this scheme was about 0.85 (see, e.g., Denisov et al., 1992).

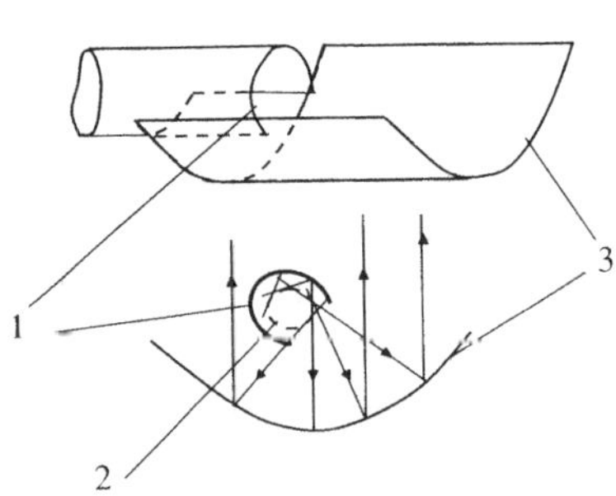

FIGURE 10.34 Quasioptical mode converter. 1, Waveguide cut; 2, caustic; 3, first mirror.

It is necessary to note that mode converters with such comparatively high transmission coefficients are not appropriate for the powerful CW gyrotrons because a diffraction RF power that is not directed through the output window ("stray RF power") overheats internal parts of the tube (basically, ceramic insulators).

A cardinal solution to this problem has been developed by Denisov et al. (1992). A regular section before the waveguide cut is replaced by a smooth perturbed (rippled-wall) waveguide. The appropriated shape of the inhomogeneous waveguide surface provides a particular distribution of the field at the G-region that optimizes the Gaussian structure of the beam and diminishes beam scattering at the cut edges (diffraction losses). In an experimental 110-GHz gyrotron, a measured transmission coefficient (by a power) after the convertor formed by a rippled-wall launcher and three-mirror line was about 95% (Denisov et al., 1992). But even in this case the fraction of the stray HF radiation is on the order of 10%. For a 1-MW gyrotron, that is 100 kW! An advanced internal converter has been designed that produces an output beam with a Gaussian mode content of 99.5%, which produces diffraction losses as small as 1 to 2% (Litvak et al., 2005). So in the 140 GHz FTZ gyrotron (Germany) power of the stray radiation was of order 1% (Dammertz et al., 2006). A synthesis technique for phase-correcting mirrors designed with reduced diffraction losses using irradiance moments developed by Shapiro et al. (2005) has been realized in a four-mirror 110-GHz gyrotron.

10.9.4 Output Windows

Development of adequate gyrotron output windows gets more and more complicated with increases in the average output power, frequency, size, and requirements concerning the quality of the output radiation. In the first small power laboratory gyrotrons ($P_{\text{out}} < 4$ kW), thin-film mica windows were applied. However, in early experiments with gyrotrons as sources of radiation for electron–cylotron heating of plasmas in a TM-3 tokamak at the Kurchatov Institute in Moscow (Alikaev et al., 1972), ceramic windows of comparatively large size were used.

The choice of construction and materials of windows for powerful gyrotrons is dictated by a combination of requirements, among which the most important are low reflection of the wave from the window, low losses in the ceramic material, high thermal conductivity, the method of mounting, mechanical strength, and even (for gyrotrons in fusion reactors) radiation insensitivity to neutron fluxes. A minimum of reflection is reached when the window thickness is equal to

$$D = n\frac{\lambda_d}{2}, \qquad n = 1, 2 \ldots \tag{10.200}$$

where $\lambda_d = \lambda/\sqrt{\varepsilon/\varepsilon_0}$ is the wavelength in a dielectric.

Over a period of development of gyrotrons with high average power in various frequency bands, a number of ceramic materials and systems of window cooling were tested. The most promising ceramics were beryllia ((BeO), boron nitride (BN), silicon nitride (Si_3N_4), sapphire (Al_2O_3), Au-doped silicon (Si + Au), and a synthetic CVD (chemical vapor deposition) diamond (see, e.g., Kartikeyan et al., 2004). The goal is to get a window for the transmission of CW power around 1 MW at a frequency on the order of 140 to 170 GHz. The frequency chosen corresponds to the condition of cyclotron resonance at the first or second harmonic in a tokamak magnetic field. A power of 1 MW corresponds to the power limitations of other gyrotron components (i.e., cavity, collector, electron gun, power supply). It turned out that only a CVD diamond satisfied the requirements indicated. The power limit of a sapphire window at a frequency of ~100 GHz at room temperature is around 0.3 to 0.4 MW. It can be used at lower frequencies and for pulsed operation, but the power criterion noted above is considerably higher. Improvements can be attained by using cooled double-disk windows and by reducing the operating temperature. However, the complicated construction and losses in cooling liquid lead to considerable difficulties.

The characteristics of CVD-diamond windows are very high. The loss tangent at frequency 145 GHz is $\sim 2 \times 10^{-5}$, approximately 10 times less than in sapphire windows, and the thermal conductivity is 50 times higher. This unique parameter of a CVD diamond allows us to use the simplest cooling method: water-edge-cooling (cooling water flows along the disk edge). Large CVD-diamond disks are manufactured in a number of countries. The leader, DeBeers Ltd. (UK), produces disks with diameter and thickness of up to 119 mm and 2.2 mm, respectively, with good reproducibility of parameters.

An interesting version of gyrotron output windows is the *Brewster window* (see, e.g., Kartikeyan et al., 2004), which has an angle $\theta_B = \tan^{-1}(\sqrt{\varepsilon/\varepsilon_0})$ between the ray and the normal to the disk surface. In a Brewster window, reflections vanish independent of the frequency for parallel polarization of a wave. These windows provide broadband operation of multifrequency or step-tunable gyrotrons (see, e.g., Kreischer and Temkin, 1987).

10.9.5 Depressed Collectors

According to Eq. (10.150), the motion of electrons in diminishing the magnetic field near a collector (Fig. 10.10) leads to decompression of the spent electron beam (i.e., to degeneration of the spent helical beam into a rectilinear beam with the corresponding spread of the longitudinal electron energy). In this case, it is possible to use a depressed collector (DC) to enhance gyrotron efficiency, similar to the use of DCs in traveling-wave tubes. The construction and electric scheme of DCs are analogous to the DC schemes used in conventional TWTs (see Section 8.6.4). The simplest single-stage DC allows for an increase of efficiency from the typical 25 to 35% to 50% and more (see, e.g., Andersen et al., 2004). Use of DCs also facilitates cooling of the collector and lowers the intensity of x-ray radiation.

10.10 GYROKLYSTRONS

10.10.1 Introduction

The gyroklystron depicted in Fig. 10.35 is analogous to a conventional klystron (Figs. 7.1 and 7.10). The analogy includes fundamental physical effects: HF modulation, inertial bunching in the drift sections, and extraction of the HF energy in the

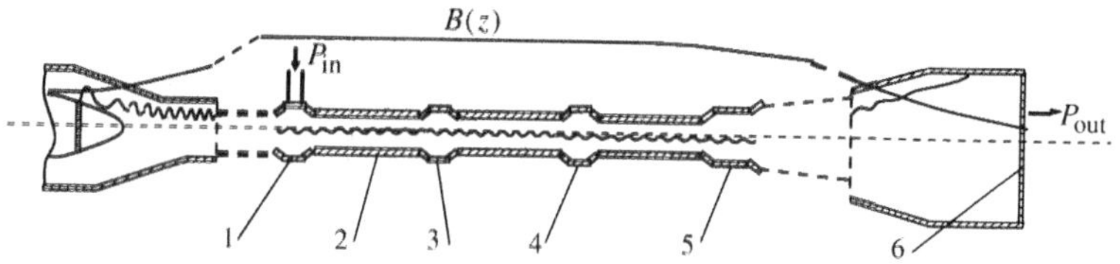

FIGURE 10.35 Four-cavity gyroklystron. 1, Input cavity; 2, drift tube; 3, intermediate cavity; 4, penultimate cavity; 5, output cavity; 6, output window.

output cavity. However, there are essential differences in the nature of the effects indicated. In particular, the energy modulation of the helical beam in the input cavity of a gyroklystron is different from the longitudinal velocity modulation of the rectilinear beam in conventional klystrons. Furthermore, the orbital (phase) bunching of electrons in gyroklystrons (in general, gyrotrons), which is an essentially relativistic effect, is not accompanied by an apparent change in the beam spatial structure, in contrast to the axial (spatial) bunching in conventional klystrons. Naturally, these differences lead to differences in the theory for both amplifiers.

The gyroklystron was invented in 1967 (Antakov et al., 1967). The idea behind the gyroklystron can also be found in early theoretical and experimental work by Wachtel and Hirshfield (1966). A detailed exposition of gyroklystron theory is beyond the scope of this book. We limit ourselves below to a description of the theoretical model, formulation of basic equations, and a summary of the most important results.

10.10.2 Basic Model

The interaction of a subrelativistic electron beam with the TE mode is assumed. Space-charge effects as well as the electron velocity spread are neglected, so the energy and phase are determined by Eqs. (10.125) and (10.129). Resonators 1, 3, 4, and 5 in the Fig. 10.35 are assumed to be short cavities where phase bunching is not developed. HF fields in drift tubes are neglected. Therefore, the electron energy w in these sections is constant, and according to Eq. (10.129), the growth rate of phase in the drift section between the $(i-1)$th and ith cavities is equal to

$$\vartheta(\zeta_i) - \vartheta(\zeta_{i-1}) = -[\Delta + w(\zeta_{i-1}) - 1](\zeta_i - \zeta_{i-1}) \tag{10.201}$$

where $w(\zeta_{i-1})$ is the electron energy at the output of the $(i-1)$th cavity, Δ is the frequency mismatch [Eq. (10.127)], and the ζ_k are dimensionless coordinates. The stagger-tuned model of the gyroklystron (Nusinovich et al., 1997) can also be taken into account, according to which the eigenfrequencies $\omega^{(k)} = \omega_{01}^{(k)} + i\omega_{02}^{(k)}$ of the cavities can be different, as well as the dimensionless frequency mismatches:

$$\delta_i = 2Q(\omega - \omega_{01}^{(i)})/\omega \tag{10.202}$$

10.10.3 Gyroklystron Equations

The analysis proceeds via successive calculations of HF fields in resonators and bunched HF currents in drift tubes beginning with the first resonator that is fed by the input power P_{in}. The resonators that follow the input cavity are excited by the bunched HF currents. The fields and the average electron energies in each resonator are determined as a result of the joint solution of the equations of motion and the equations of complex power balance.

Balance equations according to Eqs. (10.66) and (10.67) can be written as

$$\begin{aligned} 2\omega_{02}^{(1)} W_1 &= \mathrm{Re}(\overline{P}_1) + P_{in} \\ 2\omega - \omega_{01}^{(1)} W_1 &= \mathrm{Im}(\overline{P}_1) \end{aligned} \tag{10.203}$$

$$\begin{aligned} 2\omega_{02}^{(i)} W_i &= \mathrm{Re}(\overline{P}_i) \\ 2(\omega - \omega_{01}^{(i)}) W_i &= \mathrm{Im}(\overline{P}_i) \end{aligned}, \qquad i = 2, 3, \ldots, N \tag{10.204}$$

where W_i is the stored energy in the ith resonator. The subrelativistic average complex power $\overline{P}$ is determined by Eq. (10.130). Equations (10.203) and (10.204) should be solved in each cavity jointly with the averaged subrelativistic equations of motion (10.125) and (10.129) for both energy and phase. The HF fields in drift sections are suppressed. Therefore, the corresponding energies are constant and equal to the energies in the resonators at the entry of each drift section. The phase shift in the drift section that determines the bunched current is given by Eq. (10.201).

Usually, dimensionless equations are used instead of Eqs. (10.203) and (10.204). In this case the susceptibility $\chi = \chi' + i\chi'' = -i(\sigma/\varepsilon_0\omega)$ is introduced. Here σ is the complex admittance of the electron beam. The complex dimensionless susceptibility (see, e.g., Nusinovich, 2004)

$$\chi = \frac{2}{F}\frac{1}{2\pi}\int_0^{2\pi}\int_{\zeta_{in}}^{\zeta_{out}} w^{n/2} e^{i\vartheta} f^*(\zeta)\, d\zeta\, d\vartheta_0 \tag{10.205}$$

Balance equations for the input resonator can be written as

$$I_1^{(1)}\chi_1 + \frac{A}{|F|}e^{-i\psi} = i - \delta_1 \tag{10.206}$$

where the dimensionless current in the ith cavity corresponding to the nth harmonic is

$$I_n^{(i)} = 8\frac{e_0 I_b}{m_0 c^3} G^{(i)} \left(\frac{n^n}{2^n n!}\right)^2 \frac{\beta_{\perp 0}^{2(n-2)}}{\beta_{z0}} \frac{1}{\int_0^{\mu} |f^{(i)}(\zeta)|^2\, d\zeta} \tag{10.207}$$

Here I_b is the beam current and G is a structure factor determined by Eq. (10.146). The amplitude A of an input signal is assumed to be real. The complex field in the input resonator is represented as $F = |F|e^{i\psi}$. The frequency mismatch δ is determined by Eq. (10.202). It is readily found from Eq. (10.206) that

$$|F|^2 = \frac{A^2}{(1 - I_1^{(1)}\chi_1'')^2 + (\delta_1 + I_1^{(1)}\chi_1')^2} \tag{10.208}$$

$$\tan\psi = \frac{1 - I_1^{(1)}\chi_1''}{\delta_1 + I_1^{(1)}\chi_1'} \tag{10.209}$$

Balance equations for resonators other than the input resonator are

$$\left.\begin{aligned} I_n^{(i)}\chi_i'' &= 1 \\ I_n^{(i)}\chi_i' &= -\delta_i \end{aligned}\right\}, \qquad i = 2, 3, \ldots, N \tag{10.210}$$

The interaction between the bunched electron beam and the HF field in the output cavity leads to the orbital efficiency:

$$\eta_\perp = 1 - \frac{1}{2\pi}\int_0^{2\pi} w_N(\vartheta_0)\,d\vartheta_0 \tag{10.211}$$

The relations (10.208)–(10.211) should be solved jointly with subrelativistic averaged equations of motion (10.125) and (10.129) and with relation (10.201) for the phase shift.

The outlined procedure in general is rather complicated and can only be implemented numerically. Considerable simplification is attained through a linear approximation which is based on the smallness of electromagnetic fields and the corresponding perturbations of the electron energy in all resonators except the output cavity. This approximation in gyromonotron theory was considered in Section 10.7.3. Linearized equations of motion are given by Eqs. (10.133) and (10.134). The corresponding susceptibility is

$$\chi = \frac{2}{F}\frac{1}{2\pi}\int_0^{2\pi}\int_{\zeta_{\rm in}}^{\zeta_{\rm out}} f^*(\zeta)e^{i(\vartheta_0 - \Delta\zeta)}\left(1 + i\vartheta_1 + \frac{n}{2}w_1\right)d\zeta\,d\vartheta_0 \tag{10.212}$$

where w_1 and ζ_1 are small perturbations of energy and phase determined by Eqs. (10.132). In a linear approximation, the relation for the phase shift can be written in an explicit form. For example, application of Eq. (10.201) to the first drift section leads to the relation

$$\vartheta(\zeta_2) = \vartheta(\zeta_1) - (\Delta + w_1)\mu_{21} \tag{10.213}$$

where $\mu_{21} = \zeta_2 - \zeta_1$ and $w_1 = w - 1$ is a small increment of the energy [see Eq. (10.132)]. Integration of Eq. (10.133) and substitution of the value obtained for w_1 into Eq. (10.213) give

$$\vartheta(\zeta_2) = \vartheta(\zeta_1) - \Delta\mu_{21} + X\sin(\psi - \vartheta_0) \tag{10.214}$$

where the parameter $X = |F|\mu_{21}$ can be called a *bunched factor*. Comparison of Eq. (10.214) with Eqs. (7.12) and (7.13) in the theory of conventional klystrons justifies this definition.

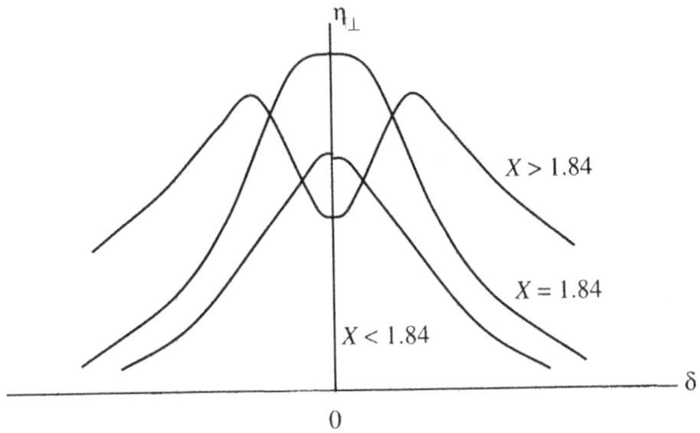

FIGURE 10.36 Typical characteristics $\eta_\perp(\delta)$ for various values of a bunched parameter.

10.10.4 Efficiency, Gain, and Bandwidth of Gyroklystrons

Two-Cavity Gyroklystrons

Efficiency Let us suppose that eigenfrequencies of both cavities are the same and their lengths are small (the point gap model). In this case the maximum efficiency $\eta_\perp \sim 34\%$ is reached for zero detuning [the frequency mismatch $\delta = 0$, as defined by Eq. (10.202)] and the bunched factor $X = 1.84$. This value corresponds to the first maximum of the function $I_1(X)$, where I_1 is a Bessel function of the first order. The value $X = 1.84$ provides a maximum efficiency $\eta = 58\%$ for a conventional klystron in a kinematic approximation (see Section 7.4.2). The maximum efficiency for $X > 1.84$ can be realized with nonzero detuning (Fig. 10.36). It is obvious that the gain is reduced if the increase of $X = |F|\mu_{21}$ is due to increased input power.

For long resonators, the maximal orbital efficiency can be increased up to 70% (Antakov et al., 1994). According to Kuraev (1979), optimization of the distribution of the HF field and use of a tapered dc magnetic field allow one to increase the efficiency up to 80% or more.

Gain In general, the gain grows with any decrease in the input power or increase in the length of the drift section. Maximum gain is attained with zero detuning.

Bandwidth Bandwidth can be defined as the frequency detuning range $\Delta\omega$ providing deviation in the gain if less than 3 dB. The order of bandwidth is defined as $\Delta\omega \sim \omega/Q$, and the normalized bandwidth is $\text{BW} = \Delta\omega/\omega$. This value can be increased substantially for slight detuning of eigenfrequencies of the cavities (stagger tuning). According to Nusinovich et al. (1997), up to a fivefold increase in bandwidth can be achieved in this case; however, it is accompanied by a reduction in gain. A useful criterion is the gain–bandwidth product.

Multicavity Gyroklystrons (Fig. 10.35) There are many adjustable constructive parameters of multicavity gyroklystrons (e.g., eigenfrequencies of cavities and their

Q-factors, lengths, operation modes, positions, number of gyrofrequency harmonics). Therefore, the choice of the optimal version that satisfies specific requirements (e.g., maximum efficiency, gain, bandwidth) is a very complicated problem, even more difficult than optimization of conventional klystrons where interaction spaces are short and the modes are fundamental. In general, this problem does not have a unique solution.

Let us outline some relevant results.

Efficiency It has been shown by Kuraev (1979) that the orbital efficiency can reach 96% in three-cavity gyroklystrons with sinusoidal distribution of HF fields in the cavities. One important condition for the realization of such high efficiency is the detuning of the penultimate cavity; its eigenfrequency should be shifted downward relative to that of the output cavity, or to the input signal if the detuning of the output cavity is zero. It is interesting that this detuning is opposite the optimal detuning upward of the penultimate cavity in conventional multicavity klystrons (see Section 7.7.2). As shown by Chu et al. (1988), this property is connected with the energy dependence of the gyrofrequency, which determines the rate of phase bunching, inverse to linear bunching in conventional klystrons. The position of the penultimate cavity is also important. According to Zasypkin et al. (1995), the highest efficiency in a three-cavity gyroklystron is achieved when the penultimate cavity is close to the output cavity. The highest gain is attained if the eigenfrequencies of all the cavities are the same and the intermediate cavity is separated equally from the input and output cavities.

Bandwidth Use of stagger-tuned gyroklystrons increases their bandwidth considerably (see, e.g., Blank et al., 2000). So, according to Nusinovich et al. (1997), the maximum bandwidth of a four-cavity gyroklystron can be 16 times larger that in the absence of stagger tuning. However, this increase leads to considerable degradation of gain. The optimal criterion for multicavity gyroklystron is thus determined by the product of bandwidth and gain (BW-G), which for proper adjustment is also increased by stagger tuning. So in a four-cavity gyroklystron, the BW-G enhancement can reach a factor of 4 to 6.

10.11 GYRO-TRAVELING-WAVE TUBES

Gyro-TWT amplifiers based on the interaction of a helical electron beam with a fast-traveling electromagnetic wave drew the attention of researchers in the earliest stages of ECM investigations (Gaponov, 1959, 1960; Pantell, 1959). However, development of gyrodevices in the USSR and after 1975 in the United States was concentrated initially on gyromonotrons and gyroklystrons as unique sources of short-wave radiation with high average power, high gain, and a comparatively large bandwidth (for gyroklystrons). Prospects for the gyro-TWT looked less attractive, because anticipated strong sensitivity to the velocity spread and low stability of these amplifiers did not promise large gain or bandwidth. However, a demand for powerful broadband amplifiers in the millimeter band (mainly for radar) has revived interest in gyro-TWTs. Intense investigations (especially in the United States) began in the late 1970s. New ideas have led to considerable progress in gyro-TWTs as powerful, stable, wideband amplifiers with high gain throughout the millimeter range. Earlier work (see, e.g., Chu et al., 1979; Seftor et al., 1979) performed with uniform interaction waveguides demonstrated power levels much higher than for short-wave conventional TWTs. However, instantaneous bandwidths at that time were on the order of only a few percentage points. Some properties of the gyro-TWT (within the scope of this book) are considered below. In particular, a short outline of the linear theory and some results of experimental studies are presented.

10.11.1 Scheme of a Gyro-TWT. Resonance Condition

A scheme of gyro-TWT is depicted in Fig. 10.37. This device, as shown in Table 10.1, is analogous to a conventional TWT (Fig. 8.1) with a distributed electrodynamic system and inertial O-type bunching (see Section 9.2.3). The magnetron-injection gun determines the required values of energy and the pitch factor $g = v_\perp / v_z$ of the helical electron beam, which interacts with the traveling electromagnetic wave in the regular waveguide.

The frequency and phase velocity of the wave required for resonance interaction of gyrating electrons with an electromagnetic field are determined by two conditions: (1) the dispersion equation of the mode in a regular waveguide [Eq. (10.47)],

$$\omega^2 = \omega_{cr}^2 + k_s^2 c^2 \tag{10.215}$$

and (2) the equation of the cyclotron resonance on the nth harmonic [Eq. (10.46)], which can be written as

$$\omega = n\omega_g + k_s v_z \tag{10.216}$$

where ω_{cr} is the cutoff frequency of the waveguide at the chosen mode and $k_s = \omega / v_{p.hs}$ is the propagation constant. The term $k_s v_z$ is the Doppler shift of the cyclotron resonance.

It is convenient to represent resonance conditions on a dispersion diagram (Fig. 10.38) as the intersection points of the hyperbola Eq. (10.215) with the straight

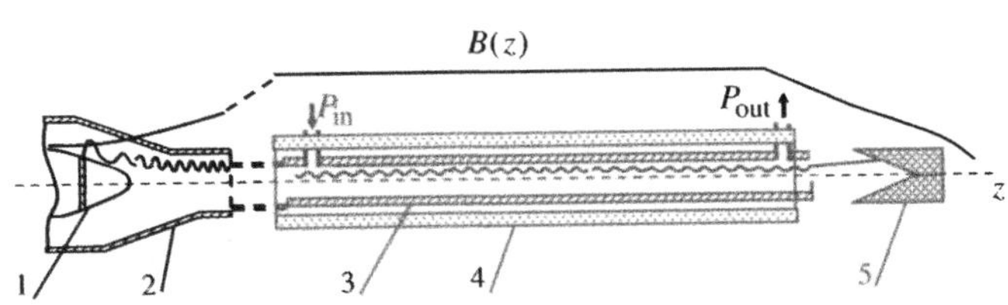

FIGURE 10.37 Gyro-TWT. 1, Cathode; 2, anode; 3, regular waveguide; 4, solenoid; 5, collector.

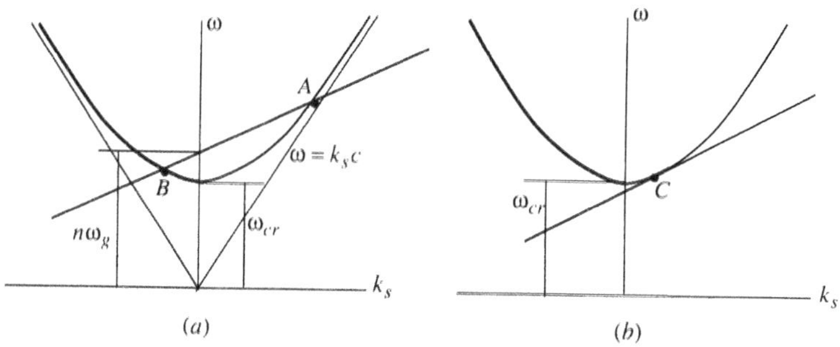

FIGURE 10.38 Dispersion diagrams of a gyro-TWT. (*a*) Two types of instability: A, forward-wave interaction; B, backward-wave interaction. (*b*) Grazing intersection: backward-wave interaction is absent. Lines $\omega = \pm k_s c$ are asymptotes of the hyperbola.

line Eq. (10.216). The case shown in Fig. 10.38*a* is definitely not preferred for a gyro-TWT. Crossing the cyclotron resonance line with the negative k_s branch of the waveguide mode hyperbola in point B corresponds to interaction with a backward wave, which can lead to an absolute instability (see, e.g., Section 8.8.2). Intersection at point A with the large Doppler shift does not provide an effective interaction with the forward wave and intensifies the velocity spread effect.

The case shown in Fig. 10.38*b* corresponds to a grazing intersection of the lines indicated above. It is readily verified that at the grazing point,

$$\omega_g = \frac{1}{n}\omega_{cr}\sqrt{1 - \left(\frac{v_z}{c}\right)^2} = \frac{\omega_{cr}}{n\gamma_z} \tag{10.217}$$

Relation (10.217) determines the gyrofrequency (the magnetic field) necessary for the grazing intersection. Point C is located near the waveguide cutoff for subrelativistic velocities. This version has several advantages: (1) the proximity of point C to the waveguide cutoff increases the interaction impedance and weakens sensitivity to the beam velocity spread; (2) there is no backward interaction; and (3) the bandwidth is enlarged. Indeed, let us estimate the rate of the cyclotron mismatch $\Delta\omega_n = \omega - n\omega_g - k_s v_z$ as a function of the signal frequency ω:

$$\frac{d\Delta\omega_n}{d\omega} = 1 - \frac{dk_s}{d\omega}v_z \tag{10.218}$$

The rate is minimal when $d\omega/dk_s = v_z$. It is the grazing intersection condition, which is provided in practically all gyro-TWTs.

10.11.2 Linear Theory of Gyro-TWTs

The linear theory of gyro-TWTs (Gaponov, 1959, 1961) is analogous to the linear theory of conventional TWTs (see, e.g., Section 8.5), although the former takes into account curvilinear electron motion, making the use of averaged equations somewhat more complicated. Let us outline the three basic steps of the theory.

1. Determination of the equation for s-mode excitation in a regular waveguide (see, e.g., Vainstein, 1956) proceeds as follows:

$$\mathbf{E}(\mathbf{r}) = C_s(z)\mathbf{E}_s(\mathbf{r}) \tag{10.219}$$

where

$$\mathbf{E}(\mathbf{r}) = \mathbf{E}(\mathbf{r}_\perp)e^{-ik_z z}, \qquad \mathbf{E}_s(\mathbf{r}) = \mathbf{E}_s(\mathbf{r}_\perp)e^{-ik_{sz} z} \tag{10.220}$$

are complex amplitudes of the excited field and the eigenfield of the s-mode of the waveguide, respectively. The amplitude C_s satisfies the equation $dC_s/dz = -(1/N_s)\int_{S_\perp} \mathbf{j}\mathbf{E}_s^*\, ds_\perp$, where N_s is the norm of the wave. The electric force in the moving frame of reference can be replaced by the Lorentz force $e_0\mathbf{E} \to e_0\mathbf{G} = e_0(\mathbf{E} + \mathbf{v}_\perp \times \mathbf{B})$ (see Section 10.6.1). Taking into account Eqs. (10.219) and (10.220), and replacing the electric fields by the Lorentz field, we obtain the thin beam approximation ($\int_{S_\perp} \mathbf{j}\, ds_\perp = -I_b(\mathbf{v}/v_{z0}) = -I_b(i\omega/v_{z0})\mathbf{r}_\omega^{(1)}$, where I_b is the beam current, $\mathbf{r}_\omega^{(1)}$ is the alternating electron's shift component), and the equation of excitation

$$(k_z - k_{sz})C_s = -\frac{\omega I_b}{v_{z0}N_s}\mathbf{r}_\omega^{(1)}\mathbf{G}_{s,\omega}^* \tag{10.221}$$

where $\mathbf{G}_{s,\omega}$ is the complex amplitude of the field, $\mathbf{G}_s = \mathbf{E}_s + \mathbf{v}_\perp \times \mathbf{B}_s$.

2. Calculation of the perturbation $\mathbf{r}_\omega^{(1)}$ of electron motion is based on the following equations of motion in cylindrical coordinates:

$$\begin{aligned}
\frac{d}{dt}\left(\gamma\frac{dr}{dt}\right) - \gamma r\dot{\theta}^2 &= -\eta[G_r + (\mathbf{v} \times \mathbf{B_0})_\mathbf{r}] \\
\frac{1}{r}\frac{d}{dt}(\gamma r^2\dot{\theta}) &= -\eta[G_\theta + (\mathbf{v} \times \mathbf{B_0})_\theta] \\
\frac{d}{dt}(\gamma\dot{z}) &= -\eta[G_z + (\mathbf{v} \times \mathbf{B_0})\mathbf{z}]
\end{aligned} \tag{10.222}$$

3. Linearization of equations of motion leads to equations for small components r_1, θ_1, and z_1 of $\mathbf{r}_\omega$. This result, combined with their representation by the Fourier series and the retention of resonance terms, gives the shift $\mathbf{r}_\omega$ as a linear function of C_s. Substitution of $\mathbf{r}_\omega$ into Eq. (10.221) and cancellation of C_s leads to the dispersion equation, which can be written as (Gaponov, 1959; Nusinovich, 2004)

$$\frac{(k_z - k_{zs})(\Delta\omega_n)^2}{\omega^3}v_{z0} = -C^3 \tag{10.223}$$

where

$$C^3 = \frac{\eta I_b f_s^2}{\gamma_0\omega^2 N_s}[(1 - \beta_{z0}^2)|G_{zn}|^2 - \beta_{\perp 0}^2|G_{\theta n}|^2 - 2\beta_{z0}\beta_{\perp 0}\mathrm{Re}(G_{\theta n}G_{zn}^*)]$$

and $\Delta\omega_n = \omega - k_z v_{z0} - n\omega_g$ is the frequency mismatch. The Fourier components G_{zn} and $G_{\theta n}$ of the Lorentz force in general depend on δ and b (Gaponov, 1959), but for TE modes and frequencies near the cutoff [see Eqs. (10.77)] they are given by Eqs. (A8.9). The dispersion equation (10.223) in this case can be written as

$$\delta^2(i\delta - b) = 1 \tag{10.224}$$

where

$$\delta = i\frac{\Delta\omega_n}{\omega C}, \qquad b = -\frac{\omega - k_{zs}v_0 - n\omega_g}{\omega C}$$

This equation coincides with the famous Pierce dispersion equation (8.52) for conventional TWTs if one neglects the space charge and uses the small nonsynchronism approximation ($Cb \ll 1$). Here the frequency mismatch of the unperturbed wave plays the role of the nonsynchronism parameter b in the theory of conventional TWTs. Further analysis of amplification properties of gyroTWTs can proceed along the lines of the design for conventional TWTs (Sections 8.5.5 and 8.5.6). In particular, the gain of gyroTWTs equals

$$G_{\mathrm{dB}} = -9.54 + 47.3CN_L \tag{10.225}$$

where $N_L = L/\beta_{z0}\lambda$. Independence of G_{dB} in Eq. (10.225) of the drive power is a result of the linearity of the small-signal gain. An increase in the input power leads to saturation effects in gain and efficiency. These effects are analogous to the nonlinear effects in conventional TWTs (Section 8.6).

10.11.3 Bandwidth of Gyro-TWTs

Equation (10.225) refers to the synchronous regime $b = 0$. In this regime the small-signal gain is maximal. The frequency shift decreases gain. An estimation of the corresponding bandwidth can be obtained from the Pierce diagrams (Fig. 8.9*a*), which refer to the case of zero space charge. However, real values of the gyro-TWT bandwidth are usually significantly lower. The limitation of the bandwidth is determined mainly by crossing the cyclotron resonance line with negative β branches of other waveguide modes on the fundamental or higher harmonics of the gyrofrequency (Fig. 10.39), which leads to absolute instability. Furthermore, absolute instability of the single operation mode is also possible near the cutoff for large beam currents, due to broadening of the cyclotron resonance band.

It turns out that the starting current for backward modes can be increased significantly by a number of methods without noticeable degradation of the saturated gain and efficiency. The most useful methods (see, e.g., Song et al., 2004) are heavy loading of the long section of the waveguide by lossy layers and limiting of the velocity spread and the factor $g = v_\perp/v_z$. So, according to the result of a large-signal simulation, a bandwidth of 5% can be reached for the powerful 94-GHz gyro-TWT (output power, 140 kW; efficiency, 28%; and saturated gain, 50 dB).

Some interesting possibilities for bandwidth widening are raised by the use of a tapered interaction waveguide with a tapered magnetic field (Fig. 10.40) (see, e.g., Park et al., 1994). The section ΔL, where condition (10.217) of the grazing

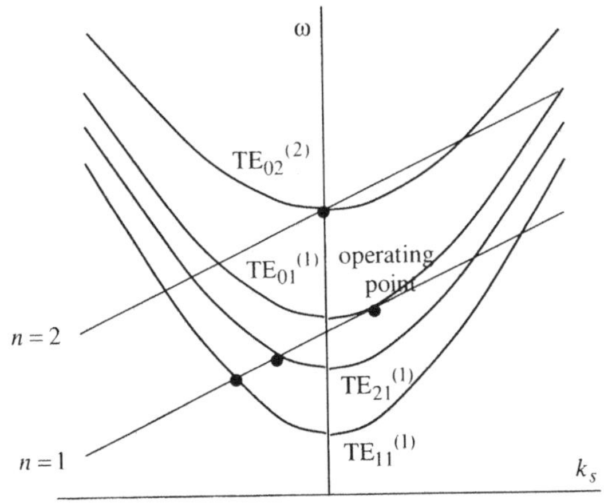

FIGURE 10.39 Dispersion diagram of the interaction of fundamental mode TE_{01} with possible oscillation modes. Lines $n = 1$ and 2 correspond to resonance on the fundamental and second harmonics of the gyrofrequency.

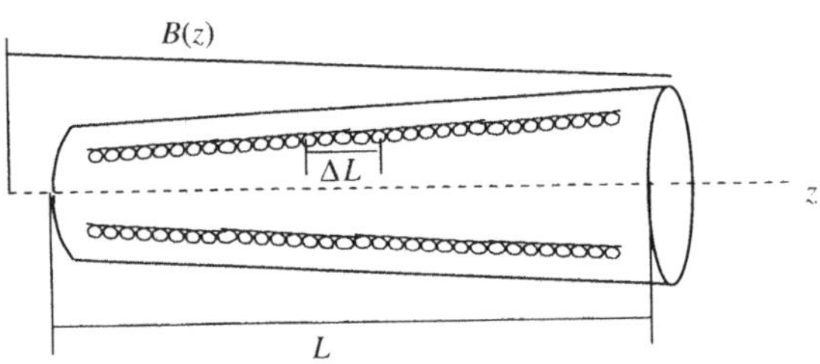

FIGURE 10.40 Interaction space in a tapered gyro-TWT.

intersection is accomplished at a given wave frequency, is much shorter than the total length L of the system. But the location of the section ΔL shifts with frequency, so the effective grazing synchronism for the adjustable tapering of the cavity and the magnetic field is achieved within a large bandwidth. The shorter effective interaction length ΔL increases the starting currents of the absolute instability and weakens the influence of velocity spread. So the main problem for these gyro-TWTs is a reflective instability (see below).

Park et al. (1994) have implemented a K_a-band tapered gyro-TWT amplifier with bandwidth 33% and gain 25 dB. Note that the tapered technique was first used for bandwidth widening of gyro-TWTs (Moiseev, 1977).

10.11.4 Reflective Instability of Gyro-TWTs

Self-excitation of oscillations can impede widening of the gyro-TWT bandwidth. This self-excitation is induced by a feedback loop as a result of the reflection of outgoing radiation from the output taper and the input cutoff narrowing. Effective means of stabilizing gyro-TWT operation by suppressing both the absolute and reflective instabilities include precise matching of the output circuit with the interaction wavegide, as well as using distributed losses in the interaction space. So a graphite-coated lossy section of length 20 cm with a full 24-cm length of interaction waveguide was used by Chu et al. (1999). Such K_a-band gyro-TWTs have demonstrated high performance (65 dB saturated gain, 80 kW output power, and bandwidth 1 GHz). Another method for suppressing reflective oscillations is the use of two-stage or, in general, multistage gyro-TWTs, similar to severed TWTs (see Section 8.7.3).

10.12 APPLICATIONS OF GYROTRONS

10.12.1 Introduction

The potential for gyrotrons (oscillators and amplifiers) as sources of high average power in the millimeter and submillimeter range has led to extraordinarily quick progress in developing these devices. Let us recall that the maximum power of conventional CW BWOs is on the order of 100 W in the W-band. The first gyromonotron (1964) produced about 200 W of CW power at a frequency of 25 GHz (Gaponov et al., 1965), but as early as 1965 CW power has reached 4 kW (Gaponov et al., 1975). It was clear that those powers were far from the limit. At present the CW output power of gyromonotrons in 140- to 170-GHz bands has been increased by more than two orders of magnitude, along with a number of new concepts enhancing the fundamental gyrotron mechanism (e.g., transfer to high-volume cavity modes, diffractive power output, CVD-diamond windows, depressed collectors, superconductive magnetic systems). Similar progress has been seen in long-pulse amplifier gyrotrons (e.g., gyroklystrons, gyro-TWTs).

The fast development of advanced gyrotrons has outpaced the dynamics of their applications. Nevertheless, at present the number of gyrotron applications is significant. Note that many corresponding fields in applied powerful high-frequency electronics were simply absent in the pre-gyrotron era. Several important gyrotron applications are considered briefly below:

- Controlled fusion experiments [electron–cyclotron resonance heating (ECRH) and electron–cyclotron current drive (ECCD)] in tokamak–stellarator plasmas
- Generation of multiply charged ions and soft x-rays; electron spin resonance spectroscopy
- Microwave procession of materials (e.g., advanced ceramics sintering, joining of ceramics)
- Millimeter radar systems (e.g., gyromonotron and gyroklystron radars)
- RF drivers for TeV electron–positron colliders

10.12.2 Electron–Cyclotron Resonance Heating and Current Drive

ECRH is based on resonance acceleration of the electron component in hot tokamak or stellarator plasmas by the electromagnetic HF field. Nonelastic collisions between accelerated electrons and ions provide additional heating of the plasma. The temperature of the latter can reach values sufficient for thermonuclear fusion if adequate power and duration of the RF radiation are achieved, and if the concentration and stability of the plasma are sufficient. The frequency of the RF drive signal should correspond to the condition of the cyclotron resonance of electrons with a toroidal magnetic field at the fundamental gyrofrequency or its harmonic (usually, the second harmonic). The toroidal magnetic field required for plasma confinement is on the order of several tesla. Therefore, in the majority of plasma fusion installations with electron-resonance heating, required frequencies are in the range 70 to 170 GHz. Powerful RF radiation penetrating the plasma induces high currents (up to 100 kA). Routine ohmic heating is also accompanied by a high current drive. However, the ohmic current is not continuous because it is induced by the alternating magnetic field. Noninductive ECCD, by contrast, is essentially a continuous effect. In comparison with other noninductive methods (e.g., neutral-beam, ion cyclotron resonance, low hybrid heating) ECRH and ECCD are characterized by highly localized deposition and flexible steering, which is especially well suited for controlled fusion applications. An important property of the ECCD is the suppression of some high plasma instabilities.

The first ECRH experiments were performed in 1972 on the tokamak TM-3 at the Kurchatov Institute in Moscow (Alikaev et al., 1972). The characteristics of the gyrotron were: wavelength, 9 mm; output power, 80 kW; and pulse duration, 0.6 ms. Early results were impressive, but considerably higher output power and pulse duration were needed. At present, practically all the world's fusion installations of tokamak and stellarator types are equipped with gyrotron units. The parameters for the gyrotrons in the two most powerful existing reactors are given in the first two lines of Table 10.6 (see Denisov, 2004). Gyrotron parameters are given for two fusion installations currently under construction: ITER (International Thermonuclear Experimental Reactor) and Stellarator W7-X, at Cadarache (France) and Greiswald (Germany), respectively. Note recent result in FZK Germany (Dammertz et al., 2006): 140 GHz gyrotron with output power 0.93 MW and pulse duration 10 min.

TABLE 10.6 Gyrotron Provided in Different Experimental Fusion Reactors

Reactor Type	Operating Mode	Frequency (GHz)	Number of Gyrotrons × Power (MW)/Pulse Duration
Tokamak DIII-D (U.S.)	$TE_{22,6}$	110	3 × 0.8/2 s; 3 × 0.6/10 s
Tokamak JT-60U (Japan)	$TE_{22,6}$	110	4 × 1.0/5 s
Tokamak ITER	$TE_{31,8}$; $TE_{25,10}$	170	24 × 1.0/CW
Stellarator W7-X (Germany)	$TE_{28,8}$	140	10 × 1.0/CW

Prospects for further increase in the output power are connected with the use of coaxial gyrotrons (Section 10.9.2). Effective mode selection in these gyrotrons, especially with corrugated coaxial inserts, allows one to use very high operational modes (e.g., $TE_{44,44}$). In this case, it is possible to obtain CW output power up to 2 MW (Piosczyk et al., 2005; Dammertz et al., 2006). Let us note the recent development of an important type of oscillator that reinforces efficiency and flexibility of ECRH–ECCD control: frequency step-tunable gyrotrons (step tuning on a few frequencies) (see, e.g., Dumbrais et al., 2001; Koppenburg et al. 2005). These gyrotrons are desired, in particular, for stabilization of neoclassical testing modes (NTM).

10.12.3 Generation of Multiply Charged Ions and Soft X-rays. Electron Spin Resonance Spectroscopy (Kartikeyan et al., 2004)

Electron–cyclotron resonance (ECR) discharge in plasma as a source of multiply charged ions can be used for accelerators and material processing (e.g., ion implantation in semiconductors). The use of such HF sources as gyrotrons significantly improves the discharge characteristics because the current of ions with fixed charge Z is proportional to the square of frequency, and an average ion charge $\overline{Z}$ in the ion current grows with frequency as $3.5 \ln f$ (Kartikeyan et al., 2004).

A gaseous source of ions is placed in a magnetic field with the configuration of an adiabatic magnetic trap (see Section 1.6.3). For example, the discharge in argon gas with a pressure of 10^{-4} torr due to radiation of 37-GHz 130-kW gyrotrons produced multiply charged ions ranging from Ar^{2+} to Ar^{18+}, with a maximum generation rate at Ar^{14+}.

ECR discharge may also be used as a source of soft x-ray radiation. According to an experiment by Golubev et al. (1996), the amount of x-ray emission power in the range of 9 nm was on the order of 7 kW with a conversion efficiency of up to 5%.

Electron spin resonance spectroscopy (ESR) is based on measurements of the resonance absorption of the gyrotron radiation in millimeter and submillimeter ranges corresponding to transitions between various spin states. Application of ESR at high frequencies and high powers offers several advantages. In particular, millimeter-wave ESR provides high resolution, so the fine-structure constants and the effects of nonlinear magnetic field interaction can be determined experimentally. Higher versatility and sensitivity is ensured by using step-tunable gyrotrons, which can cover a broad frequency range in millimeter and submillimeter wavelength bands.

10.12.4 Microwave Procession of Materials

Microwave heating, being a fast, noninertial, and easily controllable procedure, promises rich capabilities for producing materials with favorable properties. Volumetric and selective heating is a unique property of microwaves. However, application of traditional microwave installations at frequencies of 0.9 to 2.5 GHz for heating of large advanced ceramic materials is complicated by low absorption and nonuniform distribution of electromagnetic field in insufficiently multimode resonators. High-temperature material procession research at frequencies of 24 to 300 GHz has been carried out in the past few years at laboratories in Russia, the United States, Germany, and Japan (see, e.g., Denisov et al., 2005; Gaponov-Grekhov and Granatstein, 1994). Available CW drivers for this purpose include 24- to 30-GHz 15-kW gyrotrons with resistive solenoids (i.e., water and oil-cooled magnets) (e.g., Denisov et al., 2005), 30-GHz 10-kW gyrotrons with permanent magnets (Kuftin et al., 2000; Thumm, 2002), and 300-GHz 3.5-kW gyrotrons with superconducting liquid helium–free magnets (Hoshizuki et al., 2005) (see Appendix 9).

To date, gyrotron technology has been used primarily for ceramic sintering and joining of ceramic parts. Deep penetration of microwave radiation compared with that of thermal radiation in conventional heating systems allows one to realize homogeneous volumetric heating of large-scale samples. Due to microwave-induced diffusion, the temperature and time of sintering are significantly lower. Therefore, the growth of grains in ceramics during sintering is very small, so even nanocrystalline powders can be sintered without losing fine nanoscale structure (Kartikeyan et al., 2004). Advanced microwave sintering opens a path to materials with new properties and novel technological applications (e.g., sintering of metal powder compact and metal–ceramic composites).

10.12.5 Millimeter Radar Systems

According to the well-known formula, the maximum range $R_{\max}$ of a radar system can be determined as

$$R_{\max}^4 = \frac{P_r \tau A^2 \sigma}{(4\pi)^2 \lambda^2 E_{\min}} \tag{10.226}$$

were P_r is the radar radiation power, τ is the pulse duration, A is the antenna area, σ is the effective cross section of a target, and $E_{\min}$ is the minimal detectable returned energy in a pulse. Therefore, millimeter gyrotrons, which combine short wavelength λ with very high average output power $P_r\tau/T$, improve radar characteristics considerably by providing a large range and high angular resolution, thus opening new avenues for their use.

In general, propagation of millimeter waves in the atmosphere is characterized by high absorption. A comparatively lower absorption occurs in *atmospheric windows*, bands near 35, 94, and 140 GHz. A number of gyrotrons acting as drivers for millimeter-wave radars have been developed: gyromonotrons (Woo et al., 1989), powerful gyroklystrons, and gyro-TWTs in the 35- and 94-GHz bands.

A very important characteristic of any radar driver is the possibility to control its frequency and phase. There are phase-locked gyromonotrons, gyroklystrons, and gyro-TWTs. Such sources can be used for phased-array antennas. In military applications, the latter allow for the simultaneous detection of multiple targets, extension of the target-detection range, and improved stealth operation. For example, the powerful mechanically steered phased-array radar system Ruza was developed in

Russia on the basis of a gyroklystron complex (two K_a-band gyroklystrons, each with 500 kW of power). This radar was used successfully in the 1990s for tracking satellites and other ballistic objects (Tolkachev et al., 2000).

High-power millimeter-wave sources have been used in studies of atmospheric phenomena as well. Loss at the W-band for very humid conditions at sea level is about 1 dB/km. Thus, the use of powerful sources in this band is promising. In practice, the propagation distance is increased by 15 km when one changes the average power from 4 W to 4 kW. A powerful millimeter radar in the W-band can resolve multiple cloud layers (structure and extension) and give information about humidity and turbulence. Note that water clouds are too dense a medium for laser probes. Detailed studies of cloud structure were performed using the 94-GHz radar system WARLOC (Fliflet et al., 2003). Radar gyroklystrons developed for the WARLOC at NRL in the United States had 100 kW of peak power and 10 kW of average power.

An interesting (but futuristic) application of gyrotron amplifiers is related to the problem of space debris detection, because collision of a spacecraft with even a 1-cm object is very dangerous. Estimates show that the effective handling of large spacecraft is possible when a dangerous object is detected at a distance of about 1000 km. In this case the proper phased-array radar should contain about thirty 35-GHz 1-MW gyroklystrons (Kartikeyan et al., 2004).

10.12.6 Gyroklystron RF Drivers for TeV Linear Electron–Positron Colliders

Possible types of drivers for colliders include conventional klystrons, gyroklystrons, TWTs, magnicons, cyclotron autoresonance masers (CARMs), and free-electron lasers (FELs). At present linear electron–positron colliders are driven by conventional klystrons. The Stanford linear electron–positron collider (SLC) on the base of SLAC (see Section 7.8), with an energy 0.1 TeV in the center-of mass (CMS), uses 2.856-GHz, 65-MW klystrons. The next collider (NLC) should be driven by 75 MW 11.424-GHz klystrons. The designed CMS energy of particles for this collider will be 0.5 TeV for the linac length 6.3 km and 1 TeV for the length 12.8 km. An effective accelerating gradient in this accelerator is 48 MeV/m. The RF peak power and, correspondingly, the accelerated gradient are limited by a RF field breakdown. The possibility of increasing the acceleration energy and/or reducing the number of RF sources, which determines the collider price, depends on the value of the parameter $P_r\tau/\lambda^2$. Note that the same parameter enters relation (2.226) for a maximum range of radars. Thus, increasing the RF frequency is a very important means for creating multi-TeV linear colliders. However, a frequency on the order of 10 GHz is the upper boundary for effective conventional klystrons.

Gyroklystrons are promising sources to replace conventional klystrons (Granatstein and Lawson, 1996). The first two-cavity gyroklystron for a collider was designed and tested at the University of Maryland. The gyroklystron operated at a frequency of 9.85 GHz, fundamental harmonic, mode TE_{01}, voltage 425 kV, and current 200 A. The amplifier demonstrated an output power of 27 MW with a pulse duration 1.5 μs, gain of 36 dB, and efficiency of 32%. A two-cavity gyroklystron with the same input cavity but with an output cavity tuned to the second gyro-harmonic and the TE_{02} mode produced peak power of 21 MW with efficiency 21% and gain 25 dB at a frequency of ~20 GHz. A design for a 91-GHz 10-MW six-cavity gyroklystron has been presented by Lawson et al. (2001). The first two cavities operate on the TE_{011} mode at a drive frequency near 46 GHz. The remaining four cavities operate at the second harmonic (frequency, 91.4 GHz; TE_{021} mode). The high output power combined with moderate voltages requires the use of high beam currents. This requirement complicates beam formation and raises a potential depression problem if one takes into account that the beam should be located at the inner maximum of the electric field (see Section 10.9.2). Solution of this problem needs transfer to higher-order modes and a corresponding enlargement of cross sections of cavities and drift sections. But at this point, two other problems arise: mode selection and suppression of parasite coupling through drift tunnels between cavities. Proper solution of these problems depends on the use of a coaxial configuration. This design facilitates problems of mode selection, potential depression, and decoupling of cavities simultaneously. At present, a frequency-doubling coaxial gyroklystron at 70 MW and 17.12 GHz is being developed at the University of Maryland (Lawson et al., 2002).

10.13 CYCLOTRON AUTORESONANCE MASERS

10.13.1 Moderately Relativistic Gyrotrons

An increase in accelerating voltage up to relativistic values in the frame of the gyrotron mechanism meets one fundamental limitation. It can readily been seen (see also Section 10.7) that the efficiency of ultrarelativistic gyrotrons should be small. Indeed, if the input orbital energy of electrons is $E_{\perp 0} \gg E_0$ (where E_0 is the electron rest energy), the relative gyrofrequency change should be

$$\frac{\Delta\omega_g}{\omega_g} \approx \frac{\Delta E}{E} \sim 1 \qquad (10.227)$$

which obviously contradicts the gyro-synchronous condition. For the phase change of the rotating electron with respect to the RF field to remain within π, the number of electron revolutions in the interaction space should be decreased. The maximum number of revolutions is estimated as

$$N_{\max} = \frac{T}{T_g} \approx \frac{\pi}{(\omega - \omega_g)T_g} = \frac{\omega_g}{2(\omega - \omega_g)} \qquad (10.228)$$

This number, according to Eq. (10.227), should be ~1. However, the mechanism for quadratic bunching cannot work properly with this condition (see Section 10.3.3), and the efficiency would be small. As a compromise, moderately relativistic energies (200 to 500 keV) can be used. In this case, according to Zaitsev et al. (2002), an electronic orbital efficiency of 30% or even 50% can be attained for $U = 280$ kV, due to shortening of the cavity and optimization of its profile.

It is possible in principle to apply a variable along the cavity axis dc magnetic field to compensate for phase desynchronization. This method leads, however, to hard excitation of oscillations. Indeed, in the small-signal regime the mass variation is insignificant, and the nonuniform magnetic field violates the synchronism condition. As a result, the "starting" effective interaction length would be significantly less than the operational length, and as a result, the starting current would be more than the working current.

10.13.2 Operation Principle and Some Properties of CARM Oscillators

The concept of CARM was developed by Petelin (Petelin (1974) on the basis of the autoresonance effect known in the theory of accelerators (Davydovsky, 1962; Kolomensky and Lebedev, 1962). In contrast to gyrotrons, CARMs operate with traveling waves whose phase velocity is close to the speed of light, c. In this case, the transverse component of a magnetic field of a TE mode is different from zero, and the HF Lorentz force has a longitudinal component. As a result, power exchange is accompanied by a change not only of the oscillatory velocity but also of the drift velocity of particles.

Let us consider radiation of one photon by the electron oscillator in the field of an electromagnetic wave traveling along a dc magnetic field with phase velocity v_{ph}. In this case, quantum laws of energy and momentum conservation of the photon can be written as

$$\begin{aligned} \Delta mc^2 &= \hbar\,\omega \\ \Delta p_z &= \hbar\,k \end{aligned} \tag{10.229}$$

where $k = \omega/v_{\text{ph}}$. The ratio of the increments

$$\frac{\Delta mc^2}{\Delta p_z} \approx \frac{dmc^2}{dp_z} = \frac{\omega}{k} \tag{10.230}$$

does not depend on Planck's constant. Therefore, Eq. (10.230) is valid in the classical limit. We obtain the following conservation law:

$$\frac{m}{m_0} - \frac{\omega p_z}{km_0c^2} = \gamma(1 - \beta_{\text{ph}}\beta z) = G = \text{const.} \tag{10.231}$$

where $\beta_{\text{ph}} = v_{\text{ph}}/c = \omega/ck$ and $\beta_z = v_z/c$. Let us combine Eq. (10.231) with the Doppler synchronism relation

$$\omega = \frac{n\omega_g}{1 - \beta_z/\beta_{\text{ph}}} \tag{10.232}$$

assuming in both equations that $\beta_{\text{ph}} = 1$ (the exact autoresonance). Then the right-hand side of Eq. (10.232) will be constant:

$$\frac{n\omega_g}{1 - \beta_z/\beta_{\text{ph}}} = \frac{n\omega_g}{1 - \beta_z} = \frac{n\omega_{gn}}{G} \tag{10.233}$$

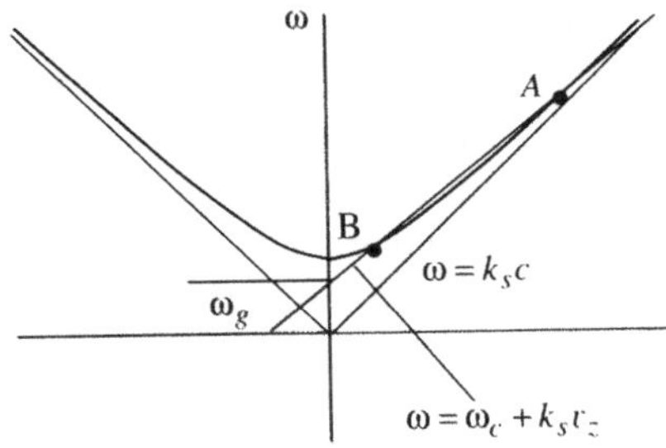

FIGURE 10.41 Dispersion characteristic of a CARM.

where $\omega_{gn} = \gamma\omega_g = \eta B$ is the nonrelativistic gyrofrequency. The synchronism takes place independent of the electron energy loss in the process of interaction with the HF field. Thus, the synchronism is maintained due to the compensation of the relativistic frequency shift and the Doppler shift stipulated by the drift velocity change.

Note that the phase of rotating electrons relative to the field phase for the strong autoresonance indicated is not changed. But then, phase bunching and stimulated radiation are absent. If, however, there is a small detuning from autoresonance ($\beta_{ph} > 1$), nonisochronism and radiation appear. The corresponding dispersion characteristic of a CARM is shown in Fig. 10.41. The intersection point A on the characteristic corresponds to the CARM regime. The low-frequency intersection point B with comparatively high phase velocity corresponds to the competing gyrotron regime.

Application of electromagnetic waves with $v_{ph} \approx c$ in a CARM leads to a significant difference between the electrodynamic system and the gyrotron resonator. In particular, the gyrotron diffraction power output turns out to be useless. There are a few substitute systems (see, e.g., Bratman and Denisov, 1992) providing a high Q-factor value for CARM resonators. In particular, Bragg resonators are used (see, e.g., Bratman and Denisov, 1992; McCowan et al., 1989). The resonator is formed by a section of the regular waveguide and a pair of Bragg reflectors (sections of corrugated waveguides) (Fig. 10.42). The geometry of Bragg reflectors provides a sufficiently high reflectivity at the frequency chosen. Effective operation of the device in the CARM regime also requires suppression of the gyrotron mode, indicated above, by raising its starting current. According to the works cited above, reaching the CARM regime requires beams with $\beta_z \sim 1$ and correspondingly reduced values of the pitch factor $g = v_\perp / v_z$. This case is favorable as well for Doppler frequency up-conversion (see Section 10.2.2), which allows for a 2γ-fold rise in frequency of the ultrarelativistic electron energy.

Small values of g do not exclude the possibility of obtaining comparatively high efficiency values. However, in contrast to the gyrotron, the large Doppler shift in a CARM leads to very stringent requirements as to velocity spread. According to Bratman and Denisov (1992), the allowable starting velocity spread is determined by the condition

$$\frac{\delta\beta_{\perp 0}}{\beta_{\perp 0}} \leq \frac{1}{4bN} \tag{10.234}$$

FIGURE 10.42 CARM Bragg resonator.

where N is the number of electron gyrorevolutions in the interaction space,

$$b = \frac{\beta_{\perp 0}^2}{2\beta_z\beta_{ph}(1 - \beta_{z0}/\beta_{ph})}$$

According to experimental data, classical magnetron injection guns do not provide proper beam quality. Kicker guns (see, e.g., Bratman et al., 1995, and Section 10.8.2) are considered more promising.

The high electronic efficiency of CARMs (26%) has been obtained in experiments by Bratman et al. Use of a kicker gun has provided a small velocity spread (the stimulated spread was ~3%). The CARM operated on the TE_{11} mode, with accelerated voltage 500 kV, beam current 100 A, and transverse particle velocity $\beta_\perp = 0.5$. The wavelength was 7.9 mm. The output Bragg reflector had a length of 83 mm and a corrugated depth 0.7 mm. Note that the efficiency indicated above was obtained by cutting a small central section with little ripples from the primary electron beam.

10.14 FREE ELECTRON LASERS

10.14.1 Introduction

Free electron lasers (FELs) originated from the Phillips ubitron (Section 10.4.3), which is an amplifier (electron oscillators excited by spatially periodic magnetic field produce a stimulated microwave radiation). The FEL era began in 1971, when J. Madey demonstrated the possibility of stimulated bresstrahlung emission in the optical range as a result of Doppler up-conversion (see Section 10.2.2) and interaction between radiation and relativistic electrons in a spatially periodic transverse magnetic field (Madey, 1971). In 1975–1977, Madey, Elias, et al. performed the first experiments with an FEL amplifier and oscillator (Elias et al., 1976) and obtained coherent emission on wavelengths corresponding to 10.2 and 3.4 μm. The next few years saw an avalanche in the amount of research devoted to FE. The number of publications up to 2005 has reached about 10^4. At present, different FELs overlap in frequency ranges from a millimeter up to the ultraviolet spectrum (wavelengths of 10 mm to 10 nm). FELs operating in long-wavelength bands usually employ nonrelativistic beams and are called free electron masers (FEMs) (Thumm, 2002). In general, the problems of FELs differ from those of microwave electronics in practice as well as in theory. However, some fundamental concepts, in particular electron velocity modulation and bunching as a necessary premise for stimulated emission, are common for FELs and other microwave devices.

Certainly, it is difficult to describe the foundations of FEL physics and technology in sufficient detail within the scope of this book. However, it is instructive to trace some principles of energy modulation and bunching in FEL. Below, the simplest version of linear FEL theory, as well as some parameters and applications of FELs are outlined.

10.14.2 Scheme of an FEL

A sketch of an FEL oscillator is depicted in Fig. 10.43. Electron gun 1 forms a thin electron beam. The guns supplied by the photocathode provide the beams with a small emittance. The electron accelerators provide high-energy pulse electron beams. The shortest-wavelength FELs use the following types of accelerators: storage rings, linear accelerators (linacs), and electrostatic accelerators (see Table 10.7). The electron energy produced by these accelerators is in the range ~5 to 100 MeV.

A spatially periodic magnetic field in the helical wiggler, depicted schematically in Fig. 10.43, is generated by bifilar current winding. The static transverse magnetic

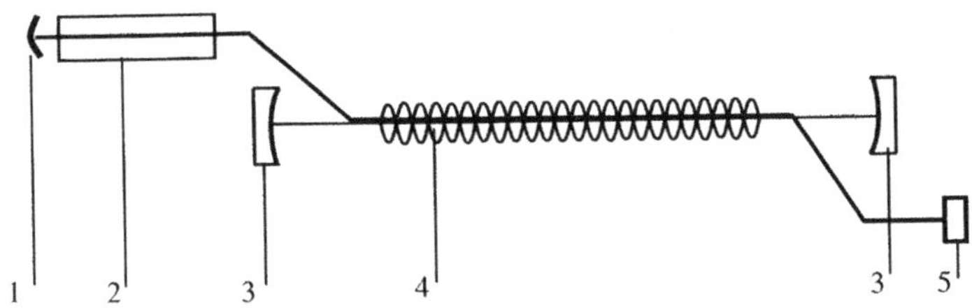

FIGURE 10.43 FEL. 1, Electron gun; 2, electron accelerator; 3, mirrors; 4, wiggler (undulator); 5, electron dump.

TABLE 10.7 Wavelength of Some Worldwide FELs

Location	Name	Wavelengths	Accelerator Type
Duke (North Carolina)	OK-4	217 nm	Storage ring
	MK-III	1.7–9.1 μm	Linac
IEFL (Japan)	1	5–22 μm	Linac
	3	230 nm–1.2 μm	Linac
Vanderbilt (Tennessee)	MK-III	2.1–9.8 μm	Linac
Stanford (California)	SCA-FEL	3–10 μm	SC-linac
University of California–Santa Barbara	30 μ-FEL	30–63 μm	Electrostatic
	MM-FEL	340 μm–2.5 mm	Electrostatic
DESY (Germany)	Tesla FEL	80 nm	SC-linac

Source: World Wide Web (2004).

field produced by this wiggler in the paraxial approximation has a simple harmonic form (see, e.g., Freund and Antonsen, 1992; and Robertson and Sprangle, 1989):

$$\mathbf{B}_w(z) = B_w(\mathbf{e}_x \cos k_w z + \mathbf{e}_y \sin k_w z) \tag{10.235}$$

where $k_w = 2\pi/\lambda_w$ and λ_w is the spatial period of the wiggler. A uniform axial magnetic field $\mathbf{B}_0 = B_0\mathbf{e}_z$ is often employed in conjugation with the helical wiggler to provide strong focusing of intense electron beams. Another widespread configuration (especially in the shortest-wavelength FELs, near-infrared, and shorter bands) is a planar wiggler that produces a linearly symmetric field. In Fig. 10.1 the design of a planar wiggler with permanent magnets is shown. A typical range for the wiggler periods is 3 to 10 cm.

The optical cavity of FEL is formed by a pair of mirrors (3). FEL requires a full overlap between electron and optical beams. Since the transverse dimensions of the electron beam are small, this implies that the transverse dimension of the optical beam must also remain small. Therefore, tolerances for the distance and the angular alignment of the mirrors are very stringent. The electromagnetic field of the optical mode is a transverse field with the magnetic vector potential

$$\mathbf{A}_R(z,t) = A_R[\cos\phi(z,t)\mathbf{e}_x + \sin\phi(z,t)e_y] = A_R(e^{i\phi(z,t)}\mathbf{e}_- + e^{-i\phi(z,t)}\mathbf{e}_+) \tag{10.236}$$

where $\phi(z,t) = kz - \omega t$, $k = 2\pi/\lambda$, and $\mathbf{e}_\pm = (\mathbf{e}_x \pm \mathbf{e}_y)/2$.

10.14.3 Linear Theory of FELs

The outline of the linear theory given below follows Freund and Neil (1999) and Robertson and Sprangle (1989). We consider the *Compton scattering regime* or *low-gain Compton regime*. This regime occurs when the electron current is sufficiently small. In this case the RF electric field is purely transverse, and as shown in the Introduction [Eq. (I.3)], it equals

$$\mathbf{E}(z,t) = -\frac{\partial \mathbf{A}_R}{\partial t} \tag{10.237}$$

In the opposite situation, when the space-charge potential is considerable, a *Raman scattering regime* is present. Note that in the first experiments of Elias et al. (1976), FELs worked in the Compton regime.

Let us write the force equation (see, e.g., Landau and Lifshitz, 1987),

$$\frac{d\mathbf{v}}{dt} = -\frac{\eta}{\gamma}\left[\mathbf{E} + \mathbf{v}\times\mathbf{B} - \frac{1}{c^2}\mathbf{v}(\mathbf{vE})\right] \tag{10.238}$$

The z-component of the force equation in the Compton regime is given by

$$\frac{dv_z}{dt} = -\frac{\eta}{\gamma}\left[(\mathbf{v} \times \mathbf{B})\mathbf{e}_z - \frac{v_z}{c^2}(vE)\right] \tag{10.239}$$

The wiggler magnetic field Eq. (10.235) can be represented as

$$\mathbf{B}_w = \text{curl}\mathbf{A} = \mathbf{e}_z \times \frac{\partial \mathbf{A}_w}{\partial z} \tag{10.240}$$

where

$$\mathbf{A}_w = A_w(e^{ik_w z}\mathbf{e}_- + e^{-ik_w z}\mathbf{e}_+) \tag{10.241}$$

Then the total magnetic field is

$$\mathbf{B} = \mathbf{B}_w + \mathbf{B}_R = \mathbf{e}_z \times \frac{\partial \mathbf{A}}{\partial z} = \mathbf{e}_z \times \frac{\partial}{\partial z}(\mathbf{A}_w + \mathbf{A}_R) \tag{10.242}$$

Using Eq. (10.242), we can write the first term in Eq. (10.239) as

$$(\mathbf{v} \times \mathbf{B})\mathbf{e}_z = \mathbf{v}\frac{\partial \mathbf{A}}{\partial z} = \mathbf{v}\frac{\partial}{\partial z}(\mathbf{A}_w + \mathbf{A}_R)$$

so Eq. (10.239) can be written

$$\frac{dv_z}{dt} = -\frac{\eta}{\gamma}\left[\mathbf{v}\frac{\partial}{\partial z}(\mathbf{A}_w + \mathbf{A}_R) + \frac{v_z}{c^2}\frac{\partial \mathbf{A}_R}{\partial t}\right] \tag{10.243}$$

The velocity can be expressed through the vector potential due to conservation of the transverse canonical momentum $d\mathbf{P}_\perp/dt = -\partial H/\partial \mathbf{r}_\perp = 0$ (H is the Hamiltonian). So $\mathbf{P}_\perp = \gamma m_0 \mathbf{v}_\perp - e_0\mathbf{A} = \text{const.} = 0$ and

$$\mathbf{v}_\perp = \frac{\eta}{\gamma}\mathbf{A} = \frac{\eta}{\gamma}(\mathbf{A}_w + \mathbf{A}_R) \approx \frac{\eta}{\gamma_0}(\mathbf{A}_w + \mathbf{A}_R) = \mathbf{v}_w + \mathbf{v}_R \tag{10.244}$$

where $\gamma_0 = \sqrt{1 + (\eta A_w/c^2)^2}\Big/\sqrt{1 - v_{z0}^2/c^2}$ and v_{z0} is the constant axial velocity. It can be shown (Robertson and Sprangle, 1989) that after substitution of Eq. (10.244) into Eq. (10.243) and linearizing of the latter while keeping terms up to the first order in the radiation field, Eq. (10.243) becomes

$$\frac{d\tilde{v}_z}{dt} = \left(\frac{\partial}{\partial z} + \frac{v_{z0}}{c^2}\frac{\partial}{\partial t}\right)\Phi_p(z,t) \tag{10.245}$$

where

$$\Phi_p = -\frac{\eta}{\gamma_0 c^2}\mathbf{A}_w\mathbf{A}_R = -\frac{\eta}{2\gamma_0 c^2}A_w A_{Re}^{i[(k+k_w)z-\omega t]} + \text{c.c.} \tag{10.246}$$

We see that the potential Φ_p that is responsible for the longitudinal acceleration of electrons is the result of wiggler and radiation field interference (i.e., it is a *beat wave*). The corresponding force is called the *pondermotive force*. Note that the phase velocity of this wave, $(v_{\rm ph})_b = \omega/(k + k_w)$, is less than the speed of light, and the beat wave can be synchronous with the electron beam. The beat effect leads to the possibility of electron bunching.

A component of the space-charge density satisfies the continuity equation. Therefore, we can write in the linear approximation

$$\frac{\partial \tilde{\rho}}{\partial t} = -\frac{\partial}{\partial z}(v_{z0}\tilde{\rho} + \rho_0 \tilde{v}) = -v_{z0}\frac{\partial \tilde{\rho}}{\partial z} - \rho_0 \frac{\partial \tilde{v}}{\partial z}$$

Then

$$\frac{d\tilde{\rho}}{dt} = \frac{\partial \tilde{\rho}}{\partial t} + v_{z0}\frac{\partial \tilde{\rho}}{\partial z} = -\rho_0 \frac{\partial \tilde{v}}{\partial z} \tag{10.247}$$

Using Eqs. (10.247) and (10.245), we obtain

$$\frac{d^2\tilde{\rho}}{dt^2} = -\frac{\rho_0 \eta}{\gamma_0}\frac{\partial}{\partial z}\left(\frac{\partial}{\partial z} + \frac{v_{z0}}{c^2}\frac{\partial}{\partial t}\right)\Phi_p(z,t) \tag{10.248}$$

Taking into account that the functions $\tilde{\rho}(z,t)$ and $\Phi_p(z,t)$ include the multiplier $e^{i[(k+k_w)z+\omega t]}$, we obtain the equation

$$[v_{z0}(k + k_w) - \omega]^2 \tilde{\rho} = \frac{\eta \varepsilon_0 \omega_b^2}{2\gamma_0^2}(k + k_w)\left(k + k_w - \frac{\omega v_{z0}}{c^2}\right)A_\omega A_R \tag{10.249}$$

where $\omega_b = \sqrt{\eta|\rho_0|/\varepsilon_0}$ is the electron plasma frequency.

The radiation field satisfies the wave equation

$$\left(\frac{\partial^2}{\partial z^2} - \frac{1}{c^2}\frac{\partial^2}{\partial t^2}\right)\mathbf{A}_R = -F\mu_0 \mathbf{j}_\perp \tag{10.250}$$

where $\mathbf{j}_\perp$ is the transverse driving current, $F = (\sigma_b/\sigma_R)$ is the filling factor associated with the radiation field, and σ_b and σ_R are the cross-sectional areas of the electron beam and the radiation field, respectively. It can be shown (Robertson and Sprangle, 1989) that

$$\mathbf{j}_\perp = \tilde{\rho}\mathbf{v}_w + \rho_0 \mathbf{v}_R \tag{10.251}$$

where $\mathbf{v}_w$ and $\mathbf{v}_R$ are defined by Eq. (10.244). Using Eqs. (10.244), (10.236), and (10.241), we obtain

$$\left(k^2 - \frac{\omega^2}{c^2} + F\frac{\omega_b^2}{\gamma_0 c^2}\right)A_R = F\mu_0 \frac{\tilde{\rho}\eta}{\gamma_0}A_w \tag{10.252}$$

Let us combine Eqs. (10.249) and (10.252). After cancelling A_R, we obtain the dispersion equation of FELs in the Compton regime:

$$\left(\frac{\omega^2 - k^2c^2 - F\omega_b^2}{\gamma_0}\right)[v_{z0}(k + k_w) - \omega]^2 = F\frac{\omega_b^2}{\gamma_0}\beta_w^2 c^2 kk_w \tag{10.253}$$

where $\beta_w = v_w/c$. In the derivation of Eq. (10.253), we used the approximation $(k + k_w)(k + k_w - v_{z0}\omega/c^2) \approx 2kk_w$. The dispersion equation leads to a gain in the Compton regime (Robertson and Sprangle, 1989):

$$G = \pi F\frac{\upsilon}{\gamma_0}\beta_w^2\frac{L^3}{r_b^2\lambda_w}\frac{\partial}{\partial\theta}\left(\frac{\sin\theta}{\theta}\right)^2 \tag{10.254}$$

where $\theta = (1 - v_{z0}/v_{\text{ph}})(L\omega/2v_{z0})$, $\upsilon = (\omega_b r_b/2c)^2 = I_b/17\beta_0$ is the Budker parameter, I_b is the beam current in kiloamperes, and r_b is the beam radius.

Intrinsic Efficiency The radiation amplitude increases together with the expense of the electron's kinetic energy until the particles are trapped in the pondermotive wave. At this point, the radiation field reaches its maximum amplitude and the electrons oscillate in a potential well. That mechanism is similar to the gain saturation effect in TWTs considered in Section 8.6.2. It can be shown that the intrinsic power efficiency

$$\eta = \frac{1}{2N} = \frac{\lambda_w}{2L} \tag{10.255}$$

The efficiency can be enhanced by the use of spatially tapered wigglers (see, e.g., Freund and Neil, 1999).

10.14.4 Parameters of FELs

Parameters for some existing FELs are listed in Table 10.7. As we see, FELs use different types of electron accelerators and overlap ranges from millimeter waves up to soft x-ray waves. SC-linac is supplied by superconducting cavities placed in cryostats. The same cryostats are used to cool superconducting quadrupole magnets. Electrostatic accelerators are based on stabilized Van de Graaft generators. Powerful FELs operate on subpicosecond and tens-of-femtosecond pulses with maximum power on the order of 1 GW. Large average power is achieved in high repetition rates. The record average power was obtained at Thomas Jefferson National Accelerator Facility in Newport News, Virginia, using a tunable laser in the range of 1 to 14 μm with an average power of 10 kW. The laser had an output power of 120 MW with pulse duration of ~1 ps and a pulse repetition frequency up to 75 MHz .

The second important path in the evolution of FELs is the shortening of the wavelength up to hard x-rays. So far, most short-wave x-ray sources have utilized spontaneous synchrotron radiation of relativistic electrons, called third-generation sources. Creation of coherent sources in x-rays on the basis of an ordinary low-gain regime is not possible because of the large losses in optical cavities. Fortunately, FELs have a high-gain regime in the self-amplified spontaneous emission (SASE) mode. SASE FELs are characterized by an exponentially growing instability, starting from noise to saturation. In the initial section of a SASE FEL, electrons radiate spontaneously. Their radiation fields have no phase correlation and the intensity of radiation is proportional to the number of electrons, N_e. After saturation, electron bunches achieve maximum modulation and all electrons radiate coherently. Then the intensity of radiation is proportional to N_e^2. SASE FELs represent the fourth generation of light sources. The peak brightness created by today's FELs exceeds the brightness of third-generation sources by up to ten orders of magnitude.

An x-ray FEL concept called a *linac coherent light source* (LCLS) was designed for a wavelength of 0.15 nm (1.5 Å) and peak power of 8 GW. Because it should operate in the single-pass regime, the crucial condition for its successful realization is a very high gain. At the same time, the important parameters of the electron beam are the energy spread and the transverse emittance. The LCLS project is scheduled to begin operation in 2008 (Pellegrini and Reiche, 2004).

10.14.5 Applications of FELs

Because of their wavelength, coherence properties, tunability, intensity, and pulse duration, FELs demonstrate unprecedented capabilities for versatile applications across a variety of fields in science, technology, and industry. The span of possible applications of FELs is already considerable. Far from complete, the current list includes such areas as pharmacology, solid-state physics, chemistry and the chemical industries, civil engineering, shipbuilding, environmental sciences, defense, space-debris orbital control, polymer surface processing, micromachining, and the photophysics of complex polyatomic molecules. Especially interesting applications of FELs are in biology and biomedicine (see, e.g., Edwards et al., 2003). The latter include surgery, photo-dynamic therapy, and mammography. Very important prospects have also been opened by the development of SASE FELs that can reach the hard x-ray range.

Certainly, only the first steps in the FEL application have been made. For FELs to become a practical radiation source at the shortest wavelengths, it will be necessary to reduce the size and cost of these devices. As a result, research is being pursued actively at several centers (see, e.g., Sprangle et al., 2004).

Appendixes

APPENDIX 1 (TO SECTION 3.4.4): PROOF OF THE 3/2 LAW FOR NONRELATIVISTIC DIODES IN THE ρ-MODE

The following conditions must be satisfied:

1. Zero potential of the cathode:

$$\varphi(\mathbf{r})_{\mathbf{r}} = \mathbf{r}_c = 0 \tag{A1.1}$$

2. Anode potential:

$$\varphi(\mathbf{r})_{\mathbf{r}=\mathbf{r}_a} = \varphi_a \tag{A1.2}$$

3. Zero initial velocities:

$$v(\mathbf{r})_{\mathbf{r}=\mathbf{r}_c} = 0 \tag{A1.3}$$

4. Zero cathode electric field (ρ-mode):

$$\left[\frac{\partial\varphi(\mathbf{r})}{\partial\mathbf{r}}\right]_{\mathbf{r}=\mathbf{r}_c} = 0 \tag{A1.4}$$

---

5. Energy integral for the static electric field; with conditions (A1.1) and (A1.3), it gives

$$v = \sqrt{2\eta\varphi} \tag{A1.5}$$

In this case the system of self-consistent equations becomes

$$\Delta\varphi = -\frac{\rho}{\varepsilon_0} \tag{A1.6}$$

$$\mathbf{j} = \rho\mathbf{v} \tag{A1.7}$$

$$\operatorname{div}\mathbf{j} = 0 \tag{A1.8}$$

Let some self-consistent flow in the diode be described by the quantities $\varphi^{(1)}$, $\mathbf{v}^{(1)}$, $\rho^{(1)}$, and $\mathbf{j}^{(1)}$. Let us perform the following operation: Increase the potential in every point of the diode n times (i.e., assume the state of the diode with $\varphi^{(2)} = n\varphi^{(1)}$) and find the change in the remaining quantities of the flow. According to the consequence of the Maupertuis principle (Section 2.2.1) the shape of the particle trajectories does not change. Therefore, the velocity directions in the each point of the flow will be the same. The values of the velocities according to Eq. (A1.5) are equal: $v^{(2)} = \sqrt{2\eta\varphi^{(2)}} = \sqrt{2\eta n\varphi^{(1)}} = \sqrt{n}v^{(1)}$. Then $\mathbf{v}^{(2)} = \sqrt{n}\,\mathbf{v}^{(1)}$. According to Eq. (A1.6), the charge density becomes $\rho^{(2)} = -\varepsilon_0\,\Delta\varphi^{(2)} = -\varepsilon_0 n\Delta\varphi^{(1)} = n\rho^{(1)}$. The current density is $\mathbf{j}^{(2)} = \rho^{(2)}\mathbf{v}^{(2)} = n^{3/2}\mathbf{j}^{(1)}$. Hence, the current in the any cross section of the beam is $I^{(2)} = \int_S j_n^{(2)} d\sigma = n^{3/2}\int_S j_n^{(1)} d\sigma = n^{3/2} I^{(1)}$ and $I^{(2)}/I^{(1)} = n^{3/2} = \left[\varphi_a^{(2)}/\varphi_a^{(1)}\right]^{3/2}$. Then

$$\frac{I^{(2)}}{(\varphi_a^{(2)})^{3/2}} = \frac{I^{(1)}}{(\varphi_a^{(1)})^{3/2}} = P = \text{const.}$$

Finally, we obtain the 3/2 law: $I = P\varphi_a^{3/2}$. The perveance P is a constant that is determined by a geometry of the diode.

APPENDIX 2 (TO SECTIONS 4.5.3; 9.2.1): SYNTHESIS OF GUNS FOR M-TYPE TWTs AND BWOs

The theory of the synthesis of these guns was developed by Kino (1960). Following this work, let us set $\theta = 0$ in Eqs. (3.144)–(3.146) for nonrelativistic electron flow in a planar magnetron. Then we obtain the flow in the (Y, Z) plane perpendicular to the magnetic field with the cathode in the plane $y = 0$ (Fig. 3.9). Assume, following Kino, that the beam is limited by the space charge $E_c = 0$ (ρ-mode). Then equations of trajectories and the potential according to Eqs. (3.144)–(3.147) acquire the following form (take into account that $M = \eta j_0/\varepsilon_0\omega_g^2$ and $N = \eta E_c/\omega_g = 0$):

$$X(\Phi) = X_0 \tag{A2.1}$$

$$Y(\Phi) = M(\Phi - \sin\Phi) \tag{A2.2}$$

$$Z(\Phi) = Z_0 + M\left(\frac{\Phi^2}{2} + \cos\Phi - 1\right) \tag{A2.3}$$

$$U(\Phi) = M^2\left(\frac{\Phi^2}{2} - \Phi\sin\Phi - \cos\Phi + 1\right) \tag{A2.4}$$

Let us also write the velocity components:

$$v_y = \frac{dY}{d\Phi} = M(1 - \cos\Phi) \tag{A2.5}$$

$$v_z = \frac{dZ}{d\Phi} = M(\Phi - \sin\Phi) \tag{A2.6}$$

The planar geometry of the beam allows us to use the Lomax–Kirstein method (Section 4.5.1). First let us continue the functions $X(\Phi)$ and $Z(\Phi)$ analytically to the upper half-plane $w = \Phi + i\Psi(\Psi > 0)$ and apply conformal transformation of the plane $v = Y + iZ$ to the plane $w = \Phi + i\Psi$:

$$Y + iZ = Y(\Phi + i\Psi) + iZ(\Phi + i\Psi) \tag{A2.7}$$

Further, we can directly apply the differential equation (4.56) of the equipotentials in the (Φ,Ψ) plane:

$$\frac{d\Psi}{d\Phi} = \frac{\mathrm{Re}[E_{t0}(w) - iE_{u0}(w)]}{\mathrm{Im}(E_{t0}(w) - iE_{u0}(w)]} \tag{A2.8}$$

where $E_{t0}(w)$ and $E_{u0}(w)$ are analytically continued tangential and normal to the trajectory components of the electric field. According to Eqs. (4.45), (4.47), and (A2.4) and Fig. 4.5 (we have replaced the x-axis by the z-axis):

$$E_{t0}(\Phi) = -\frac{\partial U}{\partial t} = \omega_g\frac{dU}{d\Phi} = \omega_g M^2\Phi(1 - \cos\Phi) \tag{A2.9}$$

$$E_{u0}(\Phi) = E_{t0}(\Phi)\tan\alpha = E_{t0}(\Phi)\frac{v_z(\Phi)}{v_y(\Phi)} = \omega_g M^2\Phi(\Phi - \sin\Phi) \tag{A2.10}$$

When integrating Eq. (A2.8), we are moving along the equipotential in the (Φ,Ψ) plane. If we replace Φ and Ψ in the each point by Y and Z from Eq. (A2.7), we immediately obtain equipotentials in the real (Y,Z) plane. This problem is easily solved numerically. According to the results of the computations (Kino, 1960), the trajectories and the electrode geometry are sufficiently simple in form; however, they can be used only until $\Phi = 2\pi$. These guns are called *short-type guns*.

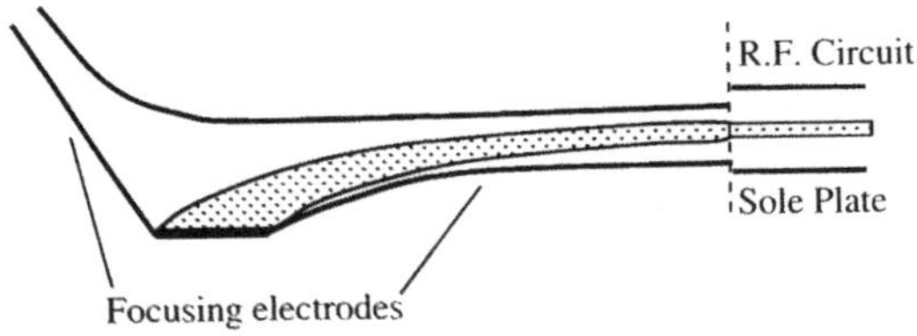

FIGURE A2.1 Long-type gun.

In another, simpler solution (solution 2), given by Kino (1960), an initial cathode velocity is assumed $(v_y)_{\Phi=0} = M$. In this case, the oscillating terms vanish in the equations above. The new equations are obtained from Eqs. (A2.2)–(A2.6) by substituting $\sin\Phi = 0$ and $\cos\Phi = 1$. We see that the trajectory acquires the form of a parabola. Note that the initial velocity $M = \eta j_0/\varepsilon_0\omega_g^2$ is usually small. The transit angle Φ of trajectories for solution 2 is not limited. Corresponding guns are called *long-type guns*. Typical trajectories and the electrode geometry of such guns are shown in Fig. A2.1.

APPENDIX 3 (TO SECTION 5.6.1): MAGNETIC FIELD IN AXIALLY SYMMETRIC SYSTEMS

Azimuthal Component of a Magnetic Field In a static approximation, Maxwell's equation [Eq. (I.5)] can be integrated around the contour in the plane perpendicular to the z-axis:

$$\int \text{curl}_z \mathbf{B}\, d\sigma = \oint B_\theta\, dl = \mu_0 \int j_z d\sigma = \mu_0 I \tag{A3.1}$$

where I is the beam current. If the contour is a circle of radius r with the center on the z-axis, we obtain from Eq. (A1.1) for axially symmetric systems

$$B_\theta(r) = \frac{\mu_0 I_r}{2\pi r} \tag{A3.2}$$

where I_r is the beam current within a circle of radius r. In a nonrelativistic approximation, we can neglect a self-magnetic field (Section 3.2.1) and assume that $I_r = 0$. Then

$$B_\theta = 0 \tag{A3.3}$$

Magnetic Vector Potential According to Eq. (1.123),

$$rA_\theta = \frac{\Psi}{2\pi} = \frac{1}{2\pi}\int B_z\, d\sigma \tag{A3.4}$$

Let us write the equation $\mathbf{B} = \text{curl}\,\mathbf{A}$ in the cylindrical coordinates

$$B_r = \frac{1}{r}\frac{\partial A_r}{\partial \theta} - \frac{\partial A_\theta}{\partial z} = -\frac{\partial A_\theta}{\partial z} \tag{A3.5}$$

$$B_\theta = \frac{\partial A_r}{\partial z} - \frac{\partial A_z}{\partial r} \tag{A3.6}$$

$$B_z = \frac{1}{r}\left[\frac{\partial (rA_\theta)}{\partial r} - \frac{\partial A_r}{\partial \theta}\right] = \frac{1}{r}\frac{\partial (rA_\theta)}{\partial r} \tag{A3.7}$$

If we take $A_r = A_z = 0$, Eq. (A3.3) will be satisfied automatically. Thus, the magnetic vector potential in the nonrelativistic approximation has only component A_θ.

In the relativistic case, the components A_r and A_z in general are not zero. In the paraxial approximation, $A_r \ll A_z$. Then, according to Eqs. (A3.6) and (A3.2),

$$\frac{\partial A_z}{\partial r} = -B_\theta = -\frac{\mu_0 I_r}{2\pi r} \tag{A3.8}$$

APPENDIX 4 (TO SECTION 8.3): DISPERSION CHARACTERISTICS OF INTERDIGITAL AND COMB STRUCTURES

Interdigital and comb structures are shown in Fig. A4.1. A rigorous analysis of the dispersion characteristics and a distribution of the electromagnetic field in these structures are complicated problems. Let us consider asymptotic solutions supposing the width of the teeth to be sufficiently small.

Interdigital (IG) Structure It is supposed that a TE wave propagates along a zigzag path. Then the phase shift on the period is equal to $\beta_0 L = \omega(L + 2h)/c$. The same phase shift would equal $\beta_0 L$, where $\beta_0 = \omega/v_{ph,0}$ is the propagation constant for the zero spatial harmonic and $v_{ph,0}$ is the phase velocity. Hence, the equality $\beta_0 = \omega(L + 2h)/cL$ is a dispersion characteristic of the zero harmonic. Taking into account that $\beta_{-1} = \beta_0 - (2\pi/L)$, we obtain the asymptotic dispersion characteristic for the -1 harmonic:

$$\frac{c}{v_{ph,-1}} = \frac{\beta_{-1} c}{\omega} = \frac{\lambda}{L} - \frac{L + 2h}{L} \tag{A4.1}$$

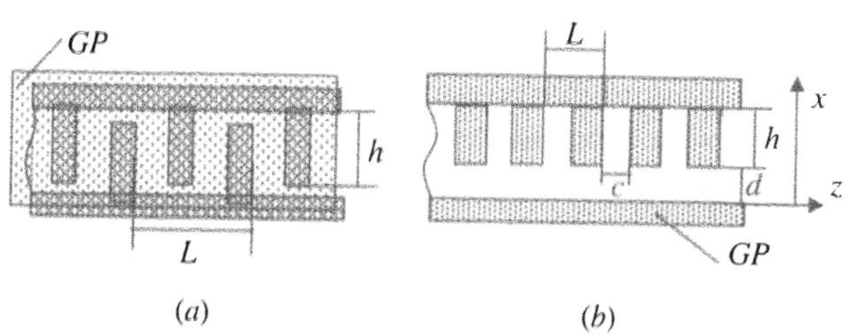

FIGURE A4.1 (*a*) Interdigital structure; (*b*) comb structure. GP, ground plate.

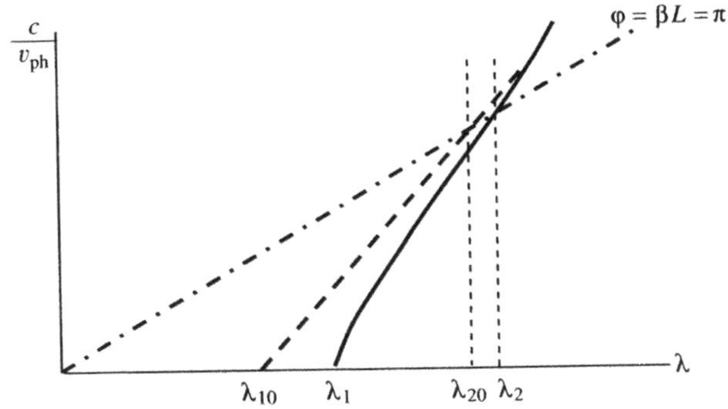

FIGURE A4.2 Dashed curve, asymptotic; solid curve, corrected. Dispersion curves of an interdigital structure.

The sign on the right side in Eq. (A4.1) is changed because $v_{ph,-1}$ should be greater than 0. We see that the delay factor $c/v_{ph,-1} \gg 1$ if $L \ll \lambda$ and $\lambda > L + 2h$. Because

$$\frac{1}{v_{gr}} = -\frac{\lambda^2}{2\pi c}\frac{\partial\beta}{\partial\lambda} \tag{A4.2}$$

we find from Eq. (A4.1) that the group velocity is negative. The dispersion curve (the straight line) corresponding to Eq. (A4.1) is depicted in Fig. A4.2. The line passes the point $\lambda_{10} = L + 2h$, where $c/v_{ph} = 0$. It is the first boundary of the bandwidth. The second boundary corresponds to the phase shift $\beta_{-1}L = \pi$. Let us take into account that the phase shift on the period

$$\phi = \beta L = \frac{c}{v_{ph}}\frac{2L}{\lambda} \quad \text{or} \quad \frac{c}{v_{ph}} = \frac{\phi}{2\pi L}\lambda$$

So the lines $\varphi =$ const. in Fig. A4.2 are straight lines intersecting the origin of coordinates. The intersection of the line $\varphi = \pi$ with the dispersion curve determines a second boundary, λ_{20}, of the bandwidth. Using Eq. (A4.1), it may readily be shown that $\lambda_{20} = 2(L + 2h)$. The delay factor for λ_{20}, $c/v_{ph} = \frac{1}{2} + (h/L) \gg 1$ when $L \ll h$. Note that according to Eq. (A4.1), the amplitude of the -1 harmonic for $\lambda > \lambda_{10} = L + 2h$ is more than the amplitude of the zero harmonic. It means that the interdigital structure has a negative dispersion (see Section 8.4.2).

A corrected analysis taking into account the presence of a grounded plate leads to a change in the dispersion curve (the solid curve in Fig. A4.2) and to a shift in the boundaries of the bandwidth to locations λ_1 and λ_2 (see, e.g., Arnaud, 1961).

Comb Structure (Fig. A4.1b) The asymptotic dispersion characteristic for the absent grounded plate and thin teeth (conditions $d \to \infty$, $L \ll h$, $kh = (2\pi/\lambda)h \ll 1$, and $r = c/L \approx 1$) is easily obtained by matching the TEM wave

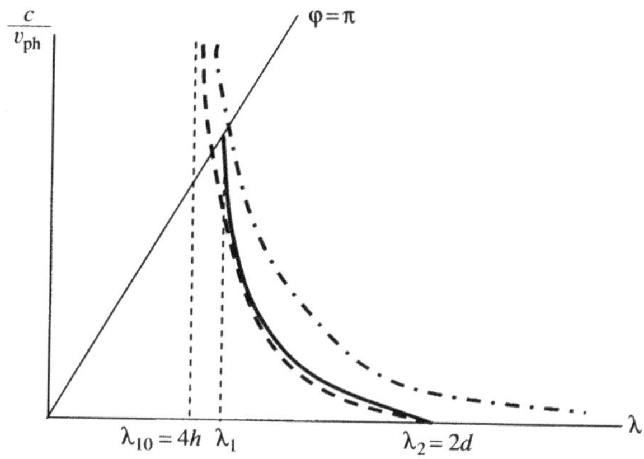

FIGURE A4.3 Dispersion characteristics of a comb structures. Dash-dotted curve, asymptotic characteristic; dashed curve, asymptotic characteristic with a grounded plate; solid curve, characteristic with finite thickness of teeth and a grounded plate.

into the comb and the delayed wave; then the delay factor for the zero harmonic is

$$\frac{c}{v_{ph}} = \frac{1}{\cos(2\pi/\lambda)h} \tag{A4.3}$$

The dispersion curve corresponding to Eq. (A4.3) is shown in Fig. A4.3. The curve stretches from $\lambda_{10} = 4h$, where the delay factor $c/v_{ph} = \infty$, to $\lambda_{20} = \infty$, where $c/v_{ph} = 1$. In latter case the wave "sees" the plane $x = d$ of the periodic structure as a smooth metal surface, and the ordinary plane wave propagates along the z-axis with velocity $v_{ph} = c$. It is obvious that the lines $\varphi = \pi$ and $\varphi = 0$ coincide with the vertical and horizontal axes, respectively. If $r = c/L < 1$, the dispersion characteristic is (Tsimring, 1957)

$$\frac{c}{v_{ph}} = \frac{\sqrt{r^2 + \cot^2 kh}}{\cot kh} \tag{A4.4}$$

When the grounded plate presents a long-wavelength boundary of the bandwidth, λ_{20} is the cutoff wavelength of the smooth waveguide $\lambda_{20} = 2d$. In this case, the delay factor $(c/v_{ph})_{\lambda_{20}} = 0$. The corresponding dispersion characteristic is shown in Fig. A4.3 as a dashed curve. Finally, finite thickness of the teeth, the short-wave boundary λ_1 grows. The resulting qualitative dispersion curve is shown in Fig. A4.3 as the solid curve. Note that the slope of the characteristics is negative. It indicates, corresponding to Eq. (A4.2), that the comb structure has a positive dispersion.

APPENDIX 5 (TO SECTION 9.2.2): ELECTROMAGNETIC FIELD IN PLANAR UNIFORM SLOW-WAVE STRUCTURES

Consider a system formed by two parallel plates (Fig. A5.1). Assume that the upper conductor is executed as a periodic structure in the z-direction and uniform in the y-direction. Let us choose some spatial harmonic that propagates in the z-direction as the wave

$$\mathbf{E}_0(x,z) = \widehat{\mathbf{E}}e^{\pm i\kappa x}e^{i(\omega t-\beta_0 z)} \tag{A5.1}$$

where $\kappa = \sqrt{k^2 - \beta_0^2}$ and $\beta_0 = \omega/v_\varphi$. For the slow harmonic, $\beta_0/k = c/v_{\rm ph} > 1$. Further, assume that the delay is large, $\beta_0 \gg k$. Then $\kappa \approx i\beta_0$ and

$$\mathbf{E}_0(x,z) = \mathbf{A}e^{\pm\beta_0 x}e^{i(\omega t-\beta_0 z)} \tag{A5.2}$$

The common solution can be represented as a linear combination of functions [Eq. (A5.2)]: that is,

$$E_{0z} = (A_1 e^{\beta_0 x} + A_2 e^{-\beta_0 x})e^{i(\omega t-\beta_0 z)} \tag{A5.3}$$

Supposing that the sole plate ($x = 0$) is an ideal conductor [i.e., $(E_{0z})_{x=0}$], we find that $A_1 = -A_2 = A/2$. Then

$$E_{0z} = A sh\beta_0 x e^{i(\omega t-\beta_0 z)} \tag{A5.4}$$

It is readily shown from the equation div $\mathbf{E}_0 = 0$ that

$$E_{0x} = iA ch\beta_0 x e^{i(\omega t-\beta_0 z)} \tag{A5.5}$$

Let us enter the potential

$$U_d = -\int_0^d |E_{0x}(x,z)|\,dx = -A\frac{sh\beta_0 d}{\beta_0} \tag{A5.6}$$

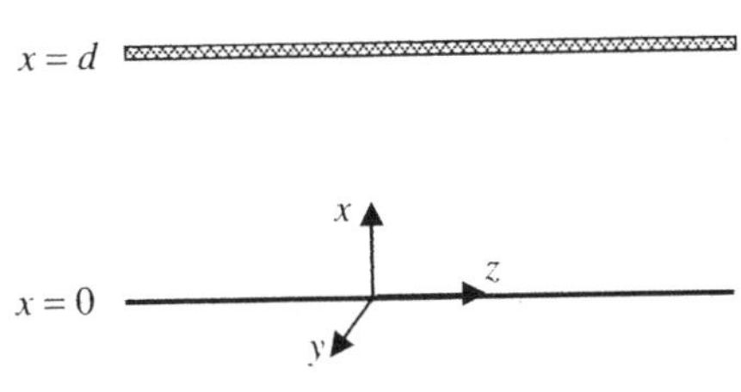

FIGURE A5.1 Uniform planar SWS.

Expressing A through U_d, we obtain the following representation for components of the electric field:

$$E_{0x} = -\beta_0 U_d \frac{ch\beta_0 x}{sh\beta_0 d} e^{i(\omega t - \beta_0 z)} \quad \text{(A5.7)}$$

$$E_{0z} = i\beta_0 U_d \frac{sh\beta_0 x}{sh\beta_0 d} e^{i(\omega t - \beta_0 z)} \quad \text{(A5.8)}$$

Equations (A5.7) and (A5.8) allow us to represent the electric field as the *gradient* of the potential:

$$U(x, z, t) = U_d \frac{sh\beta_0 x}{sh\beta_0 d} e^{i(\omega t - \beta_0 z)} \quad \text{(A5.9)}$$

As is easily seen, this is a consequence of a transition of the wave equation to a Laplace equation for the large delay (if neglect a space charge). So

$$E_{0x} = -\frac{\partial U}{\partial x}, \qquad E_{0z} = -\frac{\partial U}{\partial z} \quad \text{or} \quad \mathbf{E}_0 = -\nabla U(x,z) \quad \text{(A5.10)}$$

APPENDIX 6 (TO SECTION 10.5.2): EQUATIONS OF FREE OSCILLATIONS OF GYROTRON RESONATORS

Assume that fields and current are harmonic functions of time. Then the complex amplitude of the first current harmonic is

$$\mathbf{j}(\mathbf{r},\omega) = \frac{1}{\pi}\int_0^{2\pi} \mathbf{j}(\mathbf{r},t)e^{-i\omega t}d\omega t \quad \text{(A6.1)}$$

and the first current harmonic is

$$\mathbf{j}(\mathbf{r}, t) = \text{Re}\,\mathbf{j}(\mathbf{r},\omega)e^{i\omega t} \quad \text{(A6.2)}$$

Complex amplitudes $\mathbf{E}(\mathbf{r},\omega)$ and $\mathbf{H}(\mathbf{r},\omega)$ can be determined analogously. Maxwell's equations [Eqs. (I.5), (I.6)] via complex amplitudes assume the form

$$\text{curl } \mathbf{B} = \mu_0 \mathbf{j} + \frac{i\omega}{c^2}\mathbf{E} \quad \text{(A6.3)}$$

$$\text{curl } \mathbf{E} = -i\omega \mathbf{B} \quad \text{(A6.4)}$$

Let us consider free oscillations ($\mathbf{j} = 0$):

$$\begin{aligned} \text{curl } \mathbf{B} &= \frac{i\omega}{c^2}\mathbf{E} \\ \text{curl } \mathbf{E} &= -i\omega \mathbf{B} \end{aligned} \quad \text{(A6.5)}$$

Combinating Eqs. (A6.5) gives the Helmholtz equations:

$$(\Delta + k^2)\begin{vmatrix} \mathbf{E} \\ \mathbf{B} \end{vmatrix} = 0 \tag{A6.6}$$

where $k = \omega\sqrt{\varepsilon_0\mu_0} = \omega/c$. The Laplacian operator $\Delta = \text{curl}\,\text{curl}$ for waveguides with a rectilinear z-axis can be written as

$$\Delta = \nabla_\perp^2 + \frac{\partial^2}{\partial z^2} \tag{A6.7}$$

where $\nabla_\perp$ is an operator of the transverse gradient. This operator is

$$\nabla_\perp = \frac{\partial}{\partial r}\mathbf{i}_r + \frac{1}{r}\frac{\partial}{\partial \varphi}\mathbf{i}_\varphi$$

in cylindrical coordinates.

It is convenient to choose the axis component of the magnetic field B_z as the fundamental function for TE modes. Transverse field components are expressed simply through B_z (Jackson, 1999):

$$\mathbf{B}_\perp = -\frac{i\beta}{\kappa^2}\nabla_\perp B_z, \qquad \mathbf{E}_\perp = -\frac{i\omega}{\kappa^2}\mathbf{i}_z \times \nabla_\perp B_z \tag{A6.8}$$

where $\kappa \equiv \kappa_{mn}$ is an eigenvalue of the membrane equations (10.52) and (10.53) for the TE_{mn} mode. The Helmholtz equation in respect to B_z has the form

$$\nabla_\perp^2 B_z + \frac{\partial^2 B_z}{\partial z^2} + k^2 B_z = 0 \tag{A6.9}$$

A transverse distribution of the electromagnetic field in any cross section of a nonuniform waveguide in the frame of the transverse cross-section method (Katsenelenbaum, 1998) can be represented as an expansion into a series of modes of a regular waveguide with the same cross section (the latter is called a *comparison waveguide*). This expansion (Tsimring and Pavelyev, 1982) describes the effects of mode transformation when an incident wave is transformed in a set of modes. However, for sufficiently weak uniformity of the waveguide, the transformation can be neglected, and the incident wave is subject only to reflection. The corresponding equation of the cross-section method is pretty simple and is written as

$$B_z(\mathbf{r}_\perp, z) = C(z)B_{z\perp}(\mathbf{r}_\perp) \tag{A6.10}$$

where $B_{z\perp}$ is a membrane function of a comparison waveguide that satisfies the equation following from Eq. (A6.9):

$$\nabla_\perp^2 B_{z\perp} + \kappa^2 B_{z\perp} = 0 \tag{A6.11}$$

with the boundary condition

$$\frac{\partial B_{z\perp}(\mathbf{r}_\perp)}{\partial n} = 0 \tag{A6.12}$$

on the cross-section contour. Nonzero solutions of Eqs. (A6.11) and (A6.12) exist for discrete wave numbers κ that form a spectrum of eigenvalues of the comparison waveguide. For an axially symmetric waveguide,

$$B_z(r,\varphi,z) = C(z)I_m(\kappa r)e^{-im\varphi} \tag{A6.13}$$

where I_m is a Bessel function of the first kind of mth order. The boundary condition in this case is

$$\left.\frac{dI_m(\kappa r)}{dr}\right|_{r=R} = 0 \tag{A6.14}$$

Because the radius of the irregular waveguide in each cross section is different [i.e., $R = R(z)$], the eigenvalues $\kappa = \kappa(z)$.

Let us transform Eq. (A6.13) to Cartesian variables with the help of an integral representation of a Bessel function (Bateman and Erdelyi, 1953):

$$\mathrm{I}_m(\kappa r) = \frac{1}{2\pi}\int_0^{2\pi} e^{i(\kappa r \sin\alpha - m\alpha)}\,d\alpha = i^{-m}\frac{1}{2\pi}\int_0^{2\pi} e^{i(\kappa r\cos\alpha + m\alpha)}\,d\alpha \tag{A6.15}$$

Then the function B_z from Eq. (A6.13) can be written as

$$B_z(r,\varphi,z) = i^{-m}C(z)\frac{1}{2\pi}\int_0^{2\pi} e^{i[\kappa r\cos\alpha + m(\alpha-\varphi)]}\,d\alpha \tag{A6.16}$$

Changing the variable $\alpha = \varphi - \psi$ and using the formula for $\cos(\varphi - \psi)$, it may readily be shown that

$$B_z(r,\varphi,z) = B_z(x,y,z) = \int_0^{2\pi} A(\psi,z)e^{i\kappa(z)(x\cos\psi + y\sin\psi)}\,d\psi \tag{A6.17}$$

where

$$A(\psi,z) = i^{-m}\frac{C(z)}{2\pi}e^{-im\psi} = \frac{C(z)}{2\pi}e^{-im[(\pi/2)+\psi]} \tag{A6.18}$$

It remains to find the amplitude $C(z)$. Substituting Eqs. (A6.10) and (A6.11) into Eq. (A6.9), we obtain

$$\frac{d^2C}{dz^2} + \beta^2(z)C = 0 \tag{A6.19}$$

Equation (A6.19) is distinguished from an analogous equation for a regular waveguide for which the value $\beta^2 = k^2 - \kappa^2$ is a constant; here, $\beta^2(z) = k^2 - \kappa^2(z)$ is a function of z that is found after solution of the boundary problem equations (A6.11) and (A6.12). In mathematical physics Eq. (A6.19) is called the equation of a nonuniform vibrating string. Solution of this equation is determined by boundary conditions. Free oscillations of the resonator that is formed as a section of the nonuniform waveguide is determined by the radiation condition at the end cross sections $z = z_1$ and $z = z_2$ (Fig. 10.12). If these cross sections are localized on the sections of a uniform waveguides, the radiation conditions can be written as

$$\left.\frac{dC(z)}{dz}\right|_{z=z_1} = i\beta C(z_1), \qquad \left.\frac{dC(z)}{dz}\right|_{z=z_2} = -i\beta C(z_2) \tag{A6.20}$$

Nontrivial solutions of Eq. (A6.19) with conditions (A6.20) exist for some discrete complex eigenvalues $\omega = \omega_0 = \omega_{01} + i\omega_{02}$, which determine the spectrum of free frequencies of the resonator. Because the oscillations are damping, $\omega_{02} > 0$. The problem is reduced totally to a solution of Eqs. (A6.11) and (A6.19) with boundary conditions (A6.14) and (A6.20). Generally, numerical methods of the solution are used.

APPENDIX 7 (TO SECTION 10.5.3): DERIVATION OF EQS. (10.66) AND (10.67)

Let us write fields according to an approximation of the fixed field structure:

$$\mathbf{E} = M\mathbf{E}_0, \qquad \mathbf{B} = N\mathbf{B}_0 \tag{A7.1}$$

where M and N in general are complex constants. It is readily verified that

$$M = \frac{\omega}{\omega_0} N \tag{A7.2}$$

To obtain Eq. (A7.2), it is sufficient to substitute Eqs. (A.7.1) into the Maxwell equation $\operatorname{curl}\mathbf{E} = -i\omega\mathbf{B}$ and use the equation $\operatorname{curl}\mathbf{E}_0 = -i\omega_0\mathbf{B}_0$. Substituting Eqs. (A7.1) and (A7.2) into Eqs. (10.61) and (10.62) with $\omega_1 = \omega$ and $\omega_2 = 0$, we obtain

$$\omega|M|^2 \int_V \left(\varepsilon_0|\mathbf{E}_0|^2 - \left|\frac{\omega_0}{\omega}\right|^2 \frac{1}{\mu_0}|\mathbf{B}_0|^2\right) d\tau$$

$$= -|M|^2 \frac{\omega_{02}}{\omega}\frac{1}{\mu_0} \int_S (\mathbf{E}_0^* \times \mathbf{B}_0)_n \, d\sigma - \operatorname{Im}\int_V \mathbf{j}\mathbf{E}^* \, d\tau \tag{A7.3}$$

$$= |M|^2 \frac{\omega_{01}}{\omega}\frac{1}{\mu_0} \int_S (\mathbf{E}_0^* \times \mathbf{B}_0)_n \, d\sigma + \operatorname{Re}\int_V \mathbf{j}\mathbf{E}^* \, d\tau \tag{A7.4}$$

Using Eqs. (10.63) and (10.63*a*), we find from Eqs. (A7.3) and (A7.4) that

$$\frac{1}{2}\omega|M|^2W_0\left[1-\left|\frac{\omega_0}{\omega}\right|^2+4\left(\frac{\omega_{02}}{\omega}\right)^2\right]=-\frac{1}{2}\text{Im}\int_V \mathbf{jE}^*\,d\tau \tag{A7.5}$$

$$2|M|^2W_0\omega_{02}\frac{\omega_{01}}{\omega}=-\frac{1}{2}\text{Re}\int_V \mathbf{jE}^*d\tau \tag{A7.6}$$

Taking into account that $W=\frac{1}{2}\int_V(\varepsilon_0|\mathbf{E}|^2+(1/\mu_0)|\mathbf{B}|^2)\,d\tau$ and using Eqs. (10.63), (A7.1), and (A7.2), we obtain

$$W=\frac{1}{2}|M|^2W_0\left(1+\left|\frac{\omega_0}{\omega}\right|^2\right) \tag{A7.7}$$

Substituting Eq. (A7.7) into Eqs. (A7.5) and (A7.6) gives

$$\omega\left[\left(1+\frac{\omega_{01}}{\omega}\right)\left(1-\frac{\omega_{01}}{\omega}\right)+3\left(\frac{\omega_{02}}{\omega}\right)^2\right]\frac{2W}{1+|\omega_0/\omega|^2}=-\frac{1}{2}\text{Im}\int_V \mathbf{jE}^*\,d\tau \tag{A7.8}$$

$$\omega_{02}\frac{\omega_{01}}{\omega}\frac{4W}{1+|\omega_0/\omega|^2}=-\frac{1}{2}\text{Re}\int_V \mathbf{jE}^*\,d\pi \tag{A7.9}$$

For a high Q-factor, we can assume that $\omega_{01}/\omega\approx|\omega_0/\omega|\approx 1$ and $(\omega_{02}/\omega)^2\ll 1-\omega_{01}/\omega$. Then Eqs. (A7.8) and (A7.9) are reduced to

$$2(\omega-\omega_{01})W=\text{Im}(P) \tag{A7.10}$$

$$2\omega_{02}W=\text{Re}(P) \tag{A7.11}$$

where a complex power that the current gives up to the resonator is

$$P=-\frac{1}{2}\int_V \mathbf{jE}^*d\tau=-\frac{1}{2\pi}\int_V\int_0^{2\pi}\mathbf{j}(\mathbf{r},t)\mathbf{E}^*e^{-i\omega t}\,d\omega t\,d\tau \tag{A7.12}$$

APPENDIX 8 (TO SECTION 10.6.3): CALCULATION OF FOURIER COEFFICIENTS IN GYROTRON EQUATIONS

The averaged ("shortened") equations (10.91)–(10.94) contain Fourier coefficients of the force component: G_x, G_y, $G_\theta=G_x\sin\theta-G_y\cos\theta$, and $G_r=G_x\cos\theta+G_y\sin\theta$, where the force $\mathbf{G}$ according to Eq. (10.77) is

$$\mathbf{G}=\frac{ic}{k}\nabla_\perp\times\mathbf{i}_zB_z+\mathbf{v}_\perp\times\mathbf{i}_zB_z \tag{A8.1}$$

A common relation for the Fourier coefficients of these components is determined by Eq. (10.84) and after using Eq. (A8.1) will contain the integrals

$$\frac{1}{2\pi}\int_0^{2\pi}F_s(\theta)e^{in\theta}\,d\theta \tag{A8.2}$$

where $F_s(\theta)$ contains B_z, $\partial B_z/\partial x$, $\partial B_z/\partial y$, and trigonometrical functions of θ. Let us consider as an example, calculation of the integral $B_{zn} = 1/2\pi \int_0^{2\pi} B_z(x,y,z)e^{in\theta}\, d\theta$. Using Eqs. (A6.17) and (A6.18), we obtain

$$B_{zn} = \frac{1}{2\pi}\int_0^{2\pi}\int_0^{2\pi} A(\psi,z)e^{i\kappa(z)(x\cos\psi + y\sin\psi)+in\theta}\, d\psi\, d\theta \qquad \text{(A8.3)}$$

where $A(\psi,z) = -i^n[C(z)/2\pi]e^{-in\psi}$. The coordinates x and y of the electron on the helical trajectory are determined by Eqs. (10.78). Then the integral in Eq. (A8.3) takes the form

$$B_{zn} = \int_0^{2\pi} A(\psi,z)e^{i\kappa(z)(X\cos\psi + Y\sin\psi)}d\psi \frac{1}{2\pi}\int_0^{2\pi} e^{i\kappa(z)r_\perp \cos(\theta-\psi)+in\theta}\, d\theta \qquad \text{(A8.4)}$$

The inner integral according to Eq. (A6.15) is $i^n I_n(\kappa r_\perp)e^{in\psi}$, where I_n is a Bessel function of the first kind of order n. We obtain

$$B_{zn} = i^n I_n(\kappa r_\perp)\int_0^{2\pi} A(\psi,z)e^{i\kappa(X\cos\psi + Y\sin\psi)+in\psi}\, d\psi \qquad \text{(A8.5)}$$

According to Eq. (A6.17),

$$B_z(X,Y,z) = \int_0^{2\pi} A(\psi,z)e^{i\kappa(X\cos\psi + Y\sin\psi)}\, d\psi \qquad \text{(A8.6)}$$

Then the integral in Eq. (A8.5) can be calculated by applying the operator $D = (1/\kappa)[(\partial/\partial X) + i(\partial/\partial Y)]$ n times successively to the integral in Eq. (A8.6):

$$D^n B_z(X,Y,z) = i^n \int_0^{2\pi} A(\psi,z)e^{i\kappa(X\cos\psi + Y\sin\psi)+in\psi} \qquad \text{(A8.7)}$$

Thus,

$$B_{zn} = I_n(\kappa r_\perp)L_n(X,Y,z) \qquad \text{(A8.8)}$$

where $L_n(X,Y,z) = D^n B_z(X,Y,z)$

Other Fourier coefficients are found analogously. Finally, we obtain

$$\begin{aligned} G_{\theta n} &= -ic\frac{dI_n}{d\xi}L_n, & G_{rn} &= -\frac{c}{n}\frac{d}{d\xi}\left(\xi\frac{dI_n}{d\xi}\right)L_n \\ G_{xn} &= -\frac{v_\perp}{k}\frac{dI_n}{d\xi}\frac{\partial L_n}{\partial X}, & G_{yn} &= -\frac{v_\perp}{k}\frac{dI_n}{d\xi}\frac{\partial L_n}{\partial Y} \end{aligned} \qquad \text{(A8.9)}$$

Here $\xi = \kappa r_\perp$. The value $\beta_\perp = v_\perp/c$ and the derivation of Eq. (A8.9) was supposed equal to ξ/n. This is satisfied for the gyroresonance when $\omega = n\omega_g$ and $\kappa = k$ (i.e., for a cutoff resonator).

APPENDIX 9 (TO SECTION 10.8.2): MAGNETIC SYSTEMS OF GYROTRONS

Operational frequencies of contemporary gyrotrons cover frequency bands on the order of 3 to 300 GHz. According to the Doppler resonance condition (10.46*a*), a magnetic field in the interaction space of the gyrotrons must be close to

$$B_T \approx \frac{10.7}{\lambda_{mm}} \frac{\gamma}{n} \tag{A9.1}$$

and have values of ~0.1 to 10 T for interaction at the fundamental cyclotron resonance. Basic types of gyrotron magnets are described briefly below; our choice is determined by a gyrotron's frequency bands, construction, and operation requirements.

Resistive (Water Cooling or Oil Cooling) Solenoids These devices were the first magnets used in Russian gyrotrons, beginning in 1965. At present, these magnets are used in technological gyrotrons on frequency bands K and K_a operating basically on the second cyclotron harmonic. In this case, the magnetic field required does not exceed 0.7 T. In principle, resistive magnets are available for creating much higher magnetic fields. However, energy consumption, power supply, and cooling systems for these solenoids become very sophisticated. So a 30-T resistive magnet (the world's highest-field dc resistive magnet) required a 20-MW power supply (Mossang et al., 2005).

Liquid Helium–Cooled Superconducting Magnets The use of these solenoids was a miletone in the development of powerful short-wave gyrotrons (see, e.g., Zaitsev et al., 1974). At present, almost all gyrotrons operating in the frequency bands 40 to 250 GHz are equipped with liquid helium magnets.

The magnet is depicted in Fig. A9.1. The superconducting coil is inserted into a helium container that is, in turn, inserted into a liquid nitrogen container. The central bore is at room temperature. Most of the magnets are wound by conductors

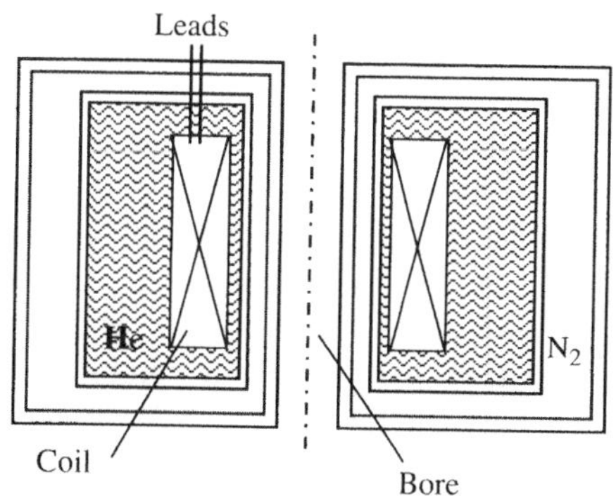

FIGURE A9.1 Liquid helium superconductive magent.

comprised of many fine filaments of NbTi alloy embedded in a copper matrix. Liquid helium has a normal boiling temperature of 4.2 K, corresponding to atmospheric pressure. In this case the critical magnetic field of an NbTi superconductor is close to 9 T. Large magnets could have a bore diameter above 100 mm for a B value of ~5 to 7 T and a length of ~1 m. Multifilamentary Nb_3Sn superconductors are now being used. The critical magnetic field of a Nb_3Sn superconductor is 10 T for a temperature of 4.2 K and 16 T for a temperature of 2.5 K. This superconductor is expensive, and the corresponding magnets cost substantially more than NbTi magnets.

The main disadvantage of liquid helium magnets is the need to in refill the liquid helium. The basic causes of helium evaporation lie in the thermal conductivity and in the electric resistance of the leads that connect the superconducting coil with a current source.

Liquid Helium–Free Superconducting Magnets (Watanabe et al., 1996) The current leads in these magnets are executed as coils of a high-temperature superconductor Bi2223 wound on ceramic cylinders. The warm ends of the leads are at liquid nitrogen temperature. The critical temperature of the Bi2223 is higher than the boiling point of liquid nitrogen at 1 atm (77 K). As a result, the heat leakage of a magnet with Bi2223 leads drops almost to zero. Now very large liquid helium–free magnets (LHFMs) are cooled by comparatively small Gifford–McMahon (GM) cryocoolers with a cooling capacity on the order of 1 to 10 W until a temperature of 4 K and lower. Using the superconductor Nb_3Sn achieves a 15-T field.

At present, LHFM magnets are produced by Japanese firms. Jastec, Inc. (Japan Superconductor Technology, Inc.) produces 12-T magnets with a room-temperature bore diameter of 100 mm. The magnet container's diameter and height are 740 and 650 mm, respectively. The magnets could operate at any angle with respect to vertical and are practically maintenance-free. A 300-GHz gyrotron with a CW output power of 3.5 kW equipped by a LHFM magnet system was used successfully for a microwave procession of materials (see Hoshizuki et al., 2005; Kuftin et al., 2000; and Section 10.12.3).

Hybrid Magnets These magnets can produce the highest dc magnetic fields. They are composed of outside coils of a low-temperature superconductor such as an Nb_3Sn superconductor and insert coils with a high critical magnetic field. Two types of hybrid systems are known. In the first system the inner coil is executed as a resistive water-cooling magnet. The high fields in this case are attained with considerable lower cost than with purely resistive magnets. The highest steady magnet field was built at the National High Magnetic Field Laboratory in Tallahasee, Florida. This 45-T hybrid system combines outer superconducting coils that produce a 14-T field and an insert 30-MW resistive magnet producing a field of 31 T (Bird, 2004).

In the second system the inner insert coil is manufactured of high-temperature superconductor Bi2223 that has an extraordinarily high upper critical magnetic field. So a hybrid liquid helium–free magnet with an 19-T field and a 52-mm bore has been designed (Kurusu et al., 2004). The outer coil is composed of Nb_3Sn and NbTi coils and delivers a maximum magnetic field of 16.5 T.

Permanent Magnets The use of permanent magnets in gyrotrons began in the 1990s (see, e.g., Kuftin et al., 1992a). At present, the most promising material for permanent magnets from the stand point of coercive force, cost saving, and temporal and temperature stability is the rare-earth alloy NdFeB. NdFeB permanent magnets used in gyrotrons provide a magnetic field of more than 1 T in a bore up to 60 mm. The frequency generated by gyrotrons with these magnets on the second harmonic of the gyrofrequency is near 56 GHz. Similar gyrotrons can be developed for CW output power up to 50 kW and are very promising technological oscillators for microwave material processing. A high harmonic gyrotron with an axis-encircling electron beam and a permanent magnet delivered on a fourth harmonic an output power of 3.5 kW in a 1-ms pulse and a frequency of 112 GHz (Idehara et al., 2004).

A principal point in a construction of a gyrotron with a permanent magnet was ensuring available distribution of the magnetic field with sufficient uniformity over the length of the interaction space and with adiabatic reduction of the field to the gun and collector regions. A solution has been found by forming an extended area in the magnet with radial magnetization (see, e.g., Kuftin et al., 2000). In Fig. A9.2, the structure of a 1 T permanent magnet is depicted.

Pulsed Magnets Pulsed magnets are divided into two categories: nondestructive magnets with a field below 100 T and a pulse duration in the range 1 ms to 1 s, and self-destruction coils with a field above 100 T and a pulse duration of

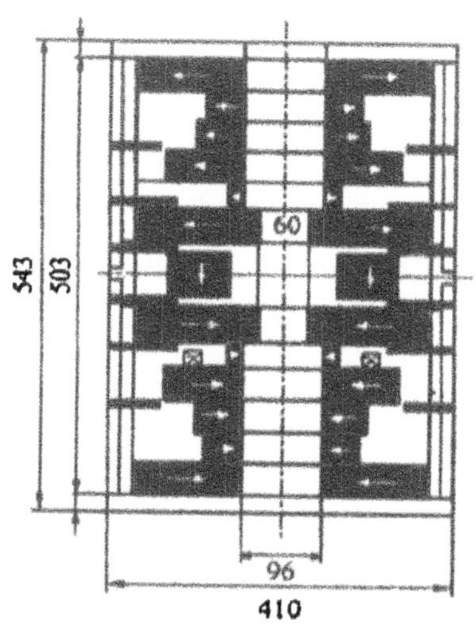

FIGURE A9.2 1-T permanent magnet. The magnetization of each element is shown by the arrows. Dimensions are in millimeters. (From Idehara et al., 2004.)

microseconds (see, e.g., Herlach, 1996). For nondestructive magnets the most important parameters of the coil material are mechanical strength and electrical conductivity. The energy density and corresponding stress due to Lorentz forces depend quadratically on the magnetic field. At 100 T the stress is up to 4 GPa. It is equal to the mechanical strength of the strongest materials presently available. By comparison, the mechanical strength of copper is on order of 0.1 GPa. Another important parameter is the size of the magnet bore. Using the most advanced materials and a sophisticated coil design obtains peak fields of about 75 T in $\frac{1}{2}$-inch bores (Herlach, 1996). Fields up to 60 T are available for routine experiments.

The use of a strong pulsed magnetic field for gyrotrons as for other cored microwave electron devices is complicated by the following contradictory requirements. On the one hand, conductivity of the metallic sheath of the cavity should be low enough and the pulse duration large enough to penetrate the magnetic field in a resonator. On the other hand, conductivity of the inner surface of the resonator should be high, to reduce ohmic losses.

In the first successful experiments with powerful submillimeter gyrotrons provided by pulsed magnets (Luchinin et al., 1983; Nusinovich, 2004), the tube body was made of thin stainless steel covered on the inner surface by a thin layer of copper (10 to 20 μm). The pulse duration of the magnet field was about 10 ms. Luchinin et al. (1983) have achieved a radiation wavelength of ~0.6 mm in a magnetic field of about 20 T on the fundamental gyroresonance. Microwave power of 100 kW at 8% efficiency for a voltage of 70 kV and 50-μs pulses has been obtained.

References

Abe, K., et al. (1998). Experimental studies of overmoded relativistic backward-wave oscillators, *IEEE Trans. Plasma Sci.*, **26**, 551.

Abramowitz, M., and I. Stegun (1972). *Handbook of Mathematical Functions with Formulas, Graphs and Mathematical Tables*, Wiley, New York.

Abubakirov, E.B., V.I. Belousov, V.N. Varganov, et al. (1983). Experimental realization of cyclotron-resonance mode selection in Cerenkov relativistic HF sources, *Sov. Tech. Phys. Lett.*, **9**(9–10).

Abubakirov, E.B., A.N. Denisenko, U.M. Fuks, et al. (2002). An X-band gigawatt amplifier, *IEEE Trans. Plasma Sci.*, **30**, 1041.

Agdur, B. (1957). On the interaction between microwave field and electrons, with special reference to the strophotron, *Ericsson Tech.*, **13**(1–3).

Aksenchik, A.V., S.V. Kolosov, A.A. Kuraev, and V.P. Shestakovich (1982). Results of optimization multicavity klystrons efficiency, *Radio Eng. Electron. Phys.*, **27**(12).

Alekseev, N., and D. Malairov (1944). Generation of high power oscillations with a magnetron in the centimeter band, *Proc. IRE*, **32**, 136.

Alfven, H., and D. Romell (1954). A new electron tube: the Strophotron, *Proc. IRE*, **42**, 1239.

Alikaev, V.V., G.A. Bobrovskiy, V.I. Poznyk, and K.A. Razumova (1972). Electron cyclotron heating at the TM-3 tokmak, *JETP Lett.*, **15**, 27.

Andersen, J.P., M.A. Shapiro, R.J. Temkin, and I. Mastovsky (2004). Operation of a 1.5 MW 110 GHz gyrotron depressed collector experiment, *Conference Digest of the Joint 29th International Conference on Infrared and Millimeter Waves and 12th Conference on Terahertz Electronics*, IEEE Press, Piscataway, NJ, p. 155.

Andronov, A.N., et al. (1995). Formation and diagnostic of helical gyrotron electron beams, in *Conference Digest of the 20th Conference on Infrared and Millimeter Waves*, Orlando, FL, pp. 141, 155.

Anon. (2003). Development of TWT for civic and military applications, *Microwave J.*, June.

Electron Beams and Microwave Vacuum Electronics. By Shulim E. Tsimring

Antakov, I.I., V.M. Bokov, R.P. Vasil'ev, and A.V. Gaponov (1960). Interaction of trochoidal beam with electromagnetic wave in a rectangular waveguide, *Izv. Vuzov's Radiofiz.*, **3**(6) (in Russian).

Antakov, I.I., A.V. Gaponov, O.V. Malygin, and V.A. Flyagin (1966). Application of induced cyclotron radiation of electrons for the generation and amplification of high-power electromagnetic waves, *Radio Eng. Electronic. Phys.*, **11**(12).

Antakov, I.I., A.V. Gaponov, V.I. Gintsburg, et al. (1967). An amplifier of electromagnetic oscillations in centimeter, millimeter, and sub-millimeter wavelength regions, Copyright 302050, June 16, *Off. Bull. KDIO SM USSR*, No. 41: 205(675).

Antakov, I.I., M.A. Moiseev, E.V. Sokolov, and E.V. Zasypkin (1994). Theoretical and experimental investigation of X-band two-cavity gyroklystrons, *Int. J. Infrared Millimeter Waves*, **15**, 873.

Antonsen, T.M., and B. Levush (1986). Influence of thermal spread on space-charge limiting current, *Int. J. Electron.*, **61**, 871.

Arnaud, J. (1961). Circuits for traveling wave crossed-field tubes, in *Crossed-Field Microwave Devices*, Vol. 1, ed. E. Okress, Academic Press, New York, p. 1.

Arnold, V.I., V.V. Kozlov, and A.T. Neistadt (1988). Mathematical aspects of classical and celestial mechanics, in *Dynamical Systems III*, Vol. 1, ed. V.I. Arnold, Springer-Verlag, New York.

Avdoshin, E.G., L.V. Nikolaev, I.N. Platonov, and Sh.E. Tsimring (1973). Experimental investigation of the velocity spread in helical electron beams, *Radiophys. Quantum Electron.*, **16**, 461.

Avdoshin, E.G., A.V. Melnikov, and Sh.E. Tsimring (1975). Influence of disturbances of axial symmetry in systems, forming helical beams, on electron velocity spread, *Elektron. Tekh., Ser. 1, Elektron. Sverch–Vysokie Chastoty (SVCh)*, **8**, 67 (in Russian).

Balaklitskii, I., I.D. Revin, B.K. Skrynik, et al. (1973). Quasi-optical generator of diffraction radiation, *Radiophys. Quantum Electron.*, **16**, 235.

Barker, R.J., and E. Schamiloglu (2001). *High Power Microwave Sources and Technologies*, IEEE Press, New York.

Barker, R.J., J.H. Booske, N.C. Luhmann, and G.S. Nusinovich (2005). *Modern Microwave and Millimeter-Wave Power Electronics*, IEEE Press, Piscataway, NJ.

Barkhausen, H., and K. Kurz (1920). Die Kurzesten, vakuumröhren herstellbaren Wellen, *Phys. Zeitung* **21**, 1 (in German).

Baryshev, Y.V., I.V. Kazarezov, E.V. Kozyrev, et al. (1994). A 100 MW electron source with extremely high beam area compression, *Nucl. Instrum. Methods Phys. Res.*, **A340**, 241.

Bateman, H., and A. Erdelyi (1953). *Higher Transcendental Functions*, Vol. 2, McGraw-Hill, New York.

Bekefi, G., and T.J. Orzechowski (1976). Giant microwave bursts emission from field emission, relativistic-electron-beam magnetron, *Phys. Rev. Lett.*, **37**, 379.

Belomytsev, S.Ya., E.A. Litvinov, G.A. Mesyats, and A.I. Fedosov (1981). Characteristics of the electron beam formed in magnetic-insulation diodes, *Sov. J. Plasma Phys.*, **7**, 48.

Benford, J. (1987). Relativistic magnetrons, in *High Power Microwave Sources*, ed. V.L. Granatstein and I. Alexev, Artech House, Norwood, MA.

Benford, J. (1998). Lowered plasma velocity with cesium iodine carbon fiber cathodes at high electric fields, *Proceedings of the 12th International Conference on High Power Particle Beams*, Haifa, Israel, Vol. 2, p. 695.

Benford, J., and G. Benford (1997). Survey of pulse shortening in high-power microwave sources, *IEEE Trans. Plasma Sci.*, **25**, 311.

Benford, J., and J. Swegle (1992). *High-Power Microwaves*, Artech House, Norwood, MA.

Benford, J., H. Sze, W. Woo, R. Smith, and B. Harteneck (1989). Phase locking in relativistic magnetrons, *Phys. Rev. Lett.*, **62**, 869.

Bernstein, M.J., and N.M. Kroll (1961). Conventional pulsed rising sun magnetrons, in *Cross-Field Microwave Devices*, Vol. 2, ed. E. Okress, Academic Press, New York, p. 220.

Beunas, A., and G. Fallon (1968). 10MW/1.5 ms L-band multi-beam klystron, *Proceedings of the Conference on Display and Vacuum Electronics*, No. 4, Barmisch-Partenkirchen, Germany.

Bevensee, R. (1964). *Electromagnetic Slow Wave Systems*, Wiley, New York.

Bird, M.D. (2004). Resistive magnet technology for hybrid inserts, *Supercond. Sci. Technol.*, **17**, 17.

Birdsall, C., and W. Bridges (1966). *Electron Dynamics of Diode Regions*, Academic Press, New York.

Black, W.M., S.H. Gold, A.W. Fliflet, et al. (1990). Megavolt, multikiloamp K_a band gyrotron oscillator experiment, *Phys. Fluids*, **B2**, 193.

Blank, M., B.G. Danly, and B. Levush (2000). Experimental demonstration of W-band gyro-klystron amplifiers with improved gain and efficiency, *IEEE Trans. Plasma Sci.*, **28**, 706.

Bogdankevich, L.S., and A.A. Rukhadze (1971). Stability of relativistic electron beams in a plasma and the problem of critical currents, *Sov. Phys. Usp.*, **14**, 163.

Boot, H., and J. Randall (1946). The cavity magnetron, *Proceedings of the IEEE Radiolocation Convention*, p. 928.

Borie, E., C. Gruber, and T. Westermann (1995). Calculation of MIG guns for gyrotrons using the BFCPIC code, *Int. J. Electron.*, **28**, 789.

Born, M., and E. Wolf (1965). *Principles of Optics*, Pergamon Press, Oxford.

Branch, G.M., and T.G. Mihran (1955). Plasma frequency reduction factors in electron beams, *IRE Trans. Electron Devices*, **2**, 3.

Bratman, V.L., and G.G. Denisov (1992). Cyclotron autoresonance masers: recent experiments and prospects, *Int. J. Electron.*, **72**, 969.

Bratman, V.L., and A.V. Savilov (2000). Space charge effects as a source of electron energy spread in efficiency degradation of gyrotrons, *IEEE Trans. Plasma Sci.*, **28**, 633.

Bratman, V.L., M.A. Moiseev, M.I. Petelin, and R.E. Institute of Applied Physics, Erm (1973). To the theory of gyrotrons with non-fixed structure of high-frequency field, *Radiophys. Quantum Electron.*, **16**, 622.

Bratman, V.L., N.S. Ginzburg, and G.S. Nusinovich (1977). Theory of the relativistic gyrotrons, *Sov. Tech. Phys. Lett.*, **3**, 395.

Bratman, V.L., M.A. Moiseev, and M.I. Petelin (1981a). Theory of gyrotrons with low Q-factors electrodynamic systems, in *Gyrotron*, ed. A. Gaponov, Institute of Applied Physics, Gorky, Russia, p. 122 (in Russian).

Bratman, V.L., N.S. Ginzburg, G.S. Nusinovich, et al. (1981b). Relativistic gyrotrons and cyclotron autoresonance masers, *Int. J. Electron.*, **51**, 541.

Bratman, V.L., G.G. Denisov, M.M. Ofitserov, et al. (1987). Millimeter-wave relativistic electron oscillators, *IEEE Trans. Plasma Sci.*, **15**, 2.

Bratman, V.L., G.G. Denisov, B.D. Kol'chugin, et al. (1995). Experimental demonstration of high-efficiency cyclotron-autoresonance-maser operation, *Phys. Rev. Lett.*, **75**, 3102.

Bratman, V.L., A.E. Fedotov, Y.K. Kalynov, et al. (1999). Moderately relativistic high-harmonic gyrotrons for millimeter/submillimeter wavelength band, *IEEE Trans. Plasma Sci.*, **27**, 456.

Breizman, B., and D. Ryutov (1974). Powerful relativistic electron beams in a plasma and in a vacuum, *Nucl. Fusion*, **14**, 873.

Brillouin, L. (1941). Theory of magnetron, *Phys. Rev.*, **60**, 385.

Brillouin, L. (1945). A theorem of Larmor and its importance for electrons in magnetic fields, *Phys. Rev.*, **67**, 260.

Brillouin, L. (1953). Wave propagation in periodic structures, in *Electric Filtres and Crystal Lattices*, Dover, New York.

Brown, W.C. (1961). The platinotron: amplitron and stabilotron, in *Crossed-Field Microwave Devices*, Vol. 2, ed. E. Okress, Academic Press, New York, p. 165.

Brown, W.C. (1965). The history of the crossed-field amplifiers, *IEEE Trans. Microwave Theory Tech. Suppl.*, 29.

Brown, W.C. (1984). The microwave magnetron and its derivatives, *IEEE Trans. Electron Devices*, **31**, 1595.

Budker, G.I., M.M. Karliner, and I.G. Makarov (1974). The gyrocon: an efficient relativistic high-power VHF generator, *Part. Accel.*, **10**, 41.

Bugaev, S.P., E.A. Litvinov, and G.A. Mesyats (1975). Explosive emission of electrons, *Sov. Phys. Usp.*, **18**, 56.

Bugaev, S.P., V.A. Cherepenin, and V.I. Kanavets (1990). Relativistic multiwave Cerenkov generators, *IEEE Trans. Plasma Sci.*, **18**, 525.

Buneman, O. (1961). Symmetric states and their breakup, in *Cross-Field Microwave Devices*, Vol. 1, ed. E. Okress, Academic Press, New York, p. 209.

Busch, H. (1926). Berechnung der Bahn von Kathodenstrahlen im axial symmetrischen elektromagnetishen Felde, *Ann. Phys. Leipzig*, **81**, 974.

Bykov, N.M., V.P. Gubanov, A.V. Gunin, et al. (1989). Relativistic carsinotron with high average power, *Sov. Phys. Tech. Phys.*, **59**, 519.

Carmel, Y., K.R. Chu, M. Read, et al. (1983). Realization of a stable and high efficient gyrotrons for controlled fusion research, *Phys. Rev. Lett.*, **50**, 112.

Caryotakis, G. (1998). The klystron: a microwave source of surprising range and endurance, *Phys. Plasmas*, **5**, 1590.

Caryotakis, G., et al. (1996). Gigawatt multibeam klystron (GMBK), *Proceedings of the 11th International Conference on High-Power Particle Beams*, Prague, Czech Republic, Vol. 1, p. 406.

Chernov, Z.S. (1956). Systems with centrifugal-electrostatic focusing of electron beams, *Radio Eng. Electron. Phys.*, **1**, 1428.

Chernov, Z.S., G.A. Bernashevsky, and V.V. Faykin (1978). Centrifugal electrostatic method of shaping electron beams, *Radio Eng. Electron. Phys.*, **23**, 147.

Chernov, Z.S., G.A. Bernasshevsky, and V.V. Faykin (1983). Electron-optic system with a thermionic cathode and with a centrifugal-electrostatic system for shaping a high-power axisymmetrical beam, *Radio Eng. Electron. Phys.*, **28**, 133.

Childs, C.D. (1911). Discharge from hot CaO, *Phys. Rev.*, **32**, 495.

Chodorow, M., and R. Craig (1957). Some new circuits for high power traveling wave tubes, *Proc. IEEE*, **45**, 1106.

Chodorow, M., and C. Susskind (1964). *Fundamentals of Microwave Electronics*, McGraw-Hill, New York.

Chodorow, M., E.L. Ginzton, I.R. Neilsen, and S. Sonkin (1953). Design and performance of high-power klystrons, *Proc. IRE*, **41**, 1584.

Chow, K., and R.H. Pantell (1960). The cyclotron resonance backward wave oscillator, *Proc. IRE*, **48**, 1865.

Chu, K.R., A. Drobot, V.L. Granatstein, and J.L. Seftor (1979). Characteristics and optimal operating parameters of a gyrotron traveling wave amplifier, *IEEE Trans. Microwave Theory Tech.*, **27**, 178.

Chu, K.R., P.E. Latham, and V.L. Granatstein (1988). Penultimate cavity tuning of the gyro-klystron amplifiers, *Int. J. Electron.*, **65**, 419.

Chu, K.R., H. Chen, C. Hung, et al. (1999). Theory and experiment of ultra-high-gain gyro-tron traveling wave amplifier, *IEEE Trans. Plasma Sci.*, **27**, 391.

Clark, M.C., B.M. Marder, and L.D. Bacon (1988). Magnetically insulated transmission line oscillator, *Appl. Phys. Lett.*, **52**, 78.

Clunie, D. et al. (1997). The design, construction, and testing of an experimental high power, short-pulse radar, strong microwave in plasmas, *Proceedings of the International Work-shop*, Nighny Novgorod, Russia, Vol. 2, p. 886.

Coleman, J. (1982). *Microwave Devices*, Reston Publishing Co., Reston, VA.

Collin, R.E. (2001). *Foundation for Microwave Engineering*, IEEE Press, Piscataway, NJ.

Collins, G. (1948). *Microwave Magnetrons*, McGraw-Hill, New York.

Cook, J.S., R. Compfner, and W.H. Yocom (1957). Slalom focusing, *Proc. IRE*, **45**, 1517.

Cutler, C.C. (1956). Instability of hollow and strip electron beam, *J. Appl. Phys.*, **27**, 1928.

Dammertz, G., S. Alberti, A. Arnold, et al. (2006). High-power gyrotron development at Forschugszentrum Karlsruhe for fusion applications, *IEEE Trans. Plasma Sci.*, **34**, p. 173.

Danly, B.G., and R.J. Temkin (1986). Generalized nonlinear harmonic gyrotron theory, *Phys. Fluids*, **29**, 561.

Davidson, R.C. (1990). *Physics of Nonneutral Plasmas*, Addison-Wesley, Reading, MA.

Davydovsky, V.Ya. (1962). On the possibility of acceleration charged particles by electromagnetic waves in a constant magnetic field, *Sov. Phys. JETP Lett.*, **43**, 886 (in Russian).

Denisov, G.G. (2004). Tendencies in the development of high-power long-pulse gyrotrons, presented at the 13th Joint Workshop on Electron Cyclotron Emission and Electron Cyclotron Resonance Heating, Institute of Applied Physics Russian Academy of Sciences (RAS), Nighny Novgorod, Russia, May.

Denisov, G.G., A.N. Kuftin, V.I. Malygin, et al. (1992). 110 GHz gyrotron with built-in high-efficiency converter, *Int. J. Electron.*, **72**, 1079.

Denisov, G.G., Yu.M. Bykov, A.G. Eremeev, et al. (2005). Development of gyrotron based technological systems at Gycom/IAP, *Conference Digest of the 30th International Con-ference on Infrared and Millimeter Waves*, Williamsburg, VA, p. 28.

DePackh, D.C., and P.P. Ulrich (1960). Brillouin flow in relativistic beams, *J. Electron. Control*, **10**, 139.

Development of TWT for civic and military applications, *Microwave Journal Magazine*, June 2003.

Doehler, O. (1961). M-carcinotron, in *Crossed-Field Microwave Devices*, Vol. 1, ed. E. Okress, Academic Press, New York, p. 1.

Dombrowski, G.E. (1959). Theory of amplitron, *IRE Trans. Electron Devices*, **6**, 419.

Drobot, A.T., and K. Kim (1981). Space-charge effects on the equilibrium of guided electron flow with gyromonotron, *Int. J. Electron.*, **51**, 351.

Dryden, V.W. (1962). Exact solutions of space charge flow in spherical coordinates with application to magnetron injection guns, *J. Appl. Phys.*, **33**, 3118.

Dumbrais, O., and J.P.T. Koponen (1999). Generalized gyrotron theory with inclusion of Electron Velocity and Energy Spreads, *Phys. Plasmas*, **6**, 2518.

Dumbrais, O., J.A. Heikkinen, and J. Zohm (2001). Electron cyclotron heating and current drive control by means of frequency step tunable gyrotrons, *Nucl. Fusion*, **41**, 927.

Dyke, W.P., and J.P. Trolan (1953). Field emission: large current densities, space charge, and the vacuum arc, *Phys. Rev.*, **89**, 799.

Eastwood, J.W., K.C. Howkins, and M.P. Hook (1998). The tapered MILO, *IEEE Trans. Plasma Sci.*, **26**, 698.

Eastwood, J.W., M.P. Hook, and K.C. Howkins (2000). Development of the tapered MILO, *Proceedings of the International Vacuum Electronics Conference*, IEEE Press, Piscataway, NJ.

Edgcombe, C.J. (1988a). Synthesis of electrodes for axisymmetric systems, *IEEE Proc.*, 135, Pt. A, 125.

Edgcombe, C.J. (1988b) Development in the design of electrodes for magnetron injection sources, *Int. J. Electron.*, **64**, 49.

Edwards, G.S., R.H. Austin, F.E. Carroll, et al. (2003). Free-electron-laser-based biophysical and biomedicine instrumentation, *Rev. Sci. Instrum.*, **74**, 3207.

Elias, L.R., W.M. Fairbank, J.M.J. Madey, et al. (1976). Observations of stimulated emission of radiation by relativistic electrons in a spatially periodic magnetic field, *Phys. Rev. Lett.*, **36**, 717.

Epsztein, B.R. (1952). Presented at the IRE conference on electron devices, Ottawa, Ontario, Canada, June.

Ergakov, V.S., M.A. Moiseev, and R.E. Erm (1980). Effect of electron velocity spread on the gyrotron characteristics, *Elektron Tekhn., Ser.1, Elektron. SVCh*, **3**, 20 (in Russian).

Fattoriny, H.V. (1983). The Cauchy problem, in *Encyclopedia of Mathematics and Its Applications*, Addison-Wesley, Reading, MA.

Fazio, M.V., W.B. Haynes, B.E. Carlsten, and R.M. Strinffield (1994). A 500 MW, 1 μs pulse length, high current relativistic klystron, *IEEE Trans. Plasma Sci.*, **22**, 740.

Fedosov, A.I., E.A. Litvinov, S.Ya. Belomytsev, and S.P. Bugaev (1977). Characteristics of electron beam formed in diodes with magnetic insulation, *Sov. Phys. J.*, **20**, 1367.

Feinstein, J., and G.S. Kino (1961). The large signal behavior of crossed-field traveling wave devices, in *Crossed-Field Microwave Devices*, Vol. 1, ed. E. Okress, Academic Press, New York, p. 528.

Fliflet, A.W., T.A. Hargreaves, R.P. Fischer, et al. (1990). Review of quasi-optical gyrotron development, *J. Fusion Energy*, **9**, 31.

Fliflet, A.W., W.H. Manheimer, K. Germain, et al. (2003). Cloud imaging using the NRL WARLOC radar, *AIP Conf. Proc.*, **691**, 329.

Flyagin, V.A., A.V. Gaponov, M.I. Petelin, and V.K. Yulpatov (1977). The gyrotron, *IEEE Trans. Microwave Theory Tech.*, **25**, 514.

Freund, H.P., and T.M. Antonsen (1992). *Principles of Free-Electron Lasers*, Chapman & Hall, New York.

Freund, H.P., and G.R. Neil (1999). Free-electron lasers: vacuum electronic generators of coherent radiation, *Proc. IEEE*, **87**, 782.

Friedman, M., and M. Uri (1970). Production and focusing of a high power relativistic annular electron beam, *Rev. Sci. Instrum.*, **41**, 1334.

Friedman, M., J. Krall, Y.Y. Lau, and V. Serlin (1990). Efficient generation of multi-GW microwave power by klystron-like amplifier, *Rev. Sci. Instrum.*, **61**, 171.

Fursei, G.N., and P.N. Vorontsov-Vel'yaminov (1968). Field-emission mechanism of vacuum arc onset, *Sov. Phys. Tech. Phys.*, **12**, 1370.

Gaiduk, V.I. (1971). Linear theory of microwave devices with curvilinear electron beams assuming space charge conditions, *Izv. Vuzov's Radioelectron.*, **14**, 17 (in Russian).

Gandhi, O., and J. Rowe (1961). Nonlinear theory of injected-beam crossed-field devices, in *Crossed-Field Microwave Devices*, Vol. 1, ed. E. Okress, Academic Press, New York, p. 439.

Ganguly, A.K., and K.R. Chu (1984). Limiting current in gyrotron, *Int. J. Infrared Millimeter Waves*, **5**, 103.

Gaponov, A.V. (1959). Interaction of irrectilinear electron beams with electromagnetic waves in transmission lines, *Radiophys. Quantum Electron.*, **2**, 450, 836.

Gaponov, A.V. (1960). Instability of a system of excited oscillators with respect to electromagnetic perturbations, *Sov. Phys. JETP*, **12**, 232.

Gaponov, A.V. (1961). Relativistic dispersion equations for waveguides with helical and trochoidal electron beams, *Izv. Vuzovs Radiofiz.*, **4**, 547 (in Russian).

Gaponov, A.V., A.L. Goldenberg, D.P. Grigoryev, et al. (1965). Induced synchronous radiation of electrons in cavity resonators, *JETP Lett.*, **2**, 267.

Gaponov, A.V., M.I. Petelin, and V.K. Yulpatov (1967). The induced radiation of excited classical oscillators and its use in high-frequency electronics, *Radiophys. Quantum Electron.*, **10**, 797.

Gaponov, A.V., A.L. Goldenberg, D.P. Grigoryev, et al. (1975). Experimental investigation of centimeter-band gyrotrons, *Radiophys. Quantum Electron.*, **18**, 204.

Gaponov, A.V., A.L. Goldenberg, M.I. Petelin, and V.K. Yulpatov (1976). A device for Cm, Mm, and Sbmm wave generation, Copyright N223931, Mar. 24, 1967, *Off. Bull. KDIO SM USSR*, No. 11, p. 200.

Gaponov, A.V., V.A. Flyagin, A.L. Goldenberg, et al. (1981). Power millimeter-wave gyrotrons, *Int. J. Electron.*, **51**, 277.

Gaponov-Grekhov, A.V., and V.L. Granatstein (1994). *Applications of High-Power Microwaves*, Artech House, Norwood, MA.

Garabedian, P.R. (1998). *Partial Differential Equations* (Chelsea) Providence, RI. American Mathematical Society.

Garabedian, P.R., and H.M. Lieberstein (1958). On the numerical calculation of detained bow shock waves and detached hypersonic flow, *Aeronaut. Sci.*, **25**, 109.

Garate E., R. McWilliams, D.E. Voss, et al. (1995). Novel cathode for field-emission applications, *Rev. Sci. Instrum*, **66**, 2528.

Gelvich, E.A., L.M. Borisov, Y.V. Zhary, et al., The new generation of high-power multiple-beam klystrons, *IEEE Trans. on Microwave Theory and Techniques*, **41**, 13.

Gilmour, A.S. (1986). *Microwave Tubes*, Artech House, Norwood, MA.

Gilmour, A.S. (1994). *Principles of Traveling Wave Tubes*, Artech House, Norwood, MA.

Ginzton, E., and A. Harrison (1946). Reflex-klystron oscillators, *Proc. IRE*, **35**, 97.

Gittins, J. (1965). *Power Traveling-Wave Tubes*, American Elsevier, New York.

Glazer, W. (1952). *Grundlagen der Elektronenoptik*, Springer-Verlag, Vienna.

Glyavin, M., A.L. Goldenberg, A.N. Kuftin, et al. (1999). Experimental studies of gyrotron electron beam systems, *IEEE Trans. Plasma Sci.*, **27**, 474.

Goebel, D., et al. (2000). Development of ultra-linear TWT for communication applications, *Abstracts of the Vacuum Electronics Conference*, Monterey, CA, May.

Gold, S.H., and G.N. Nusinovich (1997). Review of high-power microwave source research, *Rev. Sci. Instrum.*, **68**, 3945.

Goldenberg, A.L. (1976). Ph.D. dissertation, Nauchno Investigations Radiophysics Institute (NIRFI), Gorky, Russia (in Russian).

Goldenberg, A.L., and M.I. Petelin (1973). Formation of helical electron beams in adiabatic gun, *Radiophys. Quantum Electron.*, **16**, 141.

Goldstein, J.I., et al. (1992). *Scanning Electron Microscopy and X-ray Microanalysis*, Plenum Press, New York.

Golubev, S.V., Yu.A. Platonov, S.V. Razin, and V.G. Zorin (1996). Soft X-ray emission from millimeter-wave electron cyclotron resonance discharge, *J. X-ray Sci. Technol.*, **6**, 244.

Gorelov, Yu.A., J. Lohr, et al. (1999). Gyrotron performance on the 110 GHz installation at the DIII tokamak, *Conference Digest of the 24th International Conference on Infrared and Millimeter Waves*, No. TU-D8 Monterey, CA.

Gould, R. (1957). Space charge effects in beam-type magnetrons, *J. Appl. Phys.*, **28**, 599.

Granatstein, V.L., and I. Alexeff (1987). *High Power Microwave Sources*, Artech House, Norwood, MA.

Granatstein, V.L. and W.G. Lawson (1996). Gyro-amplifiers as candidate RF drivers for TeV linear colliders, *IEEE Trans. Plasma Sci.*, **24**, 648.

Graybill, S.E., and S.V. Nable (1966). Observations of magnetic self focusing in electron streams, *Appl. Phys. Lett.*, **8**, 18.

Grinberg, G.A. (1948). *Selected Problems in Mathematic Theory of Electric and Magnetic Phenomena*, Academy of Sciences, Moscow (in Russian).

Grivet, P. (1972). *Electron Optics*, Pergamon Press, Oxford.

Grow, R., and D.Watkins (1955). Backward-wave oscillator efficiency, *Proc. IRE*, **43**, 848.

Gulyaev, Yu.V., V.E. Kravchenko, and A.A. Kuraev (2004). Vavilov–Cerenkov amplifiers with irregular electrodynamic structures, *Sov. Phys. Usp.*, **174**, 639.

Haeff, A.V. (1939). Space charge effects in electron beams, *Proc. IRE*, **27**, 586.

Hahn, W.C. (1939). Small signal theory of velocity modulated electron beams, *Gen. Elec. Rev.*, **42**, 258.

Hamilton, D., J. Knipp, and J. Kuper (1966). *Klystrons and Microwave Tubes*, Dover, New York.

Hammer, J., and C. Wen (1964). Effect of high magnetic field on electron-beam noise, *RCA Rev.*, **12**, 785.

Harker, K.J. (1955). Periodic focusing of beams from partially shielding cathodes, *IRE Trans. Electron Devices*, **2**, 13.

Harker, K.J. (1960). Determination of electrode shapes for axially symmetric electron guns, *J. Appl. Phys.*, **31**, 2165.

Harker, K.J. (1963). Solution of the Cauchy problem for Laplace's equation in axially symmetric systems, *J. Math. Phys.*, **4**, 993.

Harris, L.A. (1952). Axially symmetric electron beam and magnetic-field systems, *Proc. IRE*, **40**, 700.

Harris, L.A. (1953). Instabilities in the smooth anode cylindrical magnetron, *J. Appl. Phys.*, **23**, 562; (1952), **24**, 1335.

Hartree, D.R., and P. Nicolson (1943). *Report Magazine*, **12**, 23.

Harvey, A. (1942). *High Frequency Thermoionic Tubes*, Chapman & Hall, London.

Haus, H., and F. Robinson (1955). The minimum noise figure of microwave beam amplifiers, *Proc. IRE*, **43**, 981.

Hawkes, P.W. (1970). Quadrupoles in electron lens design, *Adv. Electron. Electron Phys.*, Suppl. 7.

Hawkes, P.W. (1982). *Magnetic Electron Lenses*, Springer-Verlag, New York.

Heffner, H. (1954). Analysis of the backward-wave traveling wave tube, *Proc. IRE*, **42**, 930.

Herlach, F. (1996). Pulsed magnets for strong and ultrastrong fields, *IEEE Trans. Magn.*, **32**, 2438.

Hess, A. (1960). Ph.D. dissertation, Electrical Engineering Department, University of California, Berkley, CA.

Hirshfield, J.L., and V.L. Granatstein (1977). The electron cyclotron maser: historical survey, *IEE Trans. Microwave Theory Tech.*, **25**, 522.

Hirshfield, J.L., and J.M. Wachtel (1964). Electron cyclotron maser, *Phys. Rev. Lett.*, **12**, 533.

Hoggs, H.A. (1958). Periodic electrostatic beam focusing, *Proc. IRE*, Pt. B, Suppl. 10–12, 1106.

Hoshizuki, H., K. Matsura, S. Matsudo, et al. (2005). Development of material processing system by using a 300 GHz gyrotron, *Conference Digest of the 30th International Conference on Infrared and Millimeter Waves*, Williamsburg, VA, p. 375.

Hull, A. (1921). The motion of electrons between coaxial cylinders in a uniform magnetic field, *Phys. Rev.*, **17**, 539.

Humphries, S.J. (1990). *Charged Particle Beams*, Wiley, New York.

Hutter, R. (1965). *Beam and Wave Electronics in Microwave Tubes*, Boston Technical Pub., Cambridge, Mass.

Idehara, T., and Y. Shimizu (1994). Mode cooperation in submillimeter wave gyrotrons, *Phys. Plasmas*, **1**, 3145.

Idehara, T., I. Ogawa, S. Mitsudo, et al. (2004). A high harmonic gyrotron with an axis-encircling electron beam and a permanent magnet, *IEEE Trans. Plasma Sci.*, **32**, 903.

Ilyin, V.I., O.I. Louksha, V.E. Myasnikov, et al. (1998). Effect of emission inhomogeneity on low-frequency oscillations in gyrotron-type electron beams, *Abstracts of the 12th Conference on High-Power Particle Beams*, Haifa, Israel, p. 297.

Jackson, J. (1999). *Classical Electrodynamics*, Wiley, New York.

Jepsen, R., and H. Miller (1951). Enhance emission from magnetron cathodes, *Appl. J. Phys.*, **22**, 1196.

Johnson, H. (1955). Backward-wave oscillators, *Proc. IRE*, **43**, 684.

Kadomtsev, B.B. (1960). Instability of the electron cloud in a magnetron, *Sov. Phys. Tech. Phys.*, **4**, 753.

Kartikeyan, M.V., E. Borie, and M.K.A. Thumm (2004). *Gyrotrons: High Power Microwave and Millimeter Wave Technology*, Springer-Verlag, New York.

Katsenelenbaum, B.Z. (1998). *Theory of Non-uniform Waveguides: The Cross-Sections Method*, Institution of Electrical Engineering and Electronic Physics, London.

Kelman, V.M., and S.Ya. Yavor (1963). Elektron optica, *Izv. Akad. Nauk. USSR* (in Russian).

Kino, G.S. (1960). Design method for crossed-field electron guns, *IRE Trans. Electron Devices*, **7**, 179.

Kino, G.S., and N. Taylor (1962). Design and performance of magnetron injection guns, *IRE Trans. Electron Devices*, **9**, 1.

Kirstein, P.T. (1958). On the determination of the electrodes required to produce a given electric field distribution along a prescribed curve, *Proc. IRE*, **46**, 1716.

Kirstein, P.T., and G.S. Kino (1958). A solution to the equations of space charge flow by the method of separation of variables, *J. Appl. Phys.*, **29**, 1758.

Kirstein, P.T., G.S. Kino, and W.E. Waters (1967). *Space-Charge Flow*, McGraw-Hill, New York.

Kisel, D.V., G.S. Korablev, V.G. Paveljef, M.I. Petelin, and Sh.E. Tsimring (1974). Experimental investigation of gyrotron on second harmonic of cyclotron frequency with optimized distribution of HF field, *Radio Eng. Electron. Phys.*, **19**, 782.

Kleen, W. (1958). *Electronics of Microwave Tubes*, Academic Press, New York.

Klemperer, O., and M. Barnett (1971). *Electron Optics*, Cambridge University Press, London.

Kolomensky, A.A., and A.N. Lebedev (1962). Autoresonance motion of particle in a plane electromagnetic wave, *Sov. Phys. Dokl. Akad. Nauk USSR*, **145**, 1259 (in Russian).

Kompfner, R. (1964). *The Invention of Traveling Wave Tube*, San Francisco Press, San Francisco, CA.

Kompfner, R., and N. Williams (1952). Presented at the IRE conference on electron devices, Ottawa, Ontario, Canada, June.

Koppenburg, K., A. Arnold, G. Dammertz, et al. (2005). Recent results of the step-tunable 105–140 GHz, 1 MW gyrotron development at Forschungszentrum Karlsruhe, *Conference Digest of the 30th International Conference on Infrared and Millimeter Waves*, Williamsburg, VA, p. 291.

Korolyov, A.N., E.A. Gelvich, Y.V. Zhary, et al. (2004). Multiple-beam klystron amplifiers: performance parameters and development trends, *IEEE Trans. Plasma Sci.*, **32**, 1109.

Korovin, S.D., G.A. Mesyats, V.V. Rostov, et al. (1985). Relativistic millimeter-wave amplifier using a high-current electron miniaccelerator, *Sov. Tech. Phys. Lett.*, **11**, 445.

Korovin, S.D., I.K. Kurkan, V.V. Rostov, and E.M. Totmeninov (1992a). Relativistic backward wave oscillator with discrete resonance reflectors, *Radiophys. Quantum Electron.*, **42**, 1189.

Korovin, S.D., S.D. Polevin, A.M. Roitman, and V.V. Rostov (1992b). Relativistic backward tube with variable phase velocity, *Sov. Tech. Lett.*, **18**(7–8).

Kosmal, H. (1982). Modern multistage depressed collectors: a review, *Proc. IEEE*, **70**, 1325.

Kovalev, N.F., and A.V. Smorgonsky (1975). The theory of ultrarelativistic TWT, *Radio Eng. Electron. Phys.*, **20**(6).

Kovalev, N.F., M.I. Petelin, M.D. Raizer, et al. (1973). Generation of powerful electromagnetic radiation pulses by a beam of relativistic electrons, *JETP Lett.*, **18**, 138.

Kovalev, N., et al. (1979). O-type devices based on stimulated Cerenkov and transit radiation of relativistic electrons, in *Relativistic High Frequency Electronics*, Institute of Applied Physics, RAS, Gorky, Russia, p. 76 (in Russian).

Kovalev, N.F., E.G. Krastelev, M.I. Kuznetsov, et al. (1980). High-power relativistic 3 cm magnetron, *Sov. Tech. Phys Lett.*, **6**, 3.

Kovalev, N.F., V.E. Nechaev, and N.I. Zaitsev (1998). Scenario for output pulse shortening in microwave generators driven by relativistic electron beams, *IEEE Trans. Plasma Sci.*, **26**, 246.

Kreischer, K.E., and R.J. Temkin (1987). Single mode operation of high-power step tunable gyrotron, *Phys. Rev. Lett.*, **59**, 547.

Kreischer, K.E., R.J. Temkin, H.R. Fetterman, and W. Milligan (1984). Multimode oscillation and mode competition in high-frequency gyrotrons, *IEEE Trans. Microwave Theory Tech.*, **32**, 484.

Kruskal, B. (1960). *The Theory of Neutral Ionized Gases*, Wiley, New York.

Kuftin, A.N., V.K. Lygin, O.V. Malygin, et al. (1992a). Gyrotrons with a permanent magnet system, *Radiophys. Quantum Electron.*, **35**, 614.

Kuftin, A.N., V.K. Lygin, Sh.E. Tsimring, and V.E. Zapevalov (1992b). Numerical simulation and experimental study of magnetron-injection guns for powerful short-wave gyrotrons, *Int. J. Electron.*, **72**, 1145.

Kuftin, A.N., V.K. Lygin, V.N. Manuilov, et al. (1993). Theory of helical electron beams in gyrotrons, *Int. J. Infrared Millimeter Waves*, **14**, 792.

Kuftin, A.N., S.P. Belov, V.K. Lygin, et al. (2000). Advanced CW technological gyrotrons with permanent magnet system, presented at the 4th All-Russia Seminar on Theoretical and Applied Electron Optics, *Proc. SPIE*, **487**, 108.

Kuraev, A.A. (1979). *Theory and Optimization of Microwave Devices*, Nauka i Tekhnika, Minsk, Russia (in Russian).

Kurusu, T., M. Ono, S. Hanai, et al. (2004). A cryocooler-cooled 19T superconducting magnet with 52 mm room temperature bore, *IEEE Trans. Appl. Supercond.*, **14**, 393.

Kyhl, R.L., and H.R. Webster (1958). Instability of hollow beams, *IRE Trans. Electron Devices*, **3**, 172.

L3 Communications (2003). *Electron Devices*, L3, San Carlos, CA.

Lamb, H. (1945). *Hydrodynamics*, Dover, New York.

Landau, L.D., and E.M. Lifshitz (1976). *Mechanics*, Pergamon Press, New York.

Landau, L.D., and E.M. Lifshitz (1977). *Quantum Mechanics: Non-relativistic Theory*, Pergamon Press, New York.

Landau, L.D., and E.M. Lifshitz (1987). *The Classical Theory of Fields*, Pergamon Press, New York.

Landau, L.D., and E.M. Lifshitz (1995). *Electrodynamics of Continuous Media*, Butterworth-Heinemann, Oxford.

Langmuir, D.B. (1937). Theoretical limitations of cathode ray tubes, *Proc. Inst. Radio Eng.*, **25**, 977.

Langmuir, I., and K.B. Blogett (1923). Currents limited by space charge between coaxial cylinders, *Phys. Rev.*, Ser. 2, **22**, 347.

Langmuir, I., and K.B. Blodgett (1924). Currents limited by space charge between concentric spheres, *Phys. Rev.*, **24**, 49.

Lau, Y.Y., M. Friedman, J. Krall, and V. Serlin (1990). Relativistic klystron amplifier driven by modulated intense relativistic electron beams, *IEEE Trans. Plasma Sci.*, **18**, 553.

Lawson, J.D. (1988). *The Physics of Charged-Particle Beams*, Clarendon Press, Oxford.

Lawson, W., W.W. Destler, and C.D. Striffler (1985). High-power microwave generation from a large-orbit gyrotron in vane and hole-and-slot conducting wall geometries, *IEEE Trans. Plasma Sci.*, **13**, 444.

Lawson, W.G., R. Lawrence, M. Mizuhara, et al. (2001). Design of a 10-MW, 91.4-GHz, frequency-doubling gyroklystron for accelerator applications, *IEEE Trans. on Plasma Sci.*, **29**, 945.

Lawson, W., B. Hogan, S. Gouvela, et al. (2002). Development of band frequency doubling coaxial gyroklystrons for acceletrator applications, *Conference Abstracts of the 3rd International Vacuum Conference* **81**, Monterey, CA, Apr.

Lehnert, B. (1964). *Dynamics of Charged Particles*, North-Holland, Amsterdam, and Interscience, New York.

Lejuene, C., and J. Aubert (1980). Emittance and brightness: definitions and measurements, applied charged particle optics, *Adv. Electron. Electron Phys.*, Suppl. 13A, Pt. A.

Lemke, R.W., S.E. Calico, and M.C. Clark (1997). Investigtion of a load-limited magnetically insulated transmission line oscillator (MILO), *IEEE Trans. Plasma Sci.*, **25**, 364.

Lenci, S., and H. Bohlen (2002). Recent progress in CW klystrons at CPI, *Proc. European Particle Accelerator Conference (EPAC)*, Geneva, Switzerland, p. 2326.

Levin, Ya.G., et al. (1992). *The Clinotron*, Naukova Dumka Press, Kiev, Russia (in Russian).

Li, H., and T.M. Antonsen (1994). Space charge instabilities in cyclotron beams, *Phys. Plasmas*, **1**, 714.

Lindenblad, N.E. (1942). U.S. patent 2,300,052, filed May 4, 1940, issued Oct. 27.

Linder, E. (1938). Excess high energy electrons and electron motion in high vacuum tubes, *Proc. IRE*, **26**, 346.

Lines, A., et al. (1950). Some properties of waveguides with periodic structure, *Proc. IRE, Pt. III*, 263.

Littlejohn, R.G. (1980). *Hamiltonian Theory of Guiding Center Motion*, Lawrence Berkeley Laboratories, Berkeley, CA.

Litvak, A.G., G.G. Denisov, V.E. Myasnikov, and E.M. Tai (2005). Recent results in GYCOM/IAP developments of high-power gyrotrons for fusion installations, *Conference Digest of the 30th International Conference on Infrared and Millimeter Waves*, Williamsburg, VA, p. 233.

Llewellyn, F.B. (1935). Operation of ultra-high frequency vacuum tube, *Bell Syst. Tech. J.*, **14**, 632.

Llewellyn, F.B. (1939). *Electron Inertia Effects*, Cambridge University Press, London.

Llewellyn, F.B., and A.E. Bowen (1939). The production of ultrahigh frequency oscillation by means of diodes, *Bell Syst. Tech. J.*, **18**, 280.

Llewellyn, F.B., and L.C. Peterson (1944). Vacuum tube networks, *Proc. IRE*, **32**.

Lomax, R.J. (1957). Exact electrode system for the formation of curved space-charge beams, *J. Electron. Control*, **3**, 367.

Luchinin, A.G., O.V. Malygin, G.S. Nusinovich, and V.A. Flyagin (1983). Submillimeter gyrotron with a pulsed magnetic field, *Sov. Phys. Tech. Phys.*, **28**, 1001.

Luksha, O.I., and G.G. Sominski (1998). Study of space-charge oscillations in gyrotron trap, *Conference Digest of the 9th International Winter School, Microwave Electronics and Radiophysics*, Saratov, Russia, p. 33 (in Russian).

Lygin, V.K., V.N. Manuilov, and Sh.E. Tsimring (1987). About methods of integral equations and auxiliary sources in trajectory analysis of intense electron beams, *Electron Tekhn., Ser. 1, Elektron. SVCh*, **7**, 36 (in Russian).

Lygin, V.K., Sh.E. Tsimring, and V.E. Zapevalov (1991). About diochotron instability of helical electron beams, *Radiophys. Quantum Electron.*, **34**, 419.

Madey, J.M.J. (1971). Stimulated emission of Bremsstrahlung in periodic magnetic field, *J. Appl. Phys.*, **42**, 1906.

Mahie, C., et al. (1998). *WITEC Panel Report on Global Communications Technology and Systems*, Dec.

Malygin, S.A. (1986). Powerful gyrotron at 3rd harmonic cyclotron frequency, *Radio Eng. Electron Phys.*, **31**, 334.

Malygin, S.A., V.G. Pavelyev, and Sh.E. Tsimring (1983). Resonance transformation of modes in oversize electrodynamic structures, *Radiophys. Quantum Electron.*, **26**, 1126.

Manuilov, V.N., and Sh.E. Tsimring (1978). Synthesis of axially symmetric systems for shaping helical electron beams, *Radio Eng. Electron. Phys.*, **18**, 111.

Manuilov, V.N., Sh.E. Tsimring, and V.E. Zapevalov (1990). On the theory of helical beams with trapped electrons, *Radiophys. Quantum Electron.*, **33**, 1406.

Manuilov, V.N., B.V. Rayski, E.A. Soluyanova, and Sh.E. Tsimring (1995). Theoretical and experimental investigation of magnetron-injection guns of gyrotrons in regimes of current limited by space charge, *Commun. Tech. Electron.*, **40**, 647.

McCowan, R., A.W. Fliflet, S.H. Gold, et al. (1989). The design of a 100 GHz CARM oscillator experiment, *IEEE Trans. Plasma Sci.*, **36**, 1968.

McDermott, D.E., N.C. Luhmann, A. Kupiszevski, and H.R. Jory (1983). Small-signal theory of a large-orbit cyclotron resonance harmonic maser, *Phys. Fluids*, **26**, 1936.

Meeker, J., and J. Rowe (1962). Phase focusing in linear-beam devices, *IRE Trans. Electron Devices*, **9**, 257.

Meltzer, B. (1956). Single-component space charge flow, *J. Electron.*, **2**, 118.

Mesyats, G.A. (1991). Vacuum discharge effects in the diodes of high current electron accelerators, *IEEE Trans. Plasma Sci.*, **19**, 683.

Mesyats, G.A. (1995). Ecton or electron avalanche from metal, *Phys. Usp.*, **38**(6), 567.

Mesyats, G.A., and D.I. Proskurovskii (1968). Current growth in pulse breakdown of a short vacuum gap, *Sov. Phys. J.*, **11**, 49.

Midford, T.A., and G.S. Kino (1962). Experiments with a new type adiabatic cross-field gun, *IRE Trans. Electron Devices*, **9**, 431.

Miller, R.B. (1982). *An Introduction to the Physics of Intense Charged Particle Beams*, Plenum Press, New York.

Mizuno, K., S. Ono, and Y. Suzuki (1973). Two different mode interactions in an electron tube with a Fabry–Perot resonator: the ledatron, *IEEE Trans.* **20**, 749.

Mobius, A., and M. Thumm (1993). Gyrotron output launchers and output tapers, in *Gyrotron Oscillators: Their Principles and Practice*, ed. G. Edgcombe, Taylor & Francis, London.

Moiseev, M.A. (1977). Maximum amplification band of a CRM twystron, *Radiophys. Quantum Electron.*, **20**, 846.

Moreland, L.D., E. Schamiloglu, W. Lemke, et al. (1994). Efficiency enhancement of high power vacuum BWO's using nonuniform slow wave structures, *IEEE Trans. Plasma Sci.*, **22**, 554.

Morozov, A.I., and L.S. Solov'ev (1960). The motion of particles in a helical toroidal magnetic field, *Soviet Physics: Technical Physics*, **30**, 250.

Moss, H. (1968). *Narrow Angle Electron Guns and Cathode Ray Tubes*, Academic Press, New York.

Mossang, E., F. Debray, H. Jongbloets, et al. (2005). The Grenoble High Magnetic Field Laboratory (GHMFL), *Abstracts of the 19th International Conference on Magnet Technology*, Genova, Italy, p. 109.

Motz, H. (1951). Application of the radiation from fast electron beams, *J. Appl. Phys.*, **22**, 527.

Mourier, G. (1961). Small signal theory, in *Cross-Field Microwave Devices*, Vol. 1, ed. E. Okress, Academic Press, New York, p. 393.

Mourier, G., and E. Okress (1961). Introduction, in *Crossed-Field Microwave Devices*, Vol. 1, ed. E. Okress, Academic Press, New York, p. 1.

Muller, J. (1934). Experimentelle Untersuchungen uber Elektronenswingungen, *Hochfrequenztech. Elektnoakust.*, **43**, 195.

Myasin, E.A., M.B. Tseitlin, and L.M. Nutovich (1990). Analysis of the electron-wave interaction in the orotron on the basis of the approximate analytic nonlinear theory, *IEEE Trans. Electron Dev.*, **37**, 2634.

Nation, J.A. (1970). On the coupling of a high-current relativistic electron beam to a slow-wave structure, *Appl. Phys. Lett.*, **17**, 491.

Nezhevenko, O.A. (1994). Gyrocons and magnicons: microwave generators with circular deflection of the electron beam, *IEEE Trans. Plasma Sci.*, **22**, 766.

Nezhevenko, O.A., V.I. Yakovlev, S. Gold, and A. Kinkead (2002). High-power, high-convergence electron gun for an 11.424 GHz pulsed magnicon, *IEEE Trans. Plasma Sci.*, **30**, 1220.

Nezhevenko, O.A., M.A. LaPointe, V.P. Yakovlev, and J.L. Hirshfield (2003). 34 GHz, 45 MW pulsed magnicon: first results, *Proceedings on Particle Accelerators*, Vol. 2, IEEE Press, Piscataway, NJ, p. 1131.

Nguen, K.T., D.E. Pershing. D.K. Abe, et al. (2005). Eighteen-beam gun design for high-power, high repetition rate, broadband multiple beam klystrons, *IEEE Trans. Plasma Sci.*, **33**, 685.

Nikonov, A.G., I.M. Roife., Yu.M. Savel'ev, and V.I. Engel'ko (1983). Operation of a magnetically insulated diode at long pulse lengths, *Sov. Phys. Tech. Phys.*, **28**, 429.

Nordsiek, A. (1953). Theory of large signal behavior of traveling-wave amplifiers, *Proc. IRE*, **41**, 630.

Northrop, T.G. (1963). *The Adiabatic Motion of Charged Particles*, Interscience, New York.

Nusinovich, G.S. (1974). Methods of voltage feeds for a pulsed gyromonotron which ensure high efficiency in a single mode operation. *Elekron. Tekh., Ser. 1, Elektron. SVCh*, **3**, 44 (in Russian).

Nusinovich, G.S. (1981). Mode interaction in gyrotrons, *Int. J. Electron.*, **51**, 457.

Nusinovich, G.S. (1999). Review of the theory of mode interaction in gyrotrons, *IEEE Trans. Plasma Sci.* **27**, 313.

Nusinovich, G.S. (2004). *Introduction to the Physics of Gyrotrons*, Johns Hopkins University Press, Baltimore, MD.

Nusinovich, G.S., and R.E. Erm (1972). Efficiency of CRM monotron with a Gaussian axial distribution of the HF field, *Elektron. Tekn., Sec. 1, Elektron. SVCh*, **8**, 55 (in Russian).

Nusinovich, G.S., B.G. Danly, and B. Levush (1997). Gain and bandwidth in stagger-tuned gyroklystrons, *Phys. Plasmas*, **4**, 469.

Nusinovich, G.S., O.V. Sinitsin, L. Velikovich, et al. (2004). Startup scenarios in high-power gyrotrons, *IEEE Trans. Plasma Sci.*, **32**, 841.

Okress, E. (1961). *Crossed-Field Microwave Devices*, Academic Press, New York, Vols. 1 and 2.

Palevski, A., and G. Bekefi (1979). Microwave emission from pulsed relativistic E-beam diodes, II: The multiresonator magnetron, *Phys. Fluids*, **22**, 985.

Pantell, R.H. (1960). Electron beam interaction with fast waves, *Proceedings of the Symposium on Millimeter Waves*, Polytechnical Institute Press, Brooklyn, NY, p. 301.

Park, G.S., S.Y. Park, R.H. Kyser, et al. (1994). Broadband operation of a Ka-band tapered gyro-traveling wave amplifier, *IEEE Trans. Plasma Sci.*, **22**, 536.

Pavelyev, V.G., and Sh.E. Tsimring (1979). Open resonator. Inventor's certificate 616664, USSR, *Byull. Izobret.*, **17**, 240.

Pavelyev, V.G., Sh.E. Tsimring, and V.E. Zapevalov (1987). Coupled cavities with mode conversion in gyrotrons, *Int. J. Electron.*, **63**, 379.

Pellegrini, C., and S. Reiche (2004). The development of x-ray free electron lasers, *IEEE J. Select. Topics Quantum Electron.*, **10**, 1393.

Petelin, M.I. (1970). The similarity principle for high-frequency devices having ultrarelativistic electron beams, *Radiophy. Quantum Electron.*, **1**, 1229.

Petelin, M.I. (1974). On the theory of ultrarelativistic cyclotron self-resonance masers, *Radiophys. Quantum Electronics*, **17**, 686.

Petelin, M.I. (1999). One century of cyclotron radiation, *IEEE Trans. Plasma Sci.*, **27**, 294.

Peter, R. and J. Ruetr (1953). Influence of secondary electrons on noise factor and stability of traveling wave tube, *RCA Rev.*, **14**, 441.

Phillips, R.M. (1960). The ubitron, a high power traveling-wave tube based on periodic beam interaction in unloaded waveguide, *IRE Trans. Electron Devices*, **7**, 231.

Phillips, R.M. (1988). History of ubitron, *Nucl. Instrum. Methods*, **A272**, 1.

Pierce, J.R. (1944). Limiting currents in electron beams in presence of ions, *J. Appl. Phys.*, **15**, 721.

Pierce, J.R. (1945). Reflex oscillators, *Proc. IRE*, **34**, 112.

Pierce, J.R. (1947). Theory of the beam type traveling wave tube, *Proc. IRE*, **35**, 111.

Pierce, J.R. (1950). *Traveling Wave Tubes*, Van Nostrand, New York.

Pierce, J.R. (1954). *Theory and Design of Electron Beams*, 2nd ed., Van Nostrand, New York.

Piosczyk, B., O. Braz, G. Dammertz, et al. (1999). 165 GHz, 1.5 MW coaxial cavity gyrotron with depressed collector, *IEEE Trans. Plasma Sci.*, **27**, 484.

Piosczyk, B., T. Rzesnicki, G. Dammertz, et al. (2005). 170 GHz, 2 MW, CW coaxial cavity gyrotron: experimental verification of the design, *Conference Digest of the 30th International Conference on Infrared and Millimeter Waves*, Williamsburg, VA, p. 289.

Pohl, W.J. (1965). The design and demonstration of wide-band multiple-beam traveling wave klystron, *IEEE Trans. Electron Devices*, **12**, 351.

Posthumus, K. (1935). Oscillations in a split-anode magnetron, *Wireless Engineer*, **12**, 126.

Radley, D.E. (1958). The theory of the Pierce-type electron guns, *J. Electron. Control*, **4**, 195.

Radley, D.E. (1963). Electrodes for convergent Pierce-type electron guns, *J. Electron. Control*, **9**, 276.

Ramo, S. (1939a). Space charge and field waves in an electron beam, *Phys. Rev.*, **56**, 276.

Ramo, S. (1939b). Currents induced by electron motion, *Proc. IRE*, **27**, 584.

Rappoport, G.N. (1964). Nonlinear theory of backward oscillator with periodical slow-wave structure, *Radio Eng. Electron. Phys.*, **9**(3).

Rayski, B.V., and Sh.E. Tsimring (1996). Numerical simulation of nonstationary processes in intense helical electron beams of gyrotrons, *IEEE Trans. Plasma Sci.*, **24**, 992.

Razeghi, M., N. Sato, T. Suzuki, et al. (1985). Modified peniotron using a rectangular waveguide cavity, *Int. J. Electron.*, **59**, 533.

Regenstreif, E. (1967). Focusing with quadrupoles, doublets, and triplets, in *Focusing of Charged Particles*, Vol. 1, ed. A. Septier, Academic Press, New York.

Reimer, L. (1985). *Scanning Electron Microscopy*, Springer-Verlag, Berlin.

Reiser, M. (1977). Laminar-flow equilibria and limiting currents in magnetically focused relativistic beams, *Phys. Fluids*, **20**, 477.

Reiser, M. (1994). *Theory and Design of Charged Particle Beams*, Wiley, New York.

Robertson, C.W., and P. Sprangle (1989). A review of free-electron lasers, *Phys. Fluids*, **B1**(1–3).

Rowe J. (1965). *Nonlinear Electron-Wave Interaction Phenomena*, Academic Press, New York.

Rusin, F.S., and G.D. Bogomolov (1969). Orotron: an electronic oscillator with an open resonator and reflecting grating, *Proc. IEEE*, **57**, 720.

Rusterholz, A. (1950). *Elektronenoptik*, Birkhauser, Basel, Switzerland.

Scherzer, O. (1936a). Die Schwäche elektrische einzellinse geringster sphärischer Aberration, *Z. Phys.*, **101**, 23.

Scherzer, O. (1936b). Einige Fehler von Elektronenlinsen, *Z. Phys.*, **101**, 593.

Schneider, J. (1959). Stimulated emission of radiation by relativistic electrons in a magnetic field, *Phys. Rev. Lett.*, **2**, 504.

Schuneman, K., and D.M. Varviv (1999). Theory of clinotron, *IEEE Trans. Electron Devices.*, **46**, 2245.

Sedin, J.W. (1961). Numerical analysis of beam type crossed-field traveling tubes, in *Crossed-Field Microwave Devices*, Vol. 1, ed. E. Okress, Academic Press, New York, p. 541.

Seftor, J.L., V.L. Granatstein, K.R. Chu, et al. (1979). The electron cyclotron maser as a high-power traveling-wave amplifier of millimeter waves, *IEEE J. Quantum Electron.*, **15**, 848.

Shapiro, M.A., I.P. Anderson, and R.J. Temkin (2005). Synthesis of gyrotron phase-correcting mirrors using irradiance moments, *IEEE Trans. Plasma Sci.*, **32**, 2610.

Shiffler D., J.A. Nation, and L. Schachteret, et al. (1991). A high-power two stage traveling-wave tube amplifier, *J. Appl. Phys.*, **70**, 106.

Shiffler, D., M. Ruebush, M. Haworth, et al. (2002). Carbon velvet field-emission cathode, *Rev. Sci. Instrum.*, **73**, 4358.

Shockley, W. (1938). Currents to conductors induced by a moving charge, *J. Appl. Phys.*, **9**, 635.

Smith, L.P., and P.L. Hartman (1940). The formation and maintenance of electron beams, *J. Appl. Phys.*, **11**, 220.

Smith, S., and E. Purcell (1953). Visible light from localized surface charges moving over grating, *Phys. Rev.*, **92**, 1069.

Smith, R.R., J. Benford, B. Harteneck, and H.M. Sze (1991). Development and test of an L-band magnetron, *IEEE Trans. Plasma Sci.*, **19**, 628.

Smythe, W.R. (1950). *Static and Dynamic Electricity*, McGraw-Hill, New York.

Solntsev, V.A. (1968). Forces on an electron beam in a traveling wave tube, *Sov. Phys. Tech. Phys.*, **13**, 76.

Soluyanova, E.A., V.A. Syrovoy, Sh.E. Tsimring, and A.V. Vashkovsky (1986). External problem of synthesis for exact solutions equations of electron beams, *Radio Eng. Electron. Phys.*, **31**(4).

Song, H.H., D. McDermott, Y. Hirata, et al. (2004). Theory and experiment of a 94 GHz gyrotron traveling wave amplifier, *Phys. Plasmas*, **11**, 2935.

Sprangle, P., B. Hafizi, and J.R. Penano (2004). Design of compact, optically guided, pinched, megawatt class free-electron laser, *IEEE J. Quantum Electron.*, **40**, 1739.

Staprans, A., W. McCune, and J.A. Ruetz (1973). A high-power linear-beam tube, *Proc. IEEE*, **61**, 299.

Sterzer, F. (1958). Improvement of traveling-wave tube efficiency through collector potential depression, *IRE Trans. Electron Devices*, **5**, 300.

Sturrock, P.A. (1955). *Static and Dynamic Electron Optics*, Cambridge University Press, London.

Surin, L.A., B.S. Dumesh, F. Lewen, et al. (2001). Millimeter wave intracavity-jet orotron spectrometer for investigation of Van der Waals complexes, *Rev. Sci. Instrum.*, **72**, 2535.

Syrovoy, V.A. (1994a). Application of Riemann's method to solution of external problem in the theory of dense electron beams formation, *Radio Eng. Electron Phys.*, **39**(7).

Syrovoy, V.A. (1994b). Determination of electrode shapes in optics of axially symmetric electron beams, *Radio Eng. Electron Phys.*, **39**(4).

Syrovoy V.A. (1999). Geometrized theory of relativistic electron beams: a survey, *J. Commun. Technol. Electron.*, **44**, 601.

Syrovoy, V.A. (2004). *Theory of Intense Beams of Charged Particles*, Energoizdat, Moscow (in Russian).

Sze, H., B. Harteneck, J. Benford, and T.S.T. Young (1987). Operating characteristics of a relativistic magnetron with washer cathode, *IEEE Trans. Plasma Sci.*, **15**, 327.

Szilagyi, M. (1988). *Electron and Ion Optics*, Plenum Press, New York.

Thumm, M. (2002). State-of-the art of high power gyro-devices and free electron masers, Update, Forschungszentrum Karlsruhe (FZK) Internal Report.

Tien, P.K. (1954). Focusing of a long cylindrical electron stream by means of periodic electrostatic fields, *J. Appl. Phys.*, **25**, 1281.

Tien, P.A. (1956). A large signal theory of traveling wave amplifiers, *Bell Syst. Tech. J.*, **35**, 349.

Tolkachev, A.A., B.A. Levitan, G.K. Solovyev, et al. (2000). A megawatt power millimeter wave phased-array radar, *IEEE Trans. Aerosp. Electron. Syst.*, **25**, 25.

Trotman, R.E. (1966). *Longitudinal Space-Charge Waves*, Chapman & Hall, London.

Tseytlin, M.B., M.A. Fursaev, and O.V. Betsky (1978). *Microwave Amplifiers with Crossed Fields*, Soviet Radio, Moscow (in Russian).

Tsimring, Sh.E. (1957). A variational method of calculating waveguides with periodic inhomogeneties, *Radio Eng. Electron Phys.*, Pt.1, **2**, #1, 3; Pt.2, **2**, #8, **32**.

Tsimring, Sh.E. (1972). On the spread of velocities in helical electron beams, *Radiophys. Quantum Electron.*, **15**, 952.

Tsimring, Sh.E. (1977). Synthesis of systems for formation of helical electron beams, *Radiophys. Quantum Electron.*, **20**, 1068.

Tsimring, Sh.E. (1993). Limiting currents of helical electron beams in gyrotrons, *Int. J. Infrared Millimeter Waves*, **14**, 817.

Tsimring, Sh.E. (2001). Gyrotron electron beams: velocity and energy spread and beam instabilities, *Int. J. Infrared Millimeter Waves*, **22**, 1433.

Tsimring, Sh.E., and V.G. Pavelyev (1982). The theory of non-homogeneous waveguides, containing critical cross-sections, *Radio Eng. Electron. Phys.*, **27**(6).

Tsimring, Sh.E., and V.E. Zapevalov (1996). Experimental study of intense helical electron beams with trapped electrons, *Int. J. Electron.*, **81**, 199.

Twiss, R.Q. (1958). Radiation transfer and the possibility of negative absorption in radio astronomy, *Austral. J. Phys.*, **11**, 564.

Vaccaro, F.E. (1961). Four-cavity tunable magnetron, in *Cross-Field Microwave Devices*, Vol. 2, ed. E. Okress, Academic Press, New York, p. 135.

Vainstein, L.A. (1956). Electron waves in slow wave systems, *Sov. Phys. Tech. Phys.*, **1**, 119.

Vainstein, L.A. (1957a). The non-linear theory of traveling wave tubes, 1: Equations and laws of conservation, *Radio Eng. Electron. Phys.*, **2**, 92.

Vainstein, L.A. (1957b). The non-linear theory of traveling wave tubes, 2: Numerical results, *Radio Eng. Electron. Phys.*, **2**, 110.

Vainstein, L.A. (1988). *Electromagnetic Waves*, Radio and Svyazy, Moscow (in Russian).

Vainstein, L.A., and V.A. Solntsev (1973). *Lectures on High Frequency Electronics*, Soviet Radio, Moscow (in Russian).

Vainstein, L.A., V.A. Isaev, and D.I. Trubetskov (1983). Electron oscillator with open resonator, *Radio Eng. Electron. Phys.*, **28**(7).

Valiev, K.A. (1992). *The Physics of Submicron Lithography*, Plenum Press, New York.

Vandervoort, P.O. (1960). The relativistic motion of charged particle in an inhomogeneous electromagnetic field, *Ann. Phys.*, **10**, 401.

Varentsov, V.A., and Sh.E. Tsimring (1983). Ion compensation in helical electron beams, *Sov. Phys. Tech. Phys.*, **53**, 263.

Varian, R.H., and S.H. Varian (1939). A high frequency oscillator and amplifier, *J. Appl. Phys.*, **10**, 321.

Vasil'ev, V., I. Vinitzenko, et al. (1987). Relativistic magnetron operating in the mode of a train of pulses, *Sov. Tech. Phys. Lett.*, **13**, 762.

Vlasov, S.N., and K.M. Likin (1980). Geometrical optics theory of wave transformers in oversize waveguides, in *Gyrotrons*, ed. V. Flyagin, Institute of Applied Physics, Gorky, Russia, p. 125 (in Russian).

Vlasov, S.N., and I.M. Orlova (1974). Quasioptical transformer which transforms the waves in a waveguide having a circular cross-section into a highly-directional wave beam, *Radiophys. Quantum Electron.*, **17**, 115.

Vlasov, S.N., L.I. Zagryadskaya, and M.I. Petelin (1975). Transformation of a whispering gallery mode, propagating in a circular waveguide into beam of waves, *Radio Eng. Electron. Phys.*, **20**, 14.

Vlasov, S.N., L.I. Zagryadskaya, and I.M. Orlova (1976). Open coaxial resonators for gyrotrons, *Radio Eng. Electron. Phys.*, **21**, 96.

Wachtel, J., and J.L. Hirshfield (1966). Interference beads in pulse-stimulated cyclotron radiation, *Phys. Rev. Lett.*, **17**, 438.

Wang Hui, Zho Jin-Lin, et al. (1985). *Proceedings of the 10th International Conference on Infrared and Millimeter Waves*, Orlando, FL, p. 265.

Warnecke, R., et al. (1950). The magnetron-type traveling-wave amplifier tube, *Proc. IRE*, **38**, 486.

Warnecke, R., et al. (1955). The M-type carcinotron tube, *Proc. IRE*, **43**, 413.

Watanabe, K., S. Awaji, J. Sakuraba, et al. (1996). 11T liquid helium-free superconducting magnet, *Cryogenics*, **36**, 1019.

Waters, W.E. (1959). Azimuthal electron flow in a spherical diode, *J. Appl. Phys.*, **30**, 368.

Waters, W.E. (1960). Electron sheet beams focusing with tape ladder lines, *J. Appl. Phys.*, **31**, 1814.

Webber, S.E. (1958). Large signal analysis of the multicavity klystron. *IRE Trans. Electron Devices*, **4**, 306.

Webster, D. (1939). The theory of klystron oscillations, *J. Appl. Phys.*, **10**, 864.

Whaley, D.K., M.Q. Tran, T.M. Tran, and T.M. Antonsen (1994). Mode competition and startup in cylindrical cavity gyrotrons using high-order operating modes, *IEEE Trans. Plasma Sci.*, **22**, 850.

Whitaker, J. (1999). *Power Vacuum Tubes Handbook*, CRC Press, Boca Raton, FL.

Whitaker, J. (2000). *Video Display Engineering*, McGraw-Hill, New York.

Wilbur, D.A., and P.H. Peters (1961). The voltage-tunable magnetron, in *Cross-Field Microwave Devices*, Vol. 2, ed. E. Okress, Academic Press, New York, p. 35.

Wirth, M., A. Singh, J. Scharen, and J. Booske (2002). Third-order intermodulation reduction by harmonic injection in a TWT amplifiers, *IEEE Trans. Electron Devices*, **49**, 1052.

Woo, W., J. Benford, et al. (1989). Phase locking of high-power microwave oscillators, *J. Appl. Phys.*, **65**, 861.

World Wide Web (2004). Free electron laser researches and applications, in *Virtual Library*.

Yakovlev, V.I., and O.A. Nezhevenko (1999). Limitation of area compression of beams from Pierce guns, *AIP Conf. Proc.*, **474**, 316.

Yamanouchi, K., S. Ono, and Y. Shibaste (1964). Fast wave tubes: the double ridge wave peniotron, *Proceedings of the 5th International Congress on Microwave Tubes*, p. 96.

Yulpatov, V.K. (1981). Shortened equations of self-excited oscillations in a gyrotron, in *Gyrotron*, ed. A. Gaponov, Institute of Applied Physics, Borky, Russia, p. 26 (in Russian).

Zaitsev, N.I., T.B. Pankratova, M.I. Petelin, and V.A. Flyagin (1974). Millimeter and submillimeter-wave gyrotrons, *Radio Eng. Electron. Phys.*, **19**, 103.

Zaitsev, N.I., N.S. Ginzburg, E.V. Ilyakov, et al. (2002). X-band high efficiency relativistic gyrotron, *IEEE Trans. Plasma Sci.*, **30**, 840.

Zapevalov, V.E., and Sh.E. Tsimring (1978). Mode competition in gyrotrons on second harmonic of cyclotron frequency, *Lectures at the 4th Winter School for Engineers*, Vol. 1, Saratov, Russia, p. 66 (in Russian).

Zapevalov, V.E., and Sh.E. Tsimring (1990). Multibeam gyrotrons, *Radiophys. Quantum Electron.*, **33**, 954.

Zapevalov, V.E., G.S. Korablev, and Sh.E. Tsimring (1977). Experimental investigation of gyrotron on second harmonic of cyclotron frequency with optimized distribution of HF field, *Radio Eng. Electron. Phys.*, **22**, 1661.

Zapevalov, V.E., V.I. Kurbatov, O.V. Malygin, et al. (1989). Twin-beam gun of cyclotron-resonance maser, Inventors's certificate 897039 USSR, July 28, 1980, *Byull. Izobret.* (in Russian).

Zapevlov, V.E., V.N. Manuilov, and Sh.E. Tsimring (1991). Electron optical systems of two-beam gyrotrons, *Radiophys. Quantum Electron.*, **34**, 2051.

Zapevalov, V.E., S.A. Malygin, and Sh.E. Tsimring (1993). High-power gyrotron on second harmonic of cyclotron frequency, *Radiophys. Quantum Electron.*, **36**, 346.

Zarnitsina, I.G., and G.S. Nusinovich (1977). Competition of modes resonant with different harmonics of cyclotron frequency in gyromonotrons, *Radiophys. Quantum Electron.*, **20**, 313.

Zasypkin, E.V., M.A. Moiseev, E.V. Sokolov, and V.K. Yulpatov (1995). Effect of penultimate cavity position and tuning on three-cavity gyrolkystron amplifiers performance, *Int. J. Electron.*, **78**, 423.

Zheleznyakov, V.A. (1960). On the instability of magneto-active plasma relative to high-frequency electromagnetic perturbations, *Radiophys. Quantum Electron.*, **3**, 57.

Index

Printed and bound by CPI Group (UK) Ltd, Croydon, CR0 4YY

16/09/2026

14963457-0007